EFFICIENCY
OF RACETRACK
BETTING MARKETS

2008 Edition

World Scientific Handbook in Financial Economics Series

Series Editor: William T. Ziemba
University of British Columbia, Canada (Emeritus);
ICMA Centre, University of Reading, UK;
Visiting Professor at The Korean Advanced Institute for Science and
Technology and Sabanci University, Istanbul

Advisory Editors:

Kenneth J. Arrow
Stanford University, USA

George C. Constantinides
University of Chicago, USA

Espen Eckbo
Dartmouth College, USA

Harry M. Markowitz
University of California, USA

Robert C. Merton
Harvard University, USA

Stewart C. Myers
Massachusetts Institute of Technology,
USA

The Handbooks in Financial Economics (HIFE) are intended to be a definitive source for comprehensive and accessible information in the field of finance. Each individual volume in the series presents an accurate self-contained survey of a sub-field of finance, suitable for use by finance, economics and financial engineering professors and lecturers, professional researchers, investments, pension fund and insurance portfolio mangers, risk managers, graduate students and as a teaching supplement.

The HIFE series will broadly cover various areas of finance in a multi-handbook series. The HIFE series has its own web page that include detailed information such as the introductory chapter to each volume, an abstract of each chapter and biographies of editors. The series will be promoted by the publisher at major academic meetings and through other sources. There will be links with research articles in major journals.

The goal is to have a broad group of outstanding volumes in various areas of financial economics. The evidence is that acceptance of all the books is strengthened over time and by the presence of other strong volumes. Sales, citations, royalties and recognition tend to grow over time faster than the number of volumes published.

Published

Vol. 1   Stochastic Optimization Models in Finance (2006 Edition)
edited by William T. Ziemba & Raymond G. Vickson

Vol. 2   Efficiency of Racetrack Betting Markets (2008 Edition)
edited by Donald B. Hausch, Victor S. Y. Lo & William T. Ziemba

Vol. 3   The Kelly Capital Growth Investment Criterion: Theory and Practice
edited by Leonard C. MacLean, Edward O. Thorp & William T. Ziemba

Vol. 4   Handbook of the Fundamentals of Financial Decision Making (In 2 Parts)
edited by Leonard C. MacLean & William T. Ziemba

Vol. 5   The World Scientific Handbook of Futures Markets
edited by Anastasios G. Malliaris & William T. Ziemba

Forthcoming
The World Scientific Handbook of Insurance (To be announced)

EFFICIENCY
OF RACETRACK
BETTING MARKETS

2008 Edition

editors

Donald B Hausch
University of Wisconsin-Madison, USA

Victor SY Lo
University of British Columbia, Canada

William T Ziemba
University of British Columbia, Canada

World Scientific

NEW JERSEY · LONDON · SINGAPORE · BEIJING · SHANGHAI · HONG KONG · TAIPEI · CHENNAI

Published by

World Scientific Publishing Co. Pte. Ltd.

5 Toh Tuck Link, Singapore 596224

USA office: 27 Warren Street, Suite 401-402, Hackensack, NJ 07601

UK office: 57 Shelton Street, Covent Garden, London WC2H 9HE

Library of Congress Cataloging-in-Publication Data
Efficiency of racetrack betting markets / edited by Donald B. Hausch, Victor S.Y. Lo,
    William T. Ziemba. -- 2008 ed.
    p. cm.
    Includes bibliographical references.
    ISBN 981-281-918-5 (978-981-281-918-5) -- ISBN 981-281-919-3 (978-981-281-919-2)
    1. Horse racing--Betting. 2. Horseplayers--Psychology. I. Hausch, Donald B.
    II. Lo, Victor S. Y. III. Ziemba, W. T.
    SF331.E36 2008
    338.4'7798401--dc22

    2008015561

British Library Cataloguing-in-Publication Data
A catalogue record for this book is available from the British Library.

First published 2008 (Hardcover)
Reprinted 2016 (in paperback edition)
ISBN 978-981-3203-51-8

Printed in Singapore

To Joanne
       D.B.H.

To my parents
       V.S.Y.L.

To Rachel
       W.T.Z.

Contents

PART III. ECONOMIC AND MATHEMATICAL INSIGHTS

PART IV. EFFICIENCY OF WIN MARKETS AND THE FAVORITE-LONGSHOT BIAS

PART VII. RESEARCH IN THE COMMONWEALTH AND ASIA

Preface to 2008 Edition

Academic Press published the first edition of this volume in 1994. Our goal was to collect the important papers and many new articles for an up-to-date and comprehensive volume on racetrack efficiency. The ensuing years have pushed the volume into classic status. The volume went out of print at the same time that various hedge fund-like racing syndicates were using these ideas as an integral part of their research that led to hundreds of millions in profits for the most successful groups and about US$10 billion in total gains. When the volume went out of print, it became a cult item. Its scarcity, its content, and the dreams of new investment groups drove the price of the volume on eBay and Amazon to thousands of dollars. Hence, the volume was only available to a few. We appreciate the classic and cult status but want to make the book more accessible to students, researchers, academics and others. So, we are pleased that World Scientific is producing this 2008 edition for this purpose. Except for this Preface, the 2008 edition is identical to the original.

This 2008 edition is coming out simultaneously with the *Handbook of Sports and Lottery Markets* edited by Donald B Hausch and William T Ziemba published by Elsevier/North Holland, 2008. That volume complements this 2008 edition by providing new, original surveys of the racetrack, sports betting and lotto investment field. Together they provide an up-to-date treatment of the field of racetrack efficiency plus the key historical and current papers.

There are a number of new developments since 1994 that have altered the racetrack investment markets. These affect the academic research surveyed in this 2008 edition and the Handbook volume as well as the professional and amateur application of these results to racetrack markets across the world.

Since circa 2000, we have seen the following important changes in racetrack markets.

1. Rebates on wagers are available to essentially all bettors. These rebates began as simply a reward to large bettors for quantity purchases, similar to discounts when one buys in bulk such as a case of San Pellegrino offered at food stores and other places. However some rebaters have extended these discounts to all bettors. The tracks provide a signal of the pools to the rebaters at a low price. Then the rebaters put the bets of their clients into the track pools, with the rebaters and the bettors sharing the difference between the regular track take and the signal price. So bettors, instead of paying a track take of 13–30% at the track on various wagers, actually face a net track take of about 10%. This lower track take for syndicates disadvantages small

bettors, though, since their effective track take becomes more than the posted track take.

2. The extent of off-track betting from other racetracks and other betting sources has expanded greatly and is now about 87% of all wagers at a typical race track (see Hausch and Ziemba, 2008). There are also delays in adding these monies to the host track pools, which can lead to substantial differences between the odds posted at the time the race begins and the final odds that determine the actual payoffs. Indeed, about 50% of all bets are not shown in the pools until after it is too late to bet at the host track and the horses are running (see Ziemba, 2008). Hence, a 4-1 horse at 1 minute to post time could end up paying 5-2. This uncertainty is challenging for any system that incorporates the public's odds. An example is the Dr Z place and show system, which is discussed in two Hausch and Ziemba (one with Rubinstein) papers in Section V of this volume and which estimates a horse's win probability based on the public's win odds. Ziemba's (2008) update of this earlier work demonstrates the challenge, but also shows that, with rebates, the system may still yield profits.

3. The 1999 introduction of Betfair and other betting exchanges allows bettors to lock in prices for win, place and show and other wagers by betting not against the house but against other bettors. The house, say Betfair of London (the biggest by far), then takes a 1–5% commission on all net winning wagers. So bettors can lock in odds rather than bear the risk of changing odds in the on-track parimutuel betting system. Equally important, the betting exchanges allow short as well as long bets. Hence, bettors can wager against over-priced horses just as hedge fund managers can short over-priced stocks, stock index futures, currencies or precious metals. This feature also allows for long-short hedging akin to long-short equity hedge funds. While these exchanges are not legal in the US, they have had an impact on some important US horse races. See Snowberg, Wolfers and Zitzewitz (2008) and Smith and Vaughan Williams (2008) for a discussion of prediction markets in politics and other areas and betting on myriad sports and other events, respectively.

4. The figure below shows the return from wagers on different odds-levels. The steeper of the two lines is from Ziemba and Hausch (1986) and is based on data from the 1980s and earlier. The flatter line is from Ziemba (2004) and is based on Equiform data from 1997 to 2004. Both lines are based on data of more than 300,000 horses, and both exhibit the favorite-longshot bias where favorites are underbet relative to the true odds and longshots are

overbet. However, the extent of the bias is clearly weaker in recent years, so the expected return on longshots is higher and the expected return on favorites is lower. Further, the positive expected return on extreme favorites is no longer available. Betting exchanges appear to partly explain this recent shift. Snowberg and Wolfers (2008) show similar results based on horses in 647,903 races, which are all races run in the US from 1992 to 2001.

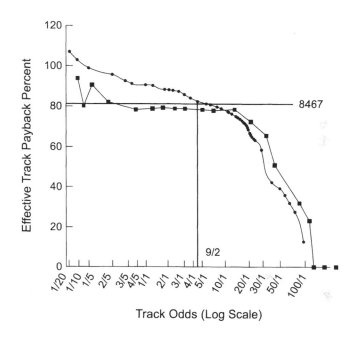

Track Odds (Log Scale)

Since the original papers were published, some have been extended. Some of these extensions are in Vaughan Williams (2003, 2005), two books we strongly recommend, and Hausch and Ziemba (2008). A few of them are now discussed.

Reference 31 in the annotated bibliography was published as Lo and Bacon-Shone (2008). They provide a discount model that is a simple approximation to the Henery and Stern models for computing ordering probabilities superior to those obtained by the Harville formulas. They show the accuracy of the approximation and the method is applied in Lo, Bacon-Shone and Bushe (1995).

The MacLean, Ziemba and Blazenko paper was extended by MacLean and Ziemba (2006) and an updated table and discussion of the good and bad properties of the Kelly criterion appear in Ziemba and Ziemba (2007).

The crosstrack betting paper by Hausch and Ziemba has evolved into a procedure for going long on horses with rebates at the host track and short on Betfair or other betting exchanges plus long/shorts across betting exchanges.

An extension of Ziemba and Hausch (1987) on the Dr Z place betting system in England arises because of the different way place pools were computed in England and other locales by splitting the net bets among the collecting horses. Hence, once about 24.5% is bet on any horse, the payoff is at its minimum. We noted this, but Jackson and Waldron (2003) extended Hausch and Ziemba's system for locks to exploit this rule. See also Edelman and O'Brian (2004) for a more general analysis of these tote arbitrages.

We hope that the readers of this 2008 edition and the Hausch and Ziemba (2008) Handbook find these papers stimulating and useful.

Donald B. Hausch, Madison
Victor S.Y. Lo, Boston
William T. Ziemba, Vancouver
February 2008

References

Edelman, D. C. and N. G. O'Brian (2004). Tote arbitrage and lock opportunities in racetrack betting. *European Journal of Finance*, **10**: 370–378.

Hausch, D.B. and W.T. Ziemba (Eds.) (2008). *Handbook of Sports and Lottery Markets.* Elsevier.

Jackson, D. and P. Waldron (2003). Parimutuel place betting in Great Britain and Ireland, in L. Vaughan Williams (Ed.), *The Economics of Gambling*, 18–29.

Lo, V. and J. Bacon-Shone (2008). Approximating the ordering probabilities of multi-entry competitions by a simple method, in D.B. Hausch and W.T. Ziemba (Eds.), *Handbook of Sports and Lottery Markets.* Elsevier.

Lo, V., J. Bacon-Shone and K. Bushe (1995). The application of ranking probability models to racetrack betting, *Management Science*, **41**: 1048–1059.

MacLean, L.C. and W.T. Ziemba (2006). Capital growth: Theory and practice, in S.A. Zenios and W.T. Ziemba (Eds.), *Handbook of Asset-Liability Management, Volume I: Theory and Methodology*, pp. 429–473. Elsevier.

Smith, M.A. and L. Vaughan Williams (2008). Betting exchanges: A technological revolution in sports betting, in D.B. Hausch and W.T. Ziemba (Eds.), *Handbook of Sports and Lottery Markets*. Elsevier.

Snowberg, E. and J. Wolfers (2008). Examining explanations of a market anomaly: Preferences or perceptions?, in D.B. Hausch and W.T. Ziemba (Eds.), *Handbook of Sports and Lottery Markets*. Elsevier.

Snowberg, E., J. Wolfers and E. Zitzewitz (2008). Prediction markets: From politics to business, in D.B. Hausch and W.T. Ziemba (Eds.), *Handbook of Sports and Lottery Markets*. Elsevier.

Vaughan Williams, L. (2003). *The Economics of Gambling*. Routledge.

Vaughan Williams, L. (2005). *Information Efficiency in Financial and Betting Markets*. Cambridge University Press.

Ziemba, W.T. (2004). Behavioral finance, racetrack betting and options and futures trading, Mathematical Finance Seminar, Stanford University, January 30.

Ziemba, W.T. (2008). Efficiency of racetrack, sports and lottery betting markets, in D.B. Hausch and W.T. Ziemba (Eds.), *Handbook of Sports and Lottery Markets*. Elsevier.

Ziemba, W.T. and D.B. Hausch (1986). *Betting at the Racetrack*. Dr Z Investments Inc, San Luis Obispo, CA.

Ziemba, W.T. and D.B. Hausch (1987). *Dr Z's Beat the Racetrack*. William Morrow.

Ziemba, R.E.S. and W.T. Ziemba (2007). *Scenarios for Risk Management and Global Investment Strategies*. Wiley.

Preface in 1994 Edition

Racetrack betting interests not just gamblers and those seeking casual entertainment, but also academics across many disciplines. Gamblers and investors certainly enjoy the sport, but making profits is their challenge. Academics are interested in this came challenge. Their approach is to determine whether the racetrack betting market is efficient and, if not, whether positive risk-adjusted returns are possible. In addition to market efficiency, active research areas include methods for analyzing racetrack data, determining probabilities of complex events, explaining gambling behavior, and developing betting strategies. Racetrack betting is of particular interest because, beyond just allowing a forum for theoretical questions of efficiency and gambling behavior, abundant and easily-accessible data from racetracks around the world permits extensive empirical consideration of these questions.

Studies related to racetrack betting have appeared in the academic literature for over 40 years. The number of articles in this area has been growing recently. A need for a book-length treatment of racing efficiency research covering its economic, psychological, financial, statistical and mathematical aspects has been felt for quite some time. This volume of collected papers is the result of collaboration with many colleagues in a variety of disciplines and from countries around the world. This volume is intended for two audiences: academics across many fields who are interested in the subject and those of the general public who wish to know more about the academic research on racetrack betting.

Following an introduction, the papers in this volume are divided into seven sections: psychological studies, utility preferences of racetrack bettors, economic and mathematical insights, efficiency of win markets and the favorite-longshot bias, place and show anomalies, efficiency of exotic wagering markets, and research in the Commonwealth and Asia.

We are pleased to be part of this distinguished series on Economic Theory, Econometrics, and Mathematical Economics, and we thank Academic Press for the opportunity to publish this volume.

Donald B. Hausch, Madison
Victor S.Y. Lo, Vancouver
William T. Ziemba, Vancouver

List of Contributors

Mukhtar M. Ali, Department of Economics, University of Kentucky, Lexington, Kentucky, USA.

> "Probability and Utility Estimates for Racetrack Bettors." *Journal of Political Economy* 83, 1977, 803-815.
>
> "Some Evidence on the Efficiency of a Speculative Market." *Econometrica* 47, 1979, 387-392.

Peter Asch, Department of Economics, Rutgers University, Newark, New Jersey, USA; Burton G. Malkiel and Richard E. Quandt, Department of Economics, Princeton University, Princeton, New Jersey, USA.

> "Racetrack Betting and Informed Behavior." *Journal of Financial Economics* 10, 1982, 187-194.
>
> "Market Efficiency in Racetrack Betting." *Journal of Business* 57, 1984, 165-174.
>
> "Market Efficiency in Racetrack Betting: Further Evidence and a Correction." *Journal of Business* 59, 1986, 157-160.

Peter Asch, Department of Economics, Rutgers University, Newark, New Jersey, USA; and Burton G. Malkiel, Department of Economics, Princeton University, Princeton, New Jersey, USA.

> "Efficiency and Profitability in Exotic Bets." *Economica* 54, 1987, 289-298.

William Benter, Hong Kong Gambling Syndicate, Central, Hong Kong.

Sandra Betton, Faculty of Commerce, University of British Columbia, Vancouver, British Columbia, Canada.

Ron Bird, TPF&C Ltd., Consulting Actuaries, Melbourne, Australia; Michael McCrae, Department of Accountancy, University of Wollongong, Wollongong, Australia.

> "Tests of the Efficiency of Racetrack Betting Using Bookmaker Odds." *Management Science* 33, 1987, 1552-1562.

Stephen R. Blough, Research Department, Federal Reserve Bank of Boston, Boston, Massachusetts, USA.

Ruth N. Bolton, GTE Laboratories Incorporated, Waltham, Massachusetts, USA; and Randall G. Chapman, Chapman and Associates, Winchester, Massachusetts, USA.
"Searching for Positive Returns at the Track: A Multinomial Logit Model for Handicapping Horse Races." *Management Science* 32, 1986, 1040-1059.

Kelly Busche and Christopher D. Hall, School of Economics, The University of Hong Kong, Hong Kong.
"An Exception to the Risk Preference Anomaly." *Journal of Business* 61, 1988, 337-346.

Brian R. Canfield and Bruce C. Fauman, Vancouver, British Columbia, Canada; William T, Ziemba, Faculty of Commerce, University of British Columbia, Vancouver, British Columbia, Canada.
"Efficient Market Adjustment of Odds Prices to Reflect Track Biases." *Management Science* 33, 1987, 1428-1439.

N.F.R. Crafts, Department of Economics, University of Warwick, Coventry, UK.

Jack Dowie, Faculty of Social Sciences, Open University, Milton Keynes, UK.
"On the Efficiency and Equity of Betting Markets." *Economica* 43, 1976, 139-150.

Stephen Figlewski, Department of Finance, Leonard N. Stern School of Business, New York University New York, New York, USA.
"Subjective Information and Market Efficiency in a Betting Model." *Journal of Political Economy* 87, 1979, 75-88.

Paul E. Gabriel, School of Business Administration, Loyola University of Chicago, Chicago, Illinois USA; and James R. Marsden, Department of Operations and Information Management, School of Business Administration, University of Connecticut, Storrs, Connecticut, USA.
"An Examination of Market Efficiency in British Racetrack Betting." *Journal of Political Economy* 98, 1990, 874-885.
"An Examination of Market Efficiency in British Racetrack Betting: Errata and Corrections." *Journal of Political Economy* 99, 1991, 657-659.

Richard M. Griffith, Veterans Administration Hospital, Lexington, Kentucky, USA.

> "Odds Adjustments by American Horse-Racing Bettors." *American Journal of Psychology* 62, 1949, 290-294.
>
> "A Footnote on Horse Race Betting." *Transactions Kentucky Academy of Science* 22, 1961, 78-81.

David A. Harville, Department of Statistics, Iowa State University, Ames, Iowa, USA.

> "Assigning Probabilities to the Outcomes of Multi-Entry Competitions." *Journal of the American Statistical Association* 68, 312-316.

Donald B. Hausch, School of Business, University of Wisconsin, Madison, Wisconsin, USA; and William T. Ziemba, Faculty of Commerce, University of British Columbia, Vancouver, British Columbia, Canada.

> "Transactions Costs, Extent of Inefficiencies, Entries and Multiple Wagers in a Racetrack Betting Model." *Management Science* 31, 1985, 381-394.
>
> "Arbitrage Strategies for Cross-Track Betting on Major Horse Races." *Journal of Business* 63, 1990, 61-78.
>
> "Locks at the Racetrack." *Interfaces* 20, 1990, 41-48.

Donald B. Hausch, School of Business, University of Wisconsin, Madison, Wisconsin, USA; William T. Ziemba, Faculty of Commerce, University of British Columbia, Vancouver, British Columbia, Canada; and Mark Rubinstein, Department of Finance, Walter A. Haas School of Business, University of California, Berkeley, California, USA.

> "Efficiency of the Market for Racetrack Betting." *Management Science* 27, 1981, 1435-1452.

Robert J. Henery, Department of Statistics and Modelling Science, University of Strathclyde, Glasgow, Scotland.

> "Permutation Probabilities as Models for Horse Races." *Journal of Royal Statistical Society B* 43, 1981, 86-91.

Rufus Isaacs, The Rand Corporation, Santa Monica, California, USA.

> "Optimal Horse Race Bets." *American Mathematical Monthly* 60, 1953, 310-315.

Hyun Song Shin, Department of Economics, University College, Oxford University, Oxford, UK.
> "Prices of State Contingent Claims with Insider Traders, and the Favourite-Longshot Bias." *The Economic Journal* 102, 1992, 426-435.

Wayne W. Snyder, Department of Finance, Sangamon State University, Sangamon, California, USA.
> "Horse Racing: Testing the Efficient Markets Model." *Journal of Finance* 33, 1978, 1109-1118.

Hal S. Stern, Department of Statistics, Harvard University, Cambridge, Massachusetts, USA.

Richard H. Thaler, Department of Economics, Johnson School of Management, Cornell University, Ithaca, New York, USA; and William T. Ziemba, Faculty of Commerce, University of British Columbia, Vancouver, British Columbia, Canada.
> "Anomalies - Parimutuel Betting Markets: Racetracks and Lotteries." *Journal of Economic Perspectives* 2, 1988, 161-174.

R.H. Tuckwell, Macquarie University, New South Wales, Australia.
> "Anomalies in the Gambling Market." *Australian Journal of Statistics* 23, 1981, 287-295.

Acknowledgments

We thank the following publishers and authors for permission to reproduce the articles listed below.

American Journal of Psychology
>
> Griffith, R.M. (1949) "Odds Adjustments by American Horse-Racing Bettors." 62, 290-294.
>
> McGlothlin, W.H. (1956) "Stability of Choices among Uncertain Alternatives." 69, 604-619.

American Mathematical Monthly
>
> Isaacs, R. (1953) "Optimal Horse Race Bets." 60, 310-315.

Australian Journal of Statistics
>
> Tuckwell, R.H (1981) "Anomalies in the Gambling Market." 23, 287-295.

Econometrica
>
> Ali, M.M. (1979) "Some Evidence on the Efficiency of a Speculative Market." 47, 387-392.

Economica
>
> Asch, P. and Quandt, R.E. (1987) "Efficiency and Profitability in Exotic Bets." 54, 289-298.
>
> Dowie, J. (1976) "On the Efficiency and Equity of Betting Markets." 43, 139-150.

Economics Letters
>
> Kallberg, J.G. and Ziemba, W.T (1979) "On the Robustness of the Arrow-Pratt Risk Aversion Measure." 2, 21-26.

Interfaces
>
> Hausch, D.B. and Ziemba, W.T (1990b) "Locks at the Racetrack." 20, 41-48.

Journal of Business

Asch, P., Malkiel, B.G. and Quandt, R.E. (1984) "Market Efficiency in Racetrack Betting." 57, 165-174.

Asch, P., Malkiel, B.G. and Quandt, R.E. (1986) "Market Efficiency in Racetrack Betting: Further Evidence and a Correction." 59, 157-160.

Busche, K. and Hall, C.D. (1988) "An Exception to the Risk Preference Anomaly." 61, 337-346.

Hausch, D.B. and Ziemba, W.T (1990a) "Arbitrage Strategies for Cross-Track Betting on Major Horse Races." 63, 61-78.

Journal of Economic Perspectives

Thaler, R. and Ziemba, W.T. (1988) "Anomalies - Parimutuel Betting Markets: Racetracks and Lotteries." 2, 161-174.

Journal of Finance

Snyder, W.W. (1978) "Horse Racing: Testing the Efficient Markets Model." 33, 1109-1118.

Losey, R.L. and Talbott, J.C., Jr. (1980) "Back on the Track with the Efficient Markets Hypothesis." 35, 1039-1043.

Journal of Financial Economics

Asch, P., Malkiel, B.G. and Quandt, R.E. (1982) "Racetrack Betting and Informed Behaviour." 10, 187-194.

Journal of Political Economy

Ali, M.M. (1977) "Probability and Utility Estimates for Racetrack Bettors." 83, 803-815.

Figlewski, S. (1979) "Subjective Information and Market Efficiency in a Betting Model." 87, 75-88.

Gabriel, P.E. and Marsden, J.R. (1990) "An Examination of Market Efficiency in British Racetrack Betting." 98, 874-885.

Gabriel, P.E. and Marsden, J.R. (1991) "An Examination of Market Efficiency in British Racetrack Betting: Errata and Corrections." 99, 657-659.

Rosett, R.N. (1965) "Gambling and Rationality." 73, 595-607.

Weitzman, M. (1965) "Utility Analysis and Group Behaviour: An Empirical Study." 73, 18-26.

Journal of Royal Statistical Society B

Henry, R.J. (1981) "Permutation Probabilities as Models for Horse Races." 43, 86-91.

Journal of the American Statistical Association

Harville, D.A. (1973) "Assigning Probabilities to the Outcomes of Multi-Entry Competitions." 68, 312-316.

Management Science

Bird, R. and McCrae, M. (1987) "Tests of the Efficiency of Racetrack Betting Using Bookmaker Odds." 33, 1552-1562.

Bolton, R.N. and Chapman, R.G. (1986) "Searching for Positive Returns at the Track: A Multinomial Logit Model for Handicapping Horse Races." 32, 1040-1059.

Canfield, B., Fauman, B.C. and Ziemba, W.T. (1987) "Efficient Market Adjustment of Odds Prices to Reflect Track Biases." 33, 1428-1439.

Hausch, D.B., Ziemba, W.T. and Rubinstein, M. (1981) "Efficiency of the Market for Racetrack Betting." 27, 1435-1452.

Hausch, D.B. and Ziemba, W.T. (1985) "Transactions Costs, Extent of Inefficiencies, Entries and Multiple Wagers in a Racetrack Betting Model." 31, 381-394.

MacLean, L.C., Ziemba, W.T. and Blazenko, G. (1992) "Growth versus Security in Dynamic Investment Analysis." 38, 1562-1585.

Psychological Reports

Metzger, M.A. (1985) "Biases in Betting: An Application of Laboratory Findings." 56, 883-888.

The Economic Journal

Shin, H.S. (1992) "Prices of State Contingent Claims with Insider Traders, and the Favourite-Longshot Bias." 102, 426-435.

The Quarterly Journal of Economics

Quandt, R.E. (1986) "Betting and Equilibrium." 101, 201-207.

Transactions Kentucky Academy of Science

Griffith, R.M. (1961) "A Footnote on Horse Race Betting." 22, 78-81.

Introduction

Gambling is decision making under risk. On the surface, it seems a simple exercise, of gaining a profit or losing one's wager. While that is, of course, the essence of the exercise, a closer inspection reveals great complexity and scope. Indeed, this simplicity veiling a complex and general process is what has attracted academics from a variety of disciplines to consider gambling as a forum for investigating matters of wider importance than simply wagering. For instance, psychologists have used gambling to illustrate and inform about fundamental behavior in the face of uncertainty, and numerous behavioral biases have seen support. Economists bring a more rational, utility-based perspective to decision making under uncertainty, and wagering has provided an opportunity to test numerous theories and has generated new theories as well. Financial experts view gambling markets as possessing many of the intricacies of financial markets; however, the repeated and short-lived reduced form of gambling markets allows cleaner analysis than is generally possible in most financial markets. Thus, starker views of aggregate investor behavior are sometimes possible, including investigations of market efficiency. Statisticians have used gambling to motivate improved estimates of complicated probabilistic events. Mathematicians and management scientists have developed useful gambling strategies that have drawn on and extended efforts in other domains (e.g., Kelly betting and information transmission). While gambling is of academic interest in its own right, it has clearly been demonstrated to enhance our understanding of more far-ranging environments.

This volume is concerned with racetrack betting. As evidenced by this collection of articles, our understanding of racetrack betting has clearly drawn from and correspondingly returned something to all the aforementioned fields of psychology, economics, finance, statistics, mathematics and management science.

It will be helpful to briefly describe the workings of the racetrack betting market. We adopt the terminology of U.S. racetracks. Typically, six to twelve horses are in a race and several wager types are available. A *win* bet requires one to name the winner of the race. A *place* (*show*) bet requires one's horse to finish at least second (third). Bets to win, place, and show involve naming just one horse and are termed *straight* bets. There are other wager types, called *exotic* bets, that require predicting the outcome of two of more horses. For example, and *quinella* bet requires one to name the horses finishing first and second, while an *exacta* bet requires naming these two horses in the right order. A *daily double* bet pays off when one correctly names the winners of two consecutive races. Separate pools are operated for each bet type.

The simplest wager is the win bet. Define W_i as the public's wager to win on horse i, $W \equiv \Sigma_i W_i$ as the public's win pool, and Q as the track payback. Then, in the event that horse i wins, QW/W_i is the payoff per dollar wagered on horse i. Define $O_i \equiv QW/W_i - 1$. The odds on horse i are expressed as O_i to 1, or O_i-1, and the return on a \$1 wager at odds of O_i-1 is the original \$1 plus another \$$O_i$.

The proportion of money bet on a horse, W_i/W, is called the win bet fraction. It is often interpreted as the public's estimate of the horse's win probability, which follows when bettors are expected profit maximizers. Several articles in this volume will show that the win bet fraction only roughly corresponds to the true win probability, though. The payoff functions for wagers other than win bets will be presented later.

The track take, $1-Q$, is the commission collected by the management of the racetrack. A portion of that amount is paid to the state government in the form of a tax. For straight bets, the track take varies from about 14 to 19 percent in the U.S. (There are other countries with much higher takes, such as the 26 percent in Japan.) Exotic bets typically have higher takes which are 20-25% or higher. The other transactions costs is *breakage* where the track rounds down returns to the nearest nickel or dime on the dollar. While breakage is not substantial for a single large payoff, it can be significant for a single small payoff and is always significant over a long sequence of wagers (see Hausch and Ziemba (1985)[1]). By comparison, breakage in Japan also involves rounding down payoffs while Hong Kong rounds payoffs to the nearest cent, which can mean rounding up or down. Thus, there is little if any cost due to breakage in Hong Kong.

There is typically about 20 minutes between races, during which time wagering occurs on the upcoming race. Most tracks publicly display the win, place and show bets of the public, and update these figures every minute or so during the betting period. The sheer quantity of numbers involved in exotic wagering (e.g. 90 exacta figures for a ten-horse race) means that these pools are often not publicly available or are displayed but only subsets at a time. While odds do change over the course of the betting period, actual payoffs in the parimutuel betting system are based on the final wagering of the public; thus, from the time one makes a wager, the odds can improve or worsen.

This volume is a compilation of recent research on various aspects of racetrack markets. The papers in this volume are contributed by researchers from different countries and various fields - Economics, Psychology, Finance, Mathematics, Statistics and Management Science/Operations Research. It is a blend of new articles produced for this volume and reprinted articles published from a wide variety of sources. The contributions cover empirical and theoretical studies of issues such as the favorite-longshot bias and efficiency of the markets to win, place, and show and markets for exotic betting. Other issues are ordering probability models and optimal betting strategies. Data is studied from racetracks around the world.

Part 1 presents some psychological studies of racetrack markets, including the earliest research in this area (Griffith (1949)[1] and McGlothlin (1956)[1]). Issues such as biases in probability estimation, the "gambler's fallacy," and framing are investigated at the racetrack.

Characteristics of the average racetrack bettor's utility function are studied in Part 2. Weitzman (1965)[1] and Ali (1977)[1] use racetrack data to empirically obtain a convex utility function, indicating that bettors are risk-lovers. With such a convex utility function, expected utility maximization generates the well-known favorite-longshot bias, i.e. bettors underbet favorites and overbet longshots.

Part 3 contains articles providing economic and mathematical insights about racetrack betting. There are three general issues in these papers: 1) optimal wagering strategies; 2) fundamental handicapping schemes; and 3) estimating ordering probabilities (the probabilities of various orders of finish of the horses). On the first of these, the earliest contributor was Isaacs (1953)[1] who developed an algorithm for optimal betting assuming our risk neutral bettor knows the true win probabilities of the horses. Kallberg and Ziemba (1994)[1] and Levin (1994)[1] extend this work to nonlinear utility and place and show betting. Thorp (1971)[1] describes the advantages of using the "Kelly criterion" originated by Kelly (1956)[2], which involves maximizing the expected logarithmic utility of capital. MacLean, Ziemba and Blazenko (1992)[1] discuss the tradeoff between return and security in gambling situations.

Like the stock market, investment strategies at the track are usually *technical* or *fundamental* in nature. Handicapping, or fundamental schemes, are described in several papers. Multinomial logit models are used with horse, jockey and race characteristics (Bolton and Chapman (1986)[1] and Chapman (1994)[1]). Benter's (1994)[1] scheme is both fundamental and technical in nature; to the aforementioned fundamental characteristics, Benter's multinomial logit analysis adds the public's win odds.

Ordering probabilities are necessary for pricing place, show and exotic bets. Three probability models of running times (Harville, Henery and Stern) are the basis for most of the research on estimating ordering probabilities. An application of these models is reported in Lo (1994a)[1].

The win market is the subject of Part 4. Using Fama's (1970)[2] definition of weak form efficiency, various papers such as Ali (1977)[1], Snyder (1978)[1] and Asch, Malkiel and Quandt (1982)[1] conclude that systematic differences exist between the public's subjective win probabilities (as measured through the odds) and the objective win probabilities (as measured by actual outcomes). The bias is such that favorites tend to be underbet and longshots tend to be overbet, hence the term "favorite-longshot bias." Quandt (1986)[1] proves that the bias implies risk-seeking bettors. The empirical studies show that the bias for favorites is insufficient to overcome the track take, except for extreme favorites, which are relatively rare (Ziemba and Hausch (1986)[2]). Thus, for practical purposes, the win market is weak form efficient despite the bias. Exceptions to the favorite-longshot bias have been documented by Busche and Hall (1988)[1] and Busche (1994)[1] for tracks in Hong Kong and Japan. Other papers apply various forms of logit model to analyze win market efficiency (see Figlewski (1979)[1], Asch, Malkiel and Quandt (1984)[1], and Lo (1994b)[1]).

Part 5 discusses the place and show markets, and the existence of the most well-documented of the racetrack's inefficiencies. Ritter (1994)[1], Asch, Malkiel and Quandt (1984,86)[1] and Hausch, Ziemba and Rubinstein (1981)[1] test various approaches for identifying inefficiencies. Hausch, Ziemba and Rubinstein (1981)[1] study an optimal growth betting strategy to exploit these inefficiencies. This strategy is extended in Hausch and Ziemba (1985[1],1990a[1],1990b[1]),

Part 6 presents studies on efficiency of exotic markets. Ali (1979)[1] compares returns from a double bet with a equivalent parlay, a double bet that is self-constructed by betting to win on two consecutive races, and concludes that the two bets are "equally priced." Asch and Quandt (1987)[1] and Lo and Busche (1994)[1], however, find that the double is more profitable than the parlay. Hausch, Lo and Ziemba (1994)[1] present methods to fairly price and thus detect mispricings of various exotic wagers. Post position bias is investigated by Canfield, Fauman and Ziemba (1987)[1] and Betton (1994)[1]. Their results indicate that such a bias does exist but that, in most circumstances, it is fully accounted for in the public's odds. Other studies on exacta and quinella bets appear in Asch and Quandt (1987)[1] and Kanto and Rosenqvist (1994)[1].

Racetrack betting in a number of the countries of the Commonwealth and in Asia are discussed in Part 7. Many of these countries use a fixed-odds system rather than a parimutuel betting system. A fixed-odds system is run by bookmakers who offer odds that, upon purchase, become fixed (unlike the parimutuel betting system). While the bookmakers can change their odds at any time, the new odds apply only to new wagers. By adjusting their odds, bookmakers attempt to balance their books so as to ensure themselves of a profit regardless of which horse wins. The papers in this section show that the favorite-longshot bias is present even in fixed-odds systems, i.e., the bookmakers' odds, in order to balance their books, reflect the public's tastes for the favorite/longshot bias.

---

[1] included in this volume
[2] cited in the Annotated Bibliography

PART I

PSYCHOLOGICAL STUDIES

Introduction to Psychological Studies

Donald B. Hausch, Victor S.Y. Lo and William T. Ziemba

Just as the racetrack market has many features that make it an interesting setting in which to study market efficiency, it likewise provides an excellent forum for a psychological investigation of investor risk behavior. As Metzger (1985)[1] notes, the racetrack offers at least four advantages over laboratory studies: 1) data is plentiful and there is considerable variation in locale, track size, race characteristics, etc.; 2) the level of dollars wagered are of economic significance to most bettors; 3) a vast amount of information is available to bettors from a variety of sources; and 4) there are several betting possibilities — straight and exotic wagering — in each race. A downside of the racetrack forum is that the odds aggregate bettors' decisions, and so individual actions are usually indiscernable; however, numerous aggregate effects have been documented that a psychological perspective provides some insights to. Foremost among these is the strong and stable favorite-longshot bias in the win market, first documented by psychologists Griffith (1949)[1] and McGlothlin (1956)[1].

Griffith (1949)[1] used the public's win odds to express, reciprocally, a psychological (subjective) probability, and used the percentage of winners in any odds category as a measure of objective probability. Taking a broad perspective on his analysis, he felt that "any consistent discrepancy between the two [probability estimates] may cast light not only on the specific topics of horse-racing betting and gambling but on the more general field of the psychology of probabilities" (pg. 290). He concluded that while final odds are, on average, accurate measures of winning, short-odds horses are systematically undervalued and long-odds horses are systematically overvalued, i.e., a favorite-longshot bias exists. Griffith (1961)[1] confirmed his earlier work on the underbetting of favorites using show wagering data on favorites; underbetting was apparent, even to the point of overcoming the track take for those with odds shorter than 1.4 to 1. Ziemba and Hausch (1986)[2] summarized additional evidence of these extreme favorites and find that at odds of 3-10 or less there are small profits.

McGlothlin (1956)[1] analyzed a larger horse-racing data set from different racetracks in California and New York states. Like Griffith, he classified his win bet data by odds and observed the favorite-longshot bias. However, he also considered results on a race-by-race basis over the racing day, finding interesting departures from the usual bias for the last two races of the day. The seventh race, the penultimate one, is usually the feature race of the day and involves horses that have received far more coverage and scrutiny than the average thoroughbred, and this is particularly so for the favorites in the race. McGlothlin observed little underbetting of these favorites. For the eighth race, the last event of the day, McGlothin calculated large and significantly positive expected returns for short-odds horses, atypically low returns for mid-range odds horses, and a small but positive expected profit for the longest-odds horses. Since the average bettor loses over the course of a racing day and the eighth race is the last opportunity to wager, an explanation of these returns is that some losers use end-game strategies that attempt to recover the day's losses, allowing them to return home a winner. Such a story leads to underbetting of favorites because of their low returns, overbetting of mid-range horses as they have adequate returns together with a reasonable probability of success, and perhaps underbetting of extreme longshots as they have a low probability of success. McGlothlin argued that his result cannot be accounted for within the usual model of subjective probability and utility maximization. Ali (1977)[1], however, finds some explanatory success with a risk-preference utility function.

Using win odds data, Metzger (1985)[1] tested other psychological hypotheses:

(i)   Gambler's fallacy: She found support for the notion that a series of wins by favorites (longshots)

7

makes less (more) favorable a wager on a favorite, which leads to underbetting (overbetting) of favorites.

(ii)   Investigation of framing of outcomes: She found a loose pattern of increased underbetting on the favorite as the racing day progresses. The importance of framing is discussed in Kahneman and Tversky (1984)[2].

(iii)   Illusion of validity for post-position bias: Classifying the odds according to number of starters and inside or outside post, she showed that the public was insensitive to contingencies, overvaluing the inside posts in accord with the expert advice.

Thaler and Ziemba (1988)[1] suggested the concept of mental accounting (Kahneman & Tversky (1984)[2]) as another psychological theory that may improve positive models of racetrack betting and other gambling behavior. Its main assumption is that bettors adopt mental accounts and act as if the funds in these accounts are not fungible. For example, suppose bettors A and B have the same betting behavior. If A just lost her $200 bet on the last horserace while B just read in the newspaper's financial section about a $200 loss in his stock portfolio, then the concept of mental accounting predicts different wagering behavior because B's loss is not related to racetrack betting.

[1] included in this volume
[2] cited in the Annotated Bibliography

ODDS ADJUSTMENTS BY AMERICAN
HORSE-RACE BETTORS

By R. M. GRIFFITH, University of Kentucky

In horse-race betting, the odds on the various horses in any race are a functioning of the proportion of the total money that is bet on each and hence are socially determined. On the other hand, the objective probability for winners from any group of horses is given a posteriori by the percentage of winners. Thus the odds express (reciprocally) a psychological probability while the percentage of winners at any odds group measures the true probability; any consistent discrepancy between the two may cast light not only on the specific topics of horse-race betting and gambling but on the more general field of the psychology of probabilities.

Becknell noted with her idealogies of probability that in many situations not the probability but the judgment of the probability is the critical variable.[1] Goodfellow[2] and Fernberger[3] have experimentally shown that such ideologies may influence ESP results. More recently Preston and Baratta have demonstrated a systematized under valuation of large probabilities and an overvaluation of small, with the indifference point falling near the geometric mean of their series in an experimental situation and have discussed certain parallels in adaptation-levels.[4] This study may help to determine the universality of their effect by submitting it to a complex, non-laboratory test.

Betting on horse-races offers per se specific but non-trivial psychological problems. Undoubtedly the most common form of gambling, it has assumed such proportions as to be a major socio-economic problem and to constitute large portions of the revenue of condoning states.[5] Gambling has been analyzed for knowledge of psychological and economic processes,[6] and pathological gambling has been clinically studied.[7] The emotionally rich background of fate and money makes this a promising field for observing variety in personality.

---

* Accepted for publication October 5, 1948. The author is indebted to Mr J A Estes, editor of *The Blood Horse*, for source material and valuable suggestions.

[1] E. A. Becknell, Probability: A function of ideology, this JOURNAL, 53, 1941, 604-609.

[2] L. D. Goodfellow, A psychological interpretation of the results of the Zenith radio experiments in telepathy, *J. Exper. Psychol.*, 23, 1938, 601-632.

[3] S. W. Fernberger, 'Extra-sensory perceptions' or instructions?, *ibid.*, 22, 1941, 602-607.

[4] M. G. Preston and P. Baratta, An experimental study of the auction-value of an uncertain outcome, this JOURNAL, 61, 1948, 183-193.

[5] See W. Steigleman, *Horseracing*, 1947, 27. In 1946 over $1,700,000,000 were bet in the pari-mutuel machines. Of this amount, nineteen states retained $80,000,000. Several times this amount is bet with illegal bookies.

[6] E. M. Riddle, Aggressive behavior in a small social group, *Arch. Psychol.*, 12, 1925 (no 78), 1-196; J. von Neuman and O. Morgenstern, *Theory of Games and Economic Behavior*, 1944, 1-641.

[7] O. Fenichel, *The Psychoanalytic Theory of Neurosis*, 1945, 372 f.

In any investigation of gambling, insight and direction may be gained from such clinical and descriptive works as *The Gambler* by Dostoevski. The briefest perusal reveals that not all forms of gambling are psychologically equivalent. Games of chance may be placed in a hierarchy of skills. The conative predictions of the fall of dice contrast with the cognitive factors involved in skillful poker-playing. Many people bet on horse-races because of an—be it as it may—uncritical, egotistical rationality as to their ability to select winners through handicapping procedures. In so far as a cognitive function is involved in any situation one may question the psychological applicability of the term 'gambling.' The most tentative list of ideologies involved in horse-race betting would, however, touch upon fundamental philosophies of probability and mathematical expectation, and few of the systems of betting neglect to give some consideration to the judgment of the odds of the selections; some employ them exclusively, as, for example, when a system permits only the favorites at odds exceeding a fixed value to be selected.

The pari-mutuel system employed at the race-tracks requires explanation. All the money bet on prospective winners of a race is pooled and a percentage (ranging from 7 to 15) known as the 'take' is deducted as revenue for the track and for state or local government. The holders of the winning tickets receive the remainder. The odds on any horse shown on the totalizator at the track are computed by subtracting the sum bet on the horse from the pool after take and dividing by the sum bet on the horse. Thus for each dollar the amount actually returned is the odds plus one, the one being the dollar bet. To preserve the relationship of the mathematical expectation as being the amount wagered divided by the odds, where the odds are the reciprocal of the probability, this value of the track-odds-plus-one will be referred to as *the odds* throughout this paper.

When betting is begun on a race, experts' opinions of the true odds on each horse are shown and the totalizator keeps the bettors informed, at about 90-second intervals, until the race starts, as to the correct odds calculated approximately from the actual money bet. No pennies, and in some states, no nickels are returned on a two-dollar bet, the stray change going into another track or state fund. In the parlance of the race-course this is termed 'breakage' to the nickel or dime.[1]

In this study, the winners and number of horses at each odds were tallied for 519 races of the 1947 spring meets of Churchill Downs, Belmont and Hialeah and for selected odds-groups (one to eleven and sixteen to twenty-one) for an additional 867 races of all tracks for the first eight days of August 1947.[2] All data were proportionately weighted

---

[1] For a more complete description of the pari-mutuel system see Steigleman, *op. cit.* 17-27.

[2] *Daily Racing Form Chart Book*, 53, 1947, nos. 1, 3, 4, and 11.

as to have arisen from 1386 races, that for odds groups 11-16 and above 21 being multiplied by the ratio 1386/519.

If the psychological odds equaled the a posteriori given by the reciprocal of the percentage winners, the product of the number of winners and their odds would equal the number of entries at each odds-group after correction of the odds had been made for loss due to breakage and take. If the product exceeds the number of entries, the psychological odds were too large: if the product is less than the number of entries, the odds were too small. Breakage, causing a larger percentage loss at the small odds than at the large, was adjusted-for by assuming continuous odds rather than odds falling into five-cent intervals. The average take, proportionate at all the odds, was computed from the data as being 13%.[10]

The arithmetic mean of the geometric means of the odds posted at the track for each of a sample of 42 races was obtained as 10.5; the theoretical minimum for the geometric mean of the track odds in a single race with nine horses—the average-size field in this study—is approximately 7.8. and both values are, of course, larger if track-odds-plus-one are used.

Results. The results are best shown by Fig. 1. Here it may be seen that the curve formed by the product of the number of winners times the actual odds falls below the curve for the number of entries throughout the series. When, however, the odds are corrected for breakage and take. the product consistently exceeds the number of entries at the short odds and is less than the entries at the longer odds. The point of equality by linear interpolation is 6.1, or at a probability value of 0.16. This indif-ference-point in track odds (to which the bettors react) is 5.1, with a reciprocal of 0.20. Thus *below* 0.16 or 0.20 the translated psychological probabilities are too large; *above,* too small. The reciprocal of the average geometric mean of the track odds of a sample of races is 0.095, and the maximum for any one nine-horse race is 0.13.

Discussion. All the factors, rational and irrational, which enter into the determination of the odds on horses tend to operate in such a manner as properly to measure the horses' chances. Errors in separate events cancel in the average so that the psychological odds almost coincide with the a posteriori odds. The near congruence of the curves of Fig. 1 attests to the skill of the handicappers and patrons at handicapping the horses.

---

[10] In any race it may be shown that the relationship exists between the odds and the take of $1/O_1 + 1/O_2 + 1/O_3 \ldots + 1/O_n = 1/(1 - y)$ where $O_1, O_2, O_3, \ldots O_n$ are the track-odds-plus-one for n horses corrected for breakage, and y is the fractional take. The average value for y for several races follows from summing both sides of the equation.

Just as necessary and psychologically as remarkable, the money wagered at each odds is such as to preserve the relationship. The final odds are the summative result of both the skill in estimating chances, *i.e.* adjusting the odds, and also the proper proportion of the patrons wagering the correct amounts at each odds to preserve them. That the number of bettors preferring each odds-group and the amount they bet are in the

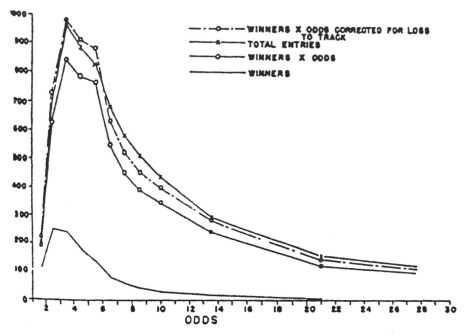

FIG. 1. NUMBER OF ENTRIES, WINNERS, AND WINNERS-TIMES-ODDS FOR EVERY ODDS-GROUP

correct proportion may be the result of several ideologies or to one group with the conscious system of betting on horses on which the track odds become more than the handicappers'.

By the nature of the odds, a greater amount is wagered on the short-odded horses than the long-odded; though the amount is greater, it is not great enough, and too much is wagered on long-odded horses. A direct comparison with the results of Preston and Baratta is made difficult by the complexity of the situation. At the same time, whether the effect is regarded as a constant error in the judgment of odds or as a preference of odds, the effect may be considered the same as Preston and Baratta's, with even the point of indifference being approximately the same. That the principle holds in this non-laboratory situation suggests its universality.

The fundamental difference of dealing with odds instead of probabilities is just one feature which made the outcome less predictable. That the point of equality should be approximately equal to that of these writers (cf. footnote p. 292) when it is translated into probability-values calls for further consideration. The nearness of this value to theirs, whereas any measure of the geometric mean is widely divergent, suggests a constancy of the point of indifference which may be relatively independent of the geometric mean.[11]

SUMMARY AND CONCLUSIONS

(1) The socially determined odds on horses in races are, on the average, correct reflections of the horses' chances.

(2) There is, however, a systematic undervaluation of the chances of short-odded horses and overvaluation of those of long-odded horses.

(3) The indifference-point occurs at odds of 6.1 or of 5.1 when track odds are used.

(4) When these values for the indifference-point are reciprocally translated into probabilities, they correspond to that found in a recent experimental work as between 0.05 and 0.25.

(5) Contrary to the study cited as relating this principle to adaptation-level theory, it is suggested that this indifference-point may not be related simply to the geometric mean of the choices.

---

[11] Since this manuscript was submitted the statistical work was repeated upon all the races of August 1934. While our primary aim was not to corroborate the results of the original investigation, that end was also served. The results were nearly as identical as could have been expected had they arisen from the same data; the curves were practically congruent and the indifference point fell at 0.18 compared to 0.16. If the theory of adaptation-level applies to this error in judgment, the amount of money the bettors have (Preston and Baratta's "endowment background") should influence the indifference-point. It was assumed that in 1934 the bettors were in a state of impecunity relative to their lush affluency of 1947 and the study was made to determine any difference in their adjustment of odds. That the point proved invariant under these diverse economic conditions lends weight to the suggestion in the original study of an unexpected constancy of the point of indifference.

STABILITY OF CHOICES AMONG UNCERTAIN ALTERNATIVES

By William H. McGlothlin, The Rand Corporation

Recently there has been an increasing interest in the theory of games and decision making, with the development of various models and strategies for determining choices. When decisions are made among alternative offers whose outcomes are unknown at the time of the choice, as in a gambling game, three main variables are involved. These are: (1) the amount wagered or risked, x; (2) the size or value of the prize, y; and (3) the objective probability of a successful outcome, P. If it is hypothesized that the individual's best strategy is to maximize the expected value, E, of his choices, it becomes a simple matter to predict uncertain decisions. Expectation, E, is defined as the summation of the products of all possible outcomes and the probability attached to each. Losses are treated as negative outcomes.

Human behavior often does not follow the above strategy, however, as in the case of buying insurance, or in accepting a gamble in which the expectation is negative. It may be assumed that the individual reacts to the psychological counterparts of x, y, and P; these are: utility of bet, $U(x)$; utility of prize, $U(y)$; and subjective probability, P'. Edwards and others have hypothesized that choices between uncertain alternatives can be predicted on the basis of maximization of 'subjectively expected utility,' SEU.[1]

$$SEU = \Sigma P'_i \, U_i$$

where U_i represents the utility of the ith possible outcome of the bet and P'_i represents the subjective probability of the outcome.

There have been some experimental attempts to measure utilities and subjective probabilities as functions of the corresponding objective scales. The validity of these functions is often in question, for it is usually nescessary to assume objective probabilities, or linear functions thereof, in order to find a subjective utility function, and to make similar assumptions about utility when deriving subjective probability. For instance, suppose that 75% of the time subjects (Ss) preferred

* Received for publication June 14, 1955. This paper is from a dissertation submitted in partial fullfillment of the requirements for the degree of Doctor of Philosophy at the University of Southern California.

[1] Ward Edwards, Probability-preferences in gambling, this JOURNAL, 66, 1953, 349-364.

a probability of 0.2 of winning $4.00 to a probability of 0.4 of winning $2.00. From these results we could conclude one of the following: (1) psychologically, an objective probability of 0.4 is less than twice the objective probability of 0.2; (2) the utility of $4.00 is more than twice the utility of $2.00; (3) neither subjective probability nor the utility of money, as valued by S, can be expressed on the corresponding objective scales.

Instead of presenting S with alternatives yielding equal expected values as in the above example, we may design the experiment such that choices are made between events yielding unequal expected values. By using some sort of competitive bidding, or by means of an information processing center such as a pari-mutuel betting machine, it is possible to have S himself establish the expected values of various combinations of probabilities and prizes. Negative expected values for a particular bet, probability, and prize would indicate at least one of the following: (1) $P' > P$; (2) $U(\$x) < x$ utiles; $U(\$y) > y$ utiles; where utiles refers to the unit of measurement for utility. Positive expected values for a given set of values would indicate the reverse relationship; i.e. (1) $P' < P$; (2) $U(\$x) > x$ utiles; (3) $U(\$y) < y$ utiles.[2]

PROBLEM

The present study assumes there is sufficient comparability among individuals to make an investigation of group risk-taking behavior meaningful. It is a statistical study of 9605 thoroughbred horse-races, mostly from California tracks, during the years 1947-1953. The primary purpose is to examine the stability of risk-taking behavior over a series of events. One approach to this question is to determine the expected values of constant-size wagers for a range of probabilities of success P. This yields an E-vs. P pattern and can be repeated for a series of risk-taking events, i.e. races. The stability of this pattern throughout the racing day allows some inferences to be made about the stability of subjective probability and utility for wager and prize over a series of events. It is also possible to obtain information about stability of risk-taking behavior that is independent of the expected-value. Variability of size of average wager over a series of events can be determined as well as preferences among wagers having equal expectations (Es) but different probabilities of success. Some limited information is available concerning differential preferences between winning and losing bettors.

PROCEDURE

Pari-mutuel wagering. Betting on horse-races is quite different from most other forms of gambling. The betting population establishes the odds, or the amount of money each horse in the race will return if successful. At a race track there is a

---

[2] For a more complete discussion of the problems involved in this type of model construction, see R. M. Thrall, C. H. Coombs, and R. L. Davis (eds.), *Decision Processes*, 1954, 255-285.

totalizator board on which the current odds-to-win on each horse appear. The odds are recalculated and flashed on the board at about 45-sec. intervals, keeping the public informed as to the amount of backing each horse has received up to that time. It is not until the final bet is made that the exact odds on each horse are established.

Betting may be for place or show in addition to win. If the horse finishes first or second the place ticket is redeemable; for the show ticket, the horse may finish first, second, or third. The amount returned on these tickets is independent of the position the horse occupies at the finish.

Win. The winning odds posted by the track are given by the formula: $a_i = [(1 - t) \cdot \Sigma A - A_i]/A_i$, where a_i = odds that the ith horse will finish first; t = proportion track takes;[*] ΣA = amount bet in the win pool on all horses for the race being considered; and A_i = amount bet on the ith horse to win.

The odds found in this manner are rounded downward to the nearest multiple of 5¢ (10¢ in some states). The odd cents so deducted are called breakage. The winning ticket pays an amount that includes these odds plus the original bet.

Place. Place odds at the track are determined by the following formula: $b_1 = [(1 - t) \cdot \Sigma B - (B_1 + B_2)]/2B_1$, where b_1 = place odds for horse finishing in the first position; ΣB = amount bet in the place pool on all horses for the race being considered; and B_1, B_2 = amount bet that the horses finishing in the first and second positions will place. The place odds for the horse finishing in the second position, b_2 are found by replacing B_1 with B_2 in the denominator of the above formula.

Show. Show odds are determined as follows: $c_1 = [(1 - t) \cdot \Sigma C - (C_1 + C_2 + C_3)]/3C_1$, where c_1 = show odds for horse finishing in the first position; ΣC = amount bet in the show pool on all horses for the race being considered; and C_1, C_2, C_3 = amount bet on the horses finishing in the first, second and third positions to show. The show odds for the horses finishing in the second and third positions, c_2 and c_3, are found by replacing C_1 with C_2 and C_3, respectively, in the denominator of the above formula.

It is apparent from the last two formulas that the place and show odds are dependent not only on the amount of money bet on the individual horse in these categories, but also on the amount bet on the other horses that appear in the numerator. While winning odds are calculated and reported on all horses in a race whether they win or not, it is practical to report the place odds only for the first two horses, and the show odds for the first three.

Range of odds. The range of odds established on the horses in any given race depends primarily on how closely the horses are matched in ability. In a typical race of 9 or 10 horses the odds-to-win range from around 2–1 on the public favorite to around 50–1 on the horse receiving the least public backing. The place odds typically range from about 1-1 to 20-1, while the show odds range from about 0.5–1 to 6–1. Thus, in a 9 horse race there are 27 possible bets with a typical range of 0.5–1 to 50–1. In the case of win-betting, good approximations of these odds are available to the bettor at the time he makes his choice. As explained above, accurate estimates of place and show odds are not available at the time of the decision making, although it is virtually certain that the place and show odds will be con-

---

[*] Track take varies from 10 to 15% in the 23 states permitting pari-mutuel wagering on thoroughbred racing. In California the figure is 13%.

siderably lower than the winning odds appearing on the totalizator board for a given horse.

Data.[4] The data used in this study were obtained from the *Daily Racing Form Chart Book* and are described in Table I.[5] The main sample consists of 1156 days or 9248 races. In view of the fact that some of the most interesting results were found in the data for the eighth race, an additional sample of 357 eighth races was analyzed to increase the reliability of the results. Whenever the eighth-race data were combined with data for other races, they were given a weight of 1156/1513.

Some tracks schedule an additional race on Saturdays, giving a total of nine races.

TABLE I
SOURCES OF DATA

Track	Years	Number of racing days
Hollywood Park, California	1947–1953	345
Santa Anita, California	1947–1953	348
Tanforan, California	1947, 1949–1951, 1953	212
Golden Gate Park, California	1947, 1949–1952	210
Bay Meadows, California	1951	41
Bay Meadows, California	1947–1950	168*
Jamaica, New York	1950	60*
Aqueduct, New York	1950	38*
Belmont Park, New York	1950	52*
Empire City, New York	1950	5*
Saratoga, New York	1950	34*
	Total	1513

* Only eighth races included.

To combine these races with the remainder of the data, the fifth race was omitted and the sixth race was used in place of the fifth-race data, and so on.

The study is of a statistical nature, and as such, lacks many of the controls found in the experimental laboratory setting. The population of bettors is not stable throughout the racing day due to late arrivals, early departures, and the fact that many bettors do not wager on every race. The amount bet by different individuals varies in an uncontrolled manner, such that persons wagering large amounts determine the size of the odds to a greater extent than do smaller bettors. Also, there is no direct way of studying differential behavior among those persons receiving reinforcement in the form of successful bets and those losing. Finally, the results found are strictly applicable only to the population from which they were derived, *i.e.* the horse-race betting public. The extent to which the results agree with other studies of this type gives some indication of their generality.

---

[4] The author wishes to express his appreciation to the public relations staff of Hollywood Park for the use of their records and office space during this study. Bill Haney, John Maluvius, James Sinnott, and Al Wesson were especially cooperative and patient.
[5] *The Daily Racing Form Chart Book*, Vols. 53-59, 1947-1953, Triangle Publication Inc., Los Angeles.

Treatment of data. The data have been handled in virtually the same manner as a similar study made by Griffith in 1948. He used data from 1386 races and divided the horses into 11 groups according to the odds established on each horse in the pari-mutuel wagering. By checking the outcome of these races, the true or objective probability (*P* = winners/entries) was found for each odds-group, and these odds were compared with the subjectively established public odds. In the present study, the total sample has been broken down into eight subsamples depending on the order of the race in the daily program. Each horse whose track odds to win (all are given as odds to one dollar) fell between 0.05 and 25.95 was placed in one of nine groups. The class intervals for the odds-groups were: 0.05–1.95; 2.00–2.95; 3.00–3.95; 4.00–4.95; 5.00–5.95; 6.00–7.95; 8.00–10.95; 11.00–15.95; and 16.00–25.95. Odds of greater than 25.95 were not recorded because the results would not have been sufficiently stable to be of use in the analysis.

The objective probability, *P*, that a horse in a particular odds-group will win the race is W/N, where W is the number of winning horses in the odds group, and N is the number of entries in that group. In discussing the expected value of bets for the various odds-groups, it is more convenient to use the expectation, *E*, found from the actual ratio of the amounts of money wagered, *i.e.* from the odds that would have prevailed had not the track take and breakage been deducted.[1] When odds are treated in this manner, positive, zero, and negative values of *E* have their conventional meaning. Expected value for a $1 bet to win in a particular track odds group is: $E = P \cdot a^* + (1 - P) \cdot -1$ where $a^* =$ mean corrected winning odds for a particular track odds group.

Reliability of the data. The approximate standard error of *E* is $\sigma_p (a^* + 1)$ where σ_p is the standard error of *P*. The standard error of a^* for a particular odds-group is so small compared to σ_p that it can be ignored. Because of the skewness of the sampling distribution for *P* at the extremes, it is usually not permissible to interpret the standard error of a proportion when *P* is as small as some of those appearing in this study, *i.e.* 0.05. The size of *N* in the present case is, however, very large (500–10,000), and under such conditions the standard error of *P* is applicable as a measure of reliability.

RESULTS

Expected values for entire sample. In Table II and Fig. 1, the expected values are given as functions of odds for the total sample of 9248 races.[3]

---

[1] R. M. Griffith, Odds adjustments by American horse-race bettors, this JOURNAL, 62, 1949, 290-294.
[2] Track odds may be corrected for track take and breakage as follows: $a_c = [(a + 1.025)/(1 - t)] - 1$, where $a_c =$ corrected odds, $a =$ track odds, and $t =$ track take. In the above equation 1.025 represents the original $1.00 bet plus the correction for breakage. With the exception of Fig. 1, expected value, *E*, always refers to values computed from *corrected* odds. The odds-groups, however, are stated in track odds.
[3] Some examples of the raw data from which the entries in Table II were computed may be helpful to the reader. In the 1156 races which occupied the first position in the day's program, there were 771 horses whose track odds-to-win were 0.05-1 to 1.95-1. Of these, 320 won and returned an average of $1.24 for each

→

TABLE II

EXPECTED VALUES OF ONE DOLLAR BETS AS A FUNCTION OF TRACK ODDS

Track odds

Position of Race	Number of Races	0.05–1.95	2.00–2.95	3.00–3.95	4.00–4.95	5.00–5.95	6.00–7.95	8.00–10.95	11.00–15.95	16.00–25.95
1	1116	.08	.04	.05	−.11	−.06	.04	−.10	.12	−.11
2	1116	.14	.13	−.05	.02	.08	−.06	−.21	−.08	−.05
3	1116	.05	.09	.06	.08	−.07	.01	−.13	−.13	−.07
4	1116	.05	.10	.04	.13	−.05	−.13	−.06	−.05	−.10
5	1116	.03	.10	−.02	−.07	.02	−.07	−.02	−.02	−.14
6	1116	.11	.03	−.01	−.05	−.06	−.07	.01	−.05	−.30
7	1116	.01	.00	.00	.08	.03	.14	−.17	−.13	−.32
8	1513	.22	.11	−.09	.04	.00	−.11	−.15	−.21	.03
1–8	9248	.08	.08	−.01	.01	.00	−.03	−.11	−.07	−.11
$(\sigma E)_{1-7}$		.048	.064	.077	.091	.108	.086	.095	.114	.131
$(\sigma E)_8$		.013	.016	.057	.069	.080	.073	.071	.082	.131
(σK) Total		.017	.022	.026	.032	.038	.031	.033	.039	.044

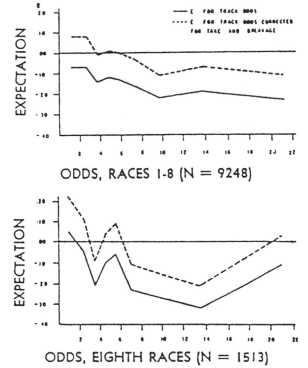

FIG. 1. EXPECTED VALUES OF ONE DOLLAR BETS AS A FUNCTION OF ODDS.

dollar bet, in addition to the original wager. The mean odds δ (a^*) that would have obtained had the track take and breakage not been deducted are: [(1.24 + 1.025)/0.87] − 1 = 1.60. From this, E = 0.415(1.60) + 0.585(−1) = 0.08. In the total sample of 9248 races over all eight positions in the racing program, there were 8781 horses entered at odds of 8.00–1 to 10.95–1. Of these, 661 won at an average odds of 9.32–1. The expected value is then −0.11.

A horizontal line is drawn on the graph at an expectation, E, of zero. Odds below 3–1 show an E of 0.08, odds of 3–1 to 6–1 have an E approximately equal to zero, and odds greater than 8–1 have an E of about −0.10. The E of 0.08 exceeds 0.00 by four standard errors; the difference between the obtained E of −0.10 and that of 0.00 is significant beyond the 5% level of confidence. The indifference point, where the graph of E is equal to 0.00, is located between odds of 3.5–1 and 5.5–1, or at the probability value of 0.15 to 0.22. This agrees well with Griffith's findings of 0.16 and 0.18 in two similar analyses.[*]

Expected values for subsamples. The primary purpose of this study was to investigate what change, if any, took place in the expected values-vs.-odds relationship as the racing day proceeded. To investigate any trend that might exist, the 9248 races were broken down into eight samples of 1156 races, depending on the position the race occupied in the daily program. Table II gives the expectations (E) for each of the eight races, and standard errors for E of each odds-group. In general, the pattern of positive E for low-odds horses and negative E for the higher odds holds for the subsamples, as it does for the total sample. The first six races all yield E-vs-odds patterns that do not differ from the pattern for the total sample by more than the sampling error.

The group of seventh races exhibits several interesting features. First, the E for odds of less than 3–1, which has been positive for all the other subsamples, is found to be roughly equal to zero. Secondly, the E for odds of 7–1 is much larger (0.19) than usual, being significantly greater (beyond the 5% level of confidence) than the corresponding value for the total sample. The third feature is the very low E (−0.32) for odds of around 21–1. This is the lowest E for any of the samples. While this pattern is quite different from those of the remainder of the data, the differences are probably not related to betting behavior on previous races. Neither of the adjacent races (sixth and eighth) shows a similar pattern. The explanation of the E-vs. odds pattern for the seventh race apparently lies in its uniqueness as the feature race of the day. The relatively low E for the low odds may be due to the increased familiarity of the public with two or three of the favorite horses in this race. These horses usually have impressive records and are highly publicized in the local newspapers. Other horses are seldom mentioned outside of charts showing entries and results.

The group of eighth races gives the most interesting results of all the subsamples. The graph of E-vs.-odds for the eighth races, in the lower half of Fig. 1, shows two significant features. First, and most outstanding, is the E of 0.22 for odds below 2–1. This is significantly above 0.00 beyond the 1/10% level of confidence and above the corresponding value for the first seven races (0.07) beyond the 2% level. The second feature of the eighth-race-graph is the sharp dip in expectation, E, at odds of 3.5–1. The E is −0.09 compared to 0.01 for the first seven races, being not quite significant at the 5% level of confidence.

In view of the relatively high E for odds of below 2–1 in the eighth races, an effort was made to investigate further this odds-group. As was explained earlier,

[*] Griffith, *op. cit.*, 290-294.

it was not possible to obtain the odds to place and show established on each horse as was the case in the win category. We may, however, categorize the data on the basis of *winning* odds as before, and then list the number of horses that finished first or second, *i.e.* placed, and the number that finished third or better, *i.e.* showed. The place and show odds are available for these horses and can be tabulated as before. The E of a place wager on horses whose odds-to-win were below 2–1 was found to be 0.24, or 6.5 standard errors above 0.00.

Unlike the seventh race, the eighth race is not unique in type. It is almost always very similar in make-up to three or four of the earlier races. The fact that the E-vs.-odds graph for the eighth race is quite different from that for the first seven races must be explained by a change in betting behavior, and this change is due to the position of the race in the daily program rather than the composition of the race. Horses with a high probability of winning, but with accompanying low pay-offs, become even more unpopular with the bettors in the last race.

Amount wagered per person as a function of position in racing meet. During the 1953 Hollywood Park season the average amount bet per person during a racing day was $72.70. The average amount paid to the track in the form of mutuel take then was $9.46 per person exclusive of breakage. These figures represent the mean amount bet. Since the distribution of bets is positively skewed, due to a few large bets, the median is undoubtedly lower than the mean. Probably the former is around $50.00 bet and $6.50 lost. There was a tendency for the amount wagered per person to increase slightly as the seasonal meet proceeded. The average amount bet per person per day for the first 10 days was $67.10 compared to $75.50 for the last 10 days. The weekday average was $76.60 per person as compared to $65.10 per person for Saturdays and holidays.

Relation between amount bet on a race and its position in the daily program. During the racing day the total amount wagered per race ordinarily increases for each succeeding event up to the eighth race. There is usually a slight decline in the total mutuel handle from the seventh to the eighth race. For the 1953 Hollywood Park data the increase from the first to the seventh race is fairly regular, with the amount bet in the latter race being about 1.8 times the amount wagered on the former. Some of this change is due to late arrivals and early departures, but the increase in sizes of wagers is clearly much more than can be accounted for by fluctuations in attendance.

Probability-preferences. Recently, Edwards, has reported several well-designed experiments using college Ss and dealing with probability-preferences in gambling.[10] These studies have held constant the E of bets and determined the preference for different probabilities by means of paired comparisons. Eight bets on a rigged pinball machine were used with probability values of 1/8, 2/8, . . . 8/8. In general, the results of these experiments have shown a definite preference for bets involving the probability of success 4/8, and a definite avoidance of the value 6/8. He found that these preferences were still distinguishable in experiments involving unequal E even though such choices violated the maximization-of-expected-value hypothesis.[11]

[10] Edwards, *op. cit.*, 349-364.

[11] Edwards, Probability-preferences among bets with differing expected values, this JOURNAL, 67, 1954, 56-67.

Furthermore, he found that the above probability-preference remained constant for different levels of expected values, thus demonstrating that the preferences exist independently of the attached utility variable.[11] Finally, Edwards carried out an experiment designed to study 'variance preferences' in gambling.[12] A conservative individual, wishing to minimize the variability of his assets over a series of risk-taking events should choose wagers with small amounts bet and high probability of success. Less conservative individuals may increase the variability of their assets, i.e. gamble on a large win at the expense of risking a large loss, by choosing to wager large amounts at low-probability values. Edwards created special situations in which the best strategy for winning bettors consisted of minimizing the variability of their assets, and the best for losing bettors consisted of maximizing the variability of their assets. The results showed that the same preferences for probabilities that had been found in earlier experiments was still the most important factor in predicting choices. Winning bettors did not change their preferences for low cost or high cost bets as the series of choices proceeded, although it would have been in their interest to do so. Losing bettors did tend to choose high-variability bets when good strategy indicated it; however, choosing bets with high variability was of less importance than the preference for a given range of probability.

The present study presents a measure of preferences for certain probabilities and asset variability in horse-race betting. In pari-mutuel wagering, the bettor may choose among win, place, and show bets. The three pools are independent, such that the amount wagered on a horse in one category has no effect on the odds on the same horse in the other two categories. The amount deducted by the track (13%) is the same for all three pools, although the factor of breakage takes a slightly larger amount from the place and show pools, since there is a higher proportion of redeemable tickets in these categories. As mentioned earlier, the typical ranges of win, place, and show odds are around 2-1 to 50-s, 1-1 to 20-1, and 0.5-1 to 6-1 respectively. Thus, the proportion of the total amount of money wagered in each pool gives a measure of preference for probability ranges among alternatives with roughly equal expected values. This is fairly analogous to Edward's measure of preference for particular probabilities. It should be noted that during a series of races these proportions may change for the group without necessarily effecting a change in the pattern of extended values discussed earlier. Fig. 2 gives the proportion of the total amount bet in the win, place, and show categories as a function of the position of the race in the daily program. The graph for the win pool shows an almost linear increase from 0.49 in the first race to 0.60 in the eighth and last race. The proportions bet in the place and show categories show corresponding decreases.

Risk-taking events and their effect on subsequent betting. While the group of bettors as a whole is always losing money due to the track-take, a proportion of them is winning on any given day. The question of differences in behavior between winners and losers has been raised. A partial answer may be obtained by determining what, if any, relationship exists between the odds that the winning horse pays and the amount of money wagered per person in the following race. If each person is assumed to bet the same amount, the proportion of winning bettors would be:

[11] Edwards, The reliability of probability preferences, this JOURNAL, 67, 1954, 68-95.
[12] Edwards, Variance preferences in gambling, this JOURNAL, 67, 1954, 441-452.

$Q = 0.87/(a_1 + 1.025)$, where $Q =$ the proportion of those persons purchasing win tickets who realize a return; and $a_1 =$ odds-to-win for horse finishing in the first position. Place and show odds were not taken into consideration. Using this measure of the proportion of the population holding successful win tickets, we tested the correlation between these values and the amount bet per person in the following race. The data from the 50-day Hollywood Park racing season were used. Saturdays and holidays were eliminated because it has been shown that these days have a smaller amount bet per person than for weekdays. This left a total of 40 racing days

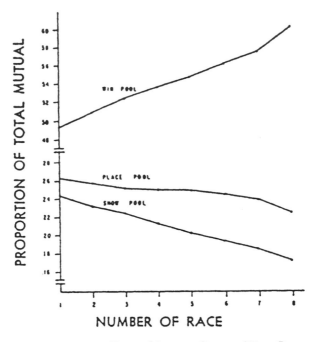

FIG. 2.   PROPORTION OF TOTAL MUTUEL BET IN WIN, PLACE, AND
SHOW POOLS AS A FUNCTION OF NUMBER OF RACE
(Data from the 50-day meet at Hollywood Park, 1953.)

and, since there were eight races a day, seven pairs of variables to be correlated. Before computing correlation coefficients, it was necessary to correct the amounts bet per person in each race for the variance contributed by the position of the racing day in the season, since this variable increases as the season proceeds. The seven coefficients ranged from −0.10 to −0.47, with one being significant beyond the 1% level of confidence, and one beyond the 5% level. Since all seven coefficients were negative, they give rather conclusive evidence that bettors increase the amount wagered more after having lost than following a successful bet.

DISCUSSION

Stability of the E-vs.-odds pattern. The general pattern is one of positive expectation, E, for low-odds wagers, high probability) and negative E for high odds (low probability), with zero E from 3.5–1 to 5.5–1 odds ($P = 0.15$ to 0.22). This pattern appears to have considerable stability inasmuch as it was found with minor variations for the first six subsamples of races. Marked variations of the E-vs.-odds relationship in the seventh race appear to be related to its uniqueness as the feature race. The sharp increase in the E for odds below 2–1 for the eighth race is probably due to certain segments of the population making decisions in accordance with their total financial losses for the day. Bettors apparently refrain from making bets which would not recoup their losses if successful. The increased popularity of odds of around 3.5–1 in the eighth race (indicated by the relatively low E of −0.09) may be due to the fact that a considerable proportion of the population has lost about three times the amount they propose to wager on the last race; however, no evidence was gathered to substantiate this speculation.

In so far as we may generalize to other populations, it appears that subjects can be expected to accept low expected values when low probability-high prize combinations are involved, while demanding higher expected values in the case of high probability-low prize combinations. The central tendency-like effect shows considerable stability over a series of risk-taking events. These findings are consistent with those of Preston and Baratta, who conducted a laboratory experiment on this problem.[14]

Variability of preferences in betting. The betting behavior of the group is such as to increase the variability of their individual assets in an almost linear fashion as the racing day proceeds. This is accomplished by increasing the amount bet per person and by choosing a higher proportion of win-category wagers in preference to place and show betting. This may be partly due to the loss of resources for the group as a whole due to the track take. There is some indication that losing bettors tend to increase the size of their wagers more than do winning bettors.

It is important to note the stability of the E-vs.-odds pattern during the first six races, in spite of the fact that during this same period, size of wagers and preference for low-probability bets (win betting) are steadily increasing. The lack of change in the E-vs.-odds pattern corresponding to an increase in size of wagers would appear to indicate that the utility scale for money in the range considered is virtually the same as the objective dollar scale, and the more important psychological variable is subjective probability. On the other hand, neither is the increasing popularity of win

---

[14] M. G. Preston and Philip Baratta, An experimental study of the auction-value of an uncertain outcome, this JOURNAL, 61, 1948, 183-193.

betting (low *P* of success) during the first six races reflected in the *E*-vs.-odds pattern. If the group's subjective evaluation of low probability-high prize wagers increases over a series of risk-taking decisions, we would expect an intensification of the negative *E* for high-odds and positive *E* for low-odds pattern. This does not occur until the last race. These results suggest the inadequacy of a model for predicting risk-taking decisions based solely on the maximization of 'subjectively expected utility.' Allais has suggested that the variances involved in the wager may also be an important factor in this type of decision making.[18] The results of the present study tend to confirm this prediction. This is not in agreement with Edwards' laboratory findings which indicated that variance-preferences were of minor importance compared to probability-preferences in gambling.[16] Edwards also found probability-preferences to be relatively stable, which is not in agreement with the present finding. Perhaps the discrepancy is due in part to the difference between college and horse-race betting populations.

SUMMARY

By means of a statistical analysis of 9605 horse-races, this study sought to obtain information about the stability of decision-making behavior over a series of risk-taking events. In general, the group tended to accept probability-prize combinations whose expected values were less for low-probability wagers than for high ones. This tendency was relatively stable over a series of decisions, and was for the most part, independent of decreasing group resources, size of average wagers, and change in group probability-preferences. The group behaved in a manner such as to increase the variability of their assets as a series of risk-taking events proceeded. This was accomplished by increasing the size of the wager and choosing lower probabilities (win bets) with accompanying prospective higher returns. There was some indication that losing bettors increased the size of their wagers more than did winning bettors.

The relatively stable *E*-vs.-odds pattern over a series of events in which sizes of wagers and preferences for probability values show consistent changes raises some questions about decision-making models. It was shown that a model making use of subjective probability and utility functions alone does not account for the results found here. In the present study, variance-preferences also play an important role in determining choices among risky alternatives.

---

[18] Maurice Allais, Le comportement de l'homme rational devant le risque: Critique des postulats et exiomes de l'ecole americaine, *Econometrica*, 21, 1953, 503-546.
[16] Edwards, Variance preferences in gambling, 441-452.

A FOOTNOTE ON HORSE RACE BETTING

RICHARD M. GRIFFITH

Veterans Administration Hospital, Lexington, Kentucky

In an early note Griffith (1949) showed that horse race bettors put too much money on horses which have little chance of winning and too little on those most likely to win. McGlothlin (1956) repeated the study, confirming the results and revealing many further potentialities in the data by considering the position of the race in the day's program. An obvious extension of the analysis—one which McGlothlin picked up but brushed over lightly—is to turn from win betting to "show" betting. The person who would like to bet on a surer thing than even the heavy favorite to win the race may bet that he will "place," that is, finish first *or* second, or, what is more likely yet, that he will "show," finish third or better. The tendency to under-bet the most probable event should appear in its most marked degree in show betting.

The reader who does not have a fundamental conception of the mechanics of pari mutuel betting may refer to one of the above references for sources. The bettor pits his skill against that of the crowd, for the pattern of their bets determines the odds. All bettors suffer under the handicap, however, that taxes and the track "take" 13-15 percent of the pool before it is divided among those with winning tickets and the amount is further reduced through "breakage," the loose change of pennies and nickels with which the track can't be bothered and which it therefore keeps. Opinions vary as to whether anyone can overcome such handicaps and win consistently.

In show betting the amounts of money bet into the show pool on the various horses also determines the amount to be returned to those bettors who win. However, the division is more complicated. Whereas the totalisator keeps the patrons informed at 45 second intervals as to the approximate amount to be returned on a horse to win, the pay-off to show cannot be easily computed because it depends on the other two horses which finish with him. After the race is run, the amount bet on each of the first three horses is removed from the total show pool, the pool first having been reduced, of course, by the take. These amounts will return to the winning bettors their original investments. The profit to the bettors on each horse is determined in a debatable way: by splitting what is left in the pool evenly three ways, irrespective of how likely or unlikely each horse had been to qualify. Furthermore, it may be realized, loss through breakage is proportionately greater at the smaller pay-offs. All of which is to say that the show bettor can have no clear notion of how much he stands to win.[1] He knows that he will generally collect less for a show bet on the favorite than on a long shot but he can only have a vague idea of how much less. His betting on horses to show, and particularly on the horses with short odds-to-win to show, can reflect only his desire to bet on the surest thing around. The present note, in the nature of a footnote to the previous paper, is aimed at measuring the extent to which bettors do this.

All horses which went to the post at odds-to-win of less than 2 to 1 (returning less than $3 for $1) for two separate months of American racing, May, 1949 and August, 1960, were considered—some 4543 of them.[2] The number which did not show was tabulated as was the pay-off to show for those which did.

As would be expected from the above discussion, the pay-off to show is quite variable; for instance, in the odds-to-win group of 1.80-1.95 to 1, the profit from a dollar bet on horses which did show ranged from .10 to $1.40. In other groups there were instances in which a horse which won the race paid more on a show ticket than he did to win.

There are two ways to compute the *post facto* "odds-to-show": one, depending on how the horses did, is the percent which did not show; the other, depending on what the bettors did, is the average profit (allowing for the horses which did not show). Fig. 1 presents the results. Below odds-to-win of 1.40 to 1, there is systematic under-betting, even to the extent of overcoming the loss to breakage and take, for which these data were not corrected.[3] The lines cross sharply in the 1.40-1.55 to 1 range due to a dip in profits in this range. Since the point of crossing was consistent between the years (and the dip, also, which may or may not mean something), only the combined results need be shown. A rough statistical test of the reliability of the results was a comparison between the two years of the directions of the differences between the two odds-to-show within each class interval: the binomial probability (7 of the 8 being in the same direction) was less than .05.

---

[1] From the viewpoint of the bettors as a whole, there is such a thing as a good bet at the race course—one which returns more money to them than they put in; oddly enough, under unusual circumstances the track can lose money. The minimum pay-off permitted by law in most states gives 10 cents profit on the dollar, the winning bettor thus being assured at least $1.10 for his $1. With only a few horses entered in a race and with one of them likely to be established as an overwhelming favorite, the track may suspend show betting to protect itself against a negative pool.

[2] As reported in the *Daily Racing Form Chart Book*, 54, No. 5, 1949 and 45 (sic.), No. 8, 1960, Triangle Publications, Inc., Chicago.

[3] *Cf.* McGlothlin, *op. cit.*, and Griffith, *op. cit.*

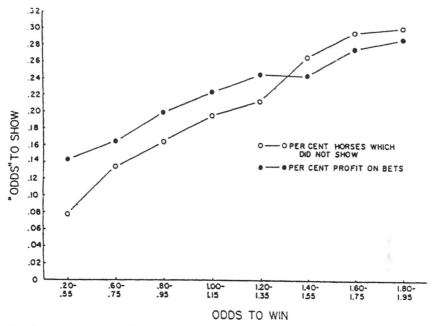

Fig. 1.—*Post facto "odds" to finish first, second, or third in relation to odds to win.*

To interpret the results in other language, a bettor would have shown a net profit had he bet on all horses to show which went to the post at odds-to-win of less than 1.40 to 1 during the periods of this study. Had it been practical to bet on every horse at the proper odds-to-win (546 of them) in May, 1949, he would have netted a profit of 6.15 cents per dollar bet; in August, 1960, his profit would have been 2.4 cents on the dollar (1496 bets).

The difference between the rates of profit in the two years was due to an increase in the number of horses which did not show in 1960. The proportion of losers in the two years differed significantly by the chi square test beyond the .01 level of confidence. The rate of return for those which *did* show rose by 2.6 cents, otherwise there would have been a net loss in 1960. (It would not be prudent with the data at hand to account for these shifts between the two years, accepting them as genuine.) Profits could be improved by other considerations, of course: the careful reader will be able to estimate that McGlothlin (1956, p. 611) found a net profit of around 10 or 11 percent on horses whose odds-to-win were less than 2 to 1 in the eighth race of the day.

Thus, as was to be expected, the tendency, which had been demonstrated with win betting, for horse race bettors to place too little money on the horses most likely to win is magnified in their even more conservative bets on the same horses to show.[4]

Literature Cited

Griffith, R. M. 1949. "Odds adjustments by American horse-race bettors." *Amer. J. Psychol.* 62:290-294.

McGlothlin, W. H. 1956. "Stability of choices among uncertain alternatives." *Amer. J. Psychol.* 69:604-615.

Accepted for publication 26 September 1961.

---

[4] We have hesitated to report the findings on the 1949 races and their confirmation in 1960. When his results are in real life, the scientist must pause to consider that some may think them practical, use or misuse them. Before one plays the fiddle he should know who will dance. (There was a strange malady epidemic in the Middle Ages (tarantism) which compelled the victim to dance, and he could not stop.) We do not court the distinction of demonstrating a method to "beat the races," even though our home in Kentucky is surrounded by the thorobred industry. As a psychologist we are gratified, however, that the discovery came about from a psychological presupposition—that the spirit of gambling is to risk little to gain much.

BIASES IN BETTING:
AN APPLICATION OF LABORATORY FINDINGS

MARY ANN METZGER

University of Maryland Baltimore County

Summary.—Three experimentally established biases away from optimality that result from misconceptions of chance, variations in the framing of outcomes, and the illusion of validity are of detectable dysfunctional consequence in the natural world of the gambler. A study of the public's betting in 12,316 horse races shows that the gambler's fallacy, increasing risk preference within a day, and insensitivity to actual contingencies contradicting expert opinion, principles derived from the results of laboratory experiments, can be meaningfully applied in the practical setting.

Systematic biases away from optimality have been shown in human judgment about quantities relevant to the evaluation of gaming propositions. These effects have been demonstrated in controlled laboratory experiments (Kahneman & Tversky, 1973, 1984; Nisbett, Krantz, Jepson, & Kunda, 1983; Tversky & Kanheman, 1974, 1981, 1983); for surveys see Einhorn and Hogarth (1981) and Kahneman, Slovic, and Tversky (1982). The experimental conditions do not simulate the natural world of the bettor, which leaves open the empirical question of to what extent the laboratory findings are of significant consequence in the natural setting (Hogarth, 1981). This study addressed that question.

Studies of parimutuel betting at racetracks have shown that some of the expected consequences of principles demonstrated in the laboratory are detectable in a natural setting. The natural setting of the parimutuel bettor differs from the laboratory setting in several ways. First, the racetrack is an attractive setting; the extent to which people participate in this activity is indicated by the fact that more than 5000 races are run each month at US Thoroughbred tracks, each with several thousand people in attendance. Second, the amounts wagered at the track are of economic significance to most bettors; the average bettor risks about $150 per day. Third, thousands of items of information on variables of high and low validity for the prediction of race outcomes are available; there are detailed lists of past performances of each runner, advice from experts, opportunities to observe the physical condition and temper of each horse just before the race, and opportunities to consult other bettors both through discussions and through observations of the tote board which is updated every minute with the precise amounts of money bet so far by the public at the track. Fourth, several independent straight or exotic betting pools are available in each race.

The field studies have shown that the public at the racetrack is well-calibrated. When the estimate of the public's subjective probability of winning for a horse in a given odds-range is taken to be the mean proportion of money in the win pool bet on horses of that class, tests have shown that the public is very accurate in the middle odds-range and tends to underestimate high probabilities and overestimate low probabilities (Ali, 1977; Fabricand, 1965; Griffith, 1949; McGlothlin, 1956; Rossett, 1966; Weitzman, 1965). In addition, the public at the track is more accurate than any individual expert or combination of expert selectors (Fabricand, 1965; Figlewski, 1979). Using the win-pool proportions as established measures of the win probabilities and assuming a constant-ratio rule for win:place or win:show probabilities, several investigators have demonstrated identifiable underestimations in the high-probability end of the place and show pools (Bond, 1980; Hausch, Ziemba, & Rubenstein, 1981; Snyder, 1978; Tuckwell, 1981).[2] McGlothlin (1949) found increased underestimations of favorites and overestimation of long shots in race 8, compared to races 1 through 8 combined. Ali (1977) found a similar effect for harness races. Calculating *delta*, a measure of risk attitude, Ali found values indicating increased risk acceptability comparing the first and second to the last race.

The purpose of this investigation was to test whether three laboratory findings are of measurable consequence in the natural setting of the parimutuel bettor. The first finding is that misconceptions of chance lead to the gambler's fallacy (Tversky & Kahneman, 1974). In particular, betting on favorites should be more acceptable after a series of races have been won by long shots than after a series of races won by favorites. The public should, therefore, overestimate the chances of favorites in the former circumstance and underestimate those chances in the latter circumstance. The second is that variations in reference points for the framing of outcomes produce variations in the acceptability of risk (Tversky & Kahneman, 1981). In particular, given that the reference point is the *status quo* at the beginning of the racing day and that the public expectation is negative, outcomes are framed increasingly in terms of getting-even vs loss rather than gain vs loss, producing fewer bets on favorites over the day. The public should increasingly underestimate the chances of favorites in later races. The third is that the illusion of validity (Tversky & Kahneman, 1974) may allow expert opinion to override the contingencies of direct experience. In particular, misleading advice about post-positions from a respected local expert (Beyer, 1978) should lead to dysfunctional biases. The misleading advice is based on the assumption that a track bias against a particular post-position is indicated by a relatively lower percentage of winners from the starters out of that post. This assumption is true only if all races have the same number of starters, which is not the case. In

---

[2]J. R. Ritter, Racetrack betting: an example of a market with efficient arbitrage. (Unpublished manuscript, University of Chicago, 1978)

this context, chance probabilities for runners from post 1 vary from .20 to .08, while those for runners from post 12 are fixed at .08. Based on the advice, inside post positions should be overestimated by the public.

METHOD

Betting patterns and winning probabilities on the public's first and second choices were obtained for 11,313 races and were used to test the first two propositions. These were all races run on nine-race cards at Thoroughbred tracks in the US in the months of May, June, and December of 1978. Similar data with respect to post-positions were obtained for 1003 sprints run on dirt tracks at Pimlico and Bowie for winter and spring of 1979 and 1980 and used to test the third proposition. The data were obtained from the Daily Racing Form monthly chart books. The public estimates of win probabilities (p) for their first (F) and second (S) choices were computed from the dollar odds in the chart using the formula $p = (1 - t) / (c + 1.05)$, where t is the track take proportion, c is the win odds posted by the track for each horse, and .05 is an adjustment for dime breakage. Ties for designation of F or S were decided by a coin flip, all candidates equally likely. Public accuracy (a) of estimation of the win probabilities for runners in a given class was computed using the formula $a = p/P$, where P is the observed proportion of winners for runners in that class.

For the investigation of the gambler's fallacy, a was classified according to the probability range and the outcomes of the two most recent previous races. The class labels are LL, SL, FL, LS, SS, FS, LF, SF, and FF. The left-hand letter indicates that the winner of the pre-previous race was F, S, or some horse (L) other than F or S. The right-hand letter indicates the winner of the immediately previous race. For the investigation of the framing of outcomes, a was classified according to the probability range and the ordinal number of the race. For the third proposition, a was classified according to the number of starters and inside (including center) or outside post.

RESULTS

The results for the first two propositions are summarized in Tables 1 and 2. The rows of both tables are labeled according to the average proportion of the win pool wagered on all runners within a given probability range. The ranges were chosen to correspond to odds ranges commonly displayed at a racetrack with a 15% take. The odds range, for example, in the first row (mean of .51) is greater than or equal to 1/9 and less than 4/5; in the second row it is greater than or equal to 4/5 and less than 9/5.

The results relevant to the evaluation of the occurrence of the gambler's fallacy are summarized in Table 1. Inspection of the table indicates that the public is measurably influenced by recent independent outcomes to over- or underestimate the win probability of their favorites. For the first choice the

TABLE 1

ACCURACY OF PUBLIC ESTIMATIONS OF PROBABILITIES I*

Public Estimate‡		Outcomes of Pre-previous and Previous Race†								
		LL	SL	FL	LS	SS	FS	LF	SF	FF
First Choice										
51 (46—77)		104	98	91	85	91	96	94	86	85
	ƒ	244§	122	189	98	55	94	209	83	145
35 (31—46)		103	109	100	109	92	97	100	97	97
	ƒ	792§	347	561	394	166	242	540	235	469
27 (25—31)		96	108	113	113	108	90	100	87	96
	ƒ	423§	168	265	181	67	124	295	130	187
24 (17—25)		98	93	115	100	159	87	89	106	89
	ƒ	454§	172	288	200	87	114	312	142	201
Second Choice										
25 (22—35)		96	96	93	100	114	119	96	104	108
	ƒ	774§	349	586	383	173	252	557	250	463
19 (17—22)		106	95	127	119	66	90	95	173	100
	ƒ	744§	307	479	331	131	197	494	223	355
15 (14—17)		100	104	99	94	88	88	150	115	89
	ƒ	395§	153	238	159	71	125	305	117	184

*Ratio (× 100) of average public estimate to true probability of winning for runners' in each odds range.   †Example: SL means the pre-previous race was won by the second choice (S) and the immediately previous race was won by a long shot (L).   L is any other than the first (F) or second choices.   ‡Mean percent of win pool wagered on runners in each odds range.   Ranges in parentheses.   §Number of observations in each row below ratios.

four outcomes which indicate a recent win by L and the most recent win by L or S (the first four columns) show 8 overestimations (entries greater than 100) of F, while the outcomes showing a recent win by F and the most recent win by F or S (the last four columns) show only 1 overestimation ($z = 2.13$, $p < .02$).   For the second choice, the analogous comparison was not significant. A further comparison over all odds ranges for the three-race sequence LLL and XXX, where X stands for either F or S, shows that favorites are underestimated 94 and 86, respectively, while second choices are underestimated 91 and over-estimated 105.   The Ns for LLL and XXX are 772 and 1186, respectively.

The results relevant to the evaluation of the proposition that betting on favorites declines over the day are summarized in Table 2.   Inspection shows that races 1 and 9, the first and last races of the day are virtually identical in their betting patterns and different from all other races.   In races 1 and 9, all classes of first choices are underestimated (mean for each race, 94), while all classes except one of second choices are overestimated (means 109 and 106 for races 1 and 9, respectively), and the exception is 100, neither over nor under. The data support an amended proposition omitting race 1 from consideration, since the first choice is underestimated in races 8 and 9 (means for each race,

TABLE 2

ACCURACY OF PUBLIC ESTIMATIONS OF PROBABILITIES II*

Public Estimate‡	Ordinal Number of Race								
	1	2	3	4	5	6	7	8	9
First Choice									
51 (46—77)	96	100	100	94	91	91	91	93	93
	106§	84	206	137	173	164	228	275	56
35 (31—46)	97	97	106	100	109	103	106	90	97
	523§	458	593	526	547	515	595	599	371
27 (25—31)	96	129	100	100	104	117	96	93	93
	279§	327	251	286	269	276	225	190	343
24 (17—25)	86	131	108	96	87	111	112	110	92
	349§	388	207	308	267	302	207	192	487
Second Choice									
25 (22—35)	109	100	100	114	100	96	93	96	104
	372§	379	569	540	563	552	569	664	330
19 (17—22)	106	106	95	100	112	100	119	112	100
	542§	576	464	489	472	457	448	402	531
15 (14—17)	115	97	107	104	87	92	125	92	116
	343§	302	224	228	221	248˙	238	190	396

*Ratio ($\times$ 100) of average public estimate to true probability of winning for runners in each odds range.
‡Mean percent of win pool wagered on runners in each odds range. Ranges in parentheses.
§Number of observations in each row below ratios.

94) compared to races 2 through 7 (over-all mean, 105, range 98 to 116). Once again, the data for the second choice show no significant pattern: the means for races 8 and 9 are 101 and 106, respectively; and the over-all mean for races 2 through 7 is 102 (range 92 to 108).

The results relevant to the third proposition show that the public was insensitive to actual contingencies, overvaluing the inside posts in accord with the expert advice. For the inside, 26 of 33 values of a were overestimations; for the outside, 13 of 30 ($\chi^2 = 6.94, p < .01$).

DISCUSSION

Over-all, the results show that for the three propositions under test here, the biases derived from experiments under artificial conditions that fail to reproduce the most obvious features of the natural world of the gambler, lead nevertheless to detectable consequences in that natural world.

The unexpected finding of identical betting patterns in the first and last races of the day invites speculation, especially since Ali's (1977) *delta* values of .22, .20, and .30 for risk acceptability in the first, second, and last races, respectively, would lead one to expect less similarity. One possibility based on an actual difference between harness and flat racing is that the field sizes are constant in harness racing whereas they vary widely in flat racing. Ali notes that he considered all of his approximately 20,000 races to be eight-horse fields, assuming the effect of the rare larger field to be negligible. In flat racing, however, fields varying from four to 12 are common within a racing day and full fields are usually run in the first and last races to support the exotic betting pools of the daily double and triple (trifecta). It could be that either the field size or the availability of exotic pools determines the betting pattern.

REFERENCES

ALI, M. M. Probability and utility estimates for racetrack bettors. *Journal of Political Economy*, 1977, 85, 803-815.

BEYER, A. *My $50,000 year at the races.* New York: Harcourt Brace, 1978.

BOND, N. A. Parimutuel gambling systems: cross-validation of behavioral and statistical models. Paper presented at the meeting of the Society for Mathematical Psychology, Providence, RI, 1980.

EINHORN, H. J., & HOGARTH, R. M. Behavioral decision theory: processes of judgment and choice. *Annual Review of Psychology*, 1981, 32, 53-88.

FABRICAND, B. P. *Horse sense.* New York: David McKay, 1965.

FIGLEWSKI, S. Subjective information and market efficiency in a betting market. *Journal of Political Economy*, 1979, 87, 75-88.

GRIFFITH, R. M. Odds adjustment by American horse-race bettors. *American Journal of Psychology*, 1949, 62, 290-294.

HAUSCH, D. B., ZIEMBA, W. T., & RUBENSTEIN, M. *Management Science*, 1981, 27, 1435-1452.

HOGARTH, R. M. Beyond discrete biases: functional and dysfunctional aspects of judgmental heuristics. *Psychological Review*, 1981, 90, 197-217.

KAHNEMAN, D., SLOVIC, P., & TVERSKY, A. (Eds.) *Judgment under uncertainty: heuristics and biases.* New York: Cambridge Univer. Press, 1982.

KAHNEMAN, D., & TVERSKY, A. On the psychology of prediction. *Psychological Review*, 1973, 80, 237-251.

KAHNEMAN, D., & TVERSKY, A. Choices, values, and frames. *American Psychologist*, 1984, 39, 341-350.

MCGLOTHLIN, W. H. Stability of choices among uncertain alternatives. *American Journal of Psychology*, 1956, 69, 604-615.

NISBETT, R. E., KRANTZ, D. H., JEPSON, C., & KUNDA, Z. The use of statistical heuristics in everyday inductive reasoning. *Psychological Review*, 1983, 90, 339-363.

ROSSETT, R. N. Gambling and rationality. *Journal of Political Economy*, 1965, 6, 595-607.

SNYDER, W. W. Horse racing: testing the efficient markets model. *Journal of Finance*, 1978, 33, 1109-1118.

TUCKWELL, R. H. Anomalies in the gambling market. *Australian Journal of Statistics*, 1981, 23, 287-295.

TVERSKY, A., & KAHNEMAN, D. Judgment under uncertainty: heuristics and biases. *Science*, 1974, 185, 1124-1131.

TVERSKY, A., & KAHNEMAN, D. The framing of decisions and the psychology of choice. *Science*, 1981, 211, 453-458.

TVERSKY, A., & KAHNEMAN, D. Extensional versus intuitive reasoning: the conjunctive fallacy in probability judgment. *Psychological Review*, 1983, 90, 293-315.

WEITZMAN, M. Utility analysis and group behavior: an empirical study. *Journal of Political Economy*, 1965, 73, 18-26.

Accepted March 27, 1985.

PART II

UTILITY PREFERENCES OF RACETRACK BETTORS

Introduction to Utility Preferences of Racetrack Bettors

Donald B. Hausch, Victor S.Y. Lo and William T. Ziemba

Racetrack betting is a classic example of decision making under uncertainty. The options available to a bettor - they types of wagers possible - are clearly understood. A bettor's probabilities are, of course, only subjective, not objective. Considerable data is available to refine one's probabilities and the parimutuel method aggregates these individual assessments to arrive at the public's assessment which, as was discussed in the previous sections, roughly approximates objective probabilities. The papers in this section attempt to describe the risk-taking behavior of the representative bettor by evaluating this bettor's utility function and by comparing reactions to different wagers that have similar risk characteristics.

A few basics first. A concave utility function offers an explanation of risk averse behavior; it explains the purchase of insurance as a rational act of paying a premium over expected loss to eliminate the risk of the loss. A concave utility function can explain risk-seeking behavior and it provides a rational basis for racetrack betting where bettors opt into a risky situation and do so despite the overall negative expected return. Simultaneous expenditures on insurance and gambling, while very common, is more difficult to explain as rational behavior. One explanation, offered by Friedman and Savage (1948)[2], proposes a utility function that is concave for low wealth, convex for intermediate levels, and concave thereafter, with current wealth being the first point of inflection. While this so-called Friedman-Savage hypothesis can explain gambling and buying insurance, Markowitz(1952)[2] noted several problems and addressed them by appending to their utility function a convex portion for low wealth.

There are many other utility functions proposed in financial literature, e.g. quadratic, exponential, logarithmic, power, etc.. Each has its own properties measured in absolute risk aversion or relative risk aversion. Kallberg and Ziemba (1979)[1] empirically examine the effect of alternative utility functions and parameter values on the optimal portfolio. Using ten NYSE securities and several rich and diverse utility function classes that spanned decreasing, constant and increasing absolute and relative risk aversion, they show that utility functions having different functional forms and parameter values but "similar" Arrow-Pratt absolute risk aversion indices produce "similar" optimal portfolios. Additional results in this area appear in Kallberg and Ziemba (1983)[2].

Examining the racetrack bettor's utility function originates with Weitzman (1965)[1]. He assumes that Mr. Avmart (average man at the race track) represents the whole population of bettors. Using simple least squares, he finds a relation between the objective win probability and the return of a win bet, then empirically determines the relationship between the utility function of Mr. Avmart and the returns. The empirical utility function is convex, matching portions of the utility functions proposed in Friedman and Savage (1948)[2] and Markowitz (1952)[2].

Rosett (1965)[1] defines some common self-constructed gambles from win bet. A martingale requires a bettor to choose n horses in different races. A high probability-low return bet is constructed by wagering on the first on the first horse in the first race. If that bet is successful, then stop. If unsuccessful, bet on the second race an amount greater than that lost in the first race. If successful then stop; if not continue this process with the remaining races. A parlay, on the other hand, is a way of constructing a low probability-high return bet. It also requires a bettor to choose n horses in different races. If the first horse wins, bet the winning proceeds on the second race, and so on. Rosett derives the relations between return and win probability under the cases that for every bet on a single horse, (i) any martingale having the same win probability as the single bet will pay the same return as the single bet (high-probability-low-return), and (ii) any parlay having the same win probability will pay the same return as the single bet (low-probability-high-return), respectively. Using Weitzman's win bet data, Rosett empirically obtains a relation which is consistent with the hypothesis that bettors are rational, sophisticated, and have strong preference for low-probability-high-return bets (i.e. in the consistent way of betting on parlay). However, his relationship is consistent with only a part of the data, and it systematically overstates the returns when the win probabilities are low.

Ali (1977)[1] assumes that bettors are expected utility maximizers, sophisticated, and behave as if the betting opportunities are limited to a single race. Following Weitzman (1965)[1], he also assumes that there is a representative bettor who establishes the final odds. Using a large win bet data set from different racetracks in the U.S., he empirically obtains a utility function of wealth that, similar to Weitzman, is convex.

Researchers using an expected utility approach to explain betting behavior have generally concluded that the bettors are risk-lovers. Such a characteristic explains the systematic discrepancy between the objective and subjective win probabilities and, in particular, the favorite-longshot bias. Other explanations are possible, too, of course. For instance, the well-documented bias, mentioned in the last section, of decision makers consistently underestimating high probabilities and overestimating low probabilities generates a similar pattern.

[1] included in this volume
[2] cited in the Annotated Bibliography

ON THE ROBUSTNESS OF THE ARROW–PRATT RISK AVERSION MEASURE *

J.G. KALLBERG
New York University, New York, NY 10003, USA

W.T. ZIEMBA
University of British Columbia, Vancouver, BC, Canada

Received November 1978

For portfolio problems with joint normally distributed asset returns, the risk aversion measure $R = -w_0(Eu''(w)/Eu'(w))$, where w_0 is initial wealth can be used to characterize optimality. Comparisons between the global measure R and local measures based on $R_A = -u''(w)/u'(w)$ are explored. Simulations for several utility function classes are described.

The Arrow–Pratt [Arrow (1971) and Pratt (1964)] risk aversion measure $R_A(w) \equiv -u''(w)/u'(w)$ (primes denote derivatives) for a utility function u defined on wealth w provides a local measure of an individual's aversion to risk. The measure is derived from Taylor series expansions assuming that the variance of final wealth is small. [See Pratt (1964) for details, regularity conditions, etc.] If the variance of wealth is infinitesimal then at initial wealth w^0, $R_A(w^0)$ essentially determines optimal behavior. For example, Samuelson (1970) proves that in the choice between a safe asset and a risky asset, the limiting optimal amount to invest in the risky asset is inversely proportional to R_A. When σ_w^2 is not infinitesimal then one must consider R_A at all w having positive mass. Exact theoretical results require very strong assumptions. For example, for two utility functions to yield the same behavior it is generally required that $R_A^1(w) = R_A^2(w)$ for all w, i.e., $u_2(w) = au_1(w) + b$, $a > 0$.

With joint normally distributed assets one has the following powerful result:

Theorem 1. Suppose $\xi \in E^n$ has a multivariate normal distribution with finite means and variances and that u_i is twice continuously differentiable with $u_i' > 0$ and $u_i'' < 0$ for $i = 1, 2$. Consider

$$\{\max E_\xi u_1(w_1 \xi^t x) \mid e^t x = 1, x \geqslant 0\}, \tag{P.1}$$

* This research was supported by Canada Council Grant No. S75-1307-R1. Thanks go to E.R. Berndt, W.E. Diewert and the editor for helpful comments on an earlier draft of this letter.

and

$$\{\max E_\xi u_2(w_2\xi^t x) \mid e^t x = 1, x \geqslant 0\},$$ (P.2)

where w_i refers to investor i's initial wealth. Suppose that all expectations are finite, x^* is an optimal solution to (P.1) and

$$\frac{w_1 E_\xi u_1''(w_1\xi^t x^*)}{E_\xi u_1'(w_1\xi^t x^*)} = \frac{w_2 E_\xi u_2''(w_2\xi^t x^*)}{E_\xi u_2'(w_2\xi^t x^*)}.$$

Then x^* is an optimal solution to (Ṗ.2).

Thus the measure

$$R \equiv \frac{-w_0 E_\xi u''(w_0\xi^t x)}{E_\xi u'(w_0\xi^t x)}$$ (1)

is the appropriate measure to use when exact portfolio results are desired. For example, suppose x^* is an optimal solution with utility function u, parameter β^* and initial wealth w^*. Then $\hat{\beta}_i$ is a solution to

$$\frac{w_i \int u_i''(\beta_i, w) \, dN(w)}{\int u_i'(\beta_i, w) \, dN(w)} = \frac{w^* \int u''(\beta^*, w) \, dN(w)}{\int u'(\beta^*, w) \, dN(w)}$$ (2)

when used as the parameter with utility function u_i when initial wealth is w_i will result in x^* being optimal for this problem as well.

The remarkable fact of Theorem 1 is that it is an exact global result independent of the variance of final wealth. With infinitesimal variances R_A approaches (1) and the reason underlying the Samuelson result is clear.

Remark.   Rubinstein (1973a) discovered the importance of the measure R in his study of capital market equilibrium. In particular the equilibrium asset prices are linearly related to R. See also Rubinstein (1973b, 1976). The extension to optimal asset proportions given in Theorem 1 was suggested to us by Rubinstein. Its statement and proof appears as an appendix to Kallberg and Ziemba (1978a). The proof utilizes the Kuhn–Tucker conditions and the result that (under the assumptions) $cov(x, u(y)) = E[u'(y)] cov(x, y)$. The gradient conditions then involve R.

In light of Theorem 1 a crucial question is: how accurate will results based on R_A be when variances are non-infinitesimal. This is especially important because R is not as rich as R_A regarding mathematical convenience, economic interpretation, qualitative results and ease of estimation. Hence despite the greater accuracy of R, use of R_A may be preferred in practical situations. To assess the adequacy of using R_A when wealth variance is not infinitesimal, we constructed an experiment using ten (assumed joint normally distributed) NYSE securities over monthly, quarterly and yearly horizons. The security variances ranged from 0.0036 (smallest monthly) to 0.3276 (largest yearly). The object was to determine if utility functions with widely varying properties but 'similar' risk aversion had 'similar' optimal portfolios. We used the familiar quadratic, exponential, logarithmic, power and negative power families plus special exponential and arctan families to provide great diversity in utility function properties and analytic convenience. Given a reference utility function u with optimal portfolio x^*, an appropriate measure of how close portfolio $\hat{x}$ is to x^* is the percentage cash equivalent difference,

$$\frac{u^{-1}[Z(x^*)] - u^{-1}[Z(\hat{x})]}{u^{-1}[Z(x^*)]}\ 100,\tag{3}$$

where

$$Z(x^*) = \max_{x \in K} E_\xi u(\xi^t x),$$

$$K \equiv \{x \mid e^t x = 1, x \geqslant 0\},$$

$$Z(\hat{x}) \equiv E_\xi u(\xi^t \hat{x}),$$

$$e \equiv (1, ..., 1)^t,$$

$$\xi \sim N(\overline{\xi}, \Sigma),$$

and u^{-1} is the inverse of u which is assumed to exist, i.e., u is strictly increasing. (In the calculations exponential u is convenient to use because of the closed form expressions: $Z(x) = E_\xi[1 - \exp(-\beta\xi^t x)] = 1 - \exp(-\beta\overline{\xi}^t x + (\beta^2/2)x'\Sigma x)$; and $u^{-1}[Z(x)] = -(1/\beta)\log(1 - Z(x))$.) If $\beta = \beta^*$ the average risk aversion of u is

$$\int R_A(\beta^*, w)\, dN(w),$$

where

$$w \equiv \xi' x^* \sim N(w) = N(\overline{\xi}^t x^*, x^{*t}\Sigma x^*).$$

Then u_i is said to have the same risk aversion as u if its average risk aversion is the same over the range of optimal final wealth. Hence a solution $\hat{\beta}_i$ to

$$\int R_A^i(\beta_i, w)\, dN(w) = \int R_A(\beta^*, w)\, dN(w)\tag{4}$$

is an appropriate value of β_i to compare u_i with u. Again exponential u simplifies the calculation because $R_A(w) = \beta^*$ for all w hence the right-hand side of (4) is the constant β^*.

Extensive calculations presented in Kallberg and Ziemba (1978a) indicate that the approximation that $\overline{R}_A^1 = \overline{R}_A^2 \Rightarrow x_1^* \simeq x_2^*$ [where $\simeq$ is measured by (3)] is very good for all the utility function classes and the three data sets for widely varying

R_A. Naturally the closeness of the approximation is in the order monthly, quarterly and yearly. For example the maximum percentage errors for these three data sets over all utility functions was 0.0004%, 0.0191% and 0.7827%, respectively. Moreover, the means, variances and optimal portfolio weights are also remarkably close, as illustrated in table 1 using quarterly data for the period January 1965–December 1969.

For most utility functions of interest (e.g. all but the arctan of those mentioned above) calculations for any given wealth level can be obtained from normalized wealth equal one calculations using a rescaled parameter. A fast and simple algorithmic procedure for portfolio calculations is described in Kallberg and Ziemba (1978b). Practical schemes for investment choice based on these results are described in Kallberg and Ziemba (forthcoming). These schemes generally use simplified methods for estimating $R_A(w)$ possibly without estimating u.

The results also yield the following conclusions:

(1)   An implication of the results for econometric research is that based only on observed portfolio data it will be extremely difficult to discriminate amongst different alternative models since these models will be observationally practically equivalent with different parameter values.

(2)   With horizons of a year or less one can substitute easily derived surrogate utility functions that are mathematically convenient for more plausible but mathematically more complicated utility functions and feel confident that the errors produced in the calculation of the optimal portfolios are at most of the order of magnitude of the errors in the data. Moreover there is a fairly well defined tradeoff of computational accuracy versus computing costs. For rough calculations one can use the measure $(-w_0 u''(w_0 \bar{\xi}^t x)/u'(w_0 \bar{\xi}^t x))$ and obtain reasonable results for small variance problems (e.g., monthly data). For the most accurate results one uses the measure (1) [with (2)]. Generally the most cost effective measure is $w_0 \bar{R}_A$ [with (4)].

(3)   The quadratic, exponential and logarithmic utility functions yield the largest range in variation of R_A and yield the safest and riskiest portfolios for extreme parameter values. Low values of R_A are easily implementable for all three utility functions. However, extremely high values of R_A are not really implementable for the exponential or logarithmic functions because of domain problems. Thus the quadratic utility function despite its two well known limitations ($u' < 0$ for large w and $R'_A > 0$) may well play a useful role as a computational surrogate for more plausible utility functions when the number of possible investment securities is large.

(4)   The results suggest that one can derive an analytically tractable utility function whose use will generate portfolio allocations nearly indistinguishable from those of a given risk averse utility function given a known distribution of returns. However if we substitute $\hat{u}$ for u when the distribution is $F(\xi)$ are the results robust if the distribution is really $\tilde{F}(\xi)$? Calculations in Kallberg and Ziemba (1978a) indicate that the maximum expected utility and optimal portfolio composition are relatively insen-

Table 1
Portfolio composition for various utility functions, using quarterly data, when $R_A = 2$. [a]

Utility function	Exponential	Quadratic	Log	Special exponential	Power	Negative power
Parameter	2.0	0.317898	−0.530359	0.086256	0.498896	0.063542
Security						
(1) Cunningham drug	0.279549	0.286024	0.273713	0.271465	0.281605	0.278865
(2) National cash	0.472942	0.477334	0.466487	0.466941	0.472449	0.457115
(3) MGM	0.104567	0.103563	0.105836	0.118487	0.093868	0.122138
(4) Gillette	*	*	*	*	*	*
(5) Household finance	0.142942	0.133079	0.153964	0.143108	0.152078	0.141881
(6) Heinz	*	*	*	*	*	*
(7) Anaconda	*	*	*	*	*	*
(8) Kaiser Al.	*	*	*	*	*	*
(9) Maytag	*	*	*	*	*	*
(10) Firestone	*	*	*	*	*	*
Mean	1.051596	1.051757	1.051418	1.051472	1.051551	1.051478
Variance	0.010109	0.010274	0.009934	0.009993	0.010070	0.010005
Expected utility	0.875441	0.875440	0.875440	0.875439	0.875390	0.875437
Error	—	0.0004	0.0004	0.0008	0.0197	0.0016

[a] *Source*: Kallberg and Ziemba (1978a). The approximation is generally more accurate for lower values of R_A because these values tend to represent utility functions with 'less curvature'. The asterisks indicate values less than 10^{-4} and the arc tangent utility function for which $\bar{R}_A = 2$ is unattainable is excluded.

sitive to errors in estimation of the variance–covariance matrix. However, errors in estimating the mean return vector do significantly change these quantities.

(5)   It is well known that an investor's risk aversion increases with R_A [Pratt (1964, Theorem 1)]. Results in Kallberg and Ziemba (1978a) provide guidance regarding the significance that particular R_A values indicate regarding the composition and change in composition of the optimal portfolio. For example, R_A values greater than about 4 always yield very risk averse portfolios with low variance, and there is little change in the optimal composition even for large changes in R_A. The range $2 \leqslant R_A \leqslant 4$ (approximately) yields moderately risk averse portfolios with a modest degree of change in the optimal portfolio composition when R_A changes. The range $0 \leqslant R_A \leqslant 2$ yields risky portfolios and there are dramatic changes in the optimal portfolio composition for even small changes in R_A. Generally if one normalizes the variance at $R_A = 0$ to be 1 then for $R_A = (0.5, 1, 2, 3, 5, 10)$ the optimal portfolio has variance about $(0.55, 0.40, 0.30, 0.25, 0.22, 0.20)$.

References

Arrow, K.J., 1971, Essays on the theory of risk bearing (Markham, Chicago, IL).

Kallberg, J.G. and W.T. Ziemba, 1978a, Comparison of alternative utility functions in portfolio selection problems, Faculty of Commerce Working Paper no. 609, Oct. (University of British Columbia, Vancouver).

Kallberg, J.G. and W.T. Ziemba, 1978b, An extended Frank–Wolfe algorithm with application to portfolio selection problems, Faculty of Commerce Working Paper no. 621, Nov. (University of British Columbia, Vancouver).

Kallberg, J.G. and W.T. Ziemba, forthcoming, A simplified procedure for portfolio selection for risk averse investors.

Pratt, J.W., 1964, Risk aversion in the small and in the large, Econometrica 32, 122–136.

Rubinstein, M.E., 1973a, The fundamental theorem of parameter-preference security valuation, Journal of Financial and Quantitative Analysis 8, 61–70.

Rubinstein, M.E., 1973b, A comparative statics analysis of risk premiums, Journal of Business 46, 605–615.

Rubinstein, M.E., 1976, The valuation of uncertain income streams and the pricing of options, Bell Journal of Economics 7, 407–425.

Samuelson, P.A., 1970, The fundamental approximation theorem of portfolio analysis in terms of means, variances and higher moments, Review of Economic Studies 37, 537–542.

UTILITY ANALYSIS AND GROUP BEHAVIOR
AN EMPIRICAL STUDY[1]

MARTIN WEITZMAN

Massachusetts Institute of Technology

I. INTRODUCTION

UTILITY analysis is a highly theoretical construct whose main function in economic theory is to serve as a link in the chain connecting human preferences with economic behavior. That few aspects of utility analysis have been satisfactorily subjected to empirical testing is unfortunate for economics because of this key role in the theory of demand.

The experiments thus far performed have primarily involved individuals or small groups in laboratory situations. Ideally a utility experiment should involve many participants in a more or less natural environment. Of course it is difficult to incorporate these features into a practical, controlled, experimental framework.

In this paper the theory and results of an altogether different approach will be presented. Extensive data of group behavior at a race track will provide the information for an analysis of several utility hypotheses. In particular an "average man at the race track" will be defined, his underlying decision-making mechanism investigated, and his indifference map between various risk combinations presented. His utility of money curve will be constructed and correlated with the theoretical literature on this subject.

[1] Thanks are due to Professor Arnold Zellner who provided insight and encouragement. This work was done at the Social Systems Research Institute, University of Wisconsin, and was supported in part by National Science Foundation Grant GS-151.

II. THE EXPECTED-UTILITY HYPOTHESIS

The theoretical content of utility analysis has been amply discussed in the literature.[2] We shall make no reference here to any but the most directly relevant aspects.

Each person is assumed to possess a subjective preference pattern among alternative situations. Generally speaking, a utility function is any function that arithmetizes the relation of preference among the situations.[3] When risk is involved, each situation is associated with a probability of occurrence.

The usefulness of the utility concept derives primarily from its behavioral implications. A fundamental postulate asserts that each individual acts in such a manner that, to a very good approximation, he is behaving *as if* he were maxi-

[2] R. Duncan Luce and Howard Raiffa, *Games and Decisions* (New York: John Wiley & Sons, 1958), chap. ii and Appendix 1; Leonard J. Savage, *The Foundations of Statistics* (New York: John Wiley & Sons, 1954), chap. v; Jerome Rothenberg, *The Measurement of Social Welfare* (Englewood Cliffs, N.J.: Prentice-Hall, Inc., 1961), Part III; Armen Alchian, "The Meaning of Utility Measurement," *American Economic Review*, XLIII (March, 1953), 26–50; Daniel Ellsberg, "Classic and Current Notions of Measurable Utility," *Economic Journal*, LXIV (September, 1954), 528–56.

[3] In this paper I am avoiding the issue between utility as a measure of preference *order* versus utility as a measure of preference *intensity*. Some strict operationalists (notably Ellsberg, *op. cit.*) regard the two concepts of utility as quite disjoint. Rothenberg (*op. cit.*, pp. 211–17) argues cogently for a high degree of likeness. A good discussion is available in Milton Friedman and L. J. Savage, "The Expected Utility Hypothesis and the Measurability of Utility," *Journal of Political Economy*, LX (December, 1952), 463–74.

mizing his utility function over the relevant range of action states.

The personal utility of the risk situation consisting of the possibility of winning m dollars with probability p will be denoted $U(p, m)$, where U is some function of p and m. In writing an individual's utility as a function of p and m alone I am abstracting from the influence of all other possible variables. Changes in these other factors might shift the utility function over time, but for any specified time we may abstract away their influence as being of fixed quality while the aforementioned two determinants serve as variables.

The expected-utility hypothesis implies, among other things, that a modification on the structure of the utility function for a risk situation can be made of the form $U(p, m) = pu(m)$, where $u(m)$ is the personal utility of the money prize, m dollars.[4] The principle of maximization of $U(p, m)$ now becomes maximization of the expected utility $pu(m)$.[5]

Friedman and Savage were the first to construct a hypothesis that explained in a quantitative way the risk behavior of low-income consumer units.[6] Their utili-

[4] We are not concerned here with the Von Neumann–Morgenstern axiomatic system underlying this hypothesis, but only with the behavioral calculus which it implies.

[5] To be more precise, the maximization of expected utility (EU) hypothesis asserts that an individual maximizes $pu(x)$, where p is the objective probability. The maximization of subjectively expected utility (SEU) hypothesis holds that an individual maximizes $p^* \cdot u(x)$, where p^* is that person's subjective probability. Strictly speaking, it is the content of the EU hypothesis that will be empirically investigated in this paper. Insofar as the audience evaluates p^* close to p, the EU and SEU hypotheses will predict almost identical behavior and, to that extent, will be operationally indistinguishable. The role of subjective probability is an important issue and shelving it at this point does not imply that it is irrelevant, but only that it is difficult to incorporate into this study.

ty of money curve contained sections of differing curvature which corresponded to qualitatively different economic levels. As a result of this construction, they were able to account for such seemingly contradictory attitudes toward risk as the simultaneous taking of insurance and participation in gambling.

Markowitz showed that the Friedman-Savage hypothesis contained certain irregularities, and presented his own amended version.[7] This utility function has no absolute domain, but is defined at

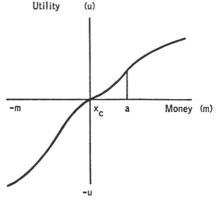

Fig. 1

any given time only with respect to an individual's "customary" or "present" wealth. Figure 1 portrays Markowitz's amended function. The origin, x_c, is not zero income but present or customary income, and is also a point of inflection. To the right of x_c (which region this study will cover), the curve is concave upward (increasing marginal utility), until a second point of inflection, a, is reached. From then on, increasing values of m yield

[6] Milton Friedman and L. J. Savage, "The Utility Analysis of Choices Involving Risk," *Journal of Political Economy*, LVI (August, 1948), 279–304.

[7] Harry Markowitz, "The Utility of Wealth," *Journal of Political Economy*, LX (April, 1952), 151–58.

decreasing marginal utility (the curve is convex upward). The exact location of a would depend on the individual's attitude toward risk. People who are less averse to bearing risk over a wide range (for example, gamblers) would be expected to possess a utility function with a located further to the right.

Markowitz uses his utility function as a device to explain and predict reactions toward risk. The shape of this utility function is consistent with many empirical generalizations about risk behavior. We shall see that the results of this study bear out Markowitz's construct for at least one section of his curve.

III. EXPERIMENTAL BACKGROUND: THE RACE TRACK AND PARIMUTUEL BETTING

Racing results at four New York race tracks (Aqueduct, Belmont, Jamaica, and Saratoga) were studied for the ten-year period 1954–63. During the racing season activity rotates among these four tracks so that at any given time only one is being used. Nine races are run each day, Monday through Saturday, and on the average nine horses are entered in each race, although the number may vary from four to fourteen. Results were obtained from May through October of each year so that during the ten-year period over 12,000 races and over 110,000 performances were investigated.

It is essential for a comprehension of the theory underlying this study that the reader understand fully the process of parimutuel betting. This system of gambling is unique in that the odds are determined by the aggregate wagering patterns of the crowd itself, rather than fixed beforehand. All the money bet on prospective winners of a race is pooled. A fixed percentage (the "take") is subtracted as revenue for the track and state government. Those people who hold the

winning tickets receive the rest. The odds on any horse are computed by subtracting the sum bet on the horse from the pool after take and dividing by the sum bet on the horse. For each dollar bet, the amount of money actually returned (the "return") is the odds *plus* one, the one being the dollar bet.

A totalizer (centrally located scoreboard) summarizes the results of previous betting by running the current odds continuously as bets pour in for a period of about twenty minutes before each race.

A rational person wagering at the race track selects a particular horse to win because he thinks that in some sense, considering his personal attitude toward risk, the horse is more of a "winner" than the odds at which he is being portrayed would seem to say. The race-track crowd is a group of people, each following his own personal motives and preferences, who collectively arrive via a market type of mechanism at the return which each horse pays. In this study we shall be interested in the general relationship between the return and the frequency of winning. The reason for the existence of this general relationship, as contrasted with the rationale behind any given bettor's specific choice, will be explained in terms of utility effects operating on the different amounts of money to be won.[8]

IV. THE EXPERIMENTAL PROBABILITY CURVE

It was found experimentally that the probability of a horse's victory could be

[8] It is possible that effects other than utility are operating. However, it is not at all clear what these other effects might be, nor does it seem, on the face of it, that a strong a priori case could be made for their relevance. In the absence of such knowledge we are forced to perform the obvious abstraction. If dynamic effects are operating, this analysis squeezes them into a static framework.

expressed as a smooth and sharply determined function of the return which it paid.[9]

After the data from all of the races were collected, each horse was classified according to the return to the dollar it would have paid had it won and subclassified within that category as a winner or a loser (values of return are figured to the nearest $0.05 at the track). An empirical probability of winning—the ratio of the number of victories divided by the number of entries—was then associated with each value of return. Weighted aggregation performed on this data yielded 257 separate points of the form (x, p), where x is the value of return to the dollar and p is the empirical probability of victory associated with that return.

The 257 points, themselves derived from over 110,000 runnings, were fitted to a curve (shown in Fig. 2) by employing a weighted least-squares method that corrected for heteroscedasticity. Various functional forms were tested in an effort to ascertain the underlying "true" function. The rectangular hyperbola, $p = A/x$, yielded fair results.[10]

The most appropriate function was judged to be the "corrected" hyperbolic form,[11]

$$p = \frac{A}{x} + \frac{B \log (1 + x)}{x}.$$

It was not understood why this particular function yielded more suitable results than others, but its superiority was evident. Neither did complicated combinations of other functions (including polynomials) make a better fit, nor did the addition of extra terms to this function significantly improve matters; furthermore, their addition could not be justified statistically.[12]

There was no trend to the residuals of the "corrected" hyperbola; they were distributed in a random fashion about their origin.[13] This also testifies to the fact that the underlying equilibrium curve is an appropriate fit. The evident conclusion is that in the empirical data there exists a pattern which appears to have a high degree of coherence, and which is aptly described by the chosen functional form.

[10] The weighted least-squares procedure yielded

$$\hat{p}_i = \underset{(0.0075)}{0.8545} \frac{1}{x_i}$$

with $R^2 = 0.9807$. The coefficient estimate is several times its standard error, and the coefficient of determination is significantly large. However, the residuals were successively correlated, indicating a systematic trend for the error term.

[11] The weighted least-squares procedure yielded

$$\hat{p}_i = \underset{(0.0190)}{1.011} \frac{1}{x_i} - \underset{(0.0099)}{0.087} \frac{\log (1 + x_i)}{x_i}$$

with $R^2 = 0.9852$. Both coefficient estimates are several times their standard errors, and the coefficient of determination is unusually large.

[12] That is, the low values computed for their t-statistics did not allow rejection of the hypothesis that the added coefficients were zero.

[13] The Durbin-Watson d-statistic was 1.94. This indicates that the null hypothesis of residual independence could be upheld against the alternative hypothesis of positively correlated successive disturbances at both the 1 and 5 per cent levels (H. Theil and A. L. Nagar, "Testing the Independence of Regression Disturbances," *Journal of the American Statistical Association*, LVI [December, 1951], 793–806).

[9] The reader may be confused at this point by usage of the word "function." It is being employed here in its strict mathematical sense; a function f of a variable x is a unique association of a number $f(x)$ with each number x in the domain of definition. No causality in the physical sense is meant to be implied by this definition. Thus it is not being maintained that the crowd-determined return *causes* the objective probability of a horse's victory—more likely it is the other way around and the crowd's estimation of the horse's probability of victory sets the return. The existence of probability as a function of return in the mathematical, not the causal, sense is under investigation.

V. FORMULATION OF THE MODEL
MR. AVMART

The study described in this paper deals with the group behavior of racing devotees in the face of uncertainty. Any conclusions relate, strictly speaking, only to the actions of the assemblage and not necessarily to the conduct of any member of it. In order to create a convenient mode of expression we will make the population homogeneous and then inquire what attitudes toward risk each hypothetical member would be required to possess and what actions he would have to perform in order to explain the data which were obtained. To avoid redundantly using a lengthy descriptive phrase, we will call a member of the artificially constructed "homogeneous" race-track crowd Mr. Avmart (average man at the race track).

Mr. Avmart's wagers are allocated among the entrants in a race in exactly the same proportion as the entire crowd apportions its money among the various horses. Of course, no real person at the track distributes his bets over a continuous spectrum of values in the manner prescribed above.[14]

Mr. Avmart bets a total of exactly $5.00 in each race. This figure is very close to the amount averaged per bettor for individual wagers placed in the win pool in the four race tracks surveyed. The exact value is not important, since it is used only in constructing a determi-

[14] Actually it is preferable for conceptual reasons to think of Avmart as using a randomizing device to select the horse he bets on. The outcomes of this randomizing device are bets on the various horses and the probabilities of these outcomes are proportional to the amount the crowd has wagered on that horse. In this manner Avmart bets on only one horse per race, the probability of betting on that horse being proportional to the aggregate amount wagered on that horse. The long-run effects are the same as if he distributed bets continuously for each race in the manner prescribed above.

nate scale for Avmart's utility of money curve and does not alter the shape of the curve. In that capacity $5.00 is the closest round figure, and a good enough approximation to the true social average.

The question of Mr. Avmart's existence is a pertinent issue. In the literal sense, of course, he is a completely fictitious entity, being nothing more than an anthropomorphic version of a race-track crowd.

As will be shown, by postulating a certain very precise utility curve for Avmart and aggregating him into a race-track crowd, the risk behavior of that crowd will be explained. The race-track crowd behaves *as if* it were a homogeneous conglomeration of a certain individual, namely, Avmart.

To a psychologist concerned with existence and reality on an individual level, this formulation of Avmart might be quite disquieting. But to an economist, interested in utility theory primarily for its implications in market behavior, attitudes toward risk, and the theory of demand, this operational definition of Avmart should coincide with that aspect of human behavior in which he is professionally interested—namely, a personal component of social behavior which when performed collectively by individuals, yields the true group action.

VI. AVMART'S INDIFFERENCE CURVE

We come now to a key theoretical derivation. The empirically acquired curve of Figure 2 with the scale of the x-axis blown up fivefold represents Avmart's indifference map between various risk situations involving money prizes and probabilities. But before this can be shown, Avmart must be more fully explained.

Mr. Avmart was constructed with the idea of a *social average* in mind. He was made to be the "most typical bettor." Of course, when he was so constructed he ceased to exist in the strict sense, but that conclusion would hold in any social averaging situation.

So far, however, nothing has been said about Avmart's motivation. Let us motivate him the way we assume any bettor at the track is motivated: to make those

fold and now called the m(money) axis. The curve of Figure 2′ is one of Avmart's utility isoquants.

Consider any two points A and B on the curve of Figure 2′. A is the point (m_1, p_1) representing the possibility that Avmart will win m_1 dollars with probability p_1, while B is the point (m_2, p_2) rep-

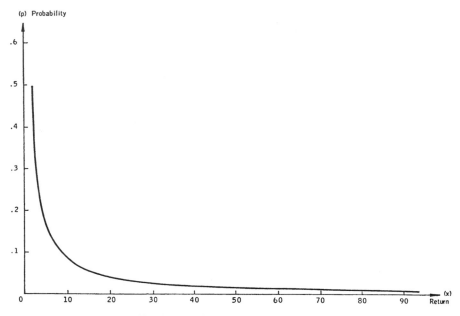

Fig. 2.—The theoretical probability curve

wagers he believes are maximizing his utility function. Avmart, average bettor that he is, apportions his $5.00 per race exactly as the entire crowd is apportioning its total money. But Avmart does not consciously or unconsciously *set out* to act in this manner. Being an individual in a risk situation, we are postulating that his behavior can be described as maximization of personal utility.

Suppose Figure 2 were (hypothetically) reproduced as Figure 2′, where the two are identical except that the scale of the x-axis of the former is magnified five-

resenting his possibility of winning m_2 dollars with probability p_2.[15] Suppose Avmart "prefers" A to B. Since he bets

[15] This option is never presented to anyone at the race track in so clear a fashion. No real bettor purchases a "possibility of winning m dollars with probability p"; but whenever he picks the horse paying m dollars the results of this study show that in the long run he can expect the proportion of victories associated with that horse to be p. In this manner we can speak of Avmart being offered the "package" (m, p)—the possibility of winning m dollars with probability p. Although, of course, no real bettor ever thinks *directly* of a horse in such a general manner, it is precisely the existence of an underlying bias toward or against certain "packages" in which we are interested.

in exactly the same ratios as the racing populace, A is *socially* preferred to B. This means that some of the money that would have been bet on B will now move to A. With more money being bet on A, the return declines and that point moves to the left while B shifts to the right as the return on it improves. These return movements cause A to become less and B more desirable. Eventually an equilibrium situation is reached where A and B are of equal desirability. But the curve of Figure 2 (and hence of Fig. 2′) is itself an equilibrium curve. Then A and B are of the same utility to Avmart, and likewise for any other pair of points on the curve of Figure 2′—that is, that curve is one of his risk-indifference contours.[16]

The fact that the rectangular hyperbola was a fairly good fit to the experimental probability curve is significant. This means that Avmart's adherence to the expected *value* hypothesis (maximization of expected *value*), which would result in a hyperbola ($mp = $ constant) for his indifference contour, is not a completely inappropriate description of his behavior. Stated more precisely, the risk-indifference contour which would result from the assumption of maximization of expected value is a good first-order ap-

proximation to the curve actually obtained. However, a more exact fit of a slightly different form could be statistically justified with a high level of confidence.

VII. AVMART'S UTILITY OF MONEY CURVE

By assuming that Avmart obeys the expected utility hypothesis, we can determine his utility of money function exactly. For every point on Avmart's indifference map of Figure 2′ the following equalities hold: $U(p, m) = pu(m) = K$, where K is the constant utility of that curve. Remembering that Figure 2′ also yields p as a function of m, $p = p(m)$ where $m = 5x$, we have that $K = p(m)u(m)$, or $u(m) = K/p(m)$. We shall make $u(\$5) = 5$ "utiles" in order to fix the utility scale. Then K is equated to $5p(\$5)$, where $p(\$5)$ is the value of the Figure 2′ function for $m = \$5$.

Figure 3 is the curve of $u(m)$, Avmart's utility of money function. Several features are noteworthy. The curve itself exhibits only slight curvature. Derived as it is from the theoretical probability curve of Figure 2, an extensive amount of raw data stands behind it (257 points, themselves averages of over 110,000 runnings). This permitted an excellent determination of the theoretical probability function which in turn was used to obtain the mathematical expression for the curve of Figure 3. These aspects of relative precision stand in marked contrast with the more usual indeterminate nature of experimentally derived utility functions.

Since the utility function was derived in analytic form, it was possible to obtain first and second derivatives.[17] The first

---

[16] Although Avmart is indifferent between any two points of Figure 2′, this does not mean that he can now carelessly place bets anywhere on that curve. Figure 2′ is an indifference map *given* the prevailing social betting patterns. If the crowd (Avmart) changes betting patterns, then Figure 2′ is altered. The situation is made somewhat clearer by considering (as we are assuming) a crowd of many Avmarts, each with identical utility functions, and each making one \$5.00 bet. If the probability versus return curve deviates momentarily from that required to give all bets equal utility, then Avmarts will line up at the ticket window to drive the probability versus return curve back to the proper shape. It is interesting to note that the existence of a "take" had no bearing on the interpretation of Figure 2′ as Avmart's risk indifference curve. The relevant question is "given the fact that Avmart pays a 'fee' for the pleasure of gambling, which gambling levels or 'packages' does he then prefer to bet on?"

[17] The calculations are not performed here. They are straightforward, but lengthy, and the reader can easily verify the results which are cited.

derivative is always positive for positive m. The second derivative is also positive for all positive values of m, but approaches zero very rapidly as m gets larger. At $m = \$500$, d^2u/dm^2 is already less than one hundredth of its value at $m = \$5.00$. It is thus approaching a point of inflection as m becomes larger, and it seems not implausible to suppose that data over a wider range would show the eventual attainment of a point of inflection, and then a change in the sign of the second derivative. Because this curve was obtained only for values of m between \$5.00 and \$500.00, the status of this conjecture unfortunately could not be tested, and it also could not be ascertained whether or not another point of inflection lay to the left of this interval. However, it does appear that the range of m for which Figure 3 has been drawn coincides with Markowitz's range of increasing marginal utility ($x_c < m < a$ in Fig. 1). This range was explicitly created in order to explain gambling behavior on a theoretical level. It is especially large

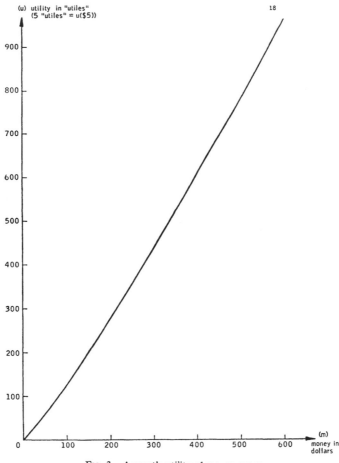

FIG. 3.—Avmart's utility of money curve

in this case ($a > \$500.00$) because Avmart represents a group that possesses greater propensity toward risk-bearing.

Because only this interval was investigated, the results do not constitute a test of the applicability of Markowitz's hypothesis to Avmart's behavior over the entire range for which it was formulated. Nevertheless, for the range examined, the results are completely in accord with what his theory predicts.[18]

VIII. CONCLUSION

A new approach to the empirical study of some facets of utility theory has been explained and presented. Instead of concentrating on individuals and trying to derive utility generalizations from their experimental behavior, more nearly the converse approach was attempted. A plethora of data concerning the collective risk actions of parimutuel bettors was employed in investigating utility aspects of the behavior of a hypothetical member of the group. Mr. Avmart was operationally defined to be an "average" constituent of the race-track populace. One of Avmart's indifference contours among risk situations was shown. Assuming Avmart's adherence to the expected utility hypothesis, his utility of money curve was derived. This curve was strikingly similar to the one proposed on theoretical grounds by Markowitz as an amendment to the Friedman-Savage hypothesis.

The conclusion is not that human beings possess, in any real sense of that word, a sharply defined utility function which they consult when making decisions. Indeed, to be of use in the context of economic theory it is sufficient only to believe that in some appropriate statistical sense people behave *as if* they possessed such a function. The results of this study show that the crowd at the race track behaves *as if* it were composed of a group of individuals each of whom possesses an identical utility function of the Markowitz variety.

[18] A study performed by R. M. Griffith ("Odds Adjustments by American Horse Race Betters," *American Journal of Psychology*, LXII [1949], 290) noted the existence of a systematic undervaluation of the chances of short-odded horses and overvaluation of long-odded horses. Griffith was primarily interested in other matters, but he noted this quantitative trend as well as a rough qualitative similarity in data from the years 1947 and 1934. This correlation appeared strange to him, since one year was a prosperous one, and the other was during the Depression. These results are neatly explained by the utility interpretation outlined above.

GAMBLING AND RATIONALITY

RICHARD N. ROSETT[1]

University of Rochester

I. INTRODUCTION

ARE gamblers rational? The sense of this question is that gamblers often accept bets with an expected value lower than some alternative, and that this seems inconsistent with the normal love of money. Friedman and Savage explained such gambling behavior by showing that it is consistent with the maximization of expected utility, given that the function relating utility to income has a certain general shape.[2]

An alternative explanation sometimes offered for particular forms of gambling is that gamblers are fooled. The possibilities for error that will lead a gambler to expect a game to pay more, on the average, than it will are probably numerous. The following two possibilities will receive consideration here:

1. The game may be so complicated that the gambler misunderstands it.

2. The gambler may not know what he can expect to win because he does not know how to perform calculations involving probabilities. This may lead to errors of judgment even though he understands the game itself and has enough information on which to base the calculations.

In order to determine whether the behavior of gamblers is consistent with these two statements, a model of pari-mutuel race-track betting is constructed on the assumption that gamblers who bet on horses understand the game—in the sense that they (1) know the factors that determine the probability that a horse will win and (2) correctly forecast the probability—and that they know how to calculate both the probability of winning a combination of simple bets and the payoff to such combinations. Observable consequences of this model are tested using race-track data. It is found that the data conform to the model.

The behavior of gamblers at the race track may not be especially interesting in itself, but as an example of rational choice between risky alternatives it is extremely interesting. A very large number of factors affect the chances that any given horse will win a race. The problem of predicting the probability that a horse will win may not require the most powerful tools of statistics, but it probably requires methods not available to even the 5–10 per cent of race-track gamblers who are the most sophisticated. Similarly, the problem of calculating the probability of winning and the distribution of

[1] This paper was written while the author was a guest at Econometric Institute, Netherlands School of Economics, Rotterdam, during his tenure as a National Science Foundation Senior Post-Doctoral Fellow. The data for the computations were made available by Martin Weitzman, who collected them in the course of his own investigation into gambling behavior ("Utility Analysis and Group Behavior: An Empirical Study," *Journal of Political Economy*, Vol. LXXIII [February, 1965]) and who generously made them available to me. Johannes Kemperman and Arthur Stone helped me to prove some of the things I strongly suspected were true, for which I thank them. J. Boas, of the Econometric Institute, and S. Beguin, of the University of Rochester, helped with the computations. Fritz Holte, Lionel McKenzie, Frederick Mosteller, and Herman Wold all read early drafts of the paper and made helpful suggestions.

[2] M. Friedman and L. J. Savage, "The Utility Analysis of Choices Involving Risk," *Journal of Political Economy*, Vol. LVI (August, 1948).

prizes, given a combination of several bets, is almost certainly beyond the capabilities of most race-track gamblers. If these gamblers behave as though they know statistical prediction methods and the probability calculus, it seems reasonable to suppose that, in a variety of other circumstances, human beings can be expected to respond appropriately to risky situations merely after having had sufficient experience with them.

II. PARI-MUTUEL BETTING

Although pari-mutuel betting actually allows bets on horses to win, to finish second, and to finish third, this discussion will be limited to bets on winners. Observations of these bets are sufficient for testing the hypothesis developed here. Observations of place (second) and show (third) bets could be used to strengthen the empirical case, but the data are not available.

Define symbols as follows:

x = the total value of bets placed on a horse race

x_i = the total value of bets placed on the ith horse in the race

k = the proportion of x that is paid out by the track to winners. The proportion $1 - k$ is retained by the track to pay taxes, expenses, and profits

R_i = the return. This is the number of dollars which will be returned by the track in exchange for a $1.00 winning ticket on the ith horse. It is equal to the odds plus 1

p_i = the probability that the ith horse will win the race.

Then

$$R_i = \frac{kx}{x_i}, \qquad (1)$$

and the expected gain of a $1.00 bet on the ith horse is given by

$$p_i R_i - 1 . \qquad (2)$$

This is almost always negative. It is important to note that the return and hence the expected gain are determined by the bettors themselves, so that an aggregate of race-track gamblers engaged in pari-mutuel betting constitute a kind of market in which it may be presumed that, except for imperfections in the operation of the market, each gambler is satisfied not to change his bet, given the returns which prevail at the moment when the race begins.[3]

III. BETTING COMBINATIONS

Simple bets on single horses can be compounded through various procedures into more complicated gambles. For instance, a gambler might choose one horse in each of eight races and bet $2.00 on each horse. Depending on how many horses won and which ones they were, it would be possible to distinguish a very large number of different prizes resulting from such a procedure. There are 2^8 distinct possible combinations of wins and losses. If the returns were chosen so as to avoid coincidences, each of those would yield a distinct prize. Bets can also be combined so as to yield only two possible outcomes, a winning outcome and a losing outcome. Suppose for some reason gamblers prefer multiple-prize gambles. At the race track these must be constructed from a number of single-prize (two-outcome) gambles. But in choosing a single-prize gamble as an element of a multiple-prize gamble, the gambler can choose either a simple bet on a single horse or a single-prize compound gamble.

The following rationality hypothesis will be made here. If a gambler must choose between two single prize bets in

[3] This may not be the case since there is no resale market for bets that have already been made. An individual may place a bet because he thinks R is high on some horse, and then he may find that the action of the market drives R down to a level such that he would like to change his bet. It is assumed here that after this happens a few times he will make his bets, taking this possibility into account.

which he risks losing $1.00, he will (1) if the probabilities of winning are equal, choose that with the greater prize; (2) if the prizes are equal, choose that with the greater probability of winning; and (3) always choose a gamble which in both prize and probability of winning is greater than the other.

Besides betting on a single horse, there are three procedures for constructing single-prize gambles at the race track. The following descriptions are of gambles in which $1.00 is risked but any amount could be risked in the same ways.

1. *The martingale.*—Select n horses each running in a different race. They pay returns $R_1, \ldots, R_n$ and will win with probability $p_1, \ldots, p_n$, where the subscripts denote the order in which the races are run. Choose sums of money $a_1, \ldots, a_n$ such that

$$a_1(R_1 - 1) = B, \qquad (3)$$

$$a_{i+1}(R_{i+1} - 1) - \sum_{j=1}^{i} a_j = B, \qquad (4)$$

$$(i = 1, \ldots, n-1)$$

$$\sum_{i=1}^{n} a_i = 1. \qquad (5)$$

Bet a_1 on the first horse. Then bet a_2 on the second horse if the first horse loses, but terminate betting if he wins. Continue betting until either n bets have been lost or one bet has been won. If one of the n horses wins, the prize will be B. If they all lose, the loss will be $1.00. The probability of winning is

$$1 - \prod_{i=1}^{n} (1 - p_i),$$

and the probability of losing is

$$\prod_{i=1}^{n} (1 - p_i).$$

It can easily be seen that the probability

of winning a martingale is greater than the probability of winning any of the bets from which it is constructed. It is therefore a method for combining low-probability–high-return bets into one bet with high probability–low return.

2. *The combination.*—Since this procedure has no widely used name, one has been provided for the purpose of reference here. It consists of choosing n horses all running in the same race. They pay returns $R_1, \ldots, R_n$ and will win with probability $p_1, \ldots, p_n$. Select sums of money $a_1, \ldots, a_n$ such that

$$a_i(R_i - 1) = B \quad (i = 1, \ldots, n) \qquad (6)$$

and

$$\sum_{i=1}^{n} a_i = 1 \qquad a_i > 0. \qquad (7)$$

Bet on all n horses, betting a_i on the ith horse. This will give a gamble in which B will be won with probability

$$\sum_{i=1}^{n} p_i$$

and $1.00 will be lost with probability

$$1 - \sum_{i=1}^{n} p_i.$$

Like the martingale, this is a procedure for combining low-probability–high-return simple bets into a high-probability–low-return compound bet.

3. *The parlay.*—Select n horses, each running in a different race, paying returns $R_1, \ldots, R_n$, and winning with probability $p_1, \ldots, p_n$, where the subscripts indicate the order in which the races will be run. Bet $1.00 on the first horse. If it wins, bet all the proceeds R_1 on the second horse. Continue this either until all the horses have won or until one of them loses. The prize

$$\$ \left(\prod_{i=1}^{n} R_i - 1 \right)$$

will be won with probability

$$\prod_{i=1}^{n} p_i.$$

This procedure can be used to combine high-probability–low-return bets into low-probability–high-return bets.

More complicated single-prize gambles can, of course, be constructed by using combinations of these procedures. A sequence of parlays could be combined into a martingale, for instance, but the three procedures described here are the only ones that can be used to combine single-prize gambles into a single-prize compound gamble.

IV. THE THEORETICAL RELATION BETWEEN RETURNS AND PROBABILITIES

If gamblers satisfy the rationality hypothesis, the possibility of combining gambles through parlays and martingales places certain constraints on the function, established in market equilibrium, which relates return to probability of winning. These constraints are as follows:

Given any probability of winning, p^*, and the return related to it, R^*,

$$[1 - (1 - p)^c]^{-1} \leq R \leq p^r$$
$$\text{for } p^* \leq p \leq 1, \quad (8)$$

where

$$c = \frac{\log[\,(R^* - 1)/R^*\,]}{\log(1 - p^*)}, \quad (9)$$

$$r = \frac{\log R^*}{\log p^*}, \quad (10)$$

$$p^r \leq R \leq [1 - (1 - p)^c]^{-1}$$
$$\text{for } 0 \leq p \leq p^*. \quad (11)$$

These limits can be derived as follows:
If, for $p^* \leq p \leq 1$, $R(p) > p^r$, it is possible to construct a parlay which is preferred to the gamble represented by (p^*, R^*). If, for $p^* \leq p \leq 1$, $R(p) <$

$[1 - (1 - p)^c]^{-1}$, it is possible to combine, through a martingale, gambles represented by (p^*, R^*) into gambles that will be preferred to those that violate the condition given by the left-hand inequality in expression (8). The analogous arguments apply for $0 \leq p \leq p^*$.

For these limits to make sense, it is necessary that

$$[1 - (1 - p)^c]^{-1} \leq p^r$$
$$\text{for } p^* \leq p \leq 1 \quad (12)$$

and

$$[1 - (1 - p)^c]^{-1} \geq p^r$$
$$\text{for } 0 \leq p \leq p^*. \quad (13)$$

If expressions (12) and (13) failed to hold, it would be possible to combine gambles (p^*, R^*) through a martingale and then combine martingales into a parlay so as to produce a gamble that would be preferred to (p^*, R^*). It is shown in Appendix A that this contradiction is not the case.

The relation

$$R = p^r \quad (14)$$

is the only function relating returns to probability for which it is true that, for every bet on a single horse, any parlay having the same probability of winning will pay the same return. This follows from the assumption that in equilibrium there cannot be two gambles offered such that one is rationally preferred to the other (see Appendix B).

The relation

$$R = [1 - (1 - p)^c]^{-1} \quad (15)$$

is the only one for which it is true that, for every bet on a single horse, any martingale having the same probability of winning as the single bet will pay the same return as the single bet (see Appendix C).

If there is strong preference for gambles with high returns and low probabilities, the relation between R and p given by equation (14) holds and r is determined by the proportion, k, retained by the track of every dollar bet as well as by the distribution of horses among different probabilities of winning.

Similarly, if there is strong preference for low-return–high-probability bets, equation (15) holds and c is determined by the same factors as determine r.

The combination possibility is not considered since it gives the upper limit

$$R \le \frac{p^* R^*}{p} \quad (p < p^*). \quad (16)$$

But this lies above the upper limit given by expression (11). Similarly the combination gives

$$R \ge \frac{p^* R^*}{p} \quad (p > p^*). \quad (17)$$

But this lies below the lower limit given by expression (8).[4]

V. RACE-TRACK DATA

The correct procedure for obtaining data to be used for the purpose of testing the hypothesis that gamblers behave rationally would be to obtain records giving, for each of a large number of horses, measures of all the factors that determine the probability of winning, whether the horse won, and the odds set by the action of the market. The relation between p and the relevant factors would be estimated, estimates of p would be obtained for each horse, and the relation between R and p would be examined.

The data used here were collected by Martin Weitzman for another purpose and hence are not perfectly suited for the purposes of this paper.

Weitzman collected observations of 110,000 horse performances at New York

State race tracks, recording R for each horse performance and whether the horse won. He arranged the horse performances in ascending order by odds. He then divided them into groups of 350, taking them in ascending order. If a group of 350 failed to include a single win, he enlarged the group, taking observations in ascending order by odds, until a win was included. These are used here, respectively, as p and R. Thus R is the average R for a group, and p is the observed average vlaue of p given the average R. Undoubtedly the use of these data to estimate the regression of R on p can be objected to on grounds of estimation bias, but no other method could be found to use these data more effective-

[4] Weitzman (op. cit.) treats the relation between R and p as if it were generated by a market of gamblers who distribute their bets among horses so as to make the marginal utility of bets on all horses equal. This would happen if the typical bettor placed bets on some horse and found that, as the amount bet increased, the marginal utility of the bet decreased until it was equal to the marginal utility of the second-best bet and so on until he found himself betting on all the horses and indifferent as to which horse he bet his next penny on. This is a possible consequence of taking a Friedman-Savage utility function to the race track, but it is much more likely that the gambler would find that the marginal utility of the most preferred gamble would fall to zero before the marginal utility of all gambles became equal.

In order to treat the relation between R and p as an indifference curve, it is necessary for Weitzman to assume, as I have done, that the gamblers are sophisticated and rational. But from these assumptions it follows that the relation between R and p is a transformation through which gambles of one sort can be combined into gambles of another sort. There may exist indifference curves which relate R and p, but we could observe them only through experiments of the sort proposed by Friedman and Savage. Under a pari-mutuel system of betting the relation between R and p will have the shape of an indifference curve only in the unlikely event that the transformation and the indifference curve are identical. If we superimpose a family of indifference curves on a transformation, we will see that the transformation will be used to acquire the highest indifference curve through the procedures described above and we will observe points, all of which lie on the transformation.

ly in connection with the problem at hand. While it is recommended that this difficulty be kept in mind while reading what follows, it seems unlikely that the conclusions reached here would be greatly affected by the replacement of these data with ideal data.

VI. THE EMPIRICAL RELATION BETWEEN RETURN AND THE PROBABILITY OF WINNING

If the gamblers who frequent the race track are rational, as hypothesized here, and sophisticated in the sense that they behave as though they understand both how to calculate the probability that a horse will win and how to combine the probabilities of combinations of events, the empirical relation between return and probability of winning should satisfy one of the following descriptions:

If there is strong preference for high-probability–low-return bets the relation should be given by equation (15). Suppose that $k = 0.9$ and that c is such as to give $R = 1.9$ when $p = .5$. From expression (1) it follows that a horse for which $p = .5$ must be bet on by slightly more than 47 cer cent of the dollars if R is to be 1.9. But suppose 60 per cent of the dollars are bet on such a horse even if R falls to 1.5. We would probably observe in such a case that equation (15) would hold because the demand for lower-probability bets would be insufficient, in itself, to bring R down to the upper limit imposed by the martingale. The additional demand for these bets would come from those who want to bet on $p = .5$ and are constructing martingales in order to do better than $R = 1.5$. These would push R for $p < .5$ just down to the upper limit imposed by the martingale.

In this case we could fit to the data the relation

$$\log (1 - R^{-1}) = c \log (1 - p). \quad (18)$$

Using the data supplied by Weitzman the function

$$\log (1 - R^{-1}) = c_0 + c_1 \log (1 - p) + c_2 \log^2 (1 - p) \quad (19)$$

was estimated. The estimated equation is

$$\log (1 - R^{-1}) = -0.0078 + 1.15 \log$$
$$(0.0216) \quad (0.02)$$
$$(1 - p) + 0.09 \log^2 (1 - p). \quad (20)$$
$$(0.03)$$

If the martingale lir 't holds throughout the observed range of p, the coefficient of the constant and squared terms in equation (20) should be zero. The constant term is not significantly different from zero, but the coefficient of the squared term is. (Standard errors are given in parentheses below the coefficients.) This suggests that the data do not support the hypothesis that the martingale limit holds over the observed range p.

If there is strong preference for low-probability–high-return gambles we would expect, from a similar argument, that the relation given by equation (14) would hold over the observed range of p. To test this the data were used to estimate the relation

$$\log R = r_0 + r_1 \log p + r_2 \log^2 p. \quad (21)$$

The estimated equation is

$$\log R = -0.365 - 1.27 \log p$$
$$(0.06) \quad (0.04)$$
$$- 0.074 \log^2 p. \quad (22)$$
$$(0.007)$$

The fact that r_0 and r_2 are significantly different from zero suggests that the parlay limit is as unsatisfactory as the martingale limit for the purpose of describing the data. Another possibility is that some intermediate level of p is in excess demand and that the function which will fit the data best is one which satisfies the martingale limit to the left of some value of p and the parlay limit to the right. It can be shown, however, that such a function would give a positive value of r_2 rather than the negative value actually obtained.

Examination of Figure 1, which gives the distribution of observed values of log R and log p, suggests that values of R associated with very low values of p are

lowest-probability horse in every race. If this proportion exceeds that which is consistent with the parlay limit, R for the lowest-probability horses is driven below what is implied by the model; and the market cannot compensate for the depression of these returns. This effect is strongest for lowest-probability horses and diminishes as p increases. This follows from the fact that a horse for which

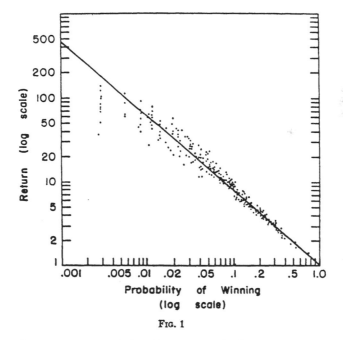

Fɪɢ. 1

too low to conform to any possible implication of the model outlined here. A further hypothesis is necessary to explain this departure from the implications of the model. Two possibilities which are consistent with rationality immediately suggest themselves:

1. A certain proportion of gamblers want the results of a day of betting at the races to be a gamble of such low probability and high return as to require that they always construct parlays using the

$p = .01$ is almost certain to be the horse with the smallest chance of winning any given race. A horse for which $p = .05$ is less likely to be the least-favored horse, and a horse for which $p = .1$ is almost certain not to be the horse least likely to win the race. Thus R will be frequently depressed for $p = .01$, less frequently for $p = .05$, and rarely for $p = .1$.

This phenomena is not inconsistent with rationality but is merely an artifact of the track's failure to provide martingale limit to the left of some value of p and the parlay limit to the right. It can be shown, however, that such a function would give a positive value of r_2 rather than the negative value actually obtained.

Examination of Figure 1, which gives the distribution of observed values of log R and log p, suggests that values of R associated with very low values of p are too low to conform to any possible implication of the model outlined here. A further hypothesis is necessary to explain this departure from the implications of the model. Two possibilities which are consistent with rationality immediately suggest themselves:

1. A certain proportion of gamblers want the results of a day of betting at the races to be a gamble of such low probability and high return as to require that they always construct parlays using the lowest-probability horse in every race. If this proportion exceeds that which is consistent with the parlay limit, R for the lowest-probability horses is driven below what is implied by the model; and the market cannot compensate for the depression of these returns. This effect is strongest for lowest-probability horses and diminishes as p increases. This follows from the fact that a horse for which

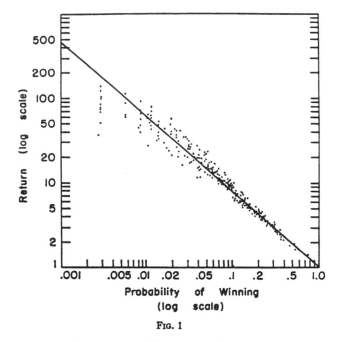

FIG. 1

$p = .01$ is almost certain to be the horse with the smallest chance of winning any given race. A horse for which $p = .05$ is less likely to be the least-favored horse, and a horse for which $p = .1$ is almost certain not to be the horse least likely to win the race. Thus R will be frequently depressed for $p = .01$, less frequently for $p = .05$, and rarely for $p = .1$.

This phenomena is not inconsistent with rationality but is merely an artifact of the track's failure to provide horses with sufficiently low probabilities of winning.

2. A certain proportion of gamblers select a horse at random (or at least on some basis which distributes their bets uniformly over horses of all probabilities of winning) so that horses for which p is low are bet on too heavily. The effect is similar to that of the first hypothesis. This too is consistent with rationality if

along with the ratios of the coefficients to their standard errors.

By the time forty observations have been dropped, the relation is almost perfectly linear and the constant term is zero. To rule out the possibility that dropping these observations might tend to make the data conform to the function given by equation (18), the same procedure was followed in re-estimating equa-

TABLE 1

No. of Observations	Coefficients			Ratios of Coefficients to Standard Errors		
	r_0	r_1	r_2	r_0/σ_{r_0}	r_1/σ_{r_1}	r_2/σ_{r_2}
257...............	−0.36	−1.27	−0.074	5.5	28.7	11.0
247...............	−0.21	−1.12	− .046	3.0	21.5	5.2
237...............	−0.09	−0.86	− .004	1.7	21.3	0.6
227...............	−0.15	−1.07	− .035	2.0	16.7	2.8
217...............	−0.05	−0.96	−0.007	0.6	13.4	0.5

TABLE 2

No. of Observations	Coefficients			Ratios of Coefficients to Standard Errors		
	c_0	c_1	c_2	c_0/σ_{c_0}	c_1/σ_{c_1}	c_2/σ_{c_2}
257.............	−0.0078	1.15	0.088	0.03	48	2.78
247.............	− .0075	1.15	.090	.03	46	2.92
237.............	− .0075	1.15	.090	.03	44	2.65
227.............	− .0075	1.15	.090	.03	42	2.56
217.............	−0.0073	1.15	0.092	0.03	40	2.50

the cost of obtaining information and estimating p is sufficiently high.

Anything which leads to an excess demand for bets on low-probability horses that cannot be satisfied by a parlay will lead to the observed effect. Irrational behavior of several types could also explain this effect.

In order to discover whether the observations for which p is very low are responsible for the significance of r_2, several regressions were estimated, dropping successively the lowest value of p, 10 at a time. The results are given in Table 1

tion (19), and the results are given in Table 2.

The observations which are dropped in estimating the coefficients given in Tables 1 and 2 are those which are inconsistent with the parlay limit but even more inconsistent with the martingale limit. The fact that c_2 remains significant when these observations are dropped strongly suggests that the martingale relation, $R = [1 − (1 − p)^c]^{-1}$, does not hold.

Assuming that the correct explanation of the observed data is that there is excess demand for gambles at very low p

which cannot be satisfied by parlays, how great must this demand be to give the observed effect? When forty observations have been dropped and the depressed values of R have been eliminated, the lowest remaining value of p is about .02. For $p = .02$ we expect R to be about 40. Assuming that the track retains about 15 per cent ($k = 0.85$) of every dollar bet, a horse for which $p = .02$ should be bet on by about 2 per cent of the bettors. If this represents the minimum proportion that will be bet on any horse under the second of the two hypotheses to explain the depressed odds for $p > .02$, and if the average race has ten horses running in it, it follows that about 20 per cent of all the dollars bet are bet by people who choose their horses at random.

Again assume that the average race has about ten horses running in it and that $k = 0.85$, but adopt the hypothesis that excess demand for low values of p springs from a demand for values of p so low and R so high as to be obtainable only through a parlay involving the horse in every race for which p is lowest. The lowest value of p observed by Weitzman was about .003. For $p = .003$, R should be about 200. This is obtained by using the line,

$$\log R = -0.8793 \log p, \quad (23)$$
$$(0.0054)$$

which is estimated from the data, and which is shown in Figure 1. The average observed value of R is about 100. If R is to be 200, about 0.425 per cent of all money bet in a given race must be bet on any horse for which $p = .003$. In order for R to fall to 100 it would be necessary for the percentage bet to rise to 0.85 per cent. This proportion is bet by people who want very low-probability–high-return bets.

VII. CONCLUSIONS

To what extent does the evidence support the hypothesis that gamblers behave as though they are both sophisticated and rational? The fact that the various functions fitted to Weitzman's data all yield correlation coefficients of about .98 suggests either that the gamblers behave as though they are sophisticated themselves or that they are following sophisticated advice. Advice that is available to race-track gamblers comes in the form of the "morning line," which is the first approximation to the odds $(R - 1)$ that will finally obtain for the horses running in each race, and predictions as to which horse in each race is most likely to win. The "morning line" is computed by a handicapper employed by the track so that odds can be posted before betting begins. After the first few minutes of betting, the "morning line" is replaced by odds based on the calculations of the pari-mutuel machine. The handicapper attempts to predict the odds which the bettors will finally achieve as a consequence of their own betting, except that he never places odds at more than 20:1 or less than 1:1. A glance at Figure 1 will certainly assure anyone that outside this range gamblers display their ability to distinguish between horses having different probabilities of winning. Thus the advice of the "morning line" is not sufficient to explain the sophisticated behavior of these gamblers. Advice in the form of prediction of favorites and best bets is confined almost exclusively to the range in which $p > .1$ and can therefore be immediately dismissed as the basis for the generally sophisticated behavior of gamblers. Of course in addition to advice, each gambler can avail himself of information on which to base his own estimates. Figure 1 suggests that somehow this information is put to good use.

With respect to rationality, the evidence strongly supports the hypothesis that a gambler will not make a simple bet when a parlay for which p is the same will yield a greater prize or when a parlay for which the prize is the same has a greater p, except for the special circumstances affecting bets on horses for which $p < .03$.

What is the practical significance of the finding that sophistication and rationality can probably be expected from people who must frequently make choices among risky alternatives? Many applications of economic theory make use of the assumption that expectations about the future are based on past experience. Managers are assumed to forecast demand on the basis of past demand and other relevant factors. In the absence of an explicit forecast an economist interested in explaining actions which depend on the forecast will assume that it was made rationally and with sophistication. Consumers are assumed to be rational and sophisticated in predicting their own future income on the basis of their past experience. Examples of studies in which observable variables on which expectations should be based are substituted for the unobservable expectations are numerous. These all make use of the assumption that the decision-makers are sophisticated and rational. The assumption is unessential when it is merely used to connect observable events in the past with observable events in the future. But when the assumption yields an otherwise unobservable quantity, for example, a subjective discount rate, it is essential, and the support which the assumption receives from the evidence presented here becomes significant.

APPENDIX A

LIMITS ON THE $p - R$ RELATION

It must be shown that for any p_0 and R_0 such that $0 \leq p_0 \leq 1$ and $1 \leq R_0 \leq 1/p_0$, the following is true:

$$[1 - (1 - p)^{c_0}]^{-1} \leq p^{r_0} \text{ for } p > p_0$$

and

$$[1 - (1 - p)^{c_0}]^{-1} \geq p^{r_0} \text{ for } p < p_0,$$

where

$$c_0 = \frac{\log(1 - R_0^{-1})}{\log(1 - p_0)}.$$

This is equivalent to

$$(R_0^{-1})^{\log p_0 / \log p}$$
$$+ (1 - R^{-1})^{\log(1-p)/\log(1-p_0)} \quad (A1)$$
$$\begin{cases} \leq 1 \text{ if } p > p_0 \\ \geq 1 \text{ if } p < p_0. \end{cases}$$

Consider the function

$$f(x) = x^a + (1 - x)^\beta,$$

where $0 \leq x \leq 1$ and $0 < a < 1 < \beta$. Evaluating $f(0)$ and $f(1)$, we find

$$f(0) = f(1) = 1.$$

Differentiating f with respect to x we obtain

$$f'(x) = ax^{a-1} + \beta(1 - x)^{\beta-1}. \quad (A2)$$

From this it follows that

$$\lim_{x \to 0} + f'(x) = +\infty > 0 \quad (A3)$$

and

$$\lim_{x \to 1} - f'(x) = a > 0.$$

From equations (A1), (A2), and (A3) and the fact that f is continuous, it follows that, for at least one x such that $0 < x < 1$, $f(x) = 1$.

Now assume that there are two values of x, x_1, and x_2 which satisfy this condition so that

$$f(0) = f(x_1) = f(x_2) = f(1) = 1 .$$

From this it follows that, for three values of x, $f'(x) = 0$, or that $ax^{a-1} = \beta(1-x)^{\beta-1}$ has three solutions all greater than zero and less than 1. This is equivalent to saying that

$$g(x) = \log a + (a-1) \log x \\ - \log \beta - (\beta - 1) \log (1-x) = 0 \quad (A4)$$

has at least three solutions such that x is

greater than zero and less than 1. It follows that

$$g'(x) = \frac{a-1}{x} + \frac{\beta-1}{1-x} = 0$$

has two solutions between zero and 1. But equation (A4) has only one solution, $x = (a-1)/(a-\beta)$. Thus there is only one x, x_0, such that

$$f(x_0) = 1 , \qquad 0 < x_0 < 1 .$$

It can be seen immediately that

$$f(x_0) > 1 , \qquad 0 < x < x_0 ,$$

and

$$f(x_0) < 1 , \qquad x_0 < x < 1 .$$

It follows immediately that (A1) holds.

APPENDIX B

THE PARLAY RELATION

It will be shown here that the function

$$R_p(p) = p^r \qquad (B1)$$

satisfies the requirement that any parlay for which the probability of winning is p_0 will yield a return $R_p(p_0)$ and that no other function has this property. It is assumed only that R_p is monotone. This corresponds to the assumption that the market behaves rationally in not allowing one gamble to be dominated by another. Select gambles such that the probabilities of winning are $p_1 \ldots p_n$ and such that

$$\prod_{i=1}^{n} p_i = p_0 .$$

A parlay consisting of bets on this sequence of gambles will, if (B1) holds, give a return which is the same as the return to a single bet which will win with probability p_0.

$$R_0 = \prod_{i=1}^{n} R_p(p_i) = \prod_{i=1}^{n} p_i^r = \left(\prod_{i=1}^{n} p_i \right)^r .$$

To show this function is unique in having this property, it will be assumed that some

other function also has the same property, and it will be shown that this assumption contradicts the assumption that the function is monotone.

Assume that $R = f(p)$ has the property

$$f(\Pi_i p_i) = \Pi_i f(p_i) ,$$

but f does not have the form given in equation (B1). It must be possible to find p_1 and p_2 such that

$$R_1 = f(p_1) = p_1^{r_1}$$

and

$$R_2 = f(p_2) = p_2^{r_2} ,$$

where $r_1 > r_2$.

For $0 < p < 1$,

$$p^{r_1} < p^{r_2} .$$

Since (p_1, R_1) and (p_2, R_2) both satisfy this function, so do all points of the form (p_1^n, R_1^n) and (p_2^m, R_2^m), where m and n are positive integers. If, for some m_0 and n_0,

$$(p_2^{m_0}, R_2^{m_0}) \geq (p_1^{n_0}, R_1^{n_0}) ,$$

in the sense that the inequality holds for at

least one element, the assumption of monotonicity (rationality) is violated.

All points of the form (p_1^n, R_1^n) lie on the curve $R = p^{r_1}$, and all points of the form (p_2^n, R_2^n) lie on the curve $R = p^{r_2}$.

Consider the graph shown in Figure 2. If

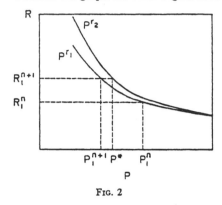

Fig. 2

points of the form (p_2^m, R_2^m) fall in intervals like that between p^* and p_1^n, where $p_1^{n+1} < p_2^m < p_1^n$ and $R_1^{n+1} > R_2^m < R_1^n$ and where $p^* = (R_1^{n+1})^{1/r_1} = p_1^{(n+1)(r_1/r_1)}$, there is no contradiction. But for all n such that

$$n > \frac{r_1/r_2}{1 - (r_1/r_2)}.$$

It follows that $p^* > p_1^n$ and that this interval vanishes. Thus for all $p > p_1^a$, where

$$a = \frac{r_1/r_2}{1 - (r_1/r_2)},$$

intervals into which points (p_2^m, R_2^m) can fall without violating the monotonicity condition vanish. Since it is possible to find values of m such that $p_1^n < p_1^a$, we find that the assumption of monotonicity is violated and the function given by (B2) is unique in having the required properties.

APPENDIX C

THE MARTINGALE RELATION

It will be shown that the function

$$R_m(p) = [1 - (1 - p)^c]^{-1},$$
$$0 < p \leqq 1, \quad (C1)$$

satisfies the requirement that any martingale for which the probability of winning is p_0 will yield a return $R_m(p_0)$ to a \$1.00 bet.

A martingale for which the probability of winning is p_0 is constructed by selecting n gambles such that

$$\prod_{i=1}^{n}(1 - p_i) = 1 - p_0,$$

where p_i is the probability of winning the ith gamble. From equation (C1) obtain R_i $(i = 1, \ldots, n)$ and solve the system of equations,

$$a_1(R_1 - 1) = k$$
$$a_i(R_i - 1) - \sum_{j=1}^{i-1} a_j = k \quad (i = 2 \ldots, n)$$
$$\sum_{i=1}^{n} a_i = 1,$$

for a_i $(i = 1, \ldots, n)$ and k. The quantity k is the amount which will be won if the martingale succeeds, and it must be shown here that, given p_0, k is independent of n and the choice of p_i if and only if (C1) holds. The quantity a_i is the proportion of a dollar to be bet on ith events.

Solving for a_i we find that

$$a_1 = \frac{k}{R_1 - 1},$$
$$a_2 = k\left(\frac{1}{R_2 - 1} + \frac{1}{(R_2 - 1)(R_1 - 1)}\right),$$

and that in general

$$a_m = k\Sigma \frac{1}{(R_m - 1)(R_{i_1} - 1)} \cdots (R_{i_v} - 1),$$

where the summation is over all distinct sets of integers

$$m > i_1 > i_2 \ldots > i_v \geq 1$$
$$\times (v = 0_1 \ldots, m - 1). \quad (C2)$$

The sum

$$\sum_{i=1}^{n} a_i = K\Sigma\Sigma \frac{1}{(R_m-1)(R_{i_1}-1)\dots(R_{i_\nu}-1)}$$

when the right-hand sum is over m as well as over all distinct sets of integers given in (C2). But this is the same as

$$k \left[\prod_{i=1}^{n}\left(1+\frac{1}{R_i-1}\right)-1 \right].$$

Since $\Pi_{i=1}^{n} a_i = 1$,

$$k = 1 \bigg/ \left[\prod_{i=1}^{n}\left(1+\frac{1}{R_i-1}\right)-1 \right].$$

From (C1),

$$1+\frac{1}{R_i-1} = (1-p_i)^{-c},$$

so that

$$k = \prod_{i=1}^{n} \frac{1}{(1-p_i)^{-c}-1} = \frac{1}{(1-p_0)^{-c}-1},$$

which is independent of n and the choice of p_i. In other words, if we construct a martingale such that the probability of winning is p_0 the prize, k, will be given by $[1-(1-p_0)^c]^{-1}-1$ regardless of how the martingale is constructed. This will be the same as the prize associated with a single bet for which the probability of winning is p_0.

The proof that this function is unique rests only on the assumption that the relation between R and p is monotone. This can be interpreted as an assumption that the market behaves rationally. The proof is similar to that given in the case of the parlay if the relation $(1-R^{-1})=(1-p)^c$ is treated in a manner similar to the parlay relation.

Probability and Utility Estimates for Racetrack Bettors

Mukhtar M. Ali

University of Kentucky

Subjective and estimated objective winning probabilities are obtained from 20,247 harness horse races. It is shown that subjectively a horse with a low winning probability is exaggerated and one with a high probability of winning is depressed. Various hypotheses characterizing the bettors' behavior to explain the observed subjective-objective probability relation are explored. Under some simplified assumptions, a utility of wealth function of a decision maker is derived, and a quantitative summary measure of his risk attitude is defined. Attitude toward risk of a representative bettor is examined. It is found that he is a risk lover and tends to take more risk as his capital dwindles.

Several laboratory experiments (Preston and Baratta 1948; Yaari 1965; Rosett 1971) suggest that in making a decision under uncertainty low probability events are overbet and high probability events are underbet. None of these experiments was conducted under a natural environment of the decision makers. In two separate studies of actual bettings in horse races, Grifith (1949, 1961) arrived at the conclusion similar to those of Preston and Baratta and Yaari. McGlothlin (1956) repeated the Grifith (1949) study with a different set of data on horse races, and among other things he confirmed Preston and Baratta's (1948) findings. None of the Grifith and McGlothlin studies considered more than 1,500 races to estimate several probabilities. Martin Weitzman (1965) analyzed over 12,000 races to obtain a relationship between the objective probability of

The author wishes to acknowledge the help and ecouragement received from his wife, Julia W. Ali. She spent hours of her time in collecting the data for the analysis. Comments on an earlier draft received from J. R. Marsden and Lawrence Fisher are highly appreciated. Discussions by the participants at the Econometric and Statistics Colloquium conducted by Arnold Zellner at the University of Chicago have been beneficial in writing this paper. The comments of the editor and the referee are highly appreciated.

TABLE 1

DATA DESCRIPTION

Race Track and Year	Racing Dates (N)	Races (N)	Average Daily Attendance	Average Bet per Race ($)*	Average Bet per Person in a Race ($)
Saratoga:					
1970	193	1,909	3,784	23,701	6.26
1971	182	1,779	3,873	25,500	6.58
1972	187	1,829	3,393	23,706	6.98
1973	189	1,834	3,486	25,350	7.27
1974	172	1,721	3,541	25,694	7.26
Roosevelt:					
1970–71† ..	192	1,698	20,426	228,884	11.21
1972	154	1,355	17,014	213,082	12.52
1973	153	1,347	17,148	214,110	12.49
1974	159	1,406	15,789	216,435	13.71
Yonkers:					
1971	155	1,381	18,025	224,289	12.44
1972	145	1,268	17,258	234,837	13.61
1973	160	1,407	15,871	225,639	14.22
1974	148	1,313	15,988	227,853	14.25

NOTE.—All tracks are in New York.
* Includes all possible betting opportunities.
† Oct.–Dec. 1970, and March–Oct. 1971. In the analysis the time period is treated as the year 1971.

a horse to win and the return for a dollar bet if the horse wins. The subjective probability defined in the Grifith or McGlothlin studies is proportional to the reciprocal of this return. The estimated relationship confirms the Grifith and McGlothlin findings.

The studies of Grifith, McGlothlin, and Weitzman were limited to the thoroughbred races. The probability estimates of these studies have serious estimation errors. The estimates obtained by Weitzman are the most reliable in which a coefficient of variation of 16 percent or higher is common.[1] This can be considered large to detect even a difference of say, 20 percent, between objective and subjective probabilities. Moreover, serious technical difficulties, though ignored, arise in their grouping of observations (see Sec. I) to estimate the objective probabilities.

The first theoretical attempt to explain the observed discrepancies between objective and subjective probabilities is by Rosett (1965). Weitzman (1965) constructed a representative utility of wealth function as an implication from the relationship between objective and subjective probabilities.

In this paper, public betting behavior in 20,247 harness horse races is analyzed. The data are described in table 1. Subjective and estimated

[1] Improved estimates can be obtained from his estimated objective probability-return relationship, but the significance of this improvement cannot be ascertained from the available data.

objective winning probabilities are obtained in Section I. It is found that subjectively a horse with a low probability of winning is overstated and that a high probability of winning is understated. Section II explores various hypotheses characterizing the bettors' behavior to explain the observed subjective-objective probability relation. Specifically, a theorem is proved showing a possible alternative explanation of these biases. Section III is devoted to studying the betting public's risk attitudes, especially as they are affected by the amount of money available to bet. A numerical measure of risk attitude is defined. It is found that the betting public behaves as risk lovers. The risk attitude seems to change with the amount of money (capital) available to bet. The representative bettor becomes more risk loving as his capital declines. Concluding remarks are in Section IV.

I. Probability Estimates

Besides the popular betting of the "place," "show," "daily double," etc., there is one regular betting opportunity in a race known as "win" bet. Betting on a horse to win pays only when the horse finishes first in the race. [2] Payoffs are according to the odds. Odds are profits per dollar bet to a successful bettor. Odds are determined from the bets made by the public, the track take, and breakage. [3] Odds for different types of bets—win, place, show, etc.—are computed separately. Suppose there are H horses numbered [4] $1, 2, \ldots, h, \ldots, H$, and let X_h be the amount of money that has been bet for horse h to win. Then,

$$W = \sum_{h=1}^{H} X_h$$

is the total win bet in the race and is called the "win pool." Let the track take and breakage be α. Then the odds on horse h are

$$a_h = [(1 - \alpha)W - X_h]/X_h = (1 - \alpha)W/X_h - 1, h = 1, 2, \ldots, H.$$

$$(1)$$

---

[2] Races with "dead heat" at any finishing position are not considered.

[3] A fixed proportion of the amount bet in a race is taken out by the track before it distributes the rest to the successful bettors. This proportion is known as the track take. The breakage arises because of the following two restrictions: (a) odds cannot be below a certain minimum, and (b) odds have to be rounded downward except when the former restriction is in effect, in which case, it is rounded upward. For the races that are analyzed, all odds are rounded to 10 cents and the minimum odds are also 10 cents.

[4] Sometimes several horses are grouped, known as "entry" or "field," for a single betting interest, and the group is assigned a single number. A bet on this number is successful if one of the horses in the group is successful. Thus, if it is a win bet, the bet is successful if one of the horses in the group finishes first. Without loss of generality, this group will be taken as a single horse.

The estimates of a_h's which are reported to the public once in every minute till the race starts fluctuate over the entire betting period. However, as fluctuation is minimum in the last few minutes of the betting period when the major amount of bets are placed, and as our data are limited to only final odds, we will proceed as if the final a_h's prevailed throughout the period (for an alternative justification of this assumption see Rosett [1965]). Our data are also limited to the odds on win bets, and therefore we will confine ourselves to the win bet only and refer to it as a bet.

Following Grifith and McGlothlin, the subjective probability of horse h to win is defined as the proportion of the win pool that is bet on it,[5] (i.e., X_h/W). This can be shown to be equal to $(1 - \alpha)/(1 + a_h)$, and from (1) it can be verified that

$$\sum_{h=1}^{H} \left(\frac{1 - \alpha}{1 + a_h} \right) = 1 \qquad (2)$$

where H is the number of horses in the race.

The objective winning probability of a horse is defined to be the proportion of times the horse wins when the race is repeated an infinitely large number of times. Thus, a race can be taken as a binomial trial with this objective probability for the winning outcome. In a race the total of the objective probabilities is one.

Both subjective and objective probabilities are different for horses in a race, and also they differ in different races. Let the subjective probability for a horse to win the ith race be ρ_i, the objective probability be π_i, and the odds be a_i. Then $\rho_i = (1 - \alpha)/(1 + a_i)$, where α is the track take and breakage. From our data on the odds, α can be computed using (2), and hence ρ_i can be obtained. However, we have only one observation to estimate π_i. Thus, no reasonably reliable estimate of π_i can be obtained. This estimation difficulty is inherent, though ignored in the studies of Grifith, McGlothlin, and Weitzman.

In our study the horses are grouped, and their average subjective probability is compared with an unbiased estimate of the corresponding average objective probability. Horses in the same group are identified with a single horse. Let there be N races competed in by a horse with ρ_i and π_i as its subjective and objective probabilities to win the ith race. Then its average subjective and objective probabilities are defined as

---

[5] Considering each dollar bet as a vote to win, the subjective probability is the total public support for the horse to win the race. It should be noted that this definition of subjective probability is not to imply expected net gain from betting a dollar on a horse is $-\alpha$, though the subjective expected net gain is $-\alpha$ and is the same for every horse. This subjective probability is also the reciprocal of the return to a dollar bet if the horse wins when there is no track take or breakage. Thus, when there is no possibility of confusion, these two terms, subjective probability and return, will be used interchangeably.

$\bar{p} = \sum p_i/N$, and $\bar{\pi} = \sum \pi_i/N$, respectively. When there is no possibility of confusion, these averages will be referred to as the respective winning probabilities of the horse. Once p_i's are computed $\bar{p}$ can be obtained. An unbiased estimate of $\bar{\pi}$ is $\hat{\pi} = \sum Y_i/N$, where $Y_i = 1$ if the horse wins the ith race and equals zero otherwise. As the races are considered independent binomial trials, it can be shown that $E(\hat{\pi}) = \bar{\pi}$, and var $(\hat{\pi}) = \bar{\pi}(1 - \bar{\pi})/N - \sigma_\pi^2/N$, where $\sigma_\pi^2 = \sum (\pi_i - \bar{\pi})^2/N$. In the absence of a reasonable estimate of σ_π^2, we neglect it in estimating var $(\hat{\pi})$ which is estimated as $\hat{\pi}(1 - \hat{\pi})/N$. Thus, the reported standard error of $\hat{\pi}$ which is the square root of this variance-estimate is exaggerated.

In grouping the horses we use the following criteria: (a) N should be as large as possible; and (b) σ_π^2 should be as small as possible. By choosing a large N, var $(\hat{\pi})$ is reduced so that $\hat{\pi}$ is a meaningful estimate of $\bar{\pi}$. Criterion (b) is necessary for a valid test of the hypothesis of subjective over- or underestimation of objective probabilities by comparing $\bar{\pi}$ with $\bar{p}$. Previous studies have grouped the horses according to their odds falling in various selected ranges. For this grouping, it is very likely to have more than one horse in a group competing in the same race. Thus, their identity with a single horse is destroyed. To avoid this possibility, such competing horses can be treated as an "entry" or "field" (see n. 4). However, the odds on this field which are to be estimated should be lower than the individual odds on these horses, and thus the field may not belong to the group from which the individual horses are drawn. Hence, not only need we estimate the odds on the field but also the groups need to be rearranged. Thus, we abandon this method of grouping.

It can be argued that the variation in the winning probabilities of the horses most likely to win the races may not be substantial, that is, σ_π^2 would be small for this group of horses. Also, for this group N would be as large as the total number of races. Hence this grouping closely satisfies criteria a and b. Similar arguments can be put forward for the horses which are second most likely to win the races. Thus, without any further information, the horses are grouped according to "favorites." The horse with the lowest odds in a race is called the first favorite; the horse with the next lowest odds is known as the second favorite, and so on. The hth favorite in a race will be called horse h,[6] and its subjective and objective winning probabilities will be denoted by $\bar{p}_h$ and $\bar{\pi}_h$, respectively.

In the following discussion, no reference is made to horses 9 and 10 because the number of races in which they competed is small. The estimates $\hat{\pi}_h$'s and their standard errors are in columns 3 and 4 of table 2, respectively. The standard errors expressed as percentages of the estimates

---

[6] A race is limited to the maximum of 10 horses, and often eight horses compete in a race. In our data, eight horses competed in 15,749 out of 20,247 races, whereas nine horses competed in only 299 races and 10 horses competed in only 71 races.

TABLE 2

PROBABILITY ESTIMATES

Horse (h) (1)	Races Competed (N) (2)	$\hat{\pi}_h$ (3)	SE $(\hat{\pi}_h)$ (4)	$100 \left(\dfrac{SE}{\hat{\pi}_h}\right)$ (5)	$\bar{p}_h$ (6)	$\dfrac{(\bar{p}_h - \hat{\pi}_h)}{SE (\hat{\pi}_h)}$ (7)
1	20,247	.3583	.0034	0.95	.3237	− 10.29
2	20,247	.2049	.0028	1.37	.2077	0.99
3	20,247	.1526	.0025	1.64	.1513	− 0.52
4	20,247	.1047	.0022	2.10	.1121	3.45
5	20,231	.0762	.0019	2.49	.0827	3.49
6	20,088	.0552	.0016	2.90	.0601	3.01
7	19,281	.0341	.0013	3.81	.0417	5.80
8	15,749	.0206	.0011	5.34	.0276	6.20
9	299	.0033	.0033	...	...	...
1C	71	.0141	.0140	...	...	...

NOTE.—Subjective probabilities for horses 9 and 10 are not shown because they are based on a very small number of observations. Corresponding objective probability estimates with their standard errors are shown to give an idea of the unreliability of these estimates.

(col. 5) vary from 0.95 percent to 5.34 percent and they can be considered small. The subjective probabilities are in column 6. Differences between subjective and estimated objective probabilities are expressed as ratios to the respective standard errors and are shown in column 7. Except for horses 2 and 3, these differences can be considered statistically significant. Every subjective probability with the exception of that of the first favorite exceeds the corresponding objective probability estimate. For the first favorite the subjective probability is significantly less than the corresponding objective probability. Thus, subjectively, a horse with a high winning probability is understated and one with a low winning probability is overstated. Qualitatively, this finding is in accord with that observed by Preston and Baratta (1948), Grifith (1949, 1961), McGlothlin (1956), and Weitzman (1965).

The present data extend over a number of years (1970–74), various tracks, and numerous race conditions. To examine the influence of some of these factors,[7] all the data were divided into 15 data sets. The previous analysis applied to each of these data sets pointed to the same conclusion as has already been reached.

II. Theoretical Explanation of the Differences between Subjective and Objective Probabilities

Despite the sampling errors and the grouping problem, the Grifith, McGlothlin, and Weitzman findings about the relationship between subjective and objective probabilities are qualitatively in accord with

[7] Some characteristics such as average daily attendance, average bet per race, average bet per race per person, the years of racing, and the racetracks are in table 1.

the present study. Grifith and McGlothlin as well as Preston and Baratta seemed to imply that the observed differences between subjective and objective probabilities are psychological. If the bettors are sophisticated the observed subjective-objective probability relation can be explained by adopting the Friedman-Savage (1948) expected utility hypothesis (EUH) if the utility functions are restricted to a properly chosen class.[8]

Rosett (1965) claims that the observed relationship is consistent with the hypothesis that bettors are rational, sophisticated, and have strong preference for low-probability–high-return bets.[9] This hypothesis is one manifestation of EUH with the properly restricted class of utility functions. To see this let us define the equilibrium betting opportunity set as the one constituting only the nondominant opportunities; that is, for any two opportunities in the set no one is preferred to the other by every bettor. Henceforth, we also maintain that an opportunity A is preferred to another opportunity B by *every* rational bettor if and only if the winning probability of A is at least as large as that of B and the return of A is larger than that of B or the winning probability of A is larger than that of B and the return of A is at least as large as that of B. Then one can prove (left to the reader):

Theorem 1.—For any two betting opportunities, A and B, A is not preferred to B by every rational bettor if and only if A is not preferred to B by every bettor with an increasing utility function who follows EUH. As can be shown from the above theorem, the equilibrium betting opportunity set as characterized by the rationality hypothesis is identical with the one obtained by adopting the EUH and restricting the utility functions to a class of increasing functions. In this case, EUH with a class of increasing utility functions is equivalent to the rationality hypothesis. The hypothesis that bettors have strong preference for low-probability–high-return bets restricts further the class of utility functions, and following Rosett (1965) it reduces the equilibrium set to the set where objective probability, π, is related to return, R, by $R = \pi^\beta$ where $\beta < 0$.

Rosett's postulated objective probability-return relationship was consistent with only a part of his data, and it systematically overstated the returns when the winning probabilities were smaller. In what follows we show that if the assumption of sophisticated bettors is replaced by the assumption that each bettor knows something about the objective winning probabilities but no one knows them exactly, then for a wide range of

---

[8] The EUH asserts that each bettor has a unique utility of wealth function and each maximizes his expected utility to choose his preferred bet. Utility functions are unique only up to positive linear transformations.

[9] Bettors are rational in the sense that no one prefers a bet with a smaller winning probability and the same or lower return, or with a lower return and the same or lower winning probability to that available to him. Bettors are sophisticated in the sense that the objective winning probabilities are known to them.

assumptions about these estimation errors the data can be consistent with the hypothesis that bettors are rational and risk neutral.

For simplicity of exposition, let us assume that there are only two horses, H_1 and H_2, in a race. Let π_h be the objective (true) probability for H_h to win the race, ρ_h be an individual's subjective estimate of π_h, and a_h be the market-established final odds on H_h. For every individual, ρ_h's are nonnegative and $\rho_1 + \rho_2 = 1$; ρ_h's vary from person to person. Let the distribution of ρ_h be $F_h(\cdot)$ so that $F_h(\rho)$ is the proportion of individual estimates of π_h not exceeding ρ. To avoid mathematical complexity, we assume that the probability of an individual estimating π_h exactly is zero. Bettors are assumed rational, and the amount of a bet for an individual is decided upon in advance and it is the same for everyone. With the above assumptions we prove:

Theorem 2.—If π_h is the median of the distribution, $F_h(\cdot)$, and the bettors are risk neutral, the subjective probability for H_h to win, $\bar{\rho}_h = (1 - \alpha)/(1 + a_h)$ where $0 \le \alpha < 1$ is the track take and breakage, exceeds the corresponding objective probability, π_h, if and only if $\pi_h < \pi_{h'} (h \ne h')$.

Proof.[10]—As bettors are risk neutral, the utility functions are linear; and as they are rational, horse H_h will be bet on if and only if

$$[(1 + a_h)\rho_h - 1] > \max \{0, [(1 + a_{h'})\rho_{h'} - 1]\}. \tag{3}$$

Without loss of generality, we assume $\alpha = 0$. As $\bar{\rho}_1 + \bar{\rho}_2 = 1$, $1/(1 + a_1) + 1/(1 + a_2) = 1$, so that from (3), the individual with subjective probability estimate ρ_h and $\rho_{h'}$ will bet on H_h if and only if $\rho_h > 1/(1 + a_h)$. Thus, the proportion of all the bettors wagering on H_h is $1 - F_h[1/(1 + a_h)] = 1 - F_h(\bar{\rho}_h)$, and this must equal $\bar{\rho}_h$ because bettors are assumed to bet the same amount. Hence, $\bar{\rho}_h$ can be obtained as a solution to the equation,

$$F_h(\bar{\rho}_h) = 1 - \bar{\rho}_h, \qquad h = 1, 2. \tag{4}$$

As $F_h(\cdot)$ is a monotonic function, the solution is unique.

If $\pi_1 = \pi_2$, then $\pi_1 = \pi_2 = \frac{1}{2}$, and as π_h is the median of the distribution, $F_h(\pi_h) = 1 - \pi_h$, so that $\bar{\rho}_h = \pi_h = \frac{1}{2}$. If $\pi_1 > \pi_2$, then $\pi_1 > \frac{1}{2}$ and $\pi_2 < \frac{1}{2}$. Consequently, $F_1(\pi_1) > 1 - \pi_1$ and $F_2(\pi_2) < 1 - \pi_2$; and as $F_h(\cdot)$s are nondecreasing functions, it follows from (4) that $\bar{\rho}_1 < \pi_1$ and $\bar{\rho}_2 > \pi_2$. Similarly, it can be shown that if $\pi_1 < \pi_2$, then $\bar{\rho}_1 > \pi_1$ and $\bar{\rho}_2 < \pi_2$. Hence the theorem.

Following the theorem, if $\pi_1 \ne \pi_2$, the larger objective probability is understated whereas the smaller objective probability is overstated. This is in accord with our empirical findings.

---

[10] Some of the assumptions and conditions of the theorem can be relaxed. For example, the condition that π_h is the median of the distribution, $F_h(.)$, can be weakened and the theorem holds if the larger of π_1 and π_2 does not fall below the median of the distribution of its subjective estimates.

If we want to maintain the hypothesis that bettors are sophisticated, the observed objective probability-return relation can be explained, as noted earlier, by EUH with a suitable restriction on the class of utility functions. One such attempt was made by Weitzman (1965) where it is assumed that bettors are homogeneous, each having the same utility function. The utility function is chosen so that the observed probability-return relation coincides with the expected utility indifference curve. It can be shown that bettors with such a utility function are not rational.[11] The source of this counterintuitive behavioral implication is the expansion possibility of the betting opportunity set through various combinations (parlay, martingale, etc.) of the simple bets available in individual races. Thus, this undesired implication can be avoided by restricting the betting opportunities to a single race. It is shown in the next section that the implied utility function is increasing with increasing rate; that is, the bettors are risk lovers. Hence, the observed discrepancies between subjective and objective probabilities may be due to estimation errors by the bettors who are risk neutral or because the bettors are sophisticated but they are risk lovers and behave as if the betting opportunities are limited to a single race.

III. Risk Attitudes

In what follows we assume that bettors are expected utility maximizers, sophisticated, and behave as if the betting opportunities are limited to a single race. Let Mr. B, like Mr. Avmart in Weitzman (1965), be a representative of all the bettors who establish the final odds. He may not be one of the bettors. Let $u(\cdot)$ be his utility of wealth function. In general, utility is also a function of various circumstantial factors. In what follows we abstract from these nuances and assume utility is exclusively a function of wealth. Let us now construct a representative race in which horses 1, 2, ..., 8 of table 2 are competing. Let the true objective probability of horse h to win a race be $\hat{\pi}_h$. By definition, horse h did not have unique odds in all the races in which it competed. The representative (estimated) odds are taken to be a weighted average of all its realized odds. The weights are chosen so that the actual average net gain per dollar bet for horse h to win over all the races it competed in is the amount that would have been obtained if this weighted average of the odds had prevailed for

---

[11] The expected utility indifference curve is estimated as

$$\pi = [1.011 - 0.087 \log_{10}(1 + R)]/R,$$

where R is the return and π is the winning probability of a bet. A parlay (for the definition, see Rosett 1965) constructed from 10 races and using a betting opportunity with $R = 10$ and π as obtained from the indifference curve would have a probability better than three times the probability of a bet from the indifference curve with the same return as this parlay. Thus, if this is an indifference curve, bettors cannot be rational.

TABLE 3

UTILITY FUNCTION AND AVERAGE AND STANDARD DEVIATION OF GAINS
FROM A DOLLAR BET

Horse (h) (1)	Estimated Odds ($\bar{a}_h$) (2)	$x = 1 + \bar{a}_h$ (3)	Utility Function $u(x)$ (4)	Average Gain ($) (5)	SD ($) (6)
1	1.5	2.5	.0574	−0.10	1.26
2	3.0	4.0	.1004	−0.19	1.63
3	4.4	5.4	.1348	−0.17	2.00
4	6.2	7.2	.1966	−0.24	2.30
5	8.7	9.7	.2699	−0.26	2.68
6	12.3	13.3	.3726	−0.27	3.22
7	17.7	18.7	.6028	−0.36	3.64
8	26.4	27.4	1.00	−0.44	4.27

every race. Let $\bar{a}_h$ be the estimated odds for horse h. These are reported in column 2 of table 3. For any race the objective as well as the subjective probabilities must add up to 1. In our case, these totals are, respectively, 1.01 and 0.99. Thus, the conditions are reasonably met. As these $\hat{\pi}_h$'s and $\bar{a}_h$'s are constructed from a number of races, the present race can be considered a representative of them.

With the choice of $\hat{\pi}_h$'s and $\bar{a}_h$'s, Mr. B's expected utility of betting M dollars on horse h is

$$E_h(u) = (1 - \hat{\pi}_h)u(X_0 - M) + \hat{\pi}_h u[(X_0 - M) + M(1 + \bar{a}_h)] \tag{5}$$

where X_0 is the initial capital of Mr. B. As the $\bar{a}_h$'s are the final odds, Mr. B must be indifferent between betting on any two of these eight horses. Thus, Mr. B being an expected utility maximizer,

$$E_1(u) = E_2(u) = \cdots = E_8(u). \tag{6}$$

Without loss of generality, we assume that $X_0 - M = 0$, and the odds are in the units of M dollars where M is the common denominator of all the bets; and as the utility function is unique only up to positive linear transformations, we assume that $u(0) = 0$ and $u(1 + \bar{a}_8) = 1$, where $\bar{a}_8$ are the highest odds. Then from (5) and (6),

$$u(1 + \bar{a}_h) = \hat{\pi}_8/\hat{\pi}_h, \qquad h = 1, 2, \ldots, 7. \tag{7}$$

Using (7), the specific values of the function, $u(\cdot)$, are computed and reported in column 4 of table 3; $u(\cdot)$ is plotted in figure 1. All the points lie below the line joining the two chosen points $[0, u(0) = 0]$ and $[1 + \bar{a}_8, u(1 + \bar{a}_8) = 1]$. Hence, the utility function of Mr. B must be

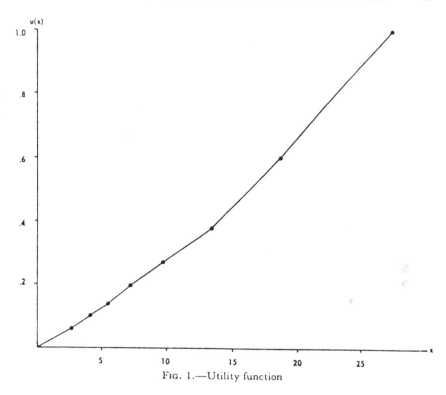

FIG. 1.—Utility function

convex and he must be a risk lover.[12] These points are joined by a broken line which is the estimated utility curve.

To obtain a mathematical function representing the estimated utility curve in figure 1, $\log_{10} u(x_h)$ is regressed on $\log_{10} x_h$ where $x_h = 1 + \bar{a}_h$. The estimated relationship is given by

$$\log_{10} u(x) = 0.2794 + 1.1784 \log_{10} x, \qquad R^2 = .9981.$$
$$\quad\quad (.0206) \quad (.0210)$$

The numbers in the parentheses are the estimated standard errors. Thus, the estimated utility function is $u(x) = 1.91x^{1.1784}$. This shows $u'(x) > 0$ and $u''(x) > 0$. Though the relative risk-aversion measure of Arrow (1971), $-xu''(x)/u'(x)$, is constant, his absolute risk-aversion measure, $-u''(x)/u'(x)$, increases with wealth, implying the representative bettor takes more risk as his capital declines. This finding supports the conjecture of Markowitz (1952) for the range of the utility function under consideration.

[12] For each of the eight horses, the comparison of the average and standard deviation of the gains from a dollar bet computed over the 20,247 races as reported in cols. 5 and 6 of table 3 points to the same conclusion.

For a further analysis, we define a measure (new) of overall risk attitude, δ, as follows:

$$\delta = 1 - \frac{\int_a^b u(x)\ dx}{(1/b - a)\int_a^b x dx}$$

where $a = 0$, $b = 1 + \bar{a}_8$, $u(a) = 0$, $u(b) = 1$. If $\delta > 0$, the bettor is said to be a risk lover; if $\delta < 0$, he is a risk averter, and if $\delta = 0$, he is a risk neutral. The absolute magnitude of δ does not exceed 1.

For our data, $\delta = .1239$. In many respects (see table 1) the races at Roosevelt are alike to those at Yonkers. But Saratoga differs considerably from either Roosevelt or Yonkers. One striking difference is that the average bet per person per race at Saratoga is about half as much as that at either Roosevelt or Yonkers. This can be interpreted as the representative bettor at Saratoga having smaller betting capital than the one at Roosevelt or Yonkers. To investigate its effects on the risk attitude, δ values are computed from the data corresponding to each of these tracks. These are respectively, .1742, .1009, and .0652. It seems the representative at Saratoga tends to take more risk than the one at Roosevelt who in turn takes more risk than the one at Yonkers. However, the difference in attitude toward risk is less pronounced between Roosevelt and Yonkers. This can be taken as extra evidence supporting the observation that the representative bettor with less capital tends to take more risk.

To investigate further the effects of capital on risk attitude, we note that out of every dollar bet, the track take and breakage are 18¢. Thus, the total capital available to bet declines through the races during a racing night. There are usually nine races in a racing night; sometimes there are 10 or even 11, but let us proceed as if there are 9 in a racing night. If we assume that 10 percent of the money available to bet at the beginning of a race is actually put down as a bet on the race, and the track take and breakage are 18 percent (average α over all the races is .180002), then the total money available to bet at the last race will be 14 percent less than it is at the beginning of the first race. We are assuming that all the bettors arrive at the first race and stay until the last. As there are a considerable number of late arrivals, we compute δ for the first two and the last. These are, respectively, .2228, .2038, and .3040; δ values are practically the same for the first two races but are considerably larger for the last. This is in accord with our previous finding about the effect of a change in capital on the representative's risk attitude.

IV. Conclusion

The subjective winning probabilities are obtained as an average of the proportion of the total money bet on a horse. Horses 1, 2, ..., 8 are defined as first, second, ..., eighth favorites in any race. The objective winning probability is computed as the proportion of times a horse finishes first out of all the races in which it competes. The data consist of 20,247 races from three tracks and over a period of 5 years. It is shown that subjectively a horse with a high objective probability of winning is understated and one with a low objective probability of winning is overstated. The pattern seems to be consistent over the years of the sample period, among the chosen tracks, and under various race conditions. Theories explaining these discrepancies between objective and subjective winning probabilities are explored. It is shown that if bettors are risk neutral but not sophisticated, the market mechanism may generate the observed objective-subjective probability relationship.

Maintaining the assumption that bettors are sophisticated, have the same utility function, and behave as if the betting opportunities are limited to a single race, the representative utility of wealth function is constructed. Identifying this utility function with a representative bettor, it is found that he is a risk lover. A measure of risk attitude, δ, is defined which varies from -1 to 1. If $\delta > 0$, the person is a risk lover; if $\delta < 0$, he is a risk averter, and $\delta = 0$ means he is a risk neutral. The representative is found to have $\delta = .1239$. The effects of capital on risk attitude are investigated, and it seems that the more capital the representative has, the less he tends to be a risk lover.

References

Arrow, K. J. *Essays in the Theory of Risk-Bearing.* Amsterdam: North-Holland, 1971.
Friedman, M., and Savage, L. J. "The Utility Analysis of Choices Involving Risk." *J.P.E.* 56, no. 4 (August 1948): 279-304.
Grifith, R. M. "Odds Adjustments by American Horse-Race Bettors." *American J. Psychology* 62 (April 1949): 290-94.
———. "A Footnote on Horse Race Betting." *Transactions Kentucky Acad. Sci.* 22 (1961): 78-81.
McGlothlin, W. H. "Stability of Choices among Uncertain Alternatives." *American J. Psychology* 69 (December 1956): 604-15.
Markowitz, H. "The Utility of Wealth." *J.P.E.* 60, no. 2 (April 1952): 151-58.
Preston, M. G., and Baratta, P. "An Experimental Study of the Auction-Value of an Uncertain Outcome." *American J. Psychology* 61 (April 1948): 183-93.
Rosett, R. N. "Gambling and Rationality." *J.P.E.* 73, no. 6 (December 1965): 595-607.
———. "Weak Experimental Verification of the Expected Utility Hypothesis." *Rev. Econ. Studies* 38 (October 1971): 481-92.
Weitzman, M. "Utility Analysis and Group Behavior: An Empirical Study." *J.P.E.* 73, no. 1 (February 1965): 18-26.
Yaari, M. E. "Convexity in the Theory of Choice under Risk." *Q.J.E.* 79 (May 1965): 278-90.

PART III

ECONOMIC AND MATHEMATICAL INSIGHTS

Introduction to Economics and Mathematical Insights

Donald B. Hausch, Victor S.Y. Lo and William T. Ziemba

The earlier sections of this book have primarily dealt with positive analysis contributions to the horseracing literature. This section begins our discussion of normative analysis. We start with the original paper in this area, Isaacs (1953)[1], where our bettor is assumed to know the true win probabilities in the horserace and wishes to maximize expected profit. Isaacs accounted for our bettor's effect on the odds and developed an ingenious algorithm to determine the optimal wagers. The remainder of the papers in this section consider three extensions of Isaacs model. First, Isaacs assumed risk neutrality while Breiman (1961)[2] (extending Kelly (1956)[2]) showed that the rate of growth of wealth is maximized, asymptotically, by assuming a logarithmic utility function. Second, Isaacs' model takes win probabilities as given; estimating these probabilities can involve *fundamental* methods (which are discussed here and in many popular trade publications) and *technical* schemes (which are considered in Section 4). Third, Isaacs' restriction to win betting can be extended to place, show and exotic wagering with ordering probability estimates. These three important extensions are now discussed in some additional detail.

The Kelly criterion is to maximize the expected logarithm of wealth. Important properties of this criterion were established by Kelly (1956)[2] for information transmission, by Breiman (1961)[2] and Thorp (1971)[2] in more general settings, and further generalized by Algoet and Cover (1988)[2]. Suppose that in each period there are k possible investments with the family of distributions of these k investments being independent and identically distributed over time. Then, three important properties of a strategy to maximize each period the expected logarithm of end-of-period wealth are: 1) it maximizes the asymptotic growth rate of capital; 2) asymptotically, it minimizes the expected time to reach a specified goal; and 3) it outperforms in the long run any other essentially different strategy almost surely. Other favorable and some unfavorable properties are discussed in MacLean, Ziemba and Blazenko (1992)[1]. With logarithmic utility introduced to Isaacs (1953)[1] model, Kallberg and Ziemba (1981)[1] described its generalized concavity properties (and those of a model allowing place and show wagering). Levin (1994)[1] extended Isaacs' optimal-wagering algorithm to the case of logarithmic utility. Empirical studies on the behavior of the optimal portfolio produced by the Kelly criterion and its comparison with the more well-known mean-variance criterion are found in Maier, Peterson and Weide (1977)[2] and Hakansson (1971)[2].

Instead of maximizing capital growth, strategies can be developed based on maximum security. For instance, ruin probability can be minimized subject to making a positive return, or confidence levels can be computed of increasing initial fortune to a given final wealth goal. To combine the ideas of capital growth and security, an alternative is a fractional Kelly criterion, i.e. compute the optimal Kelly investment but invest only a specified fraction of that amount (originated by MacLean and Ziemba (1991)[2] and MacLean, Ziemba and Blazenko (1992)[1]). Thus security can be gained at the price of growth by reducing the investment fraction. MacLean, Ziemba and Blazenko (1992)[1] developed the theoretical properties of this new criterion and provided applications in different investment and gambling areas including horse-racing and futures trading in the stock market.

Aucamp (1993)[2] determined the number of periods necessary to ensure that the Kelly criterion beats another strategy with high probability. Both theoretical and experimental evidence point to the need for a large number of periods in risky situations unless the expected logarithm of wealth between the two strategies is relatively large. Ziemba and Hausch (1986)[2] contrasted various strategies through simulations, and for some insight we summarize one example. Suppose an wagering opportunity is drawn at random from the five opportunities in Table 1. Each wager has an expected return of 1.14 and the optimal Kelly bets are indicated. Suppose these opportunities and their relative frequency of occurrence represent typical positive expectation wagers at a racetrack, and suppose over the course of a season of racing that 700 such wagers present themselves. How does the Kelly criterion compare with other strategies over this short horizon? Table 2 provides the results of 1000 trials, or racing seasons, each of 700 races, assuming an initial wealth of $1000. The Kelly criterion is compared to the fractional Kelly criterion with the fraction ½. Also considered are: 1) "fixed" bet strategies that establish a fixed bet regardless of the probability of winning, the bet's expected return, or current wealth; and 2) "proportional" bet strategies that establish a proportion of current wealth to bet regardless of the circumstances of the wager.

Probability of collecting on the wager	Mutuel Payoff on a $2 bet (including the bettor's two dollars)	Likelihood of each wager being chosen in the simulation	Optimal Kelly bet fraction
0.570	4.00	0.10	0.140
0.380	6.00	0.30	0.070
0.285	8.00	0.30	0.047
0.228	10.00	0.20	0.035
0.190	12.00	0.10	0.028

Table 1: Simulation's wagering opportunities

The simulation provides support for the Kelly system, even over a horizon as short as one racing season. Some bettors may find the distributions of final wealth from other systems may be more appealing for this short horizon, though; e.g., the fractional Kelly system may appeal to a more conservative bettor.

Isaacs (1953)[1] assumed that the bettor knows each horse's win probability. Determining win probabilities can generally take two forms: 1) technical approaches that use price information (see Section 4); and 2) fundamental methods, the second contribution of papers in this section. Bolton and Chapman (1986)[1] developed a multinomial logit model using horse, jockey and race characteristics, but found little evidence that a profitable strategy could be developed on the basis of their win estimates. Chapman (1994)[1] extended this model to analyze a larger data set from Hong Kong with more covariates, and provided evidence of positive returns. Using the same Hong Kong data, Benter (1994)[1] combined his sophisticated fundamental model's probability estimates with the public's implied probability estimates, and reported positive results for five years of actual large scale implementation. Ludlow (1993)[1] subjected his racing data to a discriminant analysis and developed several classification criteria.

A third extension of Isaacs is to allow place, show and exotic wagers (see Sections 5 and 6), which requires the estimation of ordering probabilities of the horses. Assuming one knows the win probabilities, Harville (1973)[1] proposed a very simple estimate for the probability that horses i and j finish first and second, respectively:

$$\pi_{ij} = \frac{\pi_i \pi_j}{1 - \pi_j}$$

where π_{ij} = P(horse i wins, j finishes second),
        π_i = P(horse i wins).

This estimate is based on the idea that the conditional probability horse j finishes second given horse i is first is the probability that j would have finished first had i not been in the race, which is estimated as:

$$\frac{\pi_j}{1 - \pi_i}$$

| System | Final Bankroll | | | | Number of Seasons Final Bankroll Was: (starting with $1000) | | | | | | | | |
	Min.	Max.	Mean	Median	Bankrupt	> $2	> $250	> $500	> $1000	> $5000	> $10000	> $50000	> $100000
Kelly	18	453883	48135	17269	0	1000	957	916	870	692	598	302	166
½ Kelly	145	111770	13069	8043	0	1000	999	990	954	654	430	430	1
Fixed:													
$10	307	3067	1861	1857	0	1000	1000	999	980	0	0	0	0
20	0	5377	2824	2822	9	991	990	988	978	9	0	0	0
30	0	7682	3739	3770	36	964	963	962	957	191	0	0	0
40	0	9986	4495	4685	94	906	906	906	904	432	0	0	0
50	0	12282	5213	5526	134	866	866	866	864	584	33	0	0
100	0	23747	7637	8722	349	651	651	651	651	613	425	0	0
Proportional:													
1%	435	8469	2535	2270	0	1000	1000	999	965	43	0	0	0
2%	173	57087	6628	4360	0	1000	999	991	940	443	180	7	0
3%	65	243281	15343	6799	0	1000	994	973	919	592	396	65	18
4%	49	483355	26202	8669	0	1000	979	935	882	627	459	146	61
5%	38	548382	32415	8907	0	1000	941	899	841	609	475	179	90
10%	18	364587	13662	602	0	1000	575	515	455	304	221	78	36

Table 2: Simulation's wealth results using various wagering systems.
Source: Ziemba and Hausch (1986)[2].

A similar formula holds for the probability that horses i, j and k finish 1-2-3. Harville's formulas follow if running times are independently and exponentially distributed with horse i's mean running time inversely related to π_i (see Dansie (1983)[2]). Empirical work has pointed to some problems with Harville's estimate. For instance, Hausch, Ziemba & Rubinstein (1981,p.1439)[1] observed that "no account is made of the possibility of the Silky Sullivan problem; that is, some horses generally either win or finish out-of-the-money; for these horses the formulas greatly over-estimate the true probability of finishing second or third." Harville (1973)[1] also showed that the formulas tend to generate a reverse favorite-longshot bias for second and third positions, i.e., the formulas tend to overestimate the likelihood that a favorite will finish second or third and underestimate the likelihood that a longshot will finish second or third. One natural conjecture is that a reverse favorite-longshot bias for second and third "compensates" for the favorite-longshot bias for winning, since the probabilities over all finish positions for a horse must sum to one. This explanation is, at best, a partial one since Benter (1994)[1] documented the reverse bias for second and third using Hong Kong data that does not exhibit the usual favorite-longshot bias for win. His explanation of the bias for second and third is that the Harville formulas do not adequately account for the randomness associated with finishing positions, randomness that plays an increasing role for positions that are further and further from first position. He proposed a correction method using a logistic model. Lo (1994)[1]) provides another explanation. He proved that if the true underlying distribution of running times is independent Gamma, then Harville's model overestimates the conditional probability that favorites are second and underestimates the conditional probability that longshots are second.

Others have tried different assumptions about the underlying distribution of running times. Henery (1981)[1] argued that a perhaps unrealistic feature of the Harville model is that the formula does not depend on the number of horses in the race. He suggested assuming that the running times are independent normal with unit variance, i.e. the running time of horse i is $T_i \sim N(\theta_i, 1)$ independently. The resulting ordering probabilities are the same as those of a general constant variance model. Further, the model also holds when a monotonic transformation of T_i is normally distributed. His model together with estimates of the win probabilities allows a two-stage process for the estimation of ordering probabilities: 1) the mean running times of horse i, θ_i, are determined as the solution of a system of equations for the win probabilities; and 2) with θ_i, the ordering probabilities are computed using numerical integrations. Without a closed form solution, though, computing these ordering probabilities is very time consuming. This does not present a problem if win probability estimates are available well in advance of the race, but the use of Henery's model is essentially precluded for systems that rely on the public's win odds to provide win probability estimates. Henery suggested the simplifying use of a first order Taylor series approximation, but Bacon-Shone, Lo & Busche (1992b)[2], demonstrated the inaccuracy of this approximation. Stern (1994)[1] also mentioned that this simplification does not seem to improve on the Harville model.

Another alternative is Stern's (1990)[2] extension of the underlying exponential distribution assumption of the Harville model to a Gamma running times model. Now the running time of horse i is $T_i \sim Gamma(r, \lambda_i)$ independently, where the shape parameter r is prespecified and the scale parameters λ_i can be determined given win probability estimates. This distribution is motivated by considering a more general competition in which n players, scoring points according to independent Poisson processes, are ranked according to the time until r points are scored. This model is equivalent to assuming that the running times follow the generalized extreme value distribution mentioned in Stern (1993)[1] because there is a monotonic transformation relationship (logarithmic transform) between the two kinds of random variables. The Stern model is the Harville model when $r = 1$ and it becomes the Henery model as r tends to ∞. However, again, the formulas for computing the ordering probabilities are complicated even though r is an integer. In a small empirical study, Stern analyzed 47 races and found that ordering probabilities estimated using $r = 1$ were less accurate than those estimated using $r = 2$. Using a likelihood approach and Japanese data, Lo (1994)[1] found Stern's model with r = 4 to be best. He reported that $r = \infty$ (Henery's model) best fit his Meadowlands and Hong Kong data, though. Thus, while one running time distribution model does not appear to hold universally, there is limited empirical support for Harville's model.

Lo (1994)[1] (and other empirical studies by Bacon-Shone, Lo and Busche (1992b)[2] and Lo and Bacon-Shone (1992a)[2]) demonstrated the improved ordering probability estimates of Stern's and Henery's models. Because their considerable computing requirements diminish their practicality, though, Lo and Bacon-Shone (1992b)[2] proposed and tested a simple approximation to both models. Lo, Bacon-Shone and Busche (1994)[2] analyzed the possible gains of the approximation's use in the wagering system developed by Hausch, Ziemba and Rubinstein (1981)[1] which was implemented, using a small hand held calculator (Ziemba and Hausch (1984[2], 1987[2])).

[1] included in this volume
[2] cited in the Annotated Bibliography

OPTIMAL HORSE RACE BETS

RUFUS ISAACS, The Rand Corporation

1. Introduction. The problem of placing straight win bets on a horse race so as to maximize the expected value of the profit is intellectually lucrative, although fiscally academic unless the bettor is extraordinarily opulent, has an excellent dopester, and can do elaborate computations on very short notice.

The crowd bets on the various horses. The total sum, after a fixed percentage being deducted as the track's profit, is allocated to the bettors on the winning horse in proportion to their bets. We assume that we know the true probability of each horse's winning.

There is nothing unsound or new in our basic wagering principle. It is known, in track parlance, as "overlays." It consists in taking advantage of the collective error of the crowd in appraising the probabilities as registered by the amounts they bet. What we believe is new is the exact solution.

Let us take an example. The third column is the amounts bet by the crowd, drastically scaled down from race track reality.

Horse	Probability of Winning	Amount Bet by Crowd
A	.4	$1000
B	.2	350
C	.2	300
D	.1	250
E	.1	100
	1.0	$2000

If we neglect, for the moment, the track's deduction, it is clear that E is the soundest investment; for he returns twenty times the stake and has a win probability of 0.1. But if our bet on him is large, the increment to the $100 will affect the entire picture and possibly even cause another horse to be the most favorable. We now can see that to maximize our expected profit we should place bets on a certain subset (possibly null) of horses.

Since this bagatelle first resulted from a conversation with a turf-minded friend, we have learned of its wider interpretations. It is akin to the current field of non-linear programming whose advocates have suggested alternative techniques. But in such fields, solutions by such definite means as algorithm and formula, as here obtained, are uncommon and we have been told our work should hold some exoteric interest. We acknowledge, with admiration, the suggestions of the referee who elegantly abridged our derivation.

2. The problem. Let the win probability of the jth horse be p_j, the amount wagered on him by others than ourself s_j, and our wager on him be x_j. Let the total sum wagered be multiplied by Q $(0 < Q < 1)$ prior to distribution to the winning bettors. Our problem then is:

What values of

$$x_j \ (x_j \geqq 0, \ j = 1, \cdots, n)$$

render the expected value of our profit,

$$(1) \qquad F(x_1, \cdots, x_n) = Q \left[\sum_{j=1}^{n} (x_j + s_j) \right] \sum_{j=1}^{n} \frac{p_j x_j}{x_j + s_j} - \sum_{j=1}^{n} x_j$$

a maximum, the value of the maximal F to be positive? Here Q and p_j, s_j are constants subject to

$$0 \leqq p_j \leqq 1, \qquad \sum_{j=1}^{n} p_j = 1,$$

$$s_j > 0,^* \qquad 0 < Q < 1,$$

and, of course, $n > 1$.

3. The solution. Let $c_j \ (j = 1, \cdots, n)$ be such that

$$c_j \geqq 0, \qquad \sum_{j=1}^{n} c_j = 1.$$

If we put $x_j = c_j u$ in (1), F becomes a function of u denoted by $f(u)$. It is easy to verify.

$$(2) \qquad f(0) = 0$$

$$(3) \qquad \lim_{u \to \infty} \frac{f(u)}{u} \leqq Q - 1 < 0$$

$$(4) \qquad f'(0) = \sum_{j=1}^{n} c_j \left[Q \frac{p_j}{s_j} \sum_{i=1}^{n} s_i - 1 \right].$$

Let u range from 0 to ∞. From (2) and (3) we see that f begins as 0 but ultimately becomes and remains negative. If at least one of the brackets of (4) is positive, the c_j can be chosen so that $f'(0) > 0$. Thus we state:

LEMMA 1. *There is always a non-negative maximum of F; if*

$$(5) \qquad \max_{1 \leqq j \leqq n} \frac{p_j}{s_j} > \frac{1}{Q \sum_i s_i}$$

then this maximum is positive.

Let $\overline{X} = (\bar{x}_1, \cdots, \bar{x}_n)$ furnish a maximum. If $\bar{x}_i > 0$, then

$$(6) \qquad \left(\frac{\partial F}{\partial x_i} \right)_{X - \overline{X}} = Q \sum_{j=1}^{n} \frac{p_j \bar{x}_j}{\bar{x}_j + s_j} + \frac{p_i s_i}{(\bar{x}_i + s_i)^2} Q \sum_{j=1}^{n} (\bar{x}_j + s_j) - 1 = 0.$$

Then $p_i s_i / (\bar{x}_i + s_i)^2$ has the same value for all i such that $\bar{x}_i > 0$ and we call it $1/\lambda^2$. Thus for such i

$$(7) \qquad \bar{x}_i = \lambda \sqrt{p_i s_i} - s_i.$$

---

* We will show later that there is not necessarily a solution if some s_j are zero.

LEMMA 2. *The maximal F never occurs when all the $\hat{x}_i$ are positive*; or: *Don't bet on all the horses.*

Substituting from (7) for *every* i into (6), the central member of the latter becomes $Q-1$ which is negative.

Recognizing that the higher p_j is, and the lower s_j is, the better bet Horse j is, it is clear that the ratio $\rho_j = p_j/s_j$ provides a heuristic measure of the soundness of a bet on Horse j. Let us renumber the horses, if necessary, in order of increasing value by this criterion, *i.e.*, so that $\rho_1 \leq \rho_2 \leq \cdots \leq \rho_n$. The next proposition tells us that, in maximizing our expectation, if we bet on a certain horse, then we bet on all horses which are as good or better bets in this ordering. It also asserts that the optimal amounts of wager are uniquely determined by the set of horses on which we bet.

LEMMA 3. *The only positive maxima of F are achieved by $\overline{X} = (\hat{x}_1, \cdots, \hat{x}_n)$ of the form*:

$$(8) \qquad \hat{x}_1 = \cdots = \hat{x}_{t-1} = 0, \qquad \hat{x}_t > 0, \cdots, \hat{x}_n > 0$$

where $\overline{X}$ is uniquely determined by t as follows:

$$(7) \qquad \hat{x}_i = \lambda_t \sqrt{p_i s_i} - s_i \qquad\qquad for\ i = t, \cdots, n.$$

Also

$$(9) \qquad \rho_i \leq \frac{1}{\lambda_t^2} \ for\ i = 1, \cdots, t-1$$

where

$$(10) \qquad \lambda_t^2 = Q \sum_{j=1}^{t-1} s_j \Big/ \left(1 - Q \sum_{j=t}^{n} p_j\right).$$

Remark: Lemma 2 assures us that $t > 1$, so that (10) is guaranteed a meaning.
Proof: From (6)

$$(11) \qquad \frac{1}{\lambda^2} = \frac{1 - Q \sum\limits_{j=1}^{n} \dfrac{p_j \hat{x}_j}{\hat{x}_j + s_j}}{Q \sum\limits_{j=1}^{n} (\hat{x}_j + s_j)}.$$

Remark that $\hat{x}_i > 0$ implies

$$(12) \qquad \rho_i > \frac{p_i s_i}{(\hat{x}_i + s_i)^2} = \frac{1}{\lambda^2}$$

while $\hat{x}_i = 0$ implies

$$\left(\frac{\partial F}{\partial x_i}\right)_{x=\overline{x}} \leq 0$$

and (6) becomes

$$Q \sum_{j=1}^{n} \frac{p_j \bar{x}_j}{\bar{x}_j + s_j} + p_i Q \sum_{j=1}^{n} (\bar{x}_j + s_j) - 1 \leqq 0.$$

A glance at (11) shows that now

(9)
$$p_i \leqq \frac{1}{\lambda^2}.$$

Hence (8) is the only possible form for a solution.

Suppose (8) to hold for some t, we calculate λ by substituting from (7) for $i = 1, \cdots, t-1$ into (6). The resulting λ depends on t; it is denoted by λ_t and given by (10).

LEMMA 4. *If*
$$p_t > \frac{1}{\lambda_t^2},$$

then
$$p_t > \frac{1}{\lambda_{t+1}^2} \qquad \qquad for\ t = 2, \cdots, n-1.$$

For
$$\frac{1}{\lambda_{t+1}^2} = \left[\left(1 - Q \sum_{j=t}^{n} p_j \right) + Q p_t \right] \Big/ Q \sum_{j=1}^{t} s_j$$

and
$$1 - Q \sum_{j=t}^{n} p_j = Q \sum_{j=1}^{t-1} s_j \Big/ \lambda_t^2 < Q p_t \sum_{j=1}^{t} s_j,$$

so that
$$\frac{1}{\lambda_{t+1}^2} < \left[p_t \sum_{j=1}^{t-1} s_j + p_t s_t \right] \Big/ \sum_{j=1}^{t} s_j = p_t.$$

Let us write the inequalities

(K_v)
$$p_v > 1/\lambda_v^2 = \left[q + \sum_{j=1}^{v-1} p_j \right] \Big/ \sum_{j=1}^{v-1} s_j$$

for $v = 2, \cdots, n$. Here $q = (1 - Q)/Q$, the new form of λ_v being convenient for computation.

LEMMA 5. *If any of the (K_v) hold, F has a positive maximum.*

Proof: If (K_v) holds with $v < n$, then, from Lemma 4, $p_{v+1} \geqq p_v > 1/\lambda_{v+1}^2$ so that (K_{v+1}) holds. Then (K_n) holds and it is

$$p_n > [1 - Q p_n]/Q \sum_{j=1}^{v-1} s_j = \left[\frac{1}{Q} - p_n s_n \right] \Big/ \left[\sum_{j=1}^{n} s_j - s_n \right]$$

or

$$1/Q \sum_{j=1}^{n} s_j < \rho_n = \max \frac{p_j}{s_j},$$

and we apply Lemma 1.

THEOREM. *If all of (K_v) are false, the $\bar{x}_j$ are all zero. Otherwise the t of Lemma 3 is the smallest v such that (K_v) is true.*

Proof: Suppose all the (K_v) false. If the $\bar{x}_j$ were not all zero, we would have the situation described in Lemma 3 with a $t > 1$. Then (7) for $i = t$ would imply the truth of (K_t).

If any (K_v) hold, Lemma 5 assures us that t exists. Again (7) tells us that (K_t) is true. Let $t > 2$ and suppose (K_{t-1}) true. Then $\rho_{t-1} > 1/\lambda_{t-1}^2$ and, from Lemma 4, $\rho_{t-1} > 1/\lambda_t^2$, contradicting (9) with $i = t - 1$.

The calculatory procedure is now clear: We successively test the inequalities (K_v). The first true one (if any) furnishes us t and λ_t. The $\bar{x}_j$ are computed from (7).

A Remark: We show now that if some of the s_j are zero a solution need not exist. Let us take $n = 2$, $s_1 > 0$, $s_2 = 0$, and $p_2 > 0$. Then a brief calculation shows that

$$\sup F(x_1, x_2) = F(0, 0+) = Q p_2 s_1 > 0$$

while

$$F(0, 0) = 0.$$

An example: Applying our method to the above example and taking Q as 0.9, we find we should wager

$$\$28.40 \text{ on } B$$
$$50.33 \text{ on } C$$
$$43.02 \text{ on } E.$$

4. Some ramifications. R. M. Thrall* has developed some interesting variants. The present mathematical scheme fits a problem of merchandising economics; here the restriction $Q < 1$ may be dropped. Then the bigger the x_j, the bigger F and a maximum no longer exists unless the sum of the x_j be bounded.

Reverting to the original setting, he observes that after a bettor has made an optimal wager, it may still be possible for a second bettor to do likewise with a positive expected profit. It will be small and the first bettor's profit will be severely diminished. Thrall's problem is: What is our optimal bet if we know that a succeeding bettor is to wager optimally?

---

* R. M. Thrall, Some results in non-linear programming. To be published.

Concavity Properties of Racetrack Betting Models*

J.G. Kallberg, New York University
W.T. Ziemba, University of British Columbia

Suppose in a given horserace there are n horses and the ith horse's chances of winning is q_i. Let W_i be the amount bet on i by the crowd, $W \equiv \sum W_i$ and $1-Q$ be the percentage track take. Then an investor/gambler with initial wealth w_0 will maximize his long run rate of growth of assets by choosing bets w_i to

$$\text{maximize} \qquad \phi(w_1,\ldots,w_n) = \sum_{i=1}^{n} q_i \log\{[\frac{Q(W+\sum_j w_j)-w_i-W_i}{w_i+W_i}]w_i + w_0 - \sum_{j\neq i} w_j\} \qquad (1)$$

$$\text{s.t.} \qquad \sum_{i=1}^{n} w_i \leq w_0 , \ w_i \geq 0 , \ i = 1,\ldots,n.$$

The payoff is determined as follows. Suppose horse i wins then all bettors on horse i are returned their bets ($w_i + W_i$) plus their share of the profits

$\frac{Q(W+\sum_j w_j)-w_i-W_i}{w_i+W_i}$ per dollar bet on i. Hence the expected utility of final wealth

is the objective function in (1).

The portfolio model (1) is distinguished by the fact that the return distributions are affected by the investment choices. This is natural since the amounts bet influence the odds. It is this feature that destroys the concavity of (1) even with linear utility. If one neglects this feature then one has concavity. Some concavity/generalized concavity properties of ϕ are presented in the following theorem. For convenience let

$$\phi(w) \equiv \sum_{i=1}^{n} q_i \log(f_i(w)) \text{ and } h(w) \equiv \sum_{i=1}^{n} q_i f_i(w).$$

Theorem 1

a) In general ϕ is not concave even if log utility is replaced by linear utility unless the bets by the investor are assumed not to influence the odds.

b) Each $f_i(w)$ is pseudoconcave, and

c) ϕ is concave transformable hence local maxima are global.

---

*This paper is modified from a section of "Generalized Concave Functions in Stochastic Programming and Portfolio Theory", in Generalized Concavity in Optimization and Economics, eds S. Schaible and W.T. Ziemba, Academic Press, 1980, pp. 719-763.

PROOF:

(a)   In the linear utility case suppose $n = 2$, $Q = .8$, $W_1 = 1$, $W_2 = 10$, $w_0 = 300$, $q_1 = 10/11$, $q_2 = 1/11$, $w^1 = (1, 10)$ and $w^2 = (19, 192)$. Then $\frac{1}{2}h(w^1) + \frac{1}{2}h(w^2) = 277.764 > 277.735 = h(\frac{1}{2}w^1 + \frac{1}{2}w^2)$ hence $h(w)$ is not concave. In the log utility case with the same data except $w_0 = 10^6$, ϕ is not concave because $\frac{1}{2}\phi(w^1) + \frac{1}{2}\phi(w^2) = 5.999990342 > 5.99990329 = \phi(\frac{1}{2}w^1 + \frac{1}{2}w^2)$. If the bets do not influence the odds then $\phi(w) = \sum q_i \log[g_i(w)]$ where

$g_i(w) \equiv (QW-W_i)w_i/W_i - \sum_{j \neq i} w_j + w_0$. Hence ϕ is the sum of concave functions of

linear functions which is concave. When log utility is replaced by linear utility ϕ is linear hence concave.

(b) $f_i(w) \quad = \quad \{Q(W+\sum w_j) - (w_i+W_i)\} \dfrac{w_i}{w_i+W_i} - \sum_{j \neq i} w_j + w_0$

$= \quad Qw_i \dfrac{W+\sum w_j}{w_i + W_i} - w_i - \sum w_j + w_i + w_0$

$= \quad Qw_i \dfrac{W+\sum w_j}{w_i + W_i} + (Q-1)\sum w_j - Q\sum w_j + w_0$

$= \quad Q(W+\sum w_j) - Qw_i \dfrac{W+\sum w_j}{w_i + W_i} + (Q-1)\sum w_j - Q\sum w_j + w_0$

$= \quad (Q-1)\sum w_j + QW + w_0 - Qw_i \dfrac{W+\sum w_j}{w_i + W_i}$

$= \quad (Q-1)\sum(w_j+W_j) + (1-Q)\sum W_j + Q\sum W_j + w_0 - \dfrac{QW_i\sum(w_j + W_j)}{w_i + W_i}$

$= \quad (Q-1)\sum(w_j+W_j) + \sum W_j + w_0 - \dfrac{QW_i\sum(w_j + W_j)}{w_i + W_i} \qquad (2)$

$\underbrace{\hspace{3cm}}_{\text{linear}} \quad \underbrace{\hspace{2cm}}_{\text{constant}} \underbrace{\hspace{2.5cm}}_{\text{fractional}}$

Thus (2) is of the form

$f(x) = a_0 + a^T x + b^T x/c^T x \qquad (3)$

where $a_0 \equiv W + w_0$, $x \equiv (w_1 + W_1 , \ldots , w_a + W_a)$,

$a \equiv (Q-1, \ldots , Q-1)$, $\qquad\qquad b \equiv (-QW_i , \ldots , -QW_i)$,

$c \equiv (0, \ldots, 0, \overset{\downarrow}{1}, 0, \ldots, 0)$ and $\downarrow$ indicates the ith column. In (2) $a = [\frac{1-Q}{QW_i}]b \equiv \mu b$. Since $Q \in (0, 1)$ and $W_i > 0$ it follows that $\mu > 0$. Thus

by Proposition II of Schaible (1977), (2) is quasiconcave and pseudoconcave (on an open convex set) of $a^T x \leq 0$ (i.e. $x \geq 0$ or $w_i + W_i \geq 0$).

(c)   If $x_j \equiv w_j + W_j$ then (2) becomes

$$f_i(x) = - \alpha \sum x_i + \beta_i - \gamma_i \sum x_j / x_i$$

where $\alpha \equiv (1-Q) > 0$, $\beta_i = \sum W_i + w_0 > 0$

and $\gamma_i = QW_i > 0$.

Letting $z_i \equiv \log x_i$, i.e., $x_i = e^{z_i}$ gives

$$f_i(z) = -\alpha \sum e^{z_i} + \beta_i - \gamma_i \sum e^{z_i - z_i}$$

which is concave.   Hence $\log f_i$ and ϕ are also concave.   That the transformed constraint set is still convex is verified as follows.   The constraint $w_i \geq 0$ becomes $x_i \geq W_i$ or $e^{z_i} \geq W_i$ or $z_i \geq \log W_i$.   Now the budget constraint $\sum w_i \leq w_0$ becomes $\sum x_i \leq w_0 + W$ or $\sum e^{z_i} \leq w_0 + W$, which describes

a convex set.   Thus problem (1) maps into an equivalent concave maximization problem via a 1-1 transformation.   This implies that problem (1) possesses the property that local maxima are global.

Remarks: Thus to obtain concavity one must either use the approximation that the investor's bets do not influence the odds or have rather severe conditions arise, e.g. $Q\sum w_j = w_i$, $w_i > 0$ for all i. the latter condition is unlikely to be satisfied in practice.   The result in (b) indicates that the objective function in (1) is the sum of concave functions of pseudoconcave functions.   It is not known if ϕ is generally pseudoconcave even in the linear utility function case. However, since ϕ is concave-transformable local maxima are global maxima.   The linear utility function case is Isaac's model (1953).   In this case one has an explicit solution as follows.   Assume that there is an *underbet* great enough to overcome the track take,

i.e. $\max_j \, q_j / (W_j / W) > 1/Q$ (4)

then at least one $w_i^* > 0$.   Renumber the horses so $p_1 \leq \ldots \leq p_n$ where $p_i \equiv q_i / W_i$.

Then $\quad w_i^* = \lambda_i \sqrt{q_i W_i} - W_i \quad , \quad i = t, \ldots, n$, where $\quad \lambda_i^2 \equiv Q \sum_{j=1}^{i-1} W_j / (1 - Q \sum_{j=1}^{n} q_j)$ and the

optimal solution has the form $w_i^* = \ldots = w_{t-1}^* = 0$, $w_t^* > 0$, $\ldots$, $w_n^* > 0$.

The difficulty with the model (1) is that efficiency studies, see e.g. Snyder (1978), indicate that the probability that horse i wins, q_i, is accurately estimated by the relative betting frequency of the betting public, W_i/W, except for stable tail biases[1]. Hence it is unlikely that condition (4) will be satisfied except in rare instances and even then there is no obvious simple way to estimate the q_i. However, the model (1) may be useful in cases where expert opinion or factor analytic methods yield estimates of the q_i that satisfy (4).

Hausch, Ziemba and Rubinstein (1981) suggest instead that there are more likely to be inefficiencies in the place and show[2] markets than the win market. There are a number of reasons for this, in particular because it is much more complicated to calculate the odds for place and show bets. The basic assumptions of their model are: 1) there is an *efficient* market to win so $q_i = (W_i/W$, 2) following Harville (1973) assume that the probability that i is first and j is second is $q_i q_j / (1-q_i)$ and of an ijk finish is $q_i q_j q_k / (1-q_i)(1-q_i-q_j) \equiv q^{ijk}$, and 3) the same assumptions as (1) with the place and show expressions replacing the win expressions. This yields the model

$$\text{Maximize}_{\{p_\ell\}\{s_\ell\}} \; \psi(p,s) \equiv \sum_{i=1}^{n} \sum_{\substack{j=1 \\ j \neq i}}^{n} \sum_{\substack{k=1 \\ k \neq i,j}}^{n} q_{ijk} \log \left[\begin{array}{c} \dfrac{Q(P + \sum_{\ell=1}^{n} p_\ell) - (p_i + p_j + P_i + P_j)}{2} \\[4pt] \times \left[\dfrac{p_i}{p_i + P_i} + \dfrac{p_j}{p_j + P_j} \right] \\[8pt] + \dfrac{Q(S + \sum_{\ell=1}^{n} s_\ell) - (s_i + s_j + s_k + S_i + S_j + S_k)}{3} \\[4pt] \times \left[\dfrac{s_i}{s_i + S_i} + \dfrac{s_j}{s_j + S_j} + \dfrac{s_k}{s_k + S_k} \right] \\[8pt] + w_0 - \sum_{\substack{\ell=1 \\ \ell \neq i,j,k}}^{n} s_\ell - \sum_{\substack{\ell=1 \\ \ell \neq i,j}}^{n} p_\ell \end{array} \right] \quad (5)$$

s.t. $\sum_{\ell=1}^{n} (p_\ell + s_\ell) \leqslant w_0$, $\; p_\ell \geqslant 0$, $\; s_\ell \geqslant 0$, $\qquad \ell = 1, \ldots, n$,

---

[1] Specifically longshots (high odds horses) are overbet and favorites (low odds horses) are underbet. However, even the latter bias is only about nine percent compared to the about eighteen percent track take.

[2] A horse is said to *place* if it finishes first or second, it *shows* if it finishes first, second or third. The order of finish is immaterial, however, the payoff depends on the amount bet on the other horse(s) that "finish in the money", see (5).

where Q = 1-the track take, W_i, P_j and S_k are the total dollar amounts bet to win, place and show on the indicated horse by the crowd, respectively, $q_i \equiv W_i/W$ is the theoretical probability that horse i wins, w_0 is initial wealth and p_ℓ and s_ℓ are the investor's bets to place and show on horse ℓ, respectively.

The argument of the log represents final wealth if there is an ijk finish so the expectation yields expected utility of final wealth. The payoffs are calculated [3] so that for horses in the money the betting amounts are returned and the resulting portion of the pool in excess of the track take is divided equally among the winning horses. We have the following partial results concerning the concavity, generalized concavity and solution properties of (5).

Theorem 2

(a)   *If each p_ℓ and s_ℓ is small in relation to P_ℓ and S_ℓ, (i.e. the investor's bets make an insignificant effect on the odds) and these terms are neglected accordingly when they appear with such a P_ℓ or P and S_ℓ or S, respectively then ψ is concave for logarithmic as well as linear utility.*

(b)   *In general ψ is not concave.*

(c)   *If there is linear utility and an a priori decision has been made, e.g. on the basis of expected returns, to bet on horse ℓ to place and horse m to show then ψ is concave and there is an exact closed form solution.*

PROOF:

(a) Under the assumptions

$$\psi(p,s) = \sum_i \sum_{j \neq i} \sum_{k \neq i,j} q_{ijk} \log(g_{ij}(p) + h_{ijk}(s) + w_0)$$

where

$$g_{ij}(p) \equiv \left(\frac{QP - (P_i + P_j)}{2}\right)\left(\frac{P_i}{P_i} + \frac{P_i}{P_j}\right) - \sum_{\ell \neq i,j} P_{\ell},$$

$$h_{ijk}(s) \equiv \left(\frac{QS - (S_i + S_j + S_k)}{3}\right)\left(\frac{s_i}{S_i} + \frac{s_j}{S_j} + \frac{s_k}{S_k}\right) - \sum_{\ell \neq i,j,k} s_{\ell},$$

which is the sum of concave functions of linear functions hence is concave. When log utility is replaced by linear utility ψ is linear in p,s hence concave.

This formulation omits the features of minus pools and breakage that slightly alter the exact model of the betting situation. See Hausch, Ziemba, and Rubinstein (1981) for discussion.

(b) Let $f(p,s) = \sum\limits_{i} \sum\limits_{j \neq i} \sum\limits_{k \neq i,j} q_{ijk}(g_{ij}(p) + h_{ijk}(s) + w_0)$,

$$g_{ij}(p) \equiv \left(\frac{Q(P+\Sigma p_\ell)-p_i-p_j-p_i-p_j}{2}\right)\left(\frac{P_i}{P_i+P_i} + \frac{P_i}{P_j+P_j}\right) - \sum\limits_{\ell \neq i,j} P_\ell$$

$$h_{ijk}(s) \equiv \left(\frac{Q(S+\Sigma s_\ell)-s_i-s_j-s_k-S_i-S_j-S_k}{3}\right)\left(\frac{s_i}{s_i+S_i} + \frac{s_i}{s_j+S_j} + \frac{s_k}{s_k+S_k}\right)$$

$$- \sum\limits_{\ell \neq i,j,k} s_\ell .$$

In the linear case suppose $n = 4$, $Q = .8$, $P_1 = S_1 = 97$,

$P_2 = P_3 = P_4 = S_2 = S_3 = S_4 = 1$, $w_0 = 5000$, $q_1 = .01$,

$q_2 = q_3 = q_4 = .33$, $p^1 = s^1 = (97,1,1,1)$ and $p^2 = s^2 = (1999,19,19,19)$.

Then $\frac{1}{2}f(p^1,s^1) + \frac{1}{2}f(p^2,s^2) = 4562.64 > 4557.97 = f(\frac{1}{2}p^1+\frac{1}{2}p^2, \frac{1}{2}s^1+\frac{1}{2}s^2)$,

hence f is not concave. In the log utility case with the same data

except $w_0 = 10^6$, ψ is not concave because $\frac{1}{2}\psi(p^1,s^1) + \frac{1}{2}\psi(p^2,s^2) =$

$5.9998107 > 5.9998080 = \psi(\frac{1}{2}p^1+\frac{1}{2}p^2, \frac{1}{2}s^1+\frac{1}{2}s^2)$.

(c) When there is linear utility and one bet on horse ℓ to place and one

bet on horse m to show the objective function becomes

$$\sum\limits_{\substack{i=1 \\ i \neq \ell}}^{n} \left[\hat{q}_{i\ell}\left(\frac{Q(P+p_\ell)-(p_\ell+P_\ell+P_i)}{2}\right)\left(\frac{P_\ell}{P_\ell+P_\ell}\right) + P_\ell\right]$$

$$+ \sum\limits_{\substack{i=1 \\ i \neq \ell}}^{n} \sum\limits_{\substack{j=1 \\ j \neq i,m}}^{n} \left[\tilde{q}_{ijm}\left(\frac{Q(S+s_m)-(s_m+S_m+S_i+S_j)}{3}\right)\left(\frac{s_m}{s_m+S_m}\right) + s_m\right]$$

$$- P_\ell - s_m + w_0$$

$$= \left(\frac{Q+1}{2}P_\ell\right) \sum\limits_{\substack{i=1 \\ i \neq \ell}}^{n} \hat{q}_{i\ell} + \left(\frac{P_\ell}{P_\ell+P_\ell}\right)\sum\limits_{\substack{i=1 \\ i \neq \ell}}^{n} \frac{\hat{q}_{i\ell}\left(Q\sum\limits_{\substack{k=1 \\ k \neq \ell}}^{n} P_k - P_i\right)}{2}$$

$$+ \left(\frac{Q+2}{3}s_m\right)\sum\limits_{\substack{i=1 \\ i \neq m}}^{n}\sum\limits_{\substack{j=1 \\ j \neq i,m}}^{n} \tilde{q}_{ijm} + \frac{s_m}{s_m+S_m}\sum\limits_{\substack{i=1 \\ i \neq m}}^{n}\sum\limits_{\substack{j=1 \\ j \neq i,m}}^{n} \tilde{q}_{ijm}\left(\frac{Q\sum\limits_{\substack{k=1 \\ k \neq m}}^{n} s_k - s_i - s_j}{3}\right)$$

$$- P_\ell - s_m + w_0 , \tag{6}$$

where

$$\hat{q}_{i\ell} \equiv \frac{q_i q_\ell}{1-q_i} + \frac{q_i q_\ell}{1-q_\ell} , \quad \text{and}$$

$$\tilde{q}_{ijm} \equiv \frac{q_i q_j q_m}{(1-q_i)(1-q_i-q_j)} + \frac{q_i q_i q_m}{(1-q_i)(1-q_i-q_m)} + \frac{q_i q_j q_m}{(1-q_m)(1-q_m-q_j)} .$$

Since (6) consists of sums of linear terms plus terms of the form $x/(a+x)$ for $a > 0$ it is concave. Hence the Kuhn-Tucker conditions are necessary and sufficient and yield the following solution, where $\ell=m$ is allowed:

$$p^* \equiv -P_\ell + \frac{\sqrt{P_\ell(Q\hat{q}_\ell(P-P_\ell)-\tilde{R}_\ell)}}{2-\hat{q}_\ell(Q+1)}$$

$$s_m^* \equiv -S_m + \frac{\sqrt{S_m(Q\hat{q}_m(S-S_m)-\tilde{R}_m)}}{3-\hat{q}_m(Q+2)} , \quad \text{where}$$

$$\hat{q} \equiv \sum_{i \neq \ell}\left(\frac{q_i q_\ell}{1-q_i}\right) + \sum_{j \neq \ell}\left(\frac{q_\ell q_1}{1-q_\ell}\right) \quad \text{and}$$

$$\tilde{q}_m \equiv \sum_{i \neq m}\sum_{j \neq i,m}\left(\frac{q_i q_j q_m}{(1-q_i)(1-q_i-q_j)}\right) + \sum_{i \neq m}\sum_{k \neq i,m}\left(\frac{q_i q_m q_k}{(1-q_i)(1-q_i-q_m)}\right)$$
$$+ \sum_{j \neq m}\sum_{k \neq j,m}\left(\frac{q_m q_j q_k}{(1-q_m)(1-q_m-q_j)}\right) ,$$

$$\tilde{R}_\ell \equiv \sum_{i \neq \ell}\left(\frac{q_i q_\ell}{1-q_i}\right)P_i + \sum_{j \neq \ell}\left(\frac{q_\ell q_1}{1-q_\ell}\right)P_j, \quad \text{and}$$

$$\tilde{R}_m \equiv \sum_{i \neq m}\sum_{j \neq i,m}\left(\frac{q_i q_j q_m}{(1-q_i)(1-q_i-q_j)}\right)(S_i+S_j)$$
$$+ \sum_{i \neq m}\sum_{k \neq i,m}\left(\frac{q_i q_m q_k}{(1-q_i)(1-q_i-q_m)}\right)(S_i+S_k)$$
$$+ \sum_{j \neq m}\sum_{k \neq j,m}\left(\frac{q_m q_j q_k}{(1-q_m)(1-q_m-q_j)}\right)(S_j+S_k) .$$

(i) If $p_\ell^* + s_m^* \leq w_0$, $p_\ell^* \geq 0$, and $s_m^* \geq 0$, then it is optimal to bet p_ℓ^* on horse ℓ to place and s_m^* on horse m to show.

(ii) If $p_\ell^* \leq 0$ and $s_m^* \leq 0$ bet nothing.

(iii) If $p_\ell^* \leq 0$ and $s_m^* > 0$ bet $\min(s_m^*, w_0)$ on horse m to show.

(iv) If $p_\ell^* > 0$ and $s_m^* \leq 0$ bet $\min(p_\ell^*, w_0)$ on horse ℓ to place.

(v) If $p_\ell^* + s_m^* > w_0$ and $p_\ell^* > 0$, $s_m^* > 0$ then the optimal amounts to bet are s_m^{**} and p_ℓ^{**}, from the Kuhn-Tucker conditions

$$w_0 - P_\ell^{**} - s_m^{**} \geq 0,$$
$$p_\ell^{**} \geq 0,$$
$$s_m^{**} \geq 0,$$
$$\lambda_1(w_0-p_\ell^{**}-s_m^{**}) = 0, \qquad \lambda_1,\lambda_2,\lambda_3 > 0$$
$$\lambda_2 p_\ell^{**} = 0,$$
$$\lambda_3 s_m^{**} = 0,$$
$$\frac{\partial f(p_\ell^{**},s_m^{**})}{\partial p_\ell} - \lambda_1 + \lambda_2 = 0, \quad \text{and}$$
$$\frac{\partial f(p_\ell^{**},s_m^{**})}{\partial s_m} - \lambda_1 + \lambda_3 = 0.$$

Remarks. It is not known whether or not the function $\psi(p, s)$ is pseudoconcave in general, even for linear utility, or in suitable instances that are likely to occur in practice. In fact results analogous to those in Theorem 1b and c that the *argument* is pseudoconcave and the objective function is concave-transformable are *not* obtainable using the methods used in the proof of that theorem. Hausch, Ziemba and Rubinstein (1981) solved several *thousand* models similar to (5) using GRG. In about twenty cases solutions were obtained using multiple starting values and in all instances the final solutions were identical. Hence it appears that ψ has a type of possibly local generalized concavity or connectivity, see Martin (1981), that yields optimal solutions using gradient algorithms. The linear exact solution result (c) is useful because generally it is optimal even with the exact logarithmic solution to bet on at most one horse. However one tends to bet too much with linear utility since one essentially bets until the odds are driven to be unprofitable in an expected value sense or the budget constraint is binding. The logarithmic formulation yields a more reasonable betting scheme that reflects the risks as well as the possible returns. Hausch, Ziemba and Rubinstein (1981) present results of the use of the model (5) and approximations to make it fully implementable in practice. Figure (1) illustrates typical behaviour. See also Ziemba and Hausch (1984) for a layman's account of the final system with the results of hypothetical and actual betting, etc.

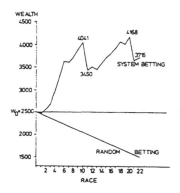

Figure 1   Results from Summer 1980 Exhibition Park Betting:
9 days 22 Races

REFERENCES

HARVILLE, D.A. (1973). "Assigning Probabilities to the Outcomes of Multi-Entry Competitions." *Journal of the American Statistical Association, 68,* 312-316.

HAUSCH, D.B., ZIEMBA, W.T. and RUBINTEIN, M.E. (1981). "Efficiency of the Market for Racetrack Betting." *Management Science.*

ISAACS, R. (1953). "Optimal Horse Race Bets." *American Mathematical Monthly,* 310 -315.

MARTIN, D.H. (1981). "Connectedness of Level Sets as a Generalization of Concavity." *Concavity Properties of Racetrack Betting Models,* 95-107.

SCHAIBLE, S. (1977). "A Note on the Sum of a Linear and a Linear-Fractional Function." *Naval Research Logistics Quarterly,* 691-693.

ZIEMBA, W.T. and HAUSCH, D.B. (1984). *Beat the Racetrack.* (Harcourt Brace Jovanovich, San Diego.)

OPTIMAL BETS IN PARI-MUTUEL SYSTEMS

NISSAN LEVIN

Faculty of management, Tel-Aviv University, Tel-Aviv, Israel

Models of optimal betting in pari-mutuel systems, using a general
utility function, are discussed. A theorem showing the relative
attractiveness of each outcome ("horse") is given. Some special cases of
the linear utility model are considered. An algorithm is presented for
the logarithmic utility. The case where the public bets are random
random variables is also analyzed.

1. Introduction

The pari-mutuel method is one of the most popular betting forms. The
bets are pooled by the management and divided among those who bet on the
winners after taxes and commissions are deducted. This system is practiced
in many countries, especially (but not only) in sports: horse racing, dog
racing, soccer and others.

Academic research has dealt primarily with horse racing. Isaacs[1953]
determined an exact closed form for the betting policy that maximizes the
expected value of the bettor's profit. Thrall [1955] showed that if there
is an infinite sequence of bettors, each using Isaacs' model, then the
expected gain of each of the bettors will be zero in the limit. Hausch et
al. [1981] and Hausch and Ziemba [1985] used the logarithmic utility function
and described a technical system for place and show betting practiced in
horse racing. Rosner [1975] also utilized logarithmic utility "... when
one's own bet is small relative to the total amount bet by the public".

The starting point of this work is Isaacs' model and various extensions
are dealt with. The general utility function problem is presented and
analyzed in Section 2. In addition to generalizing Isaacs' model, we
include two new parameters. The first is the current bettor wealth
and the second is the money remaining in the pool from previous games.
The most important result in this section is the "merit order", i.e.,
the condition specifying when one outcome ("horse") is more attractive
than another.

Section 3 analyzes the linear utility model. It gives the algorithm
determining the optimal betting policy and discusses some special situations:
A "deserted" outcome (i.e., a "horse" that no one is betting on) and a "fair
game" (i.e., a system with no taxes and management commissions).

Logarithmic utility is also analyzed in previous works, since it
should be used by bettors wishing to maximize the long-run rate of their
asset growth. The optimal solution for this case is presented in section 4.

In the current literature as well as in sections 2-4, it is assumed
that the public's bets are known. This assumption is relaxed in section 5
where we assume that these bets are random variables with known probability
distributions. This new model is referred to as the "stochastic model".

Two significant results are obtained in this section: the first is that
the expected gain is higher than the expected gain calculated using the
"deterministic model". The second is that the "merit order" calculated
by using the expected values of the public's bets is not necessarily
equal to the merit order of the stochastic case. The section concludes
with a heuristic algorithm.

    Some numerical results are given throughout the paper to demonstrate
the algorithms.

    For the sake of brevity some of the proofs are omitted and can be
found in Levin [1984].

2. The General Deterministic Model

Notation

    The following notation is used:

n - the number of possible outcomes (= the number of horses).

$b(i)$, $i=1,\cdots,n$ - the total amount bet by the public on outcome i. $b(i)$
        does not include the amount bet by our "optimal" bettor - Mr. G.
        (Without loss of generality, we assume that each ticket costs $1.00.)

$$b = \sum_{i=1}^{n} b(i)$$

$p(i)$, $i=1,\cdots,n$ - the occurrence probability of outcome i.

$1-Q$ - the percentage taken by the management.

c - the "carryover" - the amount of money left over in the pool from
        previous games. As far as we know, c has not been included in previous
        models, probably because it is not relevant in horse racing (except
        for games like "pick six" which are relatively new). But there are
        many other games where c plays a very important role. In these
        games, n - the number of possible outcomes - is very large, so in the
        frequent event that no bettor wins the prize, a carryover is added to
        the prize of the next game. Examples are soccer betting, the 6/49
        lottery (see Ziemba et al. [1985]) and as mentioned above, the
        "pick six".

$m(i)$, $i=1,\cdots,n$ - Mr. G's wager, on outcome i.

$$m = \sum_{i=1}^{n} m(i)$$

$$M = [m(1),\cdots,m(n)]$$

$$a(i) = b(i) + m(i) , \quad i=1,\cdots,n$$

```
a = b + m

A = [a(1),···,a(n)]
```

$$s(i) = (Q \cdot a + c)/a(i) , \quad i=1,\cdots,n \tag{1}$$
the share per ticket when outcome i is realized.

w   - the current wealth of Mr. G before he bets. This parameter is not
      included in Isaacs [1953] and Thrall [1955] because it is not crucial
      in the linear case. But in the nonlinear case it is necessary.
      Therefore w appears in Hausch et al. [1981]).
      When the betting budget is smaller than the bettor's wealth, w can
      represent this budget.

```
w(i) = m(i)·s(i) + w - m     i=1,···,n
```
      Mr. G's wealth after the realization of outcome i.

U(·)   - Mr. G's utility function defined on his wealth.

Assumptions

The models discussed in sections 2 - 4 assume:
a. The outcomes are mutually exclusive and totally exhaustive.
b. b(i), p(i), Q, c, and w are known parameters.
c. Each winning ticket entitles the bettor to one share as defined by (1).
d. Our optimal bettor - Mr. G - cannot borrow money.

Problem formulation

The following optimization problem is obtained:

$$\text{MAX}_{M} \quad R = \sum_{i=1}^{n} p(i) \cdot U[(w(i)])$$

s.t. $m(i) \in \{0,1,2,\cdots\}$

       $m \leq w$

If $M = (0,0,\cdots,0)$ then $w(i) = w$ and $R = U(w)$. Hence the optimal
value of R is always greater than or equal to U(w).

It is more convenient to use the a(i) as decision variables. Hence by
ignoring the integrality constraints, the problem can be reformulated as:

$$\text{MAX}_{A,a} \quad R = \sum_{i=1}^{n} p(i) \cdot U[(w(i)]) = \sum p(i) \cdot U\{[1-b(i)/a(i)] \cdot (Q \cdot a+c) + w - a +b\}$$

$$\text{s.t.} \quad \sum_{i=1}^{n} a(i) = a \tag{2}$$

$$a(i) \geq b(i) \quad i=1,\cdots,n \tag{3}$$

$$a \leq w + b \tag{4}$$

Denoting by α, $\alpha(i)$ and θ the dual variables of (2), (3) and (4) respectively, yield the Lagrangian:

$$L[A,a,\alpha,\alpha(i),\theta] = R + \alpha \cdot [a - \sum_{i=1}^{n} a(i)] + \sum \alpha(i) \cdot [a(i) - b(i)] + \theta \cdot (w+b-a)$$

The first order conditions for optimality are:

$$\partial L/\partial a(i) = p(i) \cdot b(i) \cdot (Q \cdot a + c)/a(i)^2 \cdot U'[w(i)] - \alpha + \alpha(i) = 0 \qquad (5)$$

$$\partial L/\partial a = \sum p(i) \cdot [Q - 1 - Q \cdot b(i)/a(i)] \cdot U'[w(i)] + \alpha - \theta = 0$$

where $U'(x) = dU/dx$

$$\alpha \cdot [a - \sum_{i=1}^{n} a(i)] = 0$$

$$\alpha(i) \cdot [a(i) - b(i)] = 0 \qquad i=1,\cdots,n$$

$$\theta \cdot (w + b - a) = 0 \qquad (6)$$

$$\alpha(i) \geq 0 \qquad i=1,\cdots,n$$

$$\theta \geq 0$$

Rearranging (5) yields

$$p(i) \cdot b(i)/a(i)^2 \cdot U'[w(i)] + \alpha(i)/(Q \cdot a + c) = g \qquad i=1,\cdots,n \qquad (7)$$

where $g \overset{\Delta}{=} \{\theta + \sum p(i) \cdot [1 - Q + Q \cdot b(i)/a(i)] \cdot U'(w(i))\} / (Q \cdot a + c)$ \qquad (8)

The "merit order"

Isaacs [1953] proved the following lemma for the linear utility case: If for some i and j, $p(i)/b(i) \geq p(j)/b(j)$, then if it is optimal to bet on outcome j, it is also optimal to bet on outcome i.

Lemma 1 shows that Isaacs' result is valid not only for the linear case, but for any concave utility function. In other words, if Mr. G is risk averter he should act in accordance with lemma 1.

Lemma 1: If $p(i)/b(i) \geq p(j)/b(j)$ and $a(j) > b(j)$ then $a(i) > b(i)$.

Proof : It suffices to show: $a(i) = b(i) \implies \alpha(i) < 0$

$$p(i)/b(i) \cdot U'[w(i)|a(i) = b(i)] = p(i)/b(i) \cdot U'[w(i) = w+b-a] =$$

$$= p(i)/b(i) \cdot U'[w(j)|a(j) = b(j)] \geq p(j)/b(j) \cdot U'[w(j)|a(j) = b(j)] =$$

$$= p(j) \cdot b(j)/a(j)^2 \cdot U'[w(j)] \cdot \{V[b(j)] / V[a(j)]\}$$

where $V(x) \overset{\Delta}{=} U'[w(j)|a(j) = x]/x^2$,

Now, $\alpha(j)=0$, thus by   (7) $p(j) \cdot b(j)/a(j)^2 \cdot U'[w(j)] = g$

or $p(i) \cdot b(i)/b(i)^2 \cdot U'[w(i)|a(i)=b(i)] > g \cdot V[b(j)] / V[a(j)]$

By the concavity of $U(\cdot)$, $V'<0$,   hence $V[b(j)] \geq V[a(j)]$, therefore

$p(i) \cdot b(i)/b(i)^2 \cdot U'[w(i)] > g$, and by   (7), $\alpha(i)<0$.

Using lemma 1, we rearrange the possible outcomes in order of decreasing values, i.e.,

$$p(1)/b(1) \geq p(2)/b(2) \geq \cdots \geq p(n)/b(n)$$

yielding what we refer to as the "merit order".*

Lemma 2 is a direct conclusion of lemma 1 and the merit order.

Lemma 2: The optimal solution is of the form:

   $a(i) > b(i)$ (or $m(i) > 0$) if $1 \leq i \leq r$

   $a(i) = b(i)$ (or $m(i) = 0$) if $r < i \leq n$

   where $0 \leq r \leq n$

Note that $r = 0$ means - do not bet at all, and $r = n$ means - bet on all possible outcomes.

3. Linear utility **

In this section we deal with the case of linear utility, improving Isaacs' algorithm by decreasing the computational effort and extending it by using the new parameters c and w. We also deal with some special cases: a fair game and a "deserted" outcome.

The solution procedure

Without loss of generality we assume that $U(x)=x$, thus we have from (7) and (8):

$$p(i) \cdot b(i)/a(i)^2 + \alpha(i)/(Q \cdot a+c) = g(r) \qquad\qquad i=1,\cdots,n \qquad\qquad (9)$$

----

* Hausch et al. [1981] used the logarithmic utility to deal with show and place bets in horse racing. In their numerical examples they found that generally speaking, the bets were almost always on those horses with maximum $p(i)/b(i)$, and they used this fact as an approximation on which they based their practical procedure (See also Ziemba and Hausch [1987]).

** With linear utility the analysis is mostly of theoretical interest since it yields "astronomical" with high chances of losing the full bet. This is due to the fact that using a linear utility is equivalent to basing the decisions only on the first moment of the distribution, while ignoring the higher moments (see Hausch et al. [1981]).

where $g(r) \overset{\Delta}{=} [\theta + 1 - Q \cdot P(r) + Q \cdot \sum\limits_{i=1}^{r} p(i) \cdot b(i)/a(i)] \;/\; (Q \cdot a + c)$ (10)

and $P(r) \overset{\Delta}{=} \sum\limits_{i=1}^{r} p(i)$.

By lemma 2, $a(i) > b(i)$ for $i=1,\cdots,r$, hence from (9)

$a(i) = \sqrt{p(i) \cdot b(i)/g(r)}$ $i=1,\cdots,r$ (11)

and

$a = \sum\limits_{i=1}^{n} a(i) = H(r)/\sqrt{g(r)} + b - B(r)$ (12)

where $H(r) \overset{\Delta}{=} \sum\limits_{i=1}^{r} \sqrt{p(i) \cdot b(i)}$ and $B(r) \overset{\Delta}{=} \sum\limits_{i=1}^{r} b(i)$

Now, by substituting (11) and (12) into (10), and solving for $g(r)$
we obtain:

$$g(r) = [\theta + 1 - Q \cdot P(r)] \;/\; [Q \cdot (b - B(r)) + c]$$ (13)

     If r is known, there are two possibilities:

- $\theta = 0$ and $a < w+b$:in this case $g(r)$ is given by (13). (When w
  is not included in the model (see Isaacs [1953] for example)
  it may be assumed that $w = \infty$, hence θ equals zero.)

- $a = w+b$: in this case by (12) $g(r) = [H(r)/(w+B(r))]^2$ (14)

The only remaining detail is to determine r, a task performed by the next
lemma. (Parts a and b of the lemma appear in Isaacs' paper but only for
the special case $c = 0$ and $w = \infty$. Part c is meaningless for $w = \infty$.)

Lemma 3:
a. If $p(1) < b(1)/(Q \cdot b + c)$ then $r = 0$ and $a = b$.

b. Let $\overline{r} = \text{MAX} \{i \mid p(i)/b(i) > [1 - Q \cdot P(i)]/[Q \cdot (b - B(i)) + c]\}$

   and $\overline{a} = H(\overline{r}) \cdot \sqrt{[Q \cdot (b - B(\overline{r})) + c]/[1 - Q \cdot P(\overline{r})]}$

   if $\overline{a} \le w+b$ then $r = \overline{r}$ and $a = \overline{a}$.

c. $a = w + b$ and r is the unique value satisfying

   $p(r+1)/b(r+1) < [H(r)/(w+B(r))]^2 < p(r)/b(r)$ (where $p(n+1) \overset{\Delta}{=} 0$).

     Lemma 4 facilitates the calculation effort and is of major importance
in games where the number of possible outcomes is very large.

     Lemma 4: If $p(i) > b(i)/(Q \cdot b + c)$ then $r \ge i$ (i.e. $m(i) > 0$)

     Bettor intuition suggests that if for any outcome $p(i) > b(i)/b$
(i.e. the real probability is higher than the probability perceived by the

public), then it is worthwhile to bet on this outcome. However, lemma 4 shows that this condition is neither sufficient nor necessary. The reasons for this are the existence of Q and c, and the fact that betting on one outcome affects the profitability of the bets on other outcomes.

We now combine lemmas 1-4 into an algorithm.

Step 0: renumber all the possible outcomes according to the merit order.

Step 1: If $p(1) < b(1)/(Q \cdot b+c)$ then $r = 0$ (lemma 3.a), go to step 4.

Step 2: Find $r(max) = MAX \{i \mid p(i) > b(i)/(Q \cdot b+c)\}$

Find $\bar{r}$ (such that $\bar{r} \geq r(max)$) and calculate $\bar{a}$ (lemma 3.b).

If $\bar{a} \leq w+b$ then $r = \bar{r}$, $\theta = 0$ and $g(r)$ is determined by (13).

Go to step 4.

Step 3: r is determined by lemma 3.c, $a = w+b$, $\theta = 0$ and $g(r)$ is

determined by (14).

Step 4: $a(i) = \sqrt{p(i) \cdot b(i)/g(r)}$ $i=1, \cdots, r$

$a(i) = b(i)$ $i=r+1, \cdots, n$

Stop.

Example 1:
Let $n = 5$, $Q = .9$, $c = 100$, $w = 300$ and $p(i)$ and $b(i)$ as presented in table 1.

Solution:

Step 0 : From column 7, it can be seen that the outcomes appear according to the merit order.

Step 1 : From columns 2 and 6 $p(1) > b(1)/(Q \cdot b+c)$, hence $r > 0$.
Step 2 : From column 6 $r(max) = 2$, hence $r \geq 2$.

From column 8 $\bar{r} = 4$, and by calculation, $\bar{a} = 1111.8$

$\bar{a} < 300+1000 = w + b$ therefore $r = 4$, and by (13) $g(r) = .00081$

Step 4: A = {157.3, 99.5, 344.6, 360.4, 150.0}

M = { 57.3, 19.5, 24.6, 10.4,   0.0}

W = {589.3, 403.9, 266.8, 220.0, 188.2}

$R = \sum_{i=1}^{5} p(i) \cdot w(i) = 323.0 = w + 23.0$

Expected rate of return: $(R-w)/(a-b) = 23/111.8 = 20.6\%$

Table 1: Data and calculations for example 1.

i	p(i)	b(i)	P(i)	B(i)	b(i)/ /(Q·b+c)	p(i)/b(i)	[1-QP(i)]/ /[Q(b-B(i))+c]
1	2	3	4	5	6	7	8
1	.2	100	.2	100	.10	.00200	.00090
2	.1	80	.3	180	.08	.00130	.00087
3	.3	320	.6	500	.32	.00094	.00084
4	.3	350	.9	850	.35	.00086	.00081
5	.1	150	1.0	1000	.15	.00067	.00010

For comparison, if we bet only on the first outcome (i.e. on the most
attractive outcome), the optimal solution is $m(1) = 49$ and $R = w+19.7$

Finally, if $w = 50$, lemma 3.c should be used and the solution is:
$r = 2$, $a(1) = 140.9$ and $a(2) = 89.1$

A "deserted" outcome *

     A deserted outcome is defined as an outcome that nobody has placed
a bet on, in other words - $b(i) = 0$. If there is only one deserted outcome
it occupies the first place in the merit order since if $p(i) >0$,
$p(i)/b(i) = \infty$.

Lemma 5: it is always worthwhile to bet on a deserted outcome.

Proof : Follows directly from lemma 4.

     It should be noted that lemma 5 is based on the assumption that $a(1)$
(the bettor's wager on the deserted outcome) may be less than 1. In
reality, one ticket has to be bought (nothing is gained by buying more than
one ticket), meaning that the constraint $a(1) = 1$ must be added to the
problem formulation. This "new" problem may have no feasible solution,
implying that there are situations where the optimal solution is "don't
bet", even though there is a deserted outcome.

Example 2 (6/49):
$n \approx 14,000000$ $b \approx n$, $Q = .25$ and $c \approx n$
(i.e. nobody guessed the 6 numbers in the last 4 draws).

----
* A deserted outcome is unlikely in horse racing except for games like "pick six".
But in case where there are millions of possible outcomes, the probability for such a
situation might be substantial.   A good example is the 6/49 lottery where the
participants have to pick 6 out of 49 numbers. The number of possible outcomes is
49!/6!/43! hence the occurrence probability is about .00000007 . We believe that most
players have no statistical background and they select their numbers by using an
"emotional process", therefore it is very likely that outcomes such as
"44 45 46 47 48 49" would not be picked.   Ziemba et al. [1985], discuss "unpopular"
numbers.

The expected net profit for $M = \{1, 0, \cdots, 0\}$ is:

$R = (Q \cdot a + c) \cdot p(1) + w - 1 \approx 5 \cdot Q \cdot n/n + w - 1 = w + .25 > w$

But, if $c < 3 \cdot n$, we have $R < 4 \cdot Q + w - 1 < w$, thus the optimal decision is not to bet at all.

A "fair" game

A fair game is a game where all the money invested by the bettors is redistributed among the winners, i.e., $Q = 1$. In this case, if not all the outcomes are the same, then it is worthwhile to bet, as stated in lemma 6.

Lemma 6: if $Q = 1$ and $p(1)/b(1) > p(n)/b(n)$ then $a(1) > b(1)$

Proof : According to lemma 4 it suffices to show that $p(1) > b(1)/(Q \cdot n + c)$,
       or, (since $Q = 1$ and $c \geq 0$) $p(1) > b(1)/n$.
       But if $p(1) < b(1)/n$ we have (due to the merit order):

$$\sum_{i=1}^{n} p(i) = \Sigma[p(i)/b(i)] \cdot b(i) < \Sigma p(1)/b(1) \cdot b(i) = p(1)/b(1) \cdot n < 1.$$

A very interesting situation may happen if c (the undistributed money from previous games) is so large that $r = n$, i.e., "bet on all outcomes". The solution is given in lemma 7 and it says that in such cases, Mr. G should invest all his money ! (Note that this result is valid also when Q is greater than 1).

Lemma 7: if $Q \geq 1$ and $r = n$, then $m = w$ (i.e. $a = w+b$)

Proof : From (13) $g(r) = g(n) = [\theta + 1 - Q \cdot P(n)]/[Q \cdot (n - B(n)) + c] =$

$$= [\theta - (Q-1)]/c \leq \theta/c$$

Hence $\theta > 0$, and by (6) $a = w+b$

4. Logarithmic utility

A property of the logarithmic utility is that under some assumptions (which are very reasonable in our context), maximizing its expected value, asymptotically maximizes the rate of asset growth.   Hence, if the bettor intends to participate in a series of bets and wishes to maximize the long-run rate of his asset growth (or to minimize the expected time to reach a fixed preassigned goal), he should adopt the strategy - $MAX\{E[\log(w)]\}$ (see Hausch et al. [1981], Thorp [1975] and Ziemba and Hausch [1987]).

The solution procedure is very similar to the one used in the linear case, and it is based on the merit order and lemmas 8 - 10.

Lemma 8: If $r < n$ then $a < w + b$

Lemma 9: If r < n then:

$$a(i) = [1 + \sqrt{1+p(i) \cdot \delta(r) \cdot (\delta(r)-2)/(b(i) \cdot G(r))}] \cdot b(i)/\delta(r) \quad i=1,\cdots,r \quad (15)$$

$$a(i) = b(i) \qquad i=r+1,\cdots,n$$

where $\qquad \delta(r) = 2 + 2 \cdot [w+b-A(r)]/[Q \cdot A(r)+c] \qquad\qquad (16)$

$$G(r) = [.5 \cdot \delta(r)-1] \cdot [\sum_{i=1}^{n} p(i) \cdot (1-Q-Q \cdot b(i)/(a(i)) \cdot w(i))] \qquad\qquad (17)$$

$$w(i) = [Q \cdot A(r)+c] \cdot [1-b(i)/a(i)] - A(r) + w + b \qquad\qquad (18)$$

Lemma 10: If r = n then

$$a(i) = .5 \cdot b(i) \cdot [1+\sqrt{1+4 \cdot p(i)/(b(i) \cdot G)}] \qquad\qquad i=1,\cdots,n \qquad (19)$$

where G is the solution of $\qquad 2 \cdot w + b = \sum_{i=1}^{n} b(i) \cdot \sqrt{1+4 \cdot p(i)/(b(i) \cdot G)} \qquad (20)$

Algorithm

Step 0: Let t = 0

Step 1: Increase t by 1 and find a and G(t) using (15)-(18)

Step 2: If p(t)/b(t) < G(t) go to step 5.
otherwise: if t < n and   a < w+b, go to step 1.
otherwise: if a < w+b, set t = n+1 and go to step 5.

Step 3: Use (15) to calculate a(i) for r = t-1
Use (18) and a(i) to calculate w(i).   Let R1 = $\sum p(i) \cdot U[w(i)]$
Use (19) to calculate a(i) for r = n
Use (18) and a(i) to calculate w(i).   Let R2 = $\sum p(i) \cdot U[w(i)]$

Step 4: If R2 > R1 > Ln(w) then   r = n and R = R2. Stop
If R1 > R2 > Ln(w) then r = t-1 and R = R1. Stop
Otherwise: r = 0.   Stop

Step 5: Use (15) to calculate a(i) for r = t-1
Use (18) and a(i) to calculate w(i).   Let R1 = $\sum p(i) \cdot U[w(i)]$
If R1 > Ln(w) then r = t-1 and R = R1. Stop
Otherwise: r = 0.   Stop

Example 3:

For the same data used for the linear utility (see table 1) we obtain the
following results:

From table 2, r = 4. Moreover, the total amount invested - $\sum m(i)$ =
= 56.1 - is much less than in the linear case - 111.8.
This result is due to the risk aversion of the logarithmic utility.
In other words, the solution of the linear utility can serve as an upper
bound to the solution for any risk averter.

Table 2: Results of example 3

t	G(t)	A(t)	p(t)/b(t)	m(1)	m(2)	m(3)	m(4)	m(5)	R	EXP(R)-w
1	2	3	4	5	6	7	8	9	10	11
0	-	1000.0	-	0.0	0.0	0.0	0.0	0.0	5.704	0.0
1	.00089	1019.1	.00200	19.1	0.0	0.0	0.0	0.0	5.730	7.87
2	.00086	1025.9	.00130	19.9	6.0	0.0	0.0	0.0	5.733	8.92
3	.00081	1036.0	.00094	21.0	6.8	8.2	0.0	0.0	5.734	9.28
4	.00072	1056.1	.00086	23.3	8.5	14.4	9.9	0.0	5.735	9.56
5	-	1300.0	.00067	52.1	29.1	92.4	93.9	32.6	5.713	2.94

If we change example 3 by increasing c from 100 to 1000, the
solution is: $r = 5$, $a = w+b=1300$, $M = \{52.2, 29.1, 92.3, 93.8, 32.6\}$
and $EXP(R) = w+433.1$

It should be noted that unlike the linear case, w (the current
wealth) does affect the solution. In the limit, w is so large that
$\delta \approx \delta-2 >> 2$ and we obtain - as expected - the linear solution.

5.   The   stochastic   model

Model   formulation

Until now it was assumed that the total amount of money bet by the
public on each outcome - $b(i)$ - was a given scalar.
In this section we assume that these are n random variables - $B(i)$ -
whose probability distributions are known.

The stochastic model is important for two reasons. The first is that
this actually happens in many real cases.* The second is that we cannot
solve the stochastic model deterministically by replacing $b(i)$ with $E[B(i)]$
since this might lead to incorrect decisions (see lemmas 11 and 12 below).
Since the stochastic issue introduces some mathematical complications,
only the linear utility case is discussed here.

The constraints are those given in the deterministic model (i.e.
(2)-(4)). The objective function, however, has the following form
(assuming independence between $B(i)$, $i=1,\cdots,n$):

$$RS = Q\cdot\sum_{b(1)}^{\infty}\cdots\cdots\sum_{b(n)}^{\infty}\prod_i Pr[B(i)=b(i)]\cdot\sum_{i=1}^{n}[m+c+\sum_{k=1}^{n}b(k)]\cdot p(i)\cdot m(i)/[b(i)+m(i)]-m+w$$

$$\text{or} \quad RS = Q\cdot\sum_{i=1}^{n}p(i)\cdot m(i)\cdot E\{[B(i)-\bar{B}(i)+\bar{B}+m+c]/[B(i)+m(i)]\} - m + w \qquad (21)$$

----
* One might claim that in the case of horse racing, $b(i)$ are known, but this is
not true since the decision of Mr. G should be made before the end of the betting period
(See Ritter [1987]).

where $\quad \bar{B}(i) \overset{\Delta}{=} E[B(i)], \qquad \bar{B} \overset{\Delta}{=} \overset{n}{\underset{i=1}{\Sigma}} \bar{B}(i) \quad$ and $\quad \bar{c} \overset{\Delta}{=} c/Q$

For comparison, the deterministic objective function is $\quad (B(i) = \bar{B}(i))$

$$RD = Q \cdot \underset{i=1}{\Sigma} p(i) \cdot m(i) \cdot [\bar{B}+m+\bar{c}] / [\bar{B}(i)+m(i)] - m + w \qquad (22)$$

First order conditions

Using the Lagrange multipliers defined in section 2.3 we obtain:

$$\partial L/\partial m(i) = Q \cdot p(i) \cdot E\{B(i) \cdot [B(i) - \bar{B}(i) + \bar{B} + m + \bar{c}] / [B(i)+m(i)]^2\} - \alpha + \alpha(i) = 0 \quad (23)$$

$$\partial L/\partial m = Q \cdot \Sigma p(i) \cdot m(i) \cdot E\{1/[B(i)+m(i)]\} - 1 + \alpha - \theta = 0$$

$$\alpha \cdot [m - \overset{n}{\underset{i=1}{\Sigma}} m(i)] = 0$$

$$\alpha(i) \cdot m(i) = 0 \qquad i=1, \cdots, n \qquad\qquad\qquad (24)$$

$$\theta \cdot (w - m) = 0$$

$$\alpha(i) \geq 0 \qquad\qquad\qquad i=1, \cdots, n \qquad\qquad (25)$$

$$\theta \geq 0$$

By rearranging (23) we get:

$$p(i) \cdot E\{B(i) \cdot [B(i) - \bar{B}(i) + \bar{B} + m + \bar{c}] / [B(i)+m(i)]^2\} + \alpha(i)/Q = g \qquad (26)$$

where $\quad g = 1/Q + \theta/Q - \Sigma p(i) \cdot m(i) \cdot E\{1/[B(i)+m(i)]\}$

Relation to the deterministic model

Lemma 11 deals with the relationship between RS and RD, stating that if B(i) are replaced by E[B(i)] and the solution yields RD > 0, then the real expected gain - RS - is also positive.

Lemma 11: $RS \geq RD$ for any set $M = \{m(1), \cdots, m(n)\}$

Proof : From (21) and (22) it suffices to show

$$E\{[B(i) - \bar{B}(i) + \bar{B} + m + \bar{c}] / [B(i)+m(i)]\} \geq [\bar{B}+m+\bar{c}] / [\bar{B}(i)+m(i)]$$

or $\quad E\{[\bar{B}+m+\bar{c} - (\bar{B}(i)+m(i))] / [B(i)+m(i)]\} + 1 \geq [\bar{B}+m+\bar{c}] / [\bar{B}(i)+m(i)]$

or $\quad E\{[\bar{B}+m+\bar{c} - (\bar{B}(i)+m(i))] / [B(i)+m(i)]\} \geq [B+m+c - (\bar{B}(i)+m(i))] / [\bar{B}(i)+m(i)]$

or $\quad E\{1/[B(i)+m(i)]\} \geq 1/[\bar{B}(i)+m(i)]$

or $\quad E\{1/[B(i)+m(i)]\} \cdot E\{B(i)+m(i)\} \geq 1 \qquad\qquad\qquad (27)$

To prove (27) we define $X = 1/\sqrt{[B(i)+m(i)]}$ and $Y = \sqrt{(B(i)+m(i)}$,
and use Schwarz' inequality (see Feller [1971], p. 152):
"For two arbitrary random variables X and Y defined on the same space,
$[E(X \cdot Y)]^2 \le E(X^2) \cdot E(Y^2)$, whenever these expectations exist."

The "intuition" behind lemma 11 is the convexity of the function $1/x$.
Therefore, the lower values $B(i)$ may assume, the larger is the difference
between RS and RD.

The following example demonstrates a special situation where the
optimal deterministic solution is not to bet at all, while the optimal
stochastic solution is to buy one ticket of each outcome.

Example 4: Let $n = 2$, $Q = .95$, $c = 0$, $p(1) = p(2) = .5$ and $B(i)$
        equal 0 or 1 with probabilities .5 .

Solution: $\bar{B}(1) = \bar{B}(2) = .5$ and $\bar{B} = 1$.

$p(1)/\bar{B}(1) = p(2)/\bar{B}(2) = .5/.5 = 1.0 < 1/.95 = 1/(Q \cdot \bar{B}+c)$

According to lemma 3.c the optimal deterministic solution is $m(1)=m(2)=0$.
On the other hand from (21): $RS[m(1)=m(2)=1] = .01875 > 0$.

The merit order

In the deterministic case we showed that the merit order is determined
by the value of $p(i)/b(i)$. In the stochastic case we are able to derive
an analogue order based on the following reasonable assumption:

$$E\{[B(i)-\bar{B}(i)]/B(i)\} << E\{[\bar{B}+m+\bar{c}]/B(i)\}$$ (28)

Lemma 12: If $p(i) \cdot E[1/B(i)] > p(j) \cdot E[1/B(j)]$ and $m(j) > 0$ then $m(i) > 0$.

Proof : From (24) and (26)

$g = p(j) \cdot E\{B(j) \cdot [B(j)-\bar{B}(j)+\bar{B}+m+\bar{c}]/[B(j)+m(j)]^2\}$

$\le p(j) \cdot E\{B(j) \cdot [B(j)-\bar{B}(j)+\bar{B}+m+\bar{c}]/B(j)^2\}$

$= p(j) \cdot E\{[B(j)-\bar{B}(j)+\bar{B}+m+\bar{c}]/B(j)\}$

$\approx [\bar{B}+m+\bar{c}] \cdot p(j) \cdot E[1/B(j)] < [\bar{B}+m+\bar{c}] \cdot p(i) \cdot E[1/B(i)]$

$\approx p(i) \cdot E\{B(i) \cdot [B(i)-\bar{B}(i)+\bar{B}+m+\bar{c}]/B(i)^2\}$

Now, if $m(i) = 0$ then, (from (26)) the last expression equals
$g - \alpha(i)/Q$, implying $\alpha(i) < 0$, in contradiction with (25).

The "intuitive" analogue of the deterministic merit order is
$p(i)/E[B(i)]$ which is different from $p(i) \cdot E[1/B(i)]$.

Example 5:
Let $n=2$, $Q=.9$, $c=1.45$, $p(1)=.76$, $p(2)=.24$, $B(1) \in \{4,5\}$
with probabilities .5   and $B(2) \in \{1,2\}$ with probabilities .5

Using the above data we get:

$p(1)/E[B(1)] = .169 > .160 = p(2)/E[B(2)]$

$p(1) \cdot E[1/B(1)] = .171 < .180 = p(2) \cdot E[1/B(2)]$

The expected gains of various solutions to example 5 are given in the next table.

Table 2: Results of example 5

m(1)	m(2)	RD	RS	RS/RD	RD/m	RS/m	remarks
2.54	.78	.1854	.2225	1.20	5.6%	6.7%	optimal deterministic solution
2.65	.81	.1852	.2227	1.20	5.4%	6.4%	approximated stochas. solution
3.00	1.00	.1800	.2178	1.21	4.5%	5.4%	optimal integer solution

The following observations should be made:
- the optimal bets corresponding to the deterministic model are slightly different from those obtained for the stochastic model.
- the optimal integer solution is identical for both models.
- the expected gain calculated using the stochastic model is about 20% higher than the deterministic expected gain

Guided by these observations we use the deterministic bets as an initial solution for the stochastic model. Moreover, we conjecture that if the bets are constrained to be integers and B(i) are relatively large and $\sigma[B(i)]/E[B(i)]$ is small then the deterministic solution is very close to the stochastic solution. These conditions are satisfied if the number of possible outcomes is small as in the case of horse racing. On the other hand if $Pr\{B(i)=0\}$ are significantly higher than zero (as in 6/49 or soccer betting systems), there might be a difference between the optimal solution of the models.

As an approximated solution we suggest to use the following algorithm based on the first order conditions and (28).

Algorithm

Step 0: Let $t = 0$.

Step 1: Increase t by 1 and find $m(i)$, $i=1, \cdots, n$, that satisfy:

$$p(i) \cdot E\{B(i)/[B(i)+m(i)]^2\} = G(t) \qquad (29)$$

where $G(t) \overset{\Delta}{=} \{1/Q - \sum_{i=1}^{t} p(i) \cdot m(i) \cdot E[1/[B(i)+m(i)]]\}/[\overline{B}+m+\overline{c}] \qquad (30)$

Step 2: If $m(i) \geq 0$ for $i=1, \cdots, t$ and $t < n$ then go to step 1.
        If $m(i) \geq 0$ for $i=1, \cdots, t$ and $t = n$ then stop.

Step 3: Decrease t by 1 and solve (29) and (30).
        $m(i) = 0$ for $i=t+1, \cdots, n$. Stop.

Equal probabilities systems

There are lotteries where all the outcomes have the same occurrence probability, i.e., $p(i) = 1/n$, $i=1,\cdots,n$. Moreover, in these lotteries, the $B(i)$'s have the same distribution, which will be denoted as the random variable Y. The probability that nobody wins the prize might be very large and it often happens that there is no winner for four or even five successive draws. In those cases, the accumulated prize - c - is so high that at least from a theoretical point of view, it is optimal to place bets on all possible outcomes.

Let $\bar{m} = m(i)$ be the number of tickets bought by Mr. G per outcome. Then from (29) and (30) we get

$$E[Y/(Y+\bar{m})]/n = \{1/Q - \bar{m}/n \cdot \sum_{i=1}^{n} E\{1/[Y+\bar{m}]\}/[\bar{B}+m+c]$$

and after rearranging

$$E[Y/(Y+\bar{m})^2] \cdot [E(Y)+\bar{m}+c/n] + \bar{m} \cdot E\{1/[Y+\bar{m}]\} = 1/Q \qquad (31)$$

which has only one unknown - $\bar{m}$.

Another interesting question is "what value of c makes it worthwhile to buy one ticket per outcome ?". The answer to this question is given by substituting $\bar{m} = 1$ in (31) and solving for c.

$$c = n \cdot \{1 - Q \cdot E[1/(Y+1)]\}/E[Y/(Y+1)^2] - n \cdot Q \cdot [E(Y) + 1] \qquad (32)$$

Example 6 (6/49 *):
$n \approx 14,000000$ $Q = .5$ and $Y \sim U(0,10)$
(i.e. the number of tickets sold is about 70,000000)

From (32) we get $c \approx \$49,000000$ and the expected gain is $\$2,700000$.

But, if $c \approx \$10,000000$ then the expected LOSS is about $\$4,500000$, (and if $Q = .25$ the expected LOSS is about 8,3000000), implying that for relevant values of c, we do not recommend to make such an investment !! (See also Ziemba et al. [1985].)

6. Conclusion

Several issues of pari-mutuel systems were discussed in this paper. We began by stating and proving the "general merit order" of the various outcomes, based on the ratio $p(i)/q(i)$ where $p(i)$ is the real occurrence probability of outcome i and $q(i)$ is the probability reflected by the public bets. Although this order is very "intuitive", as far as we know, it has never been proven for nonlinear utility cases.

We proceeded by extending the linear utility model, introduced by Isaacs [1953]. Unlike previous writers, we paid attention to systems with

----
* the results are based on the assumption that each ticket costs $1.00

a large number of possible outcomes. These systems have some new features like the carryover of the prize and the possibility of a deserted outcome, that offer the bettor new opportunities to be taken into consideration.,

The third issue considered was the logarithmic utility. We gave an algorithm for the optimal betting policy (here too, we included the carryover in the model).

Finally, we discussed the "stochastic model", where we assumed that the public bets were random variables. This feature, shared by all pari-mutuel systems, has been ignored by previous works. By using a reasonable approximation we were able to derive a "merit order" and to give an algorithm for the optimal betting policy.

It should be mentioned that further work is needed in two directions:
- Empirical data has to be gathered in order to evaluate the "degree of randomness" in real systems. For example, what is the behavior of the public in the last two (or one) minutes of the betting period in horse racing. (A good example of such work is Ritter's paper [1987].) Or, what is the distribution of the bets on each outcome in the 6/49 lottery.
- Theoretical research extending our results to other utility functions such as the logarithmic utility.

Acknowledgment

I wish to thank William Ziemba and Donald Hausch for their helpful suggestions.

1. Feller, W., "An Introduction to probability Theory and its Applications", Second Edition, Vol. 2, John Wiley & Sons, New York, 1971.

2. Hausch, D. B., W. T. Ziemba and M. E. Rubinstein, "Efficiency of the Market for Racetrack Betting", Management Science, Vol. 27, 1981, pp. 1435-1452.

3. Hausch, D. B. and W. T. Ziemba, "Transactions Costs, Extent of Inefficiencies, Entries and Multiple Wagers in a Racetrack Betting Model", Management Science, Vol. 31, 1985, pp. 381-394.

4. Isaacs, R., "Optimal Horse Race Bets", American Mathematical Monthly, Vol. 60, 1953, pp. 310-315.

5. Levin, N., "Optimal Bets in Pari-mutuel Systems", Working paper No. 821/84, The Israel Institute of Business Research, 1984.

6. Ritter, J. R., "Racetrack Betting - An Example of a Market with Efficient Arbitrage", to appear in this volume.

7. Rosner, B., "Optimal Allocation of Resources in Pari-mutuel Setting", Management Science, Vol. 21, 1975, pp. 997-1006.

8. Thorp, E. O., "Portfolio Choice and the Kelly Criterion", in Stochastic Optimization Models in Finance, edited by W. T. Ziemba and R. G. Vickson, Academic Press, New York, 1975, pp. 599-619.

9. Thrall, R. M., "Some Results in Non-linear Programming", Proceedings Second Symposium in Linear Programming, Vol. 2, National Bureau of Statistics, Washington D. C., January 1955, pp. 471-493.

10. Ziemba, W. T. and D. B. Hausch, "Dr. Z's Beat the Racetrack", Dr. Z Investments Inc., Box 35334, Los Angeles, CA 90035, 1987.

11. Ziemba, W. T. and D. B. Hausch, "Betting at the Racetrack", Dr. Z Investments Inc., Box 35334, Los Angeles, CA 90035, 1986.

12. Ziemba, W. T. et al, "DR. Z's 6/49 Lotto Guidebook", Dr. Z Investments Inc., Box 35334, Los Angeles, CA 90035, 1985.

GROWTH VERSUS SECURITY IN DYNAMIC INVESTMENT ANALYSIS*

L. C. MacLEAN, W. T. ZIEMBA AND G. BLAZENKO

*School of Business Administration, Dalhousie University, Halifax,
Nova Scotia, Canada B3H 1Z5*
*Faculty of Commerce, University of British Columbia, Vancouver,
British Columbia, Canada V6T 1Z2*
*School of Business Administration, Simon Fraser University,
Burnaby, British Columbia, Canada V5A 1S6*

This paper concerns the problem of optimal dynamic choice in discrete time for an investor. In each period the investor is faced with one or more risky investments. The maximization of the expected logarithm of the period by period wealth, referred to as the Kelly criterion, is a very desirable investment procedure. It has many attractive properties, such as maximizing the asymptotic rate of growth of the investor's fortune. On the other hand, instead of focusing on maximal growth, one can develop strategies based on maximum security. For example, one can minimize the ruin probability subject to making a positive return or compute a confidence level of increasing the investor's initial fortune to a given final wealth goal. This paper is concerned with methods to combine these two approaches. We derive computational formulas for a variety of growth and security measures. Utilizing fractional Kelly strategies, we can develop a complete tradeoff of growth versus security. The theory is applicable to favorable investment situations such as blackjack, horseracing, lotto games, index and commodity futures and options trading. The results provide insight into how one should properly invest in these situations.
(CAPITAL ACCUMULATION; FRACTIONAL KELLY STRATEGIES; EFFECTIVE GROWTH-SECURITY TRADEOFF; BLACKJACK: HORSERACING; LOTTO GAMES; TURN OF THE YEAR EFFECT)

This paper develops an approach to the analysis of risky investment problems for practical use by individuals. The idea is to provide simple-to-understand two-dimensional graphs that provide essential information for intelligent investment choice. Although the situation studied is multiperiod, the approach is couched in a growth versus security fashion akin to the static Markowitz mean-variance portfolio selection tradeoff. The approach is a marriage of the capital growth literature of Kelly (1956), Breiman (1961), Thorp (1966, 1975), Hakansson (1971, 1979), Algoet and Cover (1988) and others, which emphasizes maximal growth, with the maximal security literature of Ferguson (1965), Epstein (1977), and Feller (1962). In §1, we formulate a general investment model and describe three growth measures and three security measures. The measures all relate to single-valued aggregates of the investment results over multiple periods such as the mean first passage time to a particular wealth level and the probability of doubling one's wealth before halving it. In §2, we model the investment process as a random walk. Then, using familiar procedures from probability theory, we can easily generate computable quantities which are usually close approximations to the measures of interest. To generate tradeoffs of growth versus security one can utilize fractional Kelly strategies as discussed in §3. A simple but powerful result is that a complete trade-off of growth versus security for the most interesting growth and security measures is implementable simply by choosing various fractional Kelly strategies. Theoretical justification for the fractional Kelly strategies in a multiperiod context can be made using a continuous time approach and log normality assumptions (see Li et al. 1990 and Wu and Ziemba 1990).

* Accepted by Stavros A. Zenios; received March 1991.

Application of the theory to four favorable investment situations is made in §4. In each case the basic game or investment situation is unfavorable to the typical or average player. However, systems have been developed that beat the game in the sense that they have positive expected value. The question then remains how large should the wagers be and how confident is one that particular goals will be achieved. The various games, blackjack, horseracing, lotto games and the turn of the year effect differ markedly in their character. The size of the wagers vary from over half of one's fortune to less than one millionth of the fortune. The graphs that are outputted for each of these applications show how to trade off risk and return and provide crucial insight into how one should invest intelligently in these situations.

1. The Basic Investment Problem

An investor has initial wealth $y_0 \in \mathbb{R}$ and is facing n risky investments in periods 1, 2, $\ldots, t, \ldots$. The return on investments follows a stochastic process defined on the probability space (Ω, B, P) with corresponding product spaces (Ω^t, B^t, P^t). Given the realization history $\omega^{t-1} \in \Omega^{t-1}$, the investor's capital at the *beginning* of period t is $Y_{t-1}(\omega^{t-1})$. The investment in period t in each opportunity i, $i = 1, \ldots, n$, is $X_{it}(\omega^{t-1})$, $X_{it}(\omega^{t-1}) \le Y_{t-1}(\omega^{t-1})$. The investment decisions in terms of proportions are $X_{it}(\omega^{t-1}) = p_{it}(\omega^{t-1})Y_{t-1}(\omega^{t-1})$ for $i = 1, \ldots, n$, and $p_t(\omega^{t-1}) = (p_{0t}(\omega^{t-1}), \ldots, p_{nt}(\omega^{t-1}))$, where $\sum_{i=0}^{n} p_{it}(\omega^t) = 1$ and $p_{0t}(\omega^{t-1})Y_{t-1}(\omega^{t-1})$ is the investment at time t in riskless cash-like instruments. In this general form the investment strategy $p_t(\omega^{t-1})$ depends on time and the history Ω^{t-1} of the investment process. The net return per unit of capital invested in i, $i = 1, \ldots, n$, given the outcome $\omega_t \in \Omega$, in period t is given by $K_{it}(\omega_t)$, $i = 1, \ldots, n$, $t = 1, \ldots$. The return on the risk-free asset is given by $K_{0t} = 0$. It is assumed that $K_{it}(\omega_i)$ is defined by the multiplicative model

$$K_{it}(\omega_t) = \alpha_i(t)E_i(\omega_t),$$

where $\alpha_i(t) > 0$ is the average time path and $E_i(\omega_i)$ is an independent error term. Special cases result when $\alpha_i(t) = \alpha_i$ or $\alpha_i(t) = \alpha_i^t$ for $t = 1, \ldots$. The error terms in this model are not autocorrelated. A more general approach would be to consider the conditional return at time t given the history ω^{t-1}, denoted by $K_{it}(\omega_t|\omega^{t-1})$. A Bayesian analysis with such a model is theoretically tractable, but it is not as suited to computations as the multiplicative model.

An investment environment is said to be favorable if $EK_i(\omega) > 0$ for some i. We are only concerned with favorable environments. The total return on investment is $R_{it}(\omega_i) = 1 + K_{it}(\omega_t)$, and the investment decisions are $p = (p_1, p_2(\omega^1), \ldots, p_t(\omega^{t-1}), \cdots)$. Then the accumulated wealth at the end of period t is

$$Y_t(p, \omega^t) = y_0 \prod_{s=1}^{t} \left(\sum_{i=0}^{n} R_{is}(\omega_s)p_{is}(\omega^{s-1}) \right), \qquad \omega^t \in \Omega^t, \qquad t = 1, \ldots. \qquad (1)$$

For the stochastic capital accumulation process $\{Y_t(p)\}_{t=1}^{\infty}$ we are interested in characterizing the accumulation paths as the investment decisions vary. Consider the following measures of growth and security.

1.1. *Measures of Growth*

G1. $\mu_t(y_0, p) = EY_t(p)$. This is the mean accumulation of wealth at the end of time t. That is, how much do we expect to have, on average, after t periods.

G2. $\phi_t(y_0, p) = E \ln (Y_t(p)^{1/t})$. This is the mean exponential growth rate over t periods. It is a measure of how fast the investor is accumulating wealth. We also consider the long-run growth rate $\phi(p) = \lim_{t \to \infty} \phi_t(y_0, p)$.

G3. $\eta(y_0, p) = E\tau_{\{Y(p)\geq U\}}$. This is the mean first passage time τ to reach the set $[U, \infty)$. That is, how long on average does it take the investor to reach a specific level of wealth. For example, how long must the investor wait before he is a millionaire.

1.2. *Measures of Security*

S1. $\gamma_t(y_0, p) = \Pr[Y_t(p) \geq b_t]$. This is the probability that the investor will have a specific accumulated wealth $b_t \in \mathbb{R}$ at time t. That is, what are his chances of reaching a given target in a fixed amount of time. For example, what chance does the investor have of accumulating \$200,000 one hundred days from now?

S2. $\alpha(y_0, p) = \Pr[Y_t(p) \geq b_t, t = 1, \cdots]$. This is the probability that the investor's wealth is above a specified path. Our concern for an investor is that his wealth will *fall back* too much in any period. He may want protection against losing more than, say, 10% of current wealth at any point in time.

S3. $\beta(y_0, p) = \Pr[\tau_{\{Y(p)\geq U\}} < \tau_{\{Y(p)\geq L\}}]$. This is the probability of reaching a goal U which is higher than his initial wealth y_0, before falling to a wealth level L which is less than y_0. For example, if $U = 2y_0$ and $L = y_0/2$, this is the probability of doubling before halving.

The rationale for the measures selected is to *profile* the growth and security dimensions of the accumulation process. In each case there is a natural pairing of the measures: (μ_t, γ_t) for accumulated wealth at a point in time; (ϕ, α) for the behavior of growth paths; and (η, β) for first passage to terminal or stopping states. The choice of profile (pair of measures) is largely a function of personal preference and problem context. It is expected that the information contained in each of the profiles would be useful in selecting an investment strategy.

The usual criterion for evaluating a decision rule is $t\phi_t(y_0, p)$, the expected log of accumulated wealth. When $K_{it}(\omega_t) = \alpha_i(t)E_i(\omega_t)$, $i = 1, \ldots, n, t = 1, \ldots$, then the log optimal strategy is *proportional*, $\bar{p}_{it}(\omega^{t-1}) = \bar{p}_{it}$. Furthermore, if $\bar{p}_i$, $i = 1, \ldots, n$, solves the one-period problem

$$\max \left\{ E \ln \left(1 + \sum_{i=1}^{n} E_i(\omega)p_i \right) \middle| \sum_{i=1}^{n} p_i \leq 1, \quad p_i \geq 0, \quad i = 1 \cdots, n \right\},$$

then $\bar{p}_{it} = \inf \{ \alpha_i^{-1}(t)\bar{p}_i, 1 \}$ for $i = 1, \ldots, n, t = 1, \ldots$. In the case where $\alpha_i(t) = \alpha_i$, $i = 1, \ldots, n$, the log optimal strategy is a fixed fraction (independent of time), and is referred to as the Kelly (1956) strategy.

Considering the other growth measures we find that the Kelly strategy (i) maximizes the long run exponential growth rate $\phi(y_0, p)$, and (ii) minimizes the expected time to reach large goals $\eta(y_0, p)$ (Breiman 1961, Algoet and Cover 1988). The main properties of the Kelly strategy are summarized in Table 1. The performance of the Kelly strategy on the security measures is not so favorable. As with any fixed fraction rule, the Kelly strategy never risks ruin, but in general it entails a considerable risk of losing a substantial portion of wealth.

In terms of security an optimal policy is often to keep virtually all wealth in the riskless asset (Eithier 1987). We will let $p_0 = (1, 0, \ldots, 0)$ be the extreme security strategy, where all assets are held in *cash*.

Our purpose is to monitor the growth and security measures as the investment decisions vary with the goal of balancing the measures to obtain a strategy which performs well on both the growth and security dimensions. For a particular (growth, security) combination, the complete set of values is given by the *graph* (see Figure 1)

$$U_i = \{(G_i(p), S_i(p)) \mid p \text{ is a feasible investment strategy}\}.$$

TABLE 1—PART 1

Main Properties for the Kelly Criterion

Good/Bad	Property	Reference
G	Maximizing $E \log X$ asymptotically maximizes the rate of asset growth.	Breiman (1961), Algoet and Cover (1988)
G	The expected time to reach a preassigned goal is asymptotically as X increases least with a strategy maximizing $E \log X_N$.	Breiman (1961), Algoet and Cover (1988)
G	Maximizing median $\log X$.	Ethier (1987)
B	False Property: If maximizing $E \log X_N$ almost certainly leads to a better outcome then the expected utility of its outcome exceeds that of any other rule provided n is sufficiently large. Counterexample: $u(x) = X$, $\frac{1}{2} < p < 1$, Bernoulli trials, $\hat{f} = 1$ maximizes $EU(x)$ but $f^* = 2p - 1 < 1$ maximizes $E \log X_N$.	Thorp (1975)
G	The $E \log X$ bettor never risks ruin.	Hakansson and Miller (1975)
B	If the $E \log X$ bettor wins then loses or loses then wins he is behind. The order of win and loss is immaterial for one, two, . . . sets of trials.	$(1 + \gamma)(1 - \gamma)X_0 = (1 - \gamma^2)X_0 < 0$
G	The absolute amount bet is monotone in wealth.	$(\partial E \log X)/\partial W_0 > 0$
B	The bets are extremely large when the wager is favorable and the risk is very low.	Roughly the optimal wager is proportional to the edge divided by the odds. Hence for low risk situations and corresponding low odds the wager can be extremely large. For one such example, see Ziemba and Hausch (1985, pp. 159–160). There in a \$3 million race the optimal fractional wager on a 3–5 shot was 64%.

The most obvious points in U_i to identify are *efficient* points given by $(G_i(p_\alpha), S_i(p_\alpha)) \in U_i$ with

$$G_i(p_\alpha) = \max \{ G_i(p) | S_i(p) \geq \alpha, p \text{ feasible} \}.$$

Alternatively the efficient points are $(G_i(p_\beta), S_i(p_\beta))$ with

$$S_i(p_\beta) = \max \{ S_i(p) | G_i(p) \geq \beta, p \text{ feasible} \}.$$

Given the required level of security (or growth) the efficient points are undominated. The set of efficient points constitutes the *efficient frontier*.

Two key efficient points are the unconstrained optimal growth and optimal security points, respectively $(G_i(\bar{p}), S_i(\bar{p}))$ with $G_i(\bar{p}) = \max \{ G_i(p) | p \text{ feasible} \}$ and $(G_i(p_0), S_i(p_0))$ with $S_i(p_0) = \max \{ S_i(p) | p \text{ feasible} \}$. The optimal growth strategy is the Kelly strategy $\bar{p}$ and the optimal security strategy is the cash strategy p_0. These are both fixed fraction strategies. Other strategies yielding points on the efficient frontier would not typically be fixed fraction, but rather would be dynamic strategies, depending on current wealth and/or time. Gottlieb (1985) has derived such a strategy for the criteria of maximizing mean first passage time to a goal (G_3), subject to a high probability of reaching the goal (S_3).

TABLE 1—PART 2

Main Properties for the Kelly Criterion

Good/Bad	Property	Reference
B	One overbets when the problem data is uncertain.	Betting more than the optimal Kelly wager is dominated in a growth-security sense. Hence if the problem data provides probabilities, edges and odds that may be in error, then the suggested wager will be too large. This property is discussed in and largely motivates this paper.
B	The total amount wagered swamps the winnings—that is, there is much "churning."	Ethier and Tavaré (1983) and Griffin (1985) show that the Expected Gain/E Bet is arbitrarily small and converges to zero in a Bernoulli game where one wins the expected fraction p of games.
B	The unweighted average rate of return converges to half the arithmetic rate of return.	Related to property 5 this indicates that you do not seem to win as much as you expect; see Ethier and Tavare (1983) and Griffin (1985).
G	The $E \log X$ bettor is never behind any other bettor on average in 1, 2, . . . trials.	Finkelstein and Whitley (1981)
G	The $E \log X$ bettor has an optimal myopic policy. He does not have to consider prior to subsequent investment opportunities.	This is a crucially important result for practical use. Hakansson (1971b) proved that the myopic policy obtains for dependent investments with the log utility function. For independent investments and power utility a myopic policy is optimal.
G	The chance that an $E \log X$ wagerer will be ahead of any other wagerer after the first play is at least 50%.	Bell and Cover (1980)
G	Simulation studies show that the $E \log X$ bettor's fortune pulls way ahead of other strategies' wealth for reasonably-sized sequences of investments.	Ziemba and Hausch (1985)

Rather than trying to move from optimal growth to optimal security along the efficient frontier, we will consider the path generated by convex combinations of the optimal growth and optimal security strategies, namely

$$p(\lambda) = \lambda \bar{p} + (1 - \lambda)p_0, \qquad 0 \le \lambda \le 1.$$

Clearly $p(\lambda)$ is a fixed fraction strategy and we will call it *fractional Kelly* since it blends part of the Kelly with the cash strategy. The fractional Kelly strategy is n-dimensional, specifying investments in all opportunities, and the key parameter is λ, which we can consider as a *tradeoff* index.

The path traced by the fractional Kelly strategies is

$$U_i^0 = \{(G_i(p(\lambda)), S_i(p(\lambda)))|0 \le \lambda \le 1\}.$$

The fractional Kelly path is illustrated in Figure 1. This path has attractive properties. First, it is easily *computable*, as shown in the next section. When λ varies from 0 to 1 we continuously trade growth for security. Although this tradeoff is not always efficient, it is *effective* in that we see a *monotone increase in security as we decrease growth*. However, the rate of change in growth or security with respect to λ is not constant and thus the Kelly path is curvilinear. In fact, MacLean and Ziemba (1990) established that the Kelly path is above the straight line (secant) joining the optimal growth and optimal security points. These properties are discussed in §3.

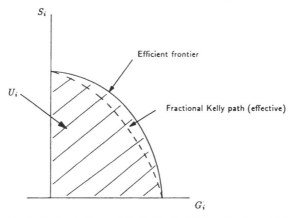

FIGURE 1.　Graph of (Growth, Security) Profile Showing Efficient Frontier and Effective Path.

2. Computation of Measures

If we select a strategy which specifies our investment in the various opportunities and we stick with that strategy, then each of the measures described in §2 is easily computable. We will now develop the computational formulae. Since most of our discussion focuses on fractional Kelly strategies, we work with them. However, more general fixed strategies could be used.

Consider the fractional Kelly strategy $p(\lambda)$ and use the transformation

$$Z_t(p(\lambda), \omega^t) = \ln Y_t(p(\lambda), \omega^t).$$

From (1),

$$Z_t(p(\lambda), \omega^t) = z_0 + \sum_{s=1}^{t} \ln \left(\sum_{i=0}^{n} R_i(\omega_s)p_i(\lambda) \right) = Z_{t-1}(p(\lambda), \omega^{t-1}) + J(p(\lambda), \omega_t),$$

where $J(p(\lambda), \omega) = \ln \left(\sum_{i=0}^{n} R_i(\omega)p_i(\lambda) \right)$ is a stationary jump process and $z_0 = \ln y_0$. Hence, Z_t, $t = 1, \ldots$, is a random walk and we can evaluate the various measures of interest using this process. An alternative approach to evaluating these measures is presented in Ethier (1987) where the discrete time (and maybe discrete state) process $Y_t(p(\lambda))$ is approximated by geometric Brownian motion.

The mean wealth accumulation measure G1 is

$$\mu_t(y_0, p(\lambda)) = EY_t(p(\lambda)) = y_0 E \prod_{s=1}^{t} \left(\sum_{i=0}^{n} R_i(\omega_s)p_i(\lambda) \right) = y_0 \left(\sum_{i=0}^{n} \bar{R}_i p_i(\lambda) \right)^t, \quad (2)$$

where $\bar{R}_i = ER_i(w)$, $i = 0, \ldots, n$.

For the mean growth rate measure G2

$$\phi_t(y_0, p(\lambda)) = E \ln \left(Y_t(p(\lambda))^{1/t} \right) = 1/t EZ_t(p(\lambda))$$

$$= E \ln \left(\sum_{i=0}^{n} R_i(\omega)p_i(\lambda) \right) + \frac{1}{t} z_0 = EJ(p(\lambda), \omega) + \frac{1}{t} z_0, \quad (3)$$

which is an easy calculation to make.

For measure S1 assume the process

$$Z_t(p(\lambda), \omega^t) = z_0 + \sum_{s=1}^{t} J(p(\lambda), \omega_s)$$

has an approximate normal distribution. Using this approximation yields

$$\gamma_t(y_0, p(\lambda)) = 1 - F\left(\frac{b_t^* - E(Z_t(p(\lambda)))}{\sigma(Z_t(p(\lambda)))}\right), \tag{4}$$

where F is the cumulative for the standard normal, $b_t^* = \ln b_t$, and $E(Z_t(p(\lambda)))$, $\sigma(Z_t(p(\lambda)))$ are the mean and standard deviation respectively.

Measure S2 is given by

$$\alpha(y_0, p(\lambda)) = \Pr[Z_t(p(\lambda)) \ge b_t^*, t = 1, \cdots] \tag{5}$$

$$= \prod_{t=1}^{\infty} \Pr[Z_t(p(\lambda)) \ge b_t^* \mid Z_s(p(\lambda)) \ge b_s^*, s = 1, \ldots, t-1]$$

$$= \prod_{t=1}^{\infty} \alpha_t(z_0, p(\lambda)). \tag{6}$$

Then with $h_t(z_t \mid Z_s \ge b_s^*, s = 1, \ldots, t-1)$ the conditional density

$$\alpha_t = \int_{z_t \ge b_t^*} h_t(z_t \mid Z_s \ge b_s^*, s = 1, \ldots, t-1) dz_t$$

$$= \int_{z_t \ge b_t^*} \left\{ \int_{z_t - x \ge b_{t-1}^*} h_{t-1}(z_t - x \mid Z_s \ge b_s^*, s = 1, \ldots, t-1) \pi(x) dx \right\} dz_t,$$

where $\pi(x)$ is the density for $J(p(\lambda))$,

$$= \int_{z_t \ge b_t^*} \left\{ \int_{x \le z_t - b_{t-1}^*} \frac{h_{t-1}(z_t - x \mid Z_s \ge b_s^*, s = 1, \ldots, t-2)}{\alpha_{t-2}} \pi(x) dx \right\} dz_t. \tag{7}$$

Equation (7) provides a sequential procedure to compute measure S2, requiring only the distribution from the previous stage and the jump probabilities.

For the other measures assume that the random variable $J(p(\lambda))$ has finite support given by the interval $[J_m(p(\lambda)), J_M(p(\lambda))]$. Consider measure S3

$$\beta(y_0, p(\lambda)) = \Pr[\tau_{\{Y(p(\lambda)) \ge U\}} < \tau_{\{Y(p(\lambda)) \le L\}} \mid y_0].$$

Transforming to an additive process with $z = \ln y$ and letting

$$\beta(z, p(\lambda)) = \Pr[\tau_{\{Z(p(\lambda)) \ge \ln U\}} < \tau_{\{Z(p(\lambda)) \le \ln L\}} \mid z]$$

yields

$$\beta(z_0, p(\lambda)) = E\{\beta(z_0 + J(p(\lambda)), p(\lambda))\}. \tag{8}$$

The expectation in (8) is over the random variable $J(p(\lambda))$. We must solve (8) for β subject to the boundary conditions

$$\beta(z, p(\lambda)) = 0 \quad \text{if} \quad z \le \ln L \tag{8a}$$

$$\beta(z, p(\lambda)) = 1 \quad \text{if} \quad z \ge \ln U. \tag{8b}$$

A solution to (8) is of the form $\beta(z, p(\lambda)) = \sum_{k=1}^{s} A_k \theta_k^z$, where A_k are chosen to satisfy (8a, 8b) and the roots $\theta_1, \ldots, \theta_s$ satisfy the equation $E\theta^{J(p(\lambda))} = 1$. It is easy to show that there are two positive roots $\theta_1 = 1$ and $\theta_2 = \theta$, where $\theta_2 < 1$ or $\theta_2 > 1$ depending on whether $EJ(p(\lambda)) > 0$ or $EJ(p(\lambda)) < 0$ respectively. Similar to the methods discussed in Feller (1962, pp. 330–334), we take the minimum $J_m(p(\lambda)) < 0$ and maximum

TABLE 2

Computational Formulas for Growth Measures G1–G3 and Security Measures S1–S3

G1 $\quad \mu_t(Y_0, p) = Y_0 \left(\sum_{i=0}^{n} \bar{R}_i p_i \right)^t$

G2 $\quad \phi(p) = E \ln \left(\sum_{i=0}^{n} R_i(\omega) p_i \right)$

G3 $\quad \bar{\eta}(Z_0, p) = \dfrac{\theta^{Z_0}}{2EJ(p)} \left(\dfrac{U_*(p)}{\theta^{U_*(p)}} + \dfrac{U^*(p)}{\theta^{U^*(p)}} \right) - \dfrac{Z_0}{EJ(p)}$

S1 $\quad \bar{\gamma}(Z_0, p) = 1 - F\left(\dfrac{b_t^* - Z_0 - tEJ(p)}{\sqrt{t\sigma(J(p))}} \right)$

S2 $\quad \alpha(Z_0, p) = \prod_{t=1}^{\infty} \alpha_t(Z_0, p),$

$\quad \alpha_t(Z_0, p) = \sum_{Z_t \geq b_t^*} \left[\sum_{J_t \geq Z_t - b_{t-1}^*} \dfrac{G_{t-1}(Z_t - J_t(p))}{\alpha_{t-1}(Z_0, p)} \cdot \pi_t \right]$

S3 $\quad \bar{\beta}(Z_0, p) = \dfrac{(\theta^{Z_0} - \theta^{U^*})(\theta^{U^*} - \theta^{L^*(p)}) + (\theta^{Z_0} - \theta^{L^*(p)})(\theta^{U^*(p)} - \theta^{L^*})}{2(\theta^{U^*(p)} - \theta^{L^*})(\theta^{U^*} - \theta^{L^*(p)})}$

$J_M(p(\lambda)) > 0$ jumps and find bounds on the solution by solving a simple system with the extreme boundary conditions satisfied as equalities. For the upper bound,

$$\beta(\ln L + J_m(p(\lambda)), p(\lambda)) = 0 \qquad \text{and} \qquad \beta(\ln U, p(\lambda)) = 1.$$

For the lower bound,

$$\beta(\ln L, p(\lambda)) = 0 \qquad \text{and} \qquad \beta(\ln U + J_M(p(\lambda)), p(\lambda)) = 1.$$

Let $L^* = \ln L$, $L^*(p(\lambda)) = \ln L + J_m(p(\lambda))$, $U^* = \ln U$, and $U^*(p(\lambda)) = \ln U + J_M(p(\lambda))$.

The bounds for measure S3 are then

$$\frac{\theta^{z_0} - \theta^{L^*}}{\theta^{U^*(p(\lambda))} - \theta^{L^*}} \leq \beta(y_0, p(\lambda)) \leq \frac{\theta^{z_0} - \theta^{L^*(p(\lambda))}}{\theta^{U^*} - \theta^{L^*(p(\lambda))}}. \tag{9}$$

For measure G3,

$$\eta(y_0, p(\lambda)) = E\tau_{\{Y(p(\lambda)) \geq U | y_0\}} = E\tau_{\{Z(p(\lambda)) \geq \ln U | z_0\}},$$

we use the recursive equation

$$\eta(z_0, p(\lambda)) = E\{\eta(z_0 + J(p(\lambda)), p(\lambda))\} + 1 \tag{10}$$

subject to the boundary condition

$$\eta(z, p(\lambda)) = 0 \qquad \text{if} \qquad z \geq \ln U. \tag{10a}$$

One solution to (10) is $\eta^*(z, p(\lambda)) = -z/(EJ(p(\lambda)))$. Any other solution can be written as $\eta(z, p(\lambda)) = \eta^*(z, p(\lambda)) + \Delta(z, p(\lambda))$ where $\Delta(z, p(\lambda))$ satisfies the system

$$\Delta(z, p(\lambda)) = E\{\Delta(z + J(p(\lambda)), p(\lambda))\}. \tag{10b}$$

Solutions of (10b) have the form $\Delta(z, p(\lambda)) = \sum A_k \theta_k^z$. Solutions of (10) are then

$$\eta(z, p) = \sum A_k \theta_k^z - z/(EJ(p(\lambda)))$$

where the A_k are chosen to satisfy the boundary conditions (10a). With $U_*(p(\lambda))$

$= \ln U - J_m(p(\lambda))$ we can solve the extreme boundary conditions as equalities, namely $\eta(U^*(p(\lambda)), p(\lambda)) = 0$ and $\eta(U_*(p(\lambda)), p(\lambda)) = 0$ yields the following bounds on G3:

$$\left[\frac{U_*(p(\lambda))}{EJ(p(\lambda))}\right]\left(\frac{\theta^{z_0}}{\theta^{U_*(p(\lambda))}}\right) - \frac{z_0}{EJ(p(\lambda))} \le \eta(y_0, p(\lambda))$$

$$\le \left[\frac{U^*(p(\lambda))}{EJ(p(\lambda))}\right]\left(\frac{\theta^{z_0}}{\theta^{U^*(p(\lambda))}}\right) - \frac{z_0}{EJ(p(\lambda))}. \quad (11)$$

The bounds in (9) and (11) are quite tight when the jumps are small relative to the initial wealth z_0.

Table 2 summarizes the formulas we can use to compute the six growth and security measures. For G3 and S3 we have bounds and the formulas given provide midpoint estimates of these bounds. A similar table appears in Ethier (1987) for the geometric Brownian motion model.

3. Effective Growth-security Tradeoff

The fact that the various growth and security measures can be easily evaluated for fractional Kelly strategies provides an opportunity to interactively investigate the performance of strategies for various measures and scenarios. The ability to evaluate various performance measures and determine a preferred strategy will depend on an *effective tradeoff* between complementary measures of growth and security. A tradeoff is effective if the loss of performance in one dimension (say growth) is compensated by a gain in the other dimension (security). In this section we establish that the tradeoff using fractional Kelly strategies is effective for each of the profile combinations. In calculating the rates of change for the various measures we will see that the tradeoff between growth and security is not constant for all λ, $0 \le \lambda \le 1$.

(i) *Rate Profile*: (ϕ, α).

LEMMA 3.1.   *Suppose $p(\lambda)$ is a fractional Kelly strategy and $\bar{p}$ is the Kelly strategy. Then*

$$\frac{\partial \phi(y_0, p(\lambda))}{\partial \lambda} > 0, \qquad 0 \le \lambda \le 1.$$

PROOF.   Consider the optimal growth strategy given by $\bar{p} = (\bar{p}_0, \ldots, \bar{p}_n)$, and the optimal security strategy $p_0 = (1, 0, \ldots, 0)$.

Then

$$p(\lambda) = (\lambda \bar{p}_0 + (1 - \lambda), \ldots, \lambda \bar{p}_n) = (p_0(\lambda), \ldots, p_n(\lambda)) \qquad \text{and}$$

$$\frac{d}{d\lambda} p_0(\lambda) < 0, \qquad \frac{d}{d\lambda} p_i(\lambda) > 0, \qquad i = 1, \ldots, n.$$

We have the optimal growth problem

$$\max \phi(p(\lambda)) = \max E \ln \left(\sum_{i=0}^{n} R_{it}(\omega) p_i(\lambda) \right).$$

Then

$$\frac{\partial}{\partial p_0} \phi(p(\lambda)) = E\left(\frac{1 - \sum_{i=1}^{n} R_{it}(\omega)}{\sum_{i=0}^{n} R_{it}(\omega) p_i(\lambda)} \right) \qquad \text{and}$$

$$\frac{\partial^2}{\partial p_0^2} \phi(p(\lambda)) = -E\left[\frac{1 - \sum_{i=1}^{n} R_{it}(\omega)}{\sum_{i=0}^{n} R_{it}(\omega) p_i(\lambda)} \right]^2 < 0.$$

So $\phi(p(\lambda))$ is concave in p_0 and with $\bar{p}_0$ such that $\partial\phi(p(\lambda))/\partial p_0 = 0$, we have

$$\frac{\partial}{\partial p_0}\,\phi(p(\lambda)) < 0, \qquad p_0 < \bar{p}_0.$$

Therefore

$$\frac{d}{d\lambda}\,\phi(p(\lambda)) = \frac{dp_0}{d\lambda}\left(\frac{\partial}{\partial p_0}\,\phi(p(\lambda))\right) > 0 \qquad \text{and } \phi(p(\lambda)) \text{ is increasing in } \lambda. \qquad \square$$

LEMMA 3.2.   *Let $p(\lambda)$ be a fractional Kelly strategy and $\ln a < 0$ be a given fallback rate. Then*

$$\frac{\partial}{\partial\lambda}\,\alpha(a, p(\lambda)) \le 0.$$

PROOF.   We have for the security measure

$$\alpha(a, p(\lambda)) = \text{Prob}\,[Y_t(p(\lambda)) \ge y_0 a^t, t = 1, \cdots]$$

$$= \prod_{t\ge 1} \text{Prob}\,[Y_t(p(\lambda)) \ge a^t \,|\, Y_s(p(\lambda)) \ge a^s, s = 1, \ldots, t-1]$$

$$= \prod_{t\ge 1} \alpha_t(a, p(\lambda)).$$

Along any path $1, y_1, \ldots, y_{t-1}$ with $y_s \ge a^s, s = 1, \ldots, t-1$.

$$\alpha_t(a, p(\lambda)) = \text{Prob}\left[\ln\left(\sum R_{jt}(\omega)p_j(\lambda)\right) > \ln a - \frac{1}{t-1}\sum_{s=1}^{t-1} z_s\right],$$

where $z_s = \ln y_s$ and $\ln a - (t-1)^{-1}\sum_{s=1}^{t-1} z_s < 0$. With $b = \ln a - (t-1)^{-1}\sum z_s$ and $R_t p(\lambda) = \sum R_{jt}(\omega)p_j(\lambda)$ we have $\alpha_t(\lambda) \equiv$

$$\alpha_t(a, p(\lambda)) = \text{Prob}\,[\ln(R_t p(\lambda)) \ge b]$$

$$= \text{Prob}\,[\ln(\lambda R_t\bar{p} + (1-\lambda)R_t p_0) \ge b]$$

$$= \text{Prob}\,[\lambda\ln(R_t\bar{p}) + (1-\lambda)\ln(R_t p_0) + \Delta \ge b]$$

$$= \text{Prob}\left[\ln(R_t\bar{p}) \ge \frac{b-\Delta}{\lambda}\right] = 1 - F_t(x(\lambda)),$$

where $x(\lambda) = (b-\Delta)/\lambda$ and F_t is the distribution function for the optimal growth process $\ln(R_t p_0)$.

Then $dx(\lambda)/d\lambda > 0$ and F nondecreasing imply $d\alpha_t(\lambda)/d\lambda \le 0$. Hence $d\alpha(a, p(\lambda))/d\lambda \le 0$. $\square$

Thus along the fractional Kelly path growth is increasing and security is decreasing and an *effective* tradeoff is possible.

(ii) *Wealth Profile*: (μ_t, γ_t).

LEMMA 3.3.   *If $p(\lambda)$ is a fractional Kelly strategy then*

$$\frac{d}{d\lambda}\,\mu_t(p(\lambda)) > 0, \qquad 0 \le \lambda \le 1.$$

PROOF. We have

$$\frac{\partial}{\partial p_0} \mu_t(p(\lambda)) = E \frac{\partial}{\partial p_0} Y_t(p(\lambda)) = E \frac{\partial}{\partial p_0} \left[y_0 \prod_{s=1}^{t} \left(\sum_{j=1}^{n} R_j(\omega_s) p_j(\lambda) \right) \right]$$

$$= E \left[y_0 \sum_{r=1}^{t} \frac{\partial}{\partial p_0} \left(\sum_{j=0}^{n} R_j(\omega_r) p_j(\lambda) \right) \prod_{s \neq r} \left(\sum_{j=0}^{n} R_j(\omega_s) p_j(\lambda) \right) \right]$$

$$= -t y_0 E[K(\omega_t) Y_{t-1}(\omega^{t-1})] = -t y_0 (EK)(EY_{t-1}) < 0.$$

Then

$$\frac{d}{d\lambda} \mu_t(p(\lambda)) = \frac{dp_0(\lambda)}{d\lambda} \frac{\partial}{\partial p_0} \mu_t(p(\lambda)) > 0. \qquad \square$$

LEMMA 3.4. *Suppose $p(\lambda)$ is a fractional Kelly strategy and consider the growth rate $\alpha = 1$. Then*

$$\frac{d}{d\lambda} \gamma_t(y_0, p(\lambda)) < 0 \qquad \text{for} \qquad 0 \leq \lambda < 1.$$

PROOF. We have

$$\gamma_t(y_0, p(\lambda)) = \text{Prob} \left[Y_t(p(\lambda)) \geq y_0 a^t \right] = \text{Prob} \left[\sum_{s=1}^{t} \ln R_s p(\lambda) \geq t \ln a \right]$$

$$= \text{Prob} \left[\sum \ln (\lambda R_s \bar{p} + (1 - \lambda) R_s p_0) \geq 0 \right]$$

$$= \text{Prob} \left[\lambda \sum \ln (R_s \bar{p}) + \sum \Delta_s \geq 0 \right] \qquad \text{where} \qquad \Delta_s > 0$$

$$= \text{Prob} \left[\sum \ln (R_s \bar{p}) \geq \frac{-\sum \Delta_s}{\lambda} \right]$$

$$= 1 - F^t(x_t(\lambda)),$$

where $x_t(\lambda) = (-\sum \Delta_s)/\lambda$ and F^t is the distribution function for $\sum_{s=1}^{t} \ln (R_s \bar{p})$. Then $(d/d\lambda) x(\lambda) > 0$ and F^t nondecreasing gives the desired result. $\square$

Hence for the wealth profile an *effective* tradeoff is possible along the fractional Kelly path.

(iii) *Stopping Rule Profile*: (η^{-1}, β). Our analysis of the stopping rule profile focuses on the computational formulas $\bar{\eta}(p(\lambda))$ and $\bar{\beta}(p(\lambda))$. Since we have upper and lower bounds for the true values, the accuracy of the approximation is easy to determine in a particular application. Our choice of strategy is based on the profile graph and that requires $\bar{\eta}$ and $\bar{\beta}$.

The formulas for $\bar{\eta}$ and $\bar{\beta}$ are based on the random walk $Z_t(p(\lambda))$ with jump process $J(p(\lambda))$. The process $J(p(\lambda))$ is *flexible* at the fractional Kelly strategy $p(\lambda)$ if the condition

$$\frac{L'^*(p(\lambda)) \theta^{L^*(p(\lambda))}}{L^*(p(\lambda)) \theta^{L^*(p(\lambda))}} < \frac{EJ'(p(\lambda)) \theta^{J(p(\lambda))}}{EJ(p(\lambda)) \theta^{J(p(\lambda))}} < \frac{U'^*(p(\lambda)) \theta^{U^*(p(\lambda))}}{U^*(p(\lambda)) \theta^{U^*(p(\lambda))}}$$

holds, where $\theta < 1$ is a positive root of the equation $E\theta(p(\lambda))^{J(p(\lambda))} = 1$, and prime denotes the derivative w.r.t. λ.

LEMMA 3.5. *Suppose $p(\lambda)$ is a fractional Kelly strategy and consider the wealth process $Y_t(p(\lambda))$, $t = 1, \cdots$ and the absorbing states $U = ky_0$, $L = k^{-1}y_0$, $k > 1$. If $J(p(\lambda))$ is flexible at $p(\lambda)$ then for $0 \leq \lambda \leq 1$*

$$\text{(i)} \quad \frac{\partial}{\partial \lambda} \bar{\eta}(y_0, p(\lambda)) < 0,$$

$$\text{(ii)} \quad \frac{\partial}{\partial \lambda} \bar{\beta}(y_0, p(\lambda)) < 0.$$

PROOF. (i) For the mean first passage time consider the lower bound

$$\eta^2(p(\lambda)) = \frac{U_*(p(\lambda))}{EJ(p(\lambda))\theta^{U_*(p(\lambda))}},$$

where without loss of generality assume $y_0 = 1$ ($z_0 = 0$). Then

$$\frac{\partial}{\partial p_0}\eta^2(p(\lambda)) = \frac{N^2(p(\lambda))}{D^2(p(\lambda))},$$

where $\quad D^2(p(\lambda)) = [EJ(p(\lambda))\theta^{U_*(p(\lambda))}]^2 \quad$ and

$$N^2(p(\lambda)) = [EJ(p(\lambda))\theta^{U_*(p(\lambda))}J'_m(p(\lambda)) - U_*(p(\lambda))EJ'(p(\lambda))\theta^{U_*(p(\lambda))}$$

$$- U_*(p(\lambda))EJ(p(\lambda))\theta^{U_*(p(\lambda))}J'_m(p(\lambda)) \ln \theta$$

$$- U_*(p(\lambda))EJ(p(\lambda))\theta^{U_*(p(\lambda))}U_*(p(\lambda))\theta'/\theta],$$

where the prime denotes the first derivative w.r.t. p_0. Since $EJ'(p(\lambda)) < 0$ and $EJ(p(\lambda)) > 0$, $J'_m(p(\lambda)) > 0$, $\theta' < 0$, we get all terms in $N^2(p(\lambda))$ positive and $(\partial/\partial p_0)\eta^2(p(\lambda)) > 0$. In the same way for the upper bound $\eta^1(p(\lambda))$ we get $(\partial/\partial p_0)\eta^1(p(\lambda)) > 0$. With $(d/d\lambda)p_0(\lambda) < 0$ the average of the bounds yields (1).

(ii) For the first passage probability we utilize the upper bound

$$\beta^1(p(\lambda)) = \frac{1 - \theta^{L^*(p(\lambda))}}{\theta^{U^*} - \theta^{L^*(p(\lambda))}}.$$

Then the numerator of $(\partial/\partial p_0)\beta^1(p(\lambda))$ is

$$N(p(\lambda)) = (\theta^{U^*} - 1)\frac{\partial}{\partial p_0}\theta^{L^*(p(\lambda))} - (1 - \theta^{L^*(p(\lambda))})\frac{\partial}{\partial p_0}\theta^{U^*}.$$

We have $E\theta(p(\lambda))^{J(p(\lambda))} = 1$ and $(\partial/\partial p_0)E\theta(p(\lambda))^{J(p(\lambda))} = 0$ from which

$$\frac{\partial}{\partial p_0}\theta(p(\lambda)) = -\theta(p(\lambda)) \ln \theta(p(\lambda))[EJ'(p(\lambda))\theta^{J(p(\lambda))}/EJ(p(\lambda))\theta^{J(p(\lambda))}]$$

$$= (-\theta(p(\lambda)) \ln \theta(p(\lambda)))\psi(p(\lambda)).$$

Then

$$\frac{\partial}{\partial p_0}\theta^{L^*(p(\lambda))} = \theta^{L^*(p(\lambda))}\left(\frac{L^*(p(\lambda))}{\theta(p(\lambda))}\frac{\partial}{\partial p_0}\theta + \ln \theta \frac{\partial}{\partial p_0}L^*(p(\lambda))\right)$$

$$= -L^*(p(\lambda))\theta^{L^*(p(\lambda))}\psi(p(\lambda)) \ln \theta + L'^*(p(\lambda))\theta^{L^*(p(\lambda))} \ln \theta,$$

and

$$\frac{\partial}{\partial p_0}\theta^{U^*} = U^*\theta^{U^*-1}\frac{\partial}{\partial p_0}\theta = -U^*\theta^{U^*}\psi(p(\lambda)) \ln \theta.$$

Substituting into $N(p(\lambda))$ with a little algebra yields

$$N(p(\lambda)) = -(\theta^{U^*} - 1)L^*(p(\lambda))\theta^{L^*(p(\lambda))}\psi(p(\lambda)) \ln \theta$$
$$+ (1 - \theta^{L^*(p(\lambda))})U^*\theta^{U^*}\psi(p(\lambda)) \ln \theta$$
$$- (\theta^{U^*} - 1)L'^*(p(\lambda))\theta^{L^*(p(\lambda))} \ln \theta.$$

With $U = ky_0$ and $L = k^{-1}y_0$ it is easy to show that

$$\frac{U^*\theta^{U^*}}{1 - \theta^{U^*}} < \frac{L(p(\lambda))\theta^{L^*(p(\lambda))}}{1 - \theta^{L^*(p(\lambda))}}$$

and therefore

$$A = (1 - \theta^{L^*(p(\lambda))})U^*\theta^{u^*} > -(\theta^{U^*} - 1)L^*(p(\lambda))\theta^{L^*(p(\lambda))} = B.$$

We have, then, substituting B for A

$$N(p(\lambda)) > 2(1 - \theta^{U^*}) \ln \theta (L^*(p(\lambda))\theta^{L^*(p(\lambda))}\psi(p(\lambda)) + L'^*(p(\lambda))\theta^{L^*(p(\lambda))}) > 0$$

since J flexible at $p(\lambda)$ implies

$$L^*(p(\lambda))\theta^{L^*(p(\lambda))}\psi(p(\lambda)) + L'^*(p(\lambda))\theta^{L^*(p(\lambda))} < 0.$$

So

$$\frac{\partial}{\partial p_0} \beta^1 > 0.$$

With the same result for the lower bound and with $(d/d\lambda)p(\lambda) < 0$ we get (ii). □

The proof of the above lemma depends on the flexibility condition but not the requirement that the strategy be fractional Kelly. The significance of the fractional Kelly strategies is that the flexibility condition is always satisfied if it is satisfied for the Kelly strategy.

For the stopping rule profile $(\bar{\eta}^{-1}, \bar{\beta})$ we have along the fractional Kelly path growth increasing and security decreasing. The inverse of expected stopping time is a measure of growth since faster growth would imply getting to the target U sooner.

In summary, for each of the profile types we have the same type of behavior. The complementary measures, growth and security, move in opposite directions as we change the blend between the Kelly and cash strategies. This enables a continuous tradeoff between growth and security.

TRADEOFF THEOREM. *Suppose $p(\lambda)$ is a fractional Kelly strategy. Then for each profile there is an effective tradeoff between growth and security by varying λ, the tradeoff index.*

Additional theoretical results concerning these tradeoffs are presented in MacLean and Ziemba (1990).

Friedman (1982) seems to have first considered fractional Kelly strategies in the context of blackjack. His results are along the lines discussed in §4.1 below. Other references on fractional Kelly strategies are Gottlieb (1984); Ethier (1987); Li, MacLean and Ziemba (1990); and Wu and Ziemba (1990).

The tradeoff theorem in this section establishes the use of growth and security measures in determining an investment strategy. There are many possible tradeoffs depending upon the choice of measures and the parameters specified in these measures. Rather than further considering the preferred measures and tradeoff in theory, we will apply the concepts to some real world examples and explore the choice of investment decisions in practice.

4. Applications

The methods described in previous sections are now applied to a variety of well-known problems in gambling and investment. Rather than concentrating on the results in the tradeoff theorem, we will explore a variety of growth and security measures to see how interactively a satisfactory decision is achieved from both perspectives. In each case we emphasize a pair of measures (growth, security). There are a few points we should make about our approach to these applications. Satisfaction with a particular investment strategy will depend on many factors. The measures of performance we have described are helpful, but they are functions of *inputs* such as initial wealth, return distribution, and performance standards. We must vary those inputs in appropriate ways to see a decision rule under different scenarios. In cases where inputs are soft we may need many diverse scenarios.

Throughout the theory and examples we use fixed strategies. This is not to imply that the particular fixed strategy is used forever, but rather a statement that if all inputs remain constant we can predict the results from the fixed strategy. Of course, the inputs are dynamic and we would regularly (continuously) update the strategy and measures of performance based on new input data. This process is facilitated by easily computable measures.

4.1. *Blackjack*: (ϕ, β)

The game of blackjack seems to have evolved from several related card games in the 19th century. It became popular in World War I and has since reached enormous popularity. It is played by millions of people in casinos around the world. Billions of dollars are lost each year by people playing the game. A relatively small number of professionals and advanced amateurs, using various methods such as card counting, are able to beat the game. The object is to reach, or be close to, twenty-one with two or more cards. Scores above twenty-one are said to bust or lose. Cards two to ten are worth their face value, Jacks, Queens, and Kings are worth ten points and Aces are worth one or eleven at the player's choice. The game is called blackjack because an Ace and a ten-valued card was paid three for two and an additional bonus accrued if the two cards were the Ace of Spades and the Jack of Spades or Clubs. While this extra bonus has been dropped by current casinos, the name has stuck. Dealers normally play a fixed strategy of hit until

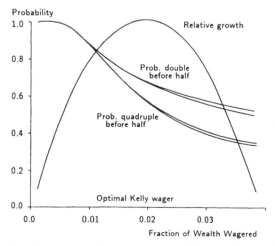

FIGURE 2.   Probability of Doubling and Quadrupling before Halving and Relative Growth Rates Versus Fraction of Wealth Wagered for Blackjack (2% Advantage, $p = 0.51$ and $q = 0.49$).

TABLE 3
Growth Rates Versus Probability of Doubling before Halving for Blackjack

	λ		$\bar{\beta}(0, p)$ = P[Doubling before Halving]		$\phi(p)$ = Relative Growth Rate
	0.1		0.999		0.19
	0.2		0.998		0.36
Range	0.3	↑	0.98	↑	0.51
for	0.4	SAFER	0.94	LESS GROWTH	0.64
Blackjack	0.5		0.89		0.75
Teams	0.6	RISKIER	0.83	MORE GROWTH	0.84
	0.7	↓	0.78	↓	0.91
	0.8		0.74		0.96
	0.9		0.70		0.99
	1.0	KELLY	0.67		1.00
	1.5		0.56		0.75
Overkill →	2.0		0.50		0.00
Too Risky					

a seventeen is reached and then stay. A variation is whether or not a soft seventeen (an ace with cards totalling six) is hit. It is slightly better for the player if the dealer stands on soft seventeen. The house has an edge of 2–10% against typical players. For example, the strategy mimicking the dealer loses about 8% because the player must hit first and busts about 28% of the time ($0.28^2 \cong 0.08$).

In general, the edge for a successful card counter varies from about −10% to +10% depending upon the favorability of the deck. By wagering more in favorable situations and less or nothing when the deck is unfavorable, an average edge weighted by the size of the bet of about 2% is reasonable. Hence, an approximation that will provide us insight into the long-run behavior of a player's fortune is to simply assume that the game is a Bernoulli trial with a probability of success $\pi = 0.51$ and probability of loss $1 - \pi = 0.49$.

We then have $\Omega = \{0, 1\}$, $K(0, p) = p$ with probability π, and $K(1, p) = -p$ with probability $1 - \pi$. The mean growth rate is

$$E \ln (1 + K(\omega, p)) = \pi \ln (1 + p) + (1 - \pi) \ln (1 - p).$$

Simple calculus gives the optimal fixed fraction strategy: $p^* = 2\pi - 1$ if $EK > 0$; $p^* = 0$ if $EK \leq 0$. (This optimal strategy may be interpreted as the edge divided by the odds (1–1 in this case).) In general, for win or lose for two outcome situations where the size of the wager does not influence the odds, the same intuitive formula holds. Hence with a 2% edge betting on a 10–1 shot, the optimal wager is 0.2% of one's fortune.) The growth rate of the investor's fortune is

$$\phi(p) = \pi \ln (1 + p) + (1 - \pi) \ln (1 - p)$$

and this is shown in Figure 2. It is nearly symmetrical around $p^* = 0.02$. Security measure S3 is also displayed in Figure 2, in terms of the probability of doubling or quadrupling before halving. The bounds, from (5), are fairly sharp. Since the growth rate and the security are both decreasing for $p > p^*$, it follows that it is never advisable to wager more than p^*. However, one may wish to trade off lower growth for more security using a fractional Kelly strategy. Table 3 illustrates the relationship between the fraction λ and growth and security. For example, a drop from $p = 0.02$ to 0.01 for a 0.5 fractional Kelly strategy, drops the growth rate by 25%, but increases the chance of doubling before halving from 67% to 89%.

In the blackjack example we see how the additional information provided by the security measure $\beta(y_0, n)$ was valuable in reaching a final investment decision. In a flexible or adaptive decision environment, competing criteria would be balanced to achieve a satisfactory path of accumulated wealth. With this approach professional blackjack teams typically use a fractional Kelly wagering strategy with the fraction $\lambda = 0.2$ to 0.8. See Gottlieb (1985) for discussion including the use of adaptive strategies.

4.2. Horseracing: (μ_t, γ_t)

Suppose we have n horses entered in a race. Only the first three finishers have positive return to the bettor. For the remaining positions, you lose your bet. Then

$$\Omega = \{(1, 2, 3), \ldots, (i, j, k), \ldots, (n - 2, n - 1, n)\}$$

is the set of all outcomes with probability π_{ijk}. We wager the fractions p_{i1}, p_{i2}, p_{i3} of our fortune Y_0 on horse i to win, place, or show, respectively. One collects on a win bet only when the horse is first, on a place bet when the horse is first or second, and on a show bet when the horse is first, second, or third. The order of finish does not matter for place and show bets. The bettors, wagering on a particular horse, share the net pool in proportion to the amount wagered, once the original amount of the bets are refunded and the winning horses share the resulting profits. Let p be the $n \times 3$ matrix of wager fractions, where $\sum_{i=1}^{n} \sum_{j=1}^{3} p_{ij} \leq 1$. The return function for a particular (i, j, k) outcome is

$$K((i, j, k), p) = (QW - w_i) \frac{p_{i1}}{w_i} + \frac{(QP - P_i - P_j)}{2} \left(\frac{p_{i2}}{P_i} + \frac{p_{j2}}{P_j} \right)$$

$$+ \left(\frac{QS - S_i - S_j - S_k}{3} \right) \left(\frac{p_{i3}}{S_i} + \frac{p_{j3}}{S_j} + \frac{p_{k3}}{S_k} \right)$$

$$- \left(\sum_{l \neq i} p_{l1} + \sum_{l \neq i, j} p_{l2} + \sum_{l \neq i, j, k} p_{l3} \right),$$

where $Q = 1$ – the track take, W_i, P_j and S_k are the total amounts bet to win, place, and show on the various horses, respectively, and $W = \sum W_i$, $P = \sum P_j$, $S = \sum S_k$.

As an example we will consider a simplified scheme where a wager is made on a single horse per race and there are five races occurring simultaneously across tracks (Hausch and Ziemba 1990). The required information on the races is given below. Each of the horses has an expected return on a $1 bet of $1.14, so this is a very favorable situation.

Race	Horse	Probability of Collecting on Wager	Odds
1	A	0.570	1–1
2	B	0.380	2–1
3	C	0.285	3–1
4	D	0.228	4–1
5	E	0.190	5–1

The Kelly strategy is

$$p^* = (p_0^*, p_1^*, p_2^*, p_3^*, p_4^*, p_5^*) = (0.8529, 0.0140, 0.0210, 0.0141, 0.0700, 0.0280).$$

So 1.4% of your fortune is bet on horse A in race 1. Using the growth measure $\mu_t(p)$ = mean accumulation of wealth, and the security measure $\gamma_t(p)$ = probability that accumulated wealth will exceed b_t, the Kelly strategy was compared to half Kelly and the fixed fraction strategies which bet 0.01, 0.05 and 0.10 respectively, distributed across

TABLE 4

$\gamma_t(Y_0, p) = $ Prob $Y_t(p)$ *Will Exceed* b_t $(t = 700, Y_0 = 1000)$

| | Strategy $- p$ | | | | |
| | | | Proportional | | |
b_t	Kelly	0.5 Kelly	1%	5%	10%
500	0.940	0.992	1.000	0.915	0.518
1,000	0.892	0.961	0.980	0.856	0.517
5,000	0.713	0.691	0.061	0.631	0.358
10,000	0.603	0.445	0.003	0.512	0.304
100,000	0.231	0.028	0.000	0.148	0.141
$\mu_t(p)$*	$17,497	$8,670	$2,394	$10,464	$1,176

* The median is used here since $Y_t(p)$ is approximately log normal (skewed) (Ethier 1987).

horses in the proportions $(0.1, 0.3, 0.3, 0.2, 0.1)$. These fixed fraction strategies are not fractional Kelly.

The results for $\mu_t(p)$ and $\gamma_t(p)$ are given in Table 4. The Kelly strategy with a 6% betting fraction and the 5% fixed fraction are similar, both having favorable growth and security. The smaller betting fractions provide slightly more security but have much less growth. Since the 10% strategy is beyond the Kelly fraction it has less growth and less security. An interesting point is that the total fraction bet is more significant than the distribution of the bets across the horses.

Implicit in the above example is the ability to identify races where there is a substantial edge in the bettor's favor. There has been considerable research into that question. Hausch et al. (1981) demonstrated the existence of anomalies in the place and show market. At thoroughbred racetracks about 2–4 profitable wagers with an edge of 10% or more exist on an average day. The profitable wagers occur mainly because: (1) the public has a distaste for the high probability-low payoff wagers that occur on short priced horses to place and show, and (2) the public's inability to properly evaluate the worth of place and show wagers because of their complexity—for example, in a ten-horse race there are 72 possible show finishes, each with a different payoff and chance of occurrence. In Hausch et al. (1981) and more fully in Hausch and Ziemba (1985) equations are developed that approximate the expected return and optimal Kelly wagers based on minimal amounts of data to make the edge operational in the limited time available at the track.

Ziemba and Hausch (1987) implement and discuss these ideas and explore various applications, extensions, simulations and results. A laymen's article discussing this appears in the May 1989 issue of *OMNI*. A survey of the academic literature on racetrack and other forms of sports betting is in Hausch and Ziemba (1992).

4.3. *Lotto Games*: (η, β)

Lotteries have been played since before the birth of Christ. Organizations and governments have long realized the enormous profits that can be made, based on the greed and hopes of the players. Lotteries tend to go through various periods of government control and when excessive abuses occur, they have often been shut down or outlawed, only to resurface later. Since 1964, there has been an unprecedented growth in lottery games in the United States and Canada. Current yearly sales are over thirty billion dollars, while net profits to various governmental bodies are well over ten billion dollars. Since the prizes are very large, the public cares little that the expected return per dollar wagered is only 40–50 cents. Typically, half the money goes to prizes, a sixth to expenses, and a third to profits. With such a low payback, it is exceedingly difficult to win at these games

TABLE 5
Lotto Games

Prizes	Probability of Winning	Case A		Case B	
		Prize	Contribution to Expected Value	Prize	Contribution to Expected Value
Jackpot	1/13,983,816	$6 M	42.9	$10 M	71.5
Bonus, 5/6⁺	1/2,330,636	$0.8	34.3	$1.2 M	51.5
5/6	1/55,492	M	9.0	$10,000	18.0
4/6	1/1,032	$5,000	14.5	$250	24.2
3/6	1/57	$150	17.6	$10	17.5
		$10			
			118.1		182.7
Edge			18.1%		82.7%
Optimal Kelly Bet			0.000,000,11		0.000,000,65
Optimal Number of Tickets Purchased per Draw with $10 M Bankroll			11		65

and the chances of winning any prize at all, let alone one of the big prizes, is very small. Is it possible to beat such a game with a scientific system? The only hope of winning is to wager on unpopular numbers in pari-mutuel lotto games. A lotto game is based on a small selection of numbers chosen by the players. Typically, games have forty to forty-nine numbers and you must choose five or six numbers. Management then draws the winning numbers and prizes typically accrue to those with three, four or more correctly matching numbers. By pari-mutuel, it is meant that the net pools for the various prizes are shared by those with winning tickets. Hence, if a small number of people, with so-called unpopular numbers, win a prize, it will be larger than if more people with popular numbers are sharing. Ziemba et al. (1986) investigated the Canadian 6/49, which played across the country, and other regional games. They found that numbers ending in nines and zeros and high numbers tended to be unpopular. Collections of unpopular numbers have a slight edge, a dollar wagered was worth more than a dollar on average, and this edge became fairly substantial when there was a large carryover. With a large carryover, that is, a jackpot pool that is steadily growing because the jackpot has not been won lately, the edge can be as high as 100% or more. Despite this promise, investors may still lose because of the very reasons that Chernoff found, essentially reverberation to the mean and gamblers ruin. Still, with such substantial edges, it is interesting to see how an investor might do playing such numbers over a long period of time.

Two realistic cases are developed in Table 5. Case A corresponds to the situation when the investor wagers only when there is a medium-sized carryover and the numbers that are drawn are quite unpopular. Case B corresponds to the situation where there is a huge carryover and the numbers drawn are the most unpopular. About one draw in every five to ten is similar to Case A and one draw in every twenty or more corresponds to Case B. Also, we have been generous in the suggested prizes. In short, we are giving the un-popular number system an at least fair chance to see if it has any hope of being a winning system in the long run. These cases correspond to the Canadian situation of paying the lotto winnings in cash up front tax free. The US situation of paying the lotto winnings over twenty years and taxing these winnings yields prizes with expected returns about one-third of those in Canada.

First we observe that the optimal Kelly wagers are miniscule. This is not surprising since for Case A, 77.2 cents of the $1.18 expected value and for Case B, $1.23 of the

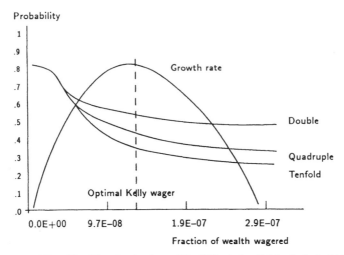

FIGURE 3.   Probability of Doubling, Quadrupling and Tenfolding before Halving for Lotto 6/49, Case A.

$1.83 expected value, respectively, is composed of less than a one in a million chance of winning the jackpot or the bonus prize. One needs a bankroll of $1 million to justify even one $1 ticket for Case A and over $150,000 for Case B. If one had a bankroll of $10 million, the optimal Kelly wagers are just 11 and 65 $1 tickets respectively. Figures 3 and 4 show the chance that the investor will double, quadruple, or tenfold his fortune before it is halved, using Kelly and fractional Kelly strategies for Cases A and B respectively. For the Kelly strategies, these chances are 0.4 to 0.6 for Case A and 0.55 to 0.80 for Case B. With fractional Kelly strategies in the range of 0.00000004 and 0.00000025 or less, the chance of tenfolding one's initial fortune before halving it is 95% or more with Cases A and B respectively. This is encouraging, but it takes an average 294 and 55 billion

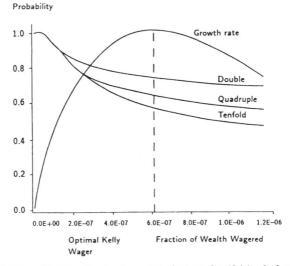

FIGURE 4.   Probability of Doubling, Quadrupling and Tenfolding before Halving for Lotto 6/49, Case B.

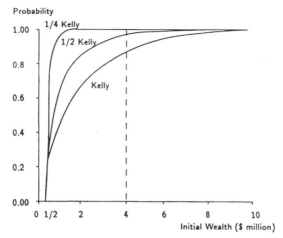

FIGURE 5. Probability of Reaching the Goal of $10 Million before Falling to $½ Million with Various Initial Wealth Levels for Kelly, ½ Kelly and ¼ Kelly Wagering Strategies for Case A.

years, respectively, to achieve this goal. These calculations assume that there are 100 draws per year.

Figure 5 shows the probability of reaching $10 million before falling to $½ million for various initial wealth levels in the range $½–$10 million for Cases A and B with full Kelly, half Kelly and quarter Kelly wagering strategies. The results are encouraging to the millionaire lotto player especially with the smaller wagers. For example, starting with $1 million there is over a 95% chance of achieving this goal with the quarter Kelly strategy for Cases A and B. Again, however, the time needed to do this is very long being 914 million years for Case A and 482 million years for Case B. More generally we have that Case A with full Kelly will take 22 million years, half Kelly 384 million years and quarter Kelly 915 million years. Case B with full Kelly will take 2.5 million years, half Kelly

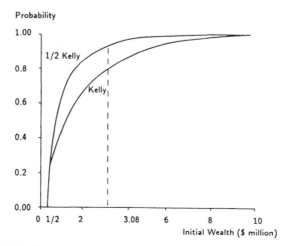

FIGURE 6. Probability of Reaching $10 million before Falling to $25,000 with Various Initial Wealth Levels for Kelly and ½ Kelly Wagering Strategies for Case B.

19.3 million years and quarter Kelly 482 million years. It will take a lot less time to merely double one's fortune rather than tenfolding it, but it is still millions of years. For Case A it will take 4.6, 2.6 and 82.3 million years for full, half, and quarter Kelly, respectively. For Case B it will take 0.792, 2.6 and 12.7 million years for full, half and quarter Kelly.

Finally, we investigate the situation for the nonmillionaire wishing to become one. First, our aspiring gambler must pool his funds with some colleagues to get enough bankroll to proceed since at least $150,000 is needed for Case B and $1 million for Case A. Such a tactic is legal in Canada and in fact highly encouraged by the lottery corporations who supply legal forms for such an arrangement. For Case A, our player needs a pool of $1 million even if the group wagers only $1 per draw. Hence the situation is well modeled by Figure 5. Our aspiring millionaire "puts up" $100,000 along with nine other friends for the $1 million bankroll and when they reach $10 million each share is worth $1 million. The pool must play full Kelly and has a chance of success of nearly 50% before disbanding if they lose half their stake. Each participant does not need to put up the whole $100,000 at the start. Indeed, the cash outflow is easy to bankroll, namely 10 cents per week per participant. However, to have a 50% chance of reaching the $1 million goal each participant (and his heirs) must have $50,000 at risk. On average it will take 22 million years to achieve the goal.

The situation is improved for Case B players. First, the bankroll needed is about $154,000 since 65 tickets are purchased per draw for a $10 million wealth level. Suppose our aspiring nouveau riche is satisfied with $500,000 and is willing to put all but $25,000/2 or $12,500 of the $154,000 at risk. With one partner he can play half Kelly strategy and buy one ticket per Case B type draw. Figure 6 indicates that the probability of success is about 0.95. On average with initial wealth of $308,000 and full Kelly it will take $\frac{3}{4}$ million years to achieve this goal. With half Kelly it will take 2.7 million years and with quarter Kelly it will take 300 million years.

The conclusion then seems to be: millionaires who play lotto games can enhance their dynasties' long-run wealth provided their wagers are sufficiently small and made only when carryovers are reasonably large; and it is not possible for non-already-rich people, except in pooled syndicates, to use the unpopular numbers in a scientific way to beat the lotto and have confidence of becoming rich; moreover these aspiring millionaires are most likely going to be residing in a cemetery when their distant heir finally reaches the goal.

4.4. *Playing the Turn of the Year Effect with Index Futures*: (ϕ, β)

Ibbotson Associates (1986) have considered the actual returns received from investments in US assets with different levels of risk during the period 1926–1985. While small stocks outperformed common stocks by more than 4 to 1, in terms of the cumulative wealth levels, their advantage is totally in the last two decades and most of the gains are in the 1974+ bull market. These returns are before taxes so that the net return after taxes adjusted for inflation for the "riskless" investments and bonds may well be negative for many investors.

One way to invest in this anomaly in light of the Roll (1983) and Ritter (1988) results is to hold long positions in a small stock index and short positions in large stock indices, because the transaction costs are less than a tenth of that of trading the corresponding basket of securities. During the time of this study the March Value Line index was a geometric average of the prices of about 1,700 securities and emphasizes the small stocks while the S&P 500 is a value weighted index of 500 large stocks. Hence the VL/S&P special makes are long in small stocks and short in big stocks at the end of the year. Each point change in the index spread is worth $500. By January 15, the biggest gains are over and the risks increase. On average, the spread drops 0.92 points in this period with a

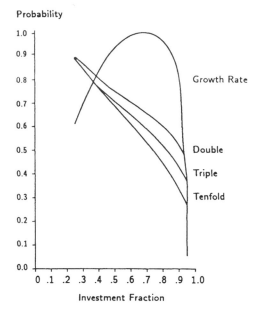

FIGURE 7. Turn of Year Effect: Relative Growth Rate and the Probability of Doubling, Tripling and Tenfolding before Halving for Various Fractional Kelly Strategies.

high variance. The projected gain from a successful trade is 0–5 points and averages 2.85 points or $1,342.50 per spread, assuming a commission of 1.5 × $55. On average, the December 15 to (−1) day gain on the spread, is 0.57 points. However, it was 1.05 in 1985 and 3.15 in 1986 which may reflect the fact that with the thin trading in the VL index, the market can be moved with a reasonably small number of players, who are learning about the success of this trade, i.e. the basis was bid up anticipating the January move. The average standard deviation of the VL/S&P spread was about 3.0. With a mean of 2.85 the following is an approximate return distribution for the trade:

Gain	7	6	5	4	3	2	1	0	−1
Probability	0.007	0.024	0.070	0.146	0.217	0.229	0.171	0.091	0.045

The optimal Kelly investment based on the return distribution is a shocking 74% of one's fortune! Such high wagers are typical for profitable situations with a small probability of loss. Given the uncertainty of the estimates involved and the volatility and margin requirements of the exchanges a much smaller wager is suggested.

Figure 7 displays the probability of doubling, tripling and tenfolding one's fortune before losing half of it, as well as the growth rates, for various fractional Kelly strategies. At fractional strategies of 25% or less the probability of tenfolding one's fortune before halving it exceeds 90% with a growth rate in excess of 50% of the maximal growth rate. Figure 8 gives the probability of reaching the distant goal of $10 million before ruining for Kelly, half Kelly and quarter Kelly strategies with wealth levels in the range of $0–$10 million. The results indicate that the quarter Kelly strategy seems very safe with a 99+% chance of achieving this goal.

These concepts were used in a $100,000 speculative account for a client of CARI Ltd., a Canadian investment management company. Five VL/S&P spreads were purchased to approximate a slightly less than 25% fractional Kelly strategy. Watching the market carefully, these were bought on December 17, 1986 at a spread of −22.18 which was

Probability

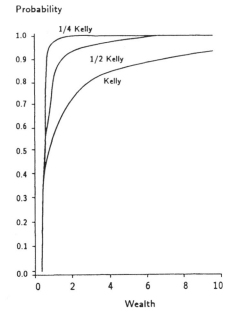

FIGURE 8.   Turn of Year Effect: Probability of Reaching $10 Million before Ruin for Kelly, $\frac{1}{2}$ Kelly and $\frac{1}{4}$ Kelly Strategies.

very close to the minimum that the spread traded at around December 15. The spread continued to gain and we cashed out at -16.47 on the 14th for a gain of 5.55 points per contract or $14,278.50 after transactions costs.

A more detailed discussion of many of the issues in this section appears in Clark and Ziemba (1987), which was updated by Van der Cruyssen and Ziemba (1992a, b).[1]

[1] This research was partially supported by Natural Sciences and Engineering Research Council of Canada grants 5-87147 and A-3152 and by the U.S. National Science Foundation. Without implicating them we would like to thank T. Cover, S. Zenios, and the referees for helpful comments on earlier drafts of this paper.

References

ALGOET, P. H. AND T. M. COVER, "Asymptotic Optimality and Asymptotic Equipartition Properties at Log-optimum Investment," *Ann. Prob.*, 16 (1988), 876–898.

BELL, R. M., AND T. M. COVER, "Competitive Optimality of Logarithmic Investment," *Math. Oper. Res.*, 5, 2 (1980), 161–166.

BREIMAN, L., "Optimal Gambling System for Favorable Games," *Proc. 4th Berkeley Symposium on Math. Statistics and Prob.*, 1 (1961), 63–68.

CHERNOFF, H., "An Analysis of the Massachusetts Numbers Game," Technical Report No. 23, Department of Math., Massachusetts Institute of Technology, Cambridge, MA, 1980 shortened version published in *Math. Intelligence*, 3 (1981), 166–172.

CLARK, R. AND W. T. ZIEMBA, "Playing the Turn of the Year with Index Futures," *Oper. Res.*, 35 (1987), 799–813.

EPSTEIN, R. A., *The Theory of Gambling and Statistical Logic*, 2nd ed., Academic Press, New York, 1977.

ETHIER, S. N., "The Proportional Bettor's Fortune," *Proc. Seventh International Conf. on Gambling and Risk Taking*, Department of Economics, University of Nevada, Reno, NV, 1987.

———, AND S. TAVARÉ, "The Proportional Bettor's Return on Investment," *J. Appl. Prob.*, 20, 563–573.

FELLER, W., *An Introduction to Probability Theory and its Applications*, Vol. 1, 2d ed., John Wiley & Sons, New York, 1962.

FERGUSON, T. S., "Betting Systems which Minimize the Probability of Ruin," *J. Soc. Appl. Math.*, 13 (1965), 795–818.

FINKELSTEIN, M. AND R. WHITLEY, "Optimal Strategies for Repeated Games," *Adv. Appl. Prob.*, 13 (1981), 415–428.

FRIEDMAN, J., "Using the Kelly Criterion to Select Optimal Blackjack Bets," Mimeo, Stanford University, 1982.

GOTTLIEB, G., "An Optimal Betting Strategy for Repeated Games," Mimeo, New York University, 1984.

——, "An Analytic Derivation of Blackjack Win Rates," *Oper. Res.*, 33 (1985), 971–988.

GRIFFIN, P., "Different Measures of Win Rate for Optimal Proportional Betting," *Management Sci.*, 30, 12 (1985), 1540–1547.

HAKANSSON, N. H. "Capital Growth and the Mean-Variance Approach to Portfolio Selection," *J. Fin. Quant. Anal.*, 6 (1971a), 517–557.

——, "On Optimal Myopic Portfolio Policies With and Without Serial Correlation," *J. Business*, 44 (1971b), 324–334.

——, "Capital Growth and the Mean-Variance Approach to Portfolio Selection," *J. Fin. Quant. Anal.*, 6 (1971c), 517–557.

——, "Optimal Multi-Period Portfolio Policies," *Portfolio Theory, TIMS Studies in the Management Sciences*, Vol. 11, E. Elton and M. J. Gruber, Eds., North-Holland, Amsterdam, 1979.

—— AND B. L. MILLER, "Compound-Return Mean-Variance Efficient Portfolios Never Risk Ruin," *Management Sci.*, 22, 4 (1975), 391–400.

HAUSCH, D. AND W. T. ZIEMBA, "Transactions Costs, Extent of Inefficiencies, Entries and Multiple Wagers in a Racetrack Betting Model," *Management Sci.*, 31, 4 (1985), 381–392.

—— AND ——, "Arbitrage Strategies for Cross-Track Betting on Major Horse Races," *J. Business*, 63, 1 (1990), 61–78.

—— AND ——, "Efficiency of Sports and Lottery Betting Markets," In *New Palgrave Dictionary on Money and Finance*, J. Eatwell, P. Newman and M. Milgatel, Eds., Macmillan, London, 1992.

——, —— AND M. RUBINSTEIN, "Efficiency of the Market for Racetrack Betting," *Management Sci.*, 27 (1981), 1435–1452.

IBBOTSON ASSOCIATES, *Stocks, Bonds, Bills and Inflation: Market Results for 1926–1985*, Ibbotson Associates, Chicago, IL, 1986.

KELLY, J., "A New Interpretation of Information Rate," *Bell System Technology J.*, 35 (1956), 917–926.

LI, Y., L. MACLEAN AND W. T. ZIEMBA, "Security Aspects of Optimal Growth Models with Minimum Expected Time Criteria," Mimeo, University of British Columbia, British Columbia, Canada, 1990.

MACLEAN, L. AND W. T. ZIEMBA, "Growth-Security Profiles in Capital Accumulation Under Risk," *Ann. Oper. Res.*, 31 (1990), 501–509.

RITTER, J. R., "The Buying and Selling Behavior of Individual Investors at the Turn of the Year: Evidence of Price Pressure Effects," *J. Finance*, 43 (1988), 701–719.

ROLL, R., "Was ist das? The Turn of the Year Effect and the Return Premia of Small Firms," *J. Portfolio Management*, 10 (1983), 18–28.

THORP, E. O., *Beat the Dealer*, 2d ed., Random House, New York, 1966.

——, "Portfolio Choice and the Kelly Criterion," In *Stochastic Optimization Models in Finance*, W. T. Ziemba and R. G. Vickson, Eds., Academic Press, New York, 1975.

VAN DER CRUYSSEN, B. AND W. T. ZIEMBA, "The Turn of the Year Effect," *Interfaces*, forthcoming (1992a).

—— AND ——, "Investing in the Turn of the Year Effect in the Futures Markets," *Interfaces*, forthcoming, (1992b).

WU, M. G. H. AND W. T. ZIEMBA, "Growth Versus Security Tradeoffs in Dynamic Investment Analysis," Mimeo, University of British Columbia, British Columbia, Canada, 1990.

ZIEMBA, W. T., "Security Market Inefficiencies: Strategies for Making Excess Profits in the Stock Market," TIMS Workshop on Investing in the Stock Market, 1986.

——, S. L. BRUMELLE, A. GAUTIER AND S. L. SCHWARTZ, *Dr. Z's 6/49 Lotto Guidebook*, Dr. Z Investments Inc., Vancouver, 1986.

—— AND D. B. HAUSCH, *Dr. Z's Beat the Racetrack*, William Morrow, New York (revised and expanded second edition of Ziemba-Hausch 1984), 1987.

—— AND ——, *Betting at the Racetrack*, Dr. Z Investments, Los Angeles, 1985.

SEARCHING FOR POSITIVE RETURNS AT THE TRACK: A MULTINOMIAL LOGIT MODEL FOR HANDICAPPING HORSE RACES

RUTH N. BOLTON AND RANDALL G. CHAPMAN

Department of Marketing & Economic Analysis, Faculty of Business, University of Alberta, Edmonton, Alberta, Canada T6G 2R6

This paper investigates fundamental investment strategies to detect and exploit the public's systematic errors in horse race wager markets. A handicapping model is developed and applied to win-betting in the pari-mutuel system. A multinomial logit model of the horse racing process is posited and estimated on a data base of 200 races. A recently developed procedure for exploiting the information content of rank ordered choice sets is employed to obtain more efficient parameter estimates. The variables in this discrete choice probability model include horse and jockey characteristics, plus several race-specific features. Hold-out sampling procedures are employed to evaluate wagering strategies. A wagering strategy that involves unobtrusive bets, with a side constraint eliminating long-shot betting, appears to offer the promise of positive expected returns, even in the presence of the typically large track take encountered at Thoroughbred racing events.
(MULTINOMIAL LOGIT MODEL; HORSE RACE WAGERING; STOCHASTIC UTILITY MODEL; RANKED CHOICE SET DATA; DISCRETE CHOICE MODELING)

Introduction

For as long as there have been horse races, bettors have searched for profitable wagering systems. The general form of any horse race wagering system involves betting against the public. If the public makes systematic and detectable errors in establishing the betting odds, it may be possible to exploit such a situation with a superior wagering strategy and make wagers with a positive expected rate of return.

Academic researchers have also searched for profitable wagering systems to evaluate the efficiency of horse race wager markets (cf., Ziemba and Hausch 1984). Such investigations have been motivated by the basic similarities of race track and stock markets, such as uncertain future returns from investments, the presence of many participants, and the availability of a variety of information concerning investments and participants. (See Copeland and Weston 1979, Fama 1970, 1975, or Rubinstein 1975 for discussions of market efficiency.)

This paper searches for a profitable wagering system to apply to win-betting in a pari-mutuel setting. Wagering systems have two components: a model of the horse race process and a wagering strategy. A model of the horse race process attempts to predict the outcome of a race. Its main output is a prediction of the probabilities of each horse winning a race. A wagering strategy then uses these probabilities as inputs to a betting algorithm which determines the amounts to wager on each horse.

Academic researchers have tended to focus primarily on devising a betting algorithm to determine the amounts to wager on each horse. These algorithms may be categorized according to whether they require knowledge of each horse's true winning probabilities. Assuming these probabilities are known with certainty, optimal wagering theorems for win-betting in a pari-mutuel setting have been developed for the expected value maximizer with infinite wealth (Isaacs 1965) and for the risk averse decision maker

(Rosner 1975). If the true winning probabilities are not known with certainty, wagering strategies may be fundamental (Arveson and Rosner 1971, 1973) or technical (Willis 1964; Harville 1973; Hausch, Ziemba, and Rubinstein 1981) in nature.

Previous research concerning the first component—modeling the horse race process— focused on the public's odds preferences, the existence of inside information, or handicapping ability. If the public consistently underestimates the true winning probabilities for horses in the low odds range (Fabricand 1965), then playing the favorites will improve upon a random betting strategy, but such a strategy may not yield positive returns (Snyder 1978; Asch, Malkiel, and Quandt 1982; Ali 1979). If horse race wager markets are weakly efficient (Dowie 1976), it may be possible to earn extraordinary returns by exploiting publicly available or inside information. Rosett (1965) and Ali (1979) found that horse race data are consistent with the efficient market hypothesis. However, handicapping systems based on publicly available information may yield positive returns (Vergin 1977).

In this paper, we focus on developing a statistical model of the horse race process. The multinomial logit model is used to analyze the horse race process since it explicitly recognizes that there are only a finite number of outcomes to a horse race, namely that one of the entered horses wins. It explicitly analyzes the effects of competition in modeling these outcomes. A recently developed approach to estimating the multinomial logit model's parameters is employed (Chapman and Staelin 1982): the rank order finishing data are exploited to yield improved statistical efficiency of the parameter estimates. Hold-out sampling is used to assess the model's predictions as inputs to various wagering strategies for pari-mutuel win-betting.

The Pari-Mutuel System

In a pari-mutuel system, bettors place wagers on a set of horses in a given race. These wagers form the betting pool from which δ, the track take, is deducted. Approximately 18% goes to the track and the various levels of government, although this amount varies across jurisdictions. The specific track take in any race is also influenced by "breakage," the practice of rounding payoffs down to the next lowest nickel or dime. (The analysis in this paper ignores breakage.) The remainder of the betting pool is allocated to the bettors on the winning horse in proportion to their bets. Hence, the final track probabilities are proportional to the amounts bet on the horses by all bettors. The pari-mutuel probability for horse h, π_h, can be written as follows:

$$\pi_h = w_h/W \tag{1}$$

where w_h is the total amount wagered on horse h by the public and W is the total size of the win-betting pool. These probabilities represent the public's consensus probabilities as reflected by their wagering patterns.

The π_h values cannot be determined until all the bettors have wagered. However, each bettor's wagering strategy depends on knowledge of the π_h values to place bets. Eisenberg and Gale (1959) have resolved this apparent contradiction by showing that a set of final track probabilities and individuals' wagers consistent both with the bettors' strategies and the pari-mutuel system do exist, and that the probabilities are unique. The bettors wager with reference to their subjective expectations of the final π_h values, in the absence of precise information about them.

A dollar bet on horse h will return r_h, if horse h wins the race:

$$r_h = \frac{(1 - \delta)W}{w_h} - 1. \tag{2}$$

From equation (1), it follows that $r_h = (1 - \delta - \pi_h)/\pi_h$. On the track toteboard, the odds typically appear in the form of $(1 - \delta - \pi_h)/\pi_h$ to 1.

How can the risk neutral bettor achieve positive returns at the race track? Define ρ_h to be the true *unknown* winning probability associated with horse h. Then the expected payoff of betting on horse h is given by $\rho_h(r_h + 1)$. Suppose that the public's consensus probabilities are equal to the true winning probabilities. In such a situation $r_h\pi_h = 1 - \delta - \pi_h$, and it follows that $\rho_h(r_h + 1) = 1 - \delta$. Therefore, it does not matter which horse the bettor wagers on, he will always expect to lose the track take, δ. In principle, then, it is only possible for a bettor to expect to discover a betting procedure that yields positive returns when the public misestimates the true winning probabilities (i.e., when $\pi_h \neq \rho_h$). Thus, positive returns at the track are only possible when $\rho_h(r_h + 1) > 1$.

A Statistical Model to Estimate Winning Probabilities

To operationalize any wagering strategy, a statistical model of the horse race process is required. It must estimate the true winning probabilities for each horse. Since accuracy is critically important, the estimates must possess good predictive ability. The results of Vergin (1977) suggest that it may be possible to develop such a model. Existing models of the horse race process are generally based on ad hoc filter rules ("don't bet on any horse that ran within the last 10 days and which lost ground in the stretch in its previous race") or regression analysis where the dependent variable is binary (a horse wins or not) conducted across many races. These models fail to account for the within-race competitive nature of the horse racing process. In addition, they have no theoretical foundation, and consequently may perform poorly. For example, Bratley (1973, p. 85) reports abandoning the search for a regression model using past performance information available before a race to predict its outcome due to lack of overall statistical significance.

A fundamental axiom of any model of the horse race process should be that the race is a probabilistic event. In this paper, this issue is recognized by developing a stochastic utility model to assess the worth of a horse. This model is parameterized in the form of the multinomial logit model. This model has been applied to a wide range of discrete choice problems in marketing and economics. Representative applications include college choice (Punj and Staelin 1978; Chapman 1979; Manski and Wise 1983), shopping center choice (Chapman 1980; Arnold, Oum, and Tigert 1983), and transportation model choice (Domencich and McFadden 1975; Hensher and Johnson 1981). In the horse racing context, Figlewski (1979) used the multinomial logit model to measure the information content of the published forecasts of professional handicappers and found that the track odds had already accounted for most of it.

A Stochastic Utility Model of the Horse Racing Process

A horse race may be thought of as an event in which a decision maker—"nature"— chooses the winning horse from among the available horses in a given race. In each race, "nature" is presented with a choice set which consists of the horses scheduled to run. Each horse h has a vector of K observed attributes (e.g., class, speed rating performance, etc.) associated with it, denoted $\mathbf{x}_h = [x_{h1}, x_{h2}, \ldots, x_{hK}]$. In addition, each horse is ridden by a jockey characterized by a vector of M attributes, $\mathbf{y}_h = [y_{h1}, y_{h2}, \ldots, y_{hM}]$.

A general specification of a statistical model of the horse racing process may be postulated as follows:

$$\rho_h = \rho(\mathbf{X}, \mathbf{Y}), \tag{3}$$

where $\mathbf{X}$ and $\mathbf{Y}$ are the relevant horse and jockey data, respectively, for all of the horses in a given race. A suitable parameterization of this choice model must be chosen so that the estimated winning probabilities satisfy the standard axioms of nonnegative probabilities and probabilities which sum to unity across all of the horses in a race. The multinomial logit model, described below, intrinsically satisfies these axioms.

Let us now assume the existence of a function which measures the worth (or "utility") of a horse h with attribute vector x_h, ridden by a jockey with attribute vector y_h in a given race. The overall worth of horse h in a race can then be written as follows:

$$U_h = U(x_h, y_h). \tag{4}$$

There is typically some measurement error in the modeling process because the attribute vectors do not capture all of the factors operating in the "choice" of a winning horse, the correct functional form for the model may not be specified, and there may be idiosyncratic aspects to any single race. Thus, the overall worth of a horse is assumed to have two parts. One part is a deterministic component, denoted $V_h = V(x_h, y_h)$. The other part is a random component, $\epsilon_h = \epsilon(x_h, y_h)$, which reflects the measurement errors in the modeling process. Assuming that the stochastic error term is independent of the deterministic component, equation (4) can be decomposed as follows:

$$U_h = V_h + \epsilon_h. \tag{5}$$

The presence of the stochastic error term in equation (5) leads to this model being described as a stochastic utility model.

Suppose that horse h^* is observed to win a race. This is equivalent to observing that nature "chose" alternative h^* from the choice set. Since nature is "rational" by definition (i.e., nature "chooses" the best horse at the time the race occurs), revealed preference implies that $U_{h*} \geq U_h$, for $h = 1, 2, \ldots, H$. Since the utility function is partly stochastic, the probability of horse h^* winning the race may be written as:

$$P_{h*} = \text{Prob}(U_{h*} \geq U_h, h = 1, 2, \ldots, H). \tag{6}$$

Further development and simplification of equation (6) requires that a joint distribution function be specified for the error terms. A natural assumption would be to invoke the normal distribution. With such an assumption, the parametric form of the model would become the multiple choice generalization of the probit model. Daganzo (1979) and Maddala (1983, pp. 62–64) may be consulted for the details of this particular discrete choice probability model. The normal distribution assumption is not without considerable cost: a formidable series of numerical integrations is required to explicitly determine the choice probabilities. Alternative error term specifications must be considered. One possible candidate is the logistic distribution. The logistic distribution assumption leads to a tractable choice probability expression, as described below. In addition to considerations of parsimony and reduced computational complexity, it may be noted that the cumulative logistic and normal distributions exhibit little numerical differences, except at the extremes. All of these considerations have led researchers to favor the logit model form over the probit model form in discrete choice modeling.

By assuming that the stochastic error terms are identically and independently distributed according to the double exponential distribution,

$$\text{Prob}(\epsilon_h \leq \epsilon) = \exp[-\exp(-\epsilon)], \tag{7}$$

then the choice probabilities assume the tractable, closed-form expression of the multinomial logit model:

$$P_{h*} = \frac{\exp(V_{h*})}{\sum\limits_{h=1}^{H} \exp(V_h)} \quad \text{for } h^* = 1, 2, \ldots, H. \tag{8}$$

To operationalize the choice probability expression in equation (8), the functional form of the deterministic component of the stochastic utility model must be specified. A linear-in-parameters specification leads to:

$$V_h = \sum_{n=1}^{N} \theta_n Z_{hn} \qquad (9)$$

where $Z_{hn} = Z_{hn}(\mathbf{x}_h, \mathbf{y}_h)$ is the measured value of attribute n for horse h in a race. The Z functions describe either the horse ($\mathbf{x}$), the jockey ($\mathbf{y}$), or both. θ_n is the relative importance of attribute n in the determination of the winning horse. The θ values in equation (9) are the parameters of the stochastic utility model that must be estimated from a sample of races.

Estimating the Parameters of the Multinomial Logit Model

The likelihood function associated with a particular sample of races can be written in the following form for the multinomial logit model:

$$\exp(L) = \prod_{j=1}^{J} P_{jh*} \qquad (10)$$

where the j subscript denotes a race ($j = 1, 2, \ldots, J$), $h*$ in equation (10) is the horse that is observed to win race j, and L refers to the log-likelihood function. Standard software packages, such as Manski (1974), are available to calculate the maximum likelihood estimates. Since maximum likelihood estimates are, in general, consistent and asymptotically normally distributed, approximate large sample confidence bounds may be constructed for parameter estimates and hypotheses may be tested in standard ways.

It is useful to describe several features of the multinomial logit model that are used extensively in this study. First, an overall goodness-of-fit measure has been proposed by McFadden (1974), which is analogous to the familiar multiple correlation coefficient in linear statistical models:

$$\tilde{R}^2 = 1 - \frac{L(\theta = \hat{\theta})}{L(\theta = \mathbf{0})}. \qquad (11)$$

To the extent that the MLE parameters, $\hat{\theta}$, explain the horse race process completely, $\tilde{R}^2$ will approach unity in value. If the vector of MLE parameters is essentially equal to $\mathbf{0}$ (implying an equal chance of each horse winning the race), then $\tilde{R}^2$ will approach zero in value. Hence, $\tilde{R}^2$ varies between zero and one depending on the "explanatory power" of $\hat{\theta}$.

Second, there is a convenient statistical test to assess whether two data subsets are characterized by the same underlying parameter vector, which would imply that the two subsets should be pooled for estimation purposes. To test the null hypothesis that $\theta^{(1)} = \theta^{(2)}$, the appropriate test statistic is:

$$-2(L(\theta = \hat{\theta}^{(1+2)}) - [L(\theta = \hat{\theta}^{(1)}) + L(\theta = \hat{\theta}^{(2)})]) \qquad (12)$$

where $\hat{\theta}^{(1+2)}$ is the MLE obtained by pooling the two data subsets, and $\hat{\theta}^{(1)}$ and $\hat{\theta}^{(2)}$ are the MLEs for the two data subsets, respectively (Watson and Westin 1975). This test statistic will be asymptotically distributed χ^2 with N degrees of freedom, the number of parameters in the model (Wald 1943).

Exploiting Rank Ordered Choice Set Data

The multinomial logit model is estimated on the basis of choice set observations of the form: nature "chooses" horse $h*$ from among all of the competing horses in a race. However, in addition to observing the winning horse in a race, it is also possible to conveniently observe the second finishing horse, the third finishing horse, and so on. Chapman and Staelin (1982) describe how the extra information inherent in such rank ordered choice sets may be exploited. The Chapman and Staelin "explosion process" is based on a Ranking Choice Theorem developed by Luce and Suppes (1965, pp. 354–355) for the class of models of which the stochastic utility model is a member.

To illustrate this procedure, suppose that a race results in the finishing order (from first to last) 4, 2, 1, and 3. By applying the Chapman and Staelin explosion process, it is possible to decompose these rank ordered data into the following three statistically independent choice sets: [horse 4 finished ahead of horses 2, 1, and 3], [horse 2 finished ahead of horses 1, and 3], and [horse 1 finished ahead of horse 3], where no ordering is implied among the "nonchosen" horses in each "choice" set.

These exploded choice sets are statistically independent, so they are equivalent to completely independent horse races. This leads to an increase in the number of independent choice sets available for analysis and, ultimately, to more precise parameter estimates. Extensive small-sample Monte Carlo results reported in Chapman and Staelin (1982) document the improved precision that can accrue by making use of the explosion process. This is valuable because it is costly to generate a sufficiently rich set of horse race data to estimate a multinomial logit model of reasonable complexity.

The rank order explosion process should only be used if it illuminates the choice process, and not if it just adds random noise. This is an important estimation issue because the reliability of the rank order finishing data may decrease for horses far behind the winner and the runners-up. The first three finishers typically receive a portion of the purse and are subject to considerable public scrutiny from track officials and bettors due to the existence of the win, place, and show betting pools. It seems reasonable to assume that those horses and their jockeys are trying and that their finishing position reflects well on their relative "worths." However, this may not be true for horses that finish out of the money.

An approach to resolving this rank order reliability issue is to only partially explode the data. Define E as the researcher-chosen depth of explosion. Then the number of independent choice set observations that may be generated from J races is defined as follows:

$$J(E) = \sum_{j=1}^{J} \min (E, d_j, H_j - 1). \tag{13}$$

$J(E)$ is only defined for nonnegative values of E and d_j represents the depth of available rank ordered choice set information for choice set (race) j.

There are three approaches to determining the appropriate depth of explosion. First, the researcher's a priori knowledge about the choice process may provide clues to the range of plausible values of E. For reasons described above, E might be as large as three in the horse race context.

Second, a heuristic approach may be utilized. This approach involves plotting values of the likelihood ratio index, $\bar{R}^2$, versus different values of E. Since this index does not depend on the number of available choice sets, the calculated $\bar{R}^2$ values should remain approximately constant as E increases, if the subsequently "exploded" observations are of equivalent reliability. If the values of $\bar{R}^2$ start to decrease substantially after some value of E, this would imply that "noisy" observations had been added and the explosion process should be terminated prior to that value of E.

Third, a formal approach to choosing the appropriate value of E involves grouping choice observations by depth of explosion and sequentially testing whether the observation groups may be pooled. Define the first subset of choice observations to consist of the $J(E)$ choice sets generated by an explosion to a depth of E. The second subset then consists of the incremental $J(E + 1) - J(E)$ choice sets generated by exploding to a depth of $E + 1$. Assuming that the $J(E + 1) - J(E)$ subset is large enough, the hypothesis that $\theta^{(E)} = \theta^{(E+1)}$ can be tested using the Watson and Westin (1975) procedure. This grouping and sequential hypothesis testing procedure can be iterated for successive values of E until the hypothesis that the subset parameter vectors are equal is rejected, or the quantity $J(E + 1) - J(E)$ yields too few exploded observations to permit meaningful maximum likelihood parameter estimates to be obtained.

Estimating a Multinomial Logit Model of the Horse Race Process

The Data Base

The horse race data base was assembled from information reported in the *Daily Racing Form*. The 200 race observations are from Aqueduct (43), Pimlico (52), Garden State (42), Keystone (32), and Suffolk Downs (31).

Each of the races satisfied the following restrictions: (i) the race was run over good or fast tracks; (ii) the race distance was in the 1–1.25 mile range; (iii) each horse in the race was a separate betting entry (i.e., there were no coupled entries); and (iv) the horses were at least three years of age. The first two restrictions were applied because the logit model only permits the direct inclusion of variables which vary across the choice set alternatives. Thus, track characteristics, which are the same for every horse in a race, cannot be directly included in the model, except possibly as interaction effects with other horse and/or jockey variables. Races were selected to hold track condition and distance roughly constant, thus obviating the need for inclusion of these variables in the model. The third restriction was made to simplify the win-betting procedure. The fourth restriction ensured that adequate past data would be available on the relevant attributes of each horse. This restriction also avoids the potential problem of dealing with high variability in performance among two-year olds.

The relevant data for each race required about one hour to assemble and code. The use of the previously described explosion process thus had a significant influence on reducing data collection costs. Instead of gathering 600 races at a corresponding cost of 600 hours, it was possible to obtain about the equivalent statistical precision in the multinomial logit model estimates by using each of the 200 races exploded to a depth of three.

Specification of the Model

The specific form of the multinomial logit model employed in this study was as follows:

$$U_h = \theta_1 \text{LIFE\%WIN}_h + \theta_2 \text{AVESPRAT}_h + \theta_3 W/\text{RACE}_h + \theta_4 \text{LSPEDRAT}_h$$
$$+ \theta_5 \text{JOCK\%WIN}_h + \theta_6 \text{JOCK\#WIN}_h + \theta_7 \text{JMISDATA}_h + \theta_8 \text{WEIGHT}_h$$
$$+ \theta_9 \text{POSTPOS}_h + \theta_{10} \text{NEWDIST}_h + \epsilon_h. \quad (14)$$

This model specification is explained in the following text.

The quality of the competing horses is presumed to be the primary determinant of horse race outcomes. The long-term quality of a horse is reflected by two aspects of its past performance: winning potential and competitive level. Current quality/ performance will also be influenced by weight, post position, whether a horse is running at a new distance, and recent workout data. The final component of the model concerns the jockey's characteristics. Each of these model components is discussed below.

A horse's quality is reflected by measures of its winning potential and competitive level. Measures of winning potential may include races won or earnings. Competitive level refers to the types of races in which a horse has previously competed. A horse which changes class is competing with horses of different quality levels. Its past performance (e.g., races won or earnings) is not directly comparable with competitors' measures. Winning potential should be adjusted by a measure of competitive level which is comparable across different past performance conditions, such as speed rating.

Overall winning potential is proxied by LIFE%WIN, the percentage of races won of those entered in the past two years. Overall competitive level is represented by AVESPRAT, an average speed rating for the last four races of each horse. (A speed rating for a horse compares its time in a race with the track record for that distance. The track record is assigned a value of 100 and a point is deducted for each one-fifth of a second that the horse's time is below that mark. The horse's raw speed rating is then adjusted by a factor to equate the track records at the various tracks used in this study, to attempt to account for differences in tracks. Thus, speed rating has been transformed to be comparable across tracks.) Recent winning potential is proxied by W/RACE, winnings per race in the current year (in $000s). The recent competitive level component of past performance is LSPEDRAT, a track-adjusted speed rating for the previous race in which the horse ran.

The effect of weight (WEIGHT) on winning probability may be positive or negative. Since weight levels are designed to handicap better horses and result in more even competition, a higher weight should result in a decrease in winning probability, ceteris paribus. However, higher weight levels are assigned to higher quality horses, so a positive effect may actually be observed because weight carried is positively correlated with a horse's quality.

An inside (lower) post position theoretically improves the probability of a horse winning because a slightly shorter race distance is involved. Higher values of POSTPOS are expected to result in a decrease in winning probability.

A horse running at a new (unfamiliar) longer distance may not perform as well initially due to the different requirements of pace, stamina, and speed. Several races may be required at the new distance before a horse is fully acclimated. Thus, running at a new distance may have a negative effect on performance. An indicator variable, NEWDIST, captures this effect. NEWDIST equals one if a horse had run three or four of its last four races at distance levels of less than one mile, and zero otherwise.

Workouts could be important in assessing a horse's current condition. If a horse has changed tracks, has not raced recently, or has experienced an injury, workout data represent an important signal as to current performance capabilities. Unfortunately, such data are difficult to interpret since a trainer's objectives for a given workout may not require that a horse perform at the maximum possible level. For this reason, workout data were considered to be of dubious value especially in comparison with the other horse quality and performance variables in the model. Thus, a workout variable was not included in the model.

Jockey characteristics may be of secondary importance in determining a horse's overall "worth." A jockey, no matter how skilled, cannot consistently win with an inferior horse. However, given horses of roughly equal quality, the more accomplished jockey may be more likely to win. There will be some positive correlation between horse and jockey quality, because owners of better horses seek to employ the better jockeys and better jockeys, in turn, prefer to ride better horses since jockeys are compensated partially on a commission system based on horses' earnings.

Jockey data on percentage of winning rides, JOCK%WIN, and number of winning rides, JOCK#WIN, over the current year were included in the model. Some jockeys' records were not available in the *Daily Racing Form*. Such missing data were accounted for by creating an indicator variable, JMISDATA, which takes on the value one when the other jockey variables are missing, and is zero otherwise. (By construction, when JMISDATA equals one, JOCK%WIN and JOCK#WIN equal zero.) Since the jockey data were missing for those who were not in the published list of leading jockeys, such missing data correspond to relatively inexperienced jockeys who are probably of lower quality than the leading jockeys. The coefficient on JMISDATA will serve as a proxy for the average values of JOCK%WIN and JOCK#WIN for such nonleading jockeys.

Results of Estimating the Base Model

The base model in equation (14) was estimated using the 200 races in the study data base. The associated empirical results are displayed in Table 1.

TABLE 1

Multinomial Logit Model Results—Base Model Estimated on 200 Races with Explosion Depths of 1, 2, and 3

Variable	Coefficient Estimates			Standard Deviation of the Variable	Standardized Coefficient Estimates
	$E = 1$	$E = 2$	$E = 3$		
LIFE%WIN	0.0143 (0.0082)	0.0076 (0.0061)	0.0066 (0.0050)	11.74	0.077
AVESPRAT	0.0789 (0.0178)	0.0615 (0.0123)	0.0546 (0.0101)	10.29	0.562
W/RACE	0.0865 (0.0506)	0.0846 (0.0398)	0.1103 (0.0356)	2.07	0.228
LSPEDRAT	0.0073 (0.0113)	0.0046 (0.0079)	0.0067 (0.0064)	14.78	0.099
JOCK%WIN	0.0205 (0.0339)	0.0297 (0.0242)	0.0236 (0.0201)	8.26	0.195
JOCK#WIN	0.0017 (0.0042)	0.0019 (0.0029)	0.0017 (0.0025)	43.41	0.076
JMISDATA	−0.0284 (0.4159)	0.1962 (0.2927)	0.1765 (0.2425)	0.48	0.086
WEIGHT	0.0227 (0.0225)	0.0076 (0.0158)	0.0030 (0.0130)	4.01	0.012
POSTPOS	−0.0804 (0.0332)	−0.0478 (0.0231)	−0.0439 (0.0189)	2.49	−0.109
NEWDIST	−0.2548 (0.2005)	−0.3902 (0.1446)	−0.3754 (0.1189)	0.46	−0.172

Summary Statistics

# of "Exploded" Choice Sets	200	400	600		
$L(\theta = 0)$	−401.4	−773.6	−1110.5		
$L(\theta = \hat{\theta})$	−364.9	−724.1	−1049.3		
$\bar{R}^2$	0.091	0.064	0.055		

Notes: [1] Asymptotic standard errors of parameter estimates are in parentheses below each coefficient estimate.

[2] The standard deviations of the variables are taken from the raw data for an "explosion" depth of 1. Thus, the "unexploded" standard deviation is used as an approximation for the "exploded" data standard deviation for each of the variables in the model.

[3] "Standardized Coefficient Estimates" are equal to the product of each variable's $E = 3$ coefficient estimate and its standard deviation.

The test of the null hypothesis that the parameter vector is equal to zero is rejected at the 0.005 level of significance. Thus, it may be concluded that the model is explaining a statistically significant amount of the variation in racing performance. Note that the overall goodness-of-fit index, $\tilde{R}^2$, has a different interpretation than the multiple correlation coefficient in linear statistical models. The standard of reference for $\tilde{R}^2$ within the multinomial logit model is the equal probability model, $\theta = 0$ (where each horse has an equal probability of winning). The multinomial logit model then attempts to explain a significant amount of the variation in the win probabilities based on the available independent variables. Furthermore, low values of $\tilde{R}^2$ should be expected in the horse racing context, since most races have a variety of constraints placed on the competing horses by the racing secretary. These constraints are designed to equalize, to some extent, the chances of the competing horses. Weight allowances, age restrictions, and specific past performance profiles (such as nonwinners in the last two months) are examples of explicit devices used to attempt to equalize the horses' win probabilities. For $E = 1$, $\tilde{R}^2$ equals about 9%, indicating that the model explains 9% more variation than the null hypothesis that all the horses have equal probabilities of winning.

The choice set data were exploded to depths of two and three. Three was chosen as the maximum depth of explosion for which prior theory would suggest that reliable rank order information might be available from the finishing order information. The results displayed in Table 1 illustrate the main value of the explosion process: the standard errors of the parameter estimates decrease when the rank ordered data are exploded to yield more choice sets for analysis. In going from $E = 1$ to $E = 2$ the average decrease in the standard errors is about 28.3%; a further 16.8% average decrease in the standard errors is achieved in going from $E = 2$ to $E = 3$. This pattern is consistent with the Monte Carlo results reported in Chapman and Staelin (1982).

The Watson and Westin (1975) sequential pooling and hypothesis testing procedure was employed to determine whether $E = 3$ was appropriate. The relevant log-likelihood values and χ^2 are described in Table 2. The null hypothesis being tested in each case is whether $\theta^{(E)} = \theta^{(E+1)}$. To assess whether a move from $E = 1$ to $E = 2$ is appropriate, the relevant χ^2 test statistic (with ten degrees of freedom, the number of variables in the model) is:

$$\chi^2 = -2(-724.13 - [(-364.94) + (-353.69)]) = 11.0.$$

Comparing this calculated test statistic with the relevant critical values leads to the conclusion that the null hypothesis should not be rejected on the basis of this sample evidence. Therefore, it is feasible to pool the observations and explode the rank ordered choice set data to a depth of two. In iterating this test to determine if an explosion

TABLE 2

Relevant Log Likelihood Values for Determining the Optimal Explosion Depth

Choice Observation Group	# of Races in This Set	Log Likelihood Value	Calculated χ^2 Value*
$J(E = 1)$	200	$L(\theta = \hat{\theta}^{(1)}) = -364.94$	
$J(E = 2) - J(E = 1)$	200	$L(\theta = \hat{\theta}^{(2)}) = -353.69$	
$J(E = 3) - J(E = 2)$	200	$L(\theta = \hat{\theta}^{(3)}) = -322.06$	
$J(E = 2)$	400	$L(\theta = \hat{\theta}^{(1+2)}) = -724.13$	11.1
$J(E = 3)$	600	$L(\theta = \hat{\theta}^{(1+2+3)}) = -1049.26$	6.0

* This is the statistic calculated to test whether the choice set data can be pooled to level E. Critical values of χ^2_{10} are 18.3 and 23.2 at the 5% and 1% levels of statistical significance, respectively. The degrees of freedom in this test are 10, corresponding to the number of variables in the model.

depth of three is appropriate, the relevant test statistic value is equal to 6.1, which again is sufficiently small that the null hypothesis should not be rejected on the basis of the sample evidence. Hence, an explosion depth of three is appropriate for these horse racing choice set data.

In examining the results reported in Table 1 for $E = 3$, it may be noted that the signs of the coefficients are consistent with a priori theoretical expectations. It is useful to attempt to measure the relative importance of each variable. This is done by calculating the coefficient estimates which would have been obtained if all variables had been standardized (to unit variance) prior to estimation. This is equivalent to assessing a variable's relative importance by taking the product of its coefficient estimate and its standard deviation. The interpretation of the standardized relative importances, also displayed in Table 1, is subject to the usual difficulties in uniquely partitioning the explained variance among any set of collinear independent variables. The results in Table 1 suggest that average speed rating (AVESPRAT) accounts for the most variation in the model. Winnings per race in the current year (W/RACE) appears to be more important than lifetime percentage wins (LIFE%WIN). This may be attributed to W/RACE taking into account high but nonwinning performances and to W/RACE being based on recent performance information. WEIGHT does not seem to be an important determinant of finishing position, given the presence of the other variables in the model. Post position (POSTPOS) and new distance (NEWDIST) appear to exhibit nontrivial effects on winning probabilities. The jockey variables appear to have less overall importance than the horse's attributes in determining winning probabilities, although this finding may be due to collinearity among the horse and jockey variables.

This model has substantial face validity on several dimensions. First, the multinomial logit model considers the competitive nature of the horse racing process. The choice probability expression explicitly includes the characteristics of each horse in comparison with all other horses in a specific race, and not relative to all horses in the universe. Second, an intuitively appealing theoretical utility maximizing (revealed preference) framework was utilized in developing the model. Third, the empirical results indicate that the model operationalization passes the usual tests of statistical significance. The empirical findings are consistent with a priori theoretical beliefs. However, it remains to be determined whether this model is sufficiently accurate to allow for the development of a superior wagering system which will earn positive returns.

Analysis of Wagering Systems: Searching for Positive Returns at the Track

In this section, the multinomial logit model of the horse race process is employed to evaluate alternative wagering strategies. Two classes of wagering strategies are considered: algorithms involving multiple bets per race and algorithms involving a single bet per race. A sequential hold-out sampling procedure was used to evaluate each wagering strategy. The model in equation (14) was estimated separately on four data subsamples drawn from the available 200 races. Each sample was a set of 150 (overlapping) choice set observations exploded to a depth of three. For each of the four estimated models, a hold-out sample of the remaining 50 races was then available to evaluate the wagering strategies. This validation approach avoids the upward bias of goodness-of-fit statistics calculated by applying a statistical model back on to the same data base from which it was originally estimated.

Strategies Involving Multiple Bets Per Race

An "optimal" set of wagers can be derived from a variety of wagering strategies based on different objective functions. For example, a wagering algorithm based on expected value maximization might be appropriate for a risk neutral bettor. Alternatively, an algorithm that maximizes expected log returns would be consistent with risk averse behavior. In addition, wagering strategies may entail either large bets whose influence on the track odds is explicitly taken into account or unobtrusive bets which do not influence the track odds.

Isaacs' Wagering Strategy. Isaacs (1953) determined the optimal amounts to wager for a risk neutral bettor with infinite wealth who has perfect estimates of the true winning probabilities. His algorithm incorporates the impact of the expected value maximizing bettor's wagers on the track odds. In operationalizing Isaacs' strategy, it is necessary to assume that the expected value maximizing bettor is the last bettor. If not, then some subsequent bettor might place wagers which would change the track odds, and thus the optimal amount the expected value maximizing bettor should have wagered. It should be recognized that there are nontrivial logistical problems associated with being the final bettor in a race, particularly if large wagers are being placed.

Isaacs' algorithm was applied to each of the four hold-out samples of 50 races. The winning probabilities were predicted using the multinomial logit model estimated on each set of 150 remaining races. The algorithm identified an average of 3.46 bets per race with expected positive returns. The average amount wagered was $958. (This was calculated as a weighted average across the four data subsets, where the weights were the total number of bets placed in each subset of 50 races.) The average return per race was −39.5%, while the weighted average return across our four hold-out samples of races was −27.8%.[1] This is considerably worse than a random betting strategy might be expected to yield. It is also much worse than Isaacs' wagering algorithm would perform if the true winning probabilities were known, rather than using fallible estimates. There was considerable variation in the returns across the four data subsets (of 50 races each). The individual 50-race subsamples had average returns of −2.6%, −65.9%, −35.6%, and −7.7%. Even allowing for sampling variation, these results suggest that the probability estimates are too imprecise to be useful in implementing Isaacs' optimal wagering strategy for the expected value maximizing bettor.

Why does Isaacs' algorithm perform so poorly? Modest errors in the estimates of the true winning probabilities could cause substantial deviations from the optimal returns of Isaacs' strategy. Isaacs' wagering algorithm determines the amounts of the wagers based on four factors: the true winning probabilities, the public's consensus probabilities (as reflected by their actual betting behavior), the size of the track take, and the size of the betting pool. The optimal amount to bet involves a trade-off between the attractiveness of wagering large amounts and the feedback effect of the resultant changes in the track odds. The bettor observes the discrepancy between the true winning probabilities and the track odds, and subsequently makes wagers which yield payoffs according to the *revised* track odds, where the revisions take into account the bettor's obtrusive wagers. The wagering of the expected value maximizing bettor results in the track consensus probabilities being driven closer to the true winning probabilities. When a fallible estimate is substituted for the true winning probability, the bettor observes the discrepancy between the estimated winning probability and the public's consensus winning probability, and then wagers in such a way as to drive the public's consensus winning probabilities toward the bettor's estimated winning probabilities. Since the estimates of the winning probabilities may be different from the true (unknown) values, the odds will be driven toward revised odds which may not necessarily yield the optimal payoffs. Therefore, Isaacs' strategy is unlikely to be

---

[1] "Return per race" is defined to be return per wager averaged across a number of races. "Return across races" is defined to be average return divided by average wager. These measures of return would, of course, be identical if wagers were of constant value.

profitable—unless the bettor has very precise estimates of the true winning probabilites—because the wagers will significantly lower the odds.[2] This finding suggests that wagering necessarily yield the optimal payoffs. Therefore, Isaacs' strategy is unlikely to be profitable unless the estimates of the true winning probabilities are sufficiently accurate.

Rosner's Wagering Strategy. Rosner (1975) determined the optimal amounts to wager for a risk averse bettor who has perfect estimates of the true winning probabilities. He derived a closed form solution for the optimal amount to bet under the assumption that the wagers have no effect on the odds. His algorithm maximizes expected log return. It has the desirable property of maximizing the long-run rate of asset growth, termed the Kelly criterion (Thorp 1975). Rosner's wagering strategy involves differential bets, where the size of the bet is a function of the attractiveness of the wager.

Unlike Isaacs' wagering strategy, Rosner's closed form solution does *not* take into account the effect of the wagers on the track odds. Consequently, if we take Rosner's suggested optimal wagers and do not correct the odds for our bettor's wager, then we will overestimate the returns to some extent. However, this simplification is progressively more reasonable as the public's wagers increase and/or our bettor's bankroll decreases. Another effect of this simplification is that the performance of the system should be less sensitive to errors in the estimation of the true winning probabilities.

Rosner's strategy was evaluated in the following way. The winning probabilities were predicted using the multinomial logit model estimated on each set of 150 races. For each of the four hold-out samples, wealth was assumed to be $1000 at the start of the first race, and then updated after each race.[3] The algorithm was applied to each of the four hold-out samples of 50 races. It identified an average of 3.48 bets per race. The average amount wagered per bet was $85. The bettor's initial wealth of $1000 decreased to $95.63 (an average across the four data subsets). In other words, the bettor had lost most of his "stake"! It is useful to examine the average return per race and average return across 50 races for Rosner's strategy, in order to compare it to Isaacs' wagering strategy. The average return per race was −14.1%. This is a substantial improvement over Isaacs' wagering strategy. However, the average return across 50 races was −37.4%. This is worse than a random wagering strategy, and worse than Isaacs' wagering strategy.

As before, there was considerable variation in the return per race within the 50 races of each data subset. (The standard error of the mean return per race is about 27.5% for each data subset.) This variation leads to large fluctuations in wealth from race to race and creates two problems. First, the bettor's wagers in a given race can be large enough to significantly affect the track odds. Since Rosner's closed form solution for the optimal amount to wager does *not* take into account the effect of the wagers on the track odds, these large wagers will not be "optimal." Second, large fluctuations in wealth from race to race lead to considerable variability in the size of the bettor's

---

[2] One of the referees commented that Isaacs' wagering strategy does poorly because it tries to "grab all the inefficiency." Since the probabilities are estimated with error, the strategy sometimes wagers incorrectly or too much. As a result, Isaacs' wagering strategy performs poorly relative to a strategy that makes a single "small" bet on a randomly chosen horse. However, it's likely that the estimated probabilities are sufficiently accurate that Isaacs' wagering strategy would perform well (in large samples) compared to a strategy that places multiple "large" bets on randomly chosen horses.

[3] The selection of $1000 for initial wealth is arbitrary. This amount was selected for calculation/ demonstration purposes. The closed form solution from Rosner's wagering strategy assumes the wagers do not affect the track odds. Hence, initial wealth of $1000 was selected with the intention that the resultant wagers from Rosner's strategy would not materially affect the existing track odds. Note that initial wealth of $1 (which would yield wagers which do not affect the existing track odds) would yield the same relative returns (on a percentage basis). As the discussion in the text indicates, the resultant wagers exhibited considerable variability in size, and this variability led to a modification of the strategy.

wagers per race, since the bettor's wagers on horses are calculated as a fraction of current wealth. Hence, the return across 50 races for a particular data subset will depend on the sequence of the races. For example, the return across 50 races will be lower in the situation where all the winning wagers occur in the "early" races, than in the situation where all the winning wagers occur in the "later" races. In other words, race sequence critically affects the average return across 50 races in the relatively small data subsets of races used in this study.

An ad hoc modification of Rosner's wagering strategy was utilized to make the wagers less obtrusive (i.e., to make the wagers have less impact on the track odds) and to remove the effect of race sequence on the return across races.[4] Both difficulties were removed by eliminating the variability in the size of the bettor's current wealth. In the modified strategy, wealth is fixed to be equal to $1000 for *each* race, and the bettor wagers some fraction of this amount.[5] The average return across 50 races was −6.4%. As before, there was substantial variation in the average returns for the four data subsets: 18.7%, −58.6%, −32.1%, and 46.4%. This is an improvement over a random wagering strategy (and over Isaacs' wagering strategy). However, this finding still indicates negative returns across races and a decrease in initial wealth.

The results of the "fixed wealth" modification provide an estimate of the returns across races that Rosner's wagering strategy should generate in the long run (i.e., when race sequence effects would "cancel out"). However, this estimate of the returns will be inaccurate for two reasons. First, returns are somewhat overestimated because they do not take into account the effect of the wagers on the track odds. Second, returns may be underestimated because the modification to Rosner's wagering strategy is suboptimal in the sense that it no longer maximizes the long run rate of asset growth. Therefore, the results of the fixed wealth modification suggest that Rosner's wagering strategy may generate returns across races of *approximately* −6.4% in the long run.

Constrained Versions of Rosner's Wagering Strategy. Rosner's wagering strategy fails to yield positive returns across races because the true winning probabilities are estimated with error. As a result, the algorithm generates some wagers with low or negative actual returns. How can the bettor avoid these wagers? Two constrained versions of Rosner's wagering strategy are considered here.

Studies of place and show betting suggest that the bettor should wager only on horses with estimated expected returns which are substantially greater than one (Harville 1973; Hausch, Ziemba, and Rubinstein 1981). That is, the bettor should wager only if $\hat{\rho}_h(r_h + 1) > \alpha$, where α is some constant exceeding one. Rosner's wagering strategy and its "fixed wealth" modification were re-evaluated utilizing the α

---

[4] An alternative solution to the problem of large wager effects is to extend Rosner's wagering strategy to take into account the effect of wagers on the track odds (e.g., Hausch, Ziemba and Rubinstein 1981). However, the results of applying Isaacs' wagering strategy indicate that a wagering strategy that alters the track odds requires very precise estimates of the true winning probabilities in order to be profitable. Hence, we chose to modify Rosner's wagering strategy to constrain the size of the wagers. An alternative solution to the problem of race sequence effects is to use a measure of wagering strategy performance which is not affected by race sequence. Since the effect of race sequence is reflected in average return across races, but not in final wealth, final wealth could be used to compare the performance of alternative wagering strategies. Final wealth is a relevant criterion for Rosner's strategy, but less useful for other wagering strategies. For example, Isaacs' wagering strategy assumes that the bettor has infinite wealth. These considerations led us to modifying Rosner's strategy by eliminating the variability in the size of the bettor's current wealth. The modification described in the text "solves" both difficulties (i.e., the effect of "large" wagers and race sequence), albeit in an ad hoc fashion.

[5] The selection of $1000 for the fixed wealth modification is arbitrary. This amount was selected for calculation/demonstration purposes. It is meant to yield wagers which are unobtrusive in the sense that they do not materially affect the existing track odds. Fixed wealth of $1 (which would not affect the existing track odds) would yield the same relative returns (on a percentage basis). See footnote 3.

constraint, for levels of α between 1.0 and 1.8 in increments of 0.1. (About 50% of the wagers are eliminated at α equals 1.8.) The average return per race and the average return across 50 races does not improve with the addition of the α constraint.[6]

It is likely that poor estimates for long shots cause the most serious errors in the calculation of expected returns and the formulation of the optimal wagering strategy. Why? First, in relative terms, errors for long shots will be larger. To illustrate the nature of this problem, a misestimate of 0.01 on a favorite whose true winning probability is 0.20 is, in percentage terms, quite small compared to a misestimate of 0.01 on a long shot whose true winning probability is 0.04. While identical in absolute size, the former represents an error of only 5% while the latter represents an error of 25%! Second, it may be easier to predict winning probabilities for favorites since their performance is likely to be more "regular", and thus more easily represented and predicted with a statistical model. This reasoning suggests the bettor may be able to avoid wagers with low or negative actual returns by wagering only on horses with an estimated probability of winning greater than some minimum value. That is, the bettor should wager only if $\hat{p}_h > p_{min}$, where p_{min} is a specified minimum winning probability estimate. It should be noted that fewer wagers are made as the value of p_{min} is raised. Hence, the race returns are calculated on correspondingly smaller sample sizes, and should be interpreted with caution.

Rosner's strategy and its "fixed wealth" modification were re-evaluated utilizing the p_{min} constraint, for levels of p_{min} between 0.00 and 0.25 in increments of 0.01. The average return per race and the average return across races are displayed in Table 3. As the prespecified value of p_{min} increases, fewer wagers are made because long shots are omitted from consideration. The average returns per race associated with $p_{min} > 0.17$ are positive for 6 of the 8 tabulated values. (The two negative values may be small sample results.) The p_{min} constraint improves the average return across races generated by Rosner's strategy and its fixed wealth modification for the majority of the tabulated values. However, the effect of the p_{min} constraint is more evident in the average return across races generated by the "fixed wealth" modification. The average return across 50 races improves for 18 of the 25 tabulated p_{min} values, and the improvement is quite large. In fact, seven of these values are positive!

The above results indicate that Rosner's wagering strategy may yield long-run positive returns when a side constraint eliminates wagers on horses for which the logit model provides poor estimates of the winning probabilities and expected returns. The bettor should wager on the horses identified by Rosner's strategy, except in the case of long shots. Long shots are horses with (estimated) winning probabilities which do not meet a p_{min} constraint of at least 0.07. (The constraint could be as high as 0.11, but apparently not higher.) Such a constraint eliminates about 17% of the wagers, and generates returns across races of about 1.3%. It is important that the p_{min} constraint not be set too high (e.g., higher than 0.11), or too many horses will be eliminated, resulting in negative returns across races. For example, at $p_{min} = 0.12$, 55% of the bets are eliminated and Rosner's differential wagers generate returns across races of -3.1%.

Multiple Unit Bets Strategy. The results obtained by applying Isaacs' and Rosner's strategies identify two features which should be incorporated in an "optimal" wagering strategy that employs fallible estimated probabilities as inputs. First, the results of

---

[6] Isaacs' strategy was evaluated with an α constraint, for α equals 1.05, 1.10, 1.15, and 1.20. The race returns for Isaacs' strategy with an α constraint do not differ substantially from the results reported in this paper. This result is not unexpected because the primary reason for the poor performance of Isaacs' strategy is the feedback effect on the track odds. The wagering strategies discussed in the remainder of this paper were also evaluated with an α constraint. The race returns do not differ substantially from the results reported.

TABLE 3

Strategies Involving Multiple Bets: Weighted Average Returns Across the Four Data Subsets***

ρ_{min}	# of Races*	Rosner's Wagering Strategy			Unit Wagers Strategy	
		Per Race	Across Races		Per Race	Across Races
			Updated W	Fixed W**		
0.00	200	−0.1411	−0.3737	−0.0638	−0.1599	−0.2178
0.01	200	−0.1411	−0.3737	−0.0638	−0.1599	−0.2178
0.02	200	−0.1411	−0.3736	−0.0635	−0.1599	−0.2138
0.03	200	−0.1402	−0.3730	−0.0617	−0.1473	−0.2050
0.04	199	−0.1324	−0.3684	−0.0516	−0.1370	−0.1818
0.05	199	−0.1265	−0.3619	−0.0386	−0.1242	−0.1479
0.06	199	−0.1197	−0.3486	−0.0169	−0.1103	−0.1040
0.07	199	−0.1019	−0.3322	0.0131	−0.0799	−0.0506
0.08	198	−0.0613	−0.3051	0.0649	−0.0282	0.0230
0.09	198	−0.1527	−0.2897	0.0442	−0.0840	0.0293
0.10	197	−0.2241	−0.2934	0.0238	−0.1671	−0.0060
0.11	191	−0.1830	−0.2495	0.0820	−0.1161	0.0833
0.12	180	−0.2277	−0.2737	−0.0307	−0.1699	0.0045
0.13	170	−0.1650	−0.3342	−0.0629	−0.1349	−0.0207
0.14	159	−0.2547	−0.4648	−0.2005	−0.2604	−0.1744
0.15	148	−0.1763	−0.4795	−0.1972	−0.1633	−0.1763
0.16	135	−0.0874	−0.4433	−0.1416	−0.0896	−0.0872
0.17	127	−0.0608	−0.4318	−0.1219	−0.0476	−0.0680
0.18	119	0.0374	−0.3889	−0.0487	0.0627	0.0393
0.19	111	0.0545	−0.3687	−0.0591	0.0757	0.0638
0.20	101	0.0780	−0.3472	−0.0401	0.0861	0.0696
0.21	93	0.0670	−0.3566	−0.0360	0.0780	0.0609
0.22	81	0.1744	−0.3074	0.0513	0.1713	0.1504
0.23	72	0.2366	−0.2888	0.0946	0.2199	0.2019
0.24	62	−0.0656	−0.3978	−0.2723	−0.0802	−0.0630
0.25	53	−0.1145	−0.4427	−0.3417	−0.1306	−0.1050

* This column describes the number of races in which at least one bet was placed for Rosner's Wagering Strategy. The number of races in which at least one bet was placed for the Unit Wagers Strategy was identical to Rosner's strategy for most values of ρ, and never differed by more than one race.

** "Updated W" denotes the pure Rosner's Wagering Strategy, in which wealth equals $1000 at the beginning of the first race, and is updated with the results of subsequent races. "Fixed W" denotes the modification of Rosner's Wagering Strategy, in which wealth is fixed to be equal to $1000 for each race.

*** The weighted averages reported in this table were calculated by using the number of races in which bets were placed (of the 50 races in a data subset) as the weight for the return from a data subset.

Isaacs' strategy indicate that wagers should be sufficiently unobtrusive that the track odds are not affected (in order to maximize average return across races and average return per race). Second, the results of Rosner's strategy indicate that the bettor should wager on horses with positive expected returns—i.e., wagers should be placed on horses when $\hat{\rho}_h(r_h + 1) > 1$. Since the logit model provides relatively poor estimates of expected returns for long shots, the results of Rosner's strategy imply that the bettor should wager on horses with estimated probabilities of winning which are greater than some minimum value. One possible ad hoc wagering strategy which satisfies these concerns is the following:

Wager one unit on each horse for which $\hat{\rho}_h(r_h + 1) > 1$ as long as $\hat{\rho}_h > \rho_{min}$, where ρ_{min} is a

specified minimum winning probability estimate and a "unit" is a dollar amount which is
sufficiently small that it does not affect the track odds.

Unlike Rosner's wagering strategy, the Multiple Unit Bets strategy does not wager
larger amounts on more attractive bets.

The results of applying the Multiple Unit Bets strategy to the four data subsets are
displayed in Table 3. The table shows the mean returns from a $1 bet on each horse
that satisfied the two conditions stipulated above. The strategy of wagering on all
horses with positive predicted expected values yields a return per race of -16.0%. The
average return across 50 races is -21.8%, somewhat worse than would be expected on
the basis of random betting. As before, this poor performance illustrates the impact
that inaccurate winning probability estimates can have on betting outcomes. As the
prespecified value of ρ_{min} increases, fewer wagers are made because long shots are
omitted from consideration. Once extreme long shots (more than 20 to 1) are removed
from consideration, the picture improves tremendously. In particular, the average
returns across races associated with ρ_{min} in excess of 0.07 are positive for 10 of the 18
tabulated values.

It is interesting to compare the Multiple Unit Bets strategy with the fixed wealth
modification of Rosner's wagering strategy. The Multiple Unit Bets strategy yields a
higher average return *per race* for almost all values of ρ_{min}, such that $0.04 < \rho_{min}$
< 0.22. However, the fixed wealth modification of Rosner's strategy yields a higher
average return *across races* for all values of $\rho_{min} < 0.11$. The pattern of dominance
then reverses, and the Multiple Unit Bets strategy dominates Rosner's strategy for all
values of $\rho_{min} \geq 0.11$.

This comparison indicates that the estimates of the winning probabilities are
sufficiently accurate to justify employing a differential betting strategy (such as Rosner's)
to maximize returns across races, rather than a unit betting strategy. Differential
betting dominates unit betting for strategies involving unconstrained multiple bets.
This finding can be explained by the observation that Rosner's strategy tends to bet
very lightly on long-odds horses, for which the winning probabilities seem to be poorly
estimated. Differential betting also yields positive returns, as well as dominates unit
betting, if a small number of misestimated long shots are eliminated from consideration.
However, if too many wagers are eliminated by the ρ_{min} constraint (i.e., at $\rho_{min} \geq 0.11$),
unit betting frequently yields positive returns, as well as dominating differential betting.
This finding cannot be completely explained by the fact that the (misestimated) long
shots—for which differential wagers are better than unit wagers—have been eliminated.

Strategies Involving a Single Bet Per Race

In the popular literature, the bettor is often advised to bet only on the horse with
the highest winning probability. Most handicapping systems are based on attempting
to identify the "best" horse, where "best" means most likely to win. This is, of course,
suboptimal since such a betting strategy does not take the expected return (the public's
wagers) into account. Thus, a superior approach for the bettor desiring to wager only
on one horse would be to bet on the horse with the maximum expected return in the
race, as long as that expected return exceeds one. The bettor could wager the same
amount in each race, or wager differential amounts across races. Strategies involving a
single bet per race could be constrained to eliminate long shots, in the same way that
strategies involving multiple bets per race were constrained.

Single Unit Bet Strategy. This strategy can be formally described as follows:

Wager one unit on the horse for which the expected return, $\hat{\rho}_h(r_h + 1)$, is a maximum, as long
as the expected return exceeds one and $\hat{\rho}_h \geq \rho_{min}$.

The results of applying this strategy to the four data subsets are reported in Table
4. Note that the average return per race and the average return across 50 races must,

TABLE 4

Single Bet Strategies: Weighted Average Returns
Across the Four Data Subsets**

ρ_{min}	# of Races	Mean Return Per Race*	Return Across Races With Differential Wagers
0.00	200	0.0310	0.0359
0.01	200	0.0310	0.0359
0.02	200	0.0310	0.0359
0.03	199	0.0362	0.0407
0.04	195	0.0574	0.0465
0.05	188	0.0968	0.0673
0.06	181	0.1392	0.0712
0.07	170	0.2129	0.0968
0.08	151	0.3656	0.1756
0.09	135	0.1919	0.0287
0.10	119	0.2009	0.0416
0.11	103	0.3874	0.1688
0.12	94	−0.1000	−0.2004
0.13	84	−0.1726	−0.2331
0.14	75	−0.0733	−0.1883
0.15	64	−0.1406	−0.2274
0.16	54	0.0185	−0.1317
0.17	44	0.2500	−0.0060
0.18	41	0.3415	0.0565
0.19	37	0.1622	−0.0282
0.20	33	0.3030	0.0506
0.21	33	0.3030	0.0506
0.22	28	0.3143	0.1608
0.23	27	0.1889	0.1756
0.24	21	−0.7333	−0.8077
0.25	20	−0.7200	−0.8079

* This is the average return per race from any strategy involving a single bet per race (i.e., a unit or differential wagering strategy). In addition, it is the average return across races for a Single Unit Bet strategy.

** The weighted averages reported in this table were calculated by using the number of races in which a bet was placed as the weight for the return from a data subset.

by definition, be identical for this wagering strategy. The results indicate that wagering on the horse with the maximum expected return yields an average return of 3.1%.

This strategy wagers on the horse with the highest expected percentage return and does not wager additional dollars on horses with lower returns. Hence, it would be expected that the average return *per race* from the Single Unit Bet strategy would dominate the average return *per race* from any of the strategies involving multiple bets per race. (This statement refers to percentage returns, not absolute returns.) In fact, the average return per race from the Single Unit Bet strategy does dominate the majority of the values in Table 3. In addition, the average return *across* 50 races is typically much higher for the Single Unit Bet strategy. In fact, 19 of the 26 tabulated values are positive! These positive values are quite large, ranging from 3.1% to 38.7%!

Single Differential Bet Strategy. A method to increase the average return across 50 races generated by the Single Unit Bet strategy is to wager different amounts of money in different races, wagering larger amounts when the betting opportunity is particularly attractive. For example, Rosner's strategy wagers smaller amounts on long shots. An ad hoc single differential bet strategy is the following:

> Wager on the horse for which the expected return, $\hat{\rho}_h(r_h + 1)$, is a maximum, as long as the expected return exceeds one and $\hat{\rho}_h \geq \rho_{min}$. The amount of the wager should be the amount that Rosner's wagering strategy recommends for that horse, assuming a current wealth of $1000.

The results of applying this strategy to the four data subsets are reported in Table 4. The average return per race is the same for both single bet wagering strategies. The average return across 50 races for the Single Differential Bet strategy (without a ρ_{min} constraint) is 3.6%. This is a slight improvement over the Single Unit Bet strategy. However, the Single Unit Bet strategy dominates for all values of ρ_{min} greater than 0.03. Apparently, a differential betting strategy cannot improve returns when long shots have already been eliminated.

Concluding Remarks

A trade-off exists between the "optimality" of a wagering strategy and the accuracy of the statistical model of the horse race process. When the true (unknown) winning probabilities are fallibly estimated, it is necessary to recognize the existence of such estimation errors within the wagering strategy. The major consequence of using estimates of the winning probabilities within a wagering strategy is that the size of the wagers and the number of wagering opportunities must be constrained.

The size of the wagers must be constrained so that the wagering strategy does not affect the track odds. Since Isaacs' wagering strategy includes feedback effects on the track odds, errors in estimating the true winning probabilities (particularly for horses which are long shots) result in an average return of -27.8% across 50 races. In contrast, Rosner's wagering strategy—which involves unobtrusive bets—improves upon a random wagering strategy. It yields an average return of -6.4% across 50 races, after adjusting for the race sequence effects that arise when this strategy is applied to small samples.

The number of wagering opportunities must be constrained so that the wagering strategy involves bets on horses which are more predictable. The multinomial logit model provides relatively poor estimates of long shots' winning probabilities and, consequently, their expected returns. Hence, the returns of most wagering strategies improve when a side constraint eliminates wagers on long shots. For example, a modification of Rosner's wagering strategy may generate positive returns once horses with predicted winning probabilities of less than about 0.07 are eliminated from consideration.

Two simple wagering strategies equal or surpass the results of Rosner's wagering strategy. The bettor can achieve comparable returns by betting identical amounts on all the favorites (winning probabilities greater than 0.19) that have positive expected returns. Or, the bettor can generate returns of 3.1% (or more) by betting a fixed amount on the horse with the highest expected return. Both these strategies involve unit bets on a limited number of horses. A strategy involving differential bets on multiple horses, such as Rosner's wagering strategy, has the potential to yield higher *long-run* returns than any strategy involving a limited number of unit bets. This study did not have a large enough data base to adequately assess the long-run rates of return across races for the various wagering strategies.

Can a horse race wagering system involving win betting yield positive returns? Given this paper's results, there appears to be room for some optimism. While this study represents a pioneering effort in statistical modeling of the determinants of horse race outcomes, a variety of avenues for followup research exist. The variability in the results across the four data subsets suggests the need for additional empirical analyses of larger samples of races to confirm or refute these findings. It is also possible that future multinomial logit modeling efforts might lead to reduced estimation errors. For example, a separate multinomial logit model could be estimated for each track, instead of pooling across tracks as was the case in this study. It would be useful to attempt to devise a complete wagering system for win, place, and show betting. Such a research effort could combine a fundamental wagering strategy similar to the one developed in this paper with a technical approach similar to Ziemba and Hausch (1984). These and other related questions will no doubt draw the future attention of researchers interested in race track wager markets.[7]

[7] The helpful comments of the Departmental Editor, an Associate Editor and the referees are gratefully acknowledged.

References

ALI, M. M., "Probability Estimates for Racetrack Betting," *J. Political Economy,* 85 (4) (1977), 803–815.

———, "Some Evidence on the Efficiency of a Speculative Market," *Econometrica,* 47 (1979), 387–392.

ARNOLD, S. J., T. H. OUM, AND D. J. TIGERT, "Determinant Attributes in Retail Patronage: Seasonal, Temporal, Regional, and International Comparisons," *J. Marketing Res.,* 20 (May 1983), 149–157.

ARVESON, J. AND B. ROSNER, "Optimal Pari-Mutuel Wagering," in S. S. Gupta and J. Yackel (Eds.), *Statistical Decision Theory and Related Topics,* Academic Press, Inc., New York, 1971.

——— AND ———, "Multivariate Analysis of Horse Race Betting," *Proc. Business and Economics Section of the American Statistical Association,* 1973, 238–240.

ASCH, P., B. MALKIEL, AND R. E. QUANDT, "Racetrack Betting and Informed Behavior," *J. Financial Economics,* 10 (2) (1982), 187–194.

BRATLEY, P., "A Day at the Races," *Infor,* 11 (2) (1973), 81–92.

CHAPMAN, R. G., "Pricing Policy and the College Choice Process," *Research in Higher Education,* 10 (1) (1979), 37–57.

———, "Retail Trade Area Analysis: Analytics and Statistics," in R. A. Leone (Ed.), *Proc.: Marketing Measurement and Analysis,* TIMS College on Marketing and The Institute of Management Sciences, Providence, R.I., 1980, 40–49.

——— AND R. STAELIN, "Exploiting Rank Ordered Choice Set Data Within the Stochastic Utility Model," *J. Marketing Res.,* 19 (August 1982), 288–301.

COPELAND, T. E. AND J. F. WESTON, *Financial Theory and Corporate Policy,* Addison-Wesley Publishing Company, Reading, Mass., 1979.

DAGANZO, C., *Multinomial Probit: The Theory and Its Application to Demand Forecasting,* Academic Press, Inc., New York, 1979.

DOMENCICH, T. AND D. McFADDEN, *Urban Travel Demand: A Behavioral Analysis,* North-Holland Publishing Company, Amsterdam, 1975.

DOWIE, J., "On the Efficiency and Equity of Betting Markets," *Economica,* 43 (May 1976), 139–150.

EISENBERG, E. AND D. GALE, "Consensus of Subjective Probabilities: The Pari-Mutuel Method," *Ann. Math. Statist.,* 30 (1959), 165–168.

FABRICAND, B. P., *Horse Sense,* David McKay Company Inc., New York, 1965.

FAMA, E. F., "Efficient Capital Markets: A Review of Theory and Empirical Work," *J. Finance,* 25 (May 1970), 383–417.

———, *Foundations of Finance,* Basic Books, New York, 1976.

FIGLEWSKI, S., "Subjective Information and Market Efficiency in a Betting Market," *J. Political Economy,* 87 (1) (1979), 75–88.

HARVILLE, D. A., "Assigning Probabilities to the Outcome in Multi-Entry Competitions," *J. Amer. Statist. Assoc.,* 68 (June 1973), 312–316.

HAUSCH, D. B., W. T. ZIEMBA, AND M. RUBINSTEIN, "Efficiency in the Market for Race Track Betting," *Management Sci.,* 27 (12) (1981), 1435–1452.

HENSHER, D. A. AND L. W. JOHNSON, *Applied Discrete Choice Modeling,* Croom Helm Ltd., London, 1981.

ISAACS, R., "Optimal Horse Race Bets," *Amer. Math. Monthly,* 60 (May 1953), 310–315.

LUCE, R. D. AND P. SUPPES, "Preference, Utility, and Subjective Probability," in R. D. Luce, R. B. Bush, and E. Galanter (Eds.), *Handbook of Mathematical Psychology. Volume 3,* John Wiley & Sons, New York, 1965, 249–410.

MADDALA, G. S., *Limited-Dependent and Qualitative Variables in Econometrics,* Cambridge University Press, Cambridge, 1983.

MANSKI, C. F., "The Conditional/Polytomous Logit Program: Instructions For Use," School of Urban and Public Affairs, Carnegie-Mellon University, 1974.

—— AND D. A. WISE, *College Choice in America,* Harvard University Press, Cambridge, Mass., 1983.

McFADDEN, D., "Conditional Logit Analysis of Qualitative Choice Behavior," in P. Zarembka (Ed.), *Frontiers in Econometrics.* Academic Press, Inc., New York, 1974, 303–328.

PUNJ, G. AND R. STAELIN, "The Choice Process For Graduate Business Schools," *J. Marketing Res.,* 15 (November 1978), 588–598.

ROSETT, R. N., "Gambling and Rationality," *J. Political Economy,* 73 (1965), 595–607.

ROSNER, B., "Optimal Allocation of Resources in a Pari-Mutuel Setting," *Management Sci.,* 21 (9) (1975), 997–1006.

RUBINSTEIN, M., "Securities Market Efficiency in an Arrow-Debreu Market," *Amer. Economic Rev.,* 65 (5) (1975), 812–824.

SNYDER, W. W., "Horse Racing: Testing the Efficient Markets Model," *J. Finance,* 33 (4) (1978), 1109–1118.

THORP, E. O., "The Portfolio Choice and the Kelly Criterion," in W. T. Ziemba and R. G. Vickson (Eds.), *Stochastic Optimization Models in Finance,* Academic Press, Inc., New York, 1975, 599–619.

VERGIN, R. C., "An Investigation of Decision Rules for Thoroughbred Race Horse Wagering," *Interfaces,* 8 (1) (1977), 34–45.

WALD, A., "Tests of Statistical Hypotheses Concerning Several Parameters, When the Number of Observations Is Large," *Trans. Amer. Math. Soc.,* 54 (1943), 426–482.

WATSON, P. L. AND R. B. WESTIN, "Transferability of Disaggregate Mode Choice Models," *Regional Sci. and Urban Economics,* 5 (2) (1975), 227–249.

WILLIS, K. E., "Optimum No Risk Strategy for Win-Place Pari-Mutuel Betting," *Management Sci.,* 10 (3) (1964), 574–577.

ZIEMBA, W. T. AND D. B. HAUSCH, *Beat the Racetrack,* Harcourt Brace Jovanovich Publishers, San Diego, Cal., 1984.

STILL SEARCHING FOR POSITIVE RETURNS AT THE TRACK: EMPIRICAL RESULTS FROM 2,000 HONG KONG RACES

Randall G. Chapman,
Chapman and Associates,
Winchester, Massachusetts

Abstract

The Bolton and Chapman (1986) multinomial logit modeling approach to handicapping horse races is extended to a new setting with much more sophisticated handicapping factors and a larger database. Applying a 20-variable pure fundamental multinomial logit handicapping model to 2,000 Hong Kong races yields clear evidence of positive returns to win betting with a simple single unit bets strategy in holdout sample predictions. By eliminating extreme longshots with estimated win probabilities less than 0.04, expected returns in excess of 20% are observed.

Introduction

Bolton and Chapman (1986) presented results that suggest it may be possible to achieve positive returns in horse race win betting. Their fundamental multinomial logit handicapping model included ten basic horse and jockey independent variables. With holdout sample predictions, they estimated that returns in excess of 5% might be possible once extreme (especially difficult-to-predict) longshots were eliminated from wagering consideration. These results were based on a modest sample of 200 races from five northeastern U.S. tracks.

This study applies the same methodology to a 2,000-race Hong Kong database. See Bolton and Chapman (1986) for details of fundamental multinomial logit horse race handicapping modeling. The 20 fundamental independent variables available in the Hong Kong database are more plentiful and much more substantive in nature than those in Bolton and Chapman (1986). They include sophisticated horse, jockey, and situational context handicapping variables. The Chapman and Staelin (1982) rank order "explosion" approach to exploiting the information content of ranked choice sets is used in this paper to identify the best possible fundamental handicapping weights. Holdout sample predictions for a single unit betting strategy are used to assess the performance of the pure fundamental multinomial logit horse race handicapping model. An augmented fundamental handicapping model, including the log of the public's win probabilities as an additional model variable, is also examined.

The Hong Kong Database

The 2,000 races in this Hong Kong database were provided by William Benter of Hong Kong. Benter and associates have extended the Bolton and Chapman (1986) modeling approach to include a much richer set of fundamental handicapping factors and have successfully applied the resulting model in Hong Kong. See Benter (1994) for some further background and details.

Table 1 contains a comparison of the Bolton and Chapman (1986) and the Hong Kong horse race databases along key dimensions. The Hong Kong database includes races from September 1985 to November 1991. These 2,000 Hong Kong horse races have at least eight runners and no first-time starters. The Hong Kong database includes relatively large fields (mean of 10) and very large wager pools (win pools greater than US$2 million are common).

Table 1: Comparison of Bolton and Chapman (1986) and Hong Kong Databases

	Bolton and Chapman (1986)	This Study
# of Races	200	2,000
Race Source	Five Northeastern U.S. Tracks	Hong Kong
Wagers	Win Wagering	Win Wagering
Tracks and Distances	"On" Tracks Over Limited Distances	"On" and "Off" Tracks Over All Distances
# of Variables	10	20
Variable Types	Basic Horse, Jockey, and Context Independent Variables	Sophisticated Horse, Workout, Jockey, Context, and Interaction Independent Variables

The overall efficiency of win betting in these 2,000 races is shown in Table 2. Traditional longshot-favorite biases are not present here. This is consistent with empirical results reported by Busche and Hall (1988) and Busche (1994). The expected win frequencies parallel the actual winning frequencies very well. On average, then, it does not seem possible to outperform the win betting public in Hong Kong. However, averages can be deceiving. Profitable win betting opportunities may still exist, if a high-performing horse race handicapping model can be developed. This is the focus of this paper.

The 20 fundamental independent variables available for analysis in this study are defined in Table 3. Horse race handicapping involves modeling (either formally or judgmentally) a horse's current performance potential based on demonstrated historical performance of both the horse and the jockey, current situational context considerations (e.g., track and distance conditions), and observable current performance signals (e.g., recent track workouts).[1] The horse race handicapping factors in Table 3 include many sophisticated horse, workout, jockey, situational context, and interaction proxies for past and current performance potential. Sophistication arises from both thoughtful variable construction (accounting for obvious and subtle handicapping principles) and variable operationalization with all past performance results. Many of these variables are based on Benter's (1994) development of a database with a complete history of each horse's past performance. While other notions of the true underlying drivers of current performance potential of horses might be posited, certainly this set of factors is substantial, sophisticated, and comprehensive.

Recency-weighted means are used to account for all past performances in operationalizing some of these variables. Recency-weighted means weight recent performance more than older performance in variable operationalization. This contrasts with Bolton and Chapman (1986) where only a few past races for each horse were used to construct the independent variables. In addition, some of these fundamental Hong Kong variables have been constructed using auxiliary regressions on all past performances of the horses or jockeys.

---

[1]This type of modeling is used in the stock market in a very similar way. See Ziemba and Schwartz (1991) for the development of such a model and its use in the Japanese stock market for hedging and other purposes.

Table 2: Overall Efficiency of Hong Kong Win Betting Pools (September 1985-November 1991)

Winning Probability	Expected Win Frequency (%)	Actual Win Frequency (%)	Sample Size
0.00-0.04	2.0	2.6	5,210
0.04-0.08	6.0	6.0	5,118
0.08-0.12	10.0	9.9	3,699
0.12-0.16	14.0	15.3	2,430
0.16-0.20	18.0	18.5	1,409
0.20-0.24	22.0	20.5	1,002
0.24-0.28	26.0	23.8	554
0.28-0.32	30.0	31.6	354
0.32-0.36	34.0	34.8	161
0.36-0.40	38.0	38.3	107

Multinomial Logit Handicapping Model Results

The 20-variable pure fundamental multinomial logit handicapping model was estimated on the 2,000-race Hong Kong database with successive explosion depths. Explosion-depth-one (first-choice data only, the winning horse versus all others in each race) and explosion-depth-two (second-choice data only, the runner-up versus third-place and lower ranked horses in each race) models were estimated separately and on a pooled basis. A Chow-test of significant differences of the separate parameter vectors provides evidence on whether the two explosion depths may be pooled for analysis purposes. If pooling is feasible (because the separate parameter vectors are not statistically different), this sequential explosion procedure is iterated to examine the appropriateness of exploding to the next possible rank order explosion depth. See Chapman and Staelin (1982) for complete details of this explosion process.

The sequential rank order explosion pooling and testing procedure yielded χ^2 values of 28.4 and 48.4, respectively, for pooling to explosion depths two and three. With a critical χ^2 value of 31.4 (for 20 d.f. at the 5% level), these results indicate that a second rank order position explosion is permissible but not a third. This contrasts with Bolton and Chapman (1986) who found empirical support for an explosion depth of three in northeastern U.S. races. In pooling the first two rank ordered choice sets for estimation purposes, the estimated standard errors of the multinomial logit model coefficients decreased by about 30% compared to using just the first choice results. This is the value of exploiting the natural rank ordered nature of horse race results — more "precise" parameter estimates result. Here, "precise" is defined in the statistical sense of reduced standard errors of coefficient estimates.

The multinomial logit coefficient estimates are reported in Table 4. Using the pooled estimates for the first and second explosion depths (a total of 4,000 Hong Kong horse race observations), the multinomial logit handicapping model empirical results may be summarized as follows:
- 13 of the 20 independent variables have t-ratios of at least 4.0 in absolute value. Of the 20 independent variables, 19 are statistically significant at the 5% level.
- The signs on the estimated fundamental horse race handicapping weights are plausible and consistent with horse race handicapping theory and principles. This is reassuring since these 20 independent variables are hardly statistically uncorrelated with each other. Indeed, variables like FIRSTCALL and

Table 3: Definitions of the 20 Pure Fundamental Horse Race Handicapping
Independent Variables Available in the 2,000-Race Hong Kong Database

Independent Variable	Definition
FINISH1	Recency-weighted mean of past normalized finishing position.
WEIGHT(REL)	Today's weight carried minus mean weight carried on all horses in this race.
WEIGHT(REL)DIST	WEIGHT(REL) times distance of this race.
FIRSTCALL	Recency-weighted mean of past first call position (an early speed factor).
FIRSTCALLDIST	FIRSTCALL times distance of this race (in combination with FIRSTCALL, this enhances the impact of early speed in short races).
WINHISTORY	Recency-weighted mean of past win history (binary coding with 1 for wins and 0 otherwise).
LENGTHSBEHIND	Recency-weighted mean of past lengths beaten, distance normalized.
DAYSSINCE	Days since last race minus median days between races.
CAREERSTARTS	Number of career starts.
FINISH2	Average past normalized finishing position.
COMPETITORS1	A complicated current-competitor factor which serves to boost horses who have actually raced with and beaten other horses in this race in a recent past race.
COMPETITORS2	Recency-weighted mean of average lengths beaten of all other horses in this horse's past races.
LASTRACE	Normalized finish of this horse in its last race times the recency weight of that race.
SPECIALDIFF	Recency-weighted mean of past officially reported "trip" difficulties this horse has experienced (e.g., "was bumped and lost two lengths in last race").
DISTANCEPREF	Preference (positive or negative) for this race's distance.
TRACKPREF	Preference for this race's track.
JOCKEYHISTORY	Recency-weighted past skill of the jockey that rode this horse in past races.
JOCKEYCURRENT	This race's jockey advantage factor (based on an auxiliary regression model).
TRACKWORK	Trackwork factor (based on an auxiliary regression model).
STRENGTH	Recency-weighted estimated strength of other horses in this horse's past races.

FIRSTCALLDIST must be highly correlated by definition, since the latter is a mathematical interaction of the former and another variable. Of course, substantive understanding of the underlying determinants of current horse race performance is not necessarily a major consideration here. Thus, the absence of

Table 4: Pure Fundamental Multinomial Logit Horse Race Handicapping Model
Coefficient Estimates (Explosion Depths One and Two Combined)

Independent Variables	Coefficient Estimate	T-Ratio
FINISH1	0.551	1.52
WEIGHT(REL)	2.444	2.02
WEIGHT(REL)DIST	-2.979	-2.48
FIRSTCALL	2.044	4.91
FIRSTCALLDIST	-1.688	-3.94
WINHISTORY	-0.666	-2.13
LENGTHSBEHIND	-6.879	-5.08
DAYSSINCE	-0.201	-4.11
CAREERSTARTS	-0.143	-9.98
FINISH2	1.062	4.91
COMPETITORS1	2.223	2.13
COMPETITORS2	9.004	5.10
LASTRACE	0.656	6.53
SPECIALDIFF	5.084	5.37
DISTANCEPREF	6.389	6.45
TRACKPREF	5.750	4.09
JOCKEYHISTORY	-1.921	-2.90
JOCKEYCURRENT	4.301	11.49
TRACKWORK	3.632	6.46
STRENGTH	3.257	7.05

plausible signs and magnitudes of coefficient estimates is not necessarily a major problem. Rather, predictive performance is key. However, it is reassuring that the results have intuitive validity.

- The log-likelihood ratio index is 8.2% for the 4,000 explosion depth two choice sets. This is a substantial improvement over the 5.5% results obtained by Bolton and Chapman (1986) for an explosion depth of three. This 50% improvement in explained variance augers well for the potential predictive performance of the 20-variable Hong Kong pure fundamental multinomial logit handicapping model. Of course, this is merely an internal measure of statistical fit. The acid test of performance comes in holdout sample predictions.

Holdout Sample Predictions of Wagering Results

The single unit bet strategy described in Bolton and Chapman (1986) is used to assess the multinomial logit horse race handicapping model's performance. A single unit bet strategy wagers a small amount (small enough so as to not influence the track odds) on the horse with the highest expected return, provided that the expected return is positive and that the estimated winning probability exceeds a specified minimum probability, p_{min}. Simulated betting results are obtained for varying levels of p_{min}, from 0.0 up to some number (perhaps 0.25) where the number of wagering opportunities is too few for reliable assessment of wagering strategy performance.

As discussed in Bolton and Chapman (1986), the heuristic of excluding extreme longshots from wagering consideration (i.e., the p_{min} rule) is an effective way of coping with the statistical modeling challenges associated with estimating precise extreme longshot winning probabilities. An error in estimated probability of 0.01 for a favorite with a true winning probability of 0.20 isn't much of a problem since it's only a relative error of 5%. An error in estimated probability of 0.01 for an extreme longshot with a true winning probability of 0.02 is a very serious problem since it's a relative error of 50%. Excluding extreme longshots tends to eliminate few apparent wagering opportunities while substantially improving expected returns.

As shown by others such as Thorp (1975), Asch and Quandt (1986), Ziemba and Hausch (1987), and McLean, Ziemba, and Blazenko (1992), more sophisticated wagering algorithms techniques usually yield greater returns than a simple single unit bet strategy. The single unit bet strategy is, however, the basic building block on which sophisticated wagering techniques are based. Thus, it establishes an important benchmark. These holdout sample predictions demonstrate that positive returns are achievable even with this very basic wagering strategy.

Holdout sampling is used to assess the out-of-sample predictive performance of the 20-variable pure fundamental multinomial logit horse race handicapping model. Five holdout samples of 400 races were constructed, with the multinomial logit horse race handicapping model being estimated on the other 1,600 races in each holdout sample prediction test.

The average results of the single unit bet strategy results across the five holdout samples using the 20-variable fundamental multinomial logit horse race handicapping model are shown in Table 5. A 17.5% track take, the current Hong Kong norm for win betting pools, is taken into account in these holdout sample prediction tests. In Table 5, "Mean" and "SD" refer, respectively, to the mean percentage return and the standard deviation of the mean percentage returns across the five 400-race holdout samples.

The most important finding reported in Table 5 is that an unconstrained single unit betting strategy ($p_{min}=0.0$) yields very attractive financial returns of 17.2% for these 2,000 Hong Kong races. It should also be noted that the standard deviations associated with these mean returns are high (31.0%). Eliminating extreme longshots (25-to-1 or greater) from consideration results in increases in expected returns from 17.2% to 25.1%. This is accompanied by no increase in the standard deviation of the mean returns. The standard deviation of returns actually decreases slightly, from 31.0% to 27.5%.

Increasing p_{min} beyond 0.06 is not desirable because the expected returns drop dramatically to near zero. There were many favorable win betting opportunities for horses in the 15-to-1 to 25-to-1 odds range in these 2,000 Hong Kong races in the September 1985-November 1991 time period. Excluding these favorable win betting opportunities has very negative consequences for overall wagering system performance.

The number of betting opportunities declines very modestly as p_{min} increases from 0.00 to 0.09. Apparently, a very small number of these 2,000 Hong Kong races include only one attractive longshot wagering opportunity (10-to-1 or greater). Few betting opportunities (less than 0.5% of these races) are sacrificed when extreme longshots (25-to-1 or greater) are eliminated from win betting consideration.

These findings agree with the pattern of the Bolton and Chapman (1986) results. However, the magnitude of the expected returns from win betting in this 2,000-race database is much higher than in Bolton and Chapman (1986). The improved predictive performance of the 20-variable pure fundamental multinomial logit handicapping model apparently translates into many very attractive win betting opportunities in Hong Kong. Of course, it is also possible that Hong Kong win betters make even larger handicapping errors than those in the Bolton and Chapman (1986) database from five northeastern U.S. tracks.

Table 5: Single Unit Bet Strategy Results Across Five Holdout 400-Race Samples Using The 20-Variable Fundamental Multinomial Logit Handicapping Model

		% Return	
ρ_{min}	# of Races	Mean	SD
0.00	1,997	17.2	31.0
0.01	1,997	17.2	31.0
0.02	1,997	17.5	30.6
0.03	1,995	19.1	33.7
0.04	1,994	25.1	27.5
0.05	1,988	29.3	32.6
0.06	1,974	15.7	19.5
0.07	1,967	3.7	20.8
0.08	1,941	-0.6	8.8
0.09	1,903	4.6	7.6
0.10	1,833	5.0	12.0
0.11	1,752	2.5	13.4
0.12	1,636	0.7	12.0
0.13	1,531	0.8	13.5
0.14	1,415	-2.2	18.2
0.15	1,304	-0.6	13.5
0.16	1,155	-1.9	20.9
0.17	1,029	-5.6	9.1
0.18	922	-8.3	13.2
0.19	812	-9.3	9.7
0.20	725	-6.3	8.8
0.21	622	-8.8	8.8
0.22	537	-13.3	14.9
0.23	464	-14.1	11.7
0.24	397	-19.6	8.7
0.25	334	-24.1	12.7

Extending Fundamental Modeling: Does The Public Have Special Insight?

This analysis has been a pure fundamental investment approach: use the public horse, jockey, and situational context (track, distance, post position, etc.) information available well ahead of race time to predict a race's outcome. However, perhaps the public knows something that is not fully captured in this pure fundamental model. Indeed, many of the papers in this volume indicate that that is the case. Whether that's true inside (non-public) information or simply model specification error is unimportant. If the public's bets themselves contain valuable insight _not already captured_ by the 20-variable pure fundamental handicapping model, that is important in and of itself.

To assess the possible presence of special public insight above and beyond that already captured by the 20-variable pure fundamental model, an additional variable was added to the model. The 21-variable augmented horse race handicapping model includes the public's win probabilities as revealed by their actual win bet wagers. As suggested by Asch and Quandt (1986, pp. 123-5) and Benter (1994), the log of the public's win probabilities was used since statistical fit slightly improves with the log transformation. This one-step reduced form approach is conceptually equivalent to a similar two-stage sequential approach suggested by Asch and Quandt (1986, pp. 123-5) for forming an augmented handicapping model.

Adding in the public's estimates of the win probabilities as proxied by the log of the public's win bets yields a log-likelihood for the 21-variable augmented model of -8,017.5. This improvement in log-likelihood (statistical fit) from -8,217.9 for the 20-variable pure fundamental model to -8,017.5 for the 21-variable augmented model is highly significant. With a critical χ^2 value of 6.6 (for 1 d.f. at the 1% level), the calculated value of 200.4 leads to a complete and unambiguous rejection of the null hypothesis that the log of the public's win probabilities do not add to the explanatory power of the pure fundamental handicapping model, given the prior presence of the 20 pure fundamental independent variables in the model.

The 20-variable pure fundamental model accounts for about 78.6% of the total explained variance in the 21-variable augmented model. The 21st variable, the log of the public's win probabilities, accounts for the remaining 21.4% of the total explained variance in the augmented model — and the last 21.4% is a highly statistically significant amount of explained variance.

Statistically, these results are absolutely clear. Even with an elaborate and sophisticated 20-variable pure fundamental horse race handicapping model, important incremental explanatory power is captured if it is possible to include the public's win probabilities as an additional predictive factor. Of course, in practical terms it would only be possible to capitalize on this if one could wager electronically at the last minute after re-calibrating the estimates of the true win probabilities with the final or virtually final public win bets.

Concluding Remarks

Six major conclusions arise from these results. First, positive returns at the track are achievable with a sophisticated pure fundamental multinomial logit horse race handicapping model. This 20-variable model with exploded choice sets applied to Hong Kong data yields very positive wagering returns (in excess of 20%), even with a simple single unit bets strategy and even in the presence of a 17.5% track take. These single unit bet strategy results from holdout predictions are so strong that it is clear that Bolton and Chapman (1986) results were not simply statistical aberrations or small sample anomalies. More elaborate wagering strategies undoubtedly would improve upon these results.

Second, many factors must be weighed to handicap horse races. Simple one- or two- or three-variable handicapping rules-of-thumb are much too simplistic to capture everything of note. Good handicapping is difficult, even for experts. There is considerable room for handicapping errors in such a complex process. Given the complexity associated with constructing and manipulating such large-scale horse racing databases, there are large barriers to entry to fundamental handicapping horse race systems. With track takes approaching 20% (or more, for exotic bets), it's not surprising that there are few very rich horse race handicappers.

Third, the inclusion of more sophisticated fundamental variables improves the statistical performance of fundamental multinomial logit handicapping models and improves the expected returns from even basic wagering strategies. In four holdout 50-race samples for a 10-variable multinomial logit horse race handicapping model, Bolton and Chapman (1986) estimated returns to be 3.1% with unconstrained single unit betting (i.e., with $p_{min}=0.0$). In this study, with five holdout 400-race samples for a 20-variable multinomial logit horse race handicapping model, returns of 17.2% exist with unconstrained single unit betting (i.e., with $p_{min}=0.0$).

Fourth, it pays to "explode" horse race results for estimation within the context of the multinomial logit model, but not too much. Ranked choice set "explosions" improve the precision of the weights in fundamental multinomial logit models horse race handicapping. These Hong Kong data suggest that an explosion depth of two is appropriate. These results are in qualitative agreement with Bolton and Chapman (1986). The important gains in estimation efficiency accrue from the first few rank order explosions.

Fifth, there are substantial variations in the returns to single unit bet strategies. Careful money management and large bankrolls are necessary to do well in applying fundamental wagering systems at the race track.

Sixth, even higher expected returns may be achievable by including the log of the public's win probabilities (as revealed by their actual win bets) as an additional variable in a multinomial logit horse race handicapping model. See Benter (1994) for more some further details of and results associated with this possibility. Of course, such an augmented horse race handicapping model form is only useful for betting purposes if it is possible to incorporate in real time the public's win betting patterns in an optimal wagering system.

The empirical results of Bolton and Chapman (1986) suggested the possibility of achieving positive returns at the track with a fundamental multinomial logit handicapping modeling approach estimated with exploded choice sets. The Bolton and Chapman (1986) results were based on a basic handicapping model applied to a modest-sized database. These results from Hong Kong provide convincing evidence of the existence of win bet positive returns with a fundamental handicapping approach.

References

Asch, Peter and Richard E. Quandt (1986), *Racetrack Betting: The Professor's Guide To Strategies* (Dover, MA: Auburn House).

Benter, William (1994), "Computer-Based Horse Race Handicapping and Wagering Systems: A Report," In this volume, 169-84.

Bolton, Ruth N. and Randall G. Chapman (1986), "Searching For Positive Returns at the Track: A Multinomial Logit Model For Handicapping Horse Races," *Management Science*, 32, 8 (August), 1040-60.

Busche, Kelly and Christopher D. Hall (1988), "An Exception to the Risk Preference Anomaly," *Journal of Business*, 61, 3 (July), 337-46.

Busche, Kelly (1994), "Efficient Market Results in an Asian Setting," In this volume, 580-1.

Chapman, Randall G. and Richard Staelin (1982), "Exploiting Rank Ordered Choice Set Data Within the Stochastic Utility Model," *Journal of Marketing Research*, XIX, 3 (August), 288-301.

McLean, Leonard C., William T. Ziemba, and George Blazenko (1992), "Growth Versus Security in Dynamic Investment Analysis," *Management Science*, 38, 11 (November), 1560-85.

Thorp, Edward O. (1975), "The Portfolio Choice and the Kelly Criterion," in W. T. Ziemba and R. G. Vickson (eds.), *Stochastic Optimization Models in Finance* (New York: Academic Press), 599-619.

Ziemba, William T. and Donald B. Hausch (1987), *Dr. Z's Beat the Racetrack*, Revised Edition (New York: William Morrow and Company, Inc.).

Ziemba, William T. and Sandra L. Schwartz (1991), *Invest Japan* (Chicago: Probus Publishers).

Computer Based Horse Race Handicapping
and Wagering Systems: A Report

William Benter

HK Betting Syndicate, Hong Kong

ABSTRACT

This paper examines the elements necessary for a practical and successful computerized horse race handicapping and wagering system. Data requirements, handicapping model development, wagering strategy, and feasibility are addressed. A logit-based technique and a corresponding heuristic measure of improvement are described for combining a fundamental handicapping model with the public's implied probability estimates. The author reports significant positive results in five years of actual implementation of such a system. This result can be interpreted as evidence of inefficiency in pari-mutuel racetrack wagering. This paper aims to emphasize those aspects of computer handicapping which the author has found most important in practical application of such a system.

INTRODUCTION

The question of whether a fully mechanical system can ever "beat the races" has been widely discussed in both the academic and popular literature. Certain authors have convincingly demonstrated that profitable wagering systems do exist for the races. The most well documented of these have generally been of the *technical* variety, that is, they are concerned mainly with the public odds, and do not attempt to predict horse performance from fundamental factors. Technical systems for place and show betting, (Ziemba and Hausch, 1987) and exotic pool betting, (Ziemba and Hausch, 1986) as well as the 'odds movement' system developed by Asch and Quandt (1986), fall into this category. A benefit of these systems is that they require relatively little preparatory effort, and can be effectively employed by the occasional racegoer. Their downside is that betting opportunities tend to occur infrequently and the maximum expected profit achievable is usually relatively modest. It is debatable whether any racetracks exist where these systems could be profitable enough to sustain a full-time professional effort.

To be truly viable, a system must provide a large number of high advantage betting opportunities in order that a sufficient amount of expected profit can be generated. An approach which does promise to provide a large number of betting opportunities is to *fundamentally* handicap each race, that is, to empirically assess each horse's chance of winning, and utilize that assessment to find profitable wagering opportunities. A natural way to attempt to do this is to develop a computer model to estimate each horse's probability of winning and calculate the appropriate amount to wager.

A complete survey of this subject is beyond the scope of this paper. The general requirements for a computer based fundamental handicapping model have been well presented by Bolton and Chapman (1986) and Brecher (1980). These two references are "required reading" for anyone interested in developing such a system. Much of what is said here has already been explained in those two works, as is much of the theoretical background which has been omitted here. What the author would hope to add, is a discussion of a few points which have not been addressed in the literature, some practical recommendations, and a report that a *fundamental* approach can in fact work in practice.

FEATURES OF THE COMPUTER HANDICAPPING APPROACH

Several features of the computer approach give it advantages over traditional handicapping. First, because of its empirical nature, one need not possess specific handicapping expertise to undertake this enterprise, as everything one needs to know can be learned from the data. Second is the testability of a computer system. By carefully partitioning data, one can develop a model and test it on *unseen* races. With this procedure one avoids the danger of overfitting past data. Using this 'holdout' technique, one can obtain a reasonable estimate of the system's real-time performance before wagering any actual money. A third positive attribute of a computerized handicapping system is its consistency. Handicapping races manually is an extremely taxing undertaking. A computer will effortlessly handicap races with the same level of care day after day, regardless of the mental state of the operator. This is a non-trivial advantage considering that a professional level betting operation may want to bet several races a day for extended periods.

The downside of the computer approach is the level of preparatory effort necessary to develop a winning system. Large amounts of past data must be collected, verified and computerized. In the past, this has meant considerable manual entry of printed data. This situation may be changing as optical scanners can speed data entry, and as more online horseracing database services become available. Additionally, several man-years of programming and data analysis will probably be necessary to develop a sufficiently profitable system. Given these considerations, it is clear that the computer approach is not suitable for the casual racegoer.

HANDICAPPING MODEL DEVELOPMENT

The most difficult and time-consuming step in creating a computer based betting system is the development of the fundamental handicapping model. That is, the model whose final output is an estimate of each horse's probability of winning. The type of model used by the author is the multinomial logit model proposed by Bolton and Chapman (1986). This model is well suited to horse racing and has the convenient property that its output is a set of probability estimates which sum to 1 within each race.

The overall goal is to estimate each horse's current performance potential. "Current performance potential" being a single overall summary index of a horse's expected performance in a particular race. To construct a model to estimate current performance potential one must investigate the available data to find those variables or *factors* which have predictive significance. The profitability of the resulting betting system will be largely determined by the predictive power of the factors chosen. The odds set by the public betting yield a sophisticated estimate of the horses' win probabilities. In order for a fundamental statistical model to be able to compete effectively, it must rival the public in sophistication and comprehensiveness. Various types of factors can be classified into groups:

Current condition:
- performance in recent races
- time since last race
- recent workout data
- age of horse

Past performance:
- finishing position in past races
- lengths behind winner in past races
- normalized times of past races

Adjustments to past performance:
- strength of competition in past races
- weight carried in past races
- jockey's contribution to past performances
- compensation for bad luck in past races
- compensation for advantageous or disadvantageous post position in past races

Present race situational factors:
- weight to be carried
- today's jockey's ability
- advantages or disadvantages of the assigned post position

Preferences which could influence the horse's performance in today's race:
- distance preference
- surface preference (turf vs dirt)
- condition of surface preference (wet vs dry)
- specific track preference

More detailed discussions of fundamental handicapping can be found in the extensive popular literature on the subject (for the author's suggested references see the list in the appendix). The data needed to calculate these factors must be entered and checked for accuracy. This can involve considerable effort. Often, multiple sources must be used to assemble complete past performance records for each of the horses. This is especially the case when the horses have run past races at many different tracks. The easiest type of racing jurisdiction to collect data and develop a model for is one with a *closed* population of horses, that is, one where horses from a single population race only against each other at a limited number of tracks. When horses have raced at venues not covered in the database, it is difficult to evaluate the elapsed times of races and to estimate the strength of their opponents. Also unknown will be the post position biases, and the relative abilities of the jockeys in those races.

In the author's experience the minimum amount of data needed for adequate model development and testing samples is in the range of 500 to 1000 races. More is helpful, but out-of-sample predictive accuracy does not seem to improve dramatically with development samples greater than 1000 races. Remember that *data for one race* means full past data on all of the runners in that race. This suggests another advantage of a *closed* racing population; by collecting the results of all races run in that jurisdiction one automatically accumulates the full set of past performance data for each horse in the population.

It is important to define factors which extract as much information as possible out of the data in each of the relevant areas. As an example, consider three different specifications of a 'distance preference' factor.

The first is from Bolton and Chapman (1986):

'NEWDIST' - this variable equals one if a horse has run three of its four previous races at a distance less than a mile, zero otherwise. (Note: Bolton and Chapman's model was only used to predict races of 1 - 1.25 miles.)

The second is from Brecher (1980):

'DOK' - this variable equals one if the horse finished in the upper 50th percentile or within 6.25 lengths of the winner in a prior race within 1/16 of a mile of today's distance, or zero otherwise

The last is from the author's current model:

'DP6A' - for each of a horse's past races, a predicted finishing position is calculated via multiple regression based on all factors except those relating to distance. This predicted finishing position in each race is then subtracted from the horse's actual finishing position. The resulting quantity can be considered to be the unexplained residual which may be due to some unknown distance preference that the horse may possess plus a certain amount of random error. To estimate the horse's preference or aversion to today's distance, the residual in each of its past races is used to estimate a linear relationship between performance and similarity to today's distance. Given the statistical uncertainty of estimating this relationship from the usually small sample of past races, the final magnitude of the estimate is standardized by dividing it by its standard error. The result is that horses with a clearly defined distance preference demonstrated over a large number of races will be awarded a relatively larger magnitude value than in cases where the evidence is less clear.

The last factor is the result of a large number of progressive refinements. The subroutines involved in calculating it run to several thousand lines of code. The author's guiding principle in factor improvement has been a combination of educated guessing and trial and error. Fortunately, the historical data makes the final decision as to which particular definition is superior. The best is the one that produces the greatest increase in predictive accuracy when included in the model. The general thrust of model development is to continually experiment with refinements of the various factors. Although time-consuming, the gains are worthwhile. In the author's experience, a model involving only simplistic specifications of factors does not provide sufficiently accurate estimates of winning probabilities. Care must be taken in this process of model development not to overfit past data. Some overfitting will always occur, and for this reason it is important to use data partitioning to maintain sets of *unseen* races for out-of-sample testing.

The complexity of predicting horse performance makes the specification of an elegant handicapping model quite difficult. Ideally, each independent variable would capture a unique aspect of the influences effecting horse performance. In the author's experience, the trial and error method of adding independent variables to increase the model's goodness-of-fit, results in the model tending to become a hodgepodge of highly correlated variables whose individual significances are difficult to determine and often counter-intuitive. Although aesthetically unpleasing, this tendency is of little consequence for the purpose which the model will be used, namely, prediction of future race outcomes. What it does suggest, is that careful and conservative statistical tests and methods should be used on as large a data sample as possible.

For example, "number of past races" is one of the more significant factors in the author's handicapping model, and contributes greatly to the overall accuracy of the predictions. The author knows of no 'common sense' reason why this factor should be important. The only reason it can be confidently included in the model is because the large data sample allows its significance to be established beyond a reasonable doubt.

Additionally, there will always be a significant amount of 'inside information' in horse racing that cannot be readily included in a statistical model. Trainer's and jockey's intentions, secret workouts, whether the horse ate its breakfast, and the like, will be available to certain parties who will no doubt take advantage of it. Their betting will be reflected in the odds. This presents an obstacle to the model developer with access to published information only. For a statistical model to compete in this environment, it must make full use of the advantages of computer modelling, namely, the ability to make complex calculations on large data sets.

CREATING UNBIASED PROBABILITY ESTIMATES

It can be presumed that valid fundamental information exists which can not be systematically or practically incorporated into a statistical model. Therefore, any statistical model, however well developed, will always be incomplete. An extremely important step in model development, and one that the author believes has been generally overlooked in the literature, is the estimation of the relation of the model's probability estimates to the public's estimates, and the adjustment of the model's estimates to incorporate whatever information can be gleaned from the public's estimates.

The public's implied probability estimates generally correspond well with the actual frequencies of winning. This can be shown with a table of estimated probability versus actual frequency of winning (Table 1).

Table 1

PUBLIC ESTIMATE VS. ACTUAL FREQUENCY

range	n	exp.	act.	Z
.000-.010	1343	.007	.007	0.0
.010-.025	4356	.017	.020	1.3
.025-.050	6193	.037	.042	2.1
.050-.100	8720	.073	.069	-1.5
.100-.150	5395	.123	.125	0.6
.150-.200	3016	.172	.173	0.1
.200-.250	1811	.222	.219	-0.3
.250-.300	1015	.273	.253	-1.4
.300-.400	716	.339	.339	0.0
> .400	312	.467	.484	0.6

races = 3198,   # horses = 32877

Table 2

FUNDAMENTAL MODEL VS. ACTUAL FREQUENCY

range	n	exp.	act.	Z
.000-.010	1173	.006	.005	-0.6
.010-.025	3641	.018	.015	-1.2
.025-.050	6503	.037	.037	-0.3
.050-.100	9642	.073	.074	0.1
.100-.150	5405	.123	.120	-0.7
.150-.200	2979	.173	.183	1.6
.200-.250	1599	.223	.232	0.9
.250-.300	870	.272	.285	0.9
.300-.400	741	.341	.320	-1.2
> .400	324	.475	.432	-1.6

races = 3198,   # horses = 32877

range = the range of estimated probabilities
n = the number of horses falling within a range
exp. = the mean expected probability
act. = the actual win frequency observed
Z = the discrepancy (+ or -) in units of standard errors

In each range of estimated probabilities, the actual frequencies correspond closely. This is not the case at all tracks (Ali, 1977) and if not, suitable corrections should be made when using the public's probability estimates for the purposes which will be discussed later. (Unless otherwise noted, data samples consist of all races run by the Royal Hong Kong Jockey Club from September 1986 through June 1993.)

A multinomial logit model using fundamental factors will also naturally produce an internally consistent set of probability estimates (Table 2). Here again there is generally good correspondence between estimated and actual frequencies. Table 2 however conceals a major, (and from a wagering point of view, disastrous) type of bias inherent in the fundamental model's probabilities. Consider the following two tables which represent roughly equal halves of the sample in Table 2. Table 3 shows the fundamental model's estimate versus actual frequency for those horses where the public's probability estimate was greater the fundamental model's. Table 4 is the same except that it is for those horses whose public estimate was less than the fundamental model's.

Table 3

Table 4

FUNDAMENTAL MODEL VS. ACTUAL FREQUENCY WHEN PUBLIC ESTIMATE IS GREATER THAN MODEL ESTIMATE

FUNDAMENTAL MODEL VS. ACTUAL FREQUENCY WHEN PUBLIC ESTIMATE IS LESS THAN MODEL ESTIMATE

range	n	exp.	act.	Z	range	n	exp.	act.	Z
.000-.010	920	.006	.005	-0.3	.000-.010	253	.007	.004	-0.6
.010-.025	2130	.017	.018	0.3	.010-.025	1511	.018	.011	-2.2
.025-.050	3454	.037	.044	2.1	.025-.050	3049	.037	.029	-2.6
.050-.100	4626	.073	.091	4.7	.050-.100	5016	.074	.058	-4.3
.100-.150	2413	.122	.147	3.7	.100-.150	2992	.123	.098	-4.2
.150-.200	1187	.172	.227	5.0	.150-.200	1792	.173	.154	-2.1
.200-.250	540	.223	.302	4.4	.200-.250	1059	.223	.196	-2.1
.250-.300	252	.270	.333	2.3	.250-.300	618	.273	.265	-0.4
.300-.400	165	.342	.448	2.9	.300-.400	576	.341	.283	-2.9
>.400	54	.453	.519	1.0	>.400	270	.480	.415	-2.1

races = 3198,   # horses = 15741   # races = 3198,   # horses = 17136

There is an extreme and consistent bias in both tables. In virtually every range the actual frequency is significantly different than the fundamental model's estimate, and always in the direction of being closer to the public's estimate. The fundamental model's estimate of the probability cannot be considered to be an unbiased estimate independent of the public's estimate. Table 4 is particularly important because it is comprised of those horses that the model would have one bet on, that is, horses whose model-estimated probability is greater than their public probability. It is necessary to correct for this bias in order to accurately estimate the advantage of any particular bet.[1]

In a sense, what is needed is a way to combine the judgements of two experts, (i.e. the fundamental model and the public). One practical technique for accomplishing this is as follows: (Asch and Quandt, 1986; pp. 123-125). See also White, Dattero and Flores, (1992).

Estimate a second logit model using the two probability estimates as independent variables. For a race with entrants $(1,2,\ldots,N)$ the win probability of horse i is given by:

$$c_i = \frac{\exp(\alpha f_i + \beta \pi_i)}{\sum \exp(\alpha f_j + \beta \pi_j)} \qquad \text{(for } j = 1 \text{ to } N)$$

(1)

f_i = log of 'out-of-sample' fundamental model probability estimate

π_i = log of public's implied probability estimate

c_i = combined probability estimate

(Natural log of probability is used rather than probability as this transformation provides a better fit)

Given a set of past races $(1,2,\ldots,R)$ for which both public probability estimates and fundamental model estimates are available, the parameters α and β can be estimated by maximizing the log likelihood function of the given set of races with respect to α and β:

$$\exp(L) = \prod c_{ji^*} \quad (j = 1 \text{ to } R) \tag{2}$$

where c_{ji^*} denotes the probability as given by equation (1) for the horse i^* observed to win race j (Bolton and Chapman, 1986 p. 1044). Equation (1) should be evaluated using fundamental probability estimates from a model developed on a separate sample of races. Use of 'out-of-sample' estimates prevents overestimation of the fundamental model's significance due to 'custom-fitting' of the model development sample. The estimated values of α and β can be interpreted roughly as the relative correctness of the model's and the public's estimates. The greater the value of α, the better the model. The probabilities that result from this model also show good correspondence between predicted and actual frequencies of winning (Table 5).

Table 5

COMBINED MODEL VS. ACTUAL FREQUENCIES

range	n	exp.	act.	Z
.000-.010	1520	.007	.005	-1.0
.010-.025	4309	.017	.018	0.1
.025-.050	6362	.037	.038	0.6
.050-.100	8732	.073	.071	-0.5
.100-.150	5119	.123	.119	-0.8
.150-.200	2974	.173	.180	1.0
.200-.250	1657	.223	.223	0.0
.250-.300	993	.272	.281	0.6
.300-.400	853	.340	.328	0.7
>.400	358	.479	.492	0.5

races = 3198,   # horses = 32877

By comparison with Tables 1 and 2, Table 5 shows that there is more *spread* in the combined model's probabilities than in either the public's or the fundamental model's alone, that is, there are more horses in both the very high and very low probability ranges. This indicates that the combined model is more informative. More important is that the new probability estimates are without the bias shown in Tables 3 and 4, and thus are suitable for the accurate estimation of betting advantage. This is borne out by Tables 6 and 7, which are analogous to Tables 3 and 4 above except that they use the combined model probabilities instead of the raw fundamental model probabilities.

Table 6

COMBINED MODEL VS. ACTUAL FREQUENCY
WHEN PUBLIC ESTIMATE IS GREATER THAN MODEL
ESTIMATE

range	n	exp.	act.	Z
.000-.010	778	.006	.005	-0.4
.010-.025	1811	.017	.015	-0.6
.025-.050	2874	.037	.035	-0.7
.050-.100	4221	.073	.073	0.0
.100-.150	2620	.123	.116	-1.0
.150-.200	1548	.173	.185	1.2
.200-.250	844	.223	.231	0.6
.250-.300	493	.272	.292	1.0
.300-.400	393	.337	.349	0.5
>.400	159	.471	.509	1.0

races = 3198,   # horses = 15741

Table 7

COMBINED MODEL VS. ACTUAL FREQUENCY
WHEN PUBLIC ESTIMATE IS LESS THAN MODEL
ESTIMATE

range	n	exp.	act.	Z
.000-.010	742	.007	.004	-0.9
.010-.025	2498	.018	.019	0.6
.025-.050	3488	.037	.041	1.4
.050-.100	4511	.072	.069	-0.7
.100-.150	2499	.123	.122	-0.1
.150-.200	1426	.173	.174	0.1
.200-.250	813	.223	.215	-0.5
.250-.300	500	.272	.270	-0.1
.300-.400	460	.342	.311	-1.4
>.400	199	.485	.477	-0.2

races = 3198,   # horses = 17136

Observe that the above tables show no significant bias one way or the other.

ASSESSING THE VALUE OF A HANDICAPPING MODEL

The log likelihood function of equation (2) can be used to produce a measure of fit analogous to the R^2 of multiple linear regression (Equation 3). This pseudo-R^2 (R^2) can be used to compare models and to assess the value of a particular model as a betting tool. Each set of probability estimates, either the public's or those of a model, achieve a certain R^2, defined as   (Bolton and Chapman, 1986)

$$R^2 = 1 - \frac{L(\text{model})}{L(1/N_j)} .$$ (3)

The R^2 value is a measure of the "explanatory power" of the model. An R^2 of 1 indicates perfect predictive ability while an R^2 of 0 means that the model is no better than random guessing. An important benchmark is the R^2 value achieved by the public probability estimate. A heuristic measure of the potential profitability of a handicapping model, borne out in practice, is the amount by which its inclusion in the combined model of equation (1) along with the public probability estimate causes the R^2 to increase over the value achieved by the public estimate alone:

$$\Delta R^2 = R^2_C - R^2_P$$ (4)

where the subscript P denotes the public's probability estimate and C stands for the combined (fundamental and public) model of equation (1) above. In a sense, ΔR^2 may be taken as a measure of the amount of information added by the fundamental model. In the case of the models which produced Tables 1,2 and 5 above these values are:

$$R^2_P = .1218 \qquad \text{(public)}$$
$$R^2_F = .1245 \qquad \text{(fundamental model)}$$
$$R^2_C = .1396 \qquad \text{(combined model)}$$

$$\Delta R^2_{C-P} = .1396 - .1218 = .0178$$

Though this value may appear small, it actually indicates that significant profits could be made with that model. The ΔR^2 value is a useful measure of the potential profitability of a particular model. It can be used to measure and compare models without the the time consuming step of a full wagering simulation. In the author's experience, greater ΔR^2 values have been invariably associated with greater wagering simulation profitability. To illustrate the point that the important criteria is the gain in R^2 in the combined model over the public's R^2, and not simply the R^2 of the handicapping model alone, consider the following two models.

The first is a logit-derived fundamental handicapping model using 9 significant fundamental factors. It achieves an out-of-sample R^2 of .1016. The second is a probability estimate derived from tallying the picks of approximately 48 newspaper tipsters. (Figlewski, 1979) The tipsters each make a selection for 1st, 2nd, and 3rd in each race. The procedure was to count the number of times each horse was picked, awarding 6 points for 1st, 3 points for 2nd, and 1 point for 3rd The point total for each horse is then divided by the total points awarded in the race (i.e. 48 * 10). This fraction of points is then taken to be the 'tipster' probability estimate. Using the log of this estimate as the sole independent variable in a logit model produces an R^2 of .1014. On the basis of their stand-alone R^2's the above two models would appear to be equivalently informative predictors of race outcome. Their vast difference appears when we perform the 'second stage' of combining these estimates with the public's. The following results were derived from logit runs on 2313 races (September 1988 to June 1993).

$$R^2_P = .1237 \qquad \text{(public estimate)}$$
$$R^2_F = .1016 \qquad \text{(fundamental model)}$$
$$R^2_T = .1014 \qquad \text{(tipster model)}$$
$$R^2_{(F\&P)} = .1327 \qquad \text{(fundamental and public)}$$
$$R^2_{(T\&P)} = .1239 \qquad \text{(tipster and public)}$$

$$\Delta R^2_{(F\&P)-P} = .1327 - .1237 = .0090$$
$$\Delta R^2_{(T\&P)-P} = .1239 - .1237 = .0002$$

As indicated by the ΔR^2 values, the tipster model adds very little to the public's estimate. The insignificant contribution of the tipster estimate to the overall explanatory power of the combined model effectively means that when there is a difference between the public estimate and the tipster estimate, then the public's estimate is superior. The fundamental model on the other hand, does contribute significantly when combined with the public's. For a player considering betting with the 'tipster' model, carrying out this 'second stage' would have saved that player from losing money; the output of the second stage model would always be virtually identical to the public estimate, thus never indicating an advantage bet.

WAGERING STRATEGY

After computing the combined and therefore unbiased probability estimates as described above, one can make accurate estimations of the advantage of any particular bet. A way of expressing advantage is as the expected return per dollar bet:

$$\text{expected return} = er = c \cdot div$$
$$\text{advantage} = er - 1$$

where c is the estimated probability of winning the bet and div is the expected dividend. For win betting the situation is straightforward. The c's are the probability estimates produced by equation (1) above, and the div's are the win dividends (as a payoff for a $1 bet) displayed on the tote board. The situation for an example race is illustrated in Table 8.

Table 8

#	c	p	er	div
1)	.021	.025	.68	33
2)	.125	.088	1.17	9.3
3)	.239	.289	.69	2.8
4)	.141	.134	.87	6.1
5)	.066	.042	1.29	19
6)	.012	.013	.75	61
7)	.107	.136	.64	6.0
8)	.144	.089	1.33	9.2
9)	.019	.014	1.18	60
10)	.067	.066	.86	12
11)	.012	.012	.83	68u
12)	.028	.047	.50	17
13)	.011	.027	.32	30
14)	.009	.019	.41	43

c = combined (second stage) probability estimate
p = public's probability estimate (1-take) / div
er = expected return on a $1 win bet
div = win dividend for a $1 bet

The 'u' after the win dividend for horse #11 stands for *unratable* and indicates that this is a horse for which the fundamental model could not produce a probability estimate. Often this is because the horse is running in its first race. A good procedure for handling such horses is to assign them the same probability as that implied by the public win odds, and renormalize the probabilities on the other horses so that the total probability for the race sums to 1. This is equivalent to saying that we have no information which would allow us to dispute the public's estimate so we will take theirs.

From Table 8 we can see that the advantage win bets are those with an *er* greater than 1. There is a positive expected return from betting on each of these horses. Given that there are several different types of wager available, it is necessary to have a strategy for determining which bets to make and in what amounts.

Kelly Betting and pool size limitations

Given the high cost in time and effort of developing a winning handicapping system, a wagering strategy which produces maximum expected profits is desirable. The stochastic nature of horse race wagering however, guarantees that losing streaks of various durations will occur. Therefore a strategy which balances the tradeoff between risk and returns is necessary. A solution to this problem is provided by the Kelly betting strategy (Kelly, 1956). The Kelly strategy specifies the fraction of total wealth to wager so as to maximize the exponential rate of growth of wealth, in situations where the advantage and payoff odds are known. As a fixed fraction strategy, it also never risks ruin. (This last point is not strictly true, as the mimimum bet limit prevents strict adherence to the strategy) For a more complete discussion of the properties of the Kelly strategy see MacLean, Ziemba and Blazenko (1992), see also Epstein (1977) and Brecher (1980).

The Kelly strategy defines the optimal bet (or set of bets) as those which maximize the expected log of wealth. In pari-mutuel wagering, where multiple bets are available in each race, and each bet effects the final payoff odds, the exact solution requires maximizing a concave logarithmic function of several variables. For a single bet, assuming no effect on the payoff odds, the formula simplifies to

$$K = \frac{(\text{advantage})}{(\text{dividend} - 1)} . \tag{5}$$

where K is the fraction of total wealth to wager. When one is simultaneously making wagers in multiple pools, further complications to the exact multiple bet Kelly solution arise due to 'exotic' bets in which one must specify the order of finish in two or more races. The expected returns from these bets must be taken into account when calculating bets for the single race pools in those races.

In the author's experience, betting the full amount recommended by the Kelly formula is unwise for a number of reasons. Firstly, accurate estimation of the advantage of the bets is critical; if one overestimates the advantage by more than a factor of two, Kelly betting will cause a negative rate of capital growth. (As a practical matter, many factors may cause one's real-time advantage to be less than past simulations would suggest, and very few can cause it to be greater. Overestimating the advantage by a factor of two is easily done in practice.) Secondly, if it is known that regular withdrawals from the betting bankroll will be made for paying expenses or taking profits, then one's effective wealth is less than their actual current wealth. Thirdly, full Kelly betting is a 'rough ride', downswings during which more than 50% of total wealth is lost are a common occurrence. For these and other reasons, a *fractional Kelly* betting strategy is advisable, that is, a strategy wherein one bets some fraction of the recommended Kelly bet (e.g. 1/2 or 1/3). For further discussion of fractional Kelly betting, and a quantitative analysis of the risk/reward tradeoffs involved, see MacLean, Ziemba and Blazenko (1992).

Another even more important constraint on betting is the effect that one's bet has on the advantage. In pari-mutuel betting markets each bet decreases the dividend. Even if the bettor possesses infinite wealth, there is a maximum bet producing the greatest expected profit, any amount beyond which lowers the expected profit. The maximum bet can be calculated by writing the equation for expected profit as a function of bet size, and solving for the bet size which maximizes expected profit. This maximum can be surprisingly low as the following example illustrates.

c	div	er
06	20	1.20

total pool size = $100,000
maximum *er* bet = $416
expected profit = $39.60

A further consideration concerns the shape of the 'expected profit versus bet size' curve when the bet size is approaching the maximum. In this example, the maximum expected profit is with a bet of $416. If one made a bet of only 2/3 the maximum, i.e. $277, the expected profit would be 35.5 dollars, or 90% of the maximum. There is very little additional gain for risking a much larger sum of money. Solving the fully formulated Kelly model (i.e. taking into account the bets' effects on the dividends) will optimally balance this tradeoff. See Kallberg and Ziemba (1994) for a discussion of the optimization properties of such formulations.

As a practical matter; given the relatively small sizes of most pari-mutuel pools, a successful betting operation will soon find that all of its bets are *pool-size-limited*. As a rule of thumb, as the bettor's wealth approaches the total pool size, the dominant factor limiting bet size becomes the effect of the bet on the dividend, not the bettor's wealth.

Exotic bets

In addition to win bets, racetracks offer numerous so-called *exotic* bets. These offer some of the highest advantage wagering opportunities. This results from the multiplicative effect on overall advantage of combining more than one advantage horse. For example, suppose that in a particular race there are two horses for which the model's estimate of the win probability is greater than the public's, though not enough so as to make them positive expectation win bets.

	c	div	p	er
1)	.115	8.3	.100	.955
2)	.060	16.6	.050	.996

By the Harville formula (Harville 1973), the estimated probability of a 1,2 or 2,1 finish is

$$C_{12,21} = (.115 * .060)/(1 - .115) + (.060 * .115)/(1 - .060) = .0151 .$$

The public's implied probability estimate is

$$P_{12,21} = (.100 * .050)/(1 - .100) + (.050 * .100)/(1 - .050) = .0108 .$$

Therefore (assuming a 17% track take) the public's rational quinella dividend should be

$$qdiv \cong (1 - .17)/.0108 = 76.85 .$$

Assuming that the estimated probability is correct the expected return of a bet on this combination is

$$er = .0151 * 76.85 = 1.16 .$$

In the above example two horses which had expected returns of less than 1 as individual win bets, in combination produce a 16% advantage quinella bet. The same principle applies, only more so, for bets in which one must specify the finishing positions of more than two horses. In *ultra-exotic* bets such as the pick-six, even a handicapping model with only modest predictive ability can produce advantage bets. The situation may be roughly summarized by stating that for a bettor in possession of accurate probability estimates which differ from the public estimates; 'the more *exotic* (i.e. specific) the bet, the higher the advantage'. Place and show bets are not considered exotic in this sense as they are less specific than normal bets. The probability differences are 'watered down' in the place and show pools.[2] Some professional players make only exotic wagers to capitalize on this effect.

First, Second, and Third

In exotic bets that involve specifying the finishing order of two or more horses in one race, a method is needed to estimate these probabilities. A popular approach is the Harville formula. (Harville, 1973):

For three horses (i, j, k) with win probabilities (π_i, π_j, π_k) the Harville formula specifies the probability that they will finish in order as

$$\pi_{ijk} = \frac{\pi_i \pi_j \pi_k}{(1 - \pi_i)(1 - \pi_i - \pi_j)} . \tag{6}$$

This formula is significantly biased, and should not be used for betting purposes, as it will lead to serious errors in probability estimations if not corrected for in some way.[3] (Henery 1981, Stern 1990, Lo and Bacon-Shone 1992). Its principle deficiency is the fact that it does not recognize the increasing randomness of the contests for second and third place. The bias in the Harville formula is demonstrated in Tables 9 and 10 which show the formula's estimated probabilities for horses to finish second and third given that the identity of the horses finishing first (and second) are known. The data set used is the same as that which produced Table 1 above.

Table 9

HARVILLE MODEL CONDITIONAL PROBABILITY OF 2ND

range	n	exp.	act.	Z
.000-.010	962	.007	.010	0.9
.010-.025	3449	.018	.030	5.3
.025-.050	5253	.037	.045	2.8
.050-.100	7682	.073	.080	2.3
.100-.150	4957	.123	.132	1.9
.150-.200	3023	.173	.161	-1.8
.200-.250	1834	.223	.195	-3.0
.250-.300	1113	.272	.243	-2.3
.300-.400	1011	.338	.317	-1.4
>.400	395	.476	.372	-4.3

races = 3198, # horses = 29679

Table 10

HARVILLE MODEL CONDITIONAL PROBABILITY OF 3RD

range	n	exp.	act.	Z
.000-.010	660	.007	.009	0.5
.010-.025	2680	.018	.033	4.3
.025-.050	4347	.037	.062	6.8
.050-.100	6646	.073	.087	4.0
.100-.150	4325	.123	.136	2.5
.150-.200	2923	.173	.178	0.7
.200-.250	1831	.223	.192	-3.4
.250-.300	1249	.273	.213	-4.9
.300-.400	1219	.341	.273	-5.3
>.400	601	.492	.333	-8.3

races = 3198, # horses = 26481

The large values of the Z-statistics show the significance of the bias in the Harville formula. The tendency is for low probability horses to finish second and third more often than predicted, and for high probability horses to finish second and third less often. The effect is more pronounced for 3rd place than for 2nd. An effective, and computationally economical way to correct for this is as follows:

Given the win probability array, $(\pi_{i \ (i=1,2,...N)})$, create a second array σ such that,

$$\sigma_i = \frac{\exp(\gamma \log(\pi_i))}{\sum \exp(\gamma \log(\pi_j))} \qquad (j = 1,2,...N) \qquad (7)$$

and a third array τ such that,

$$\tau_i = \frac{\exp(\delta \log(\pi_i))}{\sum \exp(\delta \log(\pi_j))} . \qquad (j = 1,2,...N) \qquad (8)$$

The probability of the three horses (i,j,k) finishing in order is then

$$\pi_{ijk} = \frac{\pi_i \, \sigma_j \, \tau_k}{(1 - \sigma_i)(1 - \tau_i - \tau_j)} . \qquad (9)$$

The parameters γ and δ can be estimated via maximum likelihood estimation on a sample of past races. For the above data set the maximum likelihood values of the parameters are $\gamma = .81$ and $\delta = .65$. Reproducing Tables 9 and 10 above using equations (7-9) with these parameter values substantially corrects for the Harville formula bias as can be seen in Tables 11 and 12.

Table 11

LOGISTIC MODEL CONDITIONAL PROBABILITY OF
2ND (γ = .81)

range	n	exp.	act.	Z
.000-.010	251	.008	.012	0.6
.010-.025	2282	.018	.024	1.9
.025-.050	5195	.037	.033	-1.6
.050-.100	8819	.074	.073	-0.4
.100-.150	6054	.123	.125	0.5
.150-.200	3388	.173	.176	0.5
.200-.250	1927	.222	.216	-0.6
.250-.300	973	.272	.275	0.2
.300-.400	616	.336	.349	0.7
>.400	174	.456	.397	-1.6

races = 3198,   # horses = 29679

Table 12

LOGISTIC MODEL CONDITIONAL PROBABILITY OF
3RD (δ = .65)

range	n	exp.	act.	Z
.000-.010	4	.009	.000	-0.2
.010-.025	712	.020	.010	-2.7
.025-.050	3525	.039	.035	-1.3
.050-.100	8272	.075	.073	-0.7
.100-.150	6379	.123	.130	1.7
.150-.200	3860	.172	.175	0.5
.200-.250	2075	.222	.228	0.7
.250-.300	921	.271	.268	-0.2
.300-.400	582	.337	.299	-2.0
>.400	151	.480	.450	-0.7

races = 3198,   # horses = 26481

The better fit provided by this model can be readily seen from the much smaller discrepancies between expected and actual frequencies. The parameter values used here should not be considered to be universal constants, as other authors have derived significantly different values for the parameters γ and δ using data from different racetracks (Lo, Bacon-Shone and Busche, 1994).

FEASIBILITY

A computer based handicapping and betting system could in princple be developed and implemented at most of the world's racetracks. Today's portable computers have sufficient capacity not only for real-time calculation of the bets, but for model development as well. However, several important factors should be considered in selecting a target venue, as potential profitability varies considerably among racetracks. The following are a few practical recommendations based on the author's experience.

Data availability

A reliable source of historical data must be available for developing the model and test samples. The track must have been in existence long enough, running races under conditions similar to today, in order to develop reliable predictions. Data availability in computer form is of great help, as data entry and checking are extremely time-consuming. The same data used in model development must also be available for real-time computer entry sufficient time before the start of each race. Additionally, final betting odds must be available over the development sample for the 'combined' model estimation of equation (1) as well as for wagering simulations.

Ease of operation

Having an accurate estimate of the final odds is imperative for betting purposes. Profitability will suffer greatly if the final odds are much different than the ones used to calculate the probabilities and bet sizes. The ideal venue is one which allows off-track telephone betting, and disseminates the odds electronically. This enables the handicapper to bet from the convenience of an office, and eliminates the need to take a portable computer to the track and type in the odds from the tote board at the last minute. Even given ideal circumstances, a professional effort will require several participants. Data entry and verification, general systems programming, and ongoing model development all require full-time efforts, as well as the day-to-day tasks of running a small business. Startup capital requirements are large, (mainly for research and development) unless the participants forgo salaries during the development phase.

Beatability of the opposition

Pari-mutuel wagering is a competion amongst participants in a highly negative sum game. Whether a sufficiently effective model can be developed depends on the predictability of the racing, and the level of skill of fellow bettors. If the races are largely dishonest, and the public odds are dominated by inside information then it is unlikely that a fundamental model will perform well. Even if the racing is honest, if the general public skill level is high, or if some well financed minority is skillful, then the relative advantage obtainable will be less. Particularly unfavorable is the presence of other computer handicappers. Even independently developed computer models will probably have a high correlation with each other and thus will be lowering the dividends on the same horses, reducing the profitability for all. Unfortunately, it is difficult to know how great an edge can be achieved at a particular track until one develops a model for that track and tests it, which requires considerable effort. Should that prove successful, there is still no guarantee that the future will be as profitable as past simulations might indicate. The public may become more skillful, or the dishonesty of the races may increase, or another computer handicapper may start playing at the same time.

Pool size limitations

Perhaps the most serious and inescapable limitation on profitability is a result of the finite amount of money in the betting pools. The high track take means that only the most extreme public probability mis-estimations will result in profitable betting opportunities, and the maximum bet size imposed by the bets' effects on the dividends limits the amount that can be wagered. Simulations by the author have indicated that a realistic estimate of the maximum expected profit achievable, as a percentage of total per-race turnover, is in the range of 0.25 - 0.5 per cent. This is for the case of a player with an effectively infinite bankroll. It may be true that at tracks with small pool sizes, that this percentage is higher due to the lack of sophistication of the public, but in any case, it is unlikely that this value could exceed 1.5 per cent. A more realistic goal for a start-up operation with a bankroll equal to approximately one half of the per-race turnover might be to win between 0.1 and 0.2 per cent of the total track turnover. The unfortunate implication of this is that at small volume tracks one could probably not make enough money for the operation to be viable.

Racetracks with small betting volumes also tend to have highly volatile betting odds. In order to have time to calculate and place one's wagers it is necessary to use the public odds available a few minutes before post time. The inaccuracy involved in using these volatile pre-post-time odds will decrease the effectiveness of the model.

RESULTS

The author has conducted a betting operation in Hong Kong following the principles outlined above for the past five years. Approximately five man-years of effort were necessary to organize the database and develop a handicapping model which showed a significant advantage. An additional five man-years were necessary to develop the operation to a high level of profitability. Under near-ideal circumstances, ongoing operations still require the full time effort of several persons.

A sample of approximately 2000 races (with complete past performance records for each entrant) was initially used for model development and testing. Improvements to the model were made on a continuing basis, as were regular re-estimations of the model which incorporated the additional data accumulated. A conservative fractional Kelly betting strategy was employed throughout, with wagers being placed on all positive expectation bets available in both normal and exotic pools (except place and show bets).[4] Extremely large pool sizes, (> USD $10,000,000 per race turnover) made for low volatility odds, therefore bets could be placed with accurate estimations of the final public odds. Bets were made on all available races except for races containing only *unratable* horses (~5%), resulting in approximately 470 races bet per year. The average track take was ~19% during this period.

Four of the five seasons resulted in net profits, the loss incurred during the losing season being approximately 20% of starting capital. A strong upward trend in rate of return has been observed as improvements were made to the handicapping model. Returns in the various betting pools have correlated well with theory, with the rate-of-return in exotic pools being generally higher than that in simple pools. While a precise calculation has not been made, the statistical significance of this result is evident. Following is a graph of the natural logarithm of [(wealth) / (initial wealth)] versus races bet.

RESULTS

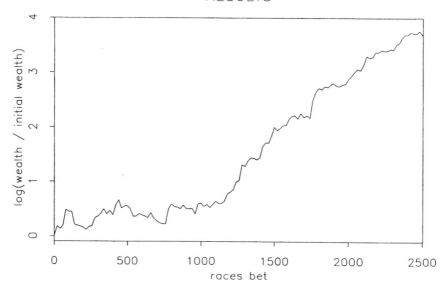

CONCLUSION

The question; "Can a system beat the races?" can surely be answered in the affirmative. The author's experience has shown that at least at some times, at some tracks, a statistically derived fundamental handicapping model can achieve a significant positive expectation. It will always remain an empirical question whether the racing at a particular track at a particular time can be beaten with such a system. It is the author's conviction that we are now experiencing the *golden age* for such systems. Advances in computer technology have only recently made portable and affordable the processing power necessary to implement such a model. In the future, computer handicappers may become more numerous, or one of the racing publications may start publishing competent computer ratings of the horses, either of which will likely cause the market to become efficient to such predictions. The profits have gone, and will go, to those who are 'in action' first with sophisticated models*+.

*An earlier version of this paper was presented at the ORSA/TIMS Joint National Meeting in Phoenix, Arizona on November 1, 1993.

+The author wishes to thank Professor George Miel (University of Nevada, Las Vegas), Paul Coladonato, Randall G. Chapman, and the editors of this volume for many helpful comments, suggestions, and corrections.

NOTES

[1]One technique to alleviate the negative consequences of biases which lead to the over-estimation of advantage is to employ a betting rule which specifies a minimum estimated advantage necessary for making a bet. In Ziemba and Hausch (1987) the authors suggest a mimimum advantage of 10% to account for the bias in their place and show betting model. Also, in their model the authors use place and show probabilities so the often present favorite-longshot win bias tends to cancel with the second and third place reverse bias. For simple probability estimations these schemes can work well, but in exotic bets whose probabilities are the products several individual win probabilities, the calculation of the correct minimum advantage becomes exceedingly complex. The author advocates the practice of correcting the probabilities first and then calculating the betting advantage.

[2]A similar calculation to the one carried out in the quinella pool example above shows that a horse with a positive expected return in the win pool will have a lower expected return as a place or show bet, given that the public bets consistently in the different pools. This effect is different than the one which produced advantages in the place and show pools for Ziemba and Hausch (1987). There the advantages arose because of inconsistencies between the public's estimated win probability for a horse, and the amount bet on that horses in the place or show pools.

[3]The bias in this formula is not as serious when used with win probabilities that show a significant favorite-longshot bias. The favorite-longshot bias often observed at racetracks (Ali, 1977) tends to cancel out the Harville formula bias in estimating second and third place probabilities.

[4]Betting off-track, the author did not have access to real-time show pool betting information. (Place betting in the North American sense is not available in Hong Kong.) Without individual horse show pool betting information, one can always achieve higher advantages by betting in 'exotic' pools such as quinella and trifecta. This follows from the above cited principle of 'the more specific the bet, the higher the advantage'. (See Note 2 above)

APPENDIX

HANDICAPPING REFERENCES*

Ainslie, Tom, *Ainslie's Complete Guide to Thoroughbred Handicapping*, (New York: Simon & Schuster, 1979)

Beyer, Andrew, *Picking Winners*, (Boston, MA: Houghton Mifflin Company, 1975)

Jones, Glendon, *Horse Racing Logic*, (New York: Vantage Press, 1989)

Quinn, J., *The ABC's of Thoroughbred Handicapping*, (New York: William Morrow and Company, 1988)

Quirin, William L., *Winning at the Races: Computer Discoveries in Thoroughbred Handicapping*, (New York: William Morrow and Company, 1979)

Scott, Don, *The Winning Way*, (Sydney: Puntwin PTY Limited, 1982)

*The following is a partial list of references which the author has found helpful in suggesting ideas for significant *factors*. A useful source for difficult to find books on handicapping is 'The Gambler's Book Club' in Las Vegas, Nevada.

REFERENCES

Ali, M., "Probability and Utility Estimates for Racetrack Betting," *Journal of Political Economy*, 85 (1977), 803-815.

Asch, P., R.E. Quandt, *Racetrack Betting: The Professors' Guide to Strategies*, (Dover, MA: Auburn House, 1986)

Bolton, Ruth N. and Randall G. Chapman, "Searching for Positive Returns at the Track: A Multinomial Logit Model for Handicapping Horse Races," *Management Science*, Vol. 32, No. 8, August (1986), 1040-1059.

Brecher, Stephen L., *Beating the Races with a Computer*, (Long Beach, CA: Software Supply, 1980)

Epstein, Richard A., *The Theory of Gambling and Statistical Logic, revised edition*, (New York, NY: Academic Press, 1977)

Figlewski, Stephen, "Subjective Information and Market Efficiency in a Betting Market," *Journal of Political Economy*, Vol. 87, No. 1, (1979), 75-88.

Harville, D.A., "Assigning Probabilities to the Outcomes of Multi-entry Competitions," *Journal of the American Statistical Association*, 68 June (1973), 312-316.

Henery, R.J., "Permutation Probabilities as Models for Horse Races,"*Journal of the Royal Statistical Society B* 43, No. 1, (1981), 86-91.

Kallberg, J.G. and W.T. Ziemba, "Pari-mutuel Betting Models," *in this volume* (1994).

Kelly, J., "A New Interpretation of Information Rate," *Bell System Technical Journal.*, 35 (1956), 917-926.

Lo, Victor S.Y. and John Bacon-Shone, "Approximating the Ordering Probabilities of Multi-entry Competions by a Simple Method," *working paper, Department of Statistics, University of Hong Kong,* (1992).

Lo, Victor S.Y., John Bacon-Shone and Kelly Busche "The Application of Ranking Probability Models to Racetrack Betting," *Management Science*, forthcoming (1994).

MacLean, L.C., W.T. Ziemba and G. Blazenko, "Growth Versus Security in Dynamic Investment Analysis," *Management Science*, Vol. 38, No. 11, November (1992), 1562-1585.

Stern, Hal, "Models for Distributions on Permutations," *Journal of the American Statistical Association*, 85, No. 410 June (1990), 558-564.

White, E.M., Ronald Dattero and Benito Flores, "Combining Vector Forecasts to Predict Thoroughbred Horse Race Outcomes," *International Journal of Forecasting* 8 (1992), 595-611.

Ziemba, William T. and Donald B. Hausch, *Betting at the Racetrack*, (Los Angeles: Dr. Z Investments, Inc., 1986)

Ziemba, William T. and Donald B. Hausch, *Dr. Z's Beat the Racetrack, revised edition*, (New York: William Morrow and Company, Inc., 1987)

An Empirical Cross-validation of Alternative Classification Strategies Applied to Harness Racing Data for Win Bets[1]

Larry H. Ludlow
Boston College

ABSTRACT

This paper presents the results of a two year cross-validation of a fundamental handicapping system. Harness race performance data from a single season's entire racing meet were subjected to a discriminant analysis and classification criteria were developed. The Year 1 discriminant function and classification criteria were applied to Year 2 data. The classification techniques are evaluated in terms of percent correct classification and return on investment.

INTRODUCTION

This study was designed to cross-validate the relative efficacy of six alternative classification techniques. In the present study the term discrimination refers to the statistical process of deriving a fundamental harness race handicapping system capable of differentiating between winners and non-winners. The term classification refers to the subsequent application of the initial rules to a second sample of races.

The data consist of performance observations of winning and losing horses racing in two seasons of harness meets (referred to as Year 1 and Year 2). It is not assumed that those two particular meets provided unique data that could not have occurred elsewhere or during a different racing season. The discrimination problem focused on whether a linear function existed that would yield significant separation between winners and non-winners based on the Year 1 data. The classification problem took the Year 1 discriminant function and classification criteria and cross-validated them upon the Year 2 data.

---

[1]Appreciation for assistance and advice is extended to Peter Tommila, Nicholas Bond, Kenneth Krueger, William Ziemba, and Donald Hausch. Correspondence should be addressed to Larry H. Ludlow, Associate Professor, Boston College, School of Education, Chestnut Hill, MA 02167-3813.

METHOD

Data

The data were obtained from two Golden Bear Raceway harness meets in Sacramento, CA. These meets ran from May through July. All races were run over a one mile distance. There were usually eight or nine horses per race.

Not all races were included in each year's analysis. First, races were included beginning at that point in the season when most of the horses had run at least three races at the Sacramento track. This was to reduce error variation due to different track conditions elsewhere that could distort the time and performace variables, i.e., Sacramento has a particulary dry and warm racing season. Second, only "claiming" races in the $1000-$24,000 range were analyzed. This was because stakes races would often include shippers brought to the track for that one race. Also, "maiden" races and conditions such as "non-winners of last three races" typically included horses with inconsistent racing histories. Since there were relatively few poor-weather racing days no effort was made to remove them from analysis.

The Year 1 data cover 255 races (N=1904 horses) and the Year 2 data cover 227 races (N=1757). Each horse was scored as a winner or a loser and the payoff for the winner was recorded in each race. The place and show finishes were not considered in this analysis.

The raw data were obtained from the Daily Racing Form. The variables and their scoring procedure were:

- a) **Stretch**-the average position at the stretch point on the track that the horse held over the last three races,
- b) **Finish**-the average finish position over the last three races,
- c) **Avet**-the average time over the last three races,
- d) **Bestt**-the best time recorded over the last three races,
- e) **Driver**-the harness driver's previous year ranking based on win, place, and show finishes (computed by the Daily Racing Form),
- f) **Pole**-the current pole position raced from,
- g) **Day**-day of the week the race was held,
- h) **Season**-time of the race season, and
- i) **Class**-claiming condition of the race.

The Stretch, Finish, and Avet data for each horse in each race were ordered and horses were assigned a rank measure ranging from best (1) to worst (as high as 9) on each variable. The Daily Racing Form driver rankings were recoded to either 1 (best), 2, or 3 (worst). Current pole positions were recoded to 1=(position 1 or 2), 2=(position 3,4,5 or 6), or 3=(position 7 or 8). This particular recoding strategy was based on the previous year's results and the Year 1 breakdowns of percent wins at post positions. Admittedly, it does not fully reflect the fact the post positions 4 and 5 may at some tracks rival position 1 for most wins.

Day of the week was coded 1 through 5 corresponding to Tuesday through Saturday, respectively. Season was coded 1 through 3 corresponding to the first, second, and third parts of the season. Class was coded as the actual claiming condition of the race.

The original continuous level data available in the Daily Racing Form (e.g. best time) were recoded to the ordinal level in order to correspond to an on-track fundamental handicapping system developed by Kenneth Krueger (personal communication). In this system one can quickly determine and sum ranks. The horse with the lowest total rank score then becomes the best bet. It is not unreasonable to assume that results based on ordinal data might by enhanced if the data were analyzed at their original continuous level.

RESULTS

Discrimination

A two-group stepwise discriminant analysis was performed (Tatsuoka, 1971). The performance data served as the predictors, winner or loser status served as the criterion. The stepwise analysis employed Rao's V (Rao, 1948) as the criterion for determining the significance of each variable's contribution to the separation of the group centroids. It is a generalized distance measure with an approximate chi-square distribution. It is used to determine if a variable is to be included in the equation and in what order of inclusion it will appear, regardless of the variable's effect on the within-group homogeneity of variance.

This initial analysis indicated that Stretch and Bestt were linear functions of Finish and Avet, respectively, and were dropped from further analysis. Although the Day, Season, and Class variables did not contribute statistically significant unique variance to the final solutions, their standardized coefficients were nevertheless extremely interesting. For each of these variables the coefficents indicated it was easier to pick a winner earlier in the week (before the better races on the weekend); easier to pick a winner earlier in the racing season (before horses and drivers had attained greater experience and better teams arrived from other tracks); and easier to pick a winner at the lower levels of claiming conditions (where the better horse and driver stood out from the lesser quality teams).

Furthermore, the structure coefficients (correlations of the variables to the discriminant function scores) revealed that the entire set of variables could be broken into three subsets. These subsets (in order of importance) were: a) Finish, Avet, Stretch, and Bestt; b) Driver and Pole; and c) Class, Day, and Season. The first subset consists of past performance variables. The second subset consists of current racing variables. The third subset consists of track variables.

Table 1
Descriptive Statistics for Year 1

	Means	
	Winners (n=255)	Losers (n=1649)
Finish	2.94	4.47
Avet	3.12	4.44
Driver	2.10	2.45
Pole	1.78	1.92

Variance - Covariance Matrix: Winners

	Finish	Avet	Driver	Pole
Finish	3.95			
Avet	1.44	3.85		
Driver	.32	.26	.86	
Pole	.11	.00	.02	.38

Variance - Covariance Matrix: Losers

	Finish	Avet	Driver	Pole
Finish	4.50			
Avet	1.72	4.69		
Driver	.22	.18	.72	
Pole	.06	.04	.06	.43

Table 1 presents descriptive statistics for the final set of statistically significant variables. As expected the winning horses had lower mean rankings than the losing horses, i.e., they tended to have better past finishes and times, and better current drivers and pole positions. They also had less variability in their covariance matrix-although Box's M test of within group covariance homogeneity (Box, 1949) was not significant (M=14.82, p>.10). This means that, given their current driver and pole position, they were more consistent in their racing performance.

Tests of multivariate normality, conducted by assessing nonnormality in the marginal distributions (Gnanadesikan, 1977), revealed that winner's distributions tended to be slightly positively skewed (some high scoring, poorly rated horses did win), while the loser's distributions tended to be slightly negatively skewed (some low scoring, highly rated horses did lose). Finally, note that the sample sizes are considerably different-a fact of considerable importance when the classifications are attempted.

Table 2
Discriminant Function Results for Year 1

	Standardized Coefficients		Unstandardized Coefficients	
Finish	.63		.29	
Avet	.44		.20	
Driver	.32		.37	
Pole	.18		.28	
		Constant	-3.53	

Centroids

Winners -.76   Losers   .12

Test Statistics

D^2	Wilks' λ	F(4,1900)	p
.791	.9159	43.62	<.01

The final discriminant analysis results are reported in Table 2. The separation of the two group centroids (Mahalanobis's D^2) is statistically significant. Although only 8.3% of the variance has been accounted for (1-Wilks' λ), this result is consistent with the 9% reported by Bolton & Chapman (1986) even though they employed a somewhat different statistical model in their attempt to build a fundamental system with positive rates of return. Under the Standardized Coefficient column the variables are ranked according to their relative contribution to the discriminant function. This ordering reveals that a horse's average last three finish positions was the best predictor of its next race performance. Next in order of importance were the average of the last three times and current driver variables. The current Pole position it raced from is least important of these four variables but it still accounts for significant variation.

The discriminant function based on the unstandardized coefficients and constant term yields the discriminant function score for each horse (hereafter denoted as y_i). As such, the resulting equation takes the following form:

$$y_i = -3.53 + .29 * Finish + .20 * Avet + .37 * Driver + .28 * Pole.$$

As an illustration the discriminant function scores are computed for one horse and driver team with the best rankings and for a team with the worst rankings. For a team with ranks of 1,1,1,1 on those four variables the resulting $y_i = -2.39$. For a team with the ranks of 9,9,3,3 the discriminant function score is $y_i = +2.83$.

Classification

Although a classification of the Year 1 data based on the Year 1 results will reflect an inflated success rate it is worthwhile because it serves as a baseline of comparison for evaluating the cross-validation results.

Bock (1975) points out that unless the calibration sample is large relative to the number of predictor variables conditional errors of misclassification will be biased downward when one employs the generalized distance function D^2. This occurs because the discriminant function maximizes the relative distance between the mean discriminant scores in the sample. The bias in error rates arises from the same source as the bias in computing a sample multiple correlation coefficient. Bock's adjustment takes the following form:

$$D^{2*} = D^2 \left(\frac{N_l + N_w - p}{N_l + N_w - 1} \right) - \frac{(p-1)(N_l + N_w)(N_l + N_w - 2)}{(N_l - N_w - 1)(N_l * N_w)},$$

where p designates the number of variables in the equation, and N_l and N_w correspond to the number of losing and winning horses, respectively. Application of his adjusted $D2^*$ formula yields $D2^* = .76$ which when compared to our original $D^2 = .79$ suggests that the Year 1 results are quite stable for prediction and classification purposes.

The following techniques were used to classify both sets of data based on the Year 1 discriminant function results. The first three techniques establish cut points along the discriminant function score continuum. The next computes distances from group centroids. The final two are probability based techniques. They were chosen because they represent different levels of ease of application and different levels of model building and testing. As classification models they differ primarily in the extent to which they utilize sample size and separate or pooled variance-covariance information. Lachenbruch (1975) provides a fuller explanation of how some of these techniques relate to one another.

1. **"Maximum likelihood"** (Overall & Klett, Eq. 9.11, 1972)

$$Y_c = \frac{\overline{X}_l + \overline{X}_w}{2} = \frac{.11825 + (-.76467)}{2} = -.32321$$

where Y_c is the cutoff score between the $\overline{X}_l$ losing group centroid and the $\overline{X}_w$ winning group centroid. The $\overline{X}_g$ centroid is defined as $D'C^{-1}B_g$ where D is a vector of mean differences between the two groups on the four variables, C^{-1} is the inverse of the pooled within groups covariance matrix, B_g is the vector of group g means on the four variables, and the group index is g=l (losers), w (winners).
Classification Rule: If y_i (the discriminant score for horse i) < -.32321, classify as a winner, otherwise classify as a loser.

2. **"Conditional Probability"** (Overall & Klett, Eq. 9.16, 1972)

$$Y_c = \frac{\overline{X}_l + \overline{X}_w}{2} + \ln p_w - \ln p_l = -1.7095$$

where $p_g = {n_g}/{N}$, (for these data $p_w = .2$, $p_l = .8$, based on population estimates of their frequency). The adjustment for sample size generally has the effect of shifting the cutoff score away from the group with the larger n_g. Since the losing group has the larger n and a positive centroid, the new cutoff becomes more negative-in effect moving to the left of the winner's centroid.
Classification Rule: If y_i < -1.7095, classify as a winner, otherwise classify as a loser.

3. **"Equal Density"** (Rulon, et al, p.218, 1967)
The univariate distribution of discriminant scores for group g with standard deviation s_g and centroid $\overline{X}_g$ may be represented as:

$$f_g(y_i) = \frac{1}{s_g \sqrt{2\pi}} \exp\left[\frac{-\frac{1}{2}\left(y_i - \overline{X}_g\right)^2}{s_g} \right].$$

The cutoff score is the point at which the density of the winner's distribution equals the density of the loser's distribution. This intersection is found by solving for the "boundary point" x_b in the following expression:

$$\left(\frac{1}{s_l^2} - \frac{1}{s_w^2}\right)x_b^2 - 2\left(\frac{\overline{X}_l}{s_l^2} - \frac{\overline{X}_w}{s_w^2}\right)x_b + \left(\frac{\overline{X}_l^2}{s_l^2} - \frac{\overline{X}_w^2}{s_w^2}\right) + 2\ln\frac{p_w s_l}{p_l s_w}$$

This "boundary point" was found through the following calculation:

a)

$$\left(\frac{1}{1.009} - \frac{1}{.94}\right)x_b^2 - 2\left(\frac{.11825}{1.009} - \frac{-.76467}{.94}\right)x_b + \left(\frac{.11825^2}{1.009} - \frac{-.76467^2}{.94}\right) + 2\ln\left(\frac{.2 * 1.005}{.8 * .969}\right) = 0$$

b) $-.0727 x_b^2 + 1.86 x_b - 3.3 = 0$

and now solving the quadratic equation for x, we have

c)

$$x = \frac{-b \pm \sqrt{b^2 - 4ac}}{2a} = \frac{-1.86 \pm \sqrt{1.86^2 - 4(-.073)(-3.3)}}{2(-.073)} = -1.925 \, or - 23.55$$

Since -23.55 is beyond the range of the discriminant function scores the only permissible root is -1.925.

Classification Rule: If $y_i < -1.925$, classify as a winner, otherwise classify as a loser.

4. "Minimum Chi-square" (Tatsuoka, Eq. 8.4, 1971)

$$\chi_{ig}^2 = X'C^{-1}X + \ln|C_g| - 2\ln p_g$$

where $|C_g|$ is the determinant of the gth group covariance matrix, X is the vector of raw score deviations from the group g means for horse i, and χ_{ig}^2 represents a generalized distance measure for horse i relative to group g. The adjustment using separate group covariance matrices has the effect of reducing the distance from a horse to the centroid of the group with the lesser covariation.

Classification Rule: Classify as a member of the group that yields the smaller χ_{ig}^2.

5. "Partial Bayes" (Day & Kerridge, Eq. 2, 1967)

$$p(H_{ig}|X) = \exp(X'B + O) / (1 + \exp(X'B + O))$$

where $p(H_{ig}|X)$ is the probability of horse H_i belonging to group g given the vector of $H_i's$ raw scores X, $B = C^{-1}(\overline{X}_l - \overline{X}_w)$ is the pooled within group covariance matrix post multiplied by the vector of mean differences between the two groups on the four variables, and

$$O = -\frac{1}{2}(\overline{X}_l - \overline{X}_w)C^{-1}(\overline{X}_l + \overline{X}_w) + \ln\frac{p_l}{p_w}.$$

Classification Rule: Classify as a winner if $p < .5$.

6. "Full Bayes" (Bock, Eq. 6.2-4, 1975)

$$B = \ln \frac{p_l}{p_w} + \frac{1}{2} \ln \left| \frac{C_l}{C_w} \right| - \frac{1}{2} \left[\left(X_i - \overline{X}_l \right)' C^{-1} \left(X_i - \overline{X}_l \right) - \left(X_i - \overline{X}_w \right)' C^{-1} \left(X_i - \overline{X}_w \right) \right].$$

This technique makes use of all sample information: distance from each centroid taking into account separate group means, variances, covariances, and sample sizes.

Classification Rule: Classify as a winner if $B < 0$.

Evaluation

The determination of percent correct classification can be defined in at least three ways. First, the total percent correct classifications across both groups can be computed but that statistic is irrelevant here because all winners could be classified as losers and the overall percent correct classification would still be 87.1. Likewise we are not interested in the success of correctly classifing each of the 255 winners since this forces us to pick a winner for each race, which may not be an efficient strategy. Rather we want the classification success rate for just those horses picked as winners who actually were winners.

The result of applying each classification technique is shown in Table 3. This table shows, in two ways, the number of correctly classified winning horses and the number of horses that were picked to be winners. The "Multiple Bets" column refers to the fact that any given method could predict more than one winner per race. Thus for each race and across all races the total number of horses meeting the respective classification criteria have been tallied. This situation, although possibly one form of betting strategy, is generally inefficient. Consequently the results under the "Single Bet" column refer to taking that horse with the largest discriminant score whenever there was a "Multiple Bet" occurrence.

The techniques differ both in the number of races in which a winner is picked and in their success rates. The "Maximum Likelihood" technique yielded a predicted winner for each race. Its classification success rate was slightly better then the track-lore standard of betting the favorite and expecting a 33% success rate. In contrast the "Equal Density" technique predicted a winner for only 50 of the 255 races but its success rate was an impressive 54%.

Table 3
Classification Results for Year 1

Technique	Percent Correctly Identified		Rate of Return on Single Bets
	Multiple Bets	Single Bets	
Maximum Likelihood	179/750= 23.9 *	98/255 = 38.4	12.9**
Conditional probability	45/ 89 = 50.6	44/ 86 = 51.2	19.1
Equal Density	28/ 54 = 51.9	27/ 50 = 54.0	9.0
Minimum chi-square	44/ 99 = 44.4	41/ 94 = 43.6	3.0
Partial Bayes	32/ 62 = 51.6	32/ 61 = 52.5	8.4
Full Bayes	26/ 50 = 52.0	26/ 50 = 52.0	9.6

*correct picks/total picks

**(total payoff - total wagered at $2.00 per wager)/total wagered

Table 3 gives the rate of return for these various classification procedures (taking into account track take and breakage). The strategy one might follow depends upon the extent of betting activity desired and the size of the return anticipated. The most active strategy ("Maximum Likelihood") does not yield the highest rate of return. The best success rate strategy ("Equal Density"), too, does not yield the best rate of return. Why, however, did the "Maximum Likelihood" technique provide as high of a return as it did? Simply because it yielded a larger average payoff ($5.88) per race due to more long-odds horses winning. It is clear that the success rate for correctly classifying winners is not necessarily the only criteria for evaluating the success of a classification technique.

Cross-validation

Unlike Bolton & Chapman (1986) who test their fundamental system by focusing on obtaining more efficient parameter estimates with an "explosion process" strategy, our purpose is to determine if the parameter estimates prove stable by significantly differentiating between winners and losers in an independent second season's sample of races. Accordingly, the Year 2 races were classified using the Year 1 criteria.

Table 4
Descriptive Statistics for Year 2

	Means	
	Winners (n=255)	Losers (n=1649)
Finish	3.25	4.53
Avet	3.53	4.49
Driver	1.58	2.02
Pole	1.87	2.01

Variance - Covariance Matrix: Winners

	Finish	Avet	Driver	Pole
Finish	4.28			
Avet	1.49	4.50		
Driver	.29	.16	.61	
Pole	-.06	.05	-.06	.46

Variance - Covariance Matrix: Losers

	Finish	Avet	Driver	Pole
Finish	4.79			
Avet	1.80	4.92		
Driver	.22	.11	.80	
Pole	-.03	-.19	-.02	.50

The descriptive statistics for the Year 2 data are given in Table 4. The pattern of winner and loser mean rankings follows the Year 1 data but note the difference between the Year 1 and Year 2 means on the Driver variable. Better drivers won relatively more races in the Year 2 data. Note also the negative covariances in the Year 2 winner's matrix. Relatively more races were won from the outside pole positions and they were won by better drivers. Like the Year 1 data Box's M was not significant but the general pattern of variation was still less for the winners than for the losing horses. The same pattern of positive skew for the winners and negative skew for the losers was seen in the marginal distributions.

Prior to applying the Year 1 classification criteria to the Year 2 data, an identical two-group stepwise discriminant analysis was performed for the sake of assessing whether the resulting discriminant function would resemble the Year 1 function. These results are presented in Table 5. The Year 2 centroids are closer than the Year 1 centroids but the most noticeable differences are in the relative magnitudes of the standardized coefficients. In particular, the Driver and Pole variables play a more important role in the Year 2 results. This means that applying Year 1 criteria to classify Year 2 data cannot result in as high of a percent correct classification rate as originally found although rates of return may change because of different payoff conditions.

Table 5
Discriminant Function Results for Year 2

	Standardized Coefficients		Unstandardized Coefficients
Finish .56			.26
Avet .33			.15
Driver .53			.45
Pole .32			.28
		Constant	-3.88

Centroids

Winners -.70   Losers   .10

Test Statistics

D^2	Wilks' λ	$F(4,1752)$	p
.653	.931	32.46	<.01

The classification success rates and return rates are reported in Table 6. Given that it is known that the relative importance of some of the variables changed from Year 1 to Year 2 it is noteworthy to observe that the Year 1 discriminant function and classification criteria still perform quite adequately. In accordance with the earlier analysis it is seen that the methods differ in the number of winners picked, number of winners correctly classified, and the rate of return. The "Full Bayes" method is particularly interesting because while it had a relatively low percent correct classification rate (based on the Year 1 criteria) it had the highest rate of return and the highest average payoff per win ($6.64).

A classification of the Year 2 data using Year 2 discriminant analysis results would, as in the case of the Year 1 data, have been more impressive. Particularly since the effect of having better drivers racing with good horses from the outside pole positions would have been taken into account. For the purpose of creating a "better" discriminant function one would combine the data from both years, rerun the discriminant analyses, recompute the classification criteria, and apply them to a third year's data. It is not inconceivable to consider that direct analysis of Daily Racing Form computer tapes could further validate, and even fine-tune, the results reported here.

Table 6
Classification Results for Year 2
(based on Year 1 criteria)

Technique	Percent Correctly Identified		Rate of Return on Single Bets
	Multiple Bets	Single Bets	
Maximum Likelihood	160/736 = 21.7 *	67/227 = 29.5	-.01 **
Conditional probability	39/99 = 39.4	38/95 = 40.0	13.6
Equal Density	18/44 = 40.9	17/43 = 39.5	13.5
Minimum chi-square	39/112 = 34.8	37/104 = 35.6	4.4
Partial Bayes	21/52 = 40.4	20/51 = 39.2	8.6
Full Bayes	11/32 = 34.4	11/32 = 34.4	14.1

*correct picks/total picks

**(total payoff - total wagered at $2.00 per wager)/total wagered

DISCUSSION

These results serve a variety of purposes. They suggest that under certain conditions a quantitative fundamental handicapping system may be successfully applied to harness racing. They demonstrate that alternative classification techniques yield different results and, more importantly, the overall determination of classification success is dependent upon the manner in which success is defined. Furthermore, one advantage of a fundamental system, such as the one presented here, may be that true probabilities of winning might be more precisely estimated. That is, given the assumption that a horse and driver's true winning probability is independent of the betting public's perception of their probability of success (as reflected in the pari-mutuel odds), it seems reasonable to assume that true win probabilities should remain unchanged regardless of how odds (and the probabilities computed from them) fluctuate prior to a race. However, it remains to be researched whether or not win probabilities computed from any of the classification strategies discussed in this paper offer an advantage over typical win probabilities (win pool/total pool) employed when computing expected values under an efficiency-of-market approach (Ziemba & Hausch, 1984) or when computing how much to bet under an optimal capital growth model approach (Hausch & Ziemba, 1985).

REFERENCES

Bock, R.D. (1975). *Multivariate Statistical Methods in Behavioral Research*. McGraw-Hill.

Bolton, R.N. & Chapman, R.G. (1986). Searching for positive returns at the track: A multinomial logit model for handicapping horse races. *Management Science, 32*, 1040-1060.

Box, G.E.O. (1949). A general distribution theory for a class of likelihood criteria. *Biometrika, 36*, 317-346.

Day, N.E. & Kerridge, O. (1967). A general maximum likelihood discriminant. *Biometrics, 23*, 313-323.

Gnanadesikan, R. (1977). *Methods for Statistical Analysis of Multivariate Observations*. Wiley.

Hausch, D.B. & Ziemba, W.T. (1985). Transactions costs, extent of inefficiencies, entries and multiple wagers in a racetrack betting model. *Management Science, 31*, 381-394.

Lachenbruch, P.A. (1975). *Discriminant Analysis*. Hefner Press.

Overall, J.E. & Klett, C.J. (1972). *Applied Multivariate Analysis*. McGraw-Hill.

Rao, C.R. (1948). Tests of significance in multivariate analysis. *Biometrika, 35*, 58-79.

Rulon, P.J., Tiedeman, D.V., Tatsuoka, M.M. & Langmuir, C.R. (1967). *Multivariate Statistics for Personnel Classification*. Wiley.

Tatsuoka, M.M. (1971). *Multivariate Analysis*. Wiley.

Ziemba, W.T. & Hausch, D.B. (1984). *Beat the Racetrack.*. Harcourt, Brace and Jovanovich.

Assigning Probabilities to the Outcomes of Multi-Entry Competitions

DAVID A. HARVILLE*

The problem discussed is one of assessing the probabilities of the various possible orders of finish of a horse race or, more generally, of assigning probabilities to the various possible outcomes of any multi-entry competition. An assumption is introduced that makes it possible to obtain the probability associated with any complete outcome in terms of only the 'win' probabilities. The results were applied to data from 335 thoroughbred horse races, where the win probabilities were taken to be those determined by the public through pari-mutuel betting.

1. INTRODUCTION

A horse player wishes to make a bet on a given horse race at a track having pari-mutuel betting. He has determined each horse's 'probability' of winning. He can bet any one of the entires to win, place (first or second), or show (first, second, or third). His payoff on a successful place or show bet depends on which of the other horses also place or show. Our horse player wishes to make a single bet that maximizes his expected return. He finds that not only does he need to know each horse's probability of winning, but that, for every pair of horses, he must also know the probability that both will place, and, for every three, he must know the probability that all three will show. Our better is unhappy. He feels that he has done a good job of determining the horses' probabilities of winning; however he must now assign probabilities to a much larger number of events. Moreover, he finds that the place and show probabilities are more difficult to assess. Our better looks for an escape from his dilemma. He feels that the probability of two given horses both placing or of three given horses all showing should be related to their probabilities of winning. He asks his friend, the statistician, to produce a formula giving the place and show probabilities in terms of the win probabilities.

The problem posed by the better is typical of a class of problems that share the following characteristics:

1. The members of some group are to be ranked in order from first possibly to last, according to the outcome of some random phenomena, or the ranking of the members has already been effected, but is unobservable.
2. The 'probability' of each member's ranking first is known or can be assessed.
3. From these probabilities alone, we wish to determine the probability that a more complete ranking of the members

* David A. Harville is research mathematical statistician, Aerospace Research Laboratories, Wright-Patterson Air Force Base, Ohio 45433. The author wishes to thank the Theory and Methods editor, an associate editor and a referee for their useful suggestions.

will equal a given ranking or the probability that it will fall in a given collection of such rankings.

Dead heats or ties will be assumed to have zero probability. For situations where this assumption is unrealistic, the probabilities of the various possible ties must be assessed separately.

We assign no particular interpretation to the 'probability' of a given ranking or collection of rankings. We assume only that the probabilities of these events satisfy the usual axioms. Their interpretation will differ with the setting.

Ordinarily, knowledge of the probabilities associated with the various rankings will be of most interest in situations like the horse player's where only the ranking itself, and not the closeness of the ranking, is important. The horse player's return on any bet is completely determined by the horses' order of finish. The closeness of the result may affect his nerves but not his pocketbook.

2. RESULTS

We will identify the n horses in the race or members in the group by the labels $1, 2, \cdots, n$. Denote by $p_k[i_1, i_2, \cdots, i_k]$ the probability that horses or members $i_1, i_2, \cdots, i_k$ finish or rank first, second, $\cdots$, kth, respectively, where $k \leq n$. For convenience, we use $p[i]$ interchangeably with $p_1[i]$ to represent the probability that horse or member i finishes or ranks first. We wish to obtain $p_k[i_1, i_2, \cdots, i_k]$ in terms of $p[1], p[2], \cdots, p[n]$, for all $i_1, i_2, \cdots, i_k$ and for $k = 2, 3, \cdots, n$. In a sense, our task is one of expressing the probabilities of elementary events in terms of the probabilities of more complex events.

Obviously, we must make additional assumptions to obtain the desired formula. Our choice is to assume that, for all $i_1, i_2, \cdots, i_k$ and for $k = 2, 3, \cdots, n$, the conditional probability that member i_k ranks ahead of members $i_{k+1}, i_{k+2}, \cdots, i_n$ given that members $i_1, i_2, \cdots, i_{k-1}$ rank first, second, $\cdots$, $(k-1)$th, respectively, equals the conditional probability that i_k ranks ahead of $i_{k+1}, i_{k+2}, \cdots, i_n$ given that $i_1, i_2, \cdots, i_{k-1}$ do not rank first. That is,

$$\frac{p_k[i_1, i_2, \cdots, i_k]}{p_{k-1}[i_1, i_2, \cdots, i_{k-1}]} \equiv \frac{p[i_k]}{q_{k-1}[i_1, i_2, \cdots, i_{k-1}]}, \quad (2.1)$$

213

where

$$q_k[i_1, i_2, \cdots, i_k] \equiv 1 - p[i_1] - p[i_2] - \cdots - p[i_k],$$

so that, for the sought-after formula, we obtain

$$p_k[i_1, i_2, \cdots, i_k]$$

$$\equiv \frac{p[i_1]p[i_2]\cdots p[i_k]}{q_1[i_1]q_2[i_1, i_2]\cdots q_{k-1}[i_1, i_2, \cdots, i_{k-1}]}. \quad (2.2)$$

In the particular case $k = 2$, the assumption (2.1) is equivalent to assuming that the event that member i_2 ranks ahead of all other members, save possibly i_1, is stochastically independent of the event that member i_1 ranks first.

The intuitive meaning and the reasonableness of the assumption (2.1) will depend on the setting. In particular, our horse player would probably not consider the assumption appropriate for every race he encounters. For example, in harness racing, if a horse breaks stride, the driver must take him to the outside portion of the track and keep him there until the horse regains the proper gait. Much ground can be lost in this maneuver. In evaluating a harness race in which there is a horse that is an 'almost certain' winner unless he breaks, the bettor would not want to base his calculations on assumption (2.1). For such a horse, there may be no such thing as an intermediate finish. He wins when he doesn't break, but finishes 'way back' when he does.

In many, though not all, cases, there is a variate (other than rank) associated with each member of the group such that the ranking is strictly determined by ordering their values. For example, associated with each horse is its running time for the race. Denote by X_i the variate corresponding to member i, $i = 1, 2, \cdots, n$. Clearly, the assumption (2.1) can be phrased in terms of the joint probability distribution of $X_1, X_2, \cdots, X_n$. It seems natural to ask whether there exist other conditions on the distribution of the X_i's which imply (2.1) or which follow from it, and which thus would aid our intuition in grasping the implications of that assumption. The answer in general seems to be no. In particular, it can easily be demonstrated by constructing a counterexample that stochastic independence of the X_i's does not in itself imply (2.1). Nor is the converse necessarily true. In fact, in many situations where assumption (2.1) might seem appropriate, it is known that the X_i's are not independent. For example, we would expect the running times of the horses to be correlated in most any horse race. An even better example is the ordering of n baseball teams according to their winning percentages over a season of play. These percentages are obviously not independent, yet assumption (2.1) might still seem reasonable.

The probability that the ranking belongs to any given collection of rankings can be readily obtained in terms of $p[1], p[2], \cdots, p[n]$ by using (2.2) to express the probability of each ranking in the collection in terms of the $p[i]$'s, and by then adding. For example, the horse player can compute the probability that both entry i and

entry j place from

$$p_2[i, j] + p_2[j, i] = \frac{p[i]p[j]}{1 - p[i]} + \frac{p[j]p[i]}{1 - p[j]}.$$

A probability of particular interest in many situations is the probability that entry or member r finishes or ranks kth or better, for which we write

$$p_k^*[r] = \sum p_k[i_1, i_2, \cdots, i_k], \quad (2.3)$$

where the summation is over all rankings $i_1, i_2, \cdots, i_k$ for which $i_u = r$ for some u. If assumption (2.1) holds, then $p_k^*[r] > p_k^*[s]$ if and only if $p[r] > p[s]$. This statement can be proved easily by comparing the terms of the right side of (2.3) with the terms of the corresponding expression for $p_k^*[s]$. Each term of (2.3), whose indices are such that $i_u = r$ and $i_v = s$ for some u, v, appears also in the second expression. Thus, it suffices to show that any term $p_k[i_1, i_2, \cdots, i_k]$, for which $i_j \neq s$, $j = 1$, $2, \cdots, k$, but $i_u = r$ for some u, is made smaller by putting $i_u = s$ if and only if $p[r] > p[s]$. That the latter assertion is true follows immediately from (2.2).

3. APPLICATION

In pari-mutuel betting, the payoffs on win bets are determined by subtracting from the win pool (the total amount bet to win by all bettors on all horses) the combined state and track take (a fixed percentage of the pool—generally about 16 percent, but varying from state to state), and by then distributing the remainder among the successful bettors in proportion to the amounts of their bets. (Actually, the payoffs are slightly smaller because of 'breakage,' a gimmick whereby the return on each dollar is reduced to a point where it can be expressed in terms of dimes.) In this section, we take the 'win probability' on each of the n horses to be in inverse proportion to what a successful win bet would pay per dollar, so that every win bet has the same 'expected return.' Note that these 'probabilities' are established by the bettors themselves and, in some sense, represent a consensus opinion as to each horse's chances of winning the race. We shall suppose that, in any sequence of races in which the number of entries and the consensus probabilities are the same from race to race, the horses going off at a given consensus probability win with a long-run frequency equal to that probability. The basis for this supposition is that, once the betting on a race has begun, the amounts bet to win on the horses are flashed on the 'tote' board for all to see and this information is updated periodically, so that, if at some point during the course of the betting the current consensus probabilities do not coincide with the bettors' experience as to the long-run win frequencies for 'similar' races, these discrepancies will be noticed and certain of the bettors will place win bets that have the effect of reducing or eliminating them.

By adopting assumption (2.1) and applying the results of the previous section, we can compute the long-run frequencies with which any given order of finish is encountered over any sequence of races having the same number

1. APPLICATION OF THEORETICAL RESULTS TO THIRD RACE OF SEPTEMBER 6, 1971, AT RIVER DOWNS RACE TRACK

Name	Amounts bet to win, place, and show as percentages of totals			Theoretical probability			Expected payoff per dollar	
	Win	Place	Show	Win	Place	Show	Place bet	Show bet
Moonlander	27.6	20.0	22.3	.275	.504	.688	1.11	1.01
E'Thon	16.5	14.2	11.1	.165	.332	.499	.94	1.06
Golden Secret	3.5	4.7	6.3	.035	.076	.126	.58	.42
Antidote	17.3	18.8	20.0	.175	.350	.521	.80	.80
Beviambo	4.0	6.2	7.8	.040	.087	.144	.51	.41
Cedar Wing	11.9	10.4	10.4	.118	.245	.382	.90	.86
Little Flitter	8.5	11.2	9.9	.085	.180	.288	.62	.68
Hot and Humid	10.7	14.4	12.2	.107	.224	.353	.62	.72

of entries and the same consensus win probabilities. In particular, we can compute the 'probability' that any three given horses in a race finish first, second, and third, respectively. As we shall now see, these probabilities are of something more than academic interest, since they are the ones needed to compute the 'expected payoff' for each place bet (a bet that a particular horse will finish either first or second) and each show bet (a bet that the horse will finish no worse than third).

Like the amounts bet to win, the amounts bet on each horse to place and to show are made available on the 'tote' board as the betting proceeds. The payoff per dollar on a successful place (show) bet consists of the original dollar plus an amount determined by subtracting from the final place (show) pool the combined state and track take and the total amounts bet to place (show) on the first two (three) finishers, and by then dividing a half (third) of the remainder by the total amount bet to place (show) on the horse in question. (Here again, the actual payoffs are reduced by breakage.) By using the probabilities computed on the basis of assumption (2.1) and the assumption that consensus win probabilities equal appropriate long-run frequencies, we can compute the expected payoff per dollar for a given place or show bet on any particular race, where the expectation is taken over a sequence of races exhibiting the same number of entries and the same pattern of win, place, and show betting. If, as the termination of betting on a given race approaches, any of the place or show bets are found to have potential expected payoffs greater than one, there is a possibility that a bettor, by making such place and show bets, can 'beat the races'. Of course, if either assumption (2.1) or the assumption that the consensus win probabilities equal long-run win frequencies for races with similar betting patterns is inappropriate, then this system will not work. It will also fail if there tend to be large last-minute adverse changes in the betting pattern, either because of the system player's own bets or because of the bets of others. However, at a track with considerable betting volume, it is not likely that such changes would be so frequent as to constitute a major stumbling block.

In Table 1, we exemplify our results by applying them to a particular race, the third race of the September 6, 1971, program at River Downs Race Track. The final win, place, and show pools were $45,071, $16,037, and $9,740, respectively. The percentage of each betting pool bet on each horse can be obtained from the table. The table also gives, for each horse, the consensus win probability, the overall probabilities of placing and showing, and the expected payoffs per dollar of place and show bets. The race was won by E'Thon who, on a per-dollar basis, paid $5.00, $3.00, and $2.50 to win, place, and show, respectively; Cedar Wing was second, paying $3.80 and $2.70 per dollar to place and show; and Beviambo finished third, returning $3.20 for each dollar bet to show.

In order to check assumption (2.1) and the assumption that the consensus win probabilities coincide with the long-run win frequencies over any sequence of races having the same number of entries and a similar betting pattern, data was gathered on 335 thoroughbred races from several Ohio and Kentucky race tracks. Data from races with finishes that involved dead heats for one or more of the first three positions were not used. Also, in the pari-mutuel system, two or more horses are sometimes lumped together and treated as a single entity for betting purposes. Probabilities and expectations for the remaining horses were computed as though these 'field' entries consisted of single horses and were included in the data, though these figures are only approximations to the 'true' figures. However, the field entries themselves were not included in the tabulations.

As one check on the correspondence between consensus win probabilities and the long-run win frequencies over races with similar patterns of win betting, the horses were divided into eleven classes according to their consensus win probabilities. Table 2 gives, for each class, the associated interval of consensus win probabilities, the average consensus win probability, the actual frequency

2. FREQUENCY OF WINNING—ACTUAL VS. THEORETICAL

Theoretical probability of winning	Number of horses	Average theoretical probability	Actual frequency of winning	Estimated standard error
.00 - .05	946	.028	.020	.005
.05 - .10	763	.074	.064	.009
.10 - .15	463	.124	.127	.016
.15 - .20	313	.175	.169	.021
.20 - .25	192	.225	.240	.031
.25 - .30	114	.272	.289	.042
.30 - .35	71	.324	.394	.058
.35 - .40	49	.373	.306	.066
.40 - .45	25	.423	.640	.096
.45 - .50	12	.464	.583	.142
.50 +	10	.554	.700	.145

3. FREQUENCY OF FINISHING SECOND— ACTUAL VS. THEORETICAL

Theoretical probability of finishing second	Number of horses	Average theoretical probability	Actual frequency of finishing second	Estimated standard error
.00 – .05	776	.030	.046	.008
.05 – .10	750	.074	.095	.011
.10 – .15	548	.124	.128	.014
.15 – .20	426	.175	.155	.018
.20 – .25	283	.223	.170	.022
.25 – .30	164	.269	.226	.033
.30 +	11	.311	.364	.145

of winners, and an estimate of the standard error associated with the actual frequency. The actual frequencies seem to agree remarkably well with the theoretical probabilities, though there seems to be a slight tendency on the part of the betters to overrate the chances of longshots and to underestimate the chances of the favorites and near-favorites. Similar results, based on an extensive amount of data from an earlier time period and from different tracks, were obtained by Fabricand [1].

Several checks were also run on the appropriateness of assumption (2.1). These consisted of first partitioning the horses according to some criterion involving the theoretical probabilities of second and third place finishes and then comparing the actual frequency with the average theoretical long-run frequency for each class. Tables 3–6 give the results when the criterion is the probability of finishing second, finishing third, placing, or showing, respectively. In general, the observed frequencies of second and third place finishes are in reasonable accord with the theoretical long-run frequencies, though there seems to be something of a tendency to overestimate the chances of a second or third place f nish for horses with high theoretical probabilities of such finishes and to underestimate the chances of those with low theoretical probabilities, with the tendency being more pronounced for third place finishes than for second place finishes. A logical explanation for the

4. FREQUENCY OF FINISHING THIRD— ACTUAL VS. THEORETICAL

Theoretical probability of finishing third	Number of horses	Average theoretical probability	Actual frequency of finishing third	Estimated standard error
.00 – .05	587	.032	.049	.009
.05 – .10	713	.074	.105	.011
.10 – .15	691	.124	.126	.013
.15 – .20	838	.175	.147	.012
.20 – .25	115	.212	.130	.031
.25 +	14	.273	.214	.110

5. FREQUENCY OF PLACING—ACTUAL VS. THEORETICAL

Theoretical probability of placing	Number of horses	Average theoretical probability	Actual frequency of placing	Estimated standard error
.00 – .05	330	.034	.036	.010
.05 – .10	526	.074	.091	.013
.10 – .15	404	.125	.121	.016
.15 – .20	358	.174	.179	.020
.20 – .25	268	.224	.257	.027
.25 – .30	240	.274	.271	.029
.30 – .35	193	.326	.306	.033
.35 – .40	175	.375	.354	.036
.40 – .45	117	.425	.359	.044
.45 – .50	109	.472	.440	.048
.50 – .55	73	.525	.425	.058
.55 – .60	51	.578	.667	.066
.60 – .65	48	.623	.625	.070
.65 – .70	29	.673	.621	.090
.70 – .75	22	.724	.909	.095
.75 +	15	.808	.867	.088

conformity of the actual place results to those predicted by the theory which is evident in Table 5 is that those horses with high (low) theoretical probabilities of finishing second generally also have high (low) theoretical

6. FREQUENCY OF SHOWING—ACTUAL VS. THEORETICAL

Theoretical probability of showing	Number of horses	Average theoretical probability	Actual frequency of showing	Estimated standard error
.00 – .05	111	.038	.045	.020
.05 – .10	316	.075	.092	.016
.10 – .15	328	.124	.180	.021
.15 – .20	266	.174	.222	.025
.20 – .25	253	.227	.257	.027
.25 – .30	243	.274	.284	.029
.30 – .35	201	.326	.303	.032
.35 – .40	196	.374	.439	.035
.40 – .45	169	.425	.426	.038
.45 – .50	150	.477	.460	.041
.50 – .55	158	.525	.468	.040
.55 – .60	137	.574	.474	.043
.60 – .65	97	.625	.577	.050
.65 – .70	100	.672	.500	.050
.70 – .75	67	.722	.627	.059
.75 – .80	67	.777	.731	.054
.80 – .85	49	.823	.816	.055
.85 – .90	30	.874	.867	.062
.90 +	20	.930	1.000	.056

7. PAYOFFS ON PLACE AND SHOW BETS— ACTUAL VS. THEORETICAL

Expected payoff per dollar	Number of different place and show bets	Average expected payoff per dollar	Average actual payoff per dollar	Estimated standard error
.00 – .25	80	.216	.088	.062
.25 – .35	214	.303	.286	.068
.35 – .45	386	.404	.609	.091
.45 – .55	628	.504	.570	.071
.55 – .65	904	.601	.730	.072
.65 – .75	980	.700	.660	.047
.75 – .85	958	.800	.947	.066
.85 – .95	819	.898	.938	.050
.95 – 1.05	546	.995	.983	.090
1.05 – 1.15	286	1.090	.989	.060
1.15 – 1.25	90	1.186	.974	.108
1.25 +	25	1.320	1.300	.258

probabilities of finishing first, so that the effects of the overestimation (underestimation) of their chances of finishing second are cancelled out by the underestimation (overestimation) of their chances of finishing first. While a similar phenomenon is operative in the show results, the cancellation is less complete and there seems to be a slight tendency to overestimate the show chances of those horses with high theoretical probabilities and to underestimate the chances of those with low theoretical probabilities.

Finally, the possible place and show bets were divided into classes according to the theoretical expected payoffs of the bets as determined from the final betting figures. The average actual payoff per dollar for each class can then be compared with the corresponding average expected payoff per dollar. The necessary figures are given in Table 7. The results seem to indicate that those place and show bets with high theoretical expected payoffs per dollar actually have expectations that are somewhat lower, giving further evidence that our assumptions are not entirely realistic, at least not for some races.

The existence of widely different expected payoffs for the various possible place and show bets implies that either the bettors 'do not feel that assumption (2.1) is entirely appropriate' or they 'believe in assumption (2.1)' but are unable to perceive its implications. Our results indicate that to some small extent the bettors are successful in recognizing situations where assumption (2.1) may not hold and in acting accordingly, but that big differences in the expected place and show payoffs result primarily from 'incorrect assessments' as to when assumption (2.1) is not appropriate or from 'ignorance as to the assumption's implications.'

A further implication of the results presented in Table 7 is that a bettor could not expect to do much better than break even by simply making place and show bets with expected payoffs greater than one.

[Received January 1972. Revised September 1972.]

REFERENCE

[1] Fabricand, Burton P., *Horse Sense*, New York: David McKay Company, Inc., 1965.

Permutation Probabilities as Models for Horse Races

By R. J. HENERY

University of Strathclyde, Scotland

SUMMARY

Some properties of models for the outcomes of races are described, these properties being consequences of a stochastic ordering of the permutations which define the outcomes of a race. Order statistics models which lead to stochastic ordering are also discussed— particular cases of these are the first-order model of Plackett (1975) and the normal model of Upton and Brook (1974). An approximation for the normal model is suggested.

Keywords: PROBABILITY; STOCHASTIC ORDER; PERMUTATIONS; ORDER STATISTICS; NORMAL; APPROXIMATION

1. INTRODUCTION

PLACKETT (1975) poses the problem of finding the probability that a horse is placed given the probabilities of a win for each of the horses in a race, and suggests the following solution. If the winning probabilities for the n horses are $p_1, p_2, ..., p_n$, then the probability that horses ijk finish 123 is

$$p_i \{p_j/(1 - p_i)\} \{p_k/(1 - p_i - p_j)\}.$$

For this solution it is easy to see that, if $p_i > p_j > p_k$, then the order ijk is more likely than jik which, in turn, is more likely than jki. In other words, the relative orders of ijk tend to be as suggested by their win probabilities. A curious, and perhaps unrealistic, feature of this solution is that the above probabilities do not depend on the number of horses in the race.

Another model which solves essentially the same problem was put forward by Upton and Brook (1974) to explain voting patterns. This model also possesses a set of preferred orders but is more complex to deal with since it requires the evaluation of a multivariate normal integral. Now all probabilities depend essentially on the number of candidates in the election and their probabilities of winning. However, we will propose an approximation in which the probability that candidates ijk finish in order 123 depends only on their respective probabilities of winning and the number of candidates.

The present paper tries to discuss the common features of these models at a general level. This leads to a stochastic ordering of permutations, i.e. the permutations of a given order become a partially ordered set.

2. STOCHASTIC ORDERING OF PERMUTATIONS

In considering possible outcomes of a race in which the runners may be ranked according to their probabilities of winning, it is plausible to postulate that the most likely outcomes are those with the runners finishing in or near their expected positions. So, for example, if we have n runners $R_1, R_2, ..., R_n$ in a race, and their respective winning probabilities are $p_1, p_2, ..., p_n$, where we take $p_1 > p_2 > ... > p_n$, then R_1 is expected to win; and it is plausible that R_n is most likely to be last and that the outcomes $\{R_1 R_2 ... R_n\}$ and $\{R_n ... R_2 R_1\}$ have the highest and lowest probabilities of all possible outcomes. It is also to be expected that, for two outcomes which differ only in the relative placings of two runners, the greater probability is associated with the outcome in which the given two runners are placed as we would expect from their win probabilities. Indeed these properties are possessed by the models of Plackett (1974) and of Upton and Brook (1974, 1975).

Since these properties reflect what one expects intuitively of a model for races we will mostly consider models with these properties. Before considering what class of models is in question, we define what may be described as the fundamental property of stochastic ordering over the outcomes or permutations.

Given two permutations D, E which differ only in the transposition of two numbers, and supposing D to be the permutation in which these two numbers are in natural order, we maintain that D has a greater probability than E and we write $D > E$. If we have a sequence of permutations $D, E, F, ..., T$ in which we proceed from one permutation to the next by transposing two numbers which were in natural order then $D > E > F > ... > T$. Not all permutations may be linked by such sequences so we say that $D > T$ if and only if there exists a sequence of "unnatural" transpositions taking D into T, and we say that D and T are stochastically ordered.

The concept is best illustrated by giving order relations for the set of permutations of order three:

$$\{1, 2, 3\} > \{2, 1, 3\} > \{2, 3, 1\} > \{3, 2, 1\}$$
$$\{1, 2, 3\} > \{1, 3, 2\} > \{3, 1, 2\} > \{3, 2, 1\}$$
$$\{2, 1, 3\} > \{3, 1, 2\}$$
$$\{1, 3, 2\} > \{2, 3, 1\}.$$

The above order relations, together with implied relations such as $\{2, 1, 3\} > \{3, 2, 1\}$, constitute a partial order. For examples of permutations which are not ordered we may take the subset of permutations with the same inversion number, e.g. $\{1, 3, 2\}$ and $\{2, 1, 3\}$ are not stochastically comparable.

As a geometric interpretation for order three we can identify the six vertices of a regular hexagon with the six permutations as shown in Fig. 1. A permutation which lies geometrically above another is stochastically greater. Permutations which are at the same height have the same inversion number (and so cannot be stochastically ordered).

In three dimensions we obtain a representation of the 24 permutations of order four by labelling the 24 vertices of the truncated octahedron (see Knuth, 1975, p. 13 for diagram) but now difference in height is no longer a guarantee of stochastic inequality—permutations are ordered only if there is a sequence joining them with all edges leading downwards from one to another. An edge is here an order relation as defined above.

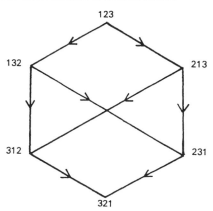

Fig. 1. Stochastic ordering of permutations of order 3.

2.1. *Consequences of Stochastic Order*

Let $p_j(k)$ be the probability that the *j*th integer occupies the *k*th position in the permutation of order n, with $1 \leqslant j \leqslant n$, $1 \leqslant k \leqslant n$. Then (i) $p_j(1)$ decreases strictly as *j* increases, (ii) $p_j(n)$ increases strictly as *j* increases, (iii) $p_1(k)$ decreases as *k* increases, (iv) $p_n(k)$ increases as *k* increases. All these properties have simple interpretations. For example, in gambling parlance property (ii) says that the horse that is most favoured to win is least favoured to be last.

We will show only how property (ii) follows from the stochastic ordering—similar arguments may be used for the others. Let us compare $p_i(n)$ and $p_j(n)$ with $i < j$. For each permutation in which *i* occupies last place we may transpose *i* and *j* to obtain a (more probable) permutation in which *j* is last. It follows immediately that $p_i(n) < p_j(n)$.

Since there are $n!$ probabilities for the permutations of order n, and these are subject only to the condition that they sum to unity, it is easily seen that properties (i) to (iv) do not imply stochastic order for $n > 2$.

3. ORDER STATISTICS MODELS

Let $X_1, X_2, ..., X_n$ be independent continuous random variables. If $p_i(1)$ is the probability that X_i is the smallest of $X_1, X_2, ..., X_n$ we assume $p_1(1) > p_2(1) > ... > p_n(1)$. The distribution function of X_i is $F(x; \theta_i)$. For example, in a race the X_i might be the times taken by the runners. The probability of the event $X_1 < X_2 < ... < X_n$ is taken as the probability of the permutation $\{1, 2, ..., n\}$ (Plackett, 1975). These models possess the important property that the relative order of any subgroup of the $X_1 ... X_n$ is independent of the relative order of the remaining X_j's. This property is an immediate consequence of the assumed independence of the X's. A special proof that exponential distributions possess this property is to be found in Upton and Brook (1975).

We will be concerned with those distributions which lead to a stochastic ordering over the permutations. The order statistics $X_{(1)}, X_{(2)}, ..., X_{(n)}$ will be written $Y_1, Y_2, ..., Y_n$. In finding the probability of a permutation $\pi = \{a_1, a_2, ..., a_n\}$ we require the probability that Y_1 comes from the a_1th distribution, Y_2 from the a_2th distribution and so on. Since we wish to compare the probabilities of two permutations π_1 and π_2, where π_2 is obtained by transposing two entries in π_1, we will consider the joint distribution of Y_j and Y_k conditional on $Y_1 = y_1$, $Y_2 = y_2, ..., Y_{j-1} = y_{j-1}, Y_{j+1} = y_{j+1}, ..., Y_{k-1} = y_{k-1}, Y_{k+1} = y_{k+1}, ..., Y_n = y_n$. For π_1 the joint probability density of Y_j and Y_k will be, assuming $y_{j-1} < y_j < y_{j+1} \leqslant y_{k-1} < y_k < y_{k+1}$,

$$\frac{f(y_j; \theta_{a_j}) f(y_k; \theta_{a_k})}{\{F(y_{j+1}; \theta_{a_{j+1}}) - F(y_{j-1}; \theta_{a_{j-1}})\} \{F(y_{k+1}; \theta_{a_{k+1}}) - F(y_{k-1}; \theta_{a_{k-1}})\}}$$

This will be recognized as equivalent to the statement that $F(Y_j; \theta_{a_j})$ and $F(Y_k; \theta_{a_k})$ are effectively uniform over their permissible ranges. For π_2 the expression for the joint density of Y_j and Y_k (conditional on all the others) is obtained by interchanging θ_{a_j} and θ_{a_k} in the above. We will ensure that $\Pr\{\pi_2\}$ is less than $\Pr\{\pi_1\}$ if we ensure that the corresponding density for π_2 is less than that for π_1.

The ratio of these two densities is

$$\frac{f(y_j; \theta_{a_j}) f(y_k; \theta_{a_k})}{f(y_k; \theta_{a_j}) f(y_j; \theta_{a_k})}$$

and we now ask for what distributions this ratio will be greater than unity for all $y_j < y_k$, provided only $a_j < a_k$. It will clearly be so if

$$\frac{f(y_j; \theta_{a_j})}{f(y_j; \theta_{a_k})}$$

is a monotonic decreasing function of y_j for $a_j < a_k$. Although the labelling of the θ_i is rather arbitrary, it will generally be the case that they are ordered, for example $\theta_1 < \theta_2 < ... < \theta_n$. Then we require $f(x; \theta_r)/(f(x; \theta_s)$ to be monotonic in x. Note the similarity with the concept of monotone likelihood ratio.

In passing we may also remark that, if independent random variables $X_1, X_2, ..., X_n$ have monotone density ratios with parameters $\theta_1 < \theta_2 < ... < \theta_n$, then the X_i are stochastically ordered in the sense that $F(x; \theta_j) \geqslant F(x; \theta_k)$ if $j < k$. By considering the Jacobian of the transformation $y = \phi(x)$, where $\phi(x)$ is a monotone increasing function, it is easily seen that $\phi(X_1), \phi(X_2), ..., \phi(X_n)$ have monotone density ratios if the X_i have, reflecting the fact that the events $\{X_1 < X_2 < ... < X_n\}$ and $\{\phi(X_1) < ... < \phi(X_n)\}$ are identical.

As examples of families of distributions which have monotone density ratios we have (a) the normal $N(\theta, \sigma^2)$ with σ^2 fixed, (b) the exponential with density $\theta^{-1} \exp(-x/\theta)$, and (c) the gamma $\Gamma(\alpha, \theta)$ or $\Gamma(\theta^{-1}, \beta)$ with α or β fixed. Of these, only in the normal case does a change in θ involve a simple shift of the distribution.

The exponential model leads to Plackett's (1975) model for permutation probabilities if we set $p_i = 1/\theta_i = \Pr\{X_i \text{ is smallest}\}$.

The normal model has been used by Upton and Brook (1974) in the analysis of voting patterns. It has the disadvantage that a multivariate normal integral is required for the probability of each permutation, although we will describe an approximation which may be used in some cases.

In horse races, especially for non-handicap races, the horses may be graded according to their inherent ability and so according to their probability of winning.

However, there are many factors involved and for the moment these models are meant to be taken qualitatively. Indeed the same is true for consumer preference and voting pattern studies where the suggested models are at best first-order approximations (Plackett, 1975).

3.1 Shifted Distributions

A natural way to generate distributions for the X_i is to make the distributions identical apart from a shift in the mean. For such variables to have monotone density ratios a necessary, but not sufficient, condition is that the distributions are unimodal. However, all shifted distribution models share properties (i) and (ii) as we now show.

Let $X_1, X_2, ..., X_n$ be independent with continuous distributions, identical apart from a simple shift, the means being $\theta_1, \theta_2, ..., \theta_n$ respectively. Then the probability that X_k is largest is

$$p_k(n) = \int_{-\infty}^{\infty} f(x - \theta_k) \prod_{i \neq k}^{n} F(x - \theta_i) \, dx$$

$$= \int_{-\infty}^{\infty} \frac{f(x)}{F(x)} \cdot \prod_{i=1}^{n} F(x + \theta_k - \theta_i) \, dx$$

and from the second form it is clear that $p_k(n)$ is (a) monotonic increasing in θ_k, (b) monotonic decreasing in θ_i for $i \neq k$, and (c) unaffected by adding the same constant to all the θ_i. Thus if $\theta_1 < \theta_2 < ... < \theta_n$ we have $p_1(n) < p_2(n) < ... < p_n(n)$, and it is similarly shown that the probabilities $p_k(1)$ that X_k is smallest must satisfy $p_1(1) > p_2(1) > ... > p_n(1)$.

As an example of an order statistics model which has properties (i) and (ii) but does not have stochastically ordered permutations, we construct a model which may be regarded as a shifted distribution model with bimodal density. Let X_i take the values i and $i+4$ with equal probability, for $1 \leqslant i \leqslant 4$. Then the permutations $\{2143\}$ and $\{4123\}$ correspond to the events $\{X_2 < X_1 < X_4 < X_3\}$ and $\{X_4 < X_1 < X_2 < X_3\}$ respectively; and these events have probabilities zero and $1/16$ respectively. However, $\{2143\}$ should be more probable than $\{4123\}$ for stochastic order of the permutations.

4. Approximate Probabilities for the Normal Case

Let $X_1, X_2, ..., X_n$ be independent normal random variables with means $\theta_1, \theta_2, ..., \theta_m$ and unit variances. The probability that $X_1 < X_2 < ... < X_n$ is

$$P\{\pi\} = \int_{-\infty}^{x} \phi(x_1 - \theta_1) \int_{x_1}^{x} \phi(x_2 - \theta_2) ... \int_{x_{n-1}}^{x} \phi(x_n - \theta_n) \, dx_1 \, dx_2 ... dx_n.$$

To find an approximation to $P\{\pi\}$ we shall develop $P\{\pi\}$ in a Taylor series in θ as far as the linear terms. This requires the evaluation of $P\{\pi\}$ and its first derivatives at $\theta_1 = \theta_2 = ... = \theta_n = 0$. The first of these quantities is obviously $1/n!$ and a typical derivative is

$$\frac{\partial P\{\pi\}}{\partial \theta_i} = \int_{-\infty}^{\infty} \phi(x_1) ... \int_{x_{i-1}}^{\infty} x_i \, \phi(x_i) \int_{x_i}^{\infty} \phi(x_{i+1}) ... \int_{x_{n-1}}^{\infty} \phi(x_n) \, dx_1 ... dx_n$$

$$= \frac{\mu_{i:n}}{n!},$$

where $\mu_{i:n}$ is the expected value of the ith order statistic in a random sample of n normal $N(0, 1)$ random variables. For small θ_i, we can approximate $P\{\pi\}$ by either

$$\frac{1}{n!} + \frac{\Sigma \theta_i \mu_{i:n}}{n!}$$

or, writing $\Phi(\xi) = 1/n!$ and $\alpha = 1/n! \, \phi(\xi)$,

$$\Phi\{\xi + \alpha \Sigma \theta_i \mu_{i:n}\}.$$

Of course the second expression has the merit of being always positive. Both approximations are unaffected if a constant is added to each of the θ_i (since $\Sigma \mu_{i:n} = 0$), so we take $\Sigma \theta_i = 0$.

For the probability that X_1 is smallest we sum $P\{\pi\}$ over all $(n-1)!$ permutations π' say in which θ_1 is in first place. Since parameters $\theta_2, ..., \theta_n$ will appear $(n-2)!$ times in each of the other positions we have

$$P\{X_1 \text{ is smallest}\} \approx \frac{1}{n} + \frac{1}{n!} \sum_{i, \pi'} \theta_{\pi_i'} \mu_{i:n}$$

$$= \frac{1}{n} + \frac{1}{n} \theta_1 \mu_{1:n} + \frac{(n-2)!}{n!} \sum_{j=2}^{n} \theta_j \cdot \sum_{i=2}^{n} \mu_{i:n}$$

$$= \frac{1}{n} + \frac{1}{n-1} \theta_1 \mu_{1:n}.$$

It should be noted that the above argument can be modified easily to cover the case of X_j being the kth smallest and we obtain

$$P\{X_j \text{ is } k\text{th smallest}\} \approx \frac{1}{n} + \frac{1}{n-1} \theta_j \mu_{k:n}.$$

Similarly for $P\{X_1 \text{ is smallest}, X_2 \text{ second smallest}, X_3 \text{ third smallest}\}$ which is approximately equal to

$$\frac{1}{n(n-1)(n-2)} \left\{ 1 + \sum_{1}^{3} \theta_i \mu_{i:n} + \frac{\sum_{1}^{3} \theta_i \sum_{1}^{3} \mu_{j:n}}{n-3} \right\}$$

where the last term in the brackets is to be taken as zero if $n = 3$.

To this order of approximation the probability $P_j(k)$ that the jth horse is placed kth may be estimated given only the probability W_j that horse j wins:

$$P_j(k) = \frac{1}{n} + \left(W_j - \frac{1}{n}\right)\frac{\mu_{k:n}}{\mu_{1:n}}.$$

However, these approximations are generally improved if we adopt the following procedure in which we assume that the win probabilities W_i are the given quantities rather than the means. Writing $z_i = \Phi^{-1}(W_i)$ and $z_0 = \Phi^{-1}(1/n)$, we approximate the mean θ_i by

$$\theta_i = \frac{(n-1)\,\phi(z_0)\,\{z_i - z_0\}}{\mu_{1:n}}.$$

For the probability that X_i is the kth smallest we use

$$\Phi\left\{z_0 + (z_i - z_0)\frac{\mu_{k:n}}{\mu_{1:n}}\right\}.$$

For the probability of a permutation we use the previously suggested approximation with the approximate values of θ_i. The approximations for other combinations of placings are obvious modifications of the above.

An approximation to $\mu_{i:n}$, due to Blom (1958) and discussed by David (1970), which is convenient and yet fairly accurate is

$$\mu_{i:n} = \Phi^{-1}\left(\frac{i - \frac{3}{8}}{n + \frac{3}{4}}\right).$$

4.1. *Example*

We may illustrate the above approximation using the data from Brook and Upton's (1974) normal model which gave win probabilities for three candidates as 0·3771, 0·3631 and 0·2598. Using the simpler approximation above, with the value $\mu_{1:3} = -0.84628$ from tables, we estimate the mean for the first as $-2(0\cdot3771 - 0\cdot3333)/0\cdot84628 = -0\cdot1034$, the other means being estimated as $-0\cdot0704$ and $+0\cdot1738$. For the permutations 123 and 321 we estimate probabilities of

$$\tfrac{1}{6} \pm \tfrac{1}{6}(0\cdot1034 + 0\cdot1738) \times 0\cdot84628,$$

i.e. 0·2058 and 0·1276.

Since the estimated means are quite small, the approximation is expected to be good. For comparison the means used by Upton and Brook were $-0\cdot1067$, $-0\cdot0762$, $+0\cdot1830$; and the exact probabilities for the permutations 123 and 321 were 0·2096 and 0·1283.

If the means θ are close to unity the more complicated approximation using the normal distribution is recommended.

Finally note that Plackett's model gives reasonable approximations to $\Pr\{123\}$ and $\Pr\{321\}$ in this and other cases, but the former probability is generally overestimated while the latter is underestimated.

REFERENCES

BLOM, G. (1958). *Statistical Estimates and Transformed Beta Variables.* New York: Wiley.
BROOK, D. and UPTON, G. J. G. (1974). Biases in Local Government elections due to position on the ballot paper. *Appl. Statist.*, **23**, 414–419.
DAVID, H. A. (1970). *Order Statistics.* New York: Wiley.
KNUTH, D. E. (1969). *The Art of Computer Programming, Vol. III*, Reading, Mass.: Addison-Wesley.
PLACKETT, R. L. (1975). The analysis of permutations. *Appl. Statist.*, **24**, 193–202.
UPTON, G. J. G. and BROOK, D. (1975). The determination of the optimum position on a ballot paper. *Appl. Statist.*, **24**, 279–287.

Estimating the Probabilities of the Outcomes of a Horse Race
(Alternatives to the Harville Formulas)

Hal S. Stern
Department of Statistics
Harvard University
Cambridge, Massachusetts 02138, USA

1. Introduction

Studies of the efficiency of the racetrack betting market usually examine the return on win bets. The conclusion of this work is that there are systematic biases in the win odds available at the track, however, these biases are insufficient to create profitable strategies using only win bets, except in rare instances with extreme favorites. To examine the return of place and show bets, estimates for the probabilities of the outcomes of a race are required. Hausch, Ziemba, and Rubinstein (1981) apply the model described by Harville (1973) to obtain such estimates. They determine that opportunities for profitable place and show betting exist. However, the Harville formulas have certain limitations describe by Hausch et al. A natural family of alternatives to the Harville model is described here. The formulas resulting from the alternative models give up the simplicity of the Harville formulas in return for slightly more accurate estimates. In the following section the racetrack betting market is described and studies of the efficiency of the market are summarized. Later sections describe the Harville formulas, alternatives to the Harville formulas, and a comparison of two different models on a small data set.

2. The racetrack betting market

In a typical race, a bettor can wager on each horse, either to win, place or show. A win bet has a positive return only if that particular horse finishes first in the race. Place bets realize a positive return if the horse finishes first or second, and show bets have a positive return if the horse finishes third or better. Of each dollar bet a fixed percentage Q, representing the track take, is not returned to winning bettors. The remaining money is divided among the winning bettors. The return on winning horses is determined by the amount of money bet on each horse. Let W_i be the amount wagered on horse i to win the race by all participants and $W = \sum_i W_i$ be the total amount bet on all horses. The return per dollar bet on horse i to win is

$$R_i = \begin{cases} \frac{(1-Q)W}{W_i} = 1 + \frac{(1-Q)W - W_i}{W_i} & \text{if } i \text{ finishes first} \\ 0 & \text{otherwise.} \end{cases}$$

The win odds reported at the race track are the last piece of this expression, $O_i = ((1-Q)W - W_i)/W_i$. Rules for determining the payoffs for place and show bets are similar. If P_j is the amount bet on horse j to place and $P = \sum_j P_j$ is the total dollar value of all place bets, then the return per dollar bet on horse j to place is

$$R_j = \begin{cases} 1 + \frac{(1-Q)P - P_i - P_j}{2P_j} & \text{if } i \text{ and } j \text{ are first and second in either order} \\ 0 & \text{otherwise.} \end{cases}$$

Note that the return on a place bet on horse j depends on which of the other horses finishes first or second. For show bets, if S_k is the amount bet on horse k to show and $S = \sum_k S_k$ is the total show pool then the return per dollar bet on horse k to show is

$$R_k = \begin{cases} 1 + \frac{(1-Q)S - S_i - S_j - S_k}{3S_k} & \text{if } i, j, k \text{ are top three finishers in any order} \\ 0 & \text{otherwise.} \end{cases}$$

This return depends on the amount bet on each of the horses that finish in the top three.

At the racetrack the payoff on each dollar bet is rounded down to the nearest nickel or dime. This is called breakage and has the effect of increasing the track take. The return for each dollar bet must be at least \$1.05. In cases where the computed return is less than this figure the track reduces the take. These "minus pools" normally occur if one horse is an extreme favorite. The effects of breakage and minus pools are not considered in this paper.

Recall that the odds on horse i to win are computed from the win pool as $O_i = [(1-Q)W - W_i]/W_i$. In the absence of breakage and minus pools, it follows that the proportion of the win pool which is bet on horse i is proportional to $1/(O_i + 1)$. If win betting at the racetrack is efficient then the proportion of the win pool bet on horse i is a good estimate of the probability that horse i wins the race. The racetrack efficiency hypothesis has been tested by many authors. A brief description of the results of such studies is provided here. More detailed summaries are provided by Hausch et al. (1981), Ziemba and Hausch (1987), and Asch and Quandt (1986).

The track take varies from racetrack to racetrack but averages 18%. Thus in an efficient market each dollar bet should have an expected return of 82 cents. Studies by Fabricand (1965), Snyder (1978), and Asch, Malkiel and Quandt (1982) demonstrate that bets on horse with odds less than two to one have an average return greater than this amount. Unfortunately, the average return is still less than one dollar. Bets on horse with extremely large odds have expected returns smaller than 82 cents. This indicates that too much money is wagered on horses that have a small probability of winning. This result is consistent with the observation that gamblers prefer low probability-high prize combinations to high probability-low prize combinations. The same phenomenon has been used to describe the popularity of government lotteries (Ziemba, Brumelle, Gautier, Schwartz, 1986).

3. The Harville Formulas

The empirical studies indicate that, despite the favorite/longshot bias, a reasonably good estimate of the probability that horse i wins a race from among n entrants is

$$q_i = \frac{W_i}{\sum_{j=1}^{n} W_j} = \frac{\frac{1}{O_i + 1}}{\sum_{j=1}^{n} \frac{1}{O_j + 1}}$$

when breakage is ignored. The q_i satisfy $q_i \geq 0$ and $\sum_i q_i = 1$. To determine whether profitable place and show bets exist, estimates for the probability that a horse finishes second or third are required. One set of estimates can be obtained directly from $q_1, \ldots, q_n$. If horse i wins the race, then the probability that horse j finishes second can be thought of as the probability that horse j wins a race among the $n - 1$ horses excluding horse i. This probability can be computed from the win pool by ignoring all bets on horse i. This leads to the conditional probability statement

$$\Pr(\, j \text{ is second given } i \text{ is first}) = \frac{W_j}{W - W_i} = \frac{q_j}{1 - q_i}.$$

Let $p(ij)$ represent the probability that horse i is first and horse j is second and let $p(ijk)$ represent the probability that horse i is first, horse j is second, and horse k is third. Then using the conditional probability result from above (and a similar result for the probability that k finishes third) leads to the Harville formulas

$$p(ij) = q_i \frac{q_j}{(1 - q_i)} \qquad \text{and} \qquad p(ijk) = q_i \frac{q_j}{(1 - q_i)} \frac{q_k}{(1 - q_i - q_j)}.$$

Harville (1973) was the first to discusses these formulas in detail in the horse racing context. The formulas are also discussed by Savage (1957), Plackett (1975), and Henery (1981).

The Harville formulas are widely used due to their simple and intuitive form. Hausch et al. (1981) use these formulas and the formulas for the payoff on place and show bets to demonstrate that some place and show bets have expected returns per dollar bet of more than one dollar. Many of these opportunities involve betting on a favorite to place or show. It seems that the betting public does not pay enough attention to the possibility that the favorite will finish second or third. Ziemba and Hausch (1987) have combined this inefficiency in the betting market with a money management scheme to create an effective betting system.

It is natural to ask whether these formulas are derived from realistic assumptions about the running times of horses. Harville (1973) argues for instance that the assumption of independent running times is not sufficient to derive these formulas. Although there do not appear to be any general conditions, like independence, that lead to the Harville formulas, there are specific probability models that do lead to the formulas. Let T_i represent the running time of the ith horse in a race. Suppose that T_i has the extreme value distribution with location parameter θ_i. The density of this distribution is

$$f_{\theta_i}(t_i) = e^{(t_i - \theta_i)} e^{-e^{(t_i - \theta_i)}} \qquad t_i \in (-\infty, \infty).$$

The parameter θ_i is called the location parameter because changing θ_i shifts the distribution to the right of left. Assume that $T_1, \ldots, T_n$ are independent random variables, having the extreme value distribution with location parameters $\theta_1, \ldots, \theta_n$. Then the probability that horse i wins the race is

$$
\begin{aligned}
p(i) &= \Pr(\, T_i = \min(T_1, \ldots, T_n)) \\
&= \int_{-\infty}^{\infty} \int_{t_i}^{\infty} \cdots \int_{t_i}^{\infty} \int_{t_i}^{\infty} \cdots \int_{t_i}^{\infty} \prod_{j=1}^{n} e^{(t_j - \theta_j)} e^{-e^{(t_j - \theta_j)}} \, dt_n \ldots dt_{i+1} dt_{i-1} \ldots dt_1 dt_i \\
&= e^{-\theta_i} \Big/ \sum_{j=1}^{n} e^{-\theta_j}.
\end{aligned}
$$

Similarly, the probability that i is first and j is second is found to be

$$p(ij) = \Pr(\, T_i < T_j < \min_{k \neq i,j} T_k) = \frac{e^{-\theta_i}}{\displaystyle\sum_{k=1}^{n} e^{-\theta_k}} \; \frac{e^{-\theta_j}}{\displaystyle\sum_{\substack{m=1 \\ m \neq i}}^{n} e^{-\theta_m}}.$$

This formula is precisely the Harville formula with $q_i = e^{-\theta_i} / \sum_{k=1}^{n} e^{-\theta_k}$. When there are $n = 2$ horses this result is closely related to the Bradley-Terry (1952) model for paired comparisons. Yellott (1977) shows that the extreme value distribution is the only location family which leads to probabilities that satisfy the Harville formulas.

There is an alternative derivation of the Harville formulas. If T_i has the exponential distribution with mean $1/\lambda_i$ and running times are again assumed independent, then the probability of a particular outcome is given by the Harville formulas with $q_i = \lambda_i / \sum_j \lambda_j$. This method for deriving the formulas is described by Henery (1981). The exponential random variable model is equivalent to the extreme value model because if X has an exponential distribution then $\ln X$ has an extreme value distribution. Further discussion of the exponential distribution and its application to estimating the probability of the order of finish are described in Stern (1987, 1990).

Neither of the two probability models that lead to the Harville formulas seem appropriate for modeling the running time of a horse race. The exponential distribution puts positive probability on times near zero. It also has a memoryless property, the distribution of the time to finish a race for a horse that has been running for two minutes is the same as distribution of the running time for a horse just starting. Of course, this is not very realistic, although it may appear this way at times to the unlucky bettor. Henery (1984) attempts to fit the times of horses to the extreme value distribution. He finds that although the extreme value distribution is not appropriate for the times of all horses, it does seem appropriate for the faster running times in a race. In the next section, alternatives to the Harville formulas are obtained by considering alternative distributions for the running times.

All of the experience to date has suggested that the Harville formulas do not correctly estimate the probability that horses will finish second or third. Horses for whom the estimated probability of finishing second is high actually finish second less frequently than predicted. Similarly, horses with low estimated probability of finishing second do so more often than expected. The same observations are true for third place finishes. This is demonstrated by Harville with data from 335 races. A possible explanation for this fact is that the information contained in the horse's failure to win is not used to adjust the probabilities (this idea is related to the memoryless property of the exponential distribution). One example illustrating the problem is the "Silky Sullivan" phenomenon described by Hausch et al (1981). Suppose horse i starts slowly and makes a late sprint to the finish line. If horse i is performing well he may pass the other horses and win the race. If not horse i is likely to finish far back in the field. It may be the case that $q_i > q_j$, where horse j is a slower but more consistent performer. When a third horse, horse k, wins the race, the Harville formulas predict that horse i is more likely to finish second than horse j because $q_i > q_j$. However, given the nature of horse i it seems that horse j is more likely to finish second in this case. The existence of such horses would explain the empirical observation that horses which have a high probability of finishing second according to the Harville formulas finish second less often than predicted. There is thus an indication that probability models which do not satisfy the Harville formulas may be better suited to estimating the probabilities of second and third place finishes. It is natural to wonder if better formulas for the probability that horse j finishes second or third would lead to improved returns.

The systematic bias in second and third place probability estimates is not as evident in place and show probability estimates. This seems to be a result of the fact that the probability that favored horses finish first is underestimated while the probability that these same horses finish second is overestimated. The two effects partially cancel each other in the place estimate. Hopefully, alternative models will improve the estimates of the probability of finishing second without damaging the estimates of place and show probabilities.

4. Alternatives to the Harville formulas

To find alternatives to the Harville formulas, the running times for horses are assumed to have distributions other than the extreme value distribution (or the equivalent exponential model). The extreme value distribution is a special case of the generalized extreme value distribution described by Mihram (1975). The generalized extreme value density is

$$g_r(t) = \frac{1}{\Gamma(r)} e^{r(t-\theta)} e^{-e^{(t-\theta)}} \qquad r > 0, \quad t \in (-\infty, \infty).$$

For $r = 1$, this is the extreme value distribution.

The extreme value distribution is the distribution of the logarithm of an exponential random variable. The generalized extreme value distribution is similarly related to the gamma distribution with shape parameter r. For larger values of r the generalized extreme value distribution is more highly peaked than the extreme value distribution with less likelihood of extremely small running times. This seems more appropriate for modeling running times than the extreme value distribution with $r = 1$.

The probability of a particular order of finish is computed under the extreme value model with parameter r (which might also be referred to as the gamma model with shape parameter r) by supposing that the running times $T_1, \ldots, T_n$ have this distribution with location parameters $\theta_1, \ldots, \theta_n$. The calculation of the probability of a particular order of finish is much more straightforward for integer values of r than for non-integer r. The non-integer values require the numerical evaluation of a multiple integral. Therefore, only integer values of r are considered here. Even for the integer values of r the resulting expression for the probability of the outcome of a race is quite complex. For example, if we take $\lambda_j = e^{-\theta_j}$ then the probability that the horses finish in the order $1, \ldots, n$ is

$$p(123 \ldots n) = \frac{\prod_{j=1}^{n} \lambda_j^r}{((r-1)!)^n} \sum_{i_n=1}^{r} \frac{(r-1)!}{(r-i_n)! \lambda_n^{i_n}} \sum_{i_{n-1}=1}^{2r-1-i_n} \frac{(2r-1-i_n)!}{(2r-i_n-i_{n-1})!(\lambda_n + \lambda_{n-1})^{i_{n-1}}}$$

$$\ldots \sum_{i_2=1}^{(n-1)r-1-\sum_{j=3}^{n} i_j} \frac{((n-1)r-1-\sum_{j=3}^{n} i_j)!}{((n-1)r-\sum_{j=2}^{n} i_j)! (\sum_{j=2}^{n} \lambda_j)^{i_2}} \frac{(nr-1-\sum_{j=2}^{n} i_j)!}{(\sum_{j=1}^{n} \lambda_j)^{nr-\sum_{j=2}^{n} i_j}}.$$

This probability is computed directly from the multiple integral expression for the probability that $T_1 < \cdots < T_n$. Alternative derivations of this result based on combinatoric arguments are given by Henery (1983) and Stern (1987). This probability expression is unchanged if each of the λ_i are multiplied by a positive constant (or equivalently if a constant is added to each of the location parameters θ_i); for convenience we often choose to have $\sum_i \lambda_i = 1$.

The probability of a partial ranking is obtained by summing the probability of all permutations that are consistent with the partial ranking or by a similar probability calculation. As an example suppose that $r = 2$ and that there are $n = 3$ horses, then

$$p(12) = \lambda_1^2 \lambda_2^2 (6\lambda_3 \lambda_{23}^{-1} + 2\lambda_{23}^{-1} + 4\lambda_3 \lambda_{23}^{-2} + \lambda_{23}^{-2} + 2\lambda_3 \lambda_{23}^{-3})$$

$$p(1) \ = \lambda_1^2 (6\lambda_2 \lambda_3 + 2\lambda_2 + 2\lambda_3 + 1)$$

with

$$\lambda_i = e^{-\theta_i}, \qquad \lambda_{23} = \lambda_2 + \lambda_3, \qquad \lambda_1 + \lambda_2 + \lambda_3 = 1.$$

These expressions are substantially more complex than the corresponding results for $r = 1$ (the Harville formulas)

$$p(1) = \lambda_1 \qquad \text{and} \qquad p(12) = \lambda_1 \frac{\lambda_2}{\lambda_2 + \lambda_3}.$$

Suppose that $\lambda_1 > \lambda_2 > \lambda_3$ (equivalent to $\theta_1 < \theta_2 < \theta_3$) so that horse 1 is the favorite and horse 3 has the longest odds. It is possible to show for this small example that when horse 2 finishes first, the probability that horse 1 finishes second is less when $r = 2$ than under the Harville formula ($r = 1$). Thus values of r larger than one lead to estimates that tend to lessen the probability that favored horses finish second. It seems reasonable to conjecture that the results demonstrated for $r = 2$ and $n = 3$ also hold for larger n and r. The generalized extreme value (or gamma) models appear to improve on the Harville formulas; the density corresponding to $r > 1$ is more likely to correspond to the distribution of race times (although this still may not be very realistic) and these models appear to improve on the empirical deficiency of the Harville formulas.

5. Comparing the models

The proposed alternatives to the Harville formulas become quite complicated for large r or large n. The results of 47 horse races with $n = 6$ horses at Bay Meadows racetrack in California were collected from the newspaper during January and February 1987. For each race the track odds and the result of the race were recorded. Races with more than six horses were not used due to the added computational burden that arises. If the win pool is efficient and breakage is ignored, then the probability that horse i wins is

$$q_i = \frac{\frac{1}{O_i + 1}}{\sum_{j=1}^{6} \frac{1}{O_j + 1}}.$$

The probability that horse j finishes second as determined by the Harville formula is

$$p_h(\cdot j) = \sum_{\substack{i=1 \\ i \neq j}} p_h(ij) = \sum_{\substack{i=1 \\ i \neq j}}^{6} q_i \frac{q_j}{1 - q_i}$$

where the dot is used to represent the fact that no first place horse has been specified. To use the alternative model with $r = 2$, we use the values of $q_1, \ldots, q_6$ to estimate the unknown parameters $\lambda_1, \ldots, \lambda_6$ that satisfy the equations,

$$q_j = \Pr(j \text{ is first in the extreme value model with } r = 2)$$

$$= \lambda_j^2 \Big(6! \prod_{\substack{i=1 \\ i \neq j}}^{6} \lambda_i + 5! \sum_{\substack{m=1 \\ m \neq j}}^{6} \Big(\prod_{\substack{i=1 \\ i \neq j, m}}^{6} \lambda_i \Big) + \cdots + 2! \sum_{\substack{m=1 \\ m \neq j}}^{6} \lambda_m + 1 \Big) \qquad j = 1, \ldots, 6.$$

This is accomplished with an iterative procedure. The probability that j finishes second under the $r = 2$ model $p_2(\cdot j)$ is found by summing the probability of all outcomes in which j is second. The probability of each such outcome is computed using the formulas of the previous section.

As an example, both estimates are computed for a sample race

Sample Horse Race Data, 2nd Race, January 9, 1987

Horse	O_i	q_i	$p_h(\cdot j)$	$p_2(\cdot j)$
1	8.9	0.084	0.107	0.113
2	19.8	0.040	0.053	0.059
3	1.1	0.395	0.280	0.266
4	3.3	0.193	0.217	0.214
5	5.2	0.134	0.162	0.165
6	4.4	0.154	0.182	0.183

The estimated probabilities of finishing second for the favorites (horse 3 and horse 4) are smaller under the $r = 2$ model than under the Harville model. The estimates for the longshots are larger.

To compare the Harville formulas and the model with $r = 2$, the number of times that horses actually finish second is compared to the expected number under each model. The total number of horses in the 47 races is 282. Each horse is placed in a category according to its estimated probability of finishing second. The expected number of second place finishes in each category is compared to the actual number in the table below.

Expected and Observed Second Place Finishes

probability of finishing 2nd	Harville Model ($r = 1$) # of horses	observed	expected	Alternative Model ($r = 2$) # of horses	observed	expected
0.00 - 0.10	66	6	4.15	58	4	3.75
0.10 - 0.15	61	7	7.69	63	9	7.94
0.15 - 0.20	49	13	8.56	57	13	9.97
0.20 - 0.25	55	12	12.54	62	13	14.10
> 0.25	51	9	14.06	42	8	11.23

In the first set of columns the same type of results that Harville observed are obtained. Horses with high probability of finishing second do so less frequently than expected. To a smaller degree, horses with small probability finish second more frequently than expected. The model with $r = 2$ provides an improvement in both of these areas. The usual chi-squared goodness-of-fit test is not appropriate for comparing the models since this is not a typical multinomial problem. Nonetheless, the chi-squared statistic for each model, which computes the sum of the squared differences between the observed and expected counts, where each squared difference is divided by the expected count, can be used as a descriptive measure. For the Harville model the observed value is 5.04 and for the $r = 2$ model the observed value is 2.10. This measures the improvement that we observe in the table. It seems that higher values of r might lead to even better fits. In fact, Bacon-Shone, Lo, and Busche (1992) find that a probability model based on the normal distribution, due to Henery (1981), which is equivalent to letting $r \to \infty$ in the generalized extreme value models, appears to fit best in a variety of settings.

6. Conclusions

The generalized extreme value distribution provides a family of models for the probabilities of the outcomes of a horse race. This family includes the Harville formulas as a special case and provides alternatives which avoid some of the problems present in the Harville formulas. Use of the alternative formulas described here involves a tradeoff between more complex formulas and improved estimates. It is difficult to embed the estimates from the model with $r = 2$ (or any value of r other than 1) in a real-time handicapping system due to the complicated calculations. An approximation based on Taylor series expansions for the probability of an outcome leads to simpler calculations. This approximation is discussed by Henery (1983) and Stern (1987); unfortunately the estimates based on this approximation does not seem to improve on the Harville formulas. A promising new approximation, the discount model, is given by Lo and Bacon-Shone (1992).

References

Asch, P., Malkiel, B. G. and Quandt, R. E. (1982). Racetrack betting and informed behavior. *Journal of Financial Economics* 10, 187-194.

Asch, P. and Quandt, R. E. (1986). *Racetrack Betting*. Dover, MA: Auburn House.

Bacon-Shone, J. H., Lo, V. S. Y. and Busche, K. (1992). Logistic analysis of complicated bets. Technical Report, Department of Statistics, University of Hong Kong.

Bradley, R. A. and Terry, M. E. (1952). Rank analysis of incomplete block designs. I. The method of paired comparisons. *Biometrika* 39, 324-345.

Fabricand, B. F. (1965). *Horse Sense*. New York: David McKay.

Harville, D. A. (1973). Assigning probabilities to the outcomes of multi-entry competitions. *Journal of the American Statistical Association* 68, 312-316.

Hausch, D. B., Ziemba, W. T., and Rubinstein, M. (1981). Efficiency of the market for racetrack betting. *Management Science* 27, 1435-1452.

Henery, R. J. (1981). Permutation probabilities as models for horse races. *Journal of the Royal Statistical Society B* **43**, 86-91.

Henery, R. J. (1983). Permutation robabilities for gamma random variables," *Journal of Applied Probability* 20, 822-834.

Henery, R. J. (1984). An extreme-value model for predicting the results of horse races, *Applied Statistics* **33**, 125-133.

Lo, V. S. Y. and Bacon-Shone, J. H. (1992). Approximating the ordering probabilities of multi-entry competitions by a simple method. Technical Report, Department of Statistics, University of Hong Kong.

Mihram, G. A. (1975). A generalized extreme-value density. *South African Statistical Journal* **9**, 153-162.

Plackett, R. L. (1975). The analysis of permutations. *Applied Statistics* **24**, 193-202.

Savage, I. R. (1957). Contributions to the theory of rank order statistics - the "trend" case. *Annals of Mathematical Statistics* **28**, 968-977.

Snyder, W. W. (1978). Horse racing: testing the efficient markets model. *Journal of Finance* **33**, 1109-1118.

Stern, H. S. (1987). Gamma processes, paired comparisons and ranking. Unpublished Ph.D. dissertation, Department of Statistics, Stanford University, Stanford, California.

Stern, H. (1990). Models for distributions on permutations. *Journal of the American Statistical Association* **85**, 558-564.

Yellott, J. I., Jr. (1977). The relationship between Luce's choice axiom, Thurstone's theory of comparative judgment, and the double exponential distribution. *Journal of Mathematical Psychology* **15**, 109-144.

Ziemba, W. T., Brumelle, S. L., Gautier, A. and Schwartz, S. L. (1986). *Dr. Z's 6/49 Lotto Guidebook*. Vancouver: Dr. Z. Investments.

Ziemba, W. T. and Hausch, D. B. (1987). *Dr. Z's Beat the Racetrack*. New York: William Morrow and Company.

Application of Running Time Distribution Models in Japan

Victor S.Y. Lo

Faculty of Commerce & Business Administration,
University of British Columiba.

Abstract

To predict the the ordering probabilities such as the probability that horse i wins and j finishes second), the Harville (1973) model has been the most popular. The model assumes the running time distribution is independent exponential. However, a recent empirical study shows that the Henery (1981) model has a better fit. In this paper, we consider the Stern (1990) model in addition to the two models above. We fit the Stern model in a Japanese data set and conclude that the Stern model with a particular value of the shape parameter is superior to the others. Under the assumption of the Stern model, we show that the Harville model has a systematic bias in predicting the ordering probabilities.

I. Introduction

Predicting the payouts of a bet in parimutuel betting of horse-racing is the dream of many bettors. Previous research in racing analysis showed that the win bet fractions (i.e. the proportions of amount bet on horses in win bet market) are quite consistent with the true win probabilities although a systematic favorite-longshot bias usually exists, i.e. the bettors underbet the favorites and overbet the longshots (e.g. Griffith (1949); McGlothlin (1956); Hoerl and Fallin (1974); Ali (1977); Snyder (1978); Hausch, Ziemba and Rubinstein (1981); Asch, Malkiel and Quandt (1982)). A survey of these papers can be found in Ziemba and Hausch (1986). Absence of the bias has also been reported in Hong Kong (Busche and Hall (1988)). We may assume that the win bet fraction is a good estimate of the win probability.

Based on the knowledge of the simple win probabilities (which are estimated by the win bet fractions), one may estimate the ordering probability, using the simple model due to Harville (1973). For example, the probability that horse i wins and j finishes second is

$$\pi_{ij} = \frac{\pi_i \, \pi_j}{1 - \pi_i} .$$

This model has been used by Hausch, Ziemba and Rubinstein (1981), Tuckwell (1981), Asch and Quandt (1987) and others. Another model originated by Henery (1981) assumes that the running times are normally distributed with different means and same variance. But there is no closed form solution for the ordering probabilities. Stern (1990) extends the exponential distribution to a Gamma

Keywords: Running time distributions; Ordering probabilities; Horse races; Win Bet Fractions; Gamma distribution.
Acknowledgements : I thank Kelly Busche, Richard Quandt and Junji Shiba for their data; John Bacon-Shone for his advice; and William Ziemba for his comments on an earlier draft of this paper.

distribution with a predetermined shape parameter, r, while the scale parameters can be estimated by the win bet fractions.

McCulloch and Zijl (1985) gave a test for the Harville model using Australian races. However, they assumed the show bet fraction was a good estimate of the show probability (i.e. P(horse i finishes first, second or third)) instead of using the observed finishing order in their analysis (the rule for computing the return for show bets in Australia is different from the other countries). Bacon-Shone, Lo and Busche (1992) report the analysis of some exotic bets - exacta, trifecta and quinella by logit models. Their findings in the Meadowlands (U.S.) and Hong Kong racetrack markets include that the Henery model is superior to the other models described above.

Here, we consider the Stern (1990) model which assumes that the running times follow the Gamma distribution with a fixed shape parameter, r. The Harville model is a special case of the Stern model. Maximum likelihood estimation of r in Japan will be reported. By using a likelihood-based argument, we will show that Stern's Gamma model with maximum likelihood estimate of r is superior to both the Harville ($r=1$) and Henery models. However, this is not true in the Meadowlands and Hong Kong data. We will theoretically show the existence of a systematic bias for predicting complicated probabilities if we use the Harville model but the true underlying model is Stern's Gamma model.

II. Fitting the Stern model

Stern's Gamma model (Stern (1990)) is motivated by considering a competition in which n players, scoring points according to independent Poisson processes, are ranked according to the time until r points are scored. Thus r should be an integer under this assumption. We can consider it as an alternative model to the Harville and Henery models. Let the running time of horse i, T_i Gamma(r, θ_i) independently or,

$$g_r(t_i | \theta_i) = \frac{1}{\Gamma(r)} \theta_i^r t_i^{r-1} \exp(-\theta_i t_i) \qquad t_i > 0.$$

where r is predetermined and θ_i can be estimated from π_i (or the bet fraction, P_i). When $r = 1$, it becomes Harville's (1973) exponential running time model. When $r = \infty$, gamma distribution tends to normal distribution. However, strictly speaking, this does not imply that the Stern model becomes Henery (1981)'s normal running time model because the variance of running time under the Stern model is varying while that under the Henery model is assumed to be a constant.

We may try to estimate this r by comparing the log likelihood :
$\sum_l \ln \pi_{[123]l}$, where $[123]l$ denotes the 3 top horses in race l,

with different values of r. The result for Japanese data is shown in Table 1. The computations are done by using Gauss-Laguerre integration for the Stern model and Gauss-Hermitian integration for the Henery model. For the Stern model, we must first find θ by solving

$$P_i = \int_0^\infty \prod_{s \neq i} [1 - G_r(t_i | \theta_s)] \, g_r(t_i | \theta_i) \, dt_i$$

where P_i = the win bet fraction for horse i, $g_r(t_s | \theta_s)$ and $G_r(t_s | \theta_s)$ are the probability density function and the cumulative distribution function of Gamma(r, θ_s). In fact, for integer r, the above P_i has a closed form in terms of θ. However, due to its complicated form, it may not be practical unless n and r are small.

From Table 1, the log likelihood is maximized when r = 4. Thus the Gamma distribution with r = 4 is a better distributional assumption of running time in Japan. The results of fitting the Stern model for the Hong Kong and Meadowlands data are shown in Table 2 and 3, respectively. The Henery model appears to be the best in both data sets. Why does not the distribution of running time universally hold ? We cannot find any sufficient reason for this. One idea is that the allocation of the purse to the top 5 horses is different in different racetracks. We do not have the purse information for Meadowlands data but we have collected this information for Hong Kong and Japanese data in Table 4.

Table 1
Log likelihood values under the Stern model
for Japanese data (1583 races)

r	log likelihood
1 (i.e. Harville)	-8977.57
2	-8954.57
3	-8950.60
4	-8950.35
5	-8950.94
6	-8951.82
7	-8952.65
8	-8953.44
Henery	-8986.88

Table 2
Log likelihood values under the Stern model
for Hong Kong (89) data (421 races)

r	log likelihood
1	-2523.37
10	-2504.55
20	-2503.58
30	-2503.44
40	-2503.72
Henery	-2502.74

Table 3
Log likelihood values under the Stern model
for Meadowlands data (510 races)

r	log likelihood
1	-2845.93
10	-2800.87
20	-2798.02
30	-2796.90
40	-2795.78
Henery	-2792.94

Table 4
Allocation of purse in percentages

Finishing horse	Hong Kong	Japan
1st	57.0	52.5
2nd	22.0	21.0
3rd	11.5	13.4
4th	6.0	7.9
5th	3.5	5.2

From Table 4, the percentages of total purse are higher for 3rd, 4th and 5th horses in Japan than in Hong Kong. We suspect that the training of horses and jockeys in Japan is not the same as in Hong Kong. More clearly, in Japan, the favourite horses may try hard to get in the top 5 positions even if they are far away from the leading horses. This will increase the conditional probabilities of finishing 3rd for favourite horses. This idea is simply a suspicion only and, of course, there may be other more reasonable factors which will affect the running time distributions of horses.

III. A theoretical comparison between Harville and Stern

In this section, under the assumption of the Stern model with parameter r, the difference in estimating the conditional probability of horse j finishing second given that horse i finishes first by the Harville and Stern models will be investigated. We study the following difference :

$$g_{j|i}^{(r)} = \pi_{j|i} - \frac{\pi_j}{1 - \pi_i} \qquad (1)$$

for $j = a$ and b, where a and b are horses chosen such that : $\theta_a = \max_{s \neq i} \theta_s$ and $\theta_b = \min_{s \neq i} \theta_s$, and π_{ij} is based on the Stern model with shape parameter r.

Note that $E(T_i) = r/\theta_i$ for $i = 1, 2, \ldots, n$ and thus, $E(T_a) = \min_{s \neq i} E(T_s)$ and $E(T_b) = \max_{s \neq i} E(T_s)$, so that a is the strongest horse and b is the weakest horse as measured by their mean running times. In the following, we use $g_r(v)$ to denote the p.d.f. of V which is a gamma random variable with shape parameter r and scale parameter 1. Its associated cumulative distribution is $G_r(v)$.

Lemma 1

Let u,v and w be non-negative functions. Suppose u and v/w are non-increasing, then

$$\frac{\int uv}{\int uw} \geq \frac{\int v}{\int w}$$

Similarly, if u is non-increasing and v/w is non-decreasing, then

$$\frac{\int uv}{\int uw} \leq \frac{\int v}{\int w}$$

Proof : See Gutmann and Maymin (1987).

Lemma 2

Define

$$K(v;\theta_j) = \frac{g_r(v\theta_j/\theta_1)\ \theta_j/\theta_1\ \prod_{w\neq 1j}\bar{G}_r(v\theta_w/\theta_1)}{\sum_{s\neq 1j} g_r(v\theta_s/\theta_1)\ \theta_s/\theta_1\ \prod_{t\neq 1s}\bar{G}_r(v\theta_t/\theta_1)} \tag{2}$$

where $\bar{G}_r(v) = 1 - G_r(v)$. Then,
$K(v;\theta_a)$ is non-increasing in v, and $K(v;\theta_b)$ is non-decreasing in v, where $\theta_a = \underset{r\neq 1}{Max}\theta_s$ and $\theta_b = \underset{r\neq 1}{Min}\theta_s$.
The proof appears in the Appendix.

Theorem 1

$$\pi_{a|1} - \frac{\pi_a}{1-\pi_1} \leq 0 \quad \text{and} \quad \pi_{b|1} - \frac{\pi_b}{1-\pi_1} \geq 0$$

where a and b are chosen such that $\theta_a = \underset{s\neq 1}{Max}\ \theta_s$ and $\theta_b = \underset{s\neq 1}{Min}\ \theta_s$

Proof:

Consider the difference (1),

$$\pi_{j|1} - \frac{\pi_j}{1-\pi_1} = \frac{1}{\pi_1}[\pi_{1j} - \frac{\pi_1\pi_j}{1-\pi_1}]\ ,$$

$$\pi_{1j} = \int_0^\infty G_r(u\theta_1/\theta_j)\ g_r(u)\prod_{s\neq 1j}[1-G_r(u\theta_s/\theta_j)]\ du$$

$$= \int_0^\infty g_r(u)\prod_{s\neq 1j}[1-G_r(u\theta_s/\theta_j)]du - \int_0^\infty g_r(u)\prod_{s\neq j}[1-G_r(u\theta_s/\theta_j)]du$$

$$= \pi_{j(1)} - \pi_j$$

where $\pi_{j(1)} = P(T_j < \underset{s\neq 1j}{Min}\{T_s\})$

i.e. the probability of horse j wins if horse i is removed from the race. Therefore,

$$\pi_{j|1} - \frac{\pi_j}{1-\pi_1} = \frac{1}{\pi_1}\left(\pi_{j(1)} - \frac{\pi_j}{1-\pi_1}\right)$$

Define $g_{j|1} = \pi_{j(1)} - \dfrac{\pi_j}{1-\pi_1}$ $\qquad\qquad\qquad\qquad\qquad$ (2)

Thus, it suffices to show that $g_{a|1} \le 0$ and $g_{b|1} \ge 0$.

$$g_{j|1} = \frac{\pi_{j(1)}\left(\sum_{s\ne 1 j}\pi_s + \pi_j\right) - \pi_j}{1-\pi_1}$$

$$= \frac{\pi_{j(1)}\sum_{s\ne 1 j}\pi_s - \pi_j(1-\pi_{j(1)})}{1-\pi_1}$$

$$= \frac{\sum_{s\ne 1 j}\pi_s \sum_{s\ne 1 j}\pi_{s(1)}}{1-\pi_1}\left[\frac{\pi_{j(1)}}{\sum_{s\ne 1 j}\pi_{s(1)}} - \frac{\pi_j}{\sum_{s\ne 1 j}\pi_s}\right]$$

Consider:

$$\frac{\pi_{j(1)}}{\sum_{s\ne 1 j}\pi_{s(1)}} - \frac{\pi_j}{\sum_{s\ne 1 j}\pi_s}$$

$$= \frac{\int_0^\infty \prod_{w\ne 1 j}\overline{G}_r(u\theta_w/\theta_j)\, g_r(u)\, du}{\sum_{s\ne 1 j}\int_0^\infty \prod_{t\ne 1 s}\overline{G}_r(u\theta_t/\theta_s)\, g_r(u)\, du}$$

$$\qquad - \frac{\int_0^\infty \overline{G}_r(u\theta_1/\theta_j)\prod_{w\ne 1 j}\overline{G}_r(u\theta_w/\theta_j)\, g_r(u)\, du}{\sum_{s\ne 1 j}\int_0^\infty \overline{G}_r(u\theta_1/\theta_s)\prod_{t\ne 1 s}\overline{G}_r(u\theta_t/\theta_s)\, g_r(u)\, du}$$

$$= \frac{\int_0^\infty g_r(v\theta_j/\theta_1)\,\theta_j/\theta_1\prod_{w\ne 1 j}\overline{G}_r(v\theta_w/\theta_1)\, dv}{\sum_{s\ne 1 j}\int_0^\infty g_r(v\theta_s/\theta_1)\,\theta_s/\theta_1\prod_{t\ne 1 s}\overline{G}_r(v\theta_t/\theta_1)\, dv}$$

$$\qquad - \frac{\int_0^\infty g_r(v\theta_j/\theta_1)\,\theta_j/\theta_1\,\overline{G}_r(v)\prod_{w\ne 1 j}\overline{G}_r(v\theta_w/\theta_1)\, dv}{\sum_{s\ne 1 j}\int_0^\infty g_r(v\theta_s/\theta_1)\,\theta_s/\theta_1\,\overline{G}_r(v)\prod_{t\ne 1 s}\overline{G}_r(v\theta_t/\theta_1)\, dv}$$

by change of variables using :

$v = u\theta_1/\theta_j$ in the numerator, and

$v = u\theta_1/\theta_s$ in the denominator.

≤ 0 when $j = a$

$\left.\rule{0pt}{28pt}\right\}$ by using Lemmas 1 and 2.

≥ 0 when $j = b$

Hence, $g_{a|1} \leq 0$ and $g_{b|1} \geq 0$ and the theorem is proved. ∎

Note that when $r=1$, the Stern model reduces to the Harville model and thus, in the above theorem, the two inequalities will become two equalities.

Theorem 1 shows that if the running times satisfy the assumption of the Stern model (with parameter $r=2,3,\ldots$), the Harville model will overestimate the conditional probability of the most favourite horse finishing second and underestimate the conditional probability of the longshot finishing second. This result is in effect a parallel result of the comparision between the Henery and Harville model reported in Lo and Bacon-Shone (1994). Similar to Theorem 1, in that paper, we have proved that under the assumption of the Henery model, the Harville model underestimate the conditional probability of the most favourite horse finishing second and underestimate the conditional probability of the longshot finishing second. Therefore, together with that theorem in Lo and Bacon-Shone (1994), we have shown that a systematic difference appears (either overestimate of underestimate the probability) when we use the Harville model to compute the ordering probability when the true running time distribution is Normal or Gamma. These two theorems are both practically and theoretically interesting since the Harville model is more commonly used in practice.

In addition, it may also be interesting to compare Stern(r) with Stern(r-1).

Conjecture

$\pi_{1a}^{(r)} \leq \pi_{1a}^{(r-1)}$ and $\pi_{1b}^{(r)} \geq \pi_{1b}^{(r-1)}$ for any i and $r = 3,4,\ldots$

where

$$\theta_a = \underset{s \neq 1}{\text{Max }} \theta_s, \quad \theta_b = \underset{s \neq 1}{\text{Min }} \theta_s,$$

$\pi_{1j}^{(t)} = P(\text{horse i wins and j finishes 2nd, Stern(t) is assumed})$

$$= \int_0^\infty G_t(u\theta_1^{(t)}/\theta_j^{(t)}) \ g_t(u) \underset{s \neq 1, j}{\prod} [1 - G_t(u\theta_s^{(t)}/\theta_j^{(t)})] du$$

for $t = r, r-1$,

and $\theta_k^{(t)}$'s are obtained by solving the following system of nonlinear equations :

$$P_k = \int_0^\infty g_t(u) \underset{s \neq k}{\prod} [1 - G_t(u\theta_s^{(t)}/\theta_k^{(t)})] du$$

for $t = r, r-1$ and $k = 1,2,\ldots,n$.

The author has not been able to prove this conjecture. But extensive numerical results (for $r=3,\ldots,8$ in Japanese data) verify that this conjecture holds. Although this conjecture is of theoretical interest, it is not very practically important since

people will not approximate Stern(r) by Stern(r-1) once they are
willing to use the Stern model.

IV. Conclusion

This paper demonstrates the application of the Stern model. The
Stern model serves as an alternative model for predicting ordering
probabilities to the Henery model. As no one model is consistently
better than the other in different racetracks, we should
investigate this issue further. There should be some reason for
this inconsistency. But we can still apply different models in
different racetracks. However, application of either Stern or
Henery model is very complicated in practice, fortunately, a
simple approximation of the Stern model in computing the
complicated ordering probabilities is proposed in Lo and
Bacon-Shone (1992). We also show that the Harville model has a
systematic bias if the running time distribution is gamma rather
than exponential.

Appendix : Proof of Lemma 2

Lemma 3

$$e'(v) \leq 1 \quad \text{where} \quad e(v) = \frac{v\,g_r(v)}{1 - G_r(v)} \quad \text{for } v > 0, \; r=1,2,3,\ldots$$

The proof, based on induction, appears in Chapter 8 of Lo (1992).

Proof of Lemma 2

Consider the derivative of $K(v;\theta_j)$ with respect to v :

$$\frac{d\,K(v;\,\theta_j)}{d\,v} \left\{ \sum_{s \neq 1\,j} g_r(v\theta_s/\theta_1)\; \theta_s/\theta_1 \prod_{t \neq 1\,s} \bar{G}_r(v\theta_t/\theta_1) \right\}^2$$

$$= \theta_j/\theta_1 \left\{ \sum_{s \neq 1\,j} g_r(v\theta_s/\theta_1)\; \theta_s/\theta_1 \prod_{t \neq 1\,s} \bar{G}_r(v\theta_t/\theta_1) \right\}$$

$$\left\{ \theta_j/\theta_1\; g_r(v\theta_j/\theta_1) \left[\frac{r-1}{v\theta_j/\theta_1} - 1 \right] \prod_{w \neq 1\,j} \bar{G}_r(v\theta_w/\theta_1) \right.$$

$$\left. + g_r(v\theta_j/\theta_1) \prod_{w \neq 1\,j} \bar{G}_r(v\theta_w/\theta_1) \sum_{w \neq 1\,j} \frac{-\theta_w/\theta_1\; g_r(v\theta_w/\theta_1)}{\bar{G}_r(v\theta_w/\theta_1)} \right\}$$

$$- \theta_j/\theta_1\; g_r(v\theta_j/\theta_1)\; \prod_{w \neq 1\,j} \bar{G}_r(v\theta_w/\theta_1)$$

$$\sum_{s \neq 1\,j} \left\{ \theta_s/\theta_1 \left[\theta_s/\theta_1\; g_r(v\theta_s/\theta_1) \left(\frac{r-1}{v\theta_s/\theta_1} - 1 \right) \prod_{t \neq 1\,s} \bar{G}_r(v\theta_t/\theta_1) \right. \right.$$

$$\left. \left. + g_r(v\theta_s/\theta_1) \prod_{t \neq 1\,s} \bar{G}_r(v\theta_t/\theta_1) \sum_{t \neq 1\,s} \frac{-\theta_t/\theta_1\; g_r(v\theta_t/\theta_1)}{\bar{G}_r(v\theta_t/\theta_1)} \right] \right\}$$

$$= \theta_j/\theta_1 \; g_r(v\theta_j/\theta_1)_{w \neq 1j} \bar{G}_r(v\theta_w/\theta_1) \left[\sum_{s \neq 1j} g_r(v\theta_s/\theta_1)\theta_s/\theta_{1t \neq 1s} \bar{G}_r(v\theta_t/\theta_1) \right.$$

$$\left[\theta_j/\theta_1 \left(\frac{r-1}{v\theta_j/\theta_1} - 1 \right) - \sum_{w \neq 1j} \frac{\theta_w/\theta_1 \; g_r(v\theta_w/\theta_1)}{\bar{G}_r(v\theta_w/\theta_1)} \right] -$$

$$\theta_j/\theta_1 \; g_r(v\theta_j/\theta_1)_{w \neq 1j} \bar{G}_r(v\theta_w/\theta_1) \sum_{s \neq 1j} \left\{ \theta_s/\theta_1 \; g_r(v\theta_s/\theta_1)_{t \neq 1s} \bar{G}_r(v\theta_t/\theta_1) \right.$$

$$\left. \left[\theta_s/\theta_1 \left(\frac{r-1}{v\theta_s/\theta_1} - 1 \right) - \sum_{t \neq 1s} \frac{\theta_t/\theta_1 \; g_r(v\theta_t/\theta_1)}{\bar{G}_r(v\theta_t/\theta_1)} \right] \right\}$$

$$= \theta_j/\theta_1 g_r(v\theta_j/\theta_1)_{w \neq 1j} \bar{G}_r(v\theta_w/\theta_1) \sum_{s \neq 1j} \left\{ \theta_s/\theta_1 \; g_r(v\theta_s/\theta_1)_{t \neq 1s} \bar{G}_r(v\theta_t/\theta_1) \right.$$

$$\left[\theta_j/\theta_1 \left(\frac{r-1}{v\theta_j/\theta_1} - 1 \right) - \theta_s/\theta_1 \left(\frac{r-1}{v\theta_s/\theta_1} - 1 \right) \right.$$

$$\left. \left. + \frac{\theta_j/\theta_1 \; g_r(v\theta_j/\theta_1)}{\bar{G}_r(v\theta_j/\theta_1)} - \frac{\theta_s/\theta_1 \; g_r(v\theta_s/\theta_1)}{\bar{G}_r(v\theta_s/\theta_1)} \right] \right\}$$

$$= \theta_j/\theta_1 g_r(v\theta_j/\theta_1)_{w \neq 1j} \bar{G}_r(v\theta_w/\theta_1) \sum_{s \neq 1j} \left\{ \theta_s/\theta_1 \; g_r(v\theta_s/\theta_1)_{t \neq 1s} \bar{G}_r(v\theta_t/\theta_1) \right.$$

$$\left. \left[\frac{\theta_s - \theta_j}{\theta_1} + \frac{e(v\theta_j/\theta_1) - e(v\theta_s/\theta_1)}{v} \right] \right\}$$

where $e(v) = \dfrac{v \; g_r(v)}{1 - G_r(v)}$

When $j = a$,

$$\frac{d \; K(v; \theta_a)}{d \; v} \left\{ \sum_{s \neq 1a} g_r(v\theta_s/\theta_1)\theta_s/\theta_{1t \neq 1s} \bar{G}_r(v\theta_t/\theta_1) \right\}^2$$

$$= \theta_a/\theta_1 g_r(v\theta_a/\theta_1)_{w \neq 1a} \bar{G}_r(v\theta_w/\theta_1) \sum_{s \neq 1a} \left\{ \theta_s/\theta_1 \; g_r(v\theta_s/\theta_1)_{t \neq 1s} \bar{G}_r(v\theta_t/\theta_1) \right.$$

$$\left. \left[\frac{\theta_s - \theta_a}{\theta_1} + v \; e'(v_0) \left(\frac{\theta_a - \theta_s}{\theta_1} \right) \frac{1}{v} \right] \right\}$$

by the mean value theorem where $v_0 \in (v\theta_s/\theta_1, v\theta_a/\theta_1)$

≤ 0 by lemma 3

On the other hand,

When $j = b$,

$$\frac{d \; K(v; \theta_b)}{d \; v} \left\{ \sum_{s \neq 1b} g_r(v\theta_s/\theta_1)\theta_s/\theta_{1t \neq 1s} \bar{G}_r(v\theta_t/\theta_1) \right\}^2$$

$$= \theta_b/\theta_1 g_r(v\theta_b/\theta_1) \underset{w\neq 1b}{\prod} \overline{G}_r(v\theta_w/\theta_1) \underset{s\neq 1b}{\sum} \left\{ \theta_s/\theta_1 \; g_r(v\theta_s/\theta_1) \underset{t\neq 1s}{\prod} \overline{G}_r(v\theta_t/\theta_1) \right.$$

$$\left. \left[\frac{\theta_s - \theta_b}{\theta_1} - v \, e'(v_0) \left(\frac{\theta_s - \theta_b}{\theta_1} \right) \frac{1}{v} \right] \right\}$$

by the mean value theorem, where $v_0 \in (v\theta_b/\theta_1, v\theta_s/\theta_1)$.

≥ 0 by lemma 3

We have shown that :

$$\frac{d \, K(v; \, \theta_a)}{d \, v} \leq 0 \text{ and } \frac{d \, K(v; \, \theta_b)}{d \, v} \geq 0$$

hence the result follows. ∎

References

Ali, M.M. (1977) "Probability and utility estimates for racetrack bettors." *Journal of Political Economy* 84 (August) :803-15.

Asch,Peter;Malkiel,B. and Quandt,R. (1982) "Racetrack betting and informed behavior." *Journal of Financial Economics* 10 (June) : 187-94.

Asch,Peter and Quandt,R. (1987) "Efficiency and profitability in exotic bets." *Economica* 54 : 289-98.

Bacon-Shone,J.H.; Lo,V.S.Y. and Busche,K. (1992) "Logistic analyses for complicated bets." Research report 11, Department of Statistics, University of Hong Kong.

Busche,K. and Hall,C.D. (1988) "An exception to the risk preference anomaly." *Journal of Business* 61 : 337-46.

Figlewski,S. (1979) "Subjective information and market efficiency in a betting model." *Journal of Political Economy* 87 (February) : 75-88.

Griffth,R.M. (1949) "Odds adjustments by American horse-racing bettors." *American Journal of Psychology* 62 : 290-294.

Gutmann,S. and Maymin,Z. (1987) "Is the selected population the best ?" *The Annals of Statistics* 15, No.1 : 456-461.

Harville,D.A. (1973) "Assigning probabilities to the outcomes of multi-entry competitions." *Journal of the American Statistical Association* 68 (June) : 312-16.

Hausch,D.B.;Ziemba,W.T. and Rubinstein,M. (1981) "Efficiency of the market for racetrack betting." *Management Science* 27 (December) : 1435-52.

Henery,R.J. (1981) "Permutation probabilities as models for

horse races." *Journal of Royal Statistical Society B* 43, No.1 : 86-91.

Hoerl,A.E.   and   Fallin,H.K.   (1974)   "Reliability   of   subjective evaluations in a high incentive situation." *Journal   of   Royal Statistical Society A* 137 : 227-230.

Lo,V.S.Y.   (1992)   "Modelling   of   gambling   probabilities."   Ph.D. thesis, Department of Statistics, University of Hong Kong.

Lo,V.S.Y.   and   Bacon-Shone,J.H.   (1992)   "An   approximation   to ordering probabilities of multi-entry competitons." Research report 16, Department of Statistics, University of Hong Kong.

Lo, V.S.Y. and Bacon-Shone, J.H. (1994) "A comparison between two models for predicting ordering probabilities in multi-entry competitions." *The Statistician*, forthcoming.

McGlothlin,W.H.   (1956)   "Stability   of   choices   among   uncertain alternatives." *American Journal of Psychology* 69 : 604-619.

McCulloch   B. and Zijl,T.V. (1985) "Direct test of Harville's mult-entry competitions model on race-track betting data." *Journal of Applied Statistics* 13, No.2 : 213-220.

Snyder, Wayne W. (1978) "Horse racing : Testing the efficient markets model." *Journal of Finance* 33 (September) : 1109-1118.

Stern,   Hal (1990) "Models for distributions on permutations." *Journal of the American Statistical Association*   85,   No.410 (June) : 558-564.

Tuckwell,R.H.   (1981)   "Anomalies   in   the   gambling   market." *Australian Journal of Statistics* 23, No.3 : 287-295.

Ziemba,W.T. and Hausch,D.B. (1986) *Betting at the racetrack*. Dr.Z Investments Inc., Los Angeles.

PART IV

EFFICIENCY OF WIN MARKETS AND THE FAVORITE-LONGSHOT BIAS

Introduction to the Efficiency of Win Markets and the Favorite-Longshot Bias

Donald B. Hausch, Victor S.Y. Lo and William T. Ziemba

Economists have long shown interest in racetrack betting as a source for investigating attitudes to risk and the efficiency of markets. Thaler and Ziemba (1988)[1], surveying research in both racing and lotteries, point that racetrack betting is an interesting application of efficient markets and rational expectations hypotheses since it possesses the usual characteristics of financial markets; namely, large numbers of investors/bettors with access to rich information sets. An advantage for testing financial theories that the racetrack has over most financial markets, though, is that there is a well-defined termination point at which final payoffs are determined.

Loosely speaking, a market is said to be efficient if no investor/bettor can obtain net profits using different sources of information. The definitions of weak, semi-strong and strong market efficiencies in financial markets are utilized by Fama (1970)[2], following earlier work by Roberts (1956)[2] and others. In weak form tests of the efficient market hypothesis, the information set is the past sequence of security price movements. Thus, weak form tests are tests of whether all information contained in historical prices is fully reflected in current prices. The random walk model for stock price was originated by Kendall(1953)[2]. Semi-strong form tests of the efficient market hypothesis are tests of whether publicly available information is fully reflected in current stock prices. Finally, strong form tests are tests of whether all information, whether public or private, is fully reflected in security prices and whether any type of investor can make an excess profit. Fama (1991)[2] surveys the more recent empirical results of different tests. Ziemba (1994)[2] surveys world wide systematic violations of equity market efficiencies.

Using analogous methods to those for studying stock market efficiency, we can examine the efficiency of racetrack markets. Most researchers have concentrated on the win market, studying whether the win bet fractions (the proportion of bets on horses) are accurate estimates of the true win probabilities; such a relation is consistent with the assumption that bettors maximize expected return.

Prior to the economists' work in this area, psychologists Griffith (1949)[1] and McGlothlin (1956)[1] analyzed win bet market data and concluded that the public systematically undervalued the win probabilities of short-odded horses (favorites) and overvalued those of long-odded horses (longshots). Such a systematic bias is consistent with risk-loving behavior. These two studies classified the data by odds levels. Ali (1977)[1] argued that horses should be classified by favorite position (ie., one for the public's favorite in the race, two for the horse with the second shortest odds, etc.) because only one horse for each favorite group exists in each race. He computed the average bet fraction and observed relative frequency of winning for each favorite-group. Simple Z-tests are used to show that the subjective probabilities (i.e. the win bet fractions) are higher (lower) than observed frequencies for longshots (favorites). This is the *favorite-longshot bias*. Given this bias, bettors improve their expected return by betting on favorites than longshots. Similar analysis was also conducted by statisticians, Hoerl and Fallin (1974)[2].

Following Fama (1970)[2]'s market efficiency concepts, Snyder (1978)[1] performed a weak form test (whether knowledge about the subjective odds assigned by bettors can be used to earn an above average return) and a strong form test (whether any special group of people such as the handicappers can outperform others). While he did not test semi-strong form efficiency, his strong form tests are perhaps better interpreted as semi-strong form tests. This is because they are concerned with returns from following the *published* predictions of experts, i.e. not information acted on privately. His weak form efficiency findings are that the favorite-longshot bias is present but it is insufficient to overcome the track take. Ziemba and Hausch (1986, chapter 6)[2] studied the case of extreme favorites where the bias may be sufficient to overcome the track take, but they occur so infrequently that for practical purposes the win market is weak form efficient. Snyder's strong form tests find that the biases produced by the official and non-official handicappers are actually larger than that generated by the public.

Figlewski (1979)[1] used a multinomial logit model to relate the observed frequency of winning to the handicappers' and final odds. He concluded that the handicappers' advice do contain considerable information but that the track odds discount almost all of it. Off-track betting systems seem not to discount the subjective handicapper information as completely, though. Losey and Talbott (1980)[1] re-analysed Snyder's data and concluded that bettors trusting the handicappers are not only unable to get above average return but they may also earn returns lower than the average. Their result strengthens Figlewski's. Following Snyder (1978)[1], Vannebo (1980)[2] argued that the skewness of rate of return may also be important in addition to the first two moments of rate of return.

Grouping by favorite position, like Ali (1977)[1], Asch, Malkiel and Quandt (1982)[1], observed the favorite-longshot bias and showed that the final parimutuel odds are more accurate than morning-line odds in estimating the probability of winning. The latter result is not surprising since the public's odds are the consensus of many bettors while the morning line odds are published by a few handicappers quite some time before the start of the race.

Busche and Hall (1988)[1] and an update by Busche (1994)[1] concluded that Hong Kong bettors do not underbet favorites and overbet longshots or vice versa, a rare exception to the favorite-longshot bias. Busche and Hall (1988)[1] also discussed the classification problem. To test the accuracy of bet fractions, previous researchers classified the data either by bet fractions or favorite positions, and then determined the win frequency within each group of "similar" horses. This may lead to an error in measuring the true winning frequencies of representative horses, though.

To address this classification issue, Bacon-Shone, Lo and Busche (1992)[2] propose to use a simple logit model to analyze the favorite-longshot bias that does not require classifying the data in a specific way. The parameter value of their model, ß, measures how serious the bias is when compared to other racetracks. Using their model, they concluded that the favorite-longshot bias exists using their U.S. data sets (which includes Ali's (1977)[1] and Asch, Malkiel and Quandt's (1984)[1] data) and a new set of Japanese data. However, consistent with Busche and Hall (1988)[1] and Busche (1994)[1], they concluded that the bias does not exist in Hong Kong. The applications of logit models to win bet data, morning line odds data and exotic bet data are summarized in Lo (1994b)[1].

The favorite-longshot bias may not be strong enough to allow a practical and profitable betting scheme, but its existence is till worthy of explanation. Two approaches based on rational betting behavior have considered risk-seeking bettors and differences of opinion. Risk-seeking bettors clearly generates the bias because bettors will demand a higher expected return for favorites which have a lower variance of return than do longshots. Weitzman (1965)[1] and Ali (1977)[1] estimated the utility function of the representative bettor and showed it to be convex. Quandt (1986)[1] proved that locally risk-seeking bettors are a necessary condition for the bias if bettors have homogeneous beliefs, since a loss to the bettors in aggregate follows from a positive track take. Ali (1977)[1] offered the second approach for generating the favorite-longshot bias. He considered races with two horses and assumed that the risk-neutral bettors hold heterogeneous beliefs about the likelihood of each horse winning, beliefs that are drawn from a distribution whose median value is the true winning probability, π. For the favorite ($\pi > 1/2$), a parimutuel market belief of π requires that a fraction π of the bettors have beliefs exceeding π. This is unlikely, though, as π is the median belief. Thus, a favorite-longshot bias will generally be exhibited.

Blough (1994)[1] extended this work to an arbitrary number of horses and allowed risk aversion. This allowed him to develop an econometric test to distinguish these two causes of the favorite-longshot bias. His data did not exhibit bias in the first place, though, and neither cause appeared to be present.

The fixed odds system in used in the U.K. also exhibits the favorite-longshot bias (Crafts (1985)[1], Henery (1985)[2] and Ziemba and Hausch (1986,1987)[2]). Shin (1991[2],1992[1]) showed that the fixed-odds system's bookmakers will construct the bias in their offered odds when there exists a fraction of "insiders" who always bet on the right horse.

[1] included in this volume
[2] cited in the Annotated Bibliography

Anomalies
Parimutuel Betting Markets:
Racetracks and Lotteries

Richard H. Thaler and William T. Ziemba

Economics can be distinguished from other social sciences by the belief that most (all?) behavior can be explained by assuming that agents have stable, well-defined preferences and make rational choices consistent with those preferences in markets that (eventually) clear. An empirical result qualifies as an anomaly if it is difficult to "rationalize," or if implausible assumptions are necessary to explain it within the paradigm. This column will present a series of such anomalies. Of course, "difficult" and "implausible" are judgments, and others might disagree with my assessment. Therefore, I invite readers to submit *brief* explanations (within the paradigm or otherwise) for any of the anomalies I report. To be considered for publication, however, proposed explanations must be falsifiable, at least in principle. Future topics for this column will come from as many fields of empirical economics as possible. Readers are invited to suggest topics by sending a note with some references to (or better yet copies of) the relevant research. The address is: Richard Thaler, c/o *Journal of Economic Perspectives*, Johnson Graduate School of Management, Malott Hall, Cornell University, Ithaca, NY 14853.

Introduction

Economists have given great attention to stock markets in their efforts to test the concepts of market efficiency and rationality. Yet wagering markets are, in one key

■ *Richard H. Thaler is the Henrietta Johnson Louis Professor of Economics at the Johnson Graduate School of Management, Cornell University, Ithaca, New York. William T. Ziemba is Alumni Professor of Management Science, Faculty of Commerce and Business Administration, University of British Columbia, Vancouver, Canada.*

respect, better suited for testing market efficiency and rationality. The advantage of wagering markets is that each asset (bet) has a well-defined termination point at which its value becomes certain. The absence of this property is one of the factors that has made it so difficult to test for rationality in the stock market. Since a stock is infinitely lived, its value today depends both on the present value of future cash flows *and* on the price someone will pay for the security tomorrow. Indeed, one can argue that wagering markets have a better chance of being efficient because the conditions (quick, repeated feedback) are those which usually facilitate learning. However, empirical research has uncovered several interesting anomalies. While there are numerous types of wagering markets, legal and otherwise, this column will concentrate on racetrack betting and lotto-type lottery games.[1]

Racetrack Betting Markets

The "market" at the racetrack convenes for about 20–30 minutes during which time participants place bets on any number of the six to twelve horses in the upcoming race. In a typical race, participants can bet on each horse, either to win, place or show (as well as "exotic" bets which depend on the combined outcomes of two or more horses). The horses that finish the race first, second or third are said to finish "in-the-money." All participants who have bet a horse to win realize a positive return on that bet only if the horse is first, while a place bet realizes a positive return if the horse is first or second, and a show bet realizes a positive return if the horse is first, second or third. There is a separate "pool" of money kept for each type of bet. Payoffs are determined in a "parimutuel" fashion, which means that the winning bets divide the money wagered on losing bets, less transactions costs.[2] The transactions costs consist of a fixed percentage t, which includes the "track take" and "breakage," the additional cost incurred because all returns per dollar bet are rounded down to the nearest five or ten cents. These transactions costs are substantial, typically in the range of 15–25 percent depending on the type of wager and the locale.

The proportion of the money in the win pool that is bet on any given horse can be interpreted as the subjective probability that this horse will win the race. By summing over many races, one can check what proportion of the horses with subjective probabilities between, say, .2 and .25 actually won races. The results of this analysis are impressive. Horses rated by the crowd as most likely to win (the "favorites") do win most often (about 1/3 of the time), and the correlation between subjective and objective probabilities is very high.[3] Apparently the bettors in these markets have considerable expertise.

---

[1] A future column may address other betting markets, such as NFL betting. Authors of recent papers on this topic are requested to send copies to Thaler.

[2] Since payoffs depend only on the final odds, bettors do not know potential payoffs when they bet. In Britain and some other places bookies accept bets on a fixed odds system where bettors are promised a certain payoff if their horse wins.

[3] See, for example, the studies by Weitzman (1965), Rosett (1965), Ali (1977), and Snyder (1978).

Does the high correlation between subjective and objective probabilities imply that the racetrack market is efficient? That depends on the definition of market efficiency. If we assume for the moment that all bettors are expected value[4] maximizers with rational expectations, then two definitions of market efficiency seem appropriate.

Market efficiency condition 1 (*weak*). No bets should have positive expected values.

Market efficiency condition 2 (*strong*). All bets should have expected values equal to $(1 - t)$ times the amount bet.

While the racetrack may be surprisingly efficient, there is substantial evidence that both of these conditions are violated. The most robust anomalous empirical regularity is called the *favorite-longshot bias*. Specifically, the expected returns per dollar bet increase monotonically with the probability of the horse winning. Favorites win more often than the subjective probabilities imply, and longshots less often. This means that favorites are much better bets than longshots. Indeed, extreme favorites, those with odds[5] of less than 3-10 (that is, with a greater than 70 percent chance to win) actually have positive expected values, in violation of condition 1.

Figure 1 (taken from Ziemba and Hausch, 1986) illustrates the favorite-longshot bias using data from most of the previously published studies (including over 50,000 races). Expected returns per dollar bet are plotted for horses at various market odds, using a transactions costs assumption of $t = 15.33$ percent, which applies in the state of California. The horizontal line indicates the point at which returns are the expected $.8467 = (1 - t)$. This occurs at odds of about 9-2 (that is, about a 15 percent probability of winning). For odds above 18-1 there is a steep drop in the expected return, with returns falling to only 13.7 cents per dollar wagered at 100-1. This implies that the typical 100-1 shot has real odds of about 730 to 1! Below 3-10 expected returns are positive, with returns of about 4-5 percent for the shortest odds horses. (This is partially explained by the $1.05 per dollar minimum payoff at nearly all U.S. tracks.) Although such overwhelming favorites are too rare to get very excited about, other profitable betting strategies are discussed below.

Another test of market efficiency is to compare the payoffs of equivalent bets. For example, most tracks offer a "daily double" bet which requires bettors to select the winners of the first two races. Suppose a bettor is considering buying a daily double ticket on horse A in the first race and B in the second race. Then an alternative betting strategy (called a parlay) would be to bet on A in the first race, and, if A wins, bet the proceeds on B to win the second race. Efficiency requires that the daily double payoff on A and B be the same as the parlay on A and B. This proposition has been tested by Ali (1979) and by Asch and Quandt (1987). The conclusion from these tests seems to be that daily double and parlay bets are priced reasonably efficiently relative to each other, though bettors should prefer the daily double because it offers lower transactions costs.

---

[4] Risk neutrality seems like a sensible initial assumption since most bettors probably wager a small portion of their total wealth. Other assumptions will be considered in the commentary section.

[5] Probabilities at the racetrack are traditionally quoted as "odds." If a horse has odds of "x to 1" then the implicit probability the horse will win is $(1 - t)/(x + 1)$.

The Effective Track Payback Less Breakage
for Various Odds Levels in California

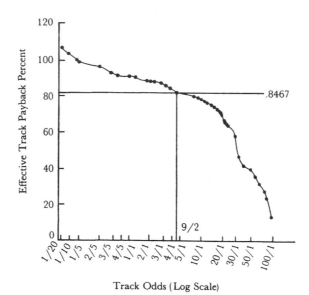

Track Odds (Log Scale)

Fig. 1. The effective track payback less breakage for various odds levels in California

A similar test is possible using exacta betting in which a bettor must correctly pick the first and second place horses in the correct order. Just as the relative amounts bet on different horses in the win pool can be used to calculate implicit forecasts of the probability of winning, similar calculations for the exacta can be made using the so-called Harville (1973) formula. If q_i is the probability that horse i wins, then it is assumed that the probability that horse i is first and horse j is second is $q_i q_j /$ $(1 - q_i)$. (Similarly, the probability that horse i is first, j is second, and k is third is $q_i q_j q_k / (1 - q_i)(1 - q_i - q_j)$.)[6] Asch and Quandt (1987) used the Harville formula to compare the subjective probability of winning implied by the betting in the win pool and the exacta pool. They found that the public did not bet in a mathematically consistent fashion. The implicit probabilities of winning for a given horse were often very different in the two pools.

Betting Strategies

Betting strategies at the track, as at the stock market, come in both fundamental and technical varieties. Fundamental strategies commonly are based on publicly available information used to "handicap" races. A bettor using a fundamental or

[6] The Harville formulas are quite accurate considering how little data they require. However, they tend to overestimate the probability that a low-odds horse will finish exactly second or third. More accurate estimation formulas are derived in Stern (1987), but these require data on all the horses in the race.

handicapping strategy attempts to determine which horses, if any, have probabilities of winning (or placing or whatever) that exceed the market-determined odds by an amount sufficient to overcome the track take.[7] Technical systems require less information and use only current betting data. Bettors using a technical system attempt to find inefficiencies in the market and bet on such "overlays" when they have positive expected value. Most academic research has concentrated on the latter strategies.[8]

Hausch, Ziemba and Rubinstein (1981) (HZR) develop and test a strategy for betting in the place and show markets. They use the amounts bet in the win pool and the Harville formulas to calculate the probabilities of placing and showing for each horse based on the betting in the win pool. Using these methods they are able to identify horses that are underbet to place or show. The basic idea is to compare the proportion of the win pool bet on horse i with the amount of the place or show pool bet on horse i. If, for example, 40 percent of the bets in the win pool are on horse i, but only 15 percent of the bets in the place pool are bet on horse i, then it is profitable to bet on horse i to place. Such profitable betting opportunities typically occur 2 to 4 times per racing day. Empirical studies on two seasons of racing data indicate that significant returns on the order of 11 percent per bet are possible in the place and show markets.[9] This violates weak market efficiency. Moreover, publication of the system does not appear to have eliminated the profitable betting opportunities. At the recent 1987 Breeders' Cup races there were five "system" bets during the racing day. As luck would have it, all five bets paid off.

Ziemba and Hausch (1986) also developed similar techniques to identify and exploit inefficiencies in the exacta markets. The most frequent profitable bets have the favorite in second position. Their plots of the probability of winning and coming in second versus odds as shown in Figure 2, show that short odds horses have a substantial probability of coming in exactly second. The betting public might easily underestimate this chance. The other common profitable wagers are derived from the extreme favorite-longshot bias. Betting extreme favorites in the first position can thus yield profitable bets. The public wagers a considerable amount on these super horses, but not as much as they should. Combinations of longshots are almost never a good exacta wager; such bets typically return 10 to 30 cents on the dollar.

Asch, Malkiel and Quandt (1984, 1986) and Asch and Quandt (1986) investigated whether a drop in the odds late in the betting period might reflect inside information and thereby point to wagers that may have positive expected returns.

---

[7]A useful and insightful book on handicapping, based on a multitude of actual data, is Quirin (1979). The current state of the art in handicapping is described in Mitchell (1987) and Quinn (1987).

[8]Asch and Quandt (1986) did investigate whether the advice of professional handicappers ("touts") or computerized handicapping systems can be used to find profitable bets. They concluded that neither were very useful. See also Mitchell (1987).

[9]For details on the system see Ziemba and Hausch (1987). Hausch and Ziemba (1985) extend the HZR results to analyze the effects of differences in transactions costs, two-horse "entries" and multiple wagers, and provide accurate regression models for varying wealth levels and track "handles" (the size of the betting pool). They also investigate how many bettors have to follow the system for the market to become efficient.

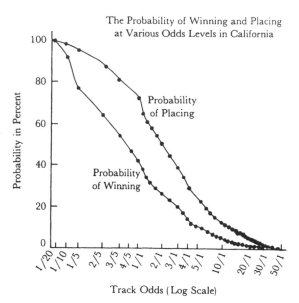

Fig. 2. The probability of winning and placing at various odds levels in California

Common racetrack folklore suggests that the smart money is bet late.[10] This is borne out by Asch, Malkiel and Quandt (1982) using data at various points in the betting cycle from 729 races at the Atlantic City Race Course. They found that for winning horses the final odds tend to be lower than the "morning line odds" (predicted odds by the track handicapper), whereas for horses finishing out of the money the final odds are much higher than the morning line odds. The later in the betting period, the more pronounced is the effect for the winners. The final odds for winners are 96 percent of the morning line odds, but for the money bet during the last eight minutes, the marginal odds are 82 percent of the morning line, and in the last five minutes they drop further to 79 percent. The final odds for losers are about 1.5 times the morning line odds. Asch and Quandt (1986) develop a logit model of the probability of winning, using the change in the odds during the last few minutes as one of the independent variables. The logit model is then used to search for profitable investment strategies. They could not find any profitable bets in the win pool, but they did find some in the place and show pools. Apparently, place and show betting on favorites whose odds have fallen in the last few minutes yields small profits. This is consistent with the Ziemba and Hausch (1987) results suggesting inefficiencies in the place and show pools.

---

[10]Asch and Quandt (1986) argue that this occurs because the smart money bettors want to withhold any information that their bets might convey. This implies that the public will bet more on a horse if the odds are *lower* than expected: an upward sloping demand curve! But if bettors have reservation odds for each horse, then they will also bet more on a horse if the odds are *higher* than expected. Probably the betting public contains a mixture of people using each strategy.

Cross Track Betting

A recent development in race track betting is the opportunity for bettors to wager at their home track on major thoroughbred races being run at another track. Cross track betting raises new and interesting questions about market efficiency. While arbitrage is made difficult by the high transactions costs and the absence of public telephones inside most race tracks, rational expectations would seem to imply that the odds at every track would be approximately the same. In fact, they frequently vary dramatically. For example, in the 1986 Kentucky Derby, the winner Ferdinand paid $16.80 for $2 at Hollywood Park in California where he had run often and was well known. He paid $37.40 at Aqueduct in New York, $79.60 at Woodbine in Toronto, $63.20 at Hialeah in Florida, and $90.00 at Evangeline in Louisiana.

While pure arbitrage may be difficult, profitable betting strategies are possible. Hausch and Ziemba (1987) have developed an optimal betting model for cross track betting under the assumption that final odds at all tracks are known in time to compute and place bets at each track. The essence of the system is to assume that the home track odds are accurate (after correcting for the favorite-longshot bias) and then to select a combination of bets at other tracks to exploit the inefficiencies. If the discrepancies in the odds at the various tracks are large enough (as they have been at some races), it is even possible to create a genuine arbitrage opportunity by betting on every horse at the track where the odds are best. Unfortunately, in the absence of a sophisticated communications system, these strategies are likely impractical (and possibly illegal). However, a Chicago commodities trader has developed and profitably used a workable one-track system using a portable television at the cross track. The bettor views the home odds when they are flashed on television, and then searches for overlays at the cross track.

Lotto Games

Lottery games date at least to biblical days. Israel was divided among the seven tribes by lot. Christ's robe was given to a lottery winner so it would not have to be cut. The Sistine chapel and its paintings were supported by lotteries. The Italian lottery has been running continuously since 1530. Lotteries are played in over 100 countries. Lotteries arrived in North America with the Pilgrims, and they were used to partially fund the new schools such as Harvard, Princeton and Yale. Later they were used to pay off debts of notables such as Thomas Jefferson. Extreme corruption led to their demise in the late 19th century, and they were banned in the United States and Canada. They resurfaced in 1964 in New Hampshire. In Canada they arrived to repay the debts from Expo 1967 in Montreal. Since then there has been explosive growth in popularity and sales. However, with an expected return of between 40 and 60 cents on the dollar, they are usually a poor investment for the rational investor.

Even with such low payout rates, it is possible to obtain positive expected value bets in lotto games. This occurs because not all numbers are equally popular with the public. The possibility of exploiting this pattern was first formalized by Chernoff

(1980) (and tested by some of his students) in the context of the Massachusetts numbers game. In this game the object is to pick a number from 0000 to 9999. If your number is drawn, then you share a portion of the total pool. A subsidiary prize is awarded if three numbers match. Chernoff found that certain numbers were unpopular: those with 0's, 9's, and to a lesser extent 8's. His theoretical analysis suggested that there were combinations with positive expected values, inducing some of his students to bet systematically on the "good" numbers. However, the students did not fare very well. First, over time the unpopular numbers became less advantageous, due to a combination of learning and simple regression. Second, they fell victim to the dreaded "gambler's ruin." The students' bankroll was not sufficient to wait out the time needed to have enough hits to generate substantial profits. Finally, they were unlucky: the unpopular numbers came up less often than would be expected.

The game that has attracted the most attention in North America is Lotto 6/49 or some similar variant. In this game one chooses six of forty-nine numbers and if they all match then one wins the jackpot. Lesser prizes are awarded for three to five matches. The probability of selecting the winning combination in this game is 1 in 13,983,816: if you play twice a week you can expect to win in 134,360 years, a long time horizon even for a rational economist!

Two features make the game interesting for the rational investor. First, as in the numbers game, some numbers are more popular than others. Second, if the grand prize is not won in a given drawing, it is carried over to the next week. Thus prizes can be enormous.[11] Ziemba and his co-workers (1986) have been studying whether these factors can produce favorable investment opportunities. Several estimation methods have been used to calculate the best numbers: simple counts of the frequency with which the numbers are picked, a regression of the log of payoffs on the winning numbers, and a sophisticated constrained maximum likelihood model. All lead to the same conclusion, namely that 15 to 20 of the numbers are quite unpopular. Moreover, the precise numbers are virtually the same from year to year. While there has been some learning over the years, so that these numbers are not quite as unpopular as they used to be, the unpopular numbers tend to remain unpopular. In fact, there are thousands of combinations of numbers that have expected returns over $1 *even when there is no carryover*. The expected value of betting the best numbers increases with the carryover and converges to about $2.25 per $1 for very large pools. The best numbers tend to be high numbers (non-birthdays) and those ending in 0's, 9's and 8's. According to the regression model, the twelve most unpopular numbers are 32, 29, 10, 30, 40, 39, 48, 12, 42, 41, 38, and 18 which tend to be 15 percent to 30 percent less popular than average. Using the marginal approach (those numbers chosen two standard deviations less than average), one finds the nineteen most unpopular

---

[11]In the U.S. lottery prizes are often announced using the accounting methods favored by university fund raisers and agents for professional athletes, namely undiscounted nominal dollars. The present value of the prize after taxes is typically about one-third of the announced value. In Canada, however, prizes are paid in cash and incur no Canadian tax liability.

numbers to be 40, 39, 20, 30, 41, 38, 42, 46, 29, 49, 48, 32, 10, 47, 1, 37, 28, 34 and 45. These numbers have edges from 26.7 percent down to 3.2 percent. The most popular number is 7, which is selected nearly 50 percent more often than the average number.

The question remains—can you make money in the lotto games playing the unpopular numbers? While you can achieve an expected value of $2 per dollar bet, there may not be any arbitrage opportunity available. Consider a hypothetical carnival game with one million spokes. You pay $1 for a number between 1 and 1 million, and you get $2 million if your choice comes up. While you have an edge, the chance of winning is so small that you probably will go bankrupt before winning the jackpot. To analyze this problem one needs to have an adequate model of growth of wealth versus security of wealth. MacLean, Ziemba and Blazenko (1987) have developed such a model to investigate questions such as: Can a dynasty enhance its long-term wealth playing lotto games? The answer is that it can. With sufficiently small wagers it can increase its initial stake of, say, $10 million by tenfold before losing $5 million with probability arbitrarily close to one, but this process takes thousands of years even if they play all over the world. For one lotto game it will take them millions of years. A more interesting question for most of us is: Can a group or single investor use the unpopular numbers to become rich? This is even more difficult, especially if one wants low risk. It is easy to bankroll, as the optimal wager can be as low as 10 cents per week for one of ten syndicate members, but these aspiring millionaires are most likely to be residing in a cemetery when their distant heirs finally reach the goal. It is still best to play unpopular numbers—they have an edge and you will win 3 to 7 times the usual prizes should you hit—but you will expect to play a very long time before winning.

One of the most attractive aspects of lotto games is that the portion of the pool designated for the jackpot is carried over to the next draw when no one wins the jackpot. Indeed it is the prospect of winning a huge jackpot that is the main driving force behind the tremendous interest and sales of lotto tickets. Does it ever pay to buy all the numbers and hence "steal the pot"? Two conditions are necessary for this to be profitable. Roughly speaking these are: (1) a large carryover (in 6/49, $7.7 million); and (2) "not very many" tickets sold. While these conditions are unlikely to obtain, there are cases that can and have arisen in minor lotto games in Canada and elsewhere where it actually would have been a reasonable idea to buy the pot. It is important to stress, however, that even if the right conditions arose, buying the pot would entail enormous transactions costs since the tickets must be bought and redeemed one by one, and, you would have to hope that no one else tried to buy the pot at the same time (see Ziemba et al., 1986, for details). Similar situations sometimes arise in exotic racetrack betting such as the "pick six" (pick the winners of six consecutive races) and related exotic bets. Substantial carryovers can exist in these pools, and making large wagers or actually buying the pot can be profitable. In fact, there are at least two major syndicates that actually try to do this, one of which made over $1 million last year.

Commentary

Racetrack betting

The racetrack betting market is surprisingly efficient. Market odds are remarkably good estimates of winning probabilities. This implies that racetrack bettors have considerable expertise, and that the markets should be taken seriously. Nevertheless, two robust anomalies are present: the favorite-longshot bias, and the inefficiencies of the place and show markets. How can these anomalies be explained?

Quandt (1986) has offered the following argument regarding the favorite-longshot bias (see also Rosett, 1965). The fact that bettors make wagers that are known to have negative expected value implies that they must be "locally" risk seeking. This implies that the usual risk-return relationship will be reversed. In equilibrium, investments (bets) with high variance will have lower average returns than investments with low variance. While this argument is logically consistent, we feel that it is not a satisfactory explanation of the observed behavior. The crucial issue is whether the inference that bettors are risk-seeking is a reasonable one to draw from the fact that they are at the track betting.

What does it mean to be "locally risk seeking"? Recognizing that most racetrack fans, including themselves, purchase insurance, Asch and Quandt (1986) suggest that the utility of wealth function may have the shape proposed by Friedman and Savage (1948), namely concave below the current wealth level and convex above it. While this assumption can explain why racetrack bettors also purchase insurance, it is surely not an adequate explanation for bettors' other behavior such as investing. We venture a guess that when it comes to retirement saving, Professors Asch and Quandt would not be willing to accept a lower mean return in order to obtain a higher level of risk. Indeed, having read their coauthor's book on the stock market (Malkiel, 1985) we guess that when it comes to investing, many racetrack bettors display concave utility functions. Thus the term "locally risk seeking" may apply to racetrack bettors, but only if the term "locally" refers to physical location rather than wealth level![12]

It is true that racetrack fans go to the track to bet—watching a horse race is just not that much fun if you do not have a rooting interest. The real question is to what extent we can explain racetrack betting with the assumptions of rational expectations, expected utility maximization, and a convex utility of wealth function. Consider some stylized facts about racetrack bettors: First, most bring a stake that represents a small portion of their wealth. The average amount bet per person in 1985 was about $150 for the day. The median is surely lower. Second, they allocate that stake over the

[12] The more basic question is whether individuals display a consistent "trait" that can be captured in an index of risk aversion or risk seeking. Psychologists have found that most such traits are highly context specific, and risk taking is no exception. As Paul Slovic (1972, p. 795) has commented: "Although knowledge of the dynamics of risk taking is still limited, there is one important aspect that has been fairly well researched–that dealing with the stability of a person's characteristic risk-taking preferences as he moves from situation to situation. Typically, a subject is tested in a variety of risk-taking tasks involving problem solving, athletic, social, vocational, and pure gambling situations. The results of close to a dozen such studies indicate little correlation, from one setting to another, in a person's preferred level of risk taking."

course of the betting day, intending to bet on nearly every race, unless they run out of money before the day ends. Third, groups of friends that attend the track together rarely bet among themselves, although they could thereby guarantee a zero sum game for the group and increase variance as much as they wanted. Are these facts consistent with maximization of a convex utility of wealth function?

Another fact that is difficult to explain within this framework is the tendency (first pointed out by McGlothlin, 1956) for the favorite-longshot bias to become more pronounced for the last couple of races of the day. Most observers (McGlothlin, 1956; Kahneman and Tversky, 1979; Asch and Quandt, 1986) seem to agree on the cause. Bettors on average are losing toward the end of the day. They would like to go home a winner, but do not want to risk losing much more money. Therefore, they bet on longshots in an attempt to break even for the day. Notice that this behavior is hard to explain within a Friedman-Savage framework. Why should a reduction in wealth increase the tendency for risk seeking?

We feel that a more promising way of modeling race track betting (and other gambling behavior) is to introduce the concept of *mental accounting* (Kahneman and Tversky, 1984; Thaler, 1985). The key assumption of mental accounting is that people adopt mental accounts and act as if the money in these accounts is not fungible. To get the feel of mental accounting, consider another thought experiment. A set of identical twins Art and Bart (with identical wealth levels) is at the race track, contemplating their bets for the last race of the day. Art has lost $100 betting so far, though he has another $100 in cash with him. Bart is even in the betting so far, but between races he read the financial page in the newspaper and discovered that a stock in which he holds 100 shares went down one point the previous day.

Notice that both twins have lost $100, and thus any wealth-based explanation of their betting behavior must predict that they will make similar bets. However, in a mental accounting formulation, Art is behind in the race track account while Bart is even; thus they might well bet differently. (See Thaler and Johnson (1986) for evidence consistent with this view.) Once the concept of mental accounting is introduced, then it becomes much easier to understand how an individual can be risk neutral or risk seeking at the racetrack but risk averse with respect to retirement savings.

As for the favorite-longshot bias, many behavioral factors are probably at work for different reasons. (1) Bettors might overestimate the chances that the long shots will win. (2) As in Kahneman and Tversky's (1979) prospect theory, bettors might overweight the small probability of winning in calculating the utility of the bet. (3) Bettors may derive utility simply from holding a ticket on a longshot. After all, $2 is a cheap thrill. (4) It is more fun to pick a long shot to win than a favorite. It is hard to claim much credit for predicting that a 1-5 favorite will win (much less place or show), but if a 20-1 longshot comes through, considerable bragging rights will have been earned. (5) Some bettors may choose horses for essentially irrational reasons, like the horse's name. Since there is no possibility of short sales, such bettors can drive the odds down on the worst horses, with the "smart money" simply taking the better bets on the favorites.

The fact that the place and show pools seem to be less efficient than the win pool is also an interesting observation. One important factor may simply be that these bets are more complicated. For example, the payoff to a bet to show depends not only on the chance the horse will be in the money, but also on which other horses are in the money and how much has been bet on each. (The greater is the share of the money that has been bet on the horses finishing in the top three positions, the smaller is the payoff.) Bettors might prefer simple bets to complicated bets,[13] or they might simply have difficulty determining when an attractive bet occurs in place and show pools.

One important conclusion we draw from this analysis is that modeling gambling behavior is complicated. Bettors' behavior seems to depend on numerous factors such as how they have done in earlier races, and which bets will yield the best stories after the fact. We should emphasize that these complications apply with equal force to investment behavior. As Merton Miller has said (1986, S467), "[to many individual investors] stocks are usually more than just the abstract 'bundle of returns' of our economic model. Behind each holding may be a story of family business, family quarrels, legacies received, divorce settlements, and a host of other considerations almost totally irrelevant to our theories of portfolio selection. That we abstract from these stories in building our models is not because the stories are uninteresting but because they may be too interesting and thereby distract us from the pervasive market forces that should be our principal concern." While we sympathize with Miller's self-control problem—we also find the stories irresistibly interesting—we feel that to understand the market forces one must enrich the models to incorporate more than the "bundle of returns." Even professional portfolio managers seem more concerned with beating the S & P index than with maximizing returns. In fact, we suspect that portfolio managers trailing the market in the 4th quarter may behave much like the racetrack bettors who bet on longshots when behind at the end of the day.

Lotteries

What can economic theory say about lotteries? Given the dreadful payout rates, one prediction might be that no one will purchase lottery tickets. However, it is easy to rationalize the purchase of a lottery ticket by saying that for a dollar purchase, the customer is paying 50 cents for a fantasy. That's a pretty good deal. The existence of popular and unpopular numbers is more difficult to rationalize. It seems that economic theory yields the following paradoxical prediction: No one will choose the most popular numbers.

To understand this phenomenon it is useful to point out that lotteries in North America did not become popular until New Jersey introduced a game which allowed players to choose their own numbers. The popularity of this feature seems to be explained by what psychologist Ellen Langer (1975) has called "the illusion of control." Even in purely chance games, players feel they have a better chance to win if they can control their own fate, rather than have it determined by purely "chance" factors. For example, Langer found that subjects in her experiments were more

---

[13] For a similar example in the finance literature, see Elton, Gruber, and Rentzler (1982).

reluctant (charged a higher price) to give up a lottery ticket they had selected themselves, than one selected at random for them.

A news story provides a vivid example of the illusion of control (and the confusion of skill and chance). One year, the winner of the Christmas drawing for the Spanish National Lottery, the "El Gordo," was interviewed on television. He was asked: "How did you do it? How did you know which ticket to buy?" Our winner replied that he had searched for a vendor who could sell him a ticket ending in 48. "Why 48?" he was asked. "Well, I dreamed of the number seven for seven nights in a row, and since seven times seven is 48 . . . "[14]

■ *The authors wish to thank, without implicating, Peter Asch, Werner De Bondt, Donald Hausch, Richard Rosett, and Richard Quandt for helpful comments on an earlier draft. Financial support was provided by the Alfred P. Sloan Foundation.*

References

Ali, Mukhtar, M., "Probability and Utility Estimates for Racetrack Bettors," *Journal of Political Economy*, 1977, *85*, 803–15.

Ali, Mukhtar, M., "Some Evidence of the Efficiency of a Speculative Market," *Econometrica*, 1979, *47*, 387–92.

Asch, Peter, and Richard E Quandt, "Efficiency and Profitability in Exotic Bets," *Economica*, August 1987, *59*, 278–98.

Asch, Peter, and Richard E. Quandt, *Racetrack Betting: The Professors' Guide to Strategies*. Dover, MA: Auburn House, 1986.

Asch, Peter, Burton G Malkiel, and Richard E. Quandt, "Market Efficiency in Racetrack Betting," *Journal of Business*, 1984, *57*, 65–75.

Asch, Peter, Burton G. Malkiel, and Richard E. Quandt, "Market Efficiency in Racetrack Betting: Further Evidence and a Correction," *Journal of Business*, 1986, *59*, 157–60.

Asch, Peter, Burton G. Malkiel, and Richard E. Quandt, "Racetrack Betting and Informed Behavior," *Journal of Financial Economics*, 1982, *10*, 187–94.

Chernoff, Herman, "An Analysis of the Massachusetts Numbers Game," Cambridge, Massachusetts, MIT Department of Mathematics, Technical Report No. 23, November 1980.

Elton, E., M. Gruber, and J. Rentzler, "Intra Day Tests of the Efficiency of the Treasury Bills Futures Market," Working paper No. CSFM-38, Columbia University Business School, October 1982.

Friedman, Milton, and L. J. Savage, "The Utility Analysis of Choices Involving Risk," *Journal of Political Economy*, August 1948, *56*, 279–304.

Harville, David A., "Assigning Probabilities to the Outcome of Multi-Entry Competitions," *Journal of the American Statistical Association*, 1973, *68*, 312–16.

Hausch, Donald B., and William T. Ziemba, "Cross Track Betting on Major Stakes Races," Vancouver: University of British Columbia, Faculty of Commerce Working Paper No. 975, June 1987.

Hausch, Donald B., and William T. Ziemba, "Transactions Costs, Extent of Inefficiencies, Entries and Multiple Wagers in a Racetrack Betting Model," *Management Science*, 1985, *31*, 381–94.

Hausch, Donald B., William T. Ziemba, and Mark Rubinstein, "Efficiency of the Market for Racetrack Betting," *Management Science*, 1981, *27*, 1435–52.

Kahneman, Daniel, and Amos Tversky, "Choices, Values, and Frames," *American Psycholo-*

[14] This example is cited in a forthcoming book on decision making by Jay Russo and Paul Schoemaker. We are grateful they shared it with us.

gist, 1984, *39*, 341–50.

Kahneman, Daniel, and Amos Tversky, "Prospect Theory: An Analysis of Decision Under Risk," *Econometrica*, 1979, *47*, 263–91.

Langer, Ellen J., "The Illusion of Control," *Journal of Personality and Social Psychology*, 1975, *32*, 311–328.

McGlothlin, William H., "Stability of Choices Among Uncertain Alternatives," *American Journal of Psychology*, 1956, *69*, 604–15.

MacLean, Leonard, William T. Ziemba, and George Blazenko, "Growth versus Security in Dynamic Investment Analysis," University of British Columbia, Faculty of Commerce and Business Administration, mimeo, 1987.

Malkiel, Burton G., *A Random Walk Down Wall Street*. New York: Norton, 1985.

Miller, Merton H., "Behavioral Rationality in Finance: The Case of Dividends," *Journal of Business*, October 1986, *59*, S451–68.

Mitchell, Dick, *A Winning Thoroughbred Strategy*. Los Angeles: Cynthia Publishing Co., 1987.

Quandt, Richard E., "Betting and Equilibrium," *Quarterly Journal of Economics*, 1986, *101*, 201–207.

Quinn, James, *The Best of Thoroughbred Handicapping 1965–1986*. New York: Morrow, 1987.

Quirin, William L., *Winning at the Races: Computer Discoveries in Thoroughbred Handicapping*. New York: Morrow, 1979.

Rosett, Richard N., "Gambling and Rationality," *Journal of Political Economy*, 1965, *73*, 595–607.

Slovic, Paul, "Psychological Study of Human Judgment: Implications for Investment Decision Making," *Journal of Finance*, 1972, *27*, 779–99.

Synder, Wayne W., "Horse Racing: Testing the Efficient Markets Model," *Journal of Finance*, 1978, *33*, 1109–18.

Stern, Hal, "Gamma Processes, Paired Comparisons and Ranking," Ph.D. dissertation, Department of Statistics, Stanford University, August 1987.

Thaler, Richard H., "Mental Accounting and Consumer Choice," *Marketing Science*, Summer 1985, *4*, 199–214.

Thaler, Richard H., and Eric Johnson, "Hedonic Framing and the Break-Even Effect," unpublished manuscript, Cornell University, Johnson Graduate School of Management, 1986.

Weitzman, Martin, "Utility Analysis and Group Behavior: An Empirical Study," *Journal of Political Economy*, 1965, *73*, 18–26.

Ziemba, William T., and Donald B. Hausch, *Betting at the Racetrack*. Vancouver and Los Angeles: Dr. Z. Investments, Inc., 1986.

Ziemba, William T., and Donald B. Hausch, *Dr. Z's Beat the Racetrack*. New York: William Morrow, 1987.

Ziemba, William T., Shelby L. Brumelle, Antoine Gautier, and Sandra L. Schwartz, *Dr. Z's 6/49 Lotto Guidebook*. Vancouver and Los Angeles: Dr. Z. Investments, Inc., 1986.

HORSE RACING: TESTING THE EFFICIENT MARKETS MODEL

WAYNE W. SNYDER*

> "It is difference of opinion
> that makes horse races."
>
> Mark Twain

SINCE THE EARLY 1960s considerable theoretical and empirical work has been published about the efficient markets model. Two important implications are: (1) prices fully reflect all available information and (2) experts do not achieve above normal profits.

If security markets and horse racing seem worlds apart, the difference is more apparent than real and can be explained largely by the social stigma society attaches to the latter. Both have common characteristics including those which form the basis for the theory of perfect competition: large numbers of participants, extensive market knowledge and ease of entry. Also, like security markets, horse racing offers an opportunity to study economic decision making under conditions of risk and uncertainty. Taken together these characteristics render horse racing both fascinating and worthwhile as an important example where the theoretical framework of the efficient markets model can be effectively applied.

The concept of "efficient markets" refers to a perfectly competitive market where prices reflect all available information. This implies that no method can be found to predict accurately future security prices or, alternatively, to detect securities which are presently undervalued by the market. In a major review of the theoretical and empirical work done in capital markets, Fama (1970) proposed three information subsets by which the efficient markets model could be appraised and tested.[1] *Weak form* tests determine whether past prices alone can predict current prices. *Semi-strong* tests use other publicly available information to predict prices. *Strong form* tests are concerned with demonstrating whether any special group is able to achieve a higher than average rate of return.

In this study I will discuss how these three tests can be applied to horse racing. I will show that horse race bettors exhibit strong and stable biases, but these are not

*Sangamon State University. An earlier version of this paper was presented December 20, 1976 at the Third Conference on Gambling, Las Vegas, Nevada. I am slightly embarrassed at the extent to which I am indebted to others for help of various kinds. Ron Sutherland suggested the framework for this study and provided helpful comments on various drafts. Nancy Jacob and Mark Rubinstein also made several recommendations. H. Fabro, J. Jung, N. Ostroot and J. Rogers gave many tireless hours collecting data. Also, I received the cooperation of the Illinois Racing Board and the *Daily Racing Form's* statistical editor Don Anderson. But my list would be totally inadequate without expressing my gratitude to J. Miller's T.A. seminar for requiring me to make a "contract" to complete this study.

1. Fama (1970) reported that a large number of studies applying all three tests generally supported the efficient markets hypothesis, but he did note some exceptions. Subsequently, Downes and Dyckman (1973) cited additional evidence that not all markets are perfectly efficient.

large enough to make it possible to earn a positive profit. Furthermore, I will demonstrate that "expert" race predictions reflect an even greater bias than the general betting public.

I. PARIMUTUEL BETTING

Horse race odds have been determined by the parimutuel system in the United States since the 1930s. The race track is authorized by the state to subtract a percentage or "take" from the total pool or "handle" bet on each race. The total take exceeds the state-authorized deduction because the track is also allowed to retain "breakage" which results from only paying odds to the nearest (and lowest) 10¢ or 20¢ on a two dollar bet. The total take is divided by rule between taxes paid to the state and the portion retained by the track to meet its operating costs, provide racing purses and profits.[2] Whatever sum is wagered, minus this total take, is paid to bettors on the winning horse. Since each horse is a potential winner, the bettor determined odds are calculated and published in "charts" for each horse in every race.[3] There is, of course, no *a priori* reason why the subjective bettor determined odds should exactly or even closely approach the empirical probabilities of winning.

II. THE WEAK FORM

A suitable weak test for horse racing is to investigate whether knowledge about the subjective odds assigned by bettors through their dollars placed at parimutuel windows can be used to earn an above average return. Perfect competition assumes that bettors will attempt to maximize profits. If the market is efficient, then the expected rate of return for all types of bets would be identical and simply equal the (negative) take. Horse racing would then be said to fully reflect the public's use of all available information and consequently the market could be called "efficient." However, if bettors systematically assign subjective odds which deviate predictably from the empirical probabilities of winning, then the rates of return will not be invariant over all possible odds. If horses are grouped by the bettor assigned odds, then the rate of return (RR) for any odds-group (O) can be calculated as:

$$RR = \frac{W(O+1)-N}{N} \tag{1}$$

where W is the *ex post* number of winning horses in each group and N is the total number of horses in each group.

Rates of return can be calculated for five previously published studies to which I have added my own collection of data.[4] Table 1 describes the six studies; the rates of return are shown in Figure 1. The horizontal axis uses a log scale for convenience of exposition to indicate the *unadjusted* track odds to one dollar. The selected odds-groups have no particular significance and were divided for convenience into eight classes.

2. Track takes have varied from a low of about 13 percent after World War II to a current high of 20 percent in some states.

3. See Fabricant (1965, pp. 22–32) for the economic arithmetic of odds calculation.

4. Too late to be included here, I learned of two other studies by Harville (1973) and Ali (1977) which also support the conclusions reached in this study.

TABLE 1

HORSE RACE STUDIES: AUTHORS, DATES AND NUMBER OF RACES

Author	Date Published	Racing Dates	No. of Races
Fabricant	1965	1955–62	10,000
Griffith	1949	1947	1,124
McGlothlin	1956	1947–53	9,248
Seligman	1975	1975	1,183
Snyder	1978	1972–74	1,730
Weitzman*	1965	1954–63	12,000

*Weitzman did not publish his data, but they were used again and published by Rosett (1965).

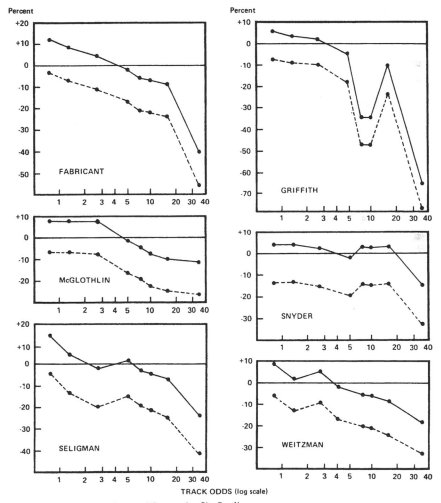

FIGURE 1.   Rates of Return for Six Studies:
Actual (dotted line) and Take Added Back (solid line)

Figure 1 also shows the rates of return with the take added back which makes them more comparable since takes have increased substantially in the thirty years since Griffith (1949) collected the earliest data. All six studies show common characteristics. Each study exhibits a clear bettor bias in that the average rate of return would have been positive on all bets which could have been placed at odds below 5 to 1, if the take were added back. Although only two of the studies (Fabricant and McGlothlin) exhibit monotonically decreasing rates of returns from the lowest to the highest odds, the rate of return tends to vary inversely with the odds in the other studies as well. There does not appear to be any clear shift in the shape of the bettor bias from the earliest study (Griffith) to the most recent (Seligman and Snyder). There is some evidence that the bettor bias is accentuated at smaller tracks where greater uncertainty exists.[5]

Differences among the studies are not very significant as can be judged from the t-values indicated in Table 2 which gives the rates of return (take added back) for each of the six studies. Although each of the studies contains several thousand observations, the rates of return are only significantly different from zero for the very largest studies (Fabricant, McGlothlin and Weitzman) and even for those studies, not all rates of return are significant. In order to show more clearly the bettor bias inherent in the six studies, they have been combined in Figure 2 where the results are summarized for over 300,000 horses which ran in more than 30,000 races held between 1947 and 1975.

TABLE 2

Rates of Return by Grouped Odds, Take Added Back

Study	Midpoint of grouped odds							
	0.75	1.25	2.5	5.0	7.5	10.0	15.0	33.0
Fabricant	11.1[a]	9.0[a]	4.6[a]	− 1.4	− 3.3	− 3.7	− 8.1	− 39.5[a]
Griffith	8.0	4.9	3.1	− 3.1	− 34.6[a]	− 34.1[a]	− 10.5	− 65.5[a]
McGlothlin	8.0[b]	8.0[a]	8.0[a]	− 0.8	− 4.6	− 7.0[b]	− 9.7	− 11.0
Seligman	14.0	4.0	− 1.0	1.0	− 2.0	− 4.0	− 7.8	− 24.2
Snyder	5.5	5.5	4.0	− 1.2	3.4	2.9	2.4	− 15.8
Weitzman	9.0[a]	3.2	6.8[a]	− 1.3	− 4.2	− 5.1	− 8.2[b]	− 18.0[a]
Combined	9.1[a]	6.4[a]	6.1[a]	− 1.2	− 5.2[a]	− 5.2[a]	− 10.2[a]	− 23.7[a]

[a] Significantly different from zero at 1% level or better.
[b] Significantly different from zero at 5% level or better.

The negative relationship in horse betting between the expected rate of return and risk is in sharp contrast with capital market theory which assumes that security buyers require a *higher* rate of return for assuming greater risk (Sharpe, 1970).[6] Another difference is that the rates of return among securities are often related, but

5. Ali (1977, p. 814) discusses this explicitly and my own analyses suggest that at the smaller tracks bettors have a stronger preference for the longer-odds horses. McGlothlin (1956, p. 610) noted that the bettor bias was smallest in the main race, presumedly because uncertainty was least then.

6. However, Miller (1977) suggests that some security buyers may prefer risky investments enough to create a *negative* relationship between the rate of return and risk.

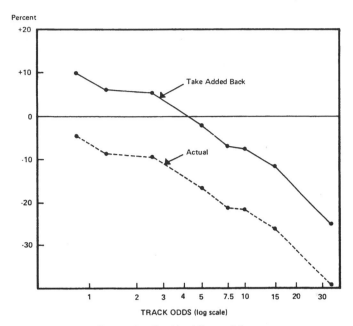

FIGURE 2.   Combined Rates of Return

there is no correlation among different horses, nor are race outcomes correlated with the market return (i.e., the negative take).[7]

The direct cause of the bettor bias is clear. In horse racing parlance, bettors create an "overlay" by betting a smaller proportion on lower-odds horses than their actual chances of winning justifies. If the direct cause is clear, there is no agreement about the basic reasons for the bettor bias. Psychologists, Griffith (1949, p. 293) and McGlothlin (1956, p. 614), concluded that the bettor bias was a reconfirmation of a general tendency for all bettors to prefer low probability-high prize combinations over high probability-low prize combinations, findings first reported in laboratory experiments by Preston and Baratta (1948). Thus, they believe that the important psychological variable is subjective probability. Economists Weitzman (1965, p. 26) and Ali (1977, p. 813) have concluded that horse race bettors behave as if they possess a utility function for money which is concave upward (i.e., increasing marginal utility).

There is, of course, no statistical technique which can disentangle the relative importance of "subjective preference for risk" and "increasing marginal utility of money." Every bet also includes, besides a potential monetary gain, the utility derived from all the factors associated with making a bet—analyzing racing forms, pitting one's predictions against others and the elements of luck. These factors are largely absent if one selects a known favorite; there are few players who can skip the challenge of trying to ferret out potential longer-odds winners. Indeed, the main reward of horse betting comes from the thrill of successfully detecting a moderately long-odds winner and thus confirming one's ability to outperform everyone else.

7. Referee's suggestion.

What then should we conclude about horse racing as an efficient market? The evidence collected for the weak form test shows that the public has a clear and strong bias which substantially affects the expected rate of return for various odds-groups, but that bias is not large enough to overcome track takes of nearly 20 percent.

III. THE SEMI-STRONG FORM

No satisfactory semi-strong test of horse racing has yet been reported. Clearly, what is needed is a general multi-information model capable of incorporating the previous performance data available for each horse and also including whatever new information is made known (e.g., new weights, jockey changes, workouts). While several models have been proposed, they have not been tested sufficiently. Dowst's *Straight, Place and Show: You can beat the races if you know how* (1945) is typical of models which fail to provide any reliability tests of their predictive ability. But there have been three attempts to develop complete models for predicting winning horses: Cohn and Stephens (1963), Fabricant (1965), and Sullivan and Adams (1974). However, only Fabricant (1965, p. 177) reported that his results were statistically significant at the one percent level of confidence. While he has presented an important study which deserves further attempts to reconfirm his results, his evidence alone is too small to conclude that the market for horse racing is not efficient by the semi-strong test.

IV. THE STRONG FORM

Horse race lore is full of persons with potential special information which would permit them to outperform the general public. Scott (1968) identifies and describes with considerable knowledge the individual roles played by owners, trainers, jockeys, grooms and clockers who all aspire to corner special information about each race's potential outcome. Thus, horse racing in one more aspect is similar to the stock market where knowledgeable "insiders" may be able to profit from their unique positions. No one has yet collected systematically the predictions of any of these "experts" and submitted them to a strong test. Fortunately, however, there are three major groups who make their predictions public knowledge: race track officials, the *Daily Racing Form*, and city newspapers.[8]

Each publishes a "line" of predicted odds for every horse entered in each race. The race track presents its line in the official track program. This is established by the track handicapper who at some tracks is also the track secretary. Since the track secretary is responsible for putting together each event, in those instances where the secretary and handicapper are the same person, the official track odds should reflect the most expert opinion about the winning chances of the various horses. Even at tracks where the two functions are assumed by different persons, a close working relationship exists between the two officials; so the handicapper's odds should reflect his special position. However, since the track secretary is

---

8. "Tip" sheets are a fourth potential source of published predictions. These, however, select only two or three horses in each race without attempting to estimate their winning probabilities. Several studies have shown that tip sheets generally produce a return that is worse than a strategy of simply betting each race's favorite.

responsible for putting together a program of similarly qualified horses, it might be predictable that the official track odds would not suggest the wide range of odds that generally prevails when the horses leave the starting gate (and as we have seen above the parimutuel odds substantially understate the true variability between horses with the most and least empirical probabilities of winning).

The *Daily Racing Form* maintains a staff of several handicappers each of whom makes a daily selection of the three horses he believes will finish as win, place and show. In addition, the *Daily Racing Form* publishes its line, a complete list of predicted odds for each horse entered in the day's events.

City newspapers also maintain handicapping experts—although often on a part-time basis only—and they too publish a line of predicted odds in addition to one or more selections of the three top horses in each event.

In order to evaluate the handicapping ability of each of the above experts, I collected and analyzed a second set of races, independent from those in Table 1. The charts for the 1975 summer meeting at Arlington Park, Chicago, were collected for the 846 races run during the 94-day meet. These results were then classified in the same manner as the data discussed in Section II above. Figure 3 shows how this sample of 7,657 entries compares with the combined data shown in Figure 2. Two differences need explaining: (1) there were too few data to warrant using a class with midpoint odds of 0.75 so the class interval was adjusted to coincide with 1.25, and (2) few experts quote odds larger than 30 to 1 so that figure was used in place of the 33 to 1 for the combined study. This new sample once again reconfirms the existence of a clear and substantial bettor bias.

In addition to the official track handicapper's odds and those of the *Daily Racing Form*, three major Chicago newspapers publish an individual line of odds for each

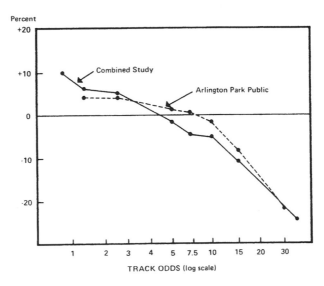

FIGURE 3. Rates of Return With Take Added Back

racing event. The predicted odds of each expert were grouped by the same midpoints used in Figure 3 for the public's parimutuel odds. The rates of return were then calculated by equation (1) using each expert's own predicted odds "as if" they were the actual pay-out odds. Figure 4 compares each expert's rate of return with the public's own.

The most important observation is that nearly all of the experts' odds diverged more from an unbiased prediction than did the general betting public's own parimutuel odds. This can be seen by the extent to which the experts' rates of return are both larger (at low odds) and smaller (at high odds) than the public's rates of return.

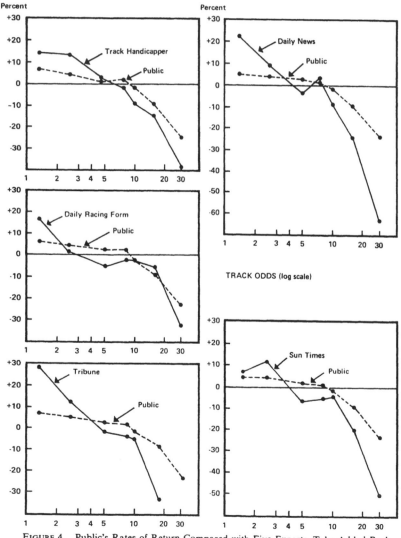

FIGURE 4. Public's Rates of Return Compared with Five Experts, Take Added Back

It might not seem too surprising that the newspaper experts as a group did substantially worse than the other two as in some instances they are part-time employees with no greater expertise than many of the well informed betting public. But it is indeed surprising that the track handicapper should exhibit a similar bias. There are several reasons why each expert's prediction exhibits an even greater bias than the general public.

My interviews with the experts revealed that they do not attempt to predict each horse's actual winning chances, rather they attempt to estimate the odds the public will create through its parimutuel betting. Figure 4 shows clearly that each expert's odds does generally approach closer the public's odds (as illustrated by the public's rate of return) than they do an unbiased estimate. But this doesn't help to explain why the bias in the experts' predictions is generally greater than the public's itself.

When I asked one of the *Daily Racing Form's* handicappers why his predictions deviated from the public's, especially at the lowest and highest odds, he offered two explanations. There was an "unwritten rule never to quote odds larger than 30 to 1." No one, however, remembered when the rule came into being or why it exists! As for the lower odds, he explained that due to his newspaper's power to influence bettor opinions, "we don't want to point the finger too clearly at a horse's winning chances." Thus, for reasons which are perhaps similar to the official track handicapper, the *Daily Racing Form's* policy is to constrain their odds predictions within a smaller range than is actually assigned by the public through its parimutuel betting.

It is possible that greater uncertainty plays a role. Each expert must make his prediction up to 30 hours before the actual race time. Other information which becomes available after their predictions may explain why the public's odds have greater variance than the experts'.[9]

The relative accuracy of each expert's predictions can be evaluated by comparing the root-mean-square differences between the public's and the expert's rates of return which are (in increasing order): official track handicapper (6.2%), *Daily Racing Form* (6.2%), *Sun-Times* (7.1%), *Daily News* (10.8%), and *Chicago Tribune* (14.8%). Thus three of the experts did about equally well while the predictions of two of the newspapers were less accurate.

These five strong tests of market efficiency have each failed to reveal any means of making a positive return after the track take is subtracted. Indeed, each expert reveals a greater bias than does the public itself.

V. CONCLUSION

I have shown through several weak and strong tests that above average and positive profits cannot be expected from horse race betting, a conclusion which is similar to that reached for security markets. While there are statistically significant differences between the subjective and empirical probabilities of winning for particular odds-groups of horses, these differences are not so large as to exceed the price of betting—the track take. Specifically, at the lower-odds the subjective probabilities significantly underestimate the empirical probabilities of horses in these groups to actually win races, whereas at the higher-odds the opposite prevails

9. Rogers offered this explanation.

as the subjective probabilities greatly exceed the empirical probabilities. Moreover, this bias exists for both the betting public and the experts. The reasons for this bias, however, are not identical. The betting public clearly prefers longer-odds to shorter-ones, and due to the parimutuel betting system this substantially decreases the (already negative) expected earnings of longer-odds bets while raising somewhat the earnings of the lower-odds horses. The expert bias arises more from perceived constraints on the range at which the experts quote entries; there appears no reason why experts should prefer longer-odds horses. As for semi-strong tests, only one model has purported to demonstrate absence of an efficient market, hence more tests are required.

REFERENCES

1. M. M. Ali. "Probability and Utility Estimates for Racetrack Bettors," *Journal of Political Economy* 85 (August 1977), pp. 803–815.
2. I. S. Cohne and G. D. Stephens. *Scientific Handicapping* (New Jersey: Prentice-Hall, Inc., 1963).
3. *Daily Racing Form* (Chicago: Triangle Publications Inc.)
4. D. Downes and T. R. Dyckman. "A Critical Look at the Efficient Market Empirical Research Literature as it Relates to Accounting Information," *The Accounting Review* (April 1973), pp. 300–317.
5. R. S. Dowst. *Straight, Place and Show* (New York: M. S. Mill Co., 1945).
6. B. F. Fabricant. *Horse Sense* (New York: David McKay Co., 1965).
7. E. F. Fama. "Efficient Capital Markets: A Review of Theory and Empirical Work," *Journal of Finance* 25 (May 1970), pp. 383–417.
8. R. M. Griffith. "Odds Adjustments by American Horse Race Bettors," *American Journal of Psychology* 62 (April 1949), pp. 290–294.
9. D. A. Harville. "Assigning Probabilities to the Outcomes of Multi-Entry Competitions," *Journal of the American Statistical Association* 68 (June 1973), pp. 312–316.
10. W. H. McGlothlin. "Stability of Choices Among Uncertain Alternatives," *American Journal of Psychology* 69 (December 1956), pp. 604–615.
11. E. M. Miller. "Risk, Uncertainty, and Divergence of Opinion," *Journal of Finance* 32 (September 1977), pp. 1151–1168.
12. M. G. Preston and P. Baratta. "An Experimental Study of the Auction-Value of an Uncertain Outcome," *American Journal of Psychology* 61 (April 1948), pp. 183–193.
13. R. H. Rosett. "Gambling and Rationality," *Journal of Political Economics* 73 (December 1965), pp. 595–607.
14. M. B. Scott. *The Racing Game* (Chicago: Aldine Publishing Co., 1968).
15. D. Seligman. "A Thinking Man's Guide to Losing at the Track," *Fortune* 92 (September 1975), pp. 81–87.
16. U. F. Sharpe. *Portfolio Theory and Capital Markets* (New York: McGraw-Hill Book Co., 1970).
17. D. Sullivan and H. Adam. *Thoroughbred Racing—Predicting the Outcome* (New York: Stafford Publishing Co., 1974).
18. M. Weitzman. "Utility Analysis and Group Behavior: An Empirical Study," *Journal of Political Economics* 73 (February 1965), pp. 18–26.

Back on the Track with the Efficient Markets Hypothesis

ROBERT L. LOSEY and JOHN C. TALBOTT, JR.*

IN A 1978 ARTICLE [10] in this journal, Wayne W. Snyder makes an analogy between security markets and pari-mutuel betting on horse races, suggesting that similarities between the two markets form a basis for the application of the theory of efficient markets to pari-mutuel betting. Snyder's analogy is apt, and his weak form test based on his own findings and similar findings by other researchers [4, 6, 7, 8, 11], is a reasonable test of the weak form efficient markets hypothesis. Furthermore, Snyder's discussion of a strong form test provides a useful basis for the formulation of a test of market efficiency; and with further refinement, it can be transformed into a more powerful test that also provides additional insight as to the influence of handicappers' quoted odds on publicly determined odds. This note further clarifies the applicability of the efficient markets model to pari-mutuel betting markets, and defines and implements a strong form test[1] of the efficiency of pari-mutuel betting markets which is a logical extension of Snyder's work.

In laying the groundwork for his strong form test, Snyder correctly argues that "owners, trainers, jockeys, grooms, and clockers ... aspire to corner special information ..." The implication is that such "insiders" are sometimes able to realize above-average profits as a result of monopolistic access to information. He bases his test of the strong-form hypothesis on the odds predicted by another group, the handicappers whose "expert opinions" are published in the *Daily Racing Form*, daily newspapers, and the track program. There is doubt as to whether handicappers can be considered insiders in the sense of having monopolistic access to information, as the odds which they quote are based primarily on information available in the *Daily Racing Form*, which is widely disseminated to bettors at the track which is the subject of Snyder's study. Furthermore, handicappers' odds are published prior to races with the express purpose of providing handicapping aid to the betting public. Thus it is reasonable to view handicappers as the race track analogue to brokerage firm and other advisory service financial analysts who "handicap" the stock market. Both race track handicappers and financial analysts are "experts" whose primary sources of information are publicly available data used in handicapping their respective markets. In this context, it seems more appropriate to consider a test of the efficiency of pari-mutuel markets

* Financial Economist, Federal Savings and Loan Insurance Corporation, and Professor of Accountancy, Wright State University, respectively.

[1] As suggested in this paper, it may be more appropriate to consider both Snyder's original efforts and the test proposed in this paper as a test of the semi-strong rather than the strong efficient market hypothesis.

279

based on handicappers' prognostications as a test of the semi-strong hypothesis rather than of the strong form.

In Snyder's section on the strong form test of the efficient markets hypothesis, the analysis is based on five separate data sets which are compilations of the public prognostications of each of five "experts". Snyder graphically illustrates the relationship of returns on hypothetical bets made at published handicappers' odds to publicly determined odds and to the empirically determined probabilities of winning. These graphs show that "experts' rates of return are both larger (at low odds) and smaller (at high odds) than the public's rates of return." Snyder then obtains a quantitative measure of the degree of this divergence by calculating the root-mean-square difference for each paired set. He then alludes to the greater bias exhibited by the experts than the public in concluding that his ... "five strong tests of market efficiency have each failed to reveal any means of making a positive return after the track take is subtracted."

While Snyder has shown that publicly determined odds and experts' odds diverge systematically at both ends of the odds continuum, he has not inquired as to how knowledge of this divergence might allow a bettor to improve his chances of finding profitable betting opportunities.[2] A logical extension of Snyder's findings is to test for the presence of a learning lag or other phenomenon which causes the expert opinions of handicappers to be imperfectly incorporated into publicly determined odds.

In constructing a test for such imperfections, we observe that studies of pari-mutuel betting consistently report that long-shots are bet down to shorter odds than their empirically determined probabilities of winning warrant, and Snyder's study reveals that handicappers' odds err even further in this direction. Thus the possibility of using comparisons of handicappers' odds to publicly determined odds in order to find profitable long-shot bets is remote.[3] However, as Snyder and the surveyed studies show, short-odds horses are underbet relative to their empirically determined probabilities; moreover, a subset of short-odds bets can be made at odds even longer than the typically generous odds quoted by handicappers.[4] Such deviations of publicly determined odds from handicappers'

---

[2] It is important to keep in mind that bets can not necessarily be made at the odds quoted by handicappers, and that handicappers typically constrain their reported odds to take account of the track take-out. In light of the latter constraint, the average return on hypothetical bets made at handicappers' odds is a loss equal to the track take-out. The average return on such bets could be positive only if handicappers followed the deceptive practice of inflating their published odds across the board.

The divergence of handicappers' odds from publicly determined odds merely indicates a(n apparent) difference of opinion, the basis of which Snyder explains in his article [10, p. 1118]. The fact that handicappers' odds diverge more from an "unbiased prediction" than publicly determined odds proves only that handicappers err more than the public in predicting a horse's probability of winning.

[3] A bettor whose strategy is to wager on all long shots with odds exceeding the published odds of handicappers would be wagering on almost every long-shot. The return from such a strategy would almost surely approximate the greater than average losses suffered by the typical long-shot bettor.

[4] The odds quoted on the shortest odds bets by two of the five experts (*Daily News* and *Tribune*) would yield a positive return after the track take, if these odds were obtainable. *Daily Racing Form* odds on such bets result in a return which approximately compensates for the track take, and the return on hypothetical bets on the shortest odds horses averaged across all five "experts", appears to allow a slight profit after the track take-out.

reported odds can be used as a basis for a test of market efficiency. The crucial question is whether such cases reflect a failure on the part of the betting public to adequately incorporate expert opinions into their estimates of a horse's chances of winning. If the market imperfectly incorporates this information, this will allow an astute bettor an opportunity to make above average returns.

In implementing our test of market efficiency, we have used data from the Arlington Park summer meet, although for 1978 rather than 1975, as the 1975 Arlington Park data used by Snyder were not readily obtainable. In order to provide a more extensive basis for our test, we have augmented the Arlington Park data with data from the Chicago area's Hawthorne meet which follows Arlington. Analytical results from both data sets were essentially identical, and were pooled in the results reported herein.

Our test makes hypothetical bets on all horses listed by the Racing Form handicapper at 3-1 or less which actually go off at longer odds than those reported by the handicapper. If we can rely on the handicapper's opinion, bets on these "handicapper's overlays" (HOs) should yield above average returns because they are underbet in the public wagering relative to their actual chances.[5] A measure of the extent of the divergence of handicapper's odds from publicly determined odds on a given horse, herein termed the overlay ratio, is obtained by dividing the final pari-mutuel (publicly determined) odds by the handicapper's odds. If we believe that handicappers are truly expert assessors of racing performance, then higher values for the overlay ratio should tend to indicate higher expected returns for the bets in question.

However, neither the average rate of return generated by hypothetical bets on all HOs, nor a further examination of returns for overlay ratios when the data are subdivided by range into three subsamples supports a belief in handicappers' acumen. In fact, the 28.4 percent rate of loss on HO system bets significantly exceeds the 17 percent track take-out, the outcome expected from a random choice strategy.[6] Moreover, it is observed that bets on horses exhibiting higher overlay ratios yield higher losses, a finding which directly contradicts expectations if handicappers' odds are indeed expertly determined.

[5] An analogous strategy in the stock market would be to purchase only those stocks that can be bought at a price less than an analyst's recommended purchase price. The return on wagers selected by the HO test tend to be biased toward above average returns (below average losses) for two reasons unrelated to the expertise of handicappers:

1) A significant majority (72%) of the bets selected by this method are at odds of 5-1 or lower. Snyder's study and the studies cited by him show that horses with publicly determined odds of 5-1 or less yield above average returns (though still negative after track take-out).

2) As pointed out by Snyder [10, p. 1117], the odds quoted by handicappers on short-odds horses tend to understate their actual chances of winning. As all bets made under our system are on horses originally quoted by handicappers at 3-1 or shorter odds, this also suggests that results should be biased toward above average returns.

The fact that results of the HO test reveal significantly greater than average losses in spite of these biases reinforces the validity of our findings.

[6] The legislated track take-out is 16%. However, because pay-offs are alsways calculated to the nearest lower $.10 on the dollar, the take-out on win-pool bets approximates 17%.

Table I
Relationship between the "overlay ratio" and return per dollar wagered for HO system bets[7]

	1.01–1.32	1.33–1.72	1.73+	
The "Overlay Ratio" (Ratio of pari-mutuel odds to handi-capper's odds)				
Mean pari-mutuel (handicap-per's) odds within range	3.05:1 (2.60:1)	4.43:1 (2.82:1)	7.28:1 (2.80:1)	
Percent return per dollar wag-ered	−21.6	−32.8	−32.7	Overall mean of HO system bets = − 28.4%*

Data consists of the 579 qualifying bets drawn from 1305 races.
* Significantly different at the 5% level from the average bettor's return of −17%.

With regard to the implications of our results for market efficiency, if we accept Snyder's implicit definition in which the absence of profit potential indicates efficiency, we conclude that we have insufficient evidence to indicate that pari-mutuel markets are inefficient. However, if we employ the more general test of market efficiency articulated by Fama [5, p. 404], and alluded to by Snyder in the first paragraph of his *Journal of Finance* article, namely that ". . . prices fully reflect all available information," the results of our test suggest that pari-mutuel markets are inefficient.[8]

REFERENCES

1. M. M. Ali. "Probability and Utility Estimates for Racetrack Bettors," *Journal of Political Economy* 85 (August 1977), pp. 803–15.

[7] As an example consider a representative wager from the middle column of overlay ratios. (The data in Table 1 is divided into the lowest, middle, and highest thirds of the overlay ratios.) Such a wager would have been listed by the handicapper at approximately 2.82:1, but closed at publicly determined odds of about 4.43:1. An average bettor wagering $100 on all available 4.43:1 shots should expect a return of *over* $83 since 4.43:1 shots win often enough on average to partially offset the 17% track take-out. However, the subset of bets clustering around 4.43:1 chosen by the HO system generates a return of $67.20 per $100 wagered ($100 less 32.8% loss of $32.80), a result that calls into question both the expertise of handicappers and the efficiency of pari-mutuel markets.

[8] The betting patterns observed in our test are easily explained if a significant part of the betting public mistakenly places its trust in handicappers' opinions and bets on HOs so as to lower the pari-mutuel odds on these supposed overlays, thereby preventing the odds from rising to a level indicative of horses' actual chances of winning.

Our analysis implies that a bettor can raise his expected return above the average by shunning the low-return bets selected by the HO system. It is not implausible to expect that an astute better may find other inefficiencies in pari-mutuel markets, each of which contributes marginally to improve the bettor's chances. On occasion, the sum of the marginal advantages may exceed the 17% track take, and thus allow bets to be made that have positive expected returns.

An interesting implication of the poor performance of handicappers in our test of market efficiency is that there is a pari-mutuel market analogue to the theory of contrary opinion often referred to by stock market analysts. (For a discussion of "contrary opinion" see [2].)

2. R. M. Bleiberg. "Contrary Opinion," *Barron's* (Feb. 25, 1980), p. 7.

3. *Daily Racing Form* (Chicago: Triangle Publications Inc.)

4. B. F. Fabricant. *Horse Sense* (New York: David McKay Co., 1965).

5. E. F. Fama. "Efficient Capital Markets: A Review of Theory and Empirical Work," *Journal of Finance* 25 (May 1970), pp. 383–417.

6. R. M. Griffith. "Odds Adjustments by American Horse Race Bettors," *American Journal of Psychology* 62 (April 1949), pp. 290–94.

7. W. H. McGlothlin. "Stability of Choices Among Uncertain Alternatives," *American Journal of Psychology* 69 (December 1956), pp. 604–15.

8. D. Seligman. "A Thinking Man's Guide to Losing at the Track," *Fortune* 92 (September 1975), pp. 81–87.

9. W. W. Snyder. "Decision-making with Risk and Uncertainty: The Case of Horse Racing," *American Journal of Psychology* 91, No. 2 (June 1978), pp. 201–209.

10. ———— "Horse Racing: Testing the Efficient Markets Model," *Journal of Finance* 33, No. 4 (September 1978), pp. 1109–18.

11. M. Weitzman. "Utility Analysis and Group Behavior: An Empirical Study," *Journal of Political Economics* 73 (February 1965), pp. 18–26.

Subjective Information and Market Efficiency in a Betting Market

Stephen Figlewski

New York University Graduate School of Business Administration

Much of the information available to participants in speculative markets is in the nature of expert opinion, analysis, professional advice, and so on. Markets discount widely held factual information very well; this paper studies market efficiency with respect to subjective information. We examine the "market" for bets on thoroughbred horse races to determine whether the published forecasts of professional handicappers are completely discounted. A multinomial logit probability model is used to measure the information content of the forecasts, and we find that they do contain considerable information but that the track odds generated by betting discount almost all of it. Within the population of bettors, those betting at the track appear to discount the handicapper information fully, but those betting through New York's off-track betting system do not.

I. Introduction

Statistical analysis of financial markets provides considerable evidence that they process publicly available information very well. The collective wisdom of the market rarely overlooks any widely held information which would allow an astute speculator to make higher-than-average profits. But while empirical tests almost always support the efficient-markets hypothesis that a speculative market fully discounts

I would like to thank the New York Racing Association and the New York Off-Track Betting Corporation for help in obtaining the data used in this study. During the research I profited greatly from conversations with Ray Kerrison of the *New York Post* and Henry Baker of the *Racing Form*. Thanks also to Steven Sheffrin, William Silber, and, especially, to Naeem Fayyaz, whose careful assistance in gathering the data was invaluable.

all publicly available information in producing a market price, it is impossible to test specifically whether a market really discounts *all* information. The simplest tests, which involve looking for statistical irregularities in the price series itself that might be exploited by a mechanical trading rule, invariably support market efficiency. Other tests examine market behavior around specific events such as stock splits, dividend changes, changes in the Federal Reserve's discount rate, and so on, and have also tended to show that once news of an event is public knowledge it is completely discounted by the market.

However, much publicly available information is of a type for which market efficiency has never really been tested. This information is what we might call "subjective information" or "expert opinion." In addition to factual data, participants in speculative markets also have access to interpretation and opinion from market analysts, brokerage houses, other investors, and other more or less qualified sources. This subjective information is a major input into the decision-making process of virtually all market participants. We should like to know whether competitive markets are as efficient in discounting it as they seem to be in dealing with factual data. The question is especially interesting since the objective of most market analysis is to produce information that has not already been discounted. A market letter that consistently said the market was efficient and all securities were correctly priced would be of no interest to investors.

Unfortunately, testing market efficiency with respect to subjective information is problematical, since such information is seldom available in an explicit enough form that it can be easily quantified and analyzed statistically. Ideally, we would like to observe a market in which, first, a number of experts made precise forecasts about the outcomes of a set of investment strategies; next, these forecasts were widely disseminated among the participants in the market; and, finally, the market generated a set of prices which could be analyzed to see if they fully discounted the experts' information. Normal financial markets do not operate in this way, but this exact situation does occur in the "market" for bets on thoroughbred horse races. Before each race at a major track a number of professional handicappers study each horse, pick the three they consider most likely to come in first, second, and third, and then publish their opinions. The choices of the handicappers writing for major newspapers are widely available to the bettors well before the race. If the market for bets is efficient, it should be true that the track odds at race time fully reflect the information contained in these handicappers' forecasts.

Data on racetrack betting have been used before to study economic behavior in a risky situation with well-defined payoffs. Rosett (1965) compared the odds on simple bets with those on more complicated strategies involving compound bets and concluded that bettors are rational and sophisticated in evaluating the various betting possibilities they have available. Weitzman (1965) and Ali (1977) both analyzed the relation between the track odds produced by betting and the objective probability that a given horse will win, in order to estimate bettors' utility functions. So far, however, the possibility of using horse-race data to observe the impact of subjective information on a competitive betting market has been overlooked.

In this paper I analyze the forecasts of 14 professional handicappers and test whether the competitive market for bets is efficient with respect to this information. It is possible to show with a very high degree of confidence that the handicappers do possess information and that much of it, though perhaps not all, is discounted by the bettors. Further, efforts at out-of-sample prediction indicate that it may be very difficult to sift any additional useful information out of the forecasts even with relatively sophisticated statistical methods. Betting on thoroughbred horse races appears to be efficient with respect to the subjective information provided by professional handicappers. However, we do find that subsets of the bettor population may differ in their treatment of information. In particular, those betting through the New York off-track betting system seem to make less efficient use of the handicapper information than do on-track bettors.

Section II describes the data and the multinomial logit model I use to analyze the handicapper forecasts. Section III presents the empirical results, and Section IV contains concluding comments.

II. The Model

Betting on horse races is a custom dating back thousands of years. (One may wonder whether the comparatively recent development of trading in corporate equity will prove to be as durable an institution.) In the present-day United States, as in many other countries, horse racing is highly institutionalized, standardized, and regulated by governmental authorities, with the result that bettors are able to place wagers on what has become a closely controlled probabilistic event, at odds that are determined by impersonal market forces in a competitive market for bets. Betting on a horse race is by its nature very different from betting on a game of chance like roulette. Roulette is a "game against nature" in which the probabilities are known ahead of time. In Knight's (1965) terminology, it is a game with risk but no uncertainty. Horse racing, on the other hand, presents the bettor with both risk and uncertainty. While each horse in a race can be thought of as having a certain probability of winning, these probabilities are not and can not be known. This changes the complexion of the game considerably. A rational bettor places a bet on a horse not simply for the pleasure of taking a risk, but because he believes the odds being offered understate the true probability that the horse will win. He is betting that his estimate of the horse's chances is more accurate than the market's estimate. Unlike a casino gambler, a good horseplayer does not count only on luck. There is a reward to gathering information to improve one's probability estimates, and the bettors with superior information and ability to analyze it will be more successful than the rest.

We will be examining data from 189 thoroughbred races run at Belmont (New York) racetrack during June and July 1977. During the racing season there are nine races a day featuring between four and 12 bettable horses. In our sample, the average number of horses per race is between eight and nine. On occasion, more than 12 horses actually compete, but only if two or more of them run "coupled." This means that because they have the same owner or trainer they are treated by the track as a single entry. In the analysis, all data for coupled horses were merged, and the entry was treated as a single horse.

Races are made up about 2 days before they are actually run, and several factors insure that all of the horses in a race are of similar ability. Entrants are restricted by age, sex, and past record. For example, a "maiden" race is only for horses which have not won within a given period. Many races are "claiming" races in which the horses can be "claimed," that is, purchased for a specified price depending on the race. This prevents an owner from entering a very fast horse in a race with those of much lower quality. Although his horse might win the race easily, there is a good chance that the horse would be claimed for a price well below its true value. Further equalizing adjustments are made by altering the weight a horse must carry. Those with poorer past records are often allowed to carry a few pounds less. Apprentice jockeys are also permitted a weight advantage to compensate for their lack of experience. Nonetheless, after all preselection and adjustments, there still remain substantial differences between the horses' probabilities of winning in a given race.

Bettors have an enormous amount of information at their disposal to evaluate these probabilities. Detailed data about each horse are readily available in the *Daily Racing Form* and from other sources; these give full descriptions of each of its previous races, its bloodline, its performance at its latest workouts, and so on. The jockey's records and data about the track are also printed. Any relevant weather or track conditions, like rain or mud, will naturally be taken into account, as will the physical appearance of the horses just before the race.

There are also the published opinions of professional handicappers who may be able to bring more and better data under consideration and to apply their personal knowledge and professional expertise in evaluating it.

We will analyze the information contained in handicappers' selections taken from the published racing columns of three major newspapers, the *Daily Racing Form*, the *New York Post*, and the *New York Daily News*. Fourteen handicappers were available for the whole sample period. The handicappers make their picks at the time the race is made up, so they do not have access to certain relevant information, such as weather conditions, on the day of the race. Nor does a handicapper normally know the selections of other handicappers when he picks his favorites. The newspapers containing these choices are not available until the day of the race, and, in any case, in personal conversations several handicappers said they made no attempt to learn the opinions of other writers before a race.

Late scratches are also a problem. In some cases, almost all of the handicappers' choices are withdrawn before the race actually takes place. To maintain a reasonably homogeneous sample, races were eliminated when a significant number of the prerace favorites were scratched.[1] When a picked horse was scratched, the lower-ranked choices of that handicapper were moved up a position. To avoid a missing data problem for the third picks, we confined the analysis to the first and second choices only. The newspapers containing the handicappers' selections were always available to the bettors well in advance of the races.

There are two ways to bet legally on a race at Belmont: either at the track just before the race, or through New York's off-track betting system (OTB). Bets made both ways are pooled together, so the overall odds are the same.[2] However, there are differences in transactions costs borne by the two sets of bettors. On-track bettors must devote an afternoon to going physically to the track and must also pay transportation costs and an entrance fee. The OTB bettors save the time and other expenses, but, in order to reduce competition with the

---

[1] This was done in the following somewhat ad hoc, but efficient, manner to avoid having to gather too many data that would ultimately prove unusable. For each race, the choices of the five *Daily Racing Form* handicappers were checked to see which picked horses had been scratched. Two points were counted for each "win" pick scratched and one point for each "place." A race with a score of five or more was eliminated.

[2] The state of Connecticut also runs OTB betting on Belmont races, but does not pool its bets with those made in New York. Thus the odds generally differ somewhat between the Connecticut OTB and the New York betting market. The situation would give rise to arbitrage possibilities in the absence of transactions costs. However, in this case, given the significant cuts both New York and Connecticut take out of the betting pools, no profitable arbitrage exists.

racetracks, New York State takes an additional 5 percent (above its normal cut) from OTB winnings.

There is also an informational difference between betting at the track and betting through the OTB. On-track betting windows open about 20 minutes before the race, and, as bets flow in, the totalizer, or "tote" board, displays a running record of the current odds, updated every 30 seconds or so. Thus, an on-track bettor has a good (though not perfect) idea of the actual payoffs on his bets, while an OTB bettor must make do with the "morning line," a set of odds forecasts made by a track employee on the morning of the race.

In a fully competitive betting market, the odds on a given bet will be bid to a level that reflects the market's best estimate of the true probability of winning it. In this case, the market odds on horse i would be given by $(T-B_i)/B_i$, where B_i is the dollar amount bet on horse i, and $T = \sum_j B_j$ is the total betting pool. These odds reflect the bettors' aggregate opinion regarding horse i's probability of winning: that is, in the eyes of the market, horse i's probability is B_i/T. The posted odds differ from these market odds because of the amounts New York State and the track withdraw from the betting pool. First, a flat 17 percent is deducted from the total. Then the payoff on a $2.00 bet is calculated, and the resulting figure is rounded down to a multiple of 20 cents. This second adjustment is known as "breakage," and it makes it impossible to calculate the market's probability estimates directly from the posted odds. For example, before breakage two horses may be bet to pay $2.61 and $2.79, but after breakage both will be posted as paying $2.60. To calculate the market's true estimate of the probabilities, the actual dollar amounts before deductions were obtained from Belmont racetrack. (Throughout the paper, the terms "market odds" and "track odds" refer to these calculated probabilities, not to the posted odds.) Separate figures were available for on-track bettors and for those betting on the same races through the OTB, making it possible to study the information processing of two distinct bettor populations. (Off-track bettors are widely regarded as being less sophisticated.) A horse may be bet to "win," "place" (finish first or second), or "show" (finish first, second, or third). More exotic bets involving two or more horses like the "daily double," "perfecta," and "trifecta" are permitted for some races. In this paper we only consider the odds on bets to win.

The odds in an informationally efficient betting market fully reflect all widely held information. If some known factor, such as a horse's post position, improved its chances of winning over that implied in the odds, astute bettors would recognize this discrepancy and bet on the horse until the market odds moved into line. Free interplay of betting should bid the odds to a level where such information as the jockey, weather conditions, a horse's past performances, and also the published selections of professional handicappers is all accurately discounted. We can write this

$$P \text{ (win} \mid \text{track odds)} = P \text{ (win} \mid \text{track odds} \\ + \text{ all publicly held information).} \quad (1)$$

Given the track odds, it should not be possible to make more accurate probability forecasts by including further treatment of other publicly available knowledge. Conversely, one can test whether the market efficiently discounts specific information by seeing whether it can be combined with the track odds to produce significantly more accurate forecasts than can be made from the odds alone. If so, the market does not accurately discount the information. Thus our strategy in testing the efficiency of this market with respect to the subjective information contained in professional handicappers' forecasts will be to fit a probability model in which their independent contribution to forecasting accuracy can be measured.

For each race in the sample, we had the dollar amount bet on every horse, the names of the two horses each of 14 handicappers considered most likely to win and to run second, and the name of the horse which actually did win. This information was transformed into numerical data in the following way. For each horse a "track-odds" variable was created by dividing the amount bet on it by the total bet for the race. Twenty-eight zero-one dummy variables were created to summarize the handicappers' picks. For example, if handicapper 6 picked horse 3 to come in second, the variable "H6 PLACE" would be set to one for horse 3 and to zero for every other horse in that race. Finally, the outcome was expressed as a dummy: one if the horse won and zero otherwise.

How can we estimate how much information each handicapper gives us about a horse's true probability when he picks it to win? One possible approach would be simply to regress the outcome dummy on the track odds and the set of handicapper dummies. The handicappers' contributions would be measured by the estimated coefficients on their choices, and the fitted value for the dependent variable could be interpreted as the estimated probability of winning. However, it is well known that ordinary least squares is not the best choice of estimator for a probability model. Among other things, it makes no use of the fact that the fitted dependent variable represents a probability and must be between zero and one. The two estimation techniques commonly used in such problems are probit and logit. Logit is far easier computationally for large problems, so it was used.

In the multinomial logit format, each race is treated as an independent drawing from a multinomial distribution in which every horse, i, has its associated probability of winning, P_i. This probability is to be evaluated and explained by a set of independent variables, Z, in this case track odds and handicapper data. The conditional probability is assumed to be related to Z by the function

$$P_i = \frac{e^{Z_i\beta}}{\sum_{j=1}^{N} e^{Z_j\beta}}, \tag{2}$$

where Z_i is the vector of explanatory variables for horse i, β is a vector of coefficients, and N is the number of horses running. This has the advantage over the least-squares formulation that estimated values of P_i all lie between zero and one and the probabilities sum to one for a race.

If we denote the estimated probability associated with the horse that actually won the kth race as P_w^k, then the joint probability of observing a whole set of outcomes in a sample of M independent races is the product of the individual win probabilities from each race,

$$L = \prod_{k=1}^{M} P_w^k. \tag{3}$$

The L is the likelihood function. Multinomial logit estimation consists of finding the parameters, β, in equation (2) such that the joint probability of the whole sample, L, is maximized. As a maximum likelihood technique, logit's large-sample theory is well developed and will allow us to make probability statements about the results. See Nerlove and Press (1973) or McFadden (1973) for a more complete treatment of multinomial logit estimation.

III. The Results

The full sample of 189 races was divided into two groups to permit in-sample estimation on 143 races and out-of-sample prediction on the remainder. The basic results did not appear to be sensitive to the manner in which the sample was split. Table 1 presents the results from the in-sample estimation. The first column represents the case in which neither the handicappers nor the bettors possess any information; β is set to zero for all of the variables, and every horse's predicted probability of winning is just $1/N$, where N is the number of horses in the race. Although this varies from race to race, the average probability of winning of a horse picked at random from the sample works out to 12.0 percent. In the next column the market odds are the only explanatory variable. The odds contribute significantly to explaining the probabilities of winning. The estimated coefficient is seven times

TABLE 1

IN-SAMPLE LOGIT ESTIMATION

	RUN					
VARIABLE	No Infor- mation	Track Odds Only	Handi- cappers Only		Track Odds and Handi- cappers	
Log likelihood	−298.8	−274.0	−264.5		−254.6	
Degrees of freedom	143	142	115		114	
Coefficients of track odds	. . .	5.198 (.73)	. . .		6.255 (1.41)	
Individual handicapper:			Win	Place	Win	Place
H1	. . .	. . .	.483 (.28)	.050 (.28)	.336 (.28)	−.039 (.29)
H2	. . .	. . .	−.211 (.31)	.210 (.26)	−.378 (.33)	.185 (.27)
H3	. . .	. . .	.194 (.28)	.119 (.27)	.066 (.28)	.054 (.27)
H4	. . .	. . .	.214 (.30)	−.359 (.30)	.038 (.31)	−.400 (.29)
H5	. . .	. . .	−.196 (.30)	.007 (.28)	−.647 (.33)	−.128 (.29)
H6	. . .	. . .	.291 (.26)	.020 (.27)	.205 (.26)	−.068 (.28)
H7	. . .	. . .	−.533 (.29)	.046 (.26)	−.560 (.30)	−.034 (.27)
H8	. . .	. . .	.073 (.26)	−.395 (.31)	−.049 (.26)	−.381 (.32)
H9	. . .	. . .	.364 (.26)	.315 (.25)	.407 (.27)	.334 (.26)
H10	. . .	. . .	.431 (.25)	.278 (.27)	.242 (.27)	.118 (.27)
H11	. . .	. . .	.226 (.29)	.629 (.25)	−.205 (.32)	.432 (.25)
H12	. .	. . .	.060 (.35)	.632 (.25)	−.009 (.35)	.605 (.25)
H13	. . .	. . .	.029 (.29)	.243 (.25)	−.239 (.30)	.103 (.25)
H14	. . .	. . .	.458 (.24)	−.061 (.28)	.409 (.25)	−.145 (.29)

NOTE.—Standard errors are given in parentheses.

its standard error. We should expect this result, since the track favorite normally wins 30.0 percent of the time. In our sample the favorite won in 29.4 percent of the races.

The next run included information from the handicappers only. As we might expect, there is considerable correlation among the handicappers' picks, so that individually the coefficients are not highly significant. Only two of them are twice their standard errors. However, as a group they have considerable explanatory power. In the sample the horse with the highest predicted probability based only on the handicappers' information won 28.7 percent of the time. The statistical significance of these variables can be shown with a χ^2 test. In maximum-likelihood estimation the quantity $2\ (\mathscr{L}_u - \mathscr{L}_r)$ is distributed as $\chi^2(k)$, where $\mathscr{L}_r$ is the maximum of the logarithm of the likelihood function estimated when there are k restrictions on the parameters, and $\mathscr{L}_u$ is the unrestricted maximum. In this case the unrestricted maximum is -264.5, and the maximum when all 28 coefficients are restricted to be zero is -298.8 from column 1. The quantity $2\ (\mathscr{L}_u - \mathscr{L}_r)$ is equal to 68.6, which is highly significant; the .001 critical value for a χ^2 distribution with 28 degrees of freedom is 56.9. The handicappers as a group clearly possess relevant information.

Looking at the individual coefficients we see a relatively wide dispersion. Coefficients differ between handicappers by more than an order of magnitude, and six of them are even negative. A negative coefficient should not be interpreted here as implying that the particular handicapper is individually worse than a random prediction. In fact, some of these variables are more accurate on an individual basis than several entering with positive coefficients. The negative coefficient arises because the variable contributes no additional information beyond what is already included in the forecasts of the other handicappers.

In the last column we combine the track odds and the handicapper picks. The significance level of the odds coefficient drops, but it is still more than four times its standard error. The important question is whether adding the handicapper information significantly improves the fit over what was obtained using the odds alone. Twice the difference in the log likelihoods from columns 4 and 2 is 38.8, a figure which is significant at the 10 percent but not at the 5 percent confidence level. While the handicappers do possess considerable information, the bulk of it is discounted in the market odds produced by the bettors. We cannot reject the hypothesis that the market is fully efficient with respect to this information at the customary 5 percent confidence level.

Next the betting pools were disaggregated into on-track betting and OTB in order to study the information processing of two distinct bettor populations. There is reason to suspect that OTB bettors as a group may be quite different from on-track bettors. The OTB is just one of a number of state-run games of chance designed to appeal to the small bettor. To at least some OTB bettors, a wager on a horse race is probably little different from a lottery ticket or a chance in one

TABLE 2

Comparison of On-Track and OTB Bettors

	RUN				
Variable	On-Track Odds Only	OTB Odds Only	On-Track Odds and Handi-cappers	OTB Odds and Handi-cappers	On-Track Odds and OTB Odds
Log likelihood	−272.1	−283.1	−253.1	−261.6	−272.0
Degrees of freedom	142	142	114	114	141
Coefficients:					
On-track odds	4.65	. . .	5.58	. . .	5.04
	(.62)		(1.18)		(1.07)
OTB odds	. . .	4.70	. . .	3.28	−.65
		(.82)		(1.35)	(1.44)

Note.—Standard errors are given in parentheses.

of a variety of legal or illegal "numbers" games. Such bettors will probably not devote much effort to actually analyzing the race. Also, since the state takes an additional 5 percent from OTB winnings, a serious horseplayer who is likely to place large bets will find it worthwhile to go to the track rather than bet via the OTB. The difference in predictive ability between these two populations shows up in the fact that OTB bettors as a group generally receive a smaller fraction of their bets back in winnings than do on-track bettors.

By disaggregating the total pools we can calculate the probabilities implied in each group's betting separately and compare their efficiency in treating handicapper data. The log likelihoods shown in the first two columns of table 2 confirm that the on-track bettors' probabilities have considerably more explanatory power than the OTB odds. This is borne out by running both together, as shown in the last column. When run along with the on-track odds, the coefficient on the OTB odds is negative and insignificant.

When the handicapper information is added to the on-track odds, the log likelihood increases by 19.0. Twice this, 38.0, is just significant at the 10 percent level for a χ^2 distribution with 28 degrees of freedom. As with the overall track odds, we are not able to reject the hypothesis that on-track bettors fully discount the handicappers' picks. For the OTB bettors, however, the situation is different. When the handicapper data are added to the OTB odds the log likelihood increases by 21.5, the coefficient on the OTB odds drops, and its standard error increases. Twice the difference in the log likelihoods is 43.0, a figure which is significant at the 5 percent level. The OTB bettors apparently do not discount the subjective handicapper infor-

TABLE 3

Oᴜᴛ-ᴏꜰ-Sᴀᴍᴘʟᴇ Eꜱᴛɪᴍᴀᴛɪᴏɴ Rᴇꜱᴜʟᴛꜱ

	RUN			
VARIABLE	No Information	Track Odds Only	Handi-cappers Only	Track Odds and Handicappers
Log likelihood	−97.6	−70.3	−90.4	−68.0
Degrees of freedom	46	45	45	44
Coefficients:				
Track odds	...	9.56 (1.49)	...	9.35 (1.50)
Handicappers alone	...	...	.72 (.19)	...
Handicappers with odds	...	...	...	−.57 (.28)

Nᴏᴛᴇ.—Standard errors are given in parentheses.

mation as accurately as do on-track bettors. This can also be seen in the predicted probabilities. Correct treatment of the information contained in published handicapper selections would allow OTB bettors to raise their average win percentage from 27.3 percent to 30.8 percent.

Table 1 shows that the track odds discount much, perhaps all, of the handicappers' information. If the logit model we fitted has in fact captured something from the handicappers' picks which was not reflected in the odds, it should be possible to predict out of sample using the coefficients from table 1. Table 3 gives the results for out-of-sample tests on the other 46 races. Column 1 is the base "no information" run, and the second column shows the results for the track odds alone. For the run labeled "handicappers only," we created a single handicapper variable calculated as the sum of the handicapper dummies multiplied by their coefficients from column 3 of table 1. (These were the maximum-likelihood coefficients from the run with the handicapper data alone.) The estimated coefficient on this composite variable is nearly four times its standard error, so the logit model does indeed forecast out of sample.

However, when a combined handicapper variable is run along with the track odds in the last column, it only contributes negatively. Here the handicapper dummies are multiplied by the coefficients from column 4 in table 1. The estimated out-of-sample coefficient on the composite is negative and even significant. The fact that the best combination of handicapper data contributes nothing in out-of-sample prediction provides further evidence that the track odds do indeed fully discount the published choices of professional handicappers.

IV. Concluding Comments

It is a cliché that differences of opinion are what makes a horse race. To a large extent differences of opinion are also responsible for the trading activity in any speculative market. When people have different information or use different techniques to analyze information it is natural for them to differ about the probabilities of uncertain events. When a market in which they may speculate on their beliefs exists, the market price that emerges embodies an aggregate opinion which to some degree incorporates all of the information available to any of them. Over time we have accumulated a great deal of evidence that competitive speculative markets are informationally efficient with respect to widely held information. That is, free interaction of agents with different beliefs results in a market price which fully and accurately discounts the information they share in common.

Much of the information available in actual markets is subjective information, the result of analysis of factual data by professional information producers. Specialization in information production is normal when data are costly to gather and analyze, especially given the inherent economies of scale. But, for an individual market participant, evaluating subjective information is potentially quite difficult, since he will not normally know the precise factual data available to the expert or his techniques for analyzing it. Further, there will normally be a variety of conflicting opinions from different experts. This poses the question of whether a competitive market will be equally efficient in discounting widely available subjective information (thereby making it valueless).

The results of this paper indicate that the market for bets on thoroughbred horse races does quite well in discounting the subjective information contained in the published predictions of professional handicappers. They did produce valuable information, but it could not be used to improve the forecast accuracy of the market odds significantly. Apparently bettors as a group are able to place appropriate weight on the handicappers' choices in making their bets. However, there is evidence that, within the population of bettors, those who bet through the OTB appear to be less good at evaluating the subjective handicapper information than on-track bettors.

References

Ali, Mukhtar. "Probability and Utility Estimates for Racetrack Bettors." *J.P.E.* 85, no. 4 (August 1977): 803–15.

Knight, Frank H. *Risk, Uncertainty and Profit*. New York: Harper Torchbooks, 1965.

McFadden, D. "Conditional Logit Analysis of Qualitative Choice Behavior." In *Frontiers in Econometrics*, edited by Paul Zarembka. New York: Academic Press, 1973.

Nerlove, M., and Press, S. "Univariate and Multivariate Log-Linear and Logistic Models." RAND Report no. R-1306-EDA/NTH, 1973.

Rosett, Richard N. "Gambling and Rationality." *J.P.E.* 73, no. 6 (December 1965): 595–607.

Weitzman, Martin. "Utility Analysis and Group Behavior: An Empirical Study." *J.P.E.* 73, no. 1 (February 1965): 18–26.

RACETRACK BETTING AND INFORMED BEHAVIOR

Peter ASCH

Rutgers University, Newark, NJ 07102, USA

Burton G. MALKIEL

Yale University, New Haven, CT 06520, USA

Richard E. QUANDT

Princeton University, Princeton, NJ 08544, USA

Received October 1981, final version received February 1982

Horse racing data permit interesting tests of attitudes toward risk. The present paper studies a new sample of racetrack results from Atlantic City, New Jersey. The questions examined are: (1) Are the market odds the best data for predicting the order of finish? (2) Do horses go off at odds that reflect their true probability of winning? (3) Is there any evidence that late bettors have better information than early bettors? It is found that market odds predict the order of finish well, but that 'favorites' are good bets and 'long shots' are poor ones. The data suggest that there does exist an 'informed' class of bettors and that bettors are on the whole neither risk neutral nor risk averse.

1. Introduction

Racetrack betting shares many similarities with investing in the stock market. In both situations future earnings are not known with certainty,[1] there are a large number of participants, there is extensive information available, professional advice abounds, and each participant has information about the activities of other bettors (investors). Horse racing data, therefore, permits interesting tests of attitudes toward risk and 'investment' behavior.

This paper studies a new sample of racetrack results from Atlantic City, New Jersey, and asks a number of questions concerned with betting rationality. These include:

(1) Are the market odds, as determined by the betting behavior of the public, the best data available for predicting the order of finish in a race?

---

[1]The dollar payoffs if the horse wins of fails to win are, of course, known at the start of the race but which of the two outcomes materializes is not.

(2) Do all horses go off at odds that reflect their true probability of winning as suggested by Baumol (1965) or is there a systematic tendency to overbet long-shots and underbet favorites as has been suggested by Rosett (1977), Ali (1977), and Snyder (1978)?

(3) Since each bet is recorded on a parimutuel tote board almost immediately after it is made, there may be some advantages for those with inside information to place their bets late in the betting period just before the race goes off. Such a strategy will minimize the time in which the signal produced by the bet will be available to other bettors. Is there then any evidence that people who bet late in the betting period have better information than other bettors?

The availability of betting information at different time periods is a unique feature of our data set. It permits the testing of somewhat different hypotheses than those that have concerned previous investigators. Section 2 discusses data and definitions. Section 3 contains the results.

2. Data and definitions

2.1. The data set

Observations are based on the entire 1978 thoroughbred racing season at the Atlantic City (NJ) Race Course, which includes 729 races and 5805 horses.[2] The data consist of:

(1) The 'morning line' odds for each horse in each race, determined by the track's professional handicapper and printed in the daily racing program. These are the handicapper's estimates of the winning probabilities for each horse that confront bettors before the start of each betting period. These odds are determined by the judgment of the handicapper and rely on such variables as a horse's past performance, class of the competition, weight carried by the horse, distance of the race, etc. This subjectively determined set of odds, as will be shown below, contains considerable information about the probability of winning for each horse.

(2) Parimutuel odds for each horse in each race at various points ('cycles') during the betting period. Twenty-four cycles are typically recorded, showing the minute-by-minute course of the actual betting for each race. Only some of the available cycles were employed in the analysis. These were chosen to represent periods during which roughly equal amounts of money were bet. Knowing the odds and amounts bet during the various cycles, we were able

---

[2]The data were provided by Eric Weiss. Thirty-six races were discarded from an initial group of 765. The most common reason for discarding was the late removal ('scratch') of an entered horse. Since scratches occur at various times prior to the beginning of a race, it was felt that the odds-behavior of the remaining horses might be distorted in an unsystematic fashion.

to calculate implicit marginal odds for various time periods within the betting period.

(3) Final parimutuel odds for each horse, given at the final betting cycle and comprising the betting pattern over the full betting period. The odds are determined by the ratio of the betting pool available for distribution to the amount bet on each horse. The racetrack is allowed to subtract a percentage or 'take' from the total betting pool to cover taxes, expenses and profits. In addition, the track enjoys 'breakage', the gain from being allowed to round payoffs downward to the nearest 10¢ or 20¢ (on a $2 bet). The total betting pool minus the 'take', including breakage, is paid on the winning horse. Only the 'win' betting pool is considered here.

(4) The outcome of each race: First, second, and third-place finishers.

2.2. Some definitions

The following definitions are employed: B is the total amount bet on the race and b_h is the amount bet on horse h, $h = 1, 2, ..., H$, where H is the number of horses in the race. It follows that

$$B = \sum_{h=1}^{H} b_h.$$

Let t be the total 'take' of the track including breakage, all expressed as a percentage amount of the total amount bet. The odds of horse h may then be stated as

$$O_h = (B(1-t) - b_h)/b_h = B(1-t)/b_h - 1.$$

We define the favorite as the horse with the lowest final odds, the second favorite as the horse with the second lowest odds, etc. We define the (bettors' subjective) probability of horse h winning the race as P_h. It follows then that

$$P_h = b_h/B = (1-t)/(1+O_h).$$

Suppose now we wish to examine the implicit or marginal odds derived from bets made late in the betting period, on the assumption that late bettors may have more or better information than early bettors. The betting period is approximately 25 minutes, which is the time between races. Define cycle 1 as the betting cycle encompassing the first 17 minutes (the first 2/3 of the betting period, and cycle 2 as the betting cycle encompassing the last eight minutes (the last 1/3 of the betting period.[3] The marginal odds on horse h in

---

[3]The two cycles used here are aggregates of the many betting cycles we recorded. One would suspect that 'informed' bettors might postpone their bets to the very end of the betting period — perhaps to the last minute or two. Our data did not permit calculation of implicit odds for this brief segment. The odds implicit in the last eight minutes of betting are thus likely to be only a rough measure of informed betting patterns.

betting cycle 1, $O_{h,1} = B_1(1-t)/b_{h,1} - 1$, where B_1 is the total amount bet up to the end of cycle 1, and $b_{h,1}$ is the amount bet on horse h up to the end of cycle 1. The final odds on horse 1 (which are also the odds on the horse derived from all betting up to and including cycle 2) are

$$O_{h,2} = B_2(1-t)/b_{h,2} - 1.$$

Define the marginal odds as the odds derived from the betting *during* a particular cycle, but *not* including previous cycles. What we specifically want are the marginal odds created by bettors in cycle 2, i.e., the bets of the 'late money', or what is hypothesized to be the 'smart money'. The marginal odds on horse h in cycle 1 are simply the regular odds, i.e.,

$$O_{h,1}^m = O_{h,1}.$$

The marginal odds on horse h in cycle 2 can be derived from the odds at the end of cycles 1 and 2, as follows:

$$O_{h,2}^m = (O_{h,2}b_{h,2} - O_{h,1}b_{h,1})/(b_{h,2} - b_{h,1}).$$

Finally, define a payoff W_h from horse h as unity plus the final odds if h wins and zero otherwise. A horse that goes off at odds of three-to-one pays $4 for each dollar bet if the horse wins. A rate of return R from betting one dollar on one horse in each of N races is

$$R = (\sum W_h - N)/N,$$

where $\sum W_h$ is the sum of payoffs in the N races.

3. Some results

How well do betting odds predict the order of finish in a race? The answer is — 'very well indeed'. Table 1 compares the subjective probability of a horse winning the race (as derived from the 'market' odds of the betting public) with the objective probabilities of winning derived from actual experience. For similar computations, see Ali (1977). In table 1, horses are grouped by the degrees to which they are favorites of the betting crowd. For example row 1 presents the results for the first favorites of bettors, i.e., for horses that go off at the lowest odds and therefore have the highest subjective probability of winning. Row 2 presents results for the second favorites, etc. The subjective probabilities of horse h winning the race are derived from the betting odds as described above.

Table 1

Subjective and objective probabilities of winning in 729 Atlantic City (NJ) races in 1978 (total number of horses = 5805).

Favorites[a] (1)	No. of races[b] (2)	Obj. prob.[c] (3)	Subj. prob. (4)	(Subj. prob. − obj. prob.)/ st. error of obj. prob.[d] (5)
1st	729	0.361	0.325	−2.119[e]
2nd	729	0.218	0.205	−0.903
3rd	729	0.170	0.145	−1.972[e]
4th	724	0.115	0.104	−0.961
5th	692	0.071	0.072	0.074
6th	598	0.050	0.048	−0.279
7th	431	0.030	0.034	0.480
8th	289	0.017	0.025	1.096
9th	165	0.006	0.018	2.095[e]

[a]Lowest odds horses.

[b]The number of races declines because many races have only a small number of entrants. It should be noted that there are numerous races in which there is a tie for which horse is the first favorite, or second favorite, etc. The pool of first favorites was taken to consist of all horses with the lowest odds, including ties, and similarly for the other positions.

[c]Note that these probabilities are the probabilities for the ith favorite conditional on there being an ith favorite in a particular race. Hence they need not sum to unity.

[d]The standard errors were computed by taking the objective probabilities as the 'true' probabilities and assuming a binomial process. Thus the standard error is $[p(1-p)/n]^{\frac{1}{2}}$ [see Ali (1977)], where p is the objective probability and n the number of races.

[e]Significant at the 0.05 level.

The objective probability of the first favorites winning the race is simply the number of times the first favorite wins divided by the total number of races. The calculation is then repeated for the second favorites,..., ith favorites, with the divisor in each case being the number of races in which there existed an ith favorite.

Note that the subjective odds of the betting public do a good job overall of predicting the true objective probabilities. The first favorites do have the highest objective probabilities of winning. They win approximately one-third of the time. The second favorites have the second highest objective probabilities of winning, and so forth. But there is also a clear tendency [as was found by Ali (1977)] that favorites tend to be underbet and long-shots overbet. The objective probability of the favorites winning (0.361) is actually higher than the subjective probabilities (0.325), as estimated by the betting crowd. On the other hand, the subjective estimates for long-shots winning (i.e., the estimates for the 8th and 9th favorites) are well above the objective probabilities of winning. The last column displays the test criterion which is distributed as $N(0, 1)$ under the null hypothesis of no difference between subjective and objective probabilities. In three cases the departure is significant at the 0.05 level. In other words, long-shots seem to go off at lower odds (higher subjective probabilities of winning) than their true objective probabilities of winning would seem to warrant.

We can shed more light on this finding by calculating rates of return from bets on horses with different odds levels in order to provide a further indication of the tendency of the betting public to overbet long-shots. With risk neutrality and in a perfect market, rates of return would tend to be equal for each odds class. In other words, the rates of return for betting horses with short odds ought to be the same as the results of betting horses with intermediate odds and with long odds. Because the track take, including breakage, is about 18.5 percent, however, all rates of return should be negative. On average, bettors should lose roughly 18.5 percent of their 'investments'.

Table 2 shows that this is not the case. There is a clear tendency for the bettor to lose less money betting short odds horses and more money betting on long shots. For example, calculating the rates of return for all races, a bettor would have lost 13.7 percent betting on all horses with odds of 2 to 1 or less and 63.7 percent betting on long shots with odds above 25 to 1. These findings are again similar to those of Ali (1979) and of Snyder (1978). These results are consistent with a view that racetrack bettors are risk lovers, a finding that supports Weitzman's (1965) results.

Table 2

Rates of return from bets on horses with different odds
levels for all races and for late races (races 8 and 9).

Odds level O (1)	Rates of return	
	All races (2)	Late races (3)
$0 \leq 2$	−0.1366	−0.0428
$2 < 0 \leq 3.5$	−0.3177	−0.3210
$3.5 < 0 \leq 5$	−0.1758	−0.0288
$5 < 0 \leq 8$	−0.2242	−0.5238
$8 < 0 \leq 14$	−0.1602	−0.1698
$14 < 0 \leq 25$	−0.3255	−0.3618
$25 < 0$	−0.6372	−0.6858

The last column of Table 2 shows the rates of return from betting results only for the last two races of the day (that is, races 8 and 9). In these races the tendency to underbet favorites and overbet long shots is even stronger. In the late races, betting on favorites produces a loss of just over 4 percent while betting on long shots produces a loss of almost 69 percent. Such findings may be related to the fact that since on average, bettors lose almost

20 percent of their capital, the total capital available to them declines during the racing day. This change in capital could be responsible for a change in risk attitude of the representative bettor. Toward the end of the racing day, bettors at the racetrack may be betting on horses with odds sufficiently long so as to give them a chance of breaking even. This phenomenon was also noted by McGlothlin (1956).

Table 3 presents statistics that shed light on the general efficiency of the final odds and on our conjecture of the behavior of 'informed' bettors with respect to the timing of their bets. Column 2 presents the ratio of final parimutuel odds to morning line odds. It will be noted that for winning horses, the final odds tend to be lower than the morning line odds whereas for horses finishing out of the money, the final odds are much higher. These data suggest that the final parimutuel odds are better predictors than the morning line odds.[4]

Table 3

Average ratios of final (O_F) and marginal odds (O_{M2} and $O_{M2'}$) to morning line odds (O_{ML}) for 729 Atlantic City (NJ) races in 1978.

Horses finishing (1)	O_F/O_{ML} (2)	$O_{M2'}/O_{ML}$ (3)	O_{M2}/O_{ML} (4)
First	0.96	0.82	0.79
Second	1.16	1.06	1.01
Third	1.22	1.17	1.07
Also rans	1.59	1.63	1.49

The last two columns of table 3 present the ratio of marginal odds to morning line odds. The marginal odds O_{M2} represent the marginal odds produced by bettors during the last third (the last eight minutes) of the betting period. The marginal odds O_{M2}, are those produced in the final five minutes of the betting period. We have suggested above that because of the potential signaling effect, bettors who feel they have inside information would prefer to bet late in the period so as to minimize the time that the signal was available to the general public. As table 3 shows, the marginal odds of the late bettors appear to be at least as good as and perhaps better than the final odds in predicting the order of finish. Horses that win have marginal odds that average 79 to 82 percent of their morning line odds (depending on

[4]A similar conclusion would be reached if table 1 is recalculated on the basis of morning line odds rather than final odds. Again, the final odds appear to be somewhat superior to handicappers' odds. It is curious that, on average, the final odds tend to be higher than the morning odds. We suspect this results because the professional handicappers are really providing a set of rankings rather than a set of odds consistent with potential payoffs after accounting for the track take.

which definition of marginal odds ·is employed). In other words, winning horses are especially favored by the late bettors.

4. Conclusions

The central conclusions of this paper are as follows:

(1) Racetrack favorites are 'good' bets and longshots are 'bad' bets — an observation that is in accord with the findings of earlier studies. Such patterns may reflect either an inefficiency in the betting market or variation in attitudes toward risk among bettors [see Losey and Talbott (1980), and Hausch, Ziemba and Rubinstein (1981) who analyze place and show bets].

(2) The data contain a suggestion that there is an 'informed' class of racetrack bettors. It is not possible, however, to define the rates of return that may be earned by this group.[5]

(3) The data suggest that racetrack bettors are on the whole, neither risk neutral nor risk averse.

(4) Despite the variation in rates of return among horses, it is not possible to devise a successful strategy based on observable betting-odds behavior.[6] Whether the variation is attributable to inefficiency or to departures from risk neutral behavior by racetrack patrons, it cannot be exploited by a risk neutral bettor. The variation is too small to overcome the unfairness of the betting terms.

---

[5]Indeed, if the 'informed' rate of return could be isolated, the informed pattern of betting might become vulnerable to imitation, in which case the information would lose its value.

[6]We shall not comment on the likelihood that this paper would have been published, had such a strategy been identified!

References

Ali, Mukhtar M., 1979, Some evidence of the efficiency of a speculative market, Econometrica 47, 387–392.

Ali, Mukhtar M., 1977, Probability and utility estimates for racetrack betting, Journal of Political Economy 85, 803–815.

Baumol, William J., 1965, The stock market and economic efficiency (Fordham University Press, New York).

Hausch, D.B., W.T. Ziemba and M. Rubinstein, 1981, Efficiency of the market for racetrack betting, Management Science 27, 1435–1452.

Losey, R.L. and J.C. Talbott, Jr., 1980, Back on the track with the efficient markets hypothesis, Journal of Finance XXXV, 1039–1043.

McGlothlin, W.H., 1956, Stability of choices among uncertain alternatives, American Journal of Psychology 69, 604–615.

Rosett, R.N., 1971, Weak experimental verification of the expected utility hypothesis, Review of Economic Studies 38, 481–492.

Snyder, Wayne N., 1978, Horse racing: Testing the efficient markets model, Journal of Finance XXXII, 1109–1118.

Weitzman, M., 1965, Utility analysis and group behavior: An empirical study, Journal of Political Economy LXXIII, 18–26.

Application of Logit Models in Racetrack Data

Victor S.Y. Lo

Faculty of Commerce and Business Administration,
University of British Columbia.

Abstract

Recent studies in racetrack market often deal with analyses of bet fractions, estimations of win probabilities and ordering probabilities. The logit model is a common way of statistical modelling of probabilities. This paper discusses and summarizes the possible applications of logit models in racetrack market analyses with win and exotic bets. In addition, we illustrate the application of logit model and how to apply the likelihood-based arguments to compare the accuracies of different sources of information using a real data set. The logit model may be more extensively used in racetrack market analyses.

I. Introduction

Analyses of racetrack data are often related to modelling of probabilities given a set of covariates. The common methods for handling binary dependent variables (such as which horse will win) and the associated probabilities (such as win probabilities) include the logit and probit models. In recent studies, some researchers still employ linear regression for modelling the probabilities due to its simplicity. For example, Asch & Quandt (1987,88) fit linear regression models of "objective probabilities" on "subjective probabilities". In their papers, objective probabilities are either aggregate relative winning frequencies or theoretical probabilities generated under some assumptions, and subjective probabilities are bet fractions. They use linear regression to analyse both win bet and exotic bet data. Although the analyses are simple, fitting a linear regression of probabilities on some covariates is unlikely to approximate the true probabilities over the middle range and completely breaks down at the extremities. In fact, one model of Asch & Quandt (1988) has a negative intercept coefficient which leads to negative probability estimates in some cases.

The probit model requires a cumulative normal distribution function for modelling the relationship between the probability and covariates. For example,
$$\pi_i = \Phi \{ \alpha + \beta x_i \}$$
where π_i = the win probability of horse i and x_i = a particular covariate associated with horse i. Due to its inconvenience of carrying the cumulative distribution function of normal, it is less commonly used in racetrack market analyses and other areas. However, Betton (1994) applies the probit analysis to

Keywords : Racetrack market; Logit models; Win bet fractions; Exotic bets; Ordering probability models; Cox's test.
Acknowledgement : I would like to thank Richard Quandt for his data and William Ziemba for his revision of the earlier draft of this paper.

post-position bias where the covariates are odds rank, post position and number of horses.

This paper discusses and summarizes the applications of logit models in racetrack market analyses. Section II introduces the multinomial logit model and its applications in win bet analysis. Possible extensions of the simple logit model are also discussed. Section III illustrates an application of the logit model using Atlantic City data. Concluding remarks appear in section IV.

II. Logit model for win bet analysis

A general form of multinomial logit model is given by Bolton and Chapman (1986), the win probability of horse i is

$$\pi_i = \frac{\exp(V_i)}{\sum_j \exp(V_j)} \qquad (1)$$

$$(i = 1, \ldots, n)$$

where $\quad V_i = \sum_{k=1}^{N} \theta_k Z_{ik}$,

n = number of horses, N = number of attributes, Z_{ik} = measured value of attribute k for horse i, e.g.

jockey characteristics, weather conditions, and θ_k is a parameter associated with attribute k.

And the log likelihood is given by : $\sum_r \log \pi_{[1],r}$,

where [1],r denotes the horse which won the race r. The win bet fractions are not used as attribute in this model. This may be because the idea is to develop a more explanatory model for prediction of win probabilities in their wagering system.

To analyse the accuracy of win bet fractions with respect to the true win probabilities, the Z-statistics analysis suggested by Ali (1977) may be used. Alternatively, we may use a particular form of the multinomial logit model

$$\pi_i = \frac{\exp(\eta P_i)}{\sum_j \exp(\eta P_j)} \qquad (2)$$

where P_i = win bet fraction of horse i and η is a parameter to be estimated. This is one of the models employed by Asch, Malkiel and Quandt (1984) and Figlewski (1979). In fact, these researchers investigate that whether the additional contributions due to morning line odds or information provided by handicappers are significant. Under the economic assumption that all the bettors are risk-neutral and expected returns maximizer, $\pi_i = P_i$. That is, the win bet fractions determined by the public will be equal to the true win probabilities. This equality is also implied by the efficient market hypothesis, namely noone can gain any benefits by betting on particular horses. However, this model does not include this simplest case, therefore we cannot judge how much better the fitted logit model is when compared to this simplest model. (In fact, Bacon-Shone, Lo and Busche (1992a) verifies empirically that the fitted logit model of (2) is inferior to the simplest model using the data set used in Asch, Malkiel and Quandt (1984)).

To overcome this difficulty, a simple way is to use the model suggested by Bacon-Shone, Lo and Busche (1992a) :

$$\pi_i = \frac{P_i^\beta}{\sum_j P_j^\beta} \qquad (3)$$

This is equivalent to replacing the P_i in (2) with $\log(P_i)$. The parameter β can be interpreted as follows : if $\beta = 1$, $\pi_i = P_i$, in economic sense, the bettors are risk-neutral; if $\beta > 1$, the bettors are risk lovers, i.e. the bettors underbet the favorites and overbet the longshots; similarly, if $\beta < 1$, the bettors are risk averse. Many empirical studies (e.g. Ali (1977); Snyder (1978); Asch, Malkiel & Quandt (1982)) conclude that the bettors underbet the favorites and overbet the longshots - the favorite-longshot bias. A survey of these can be found in Ziemba and Hausch (1985). Therefore, to test whether the bias exists or not, we can simply test whether or not $\beta = 1$. The results in Bacon-Shone, Lo & Busche (1992a) suggest that the favorite-longshot bias exists in many U.S. racetracks but not in Hong Kong. The latter empirical result is consistent with Busche & Hall (1988).

Interestingly, under the assumption of expected utility maximization, the above empirical model (3) is consistent with the utility function suggested in Ali (1977) : $U(x) = x^\beta$ where $U(.)$ is the utility function and x is the return of betting on a horse (i.e. odds plus 1). Assuming that the bettors are risk loving (i.e. the utility function is convex or $\beta > 1$) and have homogeneous beliefs, Blough (1994) also comes up with the same model in (3) using Ali's utility function. He also mentions a similar interpretation for the parameter β. In addition, Blough (1994) includes a more general case that the bettors are both risk loving and have heterogeneous beliefs. However, using a logit model with the additional consideration of heterogeneous beliefs, none of the effects of risk loving and heterogeneous beliefs appear in his data. Extension of the simple β model in (3) includes : (i) inclusion of more covariates such as morning line odds, handicapper information, weather condition and jockey characteristics, etc. in addition to the win bet fractions ; (ii) analysis of exotic bets such as exacta and trifecta, etc.

For (i), inclusion of more covariates, we may use the following extended model :

$$\pi_i = \frac{P_i^\beta Z_{1i}^{\nu(1)} \ldots Z_{mi}^{\nu(m)}}{\sum_j P_j^\beta Z_{1j}^{\nu(1)} \ldots Z_{mj}^{\nu(m)}} \qquad (4)$$

where $Z_{1i}, \ldots, Z_{mi}$ are m additional covariates for horse i and $\nu(1), \ldots, \nu(m)$ are their associated parameters. The significance of contribution made by these additional covariates can be easily assessed by the likelihood ratio test. We will illustrate this model in the next section.

For (ii), analysis of exotic bets, Bacon-Shone, Lo & Busche (1992b) suggest model the probability that horse i wins and j finishes 2nd, π_{ij}, as follows :

$$\pi_{ij} = \frac{\hat{\pi}_i^{\beta} \; \hat{\pi}_{j|1}^{\mu}}{\sum_r \hat{\pi}_r^{\beta} \; \sum_{s \neq 1} \hat{\pi}_{s|1}^{\mu}} \tag{5}$$

where $\hat{\pi}_i = \sum_{j \neq 1} \hat{\pi}_{ij}$, $\hat{\pi}_{j|1} = \hat{\pi}_{ij}/\hat{\pi}_i$, and $\hat{\pi}_{ij}$ = estimated value of π_{ij} —

one may use the win bet fractions together with some ordering probability models (Harville (1973), Henery (1981) or Stern (1990)) to estimate these probabilities. The additional parameter μ in (5) actually adjusts for both the bias due to the ordering probability model used and the favorite-longshot bias. In particular, when $\beta = \mu = 1$, we essentially ignore the favorite-longshot bias and estimate π_{ij} using a particular ordering probability model. The log likelihood for this restricted case then becomes :
$\sum_r \log \hat{\pi}_{[12],r}$ where [12] denotes the pair of horses finishing in the first two positions, and r is the race number.
We may use this log likelihood to compare the accuracies of different ordering probability models. To test this non-nested hypothesis that which one is better, we may use the non-nested test proposed by Cox (1961,62). We illustrate an application of the Cox's test in the next section.

III. Illustrations of the logit model

In this section, we illustrate an application of the simple β logit model in (3) to Atlantic City data (712 races) in 1978. The win bet fractions are obtained just before the race starts, the associated odds are called final odds. In addition, there are morning line odds (usually available in the U.S. racetracks) and earlier odds which are included in our data set.

Before the start of a betting period, the official handicapper gives estimates of the winning probabilities for each horse which are expressed in odds. These are called the morning line odds. These odds are determined by the judgment of the handicapper based on such factors as a horse's past performance, distance of the race, weight carried by the horse, and past performance of the jockey, etc.. We would like to know whether these odds are better than the final odds.

Moreover, the odds determined by the bettors are changing over time until the start of a race. We have obtained some odds data before the start of a race. I investigate whether the final odds are better than the previous odds in estimating the winning probabilities of horses.

We obtain estimates of β in Atlantic City (712 races) using the above two kinds of odds. The empirical results of estimates of β's, their estimated standard errors and log likelihood values are shown in Table 1.

Table 1

Comparisons of $\hat{\beta}$'s using different odds

odds type	$\hat{\beta}$	$\hat{se}(\hat{\beta})$	loglik(1)	loglik($\hat{\beta}$)
Morning line odds	1.5655	.08632	-1282.79	-1259.44
Odds at time before race starts :				
13 min.	1.0909	.06411	-1284.70	-1283.68
9 min.	1.2047	.06825	-1267.30	-1262.58
6 min.	1.2550	.06909	-1251.84	-1244.62
4 min.	1.2406	.06741	-1242.30	-1235.52
3 min.	1.1962	.06508	-1241.98	-1237.21
2 min.	1.1577	.06257	-1237.39	-1234.07
Final odds	1.1023	.05912	-1235.53	-1233.99

For the final odds data (i.e. the win bet fractions) above, $\hat{\beta} = 1.1023$ and is significantly greater from 1 at 5% level of significance (judging from the ratio $(\beta-1)/\hat{se}(\beta)$). This is consistent with the traditional favorite-longshot bias. But the parameter associated with the morning line odds, β, is much larger, this means the bias associated with the morning line odds is more serious.

Another observation is that the log likelihood values increase over time. The log likelihood here are simply the logarithm of joint probability for the observed values of the winning events (discrete events). Therefore it may be interpreted as a measure of how likely the associated odds are for estimating the true probabilities. To test whether the final odds are significantly better than the morning line odds and earlier odds, Cox's (1961,1962) test for non-nested hypothesis can be applied. For example, to test :

H_f : final odds reflect the true win probabilities

vs. H_g : morning line odds reflect the true win probabilities

The test involves two stages. First, evaluate :

$$T_f = \frac{(loglik_f - loglik_g) - E(loglik_f - loglik_g \mid H_f)}{SD(loglik_f - loglik_g \mid H_f)}$$

then, evaluate :

$$T_g = \frac{(loglik_g - loglik_f) - E(loglik_g - loglik_f \mid H_g)}{SD(loglik_g - loglik_g \mid H_g)}$$

where $loglik_i$ = the loglik under H_i with the estimated parameter $(i=f,g)$,

$E(. \mid H_i)$ and $SD(. \mid H_i)$ are expectation and standard deviation when H_i is true.

Under H_f, $T_f \sim N(0,1)$. Under H_g, $T_g \sim N(0,1)$.

If H_g is better than H_f, T_f will be largely negative.

If H_f is better than H_g, T_g will be largely negative.

Sometimes it is difficult to evaluate the expectations and the standard deviations and thus boostrap method is necessary (see Bacon-Shone, Lo & Busche (1992a), for example). In this case, the conditional expectations and standard deviations can be easily obtained and are given by :

$$E(loglik_f - loglik_g \mid H_f) = \sum_r \sum_i {}^f\hat{\pi}_{ir}\left(\ln \frac{{}^f\hat{\pi}_{ir}}{{}^g\hat{\pi}_{ir}}\right) \ ,$$

$$Var(loglik_f - loglik_g \mid H_f)$$

$$= \sum_r \left\{ \sum_i \left(\ln \frac{{}^f\hat{\pi}_{ir}}{{}^g\hat{\pi}_{ir}}\right)^2 Var(Y_{ir} \mid H_f) + \right.$$

$$\left. \sum_{i \neq s} \sum \left(\ln \frac{{}^f\hat{\pi}_{ir}}{{}^g\hat{\pi}_{ir}}\right) \left(\ln \frac{{}^f\hat{\pi}_{sr}}{{}^g\hat{\pi}_{sr}}\right) Cov(Y_{ir}, Y_{sr} \mid H_f) \right\},$$

$$Var(Y_{ir} \mid H_f) = {}^f\hat{\pi}_{ir}(1 - {}^f\hat{\pi}_{ir}) \ ,$$

$$Cov(Y_{ir}, Y_{sr} \mid H_f) = -{}^f\hat{\pi}_{ir} \ {}^f\hat{\pi}_{sr}.$$

Similarly for $E(loglik_g - loglik_f \mid H_g)$ and $Var(loglik_g - loglik_f \mid H_g)$, where

$loglik_k$ = loglikelihood value for the model under H_k (k=f,g)

$$= \sum_r \sum_i Y_{ir} \ln {}^k\hat{\pi}_{ir}$$

Y_{ir} = 0 or 1 according to whether horse i wins the race r

${}^k\hat{\pi}_{ir}$ = estimated winning probability of horse i in the race r under H_k (k = f,g).

The empirical results of Cox's test appear in Table 2. From Table 2, it is clear that, the final odds are more accurate than the morning line odds. This is not very surprising since the final odds are determined by a very large number of bettors but the morning line odds are determined by a few people. This result is consistent with Asch, Malkiel & Quandt (1982,1984) using numerical comparisons rather than statistical tests. Further, Cox's tests support that the final odds are better than any earlier odds above at 5% level of significance. An explanation is that the bettors who have some inside information prefer to bet as late as possible in order to reduce the time that this signal is available to the other bettors.

Next, we consider whether the contribution of morning line odds or earlier odds to the final odds in estimating the win probabilities are significant or not. This can be easily achieved using a particular form of the model (4) :

Table 2
Cox's tests for comparing final odds with other odds

H_f	H_g	T_f	T_g
Final odds	Morning line odds	-0.7131	-9.3762
	Odds at time before		
Final odds	13 min.	1.9263	-10.0018
Final odds	9 min.	1.6900	-8.0501
Final odds	6 min.	0.7170	-5.7250
Final odds	4 min.	-0.1116	-3.6721
Final odds	3 min.	0.6247	-3.6431
Final odds	2 min.	-0.2019	-1.9412

$$\pi_1 = \frac{P_1^{\beta} Z_1^{\nu}}{\sum_j P_j^{\beta} Z_j^{\nu}}$$

where Z_1 is either the subjective probability implied by the morning odds or earlier odds. The empirical results appear in Table 3.

Table 3
Test of contribution of additional covariate

odds type	$\hat{\beta}$	$\hat{\nu}$	max. loglik
Final odds	1.1023	———	-1233.99
Final odds and Morning line odds	1.1013	0.0762	-1233.44
Final odds and Odds at time before race starts :			
13 min.	1.2454	-0.1805	-1233.16
9 min.	1.2355	-0.1668	-1233.60
6 min.	1.0133	0.1078	-1233.90
4 min.	0.6797	0.4864	-1232.94
3 min.	0.9516	0.1668	-1233.90
2 min.	0.5721	0.5605	-1233.42

Table 3 shows that the likelihood ratio statistics, given by $2\{loglik(\beta,\hat{\nu}) - loglik(\beta,0)\}$, are all very small, and thus each additional parameter ν is not significantly different from 0 at any reasonable level of significance. Thus the additional information of morning line odds or earlier odds do not provide any significant contribution to the final odds in estimating the true win probabilities.

IV. Concluding remarks

The logit model for win probabilities on win bet fractions can serve two purposes : (i) detection of favorite-longshot bias by interpreting the parameter estimate appropriately; (ii) estimation of win probabilities by using appropriate covariate(s). In particular, the simple β model (3), which is consistent with some theoretical considerations in Ali (1977) and Blough (1994), can be easily extended to more complicated models (4) or (5). Further, using the log likelihood values which are by-products of the model estimations, we can compare the accuracies among different odds, bet types or ordering probability models (e.g. Bacon-Shone, Lo and Busche (1992b)). With the growing tendency of empirical studies in racetrack market analyses, it is believed that the logit models will be more widely used in this area.

References

Ali, M.M. (1977) "Probability and utility estimates for racetrack bettors." *Journal of Political Economy* 84 (August) ,803-15.

Asch,P.; Malkiel,B. and Quandt,R. (1982) "Racetrack betting and informed behavior." *Journal of Financial Economics* 10 (June) , 187-94.

Asch,P.; Malkiel,B. and Quandt,R. (1984) "Market efficiency in racetrack betting." *Journal of Business* 57 (April) , 165-74.

Asch,P. and Quandt,R. (1987) "Efficiency and profitability in exotic bets." *Economica* 54 , 289-98.

Asch, P. and Quandt,R. (1988) "Betting bias in exotic bets." *Economics Letters* 28, 215-219.

Bacon-Shone,J.H.; Lo,V.S.Y. and Busche,K. (1992a) "Modelling winning probability." Research report 10, Department of Statistics, University of Hong Kong.

Bacon-Shone,J.H.; Lo,V.S.Y. and Busche,K. (1992b) "Logistic analyses for complicated bets." Research report 11, Department of Statistics, University of Hong Kong.

Betton,S. (1994) "Post position bias : An econometric analysis of the 1987 season at Exhibition Park." In this book.

Blough,S.R. (1994) "Differences of opinion at the racetrack." In this book.

Bolton, R.N. and Chapman, R.G. (1986) "Searching for positive returns at the track , A multinomial logit model for handicapping horse races." *Management Science* 32, No.8 (August) , 1040-1059.

Cox, D.R. (1961) "Tests of separate families of hypotheses." *Proc. 4th Berkeley Symp.*, 1 , 105-123.

Cox. D.R. (1962) "Further results on tests of separate families of hypotheses." *Journal of Royal Statistical Society* B 24, No.2 , 406-424.

Figlewski,S. (1979) "Subjective information and market efficiency in a betting model." *Journal of Political Economy* 87 (February) , 75-88.

Harville,D.A. (1973) "Assigning probabilities to the outcomes of multi-entry competitions." *Journal of the American Statistical Association* 68 (June) , 312-16.

Henery,R.J. (1981) "Permutation probabilities as models for horse races." *Journal of Royal Statistical Society* B 43, No.1 , 86-91.

Snyder, Wayne W. (1978) "Horse racing : Testing the efficient markets model." *Journal of Finance* 33 (September) , 1109-1118.

Stern, Hal (1990) "Models for distributions on permutations." *Journal of the American Statistical Association* 85, No.410 (June) , 558-564.

Ziemba,W.T. and Hausch,D.B. (1985) *Betting at the racetrack.* Dr.Z Investments Inc., Los Angeles.

BETTING AND EQUILIBRIUM*

RICHARD E. QUANDT

I. INTRODUCTION

Racetrack betting is a particularly simple situation in which individuals invest money for an uncertain return. In parimutuel betting, individuals invest in "shares" of the various horses. The prices of the shares are standardized, but the payoffs depend on the amount bet on a particular horse relative to the amount bet on all horses. If b_h is the amount bet on horse h and if the racetrack retains a fraction t of all money bet for taxes, expenses, and profit, the payoff per dollar invested on horse h is $(1 - t)\Sigma_i b_i/b_h$ if the horse wins and zero otherwise.

As with the more usual financial instruments, the profitability of "investing" in a particular horse depends on objective factors (the intrinsic ability of the horse, the condition of the track, the skill of the jockey, the qualities of the other horses) and on subjective factors (what other bettors think of the horse, which influences the amount bet on it). Racetrack betting thus shares important characteristics of more general investment markets and, because of the particular simplicity of racetrack betting, permits revealing analyses of risk attitudes and equilibrium.

Various aspects of racetrack betting have been studied. Several authors have examined the relationship between the objective probability of winning a race and the probability reflected by the market odds [Baumol, 1965; Rosett, 1977; Ali, 1977; Snyder, 1978; Asch, Malkiel, and Quandt, 1982]. In general, it is found that "subjective" and objective probabilities are similar, but that favorites tend to be underbet and long shots tend to be overbet. Weitzman [1965] analyzed the extent of risk-loving behavior among bettors. Several authors have examined the question of the efficiency of the betting market [Ali, 1979; Hausch, Ziemba, Rubinstein, 1981; Losey and Talbott, 1980; Asch, Malkiel, and Quandt, 1984]. It appears that small amounts of inefficiency exist in betting "to win" and that somewhat larger, profitably exploitable, inefficiencies exist in betting "to place" or "to show" (i.e., to come in second or third).

*I am indebted to Peter Asch, Avinash Dixit, and Dwight Jaffee for very valuable assistance and advice. Financial assistance from the National Science Foundation is gratefully acknowledged.

In the present paper we investigate the empirical regularity according to which favorites are underbet and long shots are overbet. We define equilibrium and show that the empirical regularity is a consequence of equilibrium in the market.

II. THE EFFICIENT LOCUS

Assume that in a give race an amount b_h is bet on horse h. We normalize the total amount bet to unity so that

$$(1) \qquad \sum_{h=1}^{n} b_h = 1.$$

Under this normalization, the b_h's are in fact probabilities and have been interpreted as the subjective probabilities that horse h will win [Asch, Malkiel, and Quandt, 1982]. These are the quantities that have typically been compared to "objective" probabilities of winning. The odds of horse h are defined as

$$(2) \qquad O_h = (1 - t)/b_h - 1.$$

where t represents the track take including breakage.[1] The outcome of betting one dollar on horse h is either a net win of O_h dollars or a loss of one dollar. Then, denoting by p_h the objective probability that horse h wins, we obtain the mean and the variance of the outcomes as

$$(3) \qquad \mu_h = p_h O_h - (1 - p_h) = p_h \left((1 - t)/b_h\right) - 1$$

and

$$(4) \qquad \begin{aligned} \sigma_h^2 &= p_h O_h^2 + (1 - p_h) - \mu_h^2 \\ &= p_h (1 - p_h) [(1 - t)/b_h]^2. \end{aligned}$$

The set of (μ_h, σ_h^2) for $h = 1, \ldots, n$ describes the feasible choices in mean-variance space.

The track odds O_h and the proportionate amounts bet b_h are observable. The objective probabilities, however, cannot be observed (although they may be estimated from a sample of races). The question is what objective probabilities are compatible with the observed b_h and with equilibrium in the betting market.

Assume that each bettor has an expected utility function over

---

1. See Asch, Malkiel, and Quandt [1982, 1984]. Breakage represents the racetrack's ability to round payoffs to the nearest 10 or 20¢. The magnitude of t is the range of 0.175 to 0.185.

mean and variance. A typical bettor will then choose among the horses on the efficient locus so as to maximize expected utility. But the frequencies with which particular horses are chosen are immediately reflected in the odds (and the b_h's). Equilibrium in the betting market exists if and only if a set of b_h (odds O_h), which establish the efficient locus, result in horse h having a fraction b_h of the total betting pool bet on it for all h. If this were not true, then the betting pattern would result in a set of choices that implies a different efficient locus, which in turn implies a new set of choices, a further change in the efficient locus, etc.

We therefore, define equilibrium in the betting market more formally as follows. The set of numbers $\{b_h, p_h\}$ determines an efficient locus F. But given any F, expected-utility-maximizing bettors make choices that determine new b_h, say b_h'. The betting market is said to be in equilibrium relative to a set of objective probabilities $\{p_h\}$ if there exists a set $\{b_h^*\}$ that is a fixed point of the mapping $\{b_h, p_h\} \rightarrow \{b_h', p_h\}$.

The shape of the efficient locus is of considerable importance in determining the nature of the equilibrium. In order to simplify the analysis, we shall assume that a bettor never diversifies in a given race and bets on a single horse only.[2] This leads to

PROPOSITION 1. The betting market cannot be in equilibrium if $p_h = b_h$ for all $h = 1, \ldots, n$.

Proof. If $p_h = b_h$, the expected payoff for each horse is $\mu_h = -t$ from equation (3), and the variance is $\sigma_h^2 = (1 - t)(1 - b_h)/b_h$ from equation (4). This implies that the efficient locus is a vertical line in mean-variance space.[3] Since racetrack betting is an unfair game, all bettors must be (locally) risk lovers. Hence only the horse with the highest variance will be bet on. But that cannot represent an equilibrium, because it implies that all $b_h = 0$ except for one which equals 1.

It follows from Proposition 1 that equilibrium is incompatible with two or more horses in a race having $p_h = b_h$, since all but one of these horses would be dominated. We must then ask in what fashion p_h can differ from b_h so as to create the possibility of equilibrium. Equilibrium can exist only if the efficient locus has a negative slope.

In light of this, the previous assumption that bettors bet only on a single horse in a race is reasonable because the correlation

---

2. Casual observation suggests that this indeed, is common practice.
3. See also Weitzman [1965] and Asch, Malkiel, and Quandt [1982].

of returns between horses i and j is $- [p_i p_j/(1 - p_i) (1 - p_j)]^{1/2}$ < 0. Diversification is thus a variance-reducing strategy. It follows that convex combinations of horses will be optimal choices only for those bettors with very low elasticities of substitution between mean and variance. This, in particular, distinguishes racetrack betting from other asset markets in which risk-averse individuals consider negative correlations as occasions for diversification.

We now turn to showing that a necessary condition for equilibrium is to have long shots overbet and favorites underbet in the sense that $p_h > b_h$ for favorites and $p_h < b_h$ for long shots.

PROPOSITION 2. *A necessary condition for equilibrium is that* $p_i/b_i > p_j/b_j$ *imply that* $p_i > p_j$.

Proof. Equilibrium implies that the efficient locus has negative slope; i.e., that for any pair of horses such that $\mu_i > \mu_j$ we have $\sigma_i^2 < \sigma_j^2$. By equation (3) this implies that

$$(5) \qquad\qquad p_i/b_i > p_j/b_j,$$

and by equation (4) it implies that

$$(6) \qquad\qquad p_i(1 - p_i)/b_i^2 < p_j(1 - p_j)/b_j^2.$$

Dividing the two sides of equation (6) by the corresponding sides of equation (5) twice, we obtain

$$(1 - p_i)/p_i < (1 - p_j)/p_j,$$

As a corollary, we may now assume that horses are so indexed that $p_1/b_1 > p_2/b_2 > \ldots > p_n/b_n$. It follows that $\mu_1 > \mu_2 > \ldots > \mu_n$ and, if the efficient locus is negatively sloped, $\sigma_1^2 < \sigma_2^2 < \ldots < \sigma_n^2$. Then from Proposition 2, $p_1 > p_2 > \ldots > p_n$. But since $\Sigma b_i = \Sigma p_i = 1$, it is not possible that $1 > p_1/b_1$ or that $1 < p_n/b_n$ There must then exist a k such that

$$\frac{p_1}{b_1} > \ldots > \frac{p_k}{b_k} > 1 \geqq \frac{p_{k+1}}{b_{k+1}} > \ldots > \frac{p_n}{b_n}.$$

We thus have the following monotonicity rule: the ith greatest (proportionate) underbetting occurs on the horse with the ith highest objective probability of winning; moreover, horses actually underbet will generally be horses with high winning probabilities, and horses actually overbet will be those with low winning probabilities.

We show finally that a similar monotonicity condition holds for the subjective probabilities.

PROPOSITION 3. If $\mu_i > \mu_j$ and $\sigma_i^2 < \sigma_j^2$, then $b_i > b_j$.

Proof. From Proposition 2, the conditions of the present proposition imply that $p_i > p_j$. But this implies that

$$p_i(1 - p_i) - p_j(1 - p_j) = p_i - p_j - (p_i^2 - p_j^2)$$
$$= (p_i - p_j)\,[1 - (p_i + p_j)] \geqq 0.$$

Hence

(7) $$p_i(1 - p_i) \geqq p_j\,(1 - p_j).$$

If b_i were less than or equal to b_j, then dividing the left- and right-hand sides of (7) by b_i^2 and b_j^2, we obtain

$$\sigma_i^2 > \sigma_j^2,$$

contradicting the hypothesis.

An intuitive way of looking at Propositions 2 and 3 is to imagine that initially $p_i = b_i$ for all i. The efficient locus is a vertical line as in Figure I. A small perturbation of the b_h's away

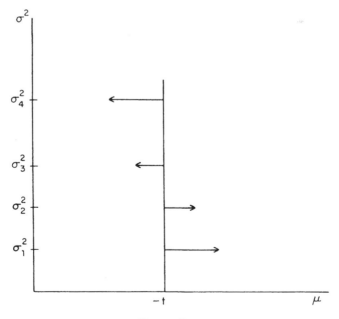

FIGURE I

from equality will change means and variances. Consider a small enough perturbation so that the order of the variances remains unaltered. Then we shall obtain a negatively sloped locus only if the lowest point moves farthest to the right, the second lowest moves second farthest, etc. This is assured by the monotonicity conditions.

It should be noted that the results rest on bettors being risk lovers in a mean-variance framework, without additional assumptions about other motives. In reality, other motives may well be present and may also explain the observed regularity. Thus, some bettors may derive utility from making successful bets on long shots [Hausch, Ziemba, Rubinstein, 1981]. Alternatively, if expected utility depends on mean, variance and on the probability of gambler's ruin, even a risk-averse individual may bet on relative long shots because the probability of gambler's ruin is not necessarily monotone in variance. It is at least conceivable that the observed regularity would manifest itself under these circumstances, although a precise description of the conditions under which it will does not seem possible.

III. CONCLUSION

The oft-encountered empirical regularity that the objective probability of winning exceeds the subjective one for favorites, with the reverse being true for long shots, is a natural consequence of requiring equilibrium to hold in the betting market with risk-loving bettors. In fact, the empirical regularity is a necessary condition for equilibrium, although it may be interpreted in slightly different ways, depending on whether "favorites" are horses with the lowest odds or with the highest objective probabilities of winning.

PRINCETON UNIVERSITY

REFERENCES

Ali, Mukhtar, "Probability and Utility Estimates for Racetrack Betting," *Journal of Political Economy*, LXXXV (1977), 803–15.
——, "Some Evidence of the Efficiency of a Speculative Market," *Econometrica*, XLVII (1979), 387–92.
Asch, P., B. G. Malkiel, and R. E. Quandt, "Racetrack Betting and Informed Behavior," *Journal of Financial Economics*, X (1982), 187–94.
——, ——, and ——, "Market Efficiency in Racetrack Betting," *Journal of Business*, LVII (1984), 165–75.

Baumol, William J., *The Stock Market and Economic Efficiency* (New York: Fordham University Press, 1965).

Hausch, D. B., W. T. Ziemba, and M. Rubenstein, "Efficiency of the Market for Racetrack Betting," *Management Science*, XXVII (1981), 1435–52.

Losey, R. L., and J. C. Talbott, Jr., "Back on the Track with the Efficient Market Hypothesis," *Journal of Finance*, XXXV (1980), 1039–43.

Rosett, R. N., "Weak Experimental Verification of the Expected Utility Hypothesis," *Review of Economic Studies*, XXXVIII (1971), 481–92.

Snyder, Wayne, N., "Horse Racing: Testing the Efficient Markets Model," *Journal of Finance*, XXXII (1978), 1109–18.

Weitzman, M., "Utility Analysis and Group Behavior: An Empirical Study," *Journal of Political Economy*, LXXII (1965), 18–26.

DIFFERENCES OF OPINION AT THE RACETRACK

Stephen R. Blough
Federal Reserve Bank of Boston

1. Introduction

Differences of opinion may be necessary to make a horserace, but in the surprisingly large economics and finance literature on the horseracing market there is little discussion of either the theoretical or the empirical importance of such differences. This paper examines the extent to which such differences can explain a well known empirical regularity in racetrack data.

Snyder [1978] summarizes results from six different data sets: his own and those examined by Griffith [1949], McGlothlin [1956], Fabricant [1965], Weitzman [1965], and Seligman [1975]. In all cases the data show a negative relationship between rates of return and track odds. That is, favored horses (those with low odds) have positive expected returns (after adjusting for track take), while longshots (horses with high odds) have negative expected returns. The phenomenon is also present in data sets used by Ali [1977], Hausch, Ziemba, and Rubinstein [1981], Asch, Malkiel, and Quandt [1982], and Henery [1985]. Ziemba and Hausch [1987] provide a survey of this literature.

The obvious explanation of this phenomenon is that bettors are risk loving. Since the variance of returns is lower for favorites (which win small amounts often) than for longshots (which win large amounts infrequently), risk loving bettors must be compensated by positive expected returns to induce them to place "boring" bets on favored horses. Ali demonstrates this result and estimates an associated utility function. In fact, Quandt [1986] argues that risk loving is a necessary condition for the horseracing market to exist: track take leads to negative expected returns in aggregate. Agents must be risk loving to make positive bets given expected losses.

However, Ali suggests an alternative explanation. He shows that, in a race between two horses, heterogeneous beliefs about the probabilities on the part of risk neutral bettors will lead to a higher expected return on the favorite. This result follows from the parimutuel system for determining the market odds, and will be explained in detail below. This case is not covered in Quandt's argument, since he assumes homogeneous beliefs on the part of bettors. This paper presents a model which allows both risk loving and heterogeneous beliefs, examines the generalization of Ali's result to races with more than two horses, and investigates empirically the relative importance of risk loving and heterogeneous beliefs in explaining racetrack data.

The possiblity of heterogeneous beliefs is important for asset markets generally. Efficient markets results rely on the assumption of homogeneous beliefs. However, heterogeneous beliefs may be important in explaining the volume of trade (Varian [1985],[1987]). Heterogeneous beliefs can lead to the "winner's curse" where each asset is held by those most optimistic about its future returns (although investors may be able to overcome this difficulty — see Milgrom and Weber [1982] regarding this auction problem). Heterogeneous beliefs also have grim implications for the ability of agents to infer from price the private information of other investors. As detailed in the Blough [1987], when there are differences of opinion and of information, agents will not be able to disentangle the influences on price of these two factors. It then makes sense for agents to invest (or bet) based on the disparity between their own evaluations and price.

Other studies of racetrack data suggest that public information is fully reflected in the market price (Figlewski [1979]), as are certain track biases (Canfield, Fauman, and Ziemba [1987]); that there is evidence of profitable private information (Dowie [1976]; Losey and Talbott [1980]; Asch, Malkiel, and Quandt [1982],[1984]; Crafts [1985]); and that there may be arbitrage possibilities between different types of bets (Harville [1973], Hausch, Ziemba, and Rubinstein [1981]; Hausch and Ziemba [1985], [1990]; Ziemba and

Hausch [1987]; Ali [1979] finds no profitable arbitrage between one pair of types of bets).

All of these results are consistent with differences of opinion. If agents do not systematically misuse public information, but rather differ in their opinions because they weight different pieces of information differently, tests of whether public information can systematically improve on market price in predicting returns are unlikely to reject market efficiency. However, differences of opinion will allow profitable insider trading to take place because outsiders cannot tell whether price movements are due to the differing information or the differing opinions of others. Similarly, arbitrage opportunities between different investments may persist if they are small compared to the profit opportunities investors perceive in the disparity between price and their own evaluations.

The next section will describe the horseracing market, assumptions about bettors, and the equilibrium condition. The third section will derive the results for races with two horses, presenting Ali's findings as special cases of a model which allows both risk loving and heterogeneous beliefs. Section 4 will consider the generalization to races with more than two horses. Section 5 will illustrate concepts introduced in section 4 using the case where beliefs can be described with a log–normal distribution. Section 6 develops a maximum likelihood procedure for distinguishing empirically the effects of risk loving and heterogeneous beliefs. Section 7 describes the data to be used, and section 8 presents results. Section 9 contains a summary and conclusions.

2. The Horseracing Market

Consider a race with J horses entered, and N agents who wish to place bets on the horses. Bettors choose their wagers in a manner to be described below. Let the total amount bet on horse j be W_j, and the total bet on all horses be W (we consider here only the win market, ignoring place, show, and "exotic" bets). Then

$$B_j \equiv W_j/W$$

is the fraction of total wagers placed on horse j.

The track keeps a fraction of the total amount bet (track take). Let this fraction be $(1-Q)$, so that Q is the payout ratio. Then an amount QW is distributed among bettors on the winning horse. The net return on a dollar bet on horse j if that horse wins is

(2.1) $$\theta_j \equiv QW/W_j - 1 = Q/B_j - 1$$

This net return θ_j is the payoff odds on horse j. Let $\underline{\theta}$ be the vector giving the odds for each of the J horses.

We can use $\underline{\theta}$ to define a vector of "market probabilities" $\underline{\pi}$:

$$\pi_j \equiv Q/(1+\theta_j)$$

or

$$\theta_j \equiv (Q-\pi_j)/\pi_j.$$

The market probabilities have the property that expected returns on each horse, evaluated using these probabilities, are equal to the aggregate net return $Q-1$:

$$\pi_j \cdot \theta_j + (1-\pi_j) \cdot (-1) = Q-1$$

Let Π_j be the true probability that horse j will win, and let $\underline{\Pi}$ be the vector of these probabilities. Just as there are market odds corresponding to the market probabilities, there are true odds corresponding to the true probabilities. Denote these by $\underline{\Theta}$, where $\Theta_j=(Q-\Pi_j)/\Pi_j$.

Having established notation, we now consider the problems of the individual bettors, followed by a description of the market equilibrium.

2.1 Behavior of Bettors

Each of the N bettors must decide on which of the J horses to bet. To simplify the exposition, we follow Ali, among others, in assuming that the amount bet is fixed in advance and the same for all bettors (the implications of this assumption are discussed in section 2.3 below). Assume then that each bettor must bet $1 on exactly one horse.

Bettors are assumed to be identical in all respects except possibly their beliefs about the vector of win probabilities. Bettors receive utility from *ex post* returns to their bets according to the utility function

$$(2.2) \qquad\qquad U(x) = (x+1)^\beta \qquad \beta > 0$$

Bettors will then be risk averse, risk neutral, or risk loving as β is less than, equal to, or greater than 1.

Each bettor has beliefs about the win probabilities of the horses. Bettor n's beliefs are given by the vector $\underline{p}_n$, so that p_{jn} gives bettor n's belief about the probability that horse j will win. Bettors are faced with "market prices" in terms of the vector $\underline{\theta}$ of payoff odds. Bettor n will evaluate the expected utility of a bet on each horse using his beliefs $\underline{p}_n$. Since the utility of a bet on a horse which loses is 0 ($x = -1$), the expected utility to bettor n of a bet on horse j is

$$(2.3) \qquad\qquad E_n[U(x_j)] = p_{jn}(\theta_j+1)^\beta$$

Bettor n will make the assumed $1 wager on a horse for which (2.3) is maximized.

2.2 Determination of Market Odds

We now turn to the determination of the equilibrium market odds $\underline{\theta}$. We have seen that these are determined by B_j, the fraction of bets placed on each horse. Since we have assumed that each bettor bets $1 on exactly one horse, B_j is also the fraction of bettors who choose to bet on horse j.

Maximization of (2.3) implies that the choice among the horses by each bettor depends on the odds offered on all the horses. Then $B_j = B_j(\underline{\theta})$, and from the definition of the odds (2.1) we have the parimutual condition for market equilibrium.

$$(2.4) \qquad\qquad \underline{\theta}^* : \quad Q/B_j(\underline{\theta}^*) - 1 = \theta_j^* \quad j = 1,...,J$$

or equivalently

$$(2.5) \qquad\qquad \underline{\pi}^* : \quad B_j(\underline{\theta}^*) = \pi_j^* \quad j = 1,...,J$$

Thus the parimutual condition requires market probabilities to be such that the fraction who bet on each horse at the associated odds be equal to the market probability on that horse. We will assume that the equilibrium odds are found by an auctioneer. The existence of a unique equilibrium vector $\underline{\pi}^*$ for general distributions of beliefs was proved by Eisenberg and Gale [1959].

The assumption that no bets are made until the equilibrium odds are known is counterfactual. The implications of betting the before final odds are known when some bettors have private information are investigated by Asch, Malkiel, and Quandt [1984].

2.3 Track Take and Optimal Bets

For the remainder of the paper, the exposition will be simplified by the assumption that $Q=1$, so that there is no track take. Track take would simply reduce the expected return on each horse by the same amount. For given market odds, track take does not affect an individual's choice of the horse which maximizes expected utility (2.3). Because we assume that the amount bet is fixed, track take therefore does not affect the market equilibrium. The fixed bet assumption means that track take is equivalent to an admission charge.

The fixed bet assumption, however, is not innocuous. It means that each bettor's beliefs have equal importance in determining the market equilibrium. If amounts bet were chosen optimally, however, an individual agent's bet would be increasing in expected return. In particular, positive track take will lead to bettors with beliefs near the market probabilities to expect losses. Optimal betting would require zero bets from these agents unless they are sufficiently risk loving (this is Quandt's argument that the racetrack market requires risk loving bettors).

Since optimal bets will increase the farther are a bettor's beliefs from the market odds, the equilibrium could be determined largely by the tails of the distribution of opinions. In this paper we choose to ignore the complications thereby introduced by fixing bets. As it happens, the results derived will not be much affected. Nothing in our assumptions about the distribution of beliefs prevents most of the mass of this distribution from being in its tails. We will however place a strong symmetry restriction on this distribution, thus ruling out *a priori* the kind of misbehaved tails which would lead to serious problems were optimal betting allowed. Perhaps this analysis will be useful in establishing more general results.

When heterogeneous beliefs are allowed, equilibrium with track take does not require bettors to be risk lovers. Optimal bets will also be positive for risk averse and risk neutral bettors whose beliefs about the probabilities are sufficiently far from the market equilibrium that they have positive expected returns (on at least one horse) after accounting for track take. Moreover, since heterogeneous beliefs will reduce expected losses, they will increase the optimal bets of risk lovers. Since bettors must be very risk loving to bet substantial amounts in the face of the roughly 20% expected losses implied by U.S. and Canadian track take, this seems an important consideration in realistic modelling of the betting market.

3. Races with 2 Horses

Ali [1977] shows that underbetting on favorites and overbetting on long shots can be explained by homogeneous, risk loving bettors who all know the true probability vector $\underline{\Pi}$. Since the variance of returns to favorites is lower than the variance of returns to longshots (favorites pay off less more often), risk loving bettors will need higher expected returns in compensation for boring bets on horses likely to win. Ali uses track returns to estimate a utility function of the form (2.2), finding $\beta = 1.1784$.

Ali also shows that in a two horse race, underbetting on favorites and overbetting on longshots can be explained by risk neutral bettors who have heterogeneous beliefs. The intuition behind this result is as follows. In a two horse race, risk neutral bettors will bet on that horse whose win probability they believe is greater than the market probability. If horse 1 shows a market win probability of $2/3$, the parimutual condition implies that, at that market probability, $2/3$ of bettors wish to bet on horse 1. Therefore, $2/3$ of bettors believe that horse 1 has a true win probability greater than $2/3$. If the median bettor is correct, the true win probability will be higher than the market probability. This will be the case whenever market odds exceed $1/2$ in this two horse race. The market win probability understates the true win probability of favorites, and favorites will have positive expected returns.

In this section these results will be shown to be special cases of the model described in section 2 which allows both risk loving and heterogeneous beliefs. We will show that underbetting on favorites will be worse when both effects are present than with only one. The next section will consider the extent to which these results generalize to races with more than two horses.

When there are two horses in the race and $Q=1$, we know that $\Pi_2 = (1-\Pi_1)$, $\pi_2 = (1-\pi_1)$, $p_{2n} = (1-p_{1n})$, and $\theta_2 = 1/\theta_1$. From equation (2.3), we know that bettor n will be willing to bet on horse 1 if

$$p_{1n}(\theta_1+1)^\beta \geq (1-p_{1n})(1/\theta_1 + 1)^\beta$$

which implies

$$\theta_1\beta \geq \frac{1-p_{1n}}{p_{1n}}$$

The bettor will bet on horse 1 if he believes the true odds are less than the market odds adjusted for risk non—neutrality. If the equalities hold in the expressions above, the bettor is indifferent between the two horses and is willing to bet on either one.

Suppose we describe bettors' beliefs about the true odds Θ_1 with a cumulative distribution function $F(x)$, where $F(x)$ gives the fraction of bettors who believe the odds are less than x. $F(x)$ is nondecreasing in x, $F(0)=0$, and $F(\infty)=1$. Since those bettors who believe the odds on horse 1 are less than $\theta_1\beta$ will bet on horse 1 at odds θ_1, we have $B_1(\theta_1) = F[(\theta_1)^\beta]$. Then the parimutual condition requires that

$$\frac{1 - F[(\theta_1)^\beta]}{F[(\theta_1)^\beta]} = \theta_1^*$$

or

(3.1) $$F[(\theta_1)^\beta] = \pi_1^*$$

Then

(3.2) $$\pi_1^* > 1/2 \text{ implies } F[(\theta_1)^\beta] \geq 1/2$$

with equality only holding in the case of homogeneous beliefs, in which case $F(x)$ is degenerate with all mass at the commonly held belief about the odds.

Equation (3.2) implies that

$$\pi_1^* > 1/2 \implies (\theta_1)^\beta \geq \text{median}\left[\frac{1 - p_{1n}}{p_{1n}}\right]$$

Suppose the distribution of beliefs about the odds is such that it has the true odds Θ_1 as its median. Then

(3.3) $$\pi_1^* > 1/2 \implies \theta_1^* \geq \Theta_1^{1/\beta}$$

where again equality holds only in the case of homogeneous beliefs.

Ali's results both follow from (3.3). If bettors have homogeneous beliefs but are risk loving, $\beta > 1$ and

$$\pi_1^* > 1/2 \implies \theta_1^* = \Theta_1^{1/\beta} > \Theta_1$$

If bettors are risk neutral but have heterogeneous beliefs, $\beta = 1$ and

$$\pi_1^* > 1/2 \implies \theta_1^* > \Theta_1$$

So we see that in a two horse race, positive returns to favorites are consistent with either risk loving or heterogeneous beliefs with risk neutrality. Further, if bettors have heterogeneous beliefs and are also risk loving,

$$\pi_1^* > 1/2 \implies \theta_1^* > \Theta_1^{1/\beta} > \Theta_1$$

so that heterogeneous beliefs will cause underbetting on favorites beyond the effect of risk loving alone.

The effect of heterogeneous beliefs will be larger the greater the dispersion of opinions. In the example presented previously, where the market probability was $2/3$ and bettors were risk neutral, the market probability will be only slightly below the true probability if beliefs are tightly distributed around the true value, because then only a small deviation will be required to persuade $2/3$ of the bettors to bet on horse 1. The larger the spread of opinion, the larger the deviation which will be necessary to meet the parimutual condition.

4. Races with J Horses

We have established that underbetting on favorites and overbetting on longshots in a two horse race is consistent with either risk loving bettors or risk neutral bettors with diverse beliefs. Let us consider how the analysis of the previous section generalizes to races with more than two horses. It will be easiest to consider risk loving and heterogeneous beliefs separately at first.

4.1 Risk Loving with Homogeneous Beliefs

If all bettors have identical beliefs, equilibrium requires that bettors be indifferent among all horses. Otherwise all bettors would bet on a single horse and the parimutual condition could not be met. If bettors' beliefs are in fact equal to the true probabilities, we have

$$\Pi_j(\theta_j{}^* + 1)^\beta = \Pi_k(\theta_k{}^* + 1)^\beta$$

for all horses j and k. The nice relationship between true and market odds for the two horse race doesn't hold in this more general formulation. However, a little algebra reveals that

(4.1) $$(\pi_j{}^*/\pi_k{}^*)^\beta = \Pi_j/\Pi_k$$

so the previous result holds as the relationship of the ratio of true probabilities to the ratio of market probabilities of any two horses in the race.

If bettors are risk lovers, the market probabilities will compress the differences in the true probabilities. Equation (4.1) implies that for $\beta > 1$, the ratio of market probabilities will be closer to 1 than the ratio of true probabilities. Then if horse 1 shows a higher market probability of winning than horse 2, the true probability that horse 1 will win exceeds the true probability that horse 2 will win by a greater amount. The market probabilities understate the true differences between horses and it follows that horses with relatively high win probabilities will have relatively high expected rates of return.

The result for the two horse race generalizes to races with more horses in the case of homogeneous beliefs. If bettors are risk lovers, horses likely to win will have higher expected rates of return than horses unlikely to win (as in Quandt [1986]).

4.2 Heterogeneous Beliefs with Risk Neutral Bettors

The extension of the result on heterogeneous beliefs to races with more than two horses is more difficult. We will find conditions on the distribution of beliefs about the true probabilities which are sufficient for the result to hold.

Assume bettors are risk neutral $(\beta=1)$ so they bet to maximize expected return. The beliefs of bettor n about a given race are given by a vector p_n, $\Sigma_j p_{jn} = 1$. From condition (2.3), bettor n bets on horse j such that

$$j = \arg \max_j \{p_{jn}(\theta_j+1)\} = \arg \max_j \{p_{jn}/\pi_j\}$$

Given a distribution of beliefs over the population of bettors, the maximization problem solved by each bettor will lead to a bet function $B_j{}'(\underline{\pi})$, j=1,...n , where we now use as arguments the market probabilities instead of the odds. The function $B_j{}'$ gives the fraction of bettors who wish to bet on horse j when the market probabilities are given by the vector $\underline{\pi}$. Recall the parimutual condition (2.5) for the n horse race

$$\underline{\pi}^* : B_j{}'(\underline{\pi}^*) = \pi_j{}^* \qquad j = 1,...,J$$

so that at equilibrium the fraction of bettors who wish to bet on each horse is equal to the market probability for that horse.

Let us now consider the distribution of beliefs $p_n = (p_{1n},...,p_{Jn})$ over the population of bettors.

If bettor n bets on horse j, it must be the case that

$$\frac{p_{jn}}{\pi_j} \geq \frac{p_{kn}}{\pi_k} \quad \text{all} \quad k$$

or

(4.2)
$$\frac{p_{kn}}{p_{jn}} \leq \frac{\pi_k}{\pi_j} \quad \text{all} \quad k$$

This suggests describing the distribution of beliefs about the probabilities as distributions of beliefs about ratios of the probabilities. In fact, the beliefs of the population of bettors can be described by any of n equivalent cumulative distribution functions

(4.3)
$$F_j\left[\frac{\pi_1}{\pi_j}, \ldots, \frac{\pi_J}{\pi_j}\right] \equiv \text{Prob}\left[\frac{p_{1n}}{p_{jn}} \leq \frac{\pi_1}{\pi_j}, \ldots, \frac{p_J}{p_{jn}} \leq \frac{\pi_J}{\pi_j}\right]$$

Any one of these functions F_j completely describes the distribution of beliefs. They are distinguished by which horse's probability is used as the denominator of the ratios which form the arguments of the function. The next section will illustrate the functions F_j and the following analysis for an example in which the ratios of probabilities follow a log—normal distribution (i.e. their logarithms have a normal distribution).

In what follows, it will be convenient to define an operator $r_j(\underline{x})$ which takes a J—vector x and transforms it as follows:

$$r_j(\underline{x}) \equiv \left[\frac{x_1}{x_j}, \ldots, \frac{x_J}{x_j}\right]$$

so that we can write (4.3) as

(4.3')
$$F_j[r_j(\underline{\pi})] \equiv \text{Prob}[r_j(\underline{p}_n) \leq r_j(\underline{\pi})]$$

It follows from (4.2) and (4.3') that we now have a convenient expression for the betting function:

$$B_j'(\underline{\pi}) = F_j[r_j(\underline{\pi})]$$

We describe the distribution of beliefs about the true probabilities using a set of cumulative distribution functions $(F_1, \ldots, F_J)$, each of which completely describes those beliefs. The function $F_j(r_j(\underline{\pi}))$ gives the fraction of bettors who will choose horse j at market probabilities $\underline{\pi}$. Then the parimutuel condition becomes

(4.4)
$$\pi_j^* = F_j[r_j(\underline{\pi}^*)] \quad \text{all} \quad j.$$

For the two horse race we needed to concern ourselves with only one parameter, the probability that horse 1 will win. We found it interesting to consider the median belief about this parameter. In a two horse race, the median belief had the property that, at market probability equal to the median belief, half of the bettors wish to bet on each of the two horses. In this sense, at the median belief the population is indifferent between the two horses. We generalize the median to the n horse race as follows:

Definition. A **consensus probability vector** $\tilde{\pi}$ is an n—vector such that

$$F_j[r_j(\tilde{\pi})] = 1/J \quad \text{all} \quad j$$

Such a consensus probability vector will exist for any continuous distribution of beliefs. As with the median, point masses in the distribution may require appropriate partitioning of the set of individual bettors who are indifferent at $\tilde{\pi}$.

Now let us specify the distributions F_j more precisely. We will assume that there is an n—vector $\underline{\lambda}$ such that

$$F_j[r_j(\underline{\pi})] = G_j\left[\ln\left[\frac{\pi_1}{\pi_j}\right] - \ln\left[\frac{\lambda_1}{\lambda_j}\right] ; \dots ; \ln\left[\frac{\pi_J}{\pi_j}\right] - \ln\left[\frac{\lambda_J}{\lambda_j}\right]\right]$$

or

(4.5) $$F_j[r_j(\underline{\pi})] = G_j[\ln(r_j(\underline{\pi})) - \ln(r_j(\underline{\lambda}))].$$

This equation simply specifies that the logs of the probability ratios follow some location distribution with location parameters $\ln(r_j(\underline{\lambda}))$. Note in particular that if the distribution is such that

$$G_j(0 , \dots , 0) = G_k(0 , \dots , 0) \quad \text{all } j, k$$

it follows from the fact that $\Sigma_j B_j{'}(\underline{\pi})=1$ that

$$G_j(0 , \dots , 0) = 1/J \quad \text{all } j$$

and therefore

$$\underline{\lambda} = \bar{\underline{\pi}}$$

so that the location vector is equal to the consensus probability vector.

We now introduce the concept of a symmetric distribution of beliefs.

Definition. The distribution of beliefs p_n about the true probability vector is symmetric iff.

$$G_j(.) = G_k(.) = G(.)$$

That is, symmetry holds if the functions are identical. This does <u>not</u> imply that $F_j[r_j(\underline{\pi})] = F_k(r_k(\underline{\pi}))$, since these have different arguments. The following section will show conditions on the log—normal distribution which give symmetry for that example.

Symmetry is a strong restriction. In the log normal case considered in the next section, we will see that symmetry requires the variance of opinion about Π_1/Π_2 to be the same as the variance of opinion about Π_3/Π_4 , and it requires that opinions about these two ratios be independent.

We can now prove the key theorem of this section.

Theorem 1. Suppose beliefs are described by a symmetric distribution of the form (4.5) , with location parameter, and thus consensus probability vector, equal to the true probability vector $\underline{\Pi}$. Then

$$\pi_j{}^* > \pi_k{}^* \quad \text{implies} \quad \frac{\pi_j{}^*}{\pi_k{}^*} < \frac{\Pi_j}{\Pi_k}$$

Proof. By symmetry, $G_j = G_k = G$. Then by (4.4)

$$\pi_j{}^* = G[\ln(r_j(\underline{\pi}^*)) - \ln(r_j(\underline{\Pi}))]$$

$$\pi_k{}^* = G[\ln(r_k(\underline{\pi}^*)) - \ln(r_k(\underline{\Pi}))]$$

$G[.]$ is a cumulative distribution function and therefore is increasing in all of its arguments. Then $\pi_j{}^* > \pi_k{}^*$ implies that

$$\frac{\pi_m{}^*}{\pi_j{}^*} \cdot \frac{\Pi_j}{\Pi_m} > \frac{\pi_m{}^*}{\pi_k{}^*} \cdot \frac{\Pi_k}{\Pi_m} \quad \text{some } m$$

The result follows immediately.   Q.E.D.

We also obtain the following useful result.

Corollary. Under the assumptions of Theorem 1, if $\pi_j^* > \pi_k^*$ then

$$\frac{\pi_m^*}{\pi_j^*} \cdot \frac{\Pi_i}{\Pi_m} > \frac{\pi_m^*}{\pi_k^*} \cdot \frac{\Pi_k}{\Pi_m} \quad \text{all } m$$

That is, all of the arguments $\ln(r_j(\underline{\pi})) - \ln(r_j(\underline{\Pi}))$ exceed the corresponding arguments $\ln(r_k(\underline{\pi})) - \ln(r_k(\underline{\Pi}))$.

Theorem 1 proves that, under a symmetry restriction on the distribution of beliefs, Ali's result for a race with two horses generalizes to a race with J horses. Underbetting on favorites and overbetting on longshots can result from heterogeneous opinions on the part of risk neutral bettors.

In the case of the two horse race, we claimed that the disparity between the true and market probabilities would be greater the greater the dispersion of opinion. We now demonstrate that this result holds for the race with J horses.

Theorem 2. Suppose the distribution of beliefs is symmetric and that the logarithms of the ratios of probabilities follow some location–scale family, with location vector equal to the true ratios and scale parameter σ, so that

$$F_j[r_j(\underline{\pi})] = H\left[\frac{\ln(\pi_1/\pi_j) - \ln(\Pi_1/\Pi_j)}{\sigma}, ..., \frac{\ln(\pi_J/\pi_j) - \ln(\Pi_J/\Pi_j)}{\sigma}\right]$$

Let $\underline{\pi}^*$ be the equilibrium market probabilities for such a population, and let $\hat{\underline{\pi}}$ be the equilibrium market probabilities for a population whose beliefs are identical except that the scale parameter is $\hat{\sigma} > \sigma$. Then if $\pi_j^*/\pi_k^* > 1$,

(4.6)
$$\frac{\hat{\pi}_i}{\hat{\pi}_k} < \frac{\pi_i^*}{\pi_k^*} < \frac{\Pi_i}{\Pi_k}$$

Proof. It follows from symmetry that either

(a)
$$\frac{\hat{\pi}_i}{\hat{\pi}_k} < \frac{\pi_i^*}{\pi_k^*} < \frac{\Pi_i}{\Pi_k}$$

or

(b)
$$\frac{\pi_i^*}{\pi_k^*} < \frac{\hat{\pi}_i}{\hat{\pi}_k} < \frac{\Pi_i}{\Pi_k}$$

for all j,k for which $\pi_j^* > \pi_k^*$. We will show that (b) leads to a contradiction. Without loss of generality, we can order the horses from lowest to highest market probability. Then $\pi_J^* > \pi_1^*$ and $\hat{\pi}_J > \hat{\pi}_1$. Further, by the Corollary to Theorem 1,

$$\ln\left[\frac{\pi_m^*}{\pi_1^*}\right] - \ln\left[\frac{\Pi_m}{\Pi_1}\right] \leq 0 \quad \text{all } m$$

$$\ln\left[\frac{\pi_m^*}{\pi_J^*}\right] - \ln\left[\frac{\Pi_m}{\Pi_J}\right] \geq 0 \quad \text{all } m$$

and the same inequalities hold when the π^*'s are replaced by $\hat{\pi}$'s.

Suppose (b) holds. Then since by assumption $\hat{\sigma} > \sigma$

$$\frac{\ln(\hat{\pi}_m/\hat{\pi}_1)-\ln(\Pi_m/\Pi_1)}{\hat{\sigma}} > \frac{\ln(\pi_m{}^*/\pi_1{}^*)-\ln(\Pi_m/\Pi_1)}{\sigma} \quad \text{all } m$$

$$\frac{\ln(\hat{\pi}_m/\hat{\pi}_j)-\ln(\Pi_m/\Pi_j)}{\hat{\sigma}} < \frac{\ln(\pi_m{}^*/\pi_j{}^*)-\ln(\Pi_m/\Pi_j)}{\sigma} \quad \text{all } m$$

But then $\hat{\pi}_1 > \pi_1{}^*$ and $\hat{\pi}_j < \pi_j{}^*$, which contradicts (b).

Q.E.D.

Thus the effect of heterogeneous opinions is greater the greater the spread of opinion, as reflected by the scale parameter σ.

4.3 Risk Loving Bettors with Heterogeneous Beliefs

We now show that when the distribution of opinions is symmetric, the result of section 3 for risk loving bettors with heterogeneous beliefs generalizes to races with J horses. Condition (4.2) for bettor n to bet on horse j is changed to

$$\frac{p_{kn}}{p_{kn}} \leq \left[\frac{\pi_k}{\pi_j}\right]^\beta \quad \text{all } k$$

When beliefs are described by a distribution function of the form (4.5), the parimutual condition becomes

$$\underline{\pi}^* : \pi_j{}^* = G_j[\beta \cdot \ln(r_j(\underline{\pi}^*)) - \ln(r_j(\underline{\Pi}))]$$

We now have the following generalization of Theorem 1.

Theorem 3. Suppose beliefs are described by a symmetric distribution function of the form (4.5). Then

$$\pi_j{}^* > \pi_k{}^* => \left[\frac{\pi_i{}^*}{\pi_k{}^*}\right]^\beta < \frac{\Pi_j}{\Pi_k}$$

Proof. Exactly analogous to the proof of Theorem 1.

Thus heterogeneous opinions lead to underbetting on favorites more severe than that implied by risk loving alone.

5. Illustration with Log–Normal Distribution

The concepts introduced in section 4 may be better understood through an example. Consider a race with three horses, and suppose that beliefs about the ratios of probabilities are distributed over the population according to a log–normal distibution. The beliefs of bettor n about the win probabilities are p_{1n}, p_{2n}, p_{3n}, and the true win probabilities are Π_1, Π_2, Π_3. In this example the logs of the beliefs follow a normal distribution:

$$x_1 \equiv \ln(p_{2n}/p_{1n}) \sim N[\ln(\Pi_2/\Pi_1), \sigma_1{}^2]$$

$$x_2 \equiv \ln(p_{3n}/p_{1n}) \sim N[\ln(\Pi_3/\Pi_1), \sigma_2{}^2]$$

$$\text{cov}(x_1,x_2) = \sigma_{12}$$

so that F_1 is given by the cumulative distribution function of a bivariate log–normal:

$$F_1\left[\frac{\pi_2}{\pi_1}, \frac{\pi_3}{\pi_1}\right] = \text{C.L.N.}\left\{\left[-\frac{\Pi_2}{\Pi_1}, -\frac{\Pi_3}{\Pi_1}\right], \Sigma_1 ; \frac{\pi_2}{\pi_1}, \frac{\pi_3}{\pi_1}\right\}$$

$$\Sigma_1 \equiv \begin{bmatrix} \sigma_1{}^2 & \sigma_{12} \\ \sigma_{12} & \sigma_2{}^2 \end{bmatrix}$$

where C.L.N. stands for cumulative log–normal. The conditions of Theorem 1 require the location parameters to equal the logs of the true probability ratios.

The function F_1, which uses the probability of horse 1 as the denominator, completely characterizes the distribution of beliefs. We can also characterize this distribution with a function F_2 which uses the probability of horse 2 as denominator. To find F_2, define

$$y_1 \equiv \ln(p_{1n}/p_{2n})$$

$$y_2 \equiv \ln(p_{3n}/p_{2n})$$

Now,

$$y_1 = -x_1 - N[\ln(\Pi_1/\Pi_1), \sigma_1{}^2]$$

$$y_2 = x_2 - x_1 - N[\ln(\Pi_3/\Pi_2), \sigma_1{}^2 + \sigma_1{}^2 - 2\sigma_{12}]$$

$$\text{cov}(y_1, y_2) = \sigma_1{}^2 - \sigma_{12}$$

So we find that

$$F_2\left[\frac{\pi_1}{\pi_2}, \frac{\pi_3}{\pi_2}\right] = \text{C.L.N.}\left\{\left[-\frac{\Pi_1}{\Pi_2}, \frac{\Pi_3}{\Pi_2}\right], \Sigma_2 ; \frac{\pi_1}{\pi_2}, \frac{\pi_3}{\pi_2}\right\}$$

$$\Sigma_2 \equiv \begin{bmatrix} \sigma_1{}^2 & \sigma_1{}^2 - \sigma_{12} \\ \sigma_1{}^2 - \sigma_{12} & \sigma_1{}^2 + \sigma_2{}^2 - 2\sigma_{12} \end{bmatrix}$$

Proceeding in the same manner, we find that

$$F_3\left[\frac{\pi_1}{\pi_3}, \frac{\pi_2}{\pi_3}\right] = \text{C.L.N.}\left\{\left[-\frac{\Pi_1}{\Pi_3}, -\frac{\Pi_2}{\Pi_3}\right], \Sigma_3 ; \frac{\pi_1}{\pi_3}, \frac{\pi_2}{\pi_3}\right\}$$

$$\Sigma_3 \equiv \begin{bmatrix} \sigma_2{}^2 & \sigma_2{}^2 - \sigma_{12} \\ \sigma_2{}^2 - \sigma_{12} & \sigma_1{}^2 + \sigma_2{}^2 - 2\sigma_{12} \end{bmatrix}$$

Each of the distributions F_1, F_2, F_3 characterizes the distribution of beliefs. If one of the distributions is known, the others can be calculated using change of variables techniques.

Now let us consider the implications of symmetry in this case. We have

$$G_i[\ln(r_i(\underline{\pi})) - \ln(r_i(\underline{\Pi}))] = \text{C.L.N.}[\underline{0}, \Sigma_i ; r_i(\underline{\pi}/\underline{\Pi})]$$

where the covariance matrices are as defined above. Suppose that $\sigma_1{}^2 = \sigma_2{}^2 = \sigma^2$ and that $\sigma_{12} = \sigma^2/2$. Then

$$\Sigma_i \equiv \begin{bmatrix} \sigma^2 & \sigma^2/2 \\ \sigma^2/2 & \sigma^2 \end{bmatrix} \quad \text{all } i$$

and the distribution of beliefs is symmetric.

6. Estimation Technique

Now let us consider how we might go about making an empirical investigation of the effects of risk loving and heterogeneous opinions. Suppose we have a sample of M races, with J_m horses in each race. The sample includes the values of the market probabilities for each horse in each race, and a measure S_m of the dispersion of opinion in each race (data will be described in the next section).

Unfortunately an exact expression for the probabilities given heterogeneous beliefs would be extremely complicated even if beliefs are assumed to be normally distributed. However, if we are willing to assume that beliefs follow a symmetric distribution, the theory of the previous section suggests the following nested model for examining the relative importance of differences of opinion and of risk loving:

$$\frac{\Pi_{jm}}{\Pi_{km}} = \left[\frac{\pi_{jm}}{\pi_{km}}\right]^{\delta S_m + \beta}$$

where Π_{jm} is the win probability of horse j in race m.

We wish to estimate the parameters δ and β. δS_m gives the contribution of differences of opinion and β the contribution of risk loving in explaining the difference between true and market probabilities. We expect to find $\delta \geq 0$ and $\beta > 0$. Bettors are risk loving if $\beta > 1$. With heterogeneous beliefs, we have no a priori reason to believe that this is the case; since individual bettors may perceive positive expected returns. However, $\delta = 0$ would require $\beta > 1$ for positive bets given track take. Even with $\delta > 0$, we might be suspicious if β were estimated to be much smaller than 1, which would imply strong risk aversion on the part of bettors.

Given an expression for the likelihood of the sample, we can proceed by the method of maximum likelihood. We have

$$\Pi_{jm} = \Pi_{km}\left[\frac{\pi_{jm}}{\pi_{km}}\right]^{\delta S_m + \beta}$$

$$1 = \sum_1^{J_m} \Pi_j = \Pi_{km}\sum_1^{J_m}\left[\frac{\pi_{jm}}{\pi_{km}}\right]^{\delta S_m + \beta}$$

$$\Pi_{km} = \left[\sum_1^{J_m}\left[\frac{\pi_{jm}}{\pi_{km}}\right]^{\delta S_m + \beta}\right]^{-1}$$

or

(6.1) $$\Pi_{km} = \left[\sum_1^{J_m} \pi_{jm}^{\delta S_m + \beta}\right]^{-1} \pi_{km}^{\delta S_m + \beta}$$

Suppose we order the horses in each race by their order of finish, so Π_{1m} is the *ex ante* probability of the horse which in fact wins. Then the likelihood of the sample is

$$L = \prod_{m=1}^{M} \Pi_{1m}$$

and we have as the log–likelihood function

(6.2) $$\ln L = \sum_{m=1}^{M}\left\{(\delta S_m + \beta)\cdot\ln(\pi_{1m}) - \ln\left[\sum_1^{J_m}\pi_{jm}^{\delta S_m + \beta}\right]^{-1}\right\}$$

For a given sample, maximization of (6.2) with respect to δ and β gives maximum likelihood estimates of these parameters.

7. Description of Data

Data were collected for the 1986 season at Golden Gate Fields in Albany, California. The original data set has 927 races, with from 5 to 12 horses running in each race. The data were taken from the San Francisco Chronicle and the Oakland Tribune. For each horse in each race, the data set includes the morning line odds $OCHRON_{jm}$ and $OTRIB_{jm}$ published by each paper on the morning of the race, and the final payoff odds $OTRACK_j$ which the newspapers reprint from the <u>Daily Racing Form</u>. The morning line odds are estimates of the odds for each horse made by a handicapper (Larry Stumes for the <u>Chronicle</u>, Jack Menges for the <u>Tribune</u>) on the basis of past performance and other information available before any betting takes place.

Several adjustment were made to the raw data to produce the data used for estimation. The probabilities directly implied by the track odds sum to more than 1 because of track take. The morning line odds are relative odds with no constraint imposed on the sum of the implied probabilities. Therefore, the probabilities used are defined by

$$PTRACK_{jm} \equiv \frac{1/(OTRACK_{jm} + 1)}{\Sigma_k 1/(OTRACK_{km} + 1)}$$

$$PCHRON_{jm} \equiv \frac{1/(OCHRON_{jm} + 1)}{\Sigma_k 1/(OCHRON_{km} + 1)}$$

$$PTRIB_{jm} \equiv \frac{1/(OTRIB_{jm} + 1)}{\Sigma_k 1/(OTRIB_{km} + 1)}$$

In some cases two horses in a race are "coupled." This means that all bets on the two horses are pooled. A bet to win on one of the coupled horses pays off if either wins. For our purposes, therefore, a pair of coupled horses is a single horse, and the finish position of the pair is the higher of the two finishes. Therefore, the lower finishing horse of these pairs was removed from the data set. In four of the 927 races a dead heat was declared and the win pool was divided among bettors on the two horses. These four races were removed from the data set leaving 923 observations for the estimation.

The morning line odds were used to construct the measure of the spread of opinion about each race. Of course the difference between the morning line of the <u>Tribune</u> and the morning line of the <u>Chronicle</u> does not directly measure the difference of opinion among bettors, but it does provide an indication of the extent of disagreement among informed individuals. Certainly the measurement would be improved if the morning lines of more handicappers were available. To the extent that disagreement between the two morning lines is a poor measure of disagreement among bettors, the estimate of δ will be biased towards 0.

The theory of section 5 requires the assumption that the distribution of opinions be symmetric. This assumption requires that the standard deviation of opinion regarding the logarithm of each probability be the same for each horse in a given race. As discussed in section 6, the theory does not give an exact expression for the measure to be used in estimation. Estimations were performed using an estimate S1 of the variance of opinion for each race, and alternatively using the associated estimate S2 of the standard deviation. The measures used are

$$S1_m \equiv \frac{1}{J_m} \Sigma_j (\ln(PCHRON_{jm}) - \ln(PTRIB_{jm}))^2$$

$$S2_m \equiv \sqrt{S1_m}$$

8. Results

Table 1 presents maximum likelihood estimates of the parameters δ and β of equation (6.2), where $\pi_{jm} = PTRACK_{jm}$ and S_m is measure alternatively as $S1_m$ and $S2_m$ as defined above, using the sample of 923 races from Golden Gate Fields. Estimation was performed using the maximum likelihood routines provided with the GAUSS statistical package.

The estimates do not support the hypothesis that differences of opinion as measured

TABLE 1

Maximum Likelihood Estimates of Equation 6.2

Spread Measure	$\hat{\beta}$	$\hat{\delta}$	$\text{corr}(\hat{\beta}, \hat{\delta})$	$-\ln(L)$
S1	1.047 (0.112)	−0.311 (0.265)	−0.746	1765.15
	L.R. stat. of $\{\beta=0, \delta=1\} = 2.83$ p−value = 0.243			
S2	1.189 (0.209)	−0.461 (0.373)	−0.935	1764.63
	L.R. stat. of $\{\beta=1, \delta=0\} = 3.87$ p−value = 0.145			

Asymptotic standard errors in parentheses.
Based on 923 observations
Source: see text.

using the morning line odds are important in explaining the relation of track probabilities to the outcome of the races. The coefficient δ on the measure of dispersion of opinion is insignificantly different from zero and has the wrong sign.

Furthermore, the coefficient β is insignificantly different from 1. The likelihood ratio test of the null hypothesis that $\beta = 1$ and $\delta = 0$ does not reject at even the 10% level for either spread measure. Despite the fact that differences of opinion appear to be unimportant, there is no indication that bettors are risk loving, either. This appears to conflict with the results found in previous studies that rates of return are higher for favorites than for longshots. The conflict is confirmed in table 2, which compares rates of return for this data set with the results reported by Snyder. Lines $1 - 7$ of that table reproduce Snyder's table 2, where he reports rates of return adjusted for track take for horses grouped by track odds. Line 8 reports rates of return for the same categories for the Golden Gate Fields data. We see that the data used in the present study do not display the phenomenon we set out to explain! We turn now to possible explanations of the discrepancy.

One possibility is of course chance. The sample of races here is small compared to other studies. Since longshots pay off large amounts, a few wins by longshots can dramatically improve the rate of return on these bets. Then the lack of a negative relationship between rate of return and odds would not imply any important differences between the bettors and market at Golden Gate Fields and those elsewhere. This possibility can be checked by examining data for a closely related market.

The Bay Meadows racetrack in Belmont, California is likely to have a population of bettors very similar to that at Golden Gate Fields. The tracks are in the same metropolitan area, their seasons have little or no overlap, and bets on races at either track can be placed at either location. Line 9 of table 2 shows rates of return by odds group for 600 races at Bay Meadows for the 1986 season through November 25. These data appear to be consistent with the Golden Gate Fields data and not with the previous data. Rates of return for Bay Meadows and Golden Gate Fields combined are shown in line 10.

The absence of a negative relationship between rate of return and odds at Golden Gate Fields does not seem to be an anomaly but rather a characteristic of the betting market in the San Francisco area.

Another possible explanation is errors in variables bias. As is usual in limited dependent variable estimation, equation (6.2) can be estimated only if the probabilities expressed in equation (6.1) are exact. If equation (6.1) omits an error term, the parameters δ and β in (6.2) are not econometrically identified.

The consequence of estimating (6.2) in the presence of errors in variables would be a bias of the exponent $\delta S_m + \beta$ towards 0. Horses with a high probability of winning

TABLE 2

Rates of Return by Grouped Odds (Take Added Back)

Study	Midpoint of Grouped Odds							
	0.75	1.25	2.5	5.0	7.5	10.0	15.0	33.0
1. Fabricant	11.1	9.0	4.6	-1.4	-3.3	-3.7	-8.1	-39.5
2. Griffith	8.0	4.9	3.1	-3.1	-34.6	-34.1	-10.5	-65.5
3. McGlothlin	8.0	8.0	8.0	-0.8	-4.6	-7.0	-9.7	-11.0
4. Seligman	14.0	4.0	-1.0	11.0	-2.0	-4.0	-7.8	-24.2
5. Snyder	5.5	5.5	4.0	-1.2	3.4	3.9	2.4	-15.8
6. Weitzman	9.0	3.2	6.8	-1.3	-4.2	-5.1	-8.2	-18.0
7. Combined	9.1	6.4	6.1	-1.2	-5.2	-5.2	-10.2	-23.7
8. GGF86	-3.1	10.4	0.6	-0.8	15.3	-2.0	9.3	5.6
9. BAYM86	-0.9	1.8	1.4	6.1	-1.9	2.7	2.9	-1.3
10. Combined	-2.5	6.4	1.0	1.9	9.0	-0.1	6.7	1.8

Source: Lines 1-7: reproduced from Snyder (1978), table 2.
Lines 8-10: see text.

according to (6.1) would tend to have a high true probability, but would also tend to be overestimated in their true probability. Thus favored horses tend to have true probabilities less than those given by (6.1), and longshots tend to have true probabilties higher (6.1) would indicate. This bias will tend to offset any negative association between rate of return and odds due to risk loving and/or heterogeneous beliefs.

The theory of section 5 required that the consensus probability vector equal the true probability vector. If this condition is violated, errors in variables bias is introduced into the estimation. The difference between this data set and others may be that the consensus probabilities at Golden Gate Fields include larger errors than those at other tracks.

This possibility is closely related to the assumption of fixed bets made in the theoretical part of the paper. With optimal betting, agents in the tails of the distribution of beliefs will have disproportionate influence on the outcome, as they will perceive the highest expected returns. We need not believe that the majority of bettors at Golden Gate Fields are making large errors, but only that the tails of the distribution are badly behaved compared to other tracks. Because of track take, there is no reason to believe that more typical bettors will bet enough to eliminate the effects of those with unusual beliefs. With variable bets, the consensus probability vector could be far from the truth even though most bettors have accurate estimates of the probabilities.

This analysis may in fact explain the negative estimated coefficient δ. Suppose that the greater the spread of opinion, the farther the consensus probability vector tends to be from the true probability vector. Then the extent of the errors in variable bias will be correlated with the measure S_m of the spread of opinion. Since errors in variables will bias the coefficient $(\delta S_m + \beta)$ towards 0, if S_m is correlated with the error a negative bias is implied for the coefficient δ. Thus the negative estimates of δ in table 1 lend support to the errors in variables explanation.

If the consensus probabilities at Golden Gate Fields include substantial errors, there may be public information available which can be used to improve on the track odds in predicting winners. In particular, we might look for additional support for the errors in variables explanation by determining whether the morning line odds can be used directly to improve on the track odds.

Table 3 reports on such an investigation. The sample used here is slightly smaller than the previous sample. Several races were dropped because of missing values for the morning line.

The first line simply reports the negative of the log likelihood of the sample when each horse is assigned a probability of $1/J_m$. The usefulness of information in predicting winners can be measured by the extent to which it reduces the negative of the log likelihood $(-\ln(L))$.

TABLE 3

Information Value of Track and Morning Line Odds

	PTRACK	PCHRON	PTRIB	$-\ln(L)$
1.	...	...	...	2003.15
2.	1	...	...	1760.69
3.	0.955 (0.160)	...	...	1760.25
4.	...	1	...	1853.14
5.	...	0.886 (0.176)	...	1850.92
6.	...	...	1	1877.98
7.	...	...	0.814 (0.176)	1872.10
8.	1.014 (0.266)	0.009 (0.270)	−0.099 (0.270)	1759.49

L.R. stat. of (8) vs. (2) = 2.4   p–value = 0.494

Asymptotic standard errors in parentheses.
Based on 919 observations.
Source: see text.

The next line reports $-\ln(L)$ when the track odds are used to determine the win probability of each horse. The following line reports $-\ln(L)$ when an adjustment to the track odds is estimated. The coefficient reported is the maximum likelihood estimate of γ from the equation

$$\Pi_{jm} = [\Sigma_k(PTRACK_{km})^\gamma]^{-1}(PTRACK_{jm})^\gamma$$

This is of course just equation (6.2) estimated when δ is constrained to equal 0.

We see that the track odds contain considerable information, reducing $-\ln(L)$ by 12%. As we expect from our previous results, allowing γ to differ from 1 yields no further improvement; the estimated value is insignificantly different from 1 which simply confirms the lack of a negative relationship between rate of return and odds.

The following four lines report similar calculations using the Chronicle and Tribune odds, respectively. We see that the morning lines do contain useful information, but not as much as the track odds. Used without adjustment, the Chronicle morning line improves $-\ln(L)$ by 7.5%, and the Tribune by 6.2%. Estimates of γ for these two sets of probabilities are somewhat, but not significantly, below 1, and allowing γ to differ from 1 allows only marginal improvements in $-\ln(L)$. Of course, the estimates may be biased downwards because of errors in variables. Note that the estimates of γ are lower for equations with higher values of $-\ln(L)$.

That the track, Chronicle, and Tribune odds contain information useful in predicting winners is not surprising. We are interested in whether the Chronicle and Tribune odds contain information not included in the track odds. If they do, that fact would lend support to the idea that the Golden Gate Fields market makes errors which account for the differences between this data set and data for other tracks.

To examine this possibility, the last line of table 3 reports maximum likelihood estimates of $\gamma 1$, $\gamma 2$, and $\gamma 3$ for the model

$$\Pi_{jm} = \frac{(PTRACK_{jm})^{\gamma 1}(PCHRON_{jm})^{\gamma 2}(PTRIB_{jm})^{\gamma 3}}{\Sigma_k(PTRACK_{km})^{\gamma 1}(PCHRON_{km})^{\gamma 2}(PTRIB_{km})^{\gamma 3}}$$

The results indicate that the information available in the morning lines is completely incorporated in the track odds (as Figlewski [1979] also found). The likelihood ratio statistic of the null hypothesis that $\left(\gamma 1=1, \gamma 2=\gamma 3=0\right)$ has a p–value of 0.494 and thus the hypothesis is accepted at any reasonable significance level.

If errors in variables, or mistakes by bettors, are the explanation for the differences between this data and the data used by other authors, such mistakes do not involve ignoring the public information available in the morning line odds. There may of course be other public information which is not fully utilized, or it may be that the public information available at this track is not as informative about the horses' chances as the information at other tracks. While the negative estimate of δ does lend some support to this explanation, it is the case that this coefficient is insignificantly different from zero.

In the absence of track take, this result would suggest that Golden Gate Fields bettors are well characterized by risk neutrality and homogeneous beliefs. Given track take, however, risk neutral bettors should not bet at all. Since there is an expected loss in aggregate, agents should bet only if they individually expect positive returns (differing beliefs) and/or are risk loving (as in Quandt). Each of these possibilities entails overbetting on longshots and underbetting on favorites, so the constant expected returns over odds classes in this data remain mysterious.

9. Conclusion

This paper examines the importance of differences of opinion at the racetrack from both a theoretical and an empirical perspective. We began with the apparently established phenomenon that rates of return are higher for favorites than for longshots. We noted that Ali suggested two possible explanations for this phenomenon — risk loving bettors, and risk neutral bettors with differing opinions.

Ali's model was generalized to allow bettors who were both risk loving and had heterogeneous beliefs. Ali's theorem about differing opinions assumed races were between only two horses. The extension of that result to races with more than two horses is problematic. We found that a strong symmetry restriction on the beliefs of bettors provided a sufficient condition for the theorem to hold for an arbitrary number of horses.

The generalized theoretical results were used to motivate an empirical model which sought to capture the effect of heterogeneous beliefs. We found that the empirical evidence does not support the hypothesis that differences of opinion are important at the racetrack. However, the data used do not display the association between rates of return and track odds which we set out to explain. Since the hypothesis of risk loving is therefore inconsistent with this data as well, the motives for betting at Golden Gate Fields are unclear. The data may differ from data used by other authors because bettors at Golden Gate Fields make larger mistakes on average than bettors elsewhere in estimating the win probabilities; however, there is no evidence that they ignore the information available in the morning line odds.

Further research on the subject should improve on this study in several respects. Optimal betting should be introduced, and the conditions on the distribution of beliefs weakened. Examination of the hypothesis that differing beliefs are important may require a better measure of those differences. The Golden Gate Fields data deserve explanation, but for testing the theory a data set which displays the overbetting/underbetting phenomenon is essential.

This paper is based on a chapter of my 1987 Ph.D. thesis at the University of California, Berkeley. I thank George Akerlof, Robert Engle, Donald Hausch, Paul Ruud, and William Ziemba for useful comments on previous versions.

10. References

1. Ali, M. M. "Probability and Utility Estimates for Racetrack Bettors." Journal of Political Economy 84 (1977) pp. 803–815.

2. _____. "Some Evidence on the Efficiency of a Speculative Market." Econometrica 47 (1979) pp. 387–392.

3. Asch, P., B. Malkiel, and R. Quandt. "Racetrack Betting and Informed Behavior." Journal of Financial Economics 10 (1982) pp. 187–194.

4. _____. "Market Efficiency in Racetrack Betting." Journal of Business 57 (1984) pp. 165–175.

5. Blough, S. "Differences of Opinion and the Information Value of Asset Prices." Working Paper #201, Johns Hopkins University, 1987.

6. Canfield, B.R., B.C. Fauman and W.T. Ziemba. "Efficient Market Adjustment of Odds Prices to Reflect Track Biases." Management Science 33 (1987) pp. 1428–1439.

7. Crafts, N.F.R. "Some Evidence of Insider Knowledge in Horse Race Betting in Britain." Economica 62 (1985) pp. 295–304.

8. Dowie, J. "On the Efficiency and Equity of Betting Markets." Economica 43 (1976) pp. 139–150.

9. Eisenberg, E. and D. Gale. "Consensus of Subjective Probabilities: the Parimutual Method." Annals of Mathematical Statistics (1959) pp.165–168.

10. Fabricant, B.F. Horse Sense. New York: David McCay, 1965.

11. Figlewski, S. "Subjective Information and Market Efficiency in a Betting Market." Journal of Political Economy 87 (1979) pp. 75–88.

12. Griffith, R.M. "Odds Adjustments by American Horse Race Bettors." American Journal of Psychology 62 (1949) pp. 290–294.

13. Harville, D.A. "Assigning Probabilities to the Outcomes of Multientry Competitions." Journal of the American Statistical Association 68 (1973) pp. 312–316.

14. Hausch, D. and W. Ziemba. "Transactions Costs, Extent of Inefficiencies, Entries and Multiple Wagers in a Racetrack Betting Model." Management Science 31 (1985) pp. 381–394.

15. _____. "Arbitrage Strategies for Cross–Track Betting on Major Horseraces." Journal of Business 63 (1990) pp. 61–78.

16. Hausch, D., W. Ziemba, and M. Rubinstein. "Efficiency of the Market for Racetrack Betting." Management Science 27 (1981) pp. 1435–1452.

17. Henery, R.J. "On the Average Probability of Losing Bets on Horses with Given Starting Price Odds." Journal of the Royal Statistical Society – A 148 (1985) pp. 342–349.

18. Losey, R. L. and J.C. Talbott Jr. "Back on the Track with the Efficient Markets Hypothesis." Journal of Finance 35 (1980) pp. 1039–1043.

19. McGlothlin, W.H. "Stability of Choices Among Uncertain Alternatives." American Journal of Psychology 69 (1956) pp. 604–615.

20.   Milgrom, P. and R.J. Weber. "A Theory of Auctions and Competitive Bidding." *Econometrica* 50 (1982) pp. 1089–1122.

21.   Quandt, R.E. "Betting and Equilibrium." *Quarter Journal of Economics* 101 (1986) pp. 201–207.

22.   Seligman, D. "A Thinking Man's Guide to Losing at the Track." *Fortune* 92 (1975) pp. 81–87.

23.   Snyder, W. W. "Horse Racing: Testing the Efficient Markets Model." *Journal of Finance* 33 (1979) pp. 1109–1118.

24.   Varian, H. "Divergence of Opinion in Complete Markets." *Journal of Finance* 40 (1985) pp. 309–317.

25.   _____. "Differences of Opinion and the Volume of Trade." Unpublished paper, University of Michigan (1987).

26.   Weitzman, M. "Utility Analysis and Group Behavior: An Empirical Study." *Journal of Political Economy* 73 (1965) pp. 18–26.

27.   Ziemba, W. T. and D.B. Hausch. *Dr. Z's Beat the Racetrack*. New York, William Morrow, 1987.

PRICES OF STATE CONTINGENT CLAIMS WITH INSIDER TRADERS, AND THE FAVOURITE-LONGSHOT BIAS*

Hyun Song Shin

The literature on the pricing of state-contingent claims has been based almost exclusively on the framework of competitive equilibrium as pioneered by Arrow (1964) and Debreu (1959). This approach rests on the classical view of markets as a place which matches buyers and sellers at a common price – the price at which supply equals demand.

In practice, however, most financial transactions take place through an intermediary, such as the market maker, who holds inventories of assets, and who sets prices for buyers and sellers. Invariably, the price facing buyers differs from that facing sellers. One source of this divergence is the cost to the market maker of maintaining the inventory. However, a more potent source of this divergence is the incidence of insider trading. The market maker faces an adverse selection problem in which a customer may be trading on the basis of superior information. In this case the bid-ask spread is determined in a trade-off between setting a large spread so as to minimise the profit of insider traders, and setting the optimal spread against the noise or liquidity traders. Bagehot (1971) hints at this source of the bid-ask spread, and subsequent papers by Copeland and Galai (1983), Glosten and Milgrom (1985) and Kyle (1985) have provided formal treatments of the problem.

This paper is an attempt to draw together these two themes in the literature. It is concerned with the pricing of contingent claims when market makers set prices in the presence of insider traders. The specific setting for our investigation is the market for bets in a horse race, in which the role of the market makers are taken by the bookmakers, and the traders are played by the potential bettors (the 'punters'). Thus, our description of betting follows the system in the United Kingdom in which bookmakers set odds rather than the system in North America in which odds are determined by the parimutuel method in which prices are proportional to amounts wagered. There are several reasons for our choice of the betting market as a vehicle for the study of asset pricing with insider trading.

Firstly, the betting market is a particularly good example of a contingent claims market. In its simplest formulation, the market for bets in an n-horse race corresponds to a market for contingent claims with n states of the world, in which the ith state corresponds to the outcome in which the ith horse wins

* Work on this paper began while visiting the University of Michigan during the academic year 1989–90. Conversations with Mark Bagnoli, Ken Binmore and Hal Varian proved valuable in clarifying my ideas, and I thank Ted Bergstrom for introducing me to the literature on horse racing. I have also benefited from the comments of Jim Mirrlees and John Vickers, and the participants at seminars in Cambridge, Southampton and Oxford. The suggestions of two referees improved the exposition, and are gratefully acknowledged.

the race. Moreover, the basic securities (Arrow-Debreu securities) which pay a dollar if a particular state obtains and nothing otherwise, have their prices determined by the betting odds. Since odds are offered on each horse, all basic securities are traded, thereby ruling out the difficulties associated with incomplete markets.

Moreover, the betting market is a particularly simple example of a financial market. The market convenes for about half an hour, at the end of which there is a definite and commonly acknowledged outcome. This is in contrast to the complex decisions faced by traders in more sophisticated financial markets in which considerations of the distant future play an integral part in current decisions.

There is also a sizeable body of evidence, both systematic and anecdotal, which points to the prevalence of insider trading in the market for bets (see, for example, Crafts (1985)). No less a body than the Committee of Inquiry of the Jockey Club acknowledges that 'trainers and their staff are insufficiently paid for their services and the majority have to resort to betting to make ends meet' (1968, p. 91). It would seem, therefore, that the activity of insider traders influences the outcome in the betting market in a significant way.

Our chief concern in this paper is to address the well-known stylised fact that the percentage mark-ups in the prices over the true probabilities is not uniform. In general, prices exhibit *favourite-longshot bias* in which, the normalised prices on the favourites of the race understate the winning chances of these horses, while the normalised prices on the longshots exaggerate their winning chances. For the betting market in the United Kingdom, Dowie (1976) confirms this bias.

Betting odds at North American race tracks are determined by the parimutuel method in which prices are proportional to amounts wagered. However, even this system cannot be regarded as being completely immune from manipulation. The setting of morning line odds, for example, will influence the early stages of betting. Nevertheless, this distinction should be borne in mind when examining North American data (see Ali (1977), Asch *et al.* (1982) and Thaler and Ziemba (1988)).

The overview of our paper is as follows. In the next section, we present our model of the betting market in an *n*-horse race. The solution of the model follows in Section II, where we derive a necessary and sufficient condition for the favourite-longshot bias. Section III investigates the robustness of this result in a more general framework. By means of a comparative statics argument, we show that the bias survives in a generalised form.

I. THE MODEL

Our model describes a market for bets in a horse race with *n* horses. The market is organised as an extensive form game in which two bookmakers compete in setting odds on the horses in anticipation of the betting behaviour to follow. There are three players in the game – the incumbent bookmaker, a potential bookmaker, and the bettor with insider information (the 'Insider'). There is also a set of *n* individuals called 'outsiders', labelled by the set $\{1, 2, ..., n\}$. The outsiders are not modelled formally as players in the game. Rather, they are mechanical traders who do not act strategically. In particular, the ith outsider attaches probability 1 to the ith horse winning the race.

However, although each outsider (taken individually) may be quite irrational, the game is designed so that the market prices which follow from outsiders' demands for bets are fully revealing, in the sense that the equilibrium prices coincide with the true probabilities. This provides a benchmark for the general case in which distortions in market prices are introduced as a result of insider trading.

There are n types of tickets which are sold in the market. The ith type of ticket pays one dollar if the ith horse wins and zero otherwise. The price of this ticket is denoted by π_i. These prices correspond to betting odds in the usual way. Odds of k to l correspond to the price of $l/(k+l)$. Negative odds are ruled out in our model, so that $0 \leqslant \pi_i \leqslant 1$ for all i. Fractional quantities of tickets may be sold.

As in any financial market, the prices quoted are not valid for all quantities. In sophisticated financial markets, the quoted price varies with the quantity traded. For our purpose, we shall assume that the bookmakers accept unit bets of one dollar. This is consistent with the convention that bookmakers' odds are the prices at which a 'substantial sum' may be wagered. The implication is that there is a limit to the size of the wager accepted at the quoted price. The assumption of unit trades is widely used in models of financial markets to bound the positions taken by insiders and risk neutral traders. Examples include Copeland and Galai (1983) and Glosten and Milgrom (1985), cited earlier. Our motivation is similar.

The game is in extensive form and describes the encounter between one bookie and one punter. It has four stages.

Stage 1 (Bidding Stage). The two bookies – the incumbent and the potential entrant – bid for monopoly rights to the betting market. Each submits a sealed bid of a positive real number. The bookie who submits the lower number wins the bid. If the bids are identical, the incumbent wins. The bookie who loses the bid gets a payoff of zero.

Stage 2 (Price-setting Stage). The bookie who wins the bid at stage 1 sets prices $\pi_1, \pi_2, \ldots, \pi_n$ for the n types of tickets subject to the constraint that the sum $\sum_i \pi_i$ does not exceed the bid submitted in stage 1 and that $0 \leqslant \pi_i \leqslant 1$, for all i.

Stage 3 (Nature's Choice of Winner and Punter). Nature then performs two experiments. In the first, a winner of the race is chosen. In the second, precisely one punter is chosen to bet against the price-setting bookie. Thus, either the Insider is chosen to play, or one of the outsiders is chosen. The experiments are governed by the probability measure μ, given as follows. Denote by O_i the event in which the ith outsider is chosen, by I the event in which the Insider is chosen, and by H_i the event in which the ith horse is chosen as the winner. The event O is the union $\bigcup_i O_i$. We shall assume that $\mu(H_i) = p_i, \mu(I) = z$, and $\mu(O_i) = (1-z)p_i$. That is, the ith horse wins with probability p_i, the Insider is chosen with probability z, and the ith outsider is chosen with probability

$(1-z)\,p_i$. We shall assume that $p_i > 0$ for all i. In keeping with the idea that the outsiders are noise traders, if an outsider is chosen, his identity is independent of the winning horse. That is, for any k and i,

$$\mu(O_k \,|\, H_i) = \mu(O_k). \tag{1}$$

In contrast, the probability that the Insider plays may not be independent of the identity of the winning horse. We denote by z_i the probability that the Insider is chosen conditional on the ith horse being the winner. That is,

$$\mu(I \,|\, H_i) = z_i. \tag{2}$$

Stage 4 (Betting Stage). The punter chosen at stage 3 meets the bookie and a bet is placed. Crucially, if the Insider is chosen to play, she is permitted to observe the identity of the winning horse, and is free to buy tickets from the bookie at the posted prices up to the value of one dollar. On the other hand, the ith outsider is an expected payoff maximiser with the belief that the ith horse wins with probability 1. Thus, the ith outsider always bets a dollar on the ith horse.

Once the betting has been completed, the race is run according to script and the horse chosen by Nature is seen to win the race. The bookie then settles with the punter in accordance with the odds offered at stage 2.

Two comments are in order concerning the model. Firstly, we could interpret the model as one in which a bookie meets a large number of punters, where the probabilities denote the proportion of each type of punter in the population. The problem for the bookie is identical. Secondly, our main interest is in the determination of prices given a zero profit condition. An alternative modelling strategy would have been to allow fully-fledged price competition between the bookies. Provided that bookies offer the full menu of bets, the outcome of price competition coincides with that of our model.

II. DERIVATION OF EQUILIBRIUM PRICES

We now proceed to the solution of the game. Our solution concept is subgame perfect equilibrium. Although there are many equilibria of our game, they are all essentially identical in the sense that the prices are identical across all equilibria. We begin by solving for the bookie's maximisation at stage two given the winning bid β at stage 1. This yields expressions for the prices $\pi_1, \ldots, \pi_n$. By substituting these expressions into the bookie's profit function, we obtain an expression for the price-setting bookie's profit in terms of the bid β. By solving for the equilibrium bid β and substituting into the expressions for $\pi_1, \ldots, \pi_n$, we obtain explicit solutions for the equilibrium prices in terms of the parameters of the model.

We denote by $\mathbf{p}$ the vector of winning probabilities $(p_1, \ldots, p_n)$ and by $\boldsymbol{\pi}$ the vector of prices $(\pi_1, \ldots, \pi_n)$. We shall denote by $V(\boldsymbol{\pi})$ the expected profit of the price-setting bookie given prices $\boldsymbol{\pi}$, and denote by $V(\boldsymbol{\pi}\,|\,I)$ the expected profit conditional on the punter being the Insider. Similarly, we denote by $V(\boldsymbol{\pi}\,|\,O)$ the expected profit conditional on the punter being an outsider. Thus,

$$V(\boldsymbol{\pi}) = z V(\boldsymbol{\pi}\,|\,I) + (1-z) V(\boldsymbol{\pi}\,|\,O). \tag{3}$$

Any punter bets precisely one dollar on a particular horse, so that if the punter bets on the ith horse, $1/\pi_i$ units of the ith ticket are sold by the bookie. Thus, if the bookie pays out, he pays out $1/\pi_i$ dollars, where i is the index of the winning horse. Since revenue is constant at 1 and the bookie always pays out to the Insider, $V(\pi \,|\, I)$ is given by $1 - \Sigma_i \mu(H_i \,|\, I)/\pi_i$. Some manipulation using (2) yields:

$$V(\pi \,|\, I) = 1 - \frac{1}{z} \sum_i \frac{z_i p_i}{\pi_i}. \tag{4}$$

Against the outsiders, the bookie pays out only in the events $\{H_i \cap O_i\}$. Thus, $V(\pi \,|\, O)$ is given by $1 - \Sigma_i \mu(H_i \cap O_i \,|\, O)/\pi_i$. From (1), we have:

$$V(\pi \,|\, O) = 1 - \sum_i \frac{p_i^2}{\pi_i}. \tag{5}$$

From (3), (4) and (5), the problem for the price-setting bookie at stage 2 of the game is to:

$$\underset{\pi}{\text{maximise}} \quad 1 - \sum_i \frac{z_i p_i + (1 - z) p_i^2}{\pi_i} \tag{6}$$

$$\text{subject to} \quad \sum_i \pi_i \leqslant \beta \quad \text{and} \quad 0 \leqslant \pi_i \leqslant 1, \quad \text{for all } i,$$

where β is the winning bid made at stage 1. The feasible set is the intersection of the halfspace $\{\pi \,|\, \Sigma_i \pi_i \leqslant \beta\}$ and the unit cube $\{\pi \,|\, 0 \leqslant \pi_i \leqslant 1, \forall i\}$. It is the intersection of two convex sets, and hence is itself convex. The objective function is strictly concave in π as can be verified from the Hessian which is a diagonal matrix with negative entries. Thus, any solution to (6) is unique. We must now consider the solution of the game for two cases – namely, the case in which the optimal π lies in the interior of the unit cube and the case in which it lies on the boundary of the unit cube.

The solution for the case in which the equilibrium π lies on the boundary of the unit cube is presented in the appendix. For the main body of the paper, we shall concentrate on the case in which bets are accepted on all horses. Since $V(\pi)$ is strictly increasing in the prices, the constraint $\Sigma_i \pi_i \leqslant \beta$ binds, and the optimal π satisfies the first-order conditions. Solving for π_i,

$$\pi_i = \frac{\beta \sqrt{[z_i p_i + (1 - z) p_i^2]}}{\sum_s \sqrt{[z_s p_s + (1 - z) p_s^2]}}. \tag{7}$$

Substituting into $V(\pi)$, we have the following expression for expected profit in stage 1 in terms of β.

$$1 - \frac{1}{\beta} \left\{ \sum_s \sqrt{[z_s p_s + (1 - z) p_s^2]} \right\}^2. \tag{8}$$

The payoff of both bookmakers must be zero in any equilibrium. It is at least zero since a bookie gets a payoff of zero by bidding a larger number than its rival. It is at most zero since the rival can undercut. Thus, in any equilibrium, the winning bid β is the number which sets (8) equal to zero. That is,

$$\beta = \left\{ \sum_s \sqrt{[z_s p_s + (1-z) p_s^2]} \right\}^2. \tag{9}$$

Substituting (9) into (7), we can solve for π_i in terms of the parameters of our model,

$$\pi_i = \sqrt{[z_i p_i + (1-z)] p_i^2} \left\{ \sum_s \sqrt{[z_s p_s + (1-z) p_s^2]} \right\}. \tag{10}$$

In the benchmark case in which $z = 0$ (when the Insider plays no part in the game), we have $\pi = p$, so that prices coincide with the true probabilities. Prices in this case mimic the determination of competitive prices in an economy consisting of the n outsiders, where the ith outsider is endowed with share p_i of the portfolio $(1, 1, ..., 1)$. Then, the total demand for the ith type of ticket is $p_i(\sum_s \pi_s)/\pi_i$, so that market clearing ensures $\pi_i = p_i \sum_s \pi_s$, and prices are proportional to the true probabilities. Notice also that the normalised price on the ith horse is identical to the proportion of the total wealth wagered on the ith horse. In other words, competitive prices are the parimutuel betting odds (see Eisenberg and Gale (1959)).

For the general case in which $z > 0$, we shall be interested in identifying the conditions under which prices exhibit favourite-longshot bias. Formally, we shall say that the equilibrium prices exhibit *favourite-longshot bias* when, $\pi_i/\pi_j < p_i/p_j$ if and only if $p_i > p_j$. In other words, the betting odds understate the winning chances of a favourite (horse i) relatively less than the winning chances of a longshot (horse j). A necessary and sufficient condition is identified in the following proposition.

PROPOSITION 1. *Suppose $z > 0$. Then, equilibrium prices exhibit favourite-longshot bias if and only if:*

$$z_i/p_i < z_j/p_j \Leftrightarrow p_i > p_j. \tag{11}$$

Proof. Since $z > 0$ and all components of $\mathbf{p}$ are non-zero, the ratio of prices in (10) is given by:

$$\frac{\pi_i}{\pi_j} = \sqrt{\left[\frac{z_i p_i + (1-z) p_i^2}{z_j p_j + (1-z) p_j^2} \right]} = \frac{p_i}{p_j} \sqrt{\left[\frac{1 - z + (z_i/p_i)}{1 - z + (z_j/p_j)} \right]}. \tag{12}$$

Thus, if (11) holds, $p_i > p_j \Leftrightarrow z_i/p_i < z_j/p_j \Leftrightarrow \pi_i/\pi_j < p_i/p_j$, which is the bias. The converse is immediate from (12).

Since $z_i = \mu(I \mid H_i)$, the condition identified in (11) rules out those cases in which the incidence of insider trading is substantially larger when a favourite is tipped to win than when a longshot is tipped to win. Since, in general, we would expect insider trading to be more prevalent given that a longshot is tipped to win, this condition accords with our intuitions. Moreover, condition (11) is satisfied even if z_i rises with p_i, provided that the ratio z_i/p_i is falling with p_i. Thus, the favourite-longshot bias would seem to be a fairly general feature of models of this kind. Notice, in particular, that (11) is satisfied when z_i is identical across all horses i. If we interpret our model as one in which the bookie meets a large number of punters, where the probabilities denote the proportion of each type of punter in the population, a low p_i would correspond to a low demand for the ith basic security from the set of outsiders. Then, (11) is satisfied if insider trading is more prevalent in 'thin' markets. Again, this accords with our intuition.

III. A MORE GENERAL ARGUMENT

The explicit solution for equilibrium prices has been obtained at the cost of attributing extreme probability beliefs to the noise traders. We shall now examine how our results are affected by a more general formulation of outsiders' beliefs.

We shall continue to assume that the outsiders are expected payoff maximisers according to their subjective probability judgements but we allow arbitrary sets of outsiders. However, we shall impose the condition that, in the absence of insider trading, prices are proportional to true probabilities. By insisting on this benchmark, any deviation from proportionality will be attributable to insider trading. We shall continue with the notation $V(\pi \mid 0)$ to denote the expected profit of the price-setting bookie against the set of outsiders. Our assumptions are:

(A1) $V(\pi \mid 0)$ is differentiable on the interior of the unit cube;

(A2) If β is the equilibrium bid of the game, then $\beta\mathbf{p}$ maximises $V(\pi \mid 0)$ subject to $\sum_i \pi_i \leqslant \beta$.

(A2) is a condition on all the subgames with bid β. Since not all of these subgames are reached in equilibrium, (A2) should be seen as a counterfactual statement concerning the set of outsiders. We note that our original model satisfies both (A1) and (A2).

Although we cannot specify the equilibrium prices without specifying $V(\pi \mid 0)$ in more detail, we have the following comparative statics result which states that, if condition (11) holds, the bookie's expected profit increases as prices move from $\beta\mathbf{p}$ toward the centre of the β-simplex. To say that prices move toward the centre of the simplex is to say that the prices on the favourites are lowered and the prices on the longshots are raised. Thus, our result can be paraphrased as saying that the introduction of favourite-longshot bias raises the bookie's expected profit.

PROPOSITION 2. *Assume* (A1) *and* (A2), *and suppose* z_i/p_i *is decreasing in* p_i. *Then, for* $\mathbf{p} \neq (1/n, ..., 1/n)$, *there exists a point* $\tilde{\pi}$ *on the line segment joining* $\beta\mathbf{p}$ *and the centre of the* β-simplex $(\beta/n, ..., \beta/n)$ *for which* $V(\tilde{\pi}) > V(\beta\mathbf{p})$.

The proof consists in showing that the directional derivative of $V(\pi)$ at $\beta\mathbf{p}$ toward the centre of the β-simplex is positive. We start by noting the following consequence of (11).

LEMMA. *If* z_i/p_i *is decreasing in* p_i, *then*

$$\sum_i \frac{z_i}{p_i}\left(\frac{1}{n} - p_i\right) > 0.$$

Proof. By hypothesis, for any $i \neq k$, $(z_i/p_i - z_k/p_k)(p_i - p_k) < 0$. In particular, for

$$p_k = \frac{1}{n}, \quad \text{we have} \quad (z_i/p_i - nz_k)\left(\frac{1}{n} - p_i\right) > 0.$$

Multiplying out and summing over i,

$$\sum_i \frac{z_i}{p_i}\left(\frac{1}{n} - p_i\right) - nz_k \sum_i \left(\frac{1}{n} - p_i\right) > 0.$$

But the second term is zero, thereby proving the claim.

Proof of Proposition 2. The directional derivative of a function f in the direction x evaluated at y is given by the inner product of the gradient ∇f at y with the unit vector $x/\|x\|$. Let $\mathbf{c} \equiv (\beta/n, \ldots, \beta/n)$ be the centre of the β-simplex. We are interested in the directional derivative of $V(\boldsymbol{\pi})$ in the direction $\mathbf{c} - \beta\mathbf{p}$ evaluated at $\beta\mathbf{p}$. This directional derivative exists by (A1), and is given by:

$$\nabla V(\beta\mathbf{p}) \frac{\mathbf{c} - \beta\mathbf{p}}{\|\mathbf{c} - \beta\mathbf{p}\|} = [(1-z)\nabla V(\beta\mathbf{p}\,|\,O) + z\nabla V(\beta\mathbf{p}\,|\,I)] \frac{\mathbf{c} - \beta\mathbf{p}}{\|\mathbf{c} - \beta\mathbf{p}\|}. \quad (13)$$

By (A2), $\beta\mathbf{p}$ maximises $V(\boldsymbol{\pi}\,|\,O)$ over the β-simplex. Since $\mathbf{c}$ lies on the β-simplex and $V(\boldsymbol{\pi}\,|\,O)$ is differentiable,

$$\nabla V(\beta\mathbf{p}\,|\,O)(\mathbf{c} - \beta\mathbf{p}) = 0. \quad (14)$$

Thus, (13) has the same sign as $\nabla V(\beta\mathbf{p}\,|\,I)(\mathbf{c} - \beta\mathbf{p})$. From the expression for $V(\boldsymbol{\pi}\,|\,I)$ given by (4),

$$z\nabla V(\beta\mathbf{p}\,|\,I)(\mathbf{c} - \beta\mathbf{p}) = \sum_i \frac{z_i p_i}{\beta^2 p_i^2}\left(\frac{\beta}{n} - \beta p_i\right) = \frac{1}{\beta}\sum_i \frac{z_i}{p_i}\left(\frac{1}{n} - p_i\right). \quad (15)$$

Thus, by the lemma, $z\nabla V(\beta\mathbf{p}\,|\,I)(\mathbf{c} - \beta\mathbf{p}) > 0$, so that from (13), (14) and (15), the directional derivative of $V(\boldsymbol{\pi})$ in the direction $\mathbf{c} - \beta\mathbf{p}$ evaluated at $\beta\mathbf{p}$ is positive. Since $V(\boldsymbol{\pi})$ is differentiable, there exists a point $\tilde{\pi}$ lying on the line $\beta\mathbf{p} + t(\mathbf{c} - \beta\mathbf{p})$, $(t \geq 0)$ for which $V(\tilde{\pi}) > V(\beta\mathbf{p})$.

This argument is limited in that it is a comparative statics argument rather than one which relies on the solution of the bookie's maximisation problem at stage 2. Indeed, it would only be by accident that the bookie's optimal π lies on the line segment joining $\beta\mathbf{p}$ and $\mathbf{c}$. Nevertheless, the result shows in an economical way the incentives at work. Also, the fact that this argument does not rely on a particular distribution of outsiders' beliefs focuses attention on the condition that z_i/p_i is a decreasing function of p_i.

IV. CONCLUDING REMARKS

A number of features make the betting market particularly tractable for formal investigation. Chief among these is the fact that prices are linear, in contrast to price schedules of more sophisticated financial markets where unit prices depend on quantities traded. Thus, betting markets form a natural bridge between the literature on linear pricing based on competitive equilibrium and that on insider trading. The necessary and sufficient condition for the favourite-longshot bias identified in this paper seems to be a natural one in this context.

Whatever one feels about the model presented in this paper, one lesson of general importance is that any departure from the classical view of markets must be accompanied by a detailed analysis of the 'market microstructure'. It is hoped that the insights gained here may contribute to the understanding of more sophisticated financial markets.

University College, Oxford

Date of receipt of final typescript: August 1991

APPENDIX

Corner Solution of π

If all the constraints $0 \leqslant \pi_i \leqslant 1$ bind, then the optimal π is given by $(1, 1, \ldots, 1)$ since $V(\pi)$ is increasing in prices. Next, if all but one of the constraints $0 \leqslant \pi_i \leqslant 1$ bind, then the optimal π is determined by the constraint $\sum_i \pi_i \leqslant \beta$ and will be of the form:

$$[1, 1, \ldots, (\beta - n + 1), 1, \ldots, 1].$$

Otherwise, we can partition the set $\{1, 2, \ldots, n\}$ into A and B where $A = \{i \mid \pi_i < 1\}$ and $B = \{i \mid \pi_i = 1\}$. Since $V(\pi)$ is increasing in prices, the constraint $\sum_i \pi_i \leqslant \beta$ binds, and any π_i, π_j where $i, j \in A$ satisfy the respective first-order conditions. Thus, for any $i, j \in A$,

$$\frac{\pi_i}{\pi_j} = \sqrt{\left[\frac{z_i p_i + (1 - z_j) p_i^2}{z_j p_j + (1 - z) p_j^2} \right]}. \tag{16}$$

Denote by $\bar{B}$ the cardinality of the set B. Since the sum of prices is β and $\pi_i = 1$ for $i \in B$, the sum $\sum_{i \in A} \pi_i$ is given by $\beta - \bar{B}$. Thus, for $i \in A$,

$$\pi_i = \frac{(\beta - \bar{B}) \sqrt{[z_i p_i + (1 - z) p_i^2]}}{\sum_{s \in A} \sqrt{[z_s p_s + (1 - z) p_s^2]}}. \tag{17}$$

By substituting (17) into $V(\pi)$, the bookie's expected profit at stage 1 is:

$$1 - \frac{1}{(\beta - \bar{B})} \left\{ \sum_{s \in A} \sqrt{[z_s p_s + (1 - z) p_s^2]} \right\}^2 - \sum_{s \in B} [z_s p_s + (1 - z) p_s^2], \tag{18}$$

which is zero in any equilibrium. Solving for $\beta - \bar{B}$ and substituting into (17), we arrive at the following solution for the equilibrium price $\pi_i (i \in A)$, in terms of the parameters of the model.

$$\pi_i = \sqrt{[z_i p_i + (1 - z) p_i^2]} \left\{ \frac{\sum_{s \in A} \sqrt{[z_s p_s + (1 - z) p_s^2]}}{1 - \sum_{s \in B} [z_s p_s + (1 - z) p_s^2]} \right\} \tag{19}$$

We note that when B is empty, (19) reduces to the interior solution (10), as we would expect.

REFERENCES

Ali, M. M. (1977). 'Probability and utility estimates for racetrack bettors.' *Journal of Political Economy*, vol. 82, pp. 803–15.

Arrow, K. (1964). 'The role of securities in the optimal allocation of risk-bearing.' *Review of Economic Studies*, vol. 31, pp. 91–6. Reprinted in *Essays in the Theory of Risk-Bearing*, North Holland, 1970.

Asch, P., Malkiel, B. and Quandt, R. (1982). 'Racetrack betting and informed behavior.' *Journal of Financial Economics*, vol. 10, pp. 187–94.

Bagehot, W. (1971). 'The only game in town.' *Financial Analysts' Journal*, vol. 22, pp. 12–4.

Copeland, T. and Galai, D. (1983). 'Information effects on the bid ask spread.' *Journal of Finance*, vol. 38, pp. 1457–69.

Crafts, N. (1985). 'Some evidence of insider knowledge in horse race betting in Britain.' *Economica*, vol. 52, pp. 295–304.

Debreu, G. (1959). *Theory of Value*. New York: Wiley.

Dowie, J. (1976). 'On the efficiency and equity of betting markets.' *Economica*, vol. 43, pp. 139–50.

Eisenberg, E. and Gale, D. (1959). 'Consensus of subjective probabilities: the parimutuel method.' *Annals of Mathematical Statistics*, vol. 30, pp. 165–8.

Glosten, L. and Milgrom, P. (1985). 'Bid, ask and transaction prices in a specialist market with heterogeneously informed traders.' *Journal of Financial Economics*, vol. 14, pp. 71–100.

Jockey Club Committee of Inquiry (1968). *The Racing Industry*. London: Trustees of the Jockey Club.

Kyle, A. (1985). 'Continuous auctions and insider trading.' *Econometrica*, vol. 53, pp. 1315–35.

Thaler, R. and Ziemba, W. (1988). 'Parimutuel betting markets: racetracks and lotteries.' *Journal of Economic Perspectives*, vol. 2, pp. 161–74.

PART V

PRICES VERSUS HANDICAPPING: PLACE AND SHOW ANOMALIES

Introduction to Prices vs. Handicapping: Place and Show Anomalies

Donald B. Hausch, Victor S.Y. Lo and William T. Ziemba

Most of the research on racetrack efficiency has focused on the win bet market, and the general conclusion has been that there exists a favorite-longshot bias but it is not sufficiently strong to allow positive profits (e.g. Snyder (1978)[1]). Extreme favorites of odds 3-10 and shorter are an exception, allowing a small positive return, but they are so uncommon that for practical purposes the win market is weak form efficient, see Ziemba and Hausch (1986)[2]. Inefficiencies in the place and show markets, however, have been reported as early as Griffith (1961)[1]. There are several explanations: 1) their pools are smaller; 2) place and show wagers are more complicated than win bets since many different payoffs are possible depending on which horses are the first two or three finishers; and 3) extrapolating the win market's bias for favorites to higher probability place and show bets on favorites suggests a potential for inefficiencies.

Asch, Malkiel and Quandt (1984[1], 1986[1]) (the second paper a correction of the first) applied a logit model estimating the true win probabilities based on the final win odds, earlier win odds and morning line odds. Comparing some filter rules for betting to win, place and show, their simulations showed returns to place and show that exceeded those to win, but in no cases were the returns positive with statistical significance.

Hausch, Ziemba and Rubinstein (1981)[1] used a more selective rule for identifying inefficiencies in the place and show markets. They first applied the Harville formulas (Harville (1973)[1]) to predict the probabilities of finishing order given the final win odds data. Then place and show probabilities were calculated followed by computation of expected returns to place and show. At this point their system involves two steps:

(i) identify those place and show bets with an expected return exceeding a specific level (a level that depends on the quality of the horses and the size of the track); and

(ii) the amount to wager is based on the "Kelly criterion", to maximize one's expected log of final wealth.

The "Kelly criterion" was originally proposed by Kelly (1956)[2] and extended by Brieman (1961)[2]. Is properties include that it maximizes the long-term capital growth rate and, asymptotically, it minimizes the time to reach a fixed wealth level. Hausch, Ziemba and Rubinstein developed regression approximations to the procedure to make possible its real-time application and, using data from Exhibition Park and Santa Anita, showed returns on the order of ten percent. Hausch and Ziemba (1985)[1] extended the wagering scheme to allow for varying track take, different initial wealth, different size tracks, multiple wagers, and multiple horse entries. Details of the betting system and more empirical results from Belmont Park and the Kentucky Derby appear in Ziemba and Hausch (1984[2], 1987[2]). For discussions of the system, see Skinner (1989)[2] and McCardell (1992)[2]. More examples and considerations of other bet types such as exotics appear in Ziemba and Hausch (1986)[2].

Ritter (1994)[1], in a revision of work predating Hausch, Ziemba and Rubinstein (1981)[1, considered similar ideas for place and show betting]. Instead of computing expected returns and employing the Kelly criterion, Ritter compared a horse's win bet fractions with those for place and show, and wagered when the difference was great enough. While he demonstrated positive profits using final odds, profits were not possible when wagers were based on odds 1½ minute from the end of betting. More recently, Lo, Bacon-Shone and Busche (1994)[2] modified Hausch, Ziemba and Rubinstein (1981)[1] by using other distributional assumptions (normal and gamma) of running times with the approximation methods proposed by Lo and Bacon-Shone (1993)[2]. More data input is required, but their simulations show improved returns.

Willis (1964)[2] developed a developed a linear programming model to exploit possible arbitrage opportunities between the win and place markets. His situations are extremely rare, though. Hausch and Ziemba (1990b)[1] studied another arbitrage possibility in the show market that may exist when there is an extreme favorite that has received most of the show betting. They showed when a "lock" exists and developed a linear program to maximized the guaranteed return. While the lock strategy guarantees a return, it is more conservative than the Kelly criterion.

Cross track betting permits bettors at their local track to wager on races being run at another track. So, in addition to exploiting possible inefficiencies in the place and show pools, cross track betting allows one to exploit inefficiencies in these markets across tracks. Hausch and Ziemba (1990a)[1] developed and tested optimal betting strategies for cross track betting. One strategy was to identify whether a risk-free hedge could be constructed by betting a sufficient amount on each horse at the track where its offered odds was longest. Examples where the variance in odds across tracks was sufficient were provided. Also analyzed was the Kelly criterion in two environments: 1) a single bettor at a cross track observing (perhaps by television) the home track's odds' and 2) a syndicate of bettors, one at each track, communicating with a central decision maker. Leong and Lim (1994)[1] also found evidence of profits using cross track betting that exists between races in Singapore and Malaysia. Both these papers showed profits but neither had sufficient data for statistically significant profits.

[1] included in this volume
[2] cited in the Annotated Bibliography

Peter Asch
Rutgers University

Burton G. Malkiel
Yale University

Richard E. Quandt
Princeton University

Market Efficiency in Racetrack Betting

I. Introduction

Betting at racetracks and investing in the stock market have several characteristics in common. In neither situation are future earnings known with certainty. Both situations are characterized by a large number of participants and by the availability of extensive information and professional advice. Horse racing data, therefore, permit interesting tests of the efficient market theory.

Numerous authors have analyzed horse racing data and various conclusions have been confirmed: (1) There is a systematic tendency to overbet long-shots and underbet favorites (Rosett 1965, 1971; Snyder 1978; and Ali 1979). (2) The tendency to overbet long-shots is strongest in late races (McGlothlin 1956; Asch, Malkiel, and Quandt 1982). (3) The utility function of bettors is convex, indicating risk love (Weitzman 1965; Asch et al. 1982). Figlewsky (1979) tested hypotheses concerning whether the betting "market" is efficient with respect to track odds and handicappers' picks of winning horses. By fitting a logit model with horses' win-loss records representing the dependent variable and track odds and handicappers' picks the independent variables, it is shown by likelihood ratio

Do market inefficiencies in racetrack betting permit profitable strategies to be devised on the basis of observed betting patterns? Logit analysis of market data indicates that profits cannot be earned in win betting, although it is possible to outperform the average bettor substantially. Surprisingly, profits net of transactions costs on place and show betting appear possible, although there are reasons to suspect that such opportunities may not be exploitable on a substantial scale. The findings do not suggest irrational behavior by bettors.

tests that handicappers' picks are substantially accounted for in the track odds. Finally, Hausch, Ziemba, and Rubinstein (1981) analyzed betting horses to place (come in first or second) or to show (first, second, or third) and found that inefficiencies exist in place and show betting.

The present paper analyzes a new sample of racetrack results from Atlantic City, New Jersey. We investigate how well the track odds and the morning line odds (rather than handicappers' picks) predict the victors of races. But there is an important further question that requires attention. If track odds and other publicly available information do very well in predicting winners, one might want to use such information in betting; yet this use of information will tend to lower the odds (and payoffs) of the horses predicted to win.

The pertinent question, then, concerns the efficiency of the race-track betting market. If efficiency is perfect, the rate of return to every betting strategy based on available information should approximate closely the track "take." For the sample of races analyzed here, the take is about 18.5%; thus investors who pursue any strategy should earn a negative return close to 18.5%. It is of particular interest to ask, however, not whether the racetrack betting market is efficient in this very stringent sense, but whether there exist departures from efficiency sufficiently large to permit profitable betting strategies. Can one, in other words, devise strategies based on observable betting patterns that imply positive rates of return? Only if such unexploited arbitrage opportunities were being consistently neglected could we conclude that the racetrack betting market was inefficient.[1]

A second question is, If one were able to predict winners by using publicly available information, would such a finding be inconsistent with the assumption of rational behavior in the racetrack market? Rosett (1965) has shown that rationality imposes requirements on the relationship between return and the probability of winning by exploiting the possibility of combination strategies, such as parlays. He found specifically that the return to low-win-probability horses appeared lower than predicted by the rationality assumption. We comment on this issue in the concluding section.

1. An analogy may be made to empirical research in the stock market. Departures from perfect efficiency have often surfaced in empirical tests, such as some findings of serial correlation and evidence of inferrable information from "insider" trading that is not immediately reflected in market prices. Nevertheless, the conclusion remains that, in general, any departures from efficiency are not large enough to permit an investor to devise an active strategy that can consistently beat a passive buy-and-hold strategy. In this sense, the empirical evidence generally supports a finding of efficiency for the stock market.

II. Data and Definitions

The Data Set

Observations are based on the entire 1978 thoroughbred racing season at the Atlantic City Race Course, which includes 712 races and 5,714 horses. The data are

1. The "morning" line odds for each horse in each race, determined by the track's professional handicapper and printed in the daily racing program. These are the handicapper's estimates of the winning probabilities for each horse that confront bettors before the start of each betting period.

2. Parimutuel odds for each horse in each race at various points ("cycles") during the betting period. Twenty-four cycles are typically recorded, showing the minute-by-minute course of the actual betting for each race. Only some of the available cycles were employed in the analysis.

3. Final parimutuel odds for each horse, given at the final betting cycle. The odds are determined by the ratio of the betting pool available for distribution to the amount bet on each horse. The racetrack subtracts a percentage or "take" from the total betting pool to cover taxes, expenses, and profits. In addition, the track enjoys "breakage," the gain from being allowed to round payoffs downward (to the nearest 10 cents or 20 cents on a $2.00 bet), or to round odds downward (to the nearest tenth). The total betting pool minus the take, including breakage, is paid on the winning horse.

4. The outcome of each race.

Some Definitions

The following definitions are employed: B = the total amount bet on the race; b_h = the amount bet on horse h, $h = 1, 2, \ldots, H$, where H is the number of horses in the race. It follows that

$$B = \sum_{h=1}^{H} b_h.$$

Let t = the total take of the track including breakage, all expressed as a percentage of the total bet. The odds of horse h may then be stated as

$$O_h = \frac{B(1-t) - b_h}{b_h} = \frac{B(1-t)}{b_h} - 1.$$

We define the (bettors' subjective) probability that horse h will win the race, P_h. It follows then that

$$P_h = \frac{b_h}{B} = (1-t)/(1 + O_h)$$

One can also define implicit or marginal odds derived from bets made late in the betting period. The betting period is approximately 25 minutes, the time between races. Define cycle 1 as the betting cycle encompassing the first 17 minutes (the first ⅔) of the betting period, and cycle 2 as the betting in the last 8 minutes (the last ⅓) of the betting period. The marginal odds on horse h in betting cycle 1, $O_{h,1} = [B_1(1 - t)/b_{h,1}] - 1$, where B_1 is the total amount bet up to the end of cycle 1, and $b_{h,1}$ is the amount bet on horse h up to the end of cycle 1. The final odds on horse 1 (which are also the odds on the horse derived from all betting up to and including cycle 2), are

$$O_{h,2} = \frac{B_2 (1 - t)}{b_{h,2}} - 1,$$

where B_2 is the total amount bet through both cycles (1 and 2) and $b_{h,2}$ is the total amount bet on horse h through both cycles (1 and 2).

Define the marginal odds as the odds derived from the betting during a particular cycle, but not including previous cycles. We seek the marginal odds created by bettors in cycle 2, namely, the bets of the "late money," or what is sometimes hypothesized to be the "smart money." The argument is that bettors with true inside information will prefer not to signal that information to the public until late in the betting cycle to minimize "following" behavior on the part of other track bettors.

The marginal odds on horse h in cycle 1 are simply the regular odds, that is:

$$O^m_{h,1} = O_{h,1}.$$

The marginal odds on horse h in cycle 2 can be derived from the odds at the end of cycles 1 and 2, as follows:

$$O^m_{h,2} = \frac{O_{h,2}b_{h,2} - O_{h,1}b_{h,1}}{b_{h,2} - b_{h,1}}.$$

Finally, define a payoff W_h from horse h as unity plus the final odds if h wins and zero otherwise. A horse that goes off at odds of three to one pays \$4.00 for each dollar bet if the horse wins. A rate of return R from betting one dollar on one horse in each of N races is $R = (\Sigma W_h - N)/N$, where ΣW_h is the sum of payoffs in the N races.

III. Models and Results

Define W_{ij} as the win function of the ith horse in the jth race, by

$$W_{ij} = \beta' x_{ij} + u_{ij},$$

where β' is a vector of coefficients, x_{ij} a vector of observable variables, and u_{ij} an error term. W_{ij} may be thought to measure the "winningness" of a horse and to depend on a systematic part ($\beta'x_{ij}$) and a stochastic part. The ith horse wins the jth race if $W_{ij} > W_{lj}, \ldots, W_{ij} > W_{n_j j}$, where n_j is the number of horses entered in the jth race. If the u_{ij} are assumed to have independent extreme value (Weibull) distributions, this leads to the well-known logit model, giving the probability that the ith horse wins the jth race as

$$P_{ij} = \frac{\exp(\beta'x_{ij})}{\displaystyle\sum_{k=1}^{n_j} \exp(\beta'x_{kj})}.$$

Estimates of the β's are obtained by maximizing the likelihood function. It is well known that the logit model possesses the independence of irrelevant alternatives property: the relative odds for horses i and k (P_{ij}/P_{kj}) depend only on the characteristics of these horses. As a first approximation this appears to be reasonable for horse racing.

Table 1 displays some basic results. All independent variables were odds converted to probabilities. Thus ML represents the win probabilities implied by the morning line odds, F the probability based on the final odds, and $M2$ the probability based on the marginal odds for cycle 2 (the last ⅓ of the betting period). The variable $M2$ is included in some of the computations to account for the possibility that the late bettors (smart money) have better information than the aggregate of all bettors. All coefficients are highly significant and, comparing runs 1, 2, and 3 with 4 and 5, respectively, the likelihood ratio test emphatically rejects the null hypothesis that the morning line odds or the final odds or the marginal odds alone provide a satisfactory explanation. In particular, we reject the null hypothesis that the morning line odds contain no information not already accounted for by the track odds. The probability of winning is positively and significantly affected by the morning line and final track probabilities as well as by the marginal probabilities. Including F and ML as well as $M2$ causes the standard errors to

TABLE 1   **Logit Results**

Variable	Logit Run				
	1	2	3	4	5
ML	9.99	. . .	. . .	4.15	4.74
	(.54)	. . .	. . .	(1.02)	(.96)
F	. . .	6.42	. . .	4.26	. . .
	. . .	(.33)	. . .	(.62)	. . .
$M2$	. . .	. . .	5.74	. . .	3.54
	. . .	. . .	(.30)	. . .	(.54)
Log likelihood	−1,280.0	−1,265.0	−1,270.0	−1,257.0	−1,258.0

NOTE.—Asymptotic standard errors in parentheses.

increase sharply, and no further significant improvement occurs in the log likelihood when both these variables are included.[2]

Simulations Using Only Win Bets

Using the results of the logit estimation one can perform the following simulations. For each race, using the coefficients in run 4 or 5 one may compute the win probability $\hat{P}_{ij}$ for each horse. We may then consider the following strategies:

Strategy 1: In each race $j(j = 1, \ldots, N)$ bet \$2.00 on the horse with the highest $\hat{P}_{ij}$.

Strategy 2: In each race bet \$2.00 on the horse with the highest $\hat{P}_{ij}$ if $\hat{P}_{ij} \geqq \gamma$, where γ is a "filter" coefficient.

Strategy 3: In each race bet \$2.00 on the horse with the highest $\hat{P}_{ij}$, if the highest $\hat{P}_{ij}$ is at least 3.5 times greater than the second highest $\hat{P}_{ij}$. Strategy 3 differs from strategy 2 in its concern, not for a high $\hat{P}_{ij}$ per se, but rather for selecting horses to bet only if the second choice by the criterion function is sufficiently inferior to the first choice.

Running such simulations on the sample from which the coefficients were obtained would bias the results. We therefore divided the sample into two halves and reestimated runs 4 and 5 from each of the two halves.[3] We then used the coefficients from each half sample for simulations on the other half sample.

For each simulation one can compute the rate of return for each of the strategies. These are displayed in table 2. The results show that all but one of the rates of return are better than the rate of return implied by the track take (-0.185) if the market is perfectly efficient, as defined above. Averaging the rates over the two half samples shows that marginally better results are obtained with run 5, that is, when the marginal odds pertaining to the last 8 minutes of betting time are used instead of the final odds. For one of the strategies the average rates of return indicate that betting can almost be a fair game. The results demonstrate clear departures from perfect efficiency but do not suggest the possibility of betting strategies that are profitable after accounting for the track take.

Simulations Using Place Bets and Show Bets

An alternative to placing win bets is to bet the selected horse to place or to show. The selection of the horse in each race was based on the same strategies discussed before but in the present simulations place or show bets were made. The rates of return are shown in table 3. All

2. The standard errors increase because of collinearity between F and M_2.
3. The coefficients from the half samples and the results of the likelihood ratio tests comparing the half-sample runs 4 and 5 with half-sample runs 1, 2, and 3 are similar to those reported in table 1.

TABLE 2   Rates of Return in Simulations Using "Win Bets"

	Estimation: First Half Simulation: Second Half		Estimation: Second Half Simulation: First Half		
	N	R	N	R	A
Run 4:					
Strategy 1	356	−.160	356	−.168	−.164
Strategy 2:					
$\gamma = .4$	132	−.111	100	−.101	−.106
$\gamma = .5$	62	−.077	57	.002	−.038
$\gamma = .6$	31	−.145	22	−.141	−.143
Strategy 3	52	−.150	51	−.047	−.098
Run 5:					
Strategy 1	356	−.186	356	−.116	−.151
Strategy 2:					
$\gamma = .4$	127	−.091	107·	−.028	−.068
$\gamma = .5$	61	−.070	58	−.052	−.061
$\gamma = .6$	28	−.114	21	−.090	−.102
Strategy 3	47	−.183	54	−.007	−.095

NOTE.—N = number of horses; R = rate of return, A = average of rates of return over the two half samples.

TABLE 3   Rates of Return in Simulations Using Place Bets and Show Bets

| | Run 4 | | | | | Run 5 | | | | |
| | Est.: First Half Sim.: Second Half | | Est.: Second Half Sim.: First Half | | | Est.: First Half Sim.: Second Half | | Est.: Second Half Sim.: First Half | | |
	N	R	N	R	A	N	R	N	R	A
A. Place bets:										
Strategy 1	356	.160	356	.127	0.143	356	.187	356	.165	.176
Strategy 2:										
$\gamma = .4$	132	.241	100	.183	0.212	127	.258	107	.269	.263
$\gamma = .5$	62	.328	57	.214	0.271	61	.301	58	.198	.249
$\gamma = .6$	31	.179	22	.527	0.353	28	.130	21	.400	.265
Strategy 3	52	.253	51	.190	0.221	47	.224	54	.287	.255
B. Show bets:										
Strategy 1	356	.090	356	.046	0.068	356	.094	356	.052	.073
Strategy 2:										
$\gamma = .4$	132	.133	100	.054	0.093	127	.140	107	.085	.112
$\gamma = .5$	62	.202	57	.016	0.109	61	.149	58	.006	.087
$\gamma = .6$	31	.100	22	.189	0.144	28	.064	21	.126	.095
Strategy 3	52	.108	51	.041	0.074	47	.102	54	.135	.118

NOTE.—See table 2 and n.

returns are positive. Betting the horse with the highest win probability to place yields a higher return than betting it to show. The number of horses on which bets are made under strategy 2 with $\gamma = .5$ or $.6$ and under strategy 3 is sufficiently small so that one cannot place great confidence in the resulting return figures. But for strategy 1, for example, involving 356 bets in each subsample, the return to place betting ranges from 0.143 to 0.176. Finally, one may note that for those strategies which allow a large number of horses to be bet on (strategy 1 and strategy 2 with $\gamma = .4$), the results of run 5, which employ the morning line and marginal odds, are slightly better.

Even simpler strategies can yield positive returns. One could, for example, predict the winner on the basis of logit runs 1, 2, or 3, employing a single explanatory variable.[4] The results are broadly comparable but somewhat less favorable. Thus, for example, strategy 1 yields an average return from place bets (over the two subsamples) of 0.081 when the morning line alone is used to predict the winner, 0.076 when the final odds are used, and 0.101 when the marginal odds are employed. This still compares favorably with placing win bets on the favorites, from which the return is -0.173.

We were not able to improve on the results of table 3 by employing more sophisticated compound strategies. The winners and runners-up can, for example, be predicted from a generalization of the logit model that explicitly takes into account not only which horse won, but which horses were second and third.[5] These predictions have been employed for combination strategies involving win, place, and show bets simultaneously.[6] The most successful of these combination strategies yielded average returns of 0.133–0.140, depending on which explanatory variables were used in the generalized logit estimation.

IV. Conclusions

The results from win betting strategies show that strategies can be developed that exceed the return from simply betting the favorite to win. The results from using place and show bets are surprising in the sense that we now find that net profits can be made. Several conjectures may be offered to explain this finding. It is possible, for example, that the tendency to overbet long shots is accentuated in place and

---

4. For strategy 1 this is equivalent to betting on the favorite.

5. The generalization follows from Harville (1973).

6. The most successful strategy requires predicting for each race three highest win probabilities denoted by $P_1 \geqq P_2 \geqq P_3$, where 1, 2, and 3 are the indices of the corresponding horses. According to this strategy, if (i) $P_1 - P_2 > .25$ and $P_2 - P_3 > .15$, bet horse 1 to win, 2 to place, 3 to show; (ii) if $P_1 - P_2 \leqq .25$ and $P_2 - P_3 > .15$, bet horse 1 to place, 2 to show; (iii) if $P_1 - P_2 > .25$, $P_2 - P_3 \leqq .15$, bet 1 to win and 1 to place; (iv) if $P_1 - P_2 \leqq .125$, $P_2 - P_3 \leqq .15$, bet 1 to show.

show betting. It is also possible that all the information contained in the win pool may not be efficiently impounded in the place and show pool. This finding confirms the broad results of Hausch et al. (1981).

Are these results incompatible with the assumption of rationality? We think not. First, we may ask whether the relationship between return and the probability of winning predicted by Rosett (1965) would still hold today. The reason is that the opportunities for low-probability parlay-type bets at racetracks have greatly increased in the last decade and there is now a plethora of exactas, trifectas, "pick fours," and so on, that are more profitable than parlays in the sense that they involve only one track take rather than n. More important, however, it is quite possible, as Rosett suggests, that even today the market fails to provide enough low-probability betting opportunities, given a substantial "taste" by the betting public for such bets. This clearly will have the effect of lowering the return on low-probability bets and is compatible with our results. Our findings that positive returns can be earned do not suggest irrational behavior; rather they suggest that the taste for certain nonpecuniary benefits associated with certain types of bets make positive profits possible for some bettors.

In other words, it is not a paradox that bets on different horses have different expected returns (and variance of returns); this is simply a consequence of bettors' having different utility functions and selecting different points among the available opportunities (characterized by different objective and subjective winning probabilities).

It should be noted, in conclusion, that these results cannot be interpreted as firm evidence of unexploited arbitrage opportunities for at least two reasons. First, for any strategy that employs more than just morning line odds, there is a serious difficulty in implementation, since final odds are typically not posted until thirty or more seconds after the betting windows close. Second, the place and show pools are usually much smaller than the win pools. Any attempt to employ the implied arbitrage strategies may prove infeasible since a relatively small volume of betting could quickly wipe out the apparent profit opportunity. Finally, it goes without saying that our simulations need to be repeated over alternative and large samples before we can have considerable confidence in the results.

References

Ali, Muktar M. 1979. Some evidence of the efficiency of a speculative market. *Econometrica* 47:387–92.

Asch, P.; Malkiel, Burton G.; and Quandt, Richard E. 1982. Racetrack betting and informed behavior. *Journal of Financial Economics* 10:187–94.

Figlewsky, Stephen. 1979. Subjective information and market efficiency in a betting market. *Journal of Political Economy* 87:75–88.

Harville, D. A. 1973. Assigning probabilities to outcomes of multi-entry competition. *Journal of the American Statistical Association* 68:312–16.

Hausch, D. B.; Ziemba, W. T.; and Rubinstein, M. 1981. Efficiency of the market for racetrack betting. *Management Science* 27:1435–52.

McGlothlin, W. H. 1956. Stability of choices among uncertain alternatives. *American Journal of Psychology* 69:604–15.

Rosett, R. N. 1965. Gambling and rationality. *Journal of Political Economy* 73:595–607.

Rosett, R. N. 1971. Weak experimental verification of the expected utility hypothesis. *Review of Economic Studies* 38:481–92.

Snyder, Wayne N. 1978. Horse racing: Testing the efficiency markets model. *Journal of Finance* 22:1109–18.

Weitzman, M. 1965. Utility analysis and group behavior: An empirical study. *Journal of Political Economy* 73:18–26.

Peter Asch
Rutgers University

Burton G. Malkiel
Yale University

Richard E. Quandt
Princeton University

Market Efficiency in Racetrack Betting: Further Evidence and a Correction*

We recently reported (Asch, Malkiel, and Quandt 1984) the results of simulated betting strategies based on logit analysis of racetrack betting data. Morning line odds, final odds, and marginal odds (defined as the odds implied by the moneys wagered during the last few minutes prior to each race) were used to predict the winning probability of each horse in a race; and a variety of strategies involving bets on the highest-probability horses was attempted. Our main conclusions were that we were unable to devise profitable strategies for win betting but that such strategies could be employed in place and show betting.

We have now run the same simulations on a new data set comprising 706 races for the summer 1984 harness meet at the Meadowlands Race-track. The logit estimations, in which winning probabilities are a function of combinations of morning line odds, final odds, and marginal odds, show coefficients remarkably similar to those of our earlier study, which was based on 1978 data for the Atlantic City Race Course. The rates of return to the place and show betting strategies, however, are substantially lower than those estimated with the 1978 data. Rerunning the simulations on the earlier data disclosed a computer programming error that had inflated the returns to place and show bets.[1]

For each data set the sample of races was divided in half. Coefficients from each half sample were used to estimate probabilities of winning. These probabilities were used in simulated betting strate-

---

* We are grateful to Bruce H. Garland, deputy director of the New Jersey Racing Commission, Joseph J. Malan, director of mutuels at the Meadowlands, and Joe LaVista, New Jersey State auditor, for their help in providing the relevant data.
1. The error in the simulations was as follows. When a successful place (or show) bet was made, the payoff credited to our hypothetical "account" was not necessarily that of the horse bet on but rather that of the *winning* horse. Since in the majority of such instances the winners were relative long shots, this error tended to inflate the rates of return.

TABLE 1   Rates of Return in Simulations Using Place Bets and Show Bets (Atlantic City)

	Run 4				Run 5			
	Estimation: First Half; Simulation: Second Half		Estimation: Second Half; Simulation: First Half		Estimation: First Half; Simulation: Second Half		Estimation: Second Half; Simulation: First Half	
	N	R	N	R	N	R	N	R
A. Place bets:								
Strategy 1	356	−.043	356	−.097	356	−.052	356	−.077
Strategy 2 ($B = .4$)	132	−.000	100	−.009	127	.021	107	.063
Strategy 2 ($B = .5$)	62	.023	57	.028	61	.017	58	.005
Strategy 2 ($B = .6$)	31	−.031	22	.109	28	−.077	22	.062
Strategy 3	52	−.049	51	.027	47	−.082	54	.106
B. Show bets:								
Strategy 1	356	−.037	356	−.102	356	−.041	356	−.100
Strategy 2 ($B = .4$)	132	.009	100	−.043	127	.019	107	−.006
Strategy 2 ($B = .5$)	62	.073	57	−.057	61	.035	58	−.053
Strategy 2 ($B = .6$)	31	−.019	22	.027	28	−.046	22	.005
Strategy 3	52	−.021	51	−.062	47	−.028	54	.016

TABLE 2   Rates of Return in Simulations Using Place Bets and Show Bets (Meadowlands)

	Run 4				Run 5			
	Estimation: First Half; Simulation: Second Half		Estimation: Second Half; Simulation: First Half		Estimation: First Half; Simulation: Second Half		Estimation: Second Half; Simulation: First Half	
	N	R	N	R	N	R	N	R
A. Place bets:								
Strategy 1	353	−.142	353	−.080	353	−.168	353	−.091
Strategy 2 ($B = .4$)	153	−.014	93	.039	150	−.041	82	.102
Strategy 2 ($B = .5$)	100	.001	54	.037	99	−.000	47	.100
Strategy 2 ($B = .6$)	57	−.028	32	−.066	63	.008	20	−.040
Strategy 3	107	−.061	70	.019	102	.006	59	.134
B. Show bets:								
Strategy 1	353	−.103	353	−.026	353	−.100	353	−.030
Strategy 2 ($B = .4$)	153	.004	93	.068	150	−.014	82	.054
Strategy 2 ($B = .5$)	100	.030	54	.027	99	−.011	47	.050
Strategy 2 ($B = .6$)	57	−.003	32	−.114	63	−.007	20	−.082
Strategy 3	107	−.015	70	.052	102	−.014	59	.162

gies applied to the other half sample. In our 1984 paper, "run 4" used morning line and final odds as predictors; "run 5" used morning line and marginal odds. The betting strategies employed were:

Strategy 1: In each race, bet $2.00 on the horse with the highest winning probability.

Strategy 2: In each race, bet $2.00 on the horse with the highest winning probability if that probability equals or exceeds a "filter" coefficient (B).

Strategy 3: In each race, bet $2.00 on the horse with the highest winning probability if that probability is at least 3.5 times as great as the probability of the horse with the second highest winning probability.

The earlier (incorrect) returns ranged from 0.127 to 0.527 on place bets and from 0.016 to 0.202 on show bets. The corrected returns, shown in table 1, range from -0.097 to 0.109 (place) and from -0.102 to 0.073 (show). The more recent (Meadowlands) rates of return, shown in table 2, are roughly comparable: the place-betting strategies yield returns from -0.168 to 0.134; for show betting, the range is from -0.114 to 0.162.

Whereas rates of return to some of the place and show betting strategies remain positive, the majority appear close to zero. Our corrected conclusions are, accordingly, as follows. Profitable strategies for win bets cannot be devised on the basis of market data. The strategies we employ get us close to a fair game (far above the average rate of return, -0.185, reflecting the track take) in place and show betting and do substantially better than the same strategies applied to win bets. The possibility of inefficiencies in place and show betting sufficiently large to permit profits remains open; but we are even more skeptical now about these possibilities than we were before. Finally, the similarities in the logit estimates over the two samples suggest that the information contained in racetrack betting data may be very much the same across times and locations.

References

Asch, Peter; Malkiel, Burton G.; and Quandt, Richard E. 1984. Market efficiency in racetrack betting. *Journal of Business* 57 (April): 165–74.

EFFICIENCY OF THE MARKET FOR RACETRACK BETTING*

DONALD B. HAUSCH, † WILLIAM T. ZIEMBA‡ AND MARK RUBINSTEIN§

Many racetrack bettors have systems. Since the track is a market similar in many ways to the stock market one would expect that the basic strategies would be either fundamental or technical in nature. Fundamental strategies utilize past data available from racing forms, special sources, etc. to "handicap" races. The investor then wagers on one or more horses whose probability of winning exceeds that determined by the odds by an amount sufficient to overcome the track take. Technical systems require less information and only utilize current betting data. They attempt to find inefficiencies in the "market" and bet on such "overlays" when they have positive expected value. Previous studies and our data confirm that for win bets these inefficiencies, which exist for underbet favorites and overbet longshots, are not sufficiently great to result in positive profits. This paper describes a technical system for place and show betting for which it appears to be possible to make substantial positive profits and thus to demonstrate market inefficiency in a weak form sense. Estimated theoretical probabilities of all possible finishes are compared with the actual amounts bet to determine profitable betting situations. Since the amount bet influences the odds and theory suggests that to maximize long run growth a logarithmic utility function is appropriate the resulting model is a nonlinear program. Side calculations generally reduce the number of possible bets in any one race to three or less hence the actual optimization is quite simple. The system was tested on data from Santa Anita and Exhibition Park using exact and approximate solutions (that make the system operational at the track given the limited time available for placing bets) and found to produce substantial positive profits. A model is developed to demonstrate that the profits are not due to chance but rather to proper identification of market inefficiencies.
(FINANCE–PORTFOLIO; GAMES–GAMBLING)

1. The Racetrack Market

For the most part,[1] academic research on security markets has bypassed an interesting and accessible market—the racetrack for thoroughbred horses—with its highly standardized form of security—the tote ticket. The racetrack shares many of the characteristics of the archtypical securities market in listed common stocks. Moreover the racetrack gains further interest from its significant differences, and more importantly, because it is inherently a more elementary market context, lacking many of the dynamic features which complicate analysis of the stock market.

The "market" at the track in North America convenes for about 20 minutes, during which participants place bets on any number of the six to twelve horses in the following race. In a typical race, participants can bet on each horse, either to win, place or show.[2] The horses that finish the race first, second or third are said to finish

---

* Accepted by Vijay S. Bawa, former Departmental Editor; received April 3, 1980. This paper has been with the authors 2 months for 1 revision.
† Northwestern University.
‡ University of British Columbia.
§ University of California, Berkeley.
[1] For surveys, see Copeland and Weston [6], Fama [9], [10] and Rubinstein [22].
[2] Other bets such as the daily double (pick the winners in the first and second race), the quinella (pick the first two finishers regardless of order in a given race) and the exacta (pick the first two finishers in exact order in a given race) as well as various combinations are possible as well. Such bets are utilized by the public to construct low probability high payoff bets. For discussion of some of the implications of such bets see Rosett [21].

"in-the-money." All participants who have bet a horse to win, realize a positive return on that bet only if the horse is first, while a place bet realizes a positive return if the horse is first or second, and a show bet realizes a positive return if the horse is first, second or third. Regardless of the outcome, all bets have limited liability. Unlike most casino games such as roulette, but like the stock market, security prices (i.e. the "odds") are jointly determined by all the participants and a rule governing transaction costs (i.e. the track "take"). To take the simplest case, for win, all bets across all horses to win are aggregated to form the win pool. If W_i represents the total amount bet by all participants on horse i to win, then $W = \sum_i W_i$ is the win pool and WQ/W_i is the payoff per dollar bet on horse l to win if and only if horse i wins, where $1 - Q$ is the percentage transaction costs.[3] If horse i does not win, the payoff per dollar bet is zero.

The track is "competitive" in the sense that every dollar bet on horse i to win, regardless of the identification of the bettor, has the same payoff. The "state contingent" price of a dollar received if and only if horse i wins is $\rho_i = (W_i/WQ)$ and the one plus riskless return is $1/\sum_i \rho_i = Q$, a number less than one. The interest rate at the track is thus negative and solely determined by the level of transactions costs. Thus, apart from an exogenously set riskless rate, the participants at the track jointly determine the security prices. For example, by betting more on one horse than another, the state-contingent price of the first horse increases relative to the second, or alternatively the payoff per dollar bet decreases on the first relative to the second.

The rules for division of the place and show pools are slightly more complex than the win pool. Let P_j be the amount bet on horse j to place and $P \equiv \sum_j P_j$ be the place pool. Similarly S_k is the amount bet on horse k to show and the show pool is $S \equiv \sum_k S_k$. The payoff per dollar bet on horse j to place is

$$1 + [PQ - P_i - P_j]/(2P_j) \quad \text{if} \quad \begin{cases} i \text{ is first and } j \text{ is second or} \\ j \text{ is first and } i \text{ is second} \end{cases}$$

$$0 \qquad\qquad\qquad\qquad\qquad \text{otherwise.}$$

Thus if horses i and j are first and second each bettor on j (and also i) to place first receives the amount of his bet back. The remaining amount in the place pool, after the track take, is then split evenly between the place bettors on i and j. The payoff to horse j to place is independent of whether j finishes first or second, but it is dependent on which horse finishes with it. A bettor on horse j to place hopes that a longshot, not a favorite, will finish with it.

The payoff per dollar bet on show is analogous

$$1 + [SQ - S_i - S_j - S_k]/(3S_k) \quad \text{if} \quad \begin{cases} k \text{ is first, second or third} \\ \text{and finishes with } i \text{ and } j \end{cases}$$

$$0 \qquad\qquad\qquad\qquad\qquad\quad \text{otherwise.}$$

In many ways the racetrack is like the stock market. A technical strategy based on discrepancies between the amounts bet on the same horses to win, place and show, is examined in this paper. Since a short position has a perfectly negative correlated outcome to the result of normal bet a given horse can be "shorted" by buying tickets on all the other horses in a race.

The racetrack also differs from the stock market in important ways. In the stock market, an investor's profit depends not only on the initial price he pays for a security,

---

[3] The actual transactions cost is more complicated and is described below.

but also on what some other investor is willing to pay him for it when he decides to sell. Thus his profit depends not only on how well the underlying firm does in terms of earnings over the time he holds its stock (i.e. supply uncertainty), but also on how other investors value that stock in the future (i.e. demand uncertainty). Given the initial price, both the nature and behavior of other market participants determine his profit. Thus current stock prices might depend not only on "fundamental" factors but also on market "psychology"—the tastes, beliefs, and endowments of other investors, etc. In contrast once all bets are placed at the track prior to a given race (i.e. initial security prices are given), the result of the race and the corresponding payoffs depend only on nature. There is no demand uncertainty at the track.

2. Previous Work on Racetrack Efficiency

A market is efficient if current security prices fully reflect all available relevant information. If this is the case, experts should not be able to achieve higher than average returns with regularity. A number of investigators have demonstrated that the New York Stock Exchange and other major security markets are efficient and so-called experts in fact achieve returns when adjusted for risk that are no higher than those that would be received from random investments (see Copeland and Weston [6], Fama [9], [10] and Rubinstein [22] for discussion, terminology, and relevant references). For an exception see Downes and Dyckman [7].

Snyder [24] provided an investigation of the efficiency of the market for racetrack bets to win. The question Snyder poses is whether or not bets at different odds levels yield the same average return. The rate of return for odds group i is

$$R_i = \frac{N_i^*(O_i + 1) - N_i}{N_i}$$

where N_i and N_i^* are the number of horses, and the number who won, respectively, at odds $O_i = (WQ/W_i - 1)$. A weakly-efficient market in Snyder's sense would set $R_i = Q$ for all i, where $1 - Q$ is the percentage track take. His results as well as those of Fabricant [8], Griffith [12], [13], McClothlin [18], Seligman [23] and Weitzman [27] suggest that there are "strong and stable biases but these are not large enough to make it possible to earn a positive profit" [24; 1110]. In particular, extreme favorites tend to be under bet and longshots overbet. The combined results of several studies comparing over 30,000 races are summarized in Table 1.

TABLE 1 (Snyder [24])

Rates of Return on Bets to Win by Grouped Odds, Take Added Back

Study	Midpoint of grouped odds							
	0.75	1.25	2.5	5.0	7.5	10.0	15.0	33.0
Fabricant	11.1[a]	9.0[a]	4.6[a]	− 1.4	− 3.3	− 3.7	− 8.1	− 39.5[a]
Griffith	8.0	4.9	3.1	− 3.1	− 34.6[a]	− 34.1[a]	− 10.5	− 65.5[a]
McGlothlin	8.0[b]	8.0[a]	8.0[a]	− 0.8	− 4.6	− 7.0[b]	− 9.7	− 11.0
Seligman	14.0	4.0	− 1.0	1.0	− 2.0	− 4.0	− 7.8	− 24.2
Snyder	5.5	5.5	4.0	− 1.2	3.4	2.9	2.4	− 15.8
Weitzman	9.0[a]	3.2	6.8[a]	− 1.3	− 4.2	− 5.1	− 8.2[b]	− 18.0[a]
Combined	9.1[a]	6.4[a]	6.1[a]	− 1.2	− 5.2[a]	− 5.2[a]	− 10.2[a]	− 23.7[a]

[a] Significantly different from zero at 1% level or better.
[b] Significantly different from zero at 5% level or better.

Since the track take averages about 18%, the net rate of return for any strategy which consistently bets within a single odds category is −9% or less. For horses with odds averaging 33 the net rate of return is about −42%.

Conventional financial theory does not explain these biases because it is usual to assume that as nondiversifiable risk (e.g. variance) rises expected return rises as well. In the win pool expected return declines as risk increases. An explanation consistent with the expected utility hypothesis is that investors (as a composite) are risk lovers and behave as if the betting opportunities are limited to a single race. Weitzman [27] and Ali [1] have estimated such utility functions. Ali's estimated utility function over wealth w is the convex function

$$u(x) = 1.91 w^{1.1784} \qquad (R^2 = .9981),$$

which has increasing absolute risk aversion. Thus by the Arrow-Pratt [3], [19] theory investors will take more risk as their wealth declines. This explains the common phenomenon that bettors, when losing, tend to bet more and more on longer odds horses in a desperate attempt to recoup earlier losses. Moreover, since u is nearly linear for large w investors are nearly risk neutral at such wealth levels.

A second explanation is that gamblers simply prefer low probability high prize combinations (i.e. longshots) to high probability low prize combinations. Besides the possible gains involved, gamblers have egos associated with analyzing racing forms and pitting one's predictions against others. Luck and entertainment as well are largely absent in betting favorites. The thrill is to successfully detect a moderate or long odds winner and thus confirm one's ability to outperform the other bettors. Such a scenario is consistent with the data and leads to the biases. Rosett [21] provides an analysis of ways to construct low probability high prize bets through parlays and other combinations that can be used to avoid the longshot tail bias problem and to take advantage of the favorite bias. In an effort to capitalize on this market many tracks feature such bets in the form of the daily double, the exacta and the quinella. However, none of these schemes appear to yield bets with positive net returns.[4]

Other studies of racetrack efficiency have been conducted by Ali [2], Figlewski [11] and Snyder [24]. Ali shows that the win market is efficient in the sense that independently derived bets with identical probabilities of winning do in fact have the same odds in a statistical sense. His analysis utilizes daily double bets and the corresponding parlays, i.e. bet the proceeds if the chosen horse wins the first race on the chosen horse in the second race, for 1089 races. Figlewski, using a multinomial logit probability model to measure the information content of the forecasts of professional handicappers and data from 189 races at Belmont in 1977, found that these forecasts do contain considerable information but the track odds generated by bettors discount almost all of it. Snyder provided strong form efficiency tests of the form: are there persons with special information that would allow them to outperform the general public? He found using data on 846 races at Arlington Park in Chicago that forecasts from three leading newspapers, the daily racing form and the official track handicapper did not lead to bets that outperformed the general public.

---

[4] For an entertaining account of an "expert" who was able to achieve positive net returns over a full racing season, see Beyer [5]. See also Vergin [26].

3. Proposed Test

The studies described in §2 examine racetrack efficiency with respect to the win pool only. In this paper our concern is with the efficiency of the place and show pools relative to the win pool and with the development of procedures to best capitalize on potential inefficiencies between these three "markets." We utilize the following definition of weak-form efficiency: the market is weakly-efficient if no individual can earn positive profits using trading rules based on historical price information. In Baumol's [4; 46] words ... "all opportunities for profit by systematic betting are eliminated". Our analysis utilizes the following two data sets: 1) Data Set 1: all dollar bets to win, place and show for the 627 races over 75 days involving 5895 horses running in the 1973/74 winter season at Santa Anita Racetrack in Arcadia, California, collected by Mark Rubinstein; and 2) Data Set 2: all dollar bets to win, place and show for the 1065 races over 110 days involving 9037 horses running in the 1978 summer season at Exhibition Park, Vancouver, British Columbia, collected by Donald B. Hausch and William T. Ziemba.

In the analysis of the efficiency of the win pool one may compare the actual frequency of winning with the theoretical probability of winning as reflected through the odds. Similar analyses are possible for the place and show pools once an estimate of the theoretical probabilities of placing and showing for all horses is available. There is no unique way to obtain these estimates. However, very reasonable estimates obtain from the natural generalization of the following simple procedure. Suppose three horses have probabilities to win of 0.5, 0.3, and 0.2, respectively. Now if horse 2 wins, a Bayesian would naturally expect that the probabilities that horses 1 and 3 place (i.e. win second place) are 0.5/0.7 and 0.2/0.7, respectively. In general if q_i ($i = 1, \ldots, n$) is the probability horse i wins, then the probability that i is first and j is second is

$$(q_i q_j)/(1 - q_i) \tag{1}$$

and the probability that i is first, j is second and k is third is

$$q_i q_j q_k /(1 - q_i)(1 - q_i - q_j). \tag{2}$$

Harville [14] gives an analysis of these formulas. Despite their apparent reasonableness they suffer from at least two flaws:

1) no account is made of the possibility of the "Silky Sullivan" problem; that is, some horses generally either win or finish out-of-the-money—for example, see footnote 5 in [15]; for these horses the formulas greatly over-estimate the true probability of finishing second or third; and

2) the formulas are not derivable from first principles involving individual horses running times; even independence of these random variables $T_1, \ldots, T_n$ is neither necessary nor sufficient to imply the formulas.

In addition to assuming (1) and (2) we assume that

$$q_i = W_i \bigg/ \sum_{i=1}^{n} W_i. \tag{3}$$

That is, the win pool is efficient. The discussion above, of course, indicates that this assumption is suspect in the tails and this is discussed below. Table 2a–c compares the actual versus theoretical probability of winning, placing and showing for data set 2.

Similar tables for data set 1 appear in King [17]; see also Harville [14]. The usual tail biases appear in Table 2a (although they are not significant at the 5% confidence level). One has reverse tail biases in the finishing second and third probabilities, see Tables 2b, c in [15]. This occurs because if the probability to win is overestimated (underestimated) and probabilities sum to one it is likely that the probabilities of finishing second and third would be underestimated (overestimated). Tables 2b, c, indicate how these biases tend to cancel when they are aggregated to form the theoretical probabilities and frequencies of placing and showing.[5]

TABLE 2

Actual vs. Theoretical Probability of Winning, Placing and Showing: Exhibition Park 1978

(a)	Theoretical Probability of Winning	Number of Horses	Average Theoretical Probability	Actual Frequency of Winning	Estimated Standard Error
	0.000 −0.025	540	0.019	0.016	0.005
	0.026 −0.050	1498	0.037	0.036	0.005
	0.051 −0.100	2658	0.073	0.079	0.005
	0.101 −0.150	1772	0.123	0.126	0.008
	0.151 −0.200	1199	0.172	0.156	0.010
	0.201 −0.250	646	0.223	0.227	0.016
	0.251 −0.300	341	0.272	0.263	0.024
	0.301 −0.350	199	0.323	0.306	0.033
	0.351 −0.400	101	0.373	0.415	0.049
	0.401 +	83	0.450	0.469	0.055
		9037			

(b)	Theoretical Probability of Placing	Number of Horses	Average Theoretical Probability	Actual Frequency of Placing	Estimated Standard Error
	0.000 −0.025	21	0.022	0.000*	0.000
	0.026 −0.050	391	0.040	0.030	0.009
	0.051 −0.100	1394	0.075	0.080	0.007
	0.101 −0.150	1335	0.124	0.152*	0.010
	0.151 −0.200	1295	0.174	0.174	0.011
	0.201 −0.250	1057	0.223	0.243	0.013
	0.251 −0.300	871	0.274	0.304	0.016
	0.301 −0.350	772	0.323	0.314	0.017
	0.351 −0.400	580	0.373	0.313*	0.019
	0.401 −0.450	420	0.424	0.395	0.024
	0.451 −0.500	321	0.472	0.457	0.028
	0.501 −0.550	202	0.523	0.415*	0.035
	0.551 −0.600	149	0.573	0.483*	0.041
	0.601 −0.650	114	0.623	0.570	0.046
	0.651 −0.700	51	0.672	0.627	0.068
	0.701 −0.750	41	0.721	0.731	0.069
	0.751 +	23	0.792	0.782	0.086
		9037			

[5]Since it is the accuracy of these probabilities rather than the q_i's that is of crucial importance in the calculations and model below this canceling provides some justification for omitting the tail biases in (3). In practice, bets are only made on horses with expected returns considerably above 1, e.g., 1.16 at Santa Anita. Modification of the q_i to include these biases might change the 1.16 to 1.14, for example. See also the discussion below and in §§4 and 5.

TABLE 2 (continued)

Actual vs. Theoretical Probability of Winning, Placing and Showing: Exhibition Park 1978

(c)	Theoretical Probability of Showing		Number of Horses	Average Theoretical Probability	Actual Frequency of Showing	Estimated Standard Error
	0.000	−0.025	0	—	—	—
	0.026	−0.050	55	0.043	0.036	0.025
	0.051	−0.100	592	0.078	0.081	0.011
	0.101	−0.150	895	0.125	0.165*	0.012
	0.151	−0.200	909	0.175	0.195	0.013
	0.201	−0.250	799	0.224	0.292*	0.016
	0.251	−0.300	885	0.275	0.289	0.015
	0.301	−0.350	794	0.324	0.346	0.017
	0.351	−0.400	703	0.374	0.398	0.018
	0.401	−0.450	655	0.425	0.433	0.019
	0.451	−0.500	617	0.475	0.452	0.020
	0.501	−0.550	542	0.524	0.477*	0.021
	0.551	−0.600	396	0.573	0.477*	0.025
	0.601	−0.650	375	0.624	0.599	0.025
	0.651	−0.700	264	0.672	0.609*	0.030
	0.701	−0.750	206	0.722	0.582*	0.034
	0.751	−0.800	154	0.773	0.655*	0.038
	0.801	−0.850	113	0.821	0.752	0.041
	0.851	−0.900	53	0.873	0.830	0.052
	0.901 +		30	0.925	0.833	0.068
			9037			

[a] Categories when the theoretical probability and the actual frequency are different at the 5% significance level are denoted by *'s. The estimated standard error is $(s^2/N)^{\frac{1}{2}}$ where the actual frequency sample variance $s^2 = N(E(X^2) - (EX)^2)/(N - 1)$. Since the X_i are either 0 or 1, $E(X^2) = EX$ and $s^2 = N(EX - (EX)^2)/(N - 1)$.

Formulas (1)–(3) can be used to develop procedures that yield net rates of return for place and show betting that are higher than expected (i.e.–18%) and indeed make positive profits. As a first step towards development of a "system" we present the results on the two data sets of \$1 bets when the theoretical expected return is α for varying α. The expected return from a \$1 bet to place on horse l is[6,7]

$$EX_l^p \equiv \sum_{\substack{j=1 \\ j \neq l}}^n \left(\frac{q_l q_j}{1 - q_l} \right) \left[1 + \frac{1}{20} \text{INT}\left\{ \left(\frac{Q(P + 1) - (1 + P_l + P_j)}{2} \right) \left(\frac{1}{1 + P_l} \right) \times 20 \right\} \right]$$

$$+ \sum_{\substack{i=1 \\ i \neq l}}^n \left(\frac{q_i q_l}{1 - q_i} \right) \left[1 + \frac{1}{20} \text{INT}\left\{ \left(\frac{Q(P + 1) - (1 + P_i + P_l)}{2} \right) \left(\frac{1}{1 + P_l} \right) \times 20 \right\} \right],$$

$$(4)$$

[6] The expressions (4) and (5) give the marginal expected return for an additional \$1 bet to place or show on horse l. To obtain the average expected rates of return one simply replaces $(1 + p_l)$ and $(1 + s_l)$ in these expressions by p_l and s_l, respectively. From a practical point of view with usual track data these quantities are virtually identical.

[7] A further complication, not reflected in (4) and (5) below, is that a \$2 winning bet must return at least \$2.10. Hence in these "minus pools" involving an extreme favorite the track's take is less than $1 - Q$.

where Q is 1 minus the track take, $P \equiv \sum P_i$ is the place pool, P_i is bet on horse i to place and $\text{INT}(Y)$ means the largest integer not exceeding Y. In (4) the quantities P and P_i are the amounts bet before an additional \$1 is bet; similarly with S and S_i in (5), below. The expressions involving INT take into account the fact that \$2 bets return payoffs rounded down to the nearest \$0.10. The two expressions represent the expected payoffs if l is first or second, respectively. Similarly the expected payoff from a \$1 bet to show on horse l is

$$
EX_l^s \equiv \sum_{\substack{j=1 \\ j \neq l}}^{n} \sum_{\substack{k=1 \\ k \neq l,j}}^{n} \frac{q_l q_j q_k}{(1 - q_l)(1 - q_l - q_j)}
$$

$$
\times \left[1 + \frac{1}{20} \text{INT} \left\{ \left(\frac{Q(S+1) - (1 + S_l + S_j + S_k)}{3} \right) \left(\frac{1}{1 + S_l} \right) \times 20 \right\} \right]
$$

$$
+ \sum_{\substack{i=1 \\ i \neq l}}^{n} \sum_{\substack{k=1 \\ k \neq i,l}}^{n} \frac{q_i q_l q_k}{(1 - q_i)(1 - q_i - q_l)}
$$

$$
\times \left[1 + \frac{1}{20} \text{INT} \left\{ \left(\frac{Q(S+1) - (1 + S_i + S_l + S_k)}{3} \right) \left(\frac{1}{1 + S_l} \right) \times 20 \right\} \right]
$$

$$
+ \sum_{\substack{i=1 \\ i \neq l}}^{n} \sum_{\substack{j=1 \\ j \neq l,i}}^{n} \frac{q_i q_j q_l}{(1 - q_i)(1 - q_i - q_j)}
$$

$$
\times \left[1 + \frac{1}{20} \text{INT} \left\{ \left(\frac{Q(S+1) - (1 + S_i + S_j + S_l)}{3} \right) \left(\frac{1}{1 + S_l} \right) \times 20 \right\} \right], \quad (5)
$$

where $S \equiv \sum S_i$ is the show pool, S_i is bet on horse i to show and the three expressions represent the expected payoffs if l is first, second and third, respectively.

Naturally one would expect that positive profits would not be obtained, given the inherent inaccuracies in assumptions (1)–(3), unless the theoretical expected return α was significantly greater than 1. However we might hope that the actual rate of return would at least increase with α and be somewhat near α. Table 3 indicates this is true for both data sets. The perverse behavior for high α in the place pool is presumably a small sample phenomenon. Additional calculations along these lines appear in Harville [14].

4. A Betting Model

The results in Table 3 give a strong indication that there are significant inefficiencies in the place and show pools and that it is possible not only to achieve above average returns but to make substantial profits. In this section we develop a model indicating not only which horses should be bet but how much should be bet taking into account investor preferences and wealth levels and the effect of bet size on the odds.

We consider an investor having initial wealth w_0 contemplating a series of bets. It is

TABLE 3

Results of Betting $1 to Place or Show on Horses with a Theoretical Expected Return of at Least α

Exhibition Park

α	Place			Show		
	Number of Bets	Total Net Profit ($)	Net Rate of Return (%)	Number of Bets	Total Net Profit ($)	Net Rate of Return (%)
1.04	225	5.10	2.3	612	33.20	5.4
1.08	126	− 10.10	− 8.0	386	53.50	13.9
1.12	69	11.10	16.1	223	40.80	18.3
1.16	40	5.10	12.8	143	26.30	18.4
1.20	18	5.30	29.4	95	21.70	22.8
1.25	11	− 2.70	− 24.5	44	11.20	25.5
1.30	3	− 3	− 100.0	27	10.80	40.0
1.50	0	0	—	3	6	200.0

Santa Anita

α	Place			Show		
	Number of Bets	Total Net Profit ($)	Net Rate of Return (%)	Number of Bets	Total Net Profit ($)	Net Rate of Return (%)
1.04	103	12.30	11.9	307	− 18.00	− 5.9
1.08	52	12.80	24.6	162	6.90	4.3
1.12	22	9.20	41.8	89	3.00	3.4
1.16	7	2.30	32.9	46	12.40	27.0
1.20	3	− 1.30	− 43.3	27	6.20	23.0
1.25	0	0	—	9	6.00	66.7
1.30	0	0	—	5	5.10	102.0
1.50	0	0	—	0	0	—

natural to suppose that the investor would wish to maximize the long run rate of asset growth and thus employ the so-called Bernoulli capital growth model; see Ziemba and Vickson [28] for references and discussion of various assumptions and results. We use the following result: if in each time period $t = 1, 2, \ldots$ there are I investment opportunities with returns per unit invested denoted by the random variables $x_{t1}, \ldots, x_{tI}$, where the x_{ti} have finitely many distinct values and for distinct t the families are independent, then maximizing $E \log \sum \lambda_{ti} x_{ti}$, s.t. $\sum \lambda_{ti} \leq w_t$, $\lambda_{ti} \geq 0$ maximizes the asymptotic rate of asset growth. The assumptions are quite reasonable in a horseracing context because there are a finite number of return possibilities and the race by race returns are likely to be nearly independent since different horses will be running (although the jockeys and trainers may not be).

The second key feature of the model is that it considers an investor's ability to influence the odds by the size of his bets.[8] This yields the following model to calculate

---

[8] The first model to include this feature seems to be Isaacs [16]. Only win bets are considered with linear utility, and he is able to determine the exact solution in closed form. His model may be useful in situations where the perfect market assumption (3) is violated or where special expertise leads one to believe their estimates of the q_i are better than those of the other bettors.

the optimal amounts to bet for place and show.

$$
\underset{\{p_l\}\{s_l\}}{\text{Maximize}} \sum_{i=1}^{n} \sum_{\substack{j=1 \\ j \neq i}}^{n} \sum_{\substack{k=1 \\ k \neq i,j}}^{n} \frac{q_i q_j q_k}{(1-q_i)(1-q_i-q_j)} \log \left[
\begin{array}{c}
\dfrac{Q(P+\sum_{l=1}^{n} p_l) - (p_i + p_j + P_i + P_j)}{2} \\[2mm]
\times \left(\dfrac{p_i}{p_i + P_i} + \dfrac{p_j}{p_j + P_j} \right) \\[2mm]
+ \dfrac{Q(S+\sum_{l=1}^{n} s_l) - (s_i + s_j + s_k + S_i + S_j + S_k)}{3} \\[2mm]
\times \left(\dfrac{s_i}{s_i + S_i} + \dfrac{s_j}{s_j + S_j} + \dfrac{s_k}{s_k + S_k} \right) \\[2mm]
+ w_0 - \sum_{\substack{l=1 \\ l \neq i,j,k}}^{n} s_l - \sum_{\substack{l=1 \\ l \neq i,j}}^{n} p_l
\end{array}
\right] \tag{6}
$$

$$
\text{s.t.} \sum_{l=1}^{n} (p_l + s_l) \leq w_0, \qquad p_l \geq 0, s_l \geq 0, l = 1, \ldots, n,
$$

where $Q = 1 -$ the track take, W_i, P_j and S_k are the total dollar amounts bet to win, place and show on the indicated horses by the crowd, respectively, $W \equiv \sum W_i$, $P \equiv \sum P_j$ and $S \equiv \sum S_k$ are the win, place and show pools, respectively, $q_i \equiv W_i/W$ is the theoretical probability that horse i wins, w_0 is initial wealth and p_l and s_l are the investor's bets to place and show on horse l, respectively.

The formulation (6) maximizes the expected logarithm of final wealth considering the probabilities and payoffs from all possible horserace finishes. It is exact except for the minor adjustment made that the rounding down to the nearest \$0.10 for a two dollar bet, see (4) and (5), is omitted.[9] For the values of $\alpha \geq 1.16$ it was observed that in a given race at most three p_l and three s_l were nonzero. When (6) is then simplified it can be solved in less than 1 second of CPU time.[10] A discussion of the generalized concavity properties of (6) will appear in a forthcoming paper by Kallberg and Ziemba.

The results are illustrated by function 1 in Figures 1 and 2 for the two data sets using an initial wealth of \$10,000. In both cases the bets produced from (6) lead to well above average returns and to positive profits.[11] These results may be contrasted with random betting; function 2 in Figures 1 and 2. Intuition suggests that Santa Anita with its larger betting pools would have more accurate estimates of the q_i than would be obtained at Exhibition Park. Hence positive profits would result from lower values of α. The results bear this out and only horses with $\alpha \geq 1.20$ for Exhibition Park were considered for possible bets. Generally speaking the bets are usually favorites and almost always on those horses with maximum $(W_i/W)/(P_i/P)$ and $(W_i/W)/(S_i/S)$

[9] It is possible to include this feature in (6) but it greatly complicates the solution procedure (e.g. differentiability is lost) with little added gain in accuracy.

[10] All calculations were made on UBC's AMDAHL 470V6 Model II computer using a code for the generalized reduced gradient algorithm.

[11] The procedure was to calculate the optimal bets to place and show in each race using (6) with the present wealth level. The results of the race and the actual payoffs that reflect the track's take and breakage are known. The payoff for our investor's bets were calculated using all the bets of the crowd plus the bets of our investor taking into account the track's take and breakage, i.e. the payoffs are thus those that would have occurred had our investor actually made his bets. The wealth level was then adjusted to reflect the race's gain or loss and the procedure continued for all races.

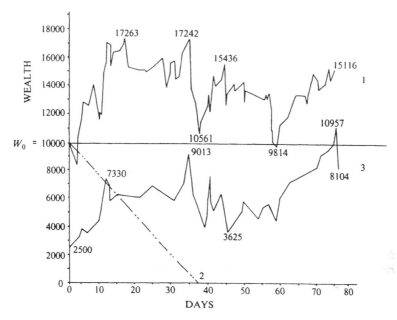

[1] Results from expected log betting to place and show when expected returns are 1.16 or better with initial wealth $10,000.

[2] Approximate wealth level history for random horse betting. Total dollars bet is as in system 1 ($116,074). Track payback is 82.5%, therefore final wealth level is $10,000 − 0.175($116.074) = −$10,313 (Note: breakage is not taken into consideration)

[3] Results from using the Exhibition Park approximate regression scheme (with initial wealth $2,500) at Santa Anita.

FIGURE 1.   Wealth Level Histories for Alternative Betting Schemes; Santa Anita: 1973/74 Season.

ratios. These ratios of the theoretical probability of winning to the track take unadjusted odds to place or show form a type of cost-benefit ratio that provides a first approximation to α. This is discussed further in §5, below. Most of the bets are to show and one tends to bet only about once per day. The numbers of bets and their size distribution are presented in Table 4. As expected, the influence on the odds made by our investor's bets is much greater at Exhibition Park hence the bets there tend to be much smaller than at Santa Anita. However, even there about 10% of the bets exceed $1000.

The log formulation has absolute risk aversion $1/w$, which for wealth around $10,000 is virtually zero. Zero absolute risk aversion is, of course, achieved by linear utility. One then will bet on the horse (or horses) with the highest α until the influence on the odds drops this horse (or horses) below another horse's α, etc., continuing until there are no favorable bets or the betting wealth has been fully utilized. The results of such linear utility betting with $w_0 = \$10,000$ are: at Exhibition Park with $\alpha \geqslant 1.20$ final wealth is $14,818; at Santa Anita with $\alpha \geqslant 1.16$ final wealth is $10,910. Such a strategy is a very risky one and leads to some very large bets. The log function has the distinct advantage that it implies negative infinite utility at zero wealth hence bets having any significant probability of yielding final wealth near zero are avoided.

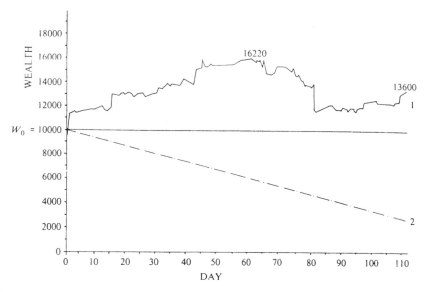

[1] Results from expected log betting to place and show when expected returns are 1.20 or better with initial wealth $10,000.

[2] Approximate wealth level history for random horse betting. Total dollars bet is as in system 2 ($8461). Track payback is 81.9% therefore final wealth level is $10,000 − 0.181($38461) = $3,035. (Note: Breakage is not taken into consideration.)

FIGURE 2.   Wealth Level Histories for Alternative Betting Schemes; Exhibition Park: 1978 Season.

TABLE 4

Size Distribution of Bets, w_0 = $10,000. Log Utility

| | Santa Anita | | | | Exhibition Park | | | |
| | Place | | Show | | Place | | Show | |
Size	% of Bets	% of $Bet	% of Bets	% of $Bet	% of Bets	% of $Bet	% of Bets	% of $Bet
0–50	7.1	0.3	2.6	0.1	29.4	3.6	17.0	1.0
51–100	0	0	1.3	0.1	23.5	6.7	13.8	2.8
101–200	0	0	3.9	0.4	5.9	2.9	22.3	9.3
201–300	21.4	6.1	5.2	1.0	5.9	4.4	14.9	10.3
301–500	7.1	3.2	9.1	2.6	23.5	30.6	8.5	9.1
501–700	14.3	10.3	13.0	5.7	0	0	6.4	10.5
701–1000	14.3	14.4	7.8	4.8	0	0	7.4	17.5
> 1001	35.8	65.7	57.1	85.3	11.8	51.8	9.7	39.5

n = 14 $11,932 n = 77 $104,142 n = 17 $4,954 n = 94[a] $33,507

Total Place Bets   Total Show Bets   Total Place Bets   Total Show Bets

Total Santa Anita Bettings = $116,074   Total Exhibition Park Betting = $38,461

[a] Two of these bets had $EX_s^l \geqslant 1.20$ and $s_l^* = 0$.

5. Making the System Operational

The calculations reported in Figures 1 and 2 were made under the assumption that the investor is free to bet once all other bettors have placed their bets. In practice one can only attempt to be one of the last bettors.[12] There is a natural tradeoff between placing a bet too early and increasing the inaccuracies and running the possibility of arriving too late at the betting window to place a bet.[13] The time just prior to the beginning of a race is crucial since many bets are typically made then including the so called "smart money" bets made very close to the beginning of the race so their impact on other bettors is minimized. It is thus extremely important that the investor be able to perform all calculations necessary to place the bet(s) very quickly. Typically, since tracks have neither public phones nor electricity, calculations can at most utilize battery operated calculators or possibly a battery operated special purpose computer. Even if computing times were negligible the very act of punching in the data needed for an exact calculation is too time-consuming since it takes more than one minute. Therefore, in practice, approximations that utilize a limited number of input data elements are required. Several types of approximations are possible such as the tabular rules of thumb developed by King [17] or the regression procedures suggested here. Our procedure indicates whether or not a bet to place or show is warranted and at what level using the following eight data inputs: w_0, W_{i*}, P_{i*}, Q_{j*}, S_{j*}, W, P and S, where $i*$ is an i for which $(W_i/W)/(P_i/P)$ is maximized and $j*$ is a j for which $(W_j/W)/(S_j/S)$ is maximized. It is easy to determine $i*$ and $j*$, particularly since $i*$ often equals $j*$, by inspection of the tote board. The approximation supposes that the only possible bets are $i*$ to place and $j*$ to show. The regressions, as given below, must be calibrated to a given track and initial wealth level. As a prelude to actual betting the regressions were calibrated at Exhibition Park for w_0 = $2500. For calculated $i*$ the expected return on a $1 bet to place is approximated by

$$E\tilde{X}^p_{i*} = 0.39445 + 0.51338 \frac{W_{i*}/W}{P_{i*}/P}, \qquad R^2 = 0.776. \tag{7}$$

If $E\tilde{X}^p_{i*} \geqslant 1.20$ then the optimal bet to place is approximated by

$$p_{i*} = -459.32 + 1715.6 q_{i*} - 0.042518 q_{i*}P - 7440.1 q^2_{i*}$$

$$+ 13791 q^3_{i*} + 0.10247 P_{i*} + 49.572 \ln w_0, R^2 = 0.954. \tag{8}$$

Similarly for $j*$ the expected return on a $1 bet to show is approximated by

$$E\tilde{X}^s_{j*} = 0.64514 + 0.32806 \frac{W_{j*}/W}{S_{j*}/S}, \qquad R^2 = 0.650. \tag{9}$$

---

[12]The model as developed in this paper utilizes an inefficiency in the place and show pools to yield positive profits. An investor's bets are determined by his wealth level as well as the profitability of one or more such bets. There may or may not exist "enough" inefficiency to provide positive profits for additional investors using a system of this nature. In the context of the Isaac's model, for win bets with linear utility, Thrall [25] showed that if there were positive profits to be made and each new investor was aware of all previous investor's bets then the profits for these various bettors are shared and in the limit become zero. It is likely that a similar result obtains for the model discussed here.

[13]At some tracks, such as Santa Anita, betting ends precisely at post time when the electric totalizator machines are shut off. At other tracks including Exhibition Park, betting ends when the horses enter the starting gate which may be 2 or even 3 minutes past the post time.

If $E\tilde{X}_j^s \geqslant 1.20$ then the optimal bet to show is approximated by

$$s_{j*} = -660.97 - 867.69 q_{j*} + 0.25933 q_{j*} S + 3715.2 q_{j*}^2$$
$$- 0.19572 S_{j*} + 77.014 \ln w_0, \qquad R^2 = 0.970. \qquad (10)$$

All the coefficients in (7)–(10) are highly significant at levels not exceeding 0.05. Utilizing the equations (7)–(10) results in a scheme in which the data input and execution time on a modern hand held programmable calculator is about 35 seconds.

The results of utilizing this method on the Exhibition Park data with initial wealth of \$2500 are shown in Figure 3, functions 1 and 2. Using the exact calculations yields a final wealth of \$5197 while the approximation scheme has an even higher final wealth of \$7698. The approximation scheme leads to 63 more bets (174 versus 111) than the exact calculation. Since these bets had a positive net return the total profit of the approximation scheme exceeds that of the exact calculation. Thus, it is clear that one would maximize profits with a cutoff rule below the conservative level of 1.20. The size distribution of bets from the exact and approximate solutions are remarkably similar; see Table 6 in [15].

For place there were 17 bets where EX^p exceeded or equalled 1.20 of which 7 (41%) were in the money. Only 2 of these 17 bets were not chosen by the regression. However 14 horses were chosen for betting by the regressions even though their "true" EX^p was less than 1.20. Most of these had "true" EX^p values close to 1.20 and were favorites. Four of these horses (29%) were in the money. For show there were 94 bets where EX^s exceeded or equalled 1.20 of which 52 (55%) were in the money. Only 8 of these were

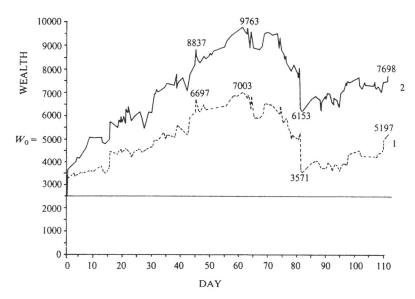

[1] Results from expected log betting to place and show when expected returns are 1.20 or better with initial wealth \$2,500.

[2] Results from using the approximate regression scheme with initial wealth \$2,500.

FIGURE 3.   Wealth Level Histories for Exact and Approximate Regression Betting Schemes; Exhibition Park: 1978 Season.

overlooked by the regressions while 59 bets with "true" EX^s values less than 1.20 were chosen by the regressions. Of these 39 (66%) were in the money.

The regression method is a simple procedure that seems to work well. For example using it on the Santa Anita data (without reestimating the coefficients) indicates that initial wealth of $2500 would yield a final wealth of $8104; see function 3 in Figure 1. Conceivably, there are many possible refinements using fundamental information that could be added. Many of these refinements as well as discussion of the results of actual betting will appear in a forthcoming book by Ziemba and Hausch. One such refinement is the supposition that the win market is more efficient under normal fast track conditions and less efficient when the track is slow, muddy, heavy, wet, sloppy, etc. Bets were made on 87 of the 110 days at Exhibition Park; of these 57 days had a fast track. Using the regressions yields "fast track" bets of $32,501 returning $38,364 for a net profit of $5863. On the nonfast days the regressions suggest bets of $17,180 returning $16,515 for a loss of $665. Hence this refinement decreases the time spent at the track and increases the net profit from $5198 to $5863.

6. Implementation and Reliability of the System

The model presented in this paper assumed that one can utilize the betting data that prevails at the end of the betting period, say t, to calculate optimal bets. In practice, however, even with the approximations given by (7)–(10), one requires about 1–1.5 minutes to physically calculate the optimal bets and place them. Thus one can only utilize betting information from τ ($\approx t - 1.5$). Hence bets that were optimal at τ may not be as profitable at t. Some evidence by Ritter [20] seems to indicate that the odds on the expost favorite at t often decrease from τ to t. In which case an optimal bet at τ may be a poor bet at t. Ritter investigated the systems: bet on i^* if $(P_{i*}/P/(W_{i*}/W) \leq 0.7$ and on j^* if $(S_{j*}/S)/(W_{j*}/W) \leq 0.7$. (For comparison equations (7) and (9) indicate the more restrictive constraints 0.6373 and 0.5912, respectively, instead of 0.7. The less restrictive Ritter constraints yield about three times as many bets as (7) and (9) indicate.) Using a sample of 229 harness races at Sportsman's Park and Hawthorne Park in Chicago he found that with $2 bets these systems gained 24% and 16%, respectively. But a 15% advantage shrunk to -1% if one uses the τ bets with a random shock over the τ to t period (for the show bets for the 95 races at Hawthorne with a 0.65 cutoff). There are some difficulties with the design of Ritter's experiment, such as the small sample, the inclusion of only $2 bets and the use of a random shock rather than an estimate of usual trends from τ to t, etc., however, his results point to the difficulty of actually making substantial profits in a racetrack setting.

An attempt was made during the summer 1980 racing season at Exhibition Park to implement the proposed system and observe the "end of betting problem." Nine racing days were attended during which 90 races were run. At $t - 2$ minutes the win, place, and show data were recorded on any horse that, through equations (7) and (9), warranted a bet. Equations (8) and (10) were then used on this data to determine the regression estimates of the optimal bets. Updated toteboard data on these horses were recorded until the end of betting to determine if the horse remained a system bet. Results of this experiment are shown in Table 5 where: 1) an initial wealth of $2500 is assumed; 2) size of bet calculations are done on data at $t - 2$ minutes; and 3) returns are based on final data. Twenty-two bets were made that yielded final wealth of $3716 for a profit of $1216. Actual betting was not done but by making the calculations two minutes before the end of betting allows 1.5 minutes to place the bet, more than

TABLE 5

Results from Summer 1980 Exhibition Park Betting

Date	Race	Regression estimate of expected return per dollar, 2 minutes before end of betting	Regression estimate of expected return per dollar, at the end of betting	Regression estimate of optimal bet 2 minutes before end of betting	Finish	Net return based on final data with consideration of our bets affecting odds	Final wealth
							$2500
July 2	9	120	122	$19,SHOW ON 4	5-6-7	− $19	2481
"	10	120	123	72,SHOW ON 8	8-1-2	72	2553
July 9	7	121	110	292,SHOW ON 1	2-7-1	131	2684
"	10	135	122	248,PLACE ON 1	1-6-2	260	2944
July 16	6	131	122	487,PLACE ON 9 ⎫		536 ⎫	
		139	117	292,SHOW ON 9 ⎬	9-8-6	146 ⎬ 682	3626
"	7	125	127	7,SHOW ON 1	5-2-8	− 7	3619
July 23	3	149	149	30,SHOW ON 2	2-10-7	92	3711
"	4	139	134	573,SHOW ON 10	6-10-4	201	3912
July 30	8	121	111	215,PLACE ON 4	4-1-5	129	4041
"	9	123	125	591,SHOW ON 6	8-1-5	− 591	3450
Aug 6	6	128	112	39,SHOW ON 4	4-3-1	59	3509
"	9	124	103	51,SHOW ON 2	4-1-3	− 51	3458
Aug 8	1	121	132	87,SHOW ON 1	1-10-4	139	3597
"	3	127	111	635,SHOW ON 3	3-4-7	127	3724
"	4	126	113	126,SHOW ON 2	2-7-1	82	3806
Aug 11	8	121	112	94,SHOW ON 8	8-6-2	113	3919
"	9	131	130	688,SHOW ON 5	5-3-4	138	4057
Aug 13	3	128	106	33,SHOW ON 2	1-6-7	− 33	4024
"	6	131	122	205,SHOW ON 5	5-8-4	144	4168
"	7	134	133	511,SHOW ON 6	8-5-9	− 511	3657
"	10	123	109	108,SHOW ON 5	3-5-1	59	3716

enough time. Note in Table 5 that many systems bets at $t - 2$ were not system bets at t, however all had regression expected returns of more than one.

An important question concerns the reliability of the results: are the results true exploitations of market inefficiencies or could they be obtained simply by chance? This question is investigated utilizing a simple model which was suggested to us by an anonymous referee. The first application is concerned with an estimate of the probability that the system's theory is vacuous and indeed the observations conform to specific favorable samples from a random betting population. The second application estimates the probability of not making a positive profit. The calculations utilize the 1980 Exhibition Park data; see Table 5.

Let π be the probability of winning a bet in each trial and

$$X_i = \begin{cases} 1 + w & \text{if the bet is won,} \\ 0 & \text{otherwise,} \end{cases}$$

be the return from a $1 bet in trial i. In n trials, the probability of winning at least $100y\%$ of the total bet is

$$\Pr\left[\frac{1}{n}\left(\sum_{i=1}^{n} X_i \right) - 1 > y \right]. \tag{11}$$

Assume that the trials are independent. Since the X_i are binomially distributed (11) can be approximated by a normal probability distribution as

$$1 - \Phi \left[\frac{\sqrt{n} \left\{ y - (1 + w)\pi + 1 \right\}}{(1 + w)\pi \sqrt{(1 - \pi)/\pi}} \right] \tag{12}$$

where Φ is the cumulative distribution function of a standard $N(0, 1)$ variable. The observed probability of winning a bet, weighted by size of bet made, yields 0.771 as an estimate of π. If the systems theory was vacuous and random betting was actually being made then $(1 + w)\pi = 0.83$ since the track's payback is approximately 83%. The 22 bets made totalled \$5304 and resulted in a profit of \$1216 for a rate of return of 22.9%. Using equation (12) with $n = 22$, gives 3×10^{-5} as the probability of making 22.9% through random betting.

Suppose that the 1980 Exhibition Park results represent typical system behavior so $\pi = 0.771$ and $(1 + w)\pi = 1.229$. In n trials the probability of making a non-positive net return is

$$\Pr \left[\frac{1}{n} \left(\sum_{i=1}^{n} X_i \right) - 1 < 0 \right] \tag{13}$$

which can be approximated as

$$\Phi \left[\frac{\sqrt{n} (1 - (1 + w)\pi)}{(1 - w)\pi \sqrt{(1 - \pi)/\pi}} \right] = \Phi(-0.342\sqrt{n}). \tag{14}$$

For $n = 22$ this probability is only 0.054 and for $n = 50$ and 100 this probability is 0.008 and 0.0003, respectively. Thus it is reasonable to suppose that the results from the 1980 Exhibition Park data (as well as the 1978 Exhibition Park and 1973/1974 Santa Anita data with their larger samples) represent true exploitation of a market inefficiency.[14]

[14] Without implicating them we would like to thank Michael Alhadeff of Longacres Racetrack in Seattle, the American Totalizator Corporation and the staff of the Jockey Club at Exhibition Park in Vancouver for helping us obtain the data used in this study, and M. J. Brennan and E. U. Choo for helpful discussions. Thanks are also due to V. S. Bawa, J. R. Ritter, and two anonymous referees for helpful comments on an earlier draft of this paper. This paper is a condensation of [15].

References

1. ALI, M. M., "Probability and Utility Estimates for Racetrack Bettors," *J. Political Economy*, Vol. 85 (1977), pp. 803–815.

2. ———, "Some Evidence of the Efficiency of a Speculative Market," *Econometrica*, Vol. 47, No. 2 (1979), pp. 387–392.

3. ARROW, K. J., *Aspects of the Theory of Risk Bearing*, Yrjö Jahnsson Foundation, Helsinki, 1965.

4. BAUMOL, W. J., *The Stock Market and Economic Efficiency*, Fordham Univ. Press, New York, 1965.

5. BEYER, A., *My \$50,000 Year at the Races*, Harcourt, Brace, Jovanovitch, New York, 1978.

6. COPELAND, J. E. AND WESTON, J. F., *Financial Theory and Corporate Policy*, Addison-Wesley, Reading, Mass., 1979.

7. DOWNES, D. AND DYCKMAN, T. R., "A Critical Look at the Efficient Market Empirical Research Literature as it Relates to Accounting Information," *Accounting Rev.* (April 1973), pp. 300–317.

8. FABRICANT, B. F., *Horse Sense*, David McKay, New York, 1965.

9. FAMA, E. F., "Efficient Capital Markets: A Review of Theory and Empirical Work," *J. Finance*, Vol. 25 (1970), pp. 383–417.

10. ———, *Foundations of Finance*, Basic Books, New York, 1976.

11. FIGLEWSKI, S., "Subjective Information and Market Efficiency in a Betting Model," *J. Political Economy*, Vol. 87 (1979), pp. 75–88.

12. GRIFFITH, R. M., "Odds Adjustments by American Horse Race Bettors," *Amer. J. Psychology*, Vol. 62 (1949), pp. 290–294.

13. ———, "A Footnote on Horse Race Betting," *Trans. Kentucky Acad. Sci.*, Vol. 22 (1961), pp. 78–81.

14. HARVILLE, D. A., "Assigning Probabilities to the Outcomes of Multi-Entry Competitions," *J. Amer. Statist. Assoc.*, Vol. 68 (1973), pp. 312–316.

15. HAUSCH, D. B., ZIEMBA, W. T. AND RUBINSTEIN, M. E., "Efficiency of the Market for Racetrack Betting," U.B.C. Faculty of Commerce, W. P. No. 712, September 1980.

16. ISAACS, R., "Optimal Horse Race Bets," *Amer. Math. Monthly* (1953), pp. 310–315.

17. KING, A. P., "Market Efficiency of a Multi-Entry Competition," MBA essay, Graduate School of Business, University of California, Berkeley, June 1978.

18. McGLOTHLIN, W. H., "Stability of Choices Among Uncertain Alternatives," *Amer. J. Psychology*, Vol. 63 (1956), pp. 604–615.

19. PRATT, J., "Risk Aversion in the Small and in the Large," *Econometrica*, Vol. 32 (1964), pp. 122–136.

20. RITTER, J. R., "Racetrack Betting: An Example of a Market with Efficient Arbitrage," mimeo, Department of Economics, University of Chicago, March 1978.

21. ROSETT, R. H., "Gambling and Rationality," *J. Political Economy*, Vol. 73 (1965), pp. 595–607.

22. RUBINSTEIN, M., "Securities Market Efficiency in an Arrow-Depreu Market," *Amer. Econom. Rev.*, Vol. 65, No. 5 (1975), pp. 812–824.

23. SELIGMAN, D., "A Thinking Man's Guide to Losing at the Track," *Fortune*, Vol. 92 (1975), pp. 81–87.

24. SNYDER, W. W., "Horse Racing: Testing the Efficient Markets Model," *J. Finance*, Vol. 33 (1978), pp. 1109–1118.

25. THRALL, R. M., "Some Results in Non-Linear Programming," *Proc. Second Sympos. in Linear Programming*, Vol. 2, National Bureau of Standards, Washington, D. C., January 27–29, 1955, pp. 471–493.

26. VERGIN, R. C., "An Investigation of Decision Rules for Thoroughbred Race Horse Wagering," *Interfaces*, Vol. 8, No. 1 (1977), pp. 34–45.

27. WEITZMAN, M., "Utility Analysis and Group Behaviour: An Empirical Study," *J. Political Economy*, Vol. 73 (1965), pp. 18–26.

28. ZIEMBA, W. T. AND VICKSON, R. G., eds., *Stochastic Optimization Models in Finance*, Academic Press, New York, 1975.

TRANSACTIONS COSTS, EXTENT OF INEFFICIENCIES, ENTRIES AND MULTIPLE WAGERS IN A RACETRACK BETTING MODEL*

DONALD B. HAUSCH AND WILLIAM T. ZIEMBA

*School of Business, University of Wisconsin,
Madison, Wisconsin 53706
Faculty of Commerce, University of British Columbia, Vancouver,
British Columbia, Canada V6T 1W5*

In a previous paper (*Management Science*, December 1981) Hausch, Ziemba and Rubinstein (HZR) developed a system that demonstrated the existence of a weak market inefficiency in racetrack place and show betting pools. The system appeared to make possible substantial positive profits. To make the system operational, given the limited time available for placing bets, an approximate regression scheme was developed for the Exhibition Park Racetrack in Vancouver for initial betting wealth between $2500 and $7500 and a track take of 17.1%. This paper: (1) extends this scheme to virtually any track and initial wealth level; (2) develops a modified system for multiple horse entries; (3) allows for multiple bets; (4) analyzes the effects of the track take and breakage on profits; (5) presents recent results using this system; and (6) considers the extent of the inefficiency, i.e., how much can be bet before the market becomes efficient?
(FINANCE–PORTFOLIO; GAMES—GAMBLING)

1. The Racetrack Market

The "market" at the track in North America convenes for about 20 minutes, during which participants make bets on any number of the six to twelve horses in the following race. In a typical race, participants can bet on each horse, either to win, place or show. All participants who have bet a horse to win realize a positive return on that bet only if the horse is first, while a place bet realizes a positive return if the horse is first or second, and a show bet realizes a positive return if the horse is first, second or third. Regardless of the outcome, all bets have limited liability. Unlike casino games such as roulette, but like the stock market, security prices (i.e. the "odds") are jointly determined by all the participants and the rules governing transactions costs (i.e. the track "take" and "breakage"). To take the simplest case, all bets across all horses to win are aggregated to form the win pool. If W_i represents the total amount bet by all participants on horse i to win, then $W = \sum_i W_i$ is the win pool and WQ/W_i is the payoff per dollar bet on horse i to win if and only if horse i wins, where Q is the track payback proportion.

The rules for division of the place and show pools are as follows. Let P_j be the amount bet on horse j to place and $P \equiv \sum_j P_j$ be the place pool. The payoff per dollar bet on horse j to place is

$$1 + [PQ - P_i - P_j]/(2P_j) \qquad \text{if} \quad \begin{cases} i \text{ is first and } j \text{ is second} \quad \text{or} \\ j \text{ is first and } i \text{ is second,} \\ \text{otherwise.} \end{cases} \qquad (1)$$
$$0$$

*Accepted by Donald G. Morrison as Special Departmental Editor; received December 13, 1983. This paper has been with the authors 2 months for 2 revisions.

Thus if horses i and j are first and second each bettor on j (and also i) to place first receives the amount of his bet back. The remaining amount in the place pool, after the track take, is then split evenly between the place bettors and i and j. The payoff to horse j to place is independent of whether j finishes first or second, but it is dependent on which horse i finishes with it. A bettor on horse j to place hopes that a longshot with a small P_i not a favorite will finish with it.

The payoff per dollar bet on horse k to show is analogous

$$
\begin{array}{ll}
1 + [SQ - S_i - S_j - S_k]/(3S_k) & \text{if} \quad \begin{cases} k \text{ is first, second or third} \\ \text{and finishes with } i \text{ and } j, \\ \text{otherwise,} \end{cases} \\
0 &
\end{array}
\tag{2}
$$

where S_k is the amount bet on horse k to show and the show pool is $S \equiv \sum_k S_k$.

Equations (1) and (2) are not quite correct as they do not account for "breakage". Breakage is discussed in §6; it is an extra commission resulting from payoffs being rounded down to the nearest 10¢ or 20¢ on a $2 bet.

2. Racetrack Efficiency

A market is efficient, see Fama (1970), if current prices fully reflect all available relevant information. In this case experts should not be able to achieve higher than average returns with regularity.

To investigate the efficiency of the racetrack's win market Snyder (1978) tested whether or not bets at different odds levels yielded the same average return. A weakly-efficient market in Snyder's sense would have the average rate of return for each odds level equal to Q, the track payback ratio which currently varies in North America from 0.852 in Ontario to 0.779 in Saskatchewan. His results suggest there are "strong and stable biases but these are not large enough to make it possible to earn a positive profit" (Snyder 1978, p. 1101). In particular, favorites tend to be underbet and longshots overbet. See Ziemba and Hausch (1984), hereafter referred to as ZH, for a survey of the literature on this bias.

To test the efficiency of the racetrack's place and show market HZR assumed:

(1) If q_i ($i = 1, \ldots, n$ horses) is the probability that i wins, then the probability that i is first and j is second is

$$
\frac{q_i q_j}{1 - q_i},
\tag{3}
$$

and the probability that i is first, j is second and k is third is

$$
\frac{q_i q_j q_k}{(1 - q_i)(1 - q_i - q_j)}.
\tag{4}
$$

Harville (1973) developed and analyzed these formulas.

(2) If W_i is the total amount bet on horse i to win and $W \equiv \sum_{i=1}^{n} W_i$ then $q_i = W_i / W$, i.e., the win market is efficient. While this assumption ignores the bias for favorites and longshots mentioned above there are reverse tail biases in the probabilities of finishing second and third. They occur because if the probability to win is overestimated (underestimated) and probabilities sum to one it is likely that the probabilities of finishing second and third are underestimated (overestimated). Tables 2b, c in HZR indicate how these biases tend to cancel when they are aggregated to form the theoretical probabilities of placing and showing which are used in the HZR model.[1]

[1] See HZR, §§4–6, for a more in-depth discussion of the model.

Using equations (1)–(4), the Bernoulli capital growth model (see Ziemba and Vickson 1975), and assuming initial wealth is w_0, the HZR model to calculate optimal amounts to bet for place ($p_l, l = 1, 2, \ldots, n$) and show ($s_l, l = 1, 2, \ldots, n$) is

$$
\text{Maximize} \sum_{i=1}^{n} \sum_{\substack{j=1 \\ j \neq i}}^{n} \sum_{\substack{k=1 \\ k \neq i,j}}^{n} \frac{q_i q_j q_k}{(1 - q_i)(1 - q_i - q_j)} \log \left[\begin{array}{l} \dfrac{Q(P + \sum_{l=1}^{n} p_l) - (p_i + p_j + P_{ij})}{2} \\[4pt] \qquad \times \left[\dfrac{p_i}{p_i + P_i} + \dfrac{p_j}{p_j + P_j} \right] \\[8pt] + \dfrac{Q(S + \sum_{l=1}^{n} s_l) - (s_i + s_j + s_k + S_{ijk})}{3} \\[4pt] \qquad \times \left[\dfrac{s_i}{s_i + S_i} + \dfrac{s_j}{s_j + S_j} + \dfrac{s_k}{s_k + S_k} \right] \\[8pt] + w_0 - \sum_{\substack{l=1 \\ l \neq i,j,k}}^{n} s_l - \sum_{\substack{l=1 \\ l \neq i,j}}^{n} p_l \end{array} \right]
\tag{5}
$$

$$
\text{s.t.} \quad \sum_{l=1}^{n} (p_l + s_l) \leqslant w_0, \quad p_l \geqslant 0, \quad s_l \geqslant 0, \quad l = 1, \ldots, n,.
$$

The formulation (5) maximizes the expected logarithm of final wealth considering the probabilities and payoffs from the possible horserace finishes. For notational simplicity $P_{ij} \equiv P_i + P_j$ and $S_{ijk} \equiv S_i + S_j + S_k$.

The generalized concavity properties of (5) are discussed in Kallberg and Ziemba (1981).

Using equations (1) and (3), the expected return on an additional \$1 bet to place on horse 1 is:

$$
\text{EX}_1^p \equiv \sum_{j=2}^{n} \left[\frac{q_1 q_j}{1 - q_1} + \frac{q_j q_1}{1 - q_j} \right] \left[1 + \frac{1}{20} \text{INT} \left[\frac{Q(P + 1) - (1 + P_1 + P_j)}{2(P_1 + 1)} \times 20 \right] \right].
\tag{6}
$$

INT[Y] means the largest integer not exceeding Y. The INT and multiplying and dividing by 20 accounts for breakage, i.e. the payoffs on a \$2 bet being rounded down to the nearest 10¢ in this instance. The $q_1 q_j/(1 - q_1)$ term is the probability that 1 is first and j is second while the $q_j q_1/(1 - q_j)$ term is the probability that j is first and 1 is second. A similar equation is available for show bets (see HZR).

One might hope that when EX_i^p or EX_i^s equals say, α, the actual average rate of return would be near α or at least increasing in α. Despite the inherent inaccuracies in assumptions (1) and (2) this is indeed the case as is borne out in Table 3 in HZR and Table 5.1 in ZH. In fact for cases of α above about 1.02, positive profits seem realizable. These profits are maximized when α is about 1.16.

An ideal model is then:

(i) for each i check EX_i^p and EX_i^s to decide whether or not to bet on horse i, and then

(ii) solve (5) to determine the optimal bet size after setting p_i and s_j to zero for horses you definitely do not want to bet on.

This analysis presumes one can bet after all other bettors have placed their bets. In practice one can only attempt to be one of the last bettors. Thus it is extremely important that one be able to perform all calculations necessary to determine the bet(s) quickly so that the bets can be placed as close to the end of the betting period as possible.

The approximations developed in HZR are regression schemes with the minimal data inputs: $w_0, W_{i^*}, P_{i^*}, W_{j^*}, S_{j^*}, W, P$ and S, where $i^* = \text{argmax}_i((W_i/W)/(P_i/P))$

and $j^* = \mathrm{argmax}_j((W_j/W)/(S_j/S))$. The ratios $(W_i/W)/(P_i/P)$ and (W_j/W) $/(S_j/S)$ may be thought of as simple measures of the inefficiency to place on i and show on j respectively. The regressions were calibrated for an initial wealth w_0 between $2500 and $7500 and a track about the size of Exhibition Park in Vancouver, B.C. (daily handle about $1.2 million) with $Q = 0.829$. Using this regression scheme on the Exhibition Part data with an initial wealth of $2500 and updating wealth over time resulted in a final wealth of $7698 at the end of the 110-day season (see Figure 3 in HZR).

We now extend equations (7)–(10) in HZR to account for different wealth levels, different size tracks, different Q's, and coupled entries.

3. Track Size and Wealth Level

Track size and wealth level do not affect the expected return per dollar bet to place or show. However both are important factors in determining the optimal amounts to bet to place and show because the larger the betting pools at the track the less our bet

TABLE 1

The Optimal Place Bet for Various Betting Wealth Levels and Place Pools Sizes

	$W_0 = \$50$	$W_0 = \$500$	$W_0 = \$2,500$	$W_0 = \$10,000$
Place Pool = \$2,000		$261q + 256q^2 + 180q^3$ $-\left(\dfrac{199qP_i}{qP_i - 0.70P_i}\right)$ [P2]	$426q + 802q^2$ $-\left(\dfrac{459qP_i}{qP - 0.60P_i}\right)$ [P5]	$487q + 901q^2$ $-\left(\dfrac{521qP_i}{qP - 0.60P_i}\right)$ [P8]
Place Pool = \$10,000	$39q + 52q^2$ $-\left(\dfrac{25qP_i}{qP - 0.75P_i}\right)$ [P1]	$375q + 525q^2$ $-\left(\dfrac{271qP_i}{qP - 0.70P_i}\right)$ [P3]	$1,307q + 1,280q^2$ $+ 902q^3$ $-\left(\dfrac{993qP_i}{qP - 0.70P_i}\right)$ [P6]	$2,497q + 1,806q^2$ $+ 2,073q^3$ $-\left(\dfrac{2,199qP_i}{qP - 0.60P_i}\right)$ [P9]
Place Pool = \$150,000		$505q + 527q^2$ $-\left(\dfrac{386qP_i}{qP - 0.60P_i}\right)$ [P4]	$2,386q + 2,668q^2$ $-\left(\dfrac{1,877qP_i}{qP - 0.60P_i}\right)$ [P7]	$7,072q + 10,470q^2$ $-\left(\dfrac{5,273qP_i}{qP - 0.70P_i}\right)$ [P10]

TABLE 2

The Optimal Show Bet for Various Betting Wealth Levels and Show Pool Sizes

	$W_0 = \$50$	$W_0 = \$500$	$W_0 = \$2.500$	$W_0 = \$10,000$
Show Pool = \$1,200		$9 + 994q^2 - 464q^3$ $-\left(\dfrac{150qS_i}{qS - 0.80S_i}\right)$ [S2]	$13 + 1.549q^2 - 901q^3$ $-\left(\dfrac{303qS_i}{qS - 0.60S_i}\right)$ [S5]	
Show Pool = \$6,000	$10 + 183q^2 - 135q^3$ $-\left(\dfrac{11S_i}{qS - 0.80S_i}\right)$ [S1]	$86 + 1,516q^2$ $- 968q^3$ $-\left(\dfrac{90.7S_i}{qS - 0.85S_i}\right)$ [S3]	$53 + 5,219q^2$ $- 2,513q^3$ $-\left(\dfrac{934qS_i}{qS - 0.70S_i}\right)$ [S6]	$58 + 7,406q^2$ $- 4.211q^3$ $-\left(\dfrac{1,359qS_i}{qS - 0.65S_i}\right)$ [S8]
Show Pool = \$100,000		$131 + 2,150q^2$ $- 1,778q^3$ $-\left(\dfrac{150S_i}{qS - 0.70S_i}\right)$ [S4]	$533 + 9,862q^2$ $- 7,696q^3$ $-\left(\dfrac{571S_i}{qS - 0.80S_i}\right)$ [S7]	$1,682 + 28.200q^2$ $- 16,880q^3$ $-\left(\dfrac{1,769S_i}{qS - 0.85S_i}\right)$ [S9]

influences the odds and as our wealth increases we tend to wager more. Therefore new regressions were calculated for most reasonable track sizes and wealth levels. The data for these regressions were the true optimal bets from the NLP model (5) over a broad range of wealth levels and track sizes. The ideal data to represent a broad range of track sizes would be a season's data from many different tracks. A more practical alternative was to multiply the Exhibition Park data by varying constants to simulate data from smaller and larger tracks. Nineteen different regressions which appear in Tables 1 and 2 were determined for place and show depending upon wealth and track size. These regressions labelled P1–P10 and S1–S9 give the optimal place or show bet for specific values of initial wealth and size of pool. For intermediate values of these variables one may determine accurate betting amounts by taking convex combinations of these basic regressions. For example, for a place bet with initial wealth $1000 and place pool of $5000 the optimal bet is α_2 (equation $[P_2]$) + α_3 (equation $[P_3]$) + α_5 (equation $[P_5]$) + α_6 (equation $[P_6]$), where

$$\alpha_2 = \tfrac{3}{4}\tfrac{5}{8} = \tfrac{15}{32}, \qquad \alpha_3 = \tfrac{1}{4}\tfrac{3}{8} = \tfrac{9}{32}, \qquad \alpha_5 = \tfrac{1}{4}\tfrac{5}{8} = \tfrac{5}{32}, \qquad \alpha_6 = \tfrac{1}{4}\tfrac{3}{8} = \tfrac{3}{32}.$$

4. Track Payback

Both the expected return per dollar bet and the optimal bet size are increasing functions of Q, the track's payback. Equations (7) to (10) in HZR were calculated with $Q = 0.829$. Modification of these equations for use at tracks with $Q \neq 0.829$ are now developed.

4.1. Adjustment of $E\tilde{X}_i^p$ and $E\tilde{X}_i^s$ for Q

The regression equivalent for EX_i^p when $Q = 0.829$ is:[2]

$$E\tilde{X}_i^p = 0.319 + 0.559 \frac{W_i / W}{P_i / P}. \tag{7}$$

How can this equation be adjusted for a track payback different from 0.829? The "true" expected return on a one dollar place bet on horse i is

$$EX_i^p = \sum_{\substack{j=1 \\ j \neq i}}^{n} \left(\frac{q_i q_j}{1 - q_i} + \frac{q_i q_j}{1 - q_j} \right)\left(1 + \frac{QP - (P_i + P_j)}{2P_i} \right). \tag{8}$$

Equation (7) is linear in Q with

$$\frac{\partial EX_i^p}{\partial Q} = \frac{q_i P}{2P_i}\left[1 + \sum_{\substack{j=1 \\ j \neq i}} \left(\frac{q_j}{1 - q_j} \right) \right].$$

Using 124 Exhibition Park races with $E\tilde{X}_i^p$ in the range 1.10 and greater, the true $\partial EX_i^p/\partial Q$ was regressed against q_i to give $\partial EX_i^p/\partial Q \approx 2.22 - 1.29q_i$ ($R^2 = 0.86$, $SE = 0.055$, both coefficients highly significant). Therefore when the track payback is Q, the expected return on a $1 place bet can be approximated by adjusting (7) to

$$E\hat{X}_i^p \equiv E\tilde{X}_i^p + (2.22 - 1.29q_i)(Q - 0.829)$$

$$= 0.319 + 0.559\left(\frac{W_i / W}{P_i / P} \right) + \left(2.22 - 1.29\left(\frac{W_i}{W} \right) \right)(Q - 0.829). \tag{9}$$

---

[2] Note that the coefficients of $E\tilde{X}_i^p$ here are different from those of equation (7) in HZR. In HZR only cases of expected return greater than 1.16 were considered. Here we may wish to bet on horses with expected returns as low as 1.10 to reflect a high quality track and high quality horses. Thus the $E\tilde{X}_i^p$ had to be recalculated to be accurate in the range 1.10 to 1.16. A similar change will be noted for $E\tilde{X}_i^s$.

A similar analysis for show yields

$$E\hat{X}_i^s \equiv E\tilde{X}_i^s + (3.60 - 2.13 q_i)(Q - 0.829)$$

$$(R^2 = 0.565, SE = 0.198, \text{ both coefficients highly significant})$$

$$= 0.543 + 0.369\left(\frac{W_i/W}{S_i/S}\right) + \left(3.60 - 2.13\left(\frac{W_i}{W}\right)\right)(Q - 0.829). \qquad (10)$$

4.2. Adjustment of $\tilde{p}^*$ and $\tilde{s}^*$ for Q

Equations (8) and (10) in HZR were calibrated for $Q = 0.829$. Since the true p^* and s^* are nondecreasing in Q which varies from track to track we must adjust $\tilde{p}^*$ and $\tilde{s}^*$ at tracks with $Q \neq 0.829$.

The exact NLP model was used on a number of Exhibition Park examples to compute the optimal place or show bets at different initial wealths, track sizes and different Q's (from 0.809 to 0.859). The results indicated that $\Delta p^*/\Delta Q$ and $\Delta s^*/\Delta Q$ are independent of Q in this range. Therefore with a ΔQ of 0.01, $\Delta p^*/\Delta Q$ and $\Delta s^*/\Delta Q$ were regressed on p^*, w_0, P_i, P and s^*, w_0, S_i, S, respectively. The analysis showed for $\Delta p^*/\Delta Q$ that p^* and w_0 were very significant independent variables but neither P_i and P were significant; similar results for $\Delta s^*/\Delta Q$ were observed. Then the $(\Delta p^*/\Delta Q, p^*, w_0)$ and $(\Delta s^*/\Delta Q, s^*, w_0)$ were aggregated (due to a small number of place data points) to give:

$$\begin{bmatrix} \Delta p^*/\Delta Q \\ \Delta s^*/\Delta Q \end{bmatrix} = [0.0316]\begin{bmatrix} \tilde{p}^* \\ \tilde{s}^* \end{bmatrix} + 0.000351 w_0$$

$$(R^2 = 0.948, SE = 2.23, n = 56, \text{ both coefficients highly significant}).$$

Therefore $\hat{p}^*$ and $\hat{s}^*$ (i.e. $\tilde{p}^*$ and $\tilde{s}^*$ adjusted for Q) are:

$$\hat{p}^* = \tilde{p}^* + (Q - 0.829)(3.16\tilde{p}^* + 0.0351 w_0) \quad \text{and} \qquad (11)$$

$$\hat{s}^* = \tilde{s}^* + (Q - 0.829)(3.16\tilde{s}^* + 0.0351 w_0). \qquad (12)$$

4.3. Example Involving $Q \neq 0.829$

May 7, 1983–Kentucky Derby at Churchill Downs, Louisville, Kentucky. The final win and show pools and the win and show bets on #8, Sunny's Halo, were:

$$W = \$3,143,669, \qquad W_8 = \$745,524,$$
$$S = \$1,099,990, \qquad S_8 = \$179,758.$$

Using just the $E\tilde{X}_i^s$ portion of equation (10) gives $E\tilde{X}_8^s = 1.08$, i.e. not enough to consider a bet if one is using a typical cutoff of 1.10 as recommended in ZH for a race like the Kentucky Derby. But Kentucky has $Q = 0.85$ and therefore it is more accurate to use equation (10) resulting in $E\hat{X}_8^s = 1.14$. Hence a show bet should be made. With an initial wealth of $1000, Table 2 gives the optimal show bet as $\tilde{s}^* = \$48$. The correction for $Q = .85$ using equation (12) yields $\hat{s}^* = \$52$. Sunny's Halo won the Derby and paid $4.00 per $2.00 bet to show so the $52 bet returned $104 for a $52 profit. Full details on this race appear in ZH.

5. Coupled Entries

Occasionally two or more horses are run as a single "coupled entry" or simply "entry" because (1) an owner or a trainer has two or more horses in the same race, or

(2) there are more horses than the toteboard can accommodate (commonly called a field). The entry wins, places or shows if just one of the horses wins, places or shows. If any two of the horses in the entry come first and second all the place pool goes to the place tickets on the entry. If two of three of the 'in-the-money' horses are the entry then typically two thirds of the show pool goes to the draw tickets on the entry (rather than the usual third).

Suppose the coupled entry has number 1 and let $q_1 = W_1/W$. Then q_1 estimates the probability that one of the horses in the entry will win the race. Suppose further that q_{1A} and q_{1B} (with $q_{1A} + q_{1B} = q_1$) are the correct winning probability estimates of the two horses in the entry. Using q_1 (i.e. thinking of the entry as a single horse) and equation (3) to calculate the probability of the entry placing gives

$$\Pr(\text{entry 1 is 1st or 2nd}) = q_1 + \sum_{i=2}^{n} \frac{q_1 q_i}{1 - q_i} . \tag{13}$$

Using q_{1A} and q_{1B} (i.e. thinking of the entry as two horses) and equation (3) to calculate the probability of the entry placing gives

$\Pr(\text{entry 1 is 1st and/or 2nd}) = \Pr(1A \text{ is 1st and any horse but } 1B \text{ is 2nd})$

$+ \Pr(1B \text{ is 1st and any horse but } 1A \text{ is 2nd})$

$+ \Pr(1A \text{ is 2nd and any horse but } 1B \text{ is 1st})$

$+ \Pr(1B \text{ is 2nd and any horse but } 1A \text{ is 1st})$

$+ \Pr(1A \text{ and } 1B \text{ are 1st and 2nd in either order})$

$$= \sum_{i=2}^{n} \frac{q_{1A} q_i}{1 - q_{1A}} + \sum_{i=2}^{n} \frac{q_{1B} q_i}{1 - q_{1B}} + \sum_{i=2}^{n} \frac{q_{1A} q_i}{1 - q_i}$$

$$+ \sum_{i=2}^{n} \frac{q_{1B} q_i}{1 - q_i} + \frac{q_{1A} q_{1B}}{1 - q_{1A}} + \frac{q_{1A} q_{1B}}{1 - q_{1B}} ,$$

which equals

$$q_{1A} + q_{1B} + \sum_{i=2}^{n} \frac{(q_{1A} + q_{1B}) q_i}{1 - q_i} ,$$

which is equation (13). Hence considering the entry as two horses does not affect our estimate of the entry's probability of placing. It does, however, affect our estimate of the expected return on a dollar bet to place since the possibility of a $1A$-$1B$ or $1B$-$1A$ finish exists and for those finishes the place payoff will be high since the whole place pool (net of the track take and breakage) goes to the holders of place tickets on the entry 1. Thus equation (9) will underestimate $E\hat{X}_1^p$. For the same reason p^*, from Table 1, will also be underestimated. We now consider the use of Tables 1 and 2 and equations (9), (11), (10) and (12) on $E\hat{X}^p$, $\hat{p}^*$, $E\hat{X}^s$ and $\hat{s}^*$, respectively, to account for an entry.

5.1. Adjustments of EX_1^p and EX_1^s for Coupled Entries

Equation (6) gives the expected return on a dollar bet to place on a horse, considering the entry 1 as a single horse. Ignoring breakage, this expected return is

$$EX_1^p = \sum_{i=2}^{n} \left(\frac{q_1 q_i}{1 - q_1} + \frac{q_1 q_i}{1 - q_i} \right) \left(1 + \frac{QP - (P_1 + P_i)}{2 P_1} \right).$$

More properly considering entry 1 as two horses, $1A$ and $1B$, the expected return on a one dollar bet to place is

$$
\begin{aligned}
EX^P_{1A,1B} = \sum_{i=2}^{n} & \left(\frac{q_{1A}q_i}{1-q_{1A}} + \frac{q_{1A}q_i}{1-q_i} \right) \left(1 + \frac{QP-(P_1+P_i)}{2P_1} \right) \left\{ \begin{array}{l} \text{horses } 1A \text{ and} \\ i \text{ place} \end{array} \right. \\
+ \sum_{i=2}^{n} & \left(\frac{q_{1B}q_i}{1-q_{1B}} + \frac{q_{1B}q_i}{1-q_i} \right) \left(1 + \frac{QP-(P_1+P_i)}{2P_1} \right) \left\{ \begin{array}{l} \text{horses } 1B \text{ and} \\ i \text{ place} \end{array} \right. \\
+ & \left(\frac{q_{1A}q_{1B}}{1-q_{1A}} + \frac{q_{1A}q_{1B}}{1-q_{1B}} \right) \left(1 + \frac{QP-P_1}{P_1} \right) \left\{ \begin{array}{l} \text{horses } 1A \text{ and} \\ 1B \text{ place}. \end{array} \right.
\end{aligned}
$$

Let $\Delta^P \equiv EX^P_{1A,1B} - EX^P_1$. It is generally the case that the two horses in the coupled entry are not of equal ability. We assume that $q_{1A} = \frac{2}{3}q_1$ and $q_{1B} = \frac{1}{3}q_1$.

Using the Exhibition Park data,[3] Δ^P was regressed on W_1/W and P_1/P, giving

$$
\tilde{\Delta}^P = 0.867 \frac{W_1}{W} - 0.857 \frac{P_1}{P} \tag{14}
$$

($R^2 = 0.996$, $SE = 0.00267$, and both coefficients highly significant).

Then using equations (7) and (18) the regression approximation for expected return on a one dollar bet on the coupled entry 1 is

$$
E\tilde{X}^P_{1A,1B} \equiv E\tilde{X}^P_1 + \tilde{\Delta}^P = 0.319 + 0.559 \left(\frac{W_1/W}{P_1/P} \right) + 0.867 \frac{W_1}{W} - 0.857 \frac{P_1}{P}. \tag{15}
$$

The same procedure for the expected return on a one dollar show bet on the coupled entry 1 yields

$$
E\tilde{X}^s_{1A,1B} \equiv E\tilde{X}^s_1 + \tilde{\Delta}^P = 0.543 + 0.369 \left[\frac{W_1/W}{S_1/S} \right] + 0.842 \frac{W_1}{W} - 0.810 \frac{S_1}{S}. \tag{16}
$$

5.2. Adjustments of p* and s* for Coupled Entries

When the possible bet is on an entry, equations (7) or (9) underestimate the expected return on an additional dollar bet to place. Therefore the optimal bet from the NLP (5) will underestimate the true optimal coupled entry place bet. To understand this phenomenon many Exhibition Park examples were solved using the exact NLP (5) assuming the entry was one horse. Then the same examples were solved supposing the entry was two horses (the formulation of the NLP was adjusted to consider the possibility of the two horses finishing first and second and then receiving a high place payoff) but the win bet on the entry was lowered, using an iterative scheme, until the optimal bet was the same as the optimal bet assuming the entry was one horse. This procedure gave pairs of $\tilde{q}^P_i$ and q_i, where $\tilde{q}^P_i$ was the probability of entry i winning in part a (thinking of the entry as one horse); and q_i was the adjusted probability that gave the same optimal bet when thinking of the entry as two horses as was observed when treating it as one horse.

Since the regression formula (13) gives the approximate optimal place bet when the horse's probability of winning is q_i, then using $\tilde{q}^P_i$ in that formula gives the approximate optimal place bet when the coupled entry's probability of winning is q_i.

---

[3] The data used were cases where the expected return (equation (7)) was $\geqslant 1.16$. i.e., the cases of interest. For instances with a low expected return the correction factor is meaningless.

The regression relating $\tilde{q}_i^p$ and q_i is

$$\tilde{q}_i^p = 0.991 q_i + 0.137 q_i^2 + 3.47 \times 10^{-7} W_0 \tag{17}$$

($R^2 = 0.9998$, $SE = 0.00161$, all coefficients highly significant).

The examples from which the data were derived spanned many wealth levels and pool sizes. While the wealth level was a very significant independent variable the pool size was found to be statistically insignificant. Therefore to compute the optimal place bet on coupled entry i the procedure is:

(0) determine that EX^p is large enough to consider a place bet,
(1) set $q_i = W_i / W$,
(2) determine $\tilde{q}_i^p$, and
(3) substitute $\tilde{q}_i^p$, w_0, P, P_i in Table 1.

The same procedure was carried out for the optimal show bet on coupled entry i:

(0) determine that EX^s is large enough to consider a show bet,
(1) set $q_i = W_i / W$,
(2) determine

$$\tilde{q}_i^s = 1.07 q_i + 4.13 \times 10^{-7} W_0 - 0.00663, \tag{18}$$

and ($R^2 = 0.999$, $SE = 0.00298$, all coefficients highly significant), and

(3) substitute $\tilde{q}_i^s$, w_0, S, and S_i, in Table 2.

Three additional questions to be considered are: (1) Will the results be different with a weighting other than $1/3$ and $2/3$ on the two horses in the entry? (2) Should there be larger adjustments for entries of three or more horses? and (3) Should there be adjustments of the equations for a single horse running against a coupled entry? The answer to all three questions is yes, but how important is it to account for these possibilities? (1) The coupled entry adjustment appears to be fairly robust to the weighting. Also each race would require handicapping to determine its more accurate weighting. Since that goes beyond the scope of this research (in fact the system described in ZH requires absolutely no handicapping), we choose to opt for no adjustments for different weightings. In the extreme case where $q_{1A} \approx q_1$ and $q_{1B} \approx 0$ it may be better to treat the entry as a single horse and ignore the entry adjustments. (2) The additional benefits of three or more horses in an entry are small beyond accounting for the entry as two horses. Thus we suggest the simplification of no further adjustment. (3) Generally the expected return equations on a single horse running against an entry will overestimate the true expected return. In most cases though the bias is small and again we suggest no further adjustment.

6. Multiple Bets

Occasionally there is more than one system bet in a given race. Since the optimal bet equations in Tables 1 and 2 were calibrated assuming only one place bet or one show bet in a race it is not correct, in a multiple betting situation, to calculate each bet individually from the tables and then wager those amounts. Often that would result in overbetting but there are also times where, for diversification reasons, that would actually result in underbetting.

We have attempted to deal with the most common multiple betting situation—a place and show bet on the same horse. Ninety-eight cases of place and show system bets on the same horse were analyzed covering a wide range of track handles, q_i's, and w_0's. Using the optimization model (5) resulted in the quadruples $(p_T^*, s_T^*, p_A^*, s_A^*)$. The p_A^* is the optimal place bet supposing it is the only good bet in the race, s_A^* is the optimal show bet supposing it is the only good bet in the race, and (p_T^*, s_T^*) is the

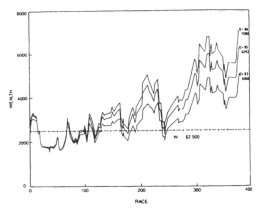

FIGURE 1. Betting Wealth Level Histories for System Bets at the 1981–82 Aqueduct Winter Meeting Using an Expected Value Cutoff of 1.14 for Track Takes of 14%, 15% and 17%.

optimal pair of place and show bets when they are considered together. The p_A^* and s_A^* are the values we should calculate from Tables 1 and 2. Then p_T^* was regressed on p_A^*, s_A^*, w_0, P_i, P and q_i. The only statistically significant independent variables were p_A^* and s_A^* leading to the regression equation

$$\tilde{p}_T^* = 1.59p_A^* - 0.639s_A^* \qquad (19)$$

($R^2 = 0.967$, $SE = 73.7$, both coefficients highly significant).

A similar procedure for s_T^* yields

$$\tilde{s}_T^* = 0.907s_A^* - 0.134p_A^* \qquad (20)$$

($R^2 = 0.992$, $SE = 72.6$, both coefficients highly significant).

7. Effect of the Track Take

The track take is a commission of 14–22% on every dollar wagered. As mentioned in §4 a change in Q, the track payback proportion, can have a substantial effect on EX_i^p and EX_i^s. A change in Q can also have a surprisingly dramatic effect on long run profit. This is illustrated in Figure 1 using data from the 1981–82 Winter Season at Aqueduct, New York[4] and supposing three different track takes: 14%, 15% and 17%. In 1981–82 the track take was 15%, earlier it had been 14% and just recently it has increased to 17%. Assuming an initial wealth of $2,500 the final wealths are $7,090, $6,292 and $5,058 at track's takes of 14%, 15% and 17% respectively. The track take increasing from 14% to 15% dropped profits by $798 (17.4%). The track take increasing from 15% to 17% dropped profits by $1,234 (32.5%).

8. Effect of Breakage

In addition to the track take, bettors must also pay an additional commission called breakage. This commission refers to the funds not returned to the betting public because the payoffs are rounded down to the nearest 10¢ or 20¢ on a $2 bet. For example, a payoff net of the track take of $6.39 would pay $6.30 or $6.20, respectively.

---

[4]The data consisted of the final win, place and show mutuels for the 43-day period December 27, 1981—March 27, 1982. During this period 3,470 horses ran in 380 races. Thanks go to Dr. Richard Van Slyke for collecting these data for us.

FIGURE 2. Wealth Level Histories at Exhibition Park (1978) with Alternative Breakage Schemes.

Breakage occurs in the win, place and show as well as other pools. We refer to rounding down to the nearest 10¢ on a $2 bet as "5¢ breakage" that is 5¢ per dollar and rounding down to the nearest 20¢ on a $2 bet as "10¢ breakage". Initially most tracks utilized a 5¢ breakage procedure. However, in recent years more and more tracks have switched to 10¢ breakage. The 10¢ breakage is never less than 5¢ breakage and usually is considerably more. As a percentage of the payoff breakage usually increases as the payoff becomes smaller unless the payoff is close to the breakage roundoff amount. An exception is the "minus pool" when $2.10[5] must be paid on a winning $2 ticket even if the payoff before breakage is only $2.08, or even $1.73.

On average bettors lose about 1.79% of the total payoff to 5¢ breakage and 3.14% to 10¢ breakage on bets using the system in ZH. Adding these amounts to the track take gives the total commission. For example, at Churchill Downs in Louisville, Kentucky the 15% track take becomes about 18.1% with their 10¢ breakage, and at Exhibition Park in Vancouver, British Columbia the 15.8% track take becomes about 17.6% with their 5¢ breakage. Thus to determine the full commission at a given racetrack one must take into account the breakage as well as the track take.

The full extent of the effect of breakage is shown in its effect on profits. Using the 1978 system bets for Exhibition Park with an initial wealth of $2500 one would have $8319 at the end of the year without breakage. With 5¢ breakage one would have $7521 and $6918 with 10¢ breakage. The effect of breakage throughout the 1978 season is shown in Figure 2. In addition to taking money away from total wealth the breakage has the effect of lowering the bet size (because of this lower wealth) thus resulting in lower future profits. These calculations indicate that 5¢ breakage averages 13.7% of profits and 10¢ breakage averages 24.1% of profits on system bets.

These calculations indicate that breakage (especially the very common 10¢ variety) is a very substantial cost. The costs are highest when one is placing bets on short odds horses. This is, unfortunately, an unavoidable aspect of the system presented in ZH.

9. Results Using the Betting System

In ZH many examples are presented showing precisely how to use the system. In Table 3 we present summary statistics on system bets made at several tracks over different seasons. The data sets for Aqueduct, Santa Anita and Exhibition Park 1978 were collected after their seasons finished. The Exhibition Park 1980 and Kentucky Derby Days data were collected race by race at the track. In all cases initial betting wealth is assumed to be $2,500. The different expected cutoff levels reflect the quality of the horses at the different tracks.

The most common system bet is to show on a favorite. Show and place bets occur about 85% and 15% of the time, respectively. The percent of bets won is about 59%

[5] In Kentucky it is $2.20.

TABLE 3

Summary Statistics on System Bets Made at Aqueduct in 1981/82, Santa Anita in 1973/74, Exhibition Park in 1978 and 1980, and at the Kentucky Derby Days 1981/82/83 with an Initial Betting Wealth of $2500

Track and Season	Number of Days	Number of Races	Track Take	Expected Value Cutoff	Number of System Bets	Number of Bets Won	Percent of Bets Won	Percent of Bets Won Weighted by Size of Bet	Total Money Wagered	Track Take	Total Profits	Average Payout Per $2 Bet	Average Rate of Return on Bets Made
Aqueduct 1981/82	43	380	15%	1.14	124	68	55%	65%	$42,686	$6,403	$3,792	$3.33	8.9%
Santa Anita 1973/74	75	627	15%	1.14	192	114	59%	69%	$51,631	$7,745	$2,837	$3.16	5.5%
Exhibition Park 1978	110	1,065	18.1%	1.20	174	97	56%	72%	$49,991	$9,048	$5,198	$3.08	10.4%
Exhibition Park 1980	10	90	17.1%	1.20	22	16	73%	77%	$5,403	$924	$1,216	$3.18	22.5%
Derby Days* 1981/82/83	3	30	15%	1.10	19	17	89%	96%	$12,766	$1,915	$5,462	$2.97	42.8%
Totals and Weighted Averages	241	2,192	—	—	531	312	59%	71%	$162,477	$26,035	$18,505	$3.12	11.4%

while the percent of bets won weighted by the size of the bet is 71%. This difference is because the bets are on shorter odds horses which finish in-the-money more often. At a track such as Santa Anita with large betting pools, our bets do not affect the odds very much and the average bet is about 7% of the betting pool. Over all these thousands of races and hundreds of system bets the total amount wagered was $162,477. The track take was $26,035. Our profit was $18,505, for an 11.4% rate of return on dollars wagered. The higher rates of return were on the races where we were at the track, so we were able to skip rainy days and reject certain horses on the basis of very simple handicapping rules. The lower rates of return, as expected, were at the tracks where we had no information other than the win, place and show mutuel pools. A simple correction which has a surprisingly large effect is removing from the Exhibition Park 1978 data the days when the track was not a fast track, i.e. rainy days when the track was slow, muddy, heavy, sloppy, etc. Doing so decreases the total money wagered from $49,991 to $32,811 but increases the profit from $5,198 to $5,863. Thus the rate of return on the "fast track" days is 17.9%, up from 10.4% over all days. It also increases the rate of return over all the racetracks from 11.4% to 13.2%.

Since the bets usually have a high EX_i^p or EX_i^s we might expect a rate of return around 16%–20%. Remember that EX_i^p and EX_i^s are on the first dollar bet. These values drop as we bet large amounts due to our bets affecting the odds.

These profits do not consider several minor "entertainment type" costs that one must incur by actual attendance at the track to apply the system. Parking, gasoline, racing program, racing form, track admission and food amount to $3–10 or more. For example, at $10 per day Aqueduct's profits of $3792 over 43 days become $3362.

Finally the average payout per $2 bet ranged from $2.97 to $3.33 at the various tracks, with an average of $3.12. This value is actually a high show return when one considers the heavy favorites the system often picks.

10. Will the Market Become Efficient?

As more and more individuals use this system the markets for place and show betting will tend to become efficient. Two important questions are: (1) How many people can play this system and still have it provide a return of 10–20%? and (2) How many people can play this system before the market becomes efficient enough that expected profits are zero?

To consider the first question we can determine how much additional money can be wagered on a particular horse to place or show before the expected value per dollar bet drops to the suggested cutoff for good betting opportunities. As an illustration Figure 3 indicates this amount to show for a cutoff of 1.14. These figures are based on a track take of 17.1% so for lower track takes more can be bet and for higher track takes less can be bet.

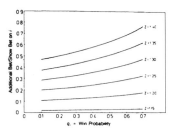

FIGURE 3. How Much Can Be Bet. B_i by System Bettors Relative to the Crowd's Show Bet, S_i, on Horse i to Lower the Expected Value to Show on the Horse i from Z to 1.14, When the Track Take is 17.1%.

An example that more directly answers questions 1 and 2 is provided below and is based on the data given in §4.3 on Sunny's Halo, the 1983 Kentucky Derby winner. Additional examples appear in ZH.

Sunny's Halo

	Betting Wealth w_0			
	$200	$500	$1000	$2000
Optimal System Bets (Using the Actual Data One Minute Before Post Time) Assuming a Betting Wealth of w_0	$11	$31	$52	$96

α	Total Amount That Can be Bet Before the Expected Return per Dollar Bet Drops to α	Number of System Bettors Needed to Drop the Expected Return per Dollar Bet to α Assuming a Betting Wealth of			
		$200	$500	$1000	$2000
1.10	$19,323	1,757	623	372	201
1.06	$41,175	3,743	1,328	792	429
1.02	$68,409	6,219	2,207	1,316	713

Our results show that $\alpha = 1.02$ is a breakeven cutoff and $\alpha = 1.06$ yields a rate of return of about 5–6%.[6]

References

FAMA, E. F., "Efficient Capital Markets: A Review of Theory and Empirical Work," *J. Finance*, 25 (1970), 383–417.

HAUSCH, D. B., W. T. ZIEMBA AND M. RUBINSTEIN, "Efficiency of the Market for Racetrack Betting," *Management Sci.*, 27 (1981), 1435–1452.

HARVILLE, D. A., "Assigning Probabilities to the Outcome of Multi-Entry Competitions," *J. Amer. Statist. Assoc.*, 68 (1973), 312–316.

KALLBERG, J. G. AND W. T. ZIEMBA, "Generalized Concave Functions in Stochastic Programming and Portfolio Theory," in *Generalized Concavity in Optimization and Economics*, S. Schaible and W. T. Ziemba (Eds.), Academic Press, New York, 1981, 719–767.

SNYDER, W. W., "Horse Racing: Testing the Efficient Markets Model," *J. Finance*, 33 (1978), 1109–1118.

ZIEMBA, W. T. AND D. B. HAUSCH, *Beat the Racetrack*, Harcourt, Brace and Jovanovich, San Diego, 1984.

—— AND R. G. VICKSON, EDS., *Stochastic Optimization Models in Finance*, Academic Press, New York, 1975.

Donald B. Hausch

University of Wisconsin—Madison

William T. Ziemba

University of British Columbia

Arbitrage Strategies for Cross-Track Betting on Major Horse Races*

I. Introduction

Racetracks and securities markets have many characteristics in common. A difference, though, is their complexity; the racetrack is really a sequence of markets that are relatively simple, short-lived, and, for the most part, independent. This "market-in-miniature" feature makes the racetrack attractive for tests of market efficiency, especially since, as Thaler and Ziemba (1988, p. 162) suggest, "one can argue that wagering markets have a better chance [than securities markets] of being efficient because the conditions (quick, repeated feedback) are those which usually facilitate learning." The many empirical racetrack studies support a weak form of efficiency for some of the available wagers, while other types of wagers seem not to be efficient. These studies are reviewed in Section II.

This article studies cross-track betting, a relatively new form of wagering. It allows bettors to wager at their track (a cross track) on a race

Cross-track betting permits bettors to place wagers at their local tracks on a race being run at another track. Since each track operates a separate betting pool, the odds can vary across the tracks. The data suggest that the odds vary, and they often vary dramatically, allowing arbitrage opportunities. This article employs a risk-free arbitrage model to demonstrate the cross-track inefficiency and recommends an optimal capital growth model for exploiting it. A simpler method is proposed for a single bettor at a single cross track. The results indicate that these methods would have worked well in practice on a number of recent Triple Crown races.

* Without implicating them, we would like to thank Bruce Fauman and Fraser Rawlinson. Also, we greatly appreciate the data supplied by a number of U.S. racetracks, and we wish to thank Victor Lespinasse for suggesting the one-track model.

405

TABLE 1   Home-Track and Cross-Track Betting, Kentucky Derby

Year	Home-Track Attendance	No. Cross Tracks	Home-Track Betting ($)	Cross-Track Betting ($)
1982	141,009	. . .	5,011,575	. . .
1983	134,444	. . .	5,546,977	. . .
1984	126,453	24	5,420,787	13,521,146
1985	108,573	32	5,770,074	14,474,555
1986	123,819	56	6,165,119	19,776,332
1987	130,532	73	6,362,673	20,829,236
1988	137,694	93	7,427,389	24,449,058

being run at another track (the home track). Since cross-track betting tends to be limited to major races, it gives the racing public an opportunity to bet on some of the world's finest racehorses. This makes it very popular with the public. Cross-track wagering can lead to increased attendance and revenues at the cross tracks and add to the revenues of the home track through a fee (usually 5% of the handle) paid by the cross tracks. Thus, all the tracks can increase profits.[1]

Separate pools for each track means the payoffs at the various tracks can differ.[2] Due to the costs of arbitrage in this setting, market efficiency across the tracks would, for practical purposes, allow some differences across the various sets of track odds. Considerable differences, however, would suggest the possibility of a market inefficiency. The data demonstrate that considerable differences do occur. For example, a $2.00 win ticket on Ferdinand, the winner of the 1986 Kentucky Derby, paid from $13.20 at Fairplex in Pomona, California, to $90.00 at Evangeline Downs in Lafayette, Louisiana.[3] Obviously, bettors would have preferred their win bets on Ferdinand to be made at

1. For an example of this effect, consider the home-track and cross-track betting on the Kentucky Derby (see table 1). The introduction in 1984 of cross-track betting on this race has greatly increased total Derby wagering. At the same time, cross-track betting seems to have had, at worst, only a minor effect on home-track wagering. The cross tracks' revenues can increase also. For instance, Illinois set a one-day pari-mutuel record on Kentucky Derby Day, 1987. Cross-track betting on the Derby accounted for $1,326,239 of the $4,534,879 wagered that day in the state. Another example is Calder Race Course in Florida. They set all-time revenue and attendance records on Kentucky Derby Day, 1985. Attendance was 23,105, and $2,775,645 was wagered, $562,453 of it on the Derby.

2. In some cases, all the wagers at the various tracks are summed. Then, on the basis of these summed values, identical payoffs are made at all the tracks. This is often called "intertrack" wagering and typically the tracks are within one state. Intertrack wagering will not be considered here.

3. These extreme payoffs are not just limited to the smaller tracks. Two large-track examples are Hollywood Park, where Ferdinand paid $16.80, and Woodbine Racetrack in Toronto where he paid $79.60. Alysheba, the 1987 Kentucky Derby winner, paid from $15.80 at Hollywood Park, California to $30.20 at Beulah Park, Ohio. Winning Colors, the 1988 derby winner, paid $7.40 at Pimlico Race Course, Maryland, and $10.40 at Beulah Park.

Evangeline Downs. The nature of the pari-mutuel betting system requires that if Ferdinand paid less at Fairplex than at Evangeline Downs, then another horse, were it to have won, would have paid more at Fairplex than at Evangeline Downs. Thus, if we are able both to learn the odds and place our bets at various tracks, it appears that significant arbitrage opportunities may exist.

Section III develops a risk-free arbitrage model to demonstrate this cross-track inefficiency. The optimal capital growth model is studied in Section IV on the general cross-track problem and, in Section V, on a simpler one-track problem. These models are tested on data from several recent Triple Crown races. A final discussion is in Section VI.

II. Efficiency of the Various Betting Markets

Among the possible wagers at the track are the so-called straight wagers to win, place, and show. They pay off when one's horse is at least first, second, or third, respectively. The "exotic" wagers include quinellas (requiring one to name the first two horses), exactors (requiring the first two horses in the correct order), trifectas (requiring the first three horses in the correct order), and daily doubles (requiring the winners of two consecutive races). Tracks have also extended the daily-double concept to picking the winners of three, four, six, and even nine consecutive races. These are very low-probability bets that can have tremendous payoffs, and they are very popular with the racing public. Before exotic wagering was offered by the tracks, bettors could use parlays and other combinations of wagers to construct low-probability/high-payoff situations. Rosett (1965) analyzed these possibilities and demonstrated that, except for extreme long shots, the bettors were rational in the sense that a simple bet would not be made if a parlay with the same probability of success had a greater return. Similarly, Ali (1973) showed that the return on a daily double is not significantly different from the return on what is an identical wager, the corresponding parlay of win bets.

Unlike typical casino games, where the odds are fixed, the odds at the track are determined by the relative amounts the bettors wager on the horses and by the track's transactions costs (the track's take and breakage). Thus, "prices" are determined at the track much like they are in securities markets. Let

n = the number of horses in a race;
T = the number of tracks accepting wagers on the race;
W_i^t = the total amount bet by the public at track t on horse i to
win ($i = 1, \ldots, n$ and $t = 1, \ldots, T$);
$W^t = \Sigma_i W_i^t$ = the win pool at track t; and
Q^t = track t's payback proportion (typically from .80 to .86).

Then the payoff per dollar bet to win on horse i at track t is

$$\begin{cases} Q^t W^t / W_i^t & \text{if horse } i \text{ wins,} \\ 0 & \text{otherwise.} \end{cases} \tag{1}$$

To determine the place payoff let P_i^t be the amount bet to place on horse i at track t and let $P^t = \Sigma_i P_i^t$ be track t's place pool. The payoff per dollar bet to place on horse i at track t is

$$\begin{cases} 1 + (Q^t P^t - P_i^t - P_j^t)/(2P_i^t) & \text{if the first two horses are } i \text{ and } j, \\ 0 & \text{if horse } i \text{ is not first or second.} \end{cases} \tag{2}$$

Thus, the track keeps $(1 - Q^t)P^t$ and the place bets on i and j are repaid. The remainder, the losing bets minus the track take, is then split evenly between those who bet on i and those who bet on j. The share for i bettors is then divided on a per-dollar-bet basis. The place payoff on i does not depend on whether i was first or second but it does depend on which horse j was the other top finisher. In a similar fashion the payoff per dollar bet to show on horse i at track t is

$$\begin{cases} 1 + (Q^t S^t - S_i^t - S_j^t - S_k^t)/(3S_i^t) & \text{if the first three horses are } i, j, \\ & \text{and } k, \\ 0 & \text{if horse } i \text{ is not at least third.} \end{cases} \tag{3}$$

Here S_i^t is the show bet on horse i by the public at track t, and $S^t = \Sigma_i S_i^t$ is track t's show pool. It is possible that $(Q^t S^t - S_i^t - S_j^t - S_k^t)/(3S_i^t)$ is less than 0.05, or even negative. In these cases, called minus pools, the track usually agrees to pay \$0.05 profit for each dollar wagered. Minus pools can also occur in the win and place markets, but are much less common. Equations (1)–(3) ignore breakage, the additional charge that results from the track rounding all payoffs down to the nearest 5 or 10 cents on the dollar.[4]

If prices reflect all available information then a market is said to be efficient (see Fama 1970). There are two conclusions that can be drawn from the many studies of win market efficiency (see, for instance, Ali [1977]; and Snyder [1978]). First, the North American public underbets favorites and overbets longshots, and this bias appears across the many years that data have been collected and across all sizes of race-track betting pools.[5] Second, despite its strength and stability, this bias is almost always less than the track take and thus it cannot be exploited to achieve positive profits. Figure 1 illustrates the favorite/longshot

4. Breakage may seem like a relatively minor cost, but we (Ziemba and Hausch 1987) demonstrate that it can have a dramatic long-run effect on a bettor's fortune.
5. Busche and Hall (1988) demonstrate an opposite bias for Hong Kong bettors. Contrary to the North American bettors, Hong Kong bettors tend to overbet favorites and underbet long shots—a bias that is consistent with risk aversion.

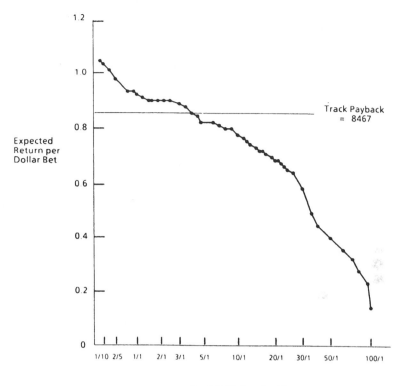

Fig. 1.—Expected return per dollar bet versus odds level: aggregation of studies involving more than 50,000 races. (Source.—Ziemba and Hausch 1986.)

bias and shows that the track take (assumed to be 15.33%, as it is in California) is sufficiently large to preclude profits. The one profitable exception is extreme favorites at odds of 3–10 or less. However, they are relatively rare. Thus, if we define a market to be weakly efficient (see Fama 1970), if no one can devise a profitable trading rule based on historical price information, the win market is, for practical purposes, weakly efficient.

Hausch, Ziemba, and Rubinstein (1981) tested the efficiency of the place-and-show markets. They assumed:

ASSUMPTION 1. If q_i is the probability that horse i wins, then the probability that i is first and j is second is $q_i q_j / (1 - q_i)$, and the probability that i is first, j is second, and k is third is

$$q_i q_j q_k / [(1 - q_i)][(1 - q_i - q_j)].$$

(These formulas were developed and tested by Harville [1973].)[6]

---

6. Henery (1981) and Stern (1987) show that these equations can be derived by associating with each horse an independent exponential random variable with a scale

ASSUMPTION 2.   The win market is efficient, so the win odds can be used to estimate q_i.

Using these two assumptions and equations (2) and (3), Hausch, Ziemba, and Rubinstein (1981) were able to identify horses that were underbet to place or show. The optimal capital growth model then determined the place-and-show wagers that maximized the expected rate of growth of one's bankroll. Since the exact model is a complicated nonlinear optimization problem that is difficult to solve at the track, Hausch, Ziemba, and Rubinstein (1981) developed simple regression approximations with quite minimal data-entry requirements. Their empirical studies on two seasons of racing data indicated that significant returns on the order of 11% were possible in the place-and-show markets. We (Hausch and Ziemba 1985; Ziemba and Hausch 1987) extended Hausch, Ziemba, and Rubinstein's (1981) results to provide further evidence of the place-and-show inefficiency. We (Ziemba and Hausch 1986) and Asch and Quandt (1987) studied inefficiency of the exotic markets. Asch, Malkiel, and Quandt (1984, 1986) investigated whether a drop in the odds late in the betting period might reflect inside information and thereby point to wagers that may have positive expected returns. Their results suggest that this is not the case, however. A more thorough literature survey is in Thaler and Ziemba (1988) and Hausch and Ziemba (1990). The latter also studies racing outside of North America.

III.   Inefficiency of the Win Market and the Risk-free Hedging Model

The literature has demonstrated weak efficiency of the win market at a single track, despite a favorite/longshot bias. To test whether this weak efficiency is maintained across the win markets with cross-track betting, data were collected on several recent Triple Crown races. Although cross-track betting is becoming more popular, it tends to be restricted to major races. The best known of these are the Triple Crown races: the Kentucky Derby at Churchill Downs on the first Saturday in May, the Preakness Stakes at Pimlico Race Course 2 weeks later, and the Belmont Stakes at Belmont Park 3 weeks after that.

A simple problem is considered first. Allowing only win betting, what is the minimum amount that our bettor must wager to ensure the return of $1.00 regardless of which horse wins the race? The solution to this problem is, for each horse, to identify the track that has it at the longest odds and bet just enough to receive $1.00 if it wins. The solution involves no estimation of the horses' win probabilities. If this

---

parameter equal to the inverse of its win probability. Then, any ordering of the random variables is just the Harville formulas. Stern also develops alternative ordering formulas using gamma distributions that are more accurate but more complicated.

TABLE 2   **Projected Win Payoffs, 1983 Preakness**

Horse No.	Highest Win Return (on a $1 Bet)	Track	$ Amount of Wager That Will Return $1
1	29.40	Louisiana Downs	.0340
2	12.70	Louisiana Downs	.0787
3	34.60	Los Alamitos	.0289
4	169.90	Hollywood	.0059
5	56.90	Louisiana Downs	.0176
6	5.70	Louisiana Downs	.1754
7	10.60	Pimlico	.0943
8	76.60	Louisiana Downs	.0131
9	116.10	Hollywood	.0086
10	2.20	Los Alamitos	.4545
11	40.60	Los Alamitos	.0246
Total			.9356

solution were employed at one track, then our bettor would have to pay $1/Q$ dollars (an amount greater than $1.00) to ensure a return of $1.00. With the opportunity of betting at several tracks, each with a different set of odds, we may be able to lower this minimum amount to below $1.00. To see how this system works, consider the 1983 Preakness. The final win odds were collected from 11 tracks that allowed wagering on this race.[7] The highest of the 11 win payoffs on horse number 1 was $29.40 per dollar wagered at Louisiana Downs. Thus, a win bet of $0.0340 there would have returned $1.00 when, in fact, horse number 1, Deputed Testamony, won the race. This wager and those on the other horses are presented in table 2.

Thus, by wagering $0.9356, our bettor is guaranteed $1.00 regardless of who wins the race. This is a certain profit of $0.0644 per $0.9356 wagered, or a guaranteed return of 6.9% rate of return in a 2-minute race.

Obviously, "risk-free" arbitrage is not possible. Implementing the system requires that the track odds be sent to a central decision maker several minutes before the end of betting, and during that time the odds can change. The system does, however, demonstrate the large discrepancies in betting across the tracks and shows how simple it can be to take advantage of them. Table 3 presents the results of applying the system to the data from other Triple Crown races.

Three of the races had insufficient variance in the win odds across

7. The number of outlets involved in cross-track betting has increased over the years. The 1985 Kentucky Derby had betting at 32 outlets, including New York City's off-track betting (OTB). The 1985 Preakness and Belmont had 28 and 41 outlets, respectively. For 1988, the number of outlets for these three races were, respectively, 93, 87, and 76. We requested the final win, place, and show data from the outlets that we knew had accepted wagers. Response rates tended to be low and, in some cases, the data were no longer available, having been stored for only a short period. All the data received were used in the analysis.

TABLE 3   Risk-free Hedging Model Results

Race	No. Horses in Race	No. Tracks	% Profit
Preakness:			
1982	7	5	.0
1983	11	11	6.9
1984	10	4	.0
1985	11	7	2.5
Belmont:			
1982	11	5	13.6
1983	12	9	8.5
1984	11	2	.0
1985	9	11	5.0
Kentucky Derby:			
1984*	12	7	.1
1985	12	6	10.1
Average			4.7

* Cross-track betting on the Kentucky Derby did not begin until 1984.

the tracks to allow a risk-free profit. This is not surprising for the 1984 Belmont because we had data from only two tracks. Also, the results for the 1982 and 1984 Preakness are based on only five and four tracks, respectively. In these three races our bettor would obviously make no wagers. These three races giving a 0% return together with the other seven races have an average risk-free profit of 4.7%. This rate of profit is for the certain return of $1.00. As the required certain return increases, then the wagers begin to affect the odds and this rate of profit will decrease.[8]

8. This "certain return" scheme can be extended to include place-and-show betting. Let R_i^t be the return on a $1.00 win bet on horse i at track t when i wins. Similarly, let R_{ij}^t be the return on a $1.00 place bet on i at track t when i and j are the top two finishers, and let R_{ijk}^t be the return on a $1.00 show bet on i at track t when i, j, and k are the top three finishers. Formulas (1)–(3), corrected for breakage, determine R_i^t, R_{ij}^t and R_{ijk}^t. The decision variables are amounts to wager to win, place, and show on horse i at track t and are represented, respectively, by w_i^t, p_i^t, and s_i^t. There is a constraint for each ijk finish that requires a return of $1.00, should that finish occur. The following formulation ignores our effect on the odds and, while this effect is negligible for a return of $1.00, it should be included for larger required returns. The formulation that determines the minimum expenditure for a certain return of at least $1.00 is

$$\text{minimize} \sum_{t=1}^{T} \sum_{i=1}^{n} (w_i^t + p_i^t + s_i^t),$$

subject to

$$\sum_{t=1}^{T} (R_i^t w_i^t + R_{ij}^t p_i^t + R_{ji}^t p_j^t + R_{ijk}^t s_i^t + R_{jik}^t s_j^t + R_{kij}^t s_k^t) \geq 1 \quad \text{for each } i, j, k,$$

$$w_i^t, p_i^t, s_i^t \geq 0 \quad \text{for all } i = 1, \ldots, n \text{ and } t = 1, \ldots, T.$$

This linear program has a large number of constraints; with n horses there are $n(n-1)(n-2)$ possible ijk finishes and, consequently, $n(n-1)(n-2)$ constraints. (The dual program has $3Tn$ constraints, however.) Interestingly, even with the possibility of place-and-show betting, betting only to win, as in Sec. III, may be optimal. The 1982

IV. The Optimal Capital Growth Model

Section III demonstrated an inefficiency in cross-track betting. We now propose and test an "optimal" wagering strategy for this inefficiency. Bettors at the racetrack appear to have many different objectives and, therefore, employ many different wagering strategies. Two quite reasonable long-term objectives, however, are (1) to maximize the expected rate of growth of one's bankroll, and (2) to minimize the expected time to reach some specified large wealth level. Breiman (1961) proved that both of these objectives are asymptotically satisfied by maximizing, on a myopic period-by-period basis, the expected logarithm of one's final wealth. This approach is termed the optimal capital growth model, and its theoretical justification for the logarithmic utility function has been well studied (see Ziemba and Vickson [1975] for references and a discussion of its assumptions and results). A simulation by us (Ziemba and Hausch 1986) suggests that this strategy not only performs well asymptotically but over a year of wagering it can also be expected to outperform other commonly used betting strategies. Other attractive features of the capital growth model are that one's effect on the odds can be accounted for and bet size is monotone in wealth. On the negative side, though, the recommended bets can be very large. Indeed, the Arrow-Pratt absolute risk-aversion index is wealth^{-1}, which is close to zero for large wealth.

Let w_0 be our bettor's initial wealth and q_{ijk} the probability that i is first, j is second, and k is third. Also, let $P_{ij}^t = P_i^t + P_j^t$ and $S_{ijk}^t = S_i^t + S_j^t + S_k^t$. The optimal capital growth model is[9]

$$
\max_{\{w_\ell^t, p_\ell^t, s_\ell^t\}} \sum_{\substack{i=1}}^{n} \sum_{\substack{j=1 \\ j \neq i}}^{n} \sum_{\substack{k=1 \\ k \neq i, j}}^{n} q_{ijk} \log \left[w_0 \right.
$$

$$
+ \sum_{t=1}^{T} \left\{ \left[Q^t \left(W^t + \sum_{\ell=1}^{n} w_\ell^t \right) - (W_i^t + w_i^t) \right] \left(\frac{w_i^t}{W_i^t + w_i^t} \right) \right.
$$

$$
\left. + \frac{\left[Q^t \left(P^t + \sum_{\ell=1}^{n} p_\ell^t \right) - (P_{ij}^t + p_i^t + p_j^t) \right]}{2}
$$

Preakness (with $7 \times 6 \times 5 = 210$ constraints) is one example of this. The reason for no place-or-show betting at optimum seems to be a coordination problem. Win bets return one and only one positive amount and these payoffs are mutually exclusive. Thus, a return of $1.00 can be efficiently guaranteed with win bets. Place-and-show bets, however, tend to have many different possible payoffs, and collecting on them is not mutually exclusive. These differences make it difficult to efficiently choose place-and-show bets over the win bets.

9. It is very simple to include exotic wagering in this capital growth model. It was formulated to consider only win, place, and show betting, though, because that was the only available data for testing the model.

$$\times \left(\frac{p_i^t}{P_i^t + p_i^t} + \frac{p_j^t}{P_j^t + p_j^t} \right)$$

$$+ \frac{\left[Q^t \left(S^t + \sum_{\ell=1}^{n} s_\ell^t \right) - \left(S_{ijk}^t + s_i^t + s_j^t + s_k^t \right) \right]}{3}$$

$$\times \left(\frac{s_i^t}{S_i^t + s_i^t} + \frac{s_j^t}{S_j^t + s_j^t} + \frac{s_k^t}{S_k^t + s_k^t} \right)$$

$$- \left(\sum_{\substack{\ell=1 \\ \ell \neq i}}^{n} w_\ell^t + \sum_{\substack{\ell=1 \\ \ell \neq i,j}}^{n} p_\ell^t + \sum_{\substack{\ell=1 \\ \ell \neq i,j,k}}^{n} s_\ell^t \right) \Bigg] \Bigg\} ,$$

subject to

$$\sum_{t=1}^{T} \sum_{\ell=1}^{n} (w_\ell^t + p_\ell^t + s_\ell^t) \leq w_0,$$

$$w_\ell^t \geq 0, \ p_\ell^t \geq 0, \ s_\ell^t \geq 0, \quad t = 1, \ldots, T, \ell = 1, \ldots, n.$$

This model requires estimates of the win probabilities for each horse. The efficiency studies of the win market have indicated that the public's win odds, adjusted for the favorite/longshot bias, provide good estimates of these probabilities. However, with cross-track wagering we not only have a different set of win odds for each participating track but, as demonstrated by the Ferdinand example in Section I and the risk-free hedging model in Section III, these odds can vary considerably. Rather than take a weighted average of all the tracks' odds to arrive at a set of win probabilities, we decided to use only the home track's odds.[10] This decision was based on a perceived informational advantage that the home-track public has over the bettors at the cross tracks, an advantage that results from several factors: (1) since these races were run near the end of the day's racing, the home-track public had watched the jockeys perform in, perhaps, several races already, they had observed the condition of the track and possibly noted any track biases, and they saw the horses in the paddock and in the parade to post; (2) the home crowd knows better if their track tends to favor front-runners or late chargers; and (3) since the home track is usually a larger track that has many major races, its public is more likely to have seen some of these horses race earlier in the season. Further, the studies supporting win-market efficiency have all involved home-track odds. Thus, we assume win-market efficiency at the home track and,

10. This decision was not the result of any analysis of the data. There are only a few cross-track races each year and the required data for each race are the final tote-board figures from several of the racetracks permitting this betting. These difficulties led to data on only 10 races being collected, much less than would be required for any analysis.

TABLE 4   **1982 Preakness**

Horse	Finish	Pimlico's Win Odds	Win Probability
1. Reinvested	. . .	7.6–1	.090
2. Cut Away	3d	41.6–1	.011
3. Water Bank	. . .	12.0–1	.056
4. Bold Style	. . .	26.2–1	.023
5. Laser Light	. . .	5.3–1	.125
6. Linkage	2d	.5–1	.597
7. Aloma's Ruler	1st	6.9–1	.098

after adjusting for the favorite/longshot bias in figure 1, use the home-win odds to estimate win probabilities. Since this win-market efficiency is not necessarily maintained at the cross tracks, the model can then include possible win betting at the cross tracks. The 1982 Preakness Stakes will be used to illustrate this model. Table 4 lists the race's entrants and their odds.

Aloma's Ruler's 6.9–1 win odds meant that, when he won, each dollar wagered on him was returned with an additional $6.90 profit. Linkage was the crowd's favorite. His 0.5–1 win odds meant that, had he won, he would have returned $1.50 per dollar wagered on him. After adjusting the win probabilities for the favorite/longshot bias, the probabilities were then normalized to sum to one. These probabilities and equation (1) adjusted for breakage allow the calculation of the expected return on an additional dollar bet to win for each horse i at each track t. Similarly, the Harville formulas in assumption 1 and equations (2) and (3), also adjusted for breakage, allow the calculation of the expected returns on place-and-show bets (see Hausch, Ziemba, and Rubinstein 1981). We received final win, place, and show figures from four of the tracks that allowed betting on the 1982 Preakness: Golden Gate and Los Alamitos in California, Centennial in Colorado, and Penn National in Pennsylvania. The calculated expected returns on the various bets at the four tracks are presented in table 5.

If a track has a payback proportion of, say, 0.85 then the average return on $1.00 bets, ignoring breakage, will be $0.85. With breakage, it will be somewhat less than $0.85. For individual horses the expected returns from table 5 vary from 0.181 (show bet on horse 2 at Golden Gate) to 1.336 (place bet on horse 6 at Los Alamitos). Of particular interest are the expected returns exceeding 1.00 since those indicate the wagers that have a positive expected profit. We restricted our attention to wagers with expected returns of at least 1.10. They are the highlighted expected returns in table 5. If the probability estimates of this model were exact, then this 1.10 cutoff could result in suboptimal wagering, particularly since diversification can even lead to the inclusion of wagers with negative expected profits. Hausch, Ziemba, and

TABLE 5   Expected Returns, Cross-Track Betting on 1982 Preakness

				Horse			
	1	2	3	4	5	6	7
Finish	...	3	...	...	...	2	1
Win probability	.090	.011	.056	.023	.125	.597	.098
Expected return on a $1 bet to win:							
Golden Gate	.837	.246	.437	.708	.788	.955	1.000
Centennial	.900	.389	.370	.570	.825	.836	1.147*
Los Alamitos	.666	.391	.347	.777	.713	1.015	1.196*
Penn National	.954	.484	.588	1.109*	.650	.896	.755
Expected return on a $1 bet to place:							
Golden Gate	.749	.193	.349	.582	.669	1.149*	.880
Centennial	.769	.238	.260	.329	.719	1.084	1.120*
Los Alamitos	.794	.233	.277	.556	.778	1.336*	.888
Penn National	.673	.399	.391	.737	.586	1.101*	.731
Expected return on a $1 bet to show:							
Golden Gate	.837	.181	.405	.413	.817	1.153*	.996
Centennial	.747	.197	.340	.252	.803	1.008	1.138*
Los Alamitos	.890	.200	.392	.341	1.180*	1.293*	.873
Penn National	.710	.235	.451	.388	.769	1.099	.793

NOTE.—In this table, horses 2, 6, and 7 are Cut Away, Linkage, and Alomá's Ruler, respectively.
*Highlighted expected returns (returns of at least 1.10).

Rubinstein (1981) found, however, that due to the approximations in the model it was more profitable to wager only if the expected return was well in excess of 1.00. For large tracks and races of the quality of these Triple Crown races, a minimum expected return of 1.10 is reasonable.

The capital growth model was run on the data from these cross tracks. Table 6 presents the optimal portfolio of wagers assuming a 1.10 expected return cutoff and an initial wealth of $2,500. This portfolio has wagers totaling $2,437 at all four tracks and its certainty equivalent can be calculated as $432. Since the first three finishers of this race were horses 7, 6, and 2, our bettor would collect on the win, place, and show bets on horse 7 and the place and show bets on horse 6, for a return of $3,716.90 and a profit of $1,279.90. The payoffs in this table account for our bettor's effect on the odds. For example, the actual payoff per $1.00 bet to show on horse 7 at Centennial was $2.70. However, our bettor's $83 show bet at this track would have lowered this payoff to $2.50. The latter payoff was used to determine our bettor's return.

Table 5 supposes only $1.00 is wagered. Thus, even though one track may have a higher expected return to, say, show on some horse than another track has, it can be that our bettor will make show wagers at

TABLE 6   Optimal Capital Growth Wagers, Cross-Track Betting on 1982 Preakness

Horse	Bet	Track	Expected Return	$ Bet	$ Payoff per $1 Bet	$ Total Return	$ Profit
4	Win	P.N.	1.109	14	. . .	. . .	−14.00
7	Win	L.A.	1.196	40	12.00	480.00	440.00
6	Place	L.A.	1.336	855	1.50	1282.50	427.50
7	Place	Cen.	1.120	46	3.30	151.80	105.80
5	Show	L.A.	1.180	172	. . .	. . .	−172.00
6	Show	G.G.	1.153	571	1.30	742.30	171.30
6	Show	L.A.	1.293	656	1.30	852.80	196.80
7	Show	Cen.	1.138	83	2.50	207.50	124.50
Totals				2,437		3,716.90	1,279.90

NOTE.—P.N. = Penn National; L.A. = Los Alamitos; Cen. = Centennial; G.G. = Golden Gate.

both tracks because of the bettor's effect on the odds. This happened at Golden Gate and Los Alamitos. Horse 6 had expected returns of 1.153 and 1.293 at these tracks and the optimal show bets were $571 and $656 at them, respectively. With these wagers, it must be that the addition to expected utility from an additional dollar bet to show will be the same at the two tracks.[11]

The capital growth model was studied with cross-track data on the other Triple Crown races. In each case it was assumed that the initial wealth was $2,500. The results of these races and the 1982 Preakness are given in table 7.

Despite losses on two of the races, including a huge loss on the 1984 Preakness, there was a total profit of $2,647.80 for a 15% return on money wagered. Average profits were $264.78. The standard error of the mean is $342, so no statistically significant statements can be made about positive profits on the basis of these results. Studying the expected returns suggests that many of the profitable overlays occur because of regional biases. For instance, Conquistador Cielo, the winner of the 1982 Belmont Stakes, raced his entire career on the East Coast. The West Coast bettors were less familiar with him, while on the East Coast he was, for many, a sentimental favorite. In the Belmont Stakes he tended to be sent off at lower odds at the East Coast tracks. When he won he paid from $5.80 to win at Commodore Downs in Pennsylvania to $15.40 at Los Alamitos in California. Another example is Tolomeo, the English colt that won the 1983 Arlington Million. His odds were 4–1 in England and 38–1 at Arlington Park in

---

11. This marginal utility can be complicated because a component of the expected return on an additional dollar bet to show on horse 6 at Los Alamitos is the positive effect that it has on the return to the show bet on horse 5 at Los Alamitos if 6 finishes out of the money.

TABLE 7   Results of Optimal Capital Growth Model on Triple Crown Races

Race	No. Horses in Race	No. Tracks	No. Wagers	Total Wagers ($)	Certainty Equivalent ($)	No. Wagers Won	Return ($)	Profit ($)
Preakness:								
1982	7	4	8	2,437	432	6	3,716.90	1,279.90
1983	11	8	13	1,949	325	1	1,647.30	−301.70
1984	10	3	8	2,282	371	0	0.00	−2,282.00
1985	11	6	11	1,817	325	4	2,014.40	197.40
Belmont:								
1982	11	4	20	942	305	5	1,880.40	938.40
1983	12	7	20	2,452	469	6	2,578.80	126.80
1984	11	2	3	1,371	317	1	2,331.00	960.00
1985	9	9	15	471	73	1	78.30	−392.70
Kentucky Derby:								
1984	12	6	13	2,027	340	5	3,027.20	1,000.20
1985	12	6	7	1,973	350	3	3,094.50	1,121.50
Totals			118	17,721	3,307	32	20,368.80	2,647.80

Illinois. This observation and table 7 suggest that a high expected return is possible even with only a few carefully selected cross tracks.

The Triple Crown races are the most widely publicized of the North American horse races. Thus, one might expect that other less publicized cross-track races would have even more divergence in the odds across the tracks. If so, they should have an even greater potential for expected profits.

A major assumption of this model is that after all other bettors have made their wagers our bettor (1) learns the tote-board information at each track, (2) runs the capital growth model, and (3) communicates the optimal wagers to agents at each track, who then make the wagers. Obviously this is extremely difficult, if not impossible, in practice, never mind any legal concerns. Just learning the tote-board information at each track is difficult because there are no pay phones inside the racetrack grounds. A central decision maker and the agents communicating with cellular phones is feasible but still requires a significant amount of time. Unfortunately, though, odds may change in the last few minutes of betting, and profitable bets a few minutes before the end of betting may not be profitable based on the final odds. Hausch, Ziemba, and Rubinstein (1981) studied the odds changes in the last 2 minutes of betting and found that expected returns did change somewhat but profitable place-and-show bets 2 minutes from the end tended to remain profitable based on final odds. However, the agents in this model, because of its extra complications, would probably have to report odds more premature than those 2 minutes before the end of betting. The results given in this article, then, may overestimate the profits possible in practice. Additionally, our profit figures do not account for the costs of implementing this procedure—the agents and other costs at each cross track, the long-distance phone calls, and the computer time.

V.   Testing the One-Track Capital Growth Model

Implementing the previous section's capital growth model is certainly difficult. However, when the race is televised from the home track there are simpler versions of this scheme possible for one bettor at one track. Our bettor, with a portable television at the cross track, can view the home odds when they are shown on TV. With these odds giving "true" win probabilities, our bettor can search for overlays at the cross track. A flat bet to win could be made on horses going off at longer odds than at the home track, or a sophisticated bettor could also bring a portable computer to the track and run the capital growth model described in Section IV for one track, that is, $T = 1$. This latter scheme is tested here with the Triple Crown race data.

As an example, consider the 1984 Kentucky Derby simulcast at

TABLE 8   **Optimal Wagers Based on One-Track Capital Growth Model, 1984 Kentucky Derby**

Wager Type	Horse	Expected Return	$ Optimal Bet	$ Payoff on a $1 Bet	$ Realized Return	$ Profit
Win	2	1.236	15	. . .	. . .	− 15.00
Win	10	1.159	. . .	5.60	. . .	. . .
Place	2	1.153	. . .	. . .	. . .	. . .
Place	10	1.410	249	3.70	921.30	672.30
Show	2	1.353	261	. . .	. . .	− 261.00
Show	10	1.275	295	2.50	737.50	442.50
Totals			820		1,658.80	838.80

Golden Gate Fields in Albany, California. Using the win odds televised from Churchill Downs, Golden Gate had six wagers with expected returns exceeding 1.10. With a wealth of $2,500, the capital growth model yields the optimal wagers shown in table 8.

This portfolio has four wagers totaling $820. The win bet on horse number 10 and the place bet on horse number 2 are zero even though they have expected returns exceeding 1.10. This is because the possibility of number 2 doing well in the race is better accounted for with the higher returning win-and-show bets on him. Also, the possibility of number 10 doing well is better accounted for with the higher returning place-and-show bets on him. Swale, number 10, won the 1984 Derby, followed by Coax Me Chad and At The Threshold for a 10-12-9 finish. Therefore, only the place-and-show bets on Swale paid off for a return of $1,658.80 and a profit of $838.80.

Table 9 presents the results of this one-track model on the other races. The average wagers on a race varied from $78.33 to $1,173.25 and the average profits varied from − $868.67 to $824.10. The average of these 10 average profits was $69.97 or 9.2% on the money wagered. Again, there is such variability in the profits that, without additional data, no statistically significant statements can be made about positive expected profits.

VI. Final Discussion

There is considerable evidence that the win market at the racetrack is weakly efficient. A risk-free arbitrage model is presented that demonstrates that this is not the case with cross-track wagering. The mispricing by the public at the cross tracks may be due partly to their more limited access to information relative to those attending the home track. Also, some of the variance in the odds across the tracks may be due to certain horses being more familiar to bettors in certain regions of North America. The optimal capital growth model suggests that the

TABLE 9   Results of One-Track Capital Growth Model on Triple Crown Races

Race	No. of Cross Tracks	Average Cross-Track Wager ($)	Average Cross-Track Realized Return ($)	Average Cross-Track Profit ($)
Preakness:				
1982	4	1,173.25	1,724.80	551.55
1983	8	909.25	353.79	−555.46
1984	3	868.67	.00	−868.67
1985	6	799.67	789.37	−10.30
Belmont:				
1982	4	407.25	639.02	231.78
1983	7	532.71	488.63	−44.08
1984	2	1,158.00	1,982.10	824.10
1985	9	78.33	7.80	−70.53
Kentucky Derby:				
1984	6	520.67	884.92	364.25
1985	6	1,161.33	1,438.42	277.08
Average	5.5	760.91	830.88	69.97

discrepancies across the tracks can allow profits, but further work is needed to demonstrate significant profits. A simpler version of the optimal capital growth model for one bettor at one cross track also demonstrates the possibility of profit.

References

Ali, M. M. 1977. Probability and utility estimates for racetrack bettors. *Journal of Political Economy* 85 (August): 803–15.

Ali, M. M. 1979. Some evidence of the efficiency of a speculative model. *Econometrica* 47 (March):387–92.

Asch, P.; Malkiel, B. G.; and Quandt, R. E. 1984. Market efficiency in racetrack betting. *Journal of Business* 57 (April): 165–75.

Asch, P.; Malkiel, B. G.; and Quandt, R. E. 1986. Market efficiency in racetrack betting: Further evidence and a correction. *Journal of Business* 59 (January): 157–60.

Asch, P., and Quandt, R. E. 1987. Efficiency and profitability in exotic bets. *Economica* 59 (August): 289–98.

Breiman, L. 1961. Optimal gambling systems for favorable games. In *Proceedings of the Fourth Berkeley Symposium,* pp. 65–68. Berkeley: University of California Press.

Busche, K., and Hall, C. D. 1988. An exception to the risk preference anomaly. *Journal of Business* 61 (July): 337–46.

Fama, E. F. 1970. Efficient capital markets: A review of theory and empirical work. *Journal of Finance* 25 (May): 383–417.

Harville, D. A. 1973. Assigning probabilities to the outcomes of multi-entry competitions. *Journal of the American Statistical Association* 68 (June): 312–16.

Hausch, D. B., and Ziemba, W. T. 1985. Transactions costs, extent of inefficiencies, entries, and multiple wagers in a racetrack betting model. *Management Science* 31 (April): 381–94.

Hausch, D. B., and Ziemba, W. T. (eds.). 1990, in press. *Efficiency of Racetrack Betting Markets.* New York: Academic Press.

Hausch, D. B.; Ziemba, W. T.; and Rubinstein, M. 1981. Efficiency of the market for racetrack betting. *Management Science* 27 (December): 1435–52.

Henery, R. J. 1981. Permutation probabilities as models for horse races. *Journal of the Royal Statistical Society,* ser. B, 43:86–91.

Rosett, R. H. 1965. Gambling and rationality. *Journal of Political Economy* 73 (December): 595–607.

Snyder, W. W. 1978. Horse racing: Testing the efficient markets model. *Journal of Finance* 33 (September): 1109–18.

Stern, H. S. 1987. Gamma processes, paired comparisons and ranking. Ph.D. dissertation. Stanford, Calif.: Stanford University, Department of Statistics.

Thaler, R., and Ziemba, W. T. 1988. Parimutuel betting markets: Racetracks and lotteries. *Journal of Economic Perspectives* 2 (Spring): 161–74.

Ziemba, W. T., and Hausch, D. B. 1986. *Betting at the Racetrack.* New York: Norris M. Strauss.

Ziemba, W. T., and Hausch, D. B. 1987. *Dr. Z's Beat the Racetrack.* New York: Morrow.

Ziemba, W. T., and Vickson, R. G. (eds.). 1975. *Stochastic Optimization Models in Finance.* New York: Academic Press.

Locks at the Racetrack

Donald B. Hausch

School of Business
University of Wisconsin
Madison, Wisconsin 53706

William T. Ziemba

Faculty of Commerce and Business Administration
University of British Columbia
Vancouver, BC
Canada V6T 1Y8

The folklore of investment is replete with stories of arbitrage opportunities where profits can be made without risk. Such a "lock" exists at the racetrack. A simple model provides a criterion for existence of a set of bets to create the arbitrage plus the size of the various investments.

Racetracks return only about 82 percent of the money wagered to the winning bettors. Thus, the average bettor loses about 18 percent of his or her wagers. There are two parts to this loss: (1) the track take is the predetermined percentage of the betting pool that is kept by the track; and (2) breakage is the process of rounding down the payoffs on winning bets to common payoff amounts. Handicappers have proposed many systems for beating the track but an 18 percent disadvantage is a formidable hurdle to overcome. Hausch, Ziemba, and Rubinstein [1981] and Hausch and Ziemba [1985] devised a method that allows the average player to win by utilizing the betting bias

of the players along with an investment-decision model to determine when and how much to wager. The method works well, but it is far from risk free. This is also true of alternative approaches that are discussed by Asch, Malkiel and Quandt [1984, 1986], Thaler and Ziemba [1988] and Hausch and Ziemba [forthcoming a]. However, there are situations at the track where it is possible to construct a risk-free hedge, or in the racing vernacular, a *lock*. The only publications we know of that concern locks are Willis [1964], Leong and Lim [1989], and Hausch and Ziemba [forthcoming b]. The latter two deal with cross-track betting where a number of tracks each offer odds, possibly very different, on the same race. While it may be possible to make risk-free profits with cross-track betting, it involves considerable overhead at the various locations, and the bettors may encounter legal problems with communicating the information. Willis [1964] discusses a much cleaner situation with all activity concentrated on one race at one track. When the public's wagers in the win market (one collects if one's horse is first) and the place market (one collects if one's horse is first or second) are very different, Willis shows how arbitrage between the two markets may be possible. However, it would be extremely rare for the odds in the two markets to vary sufficiently. In fact, Willis supplies no actual examples. Our lock concentrates on another market, the show market.

One collects on a show bet if the horse finishes first, second, or third. Let S_i represent the public's show bet on horse i and let Q be the track's payback proportion (about 0.80 to 0.86). Finally, let n be the number of horses and let $S = \sum_{i=1}^{n} S_i$ be the show pool. Then if horses i, j, and k finish 1-2-3 in any order, the payoff per dollar wagered on i is

$$1 + \frac{QS - S_i - S_j - S_k}{3S_i}. \tag{1}$$

The ticket holders of the first three horses are repaid their original bets plus a share of the profits. In practice, breakage rounds this payoff down to the nearest 5 cents or, more commonly, 10 cents. With $INT[Y]$ giving the largest integer not exceeding Y, the formula for the payoff including the effect of breakage is

$$1 + \frac{1}{N} INT \left[\frac{N[QS - S_i - S_j - S_k]}{3S_i} \right] \tag{2}$$

where $N = 10$ or 20 for 10 cents or 5 cents breakage, respectively. If this payoff is less than $1.05 for each dollar wagered, then most tracks guarantee a return of $1.05, that is a five percent profit.

Situations where the track must honor this guarantee of a five percent minimum payoff are called minus pools. They result in the track collecting less than $(1 - Q)S$, and hence, tracks try to avoid them whenever possible. However, they are not

Our lock concentrates on the show market.

uncommon. We focus here on a particular type of minus pool, one where a heavy favorite has about 95 percent of the show pool bet on it, that is where $S_i \geq 0.95S$. The crowd figures that this horse, or group of horses if it is a betting entry, is so good that it almost surely will finish at least third. In such a situation, one can construct a lock in the show pool.

Locks were described in articles in *Sports Illustrated* [Gelband 1979] and *Fortune* [Seligman 1979] using the example in Table 1, the Alabama Stakes at Saratoga on August 11, 1979.

Davona Dale's 95.5 percent share of the show pool is unusually high and has created a minus pool. If Davona Dale finishes in the money, that is, at least third, then the show payoffs on the first three finishers will be the minimum $1.05 per dollar bet. Here is how a lock can be devised: If Davona Dale is in the money, then we will receive five percent on the money we wagered on her plus five percent on the money we wagered on the

Horse	Win Odds to $1	Win Probability	Show Bet	% of Show Pool
Davona Dale	0.30	0.661	$435,825	95.5
It's in the Air	3.10	0.210	7,901	1.7
Mairzy Doates	13.40	0.050	4,518	1.0
Poppycock	17.50	0.037	4,417	1.0
Croquis	15.40	0.042	3,873	0.8
		1.000	$456,534	100.0

Table 1: The public's win odds and show bets are given for the 1979 Alabama Stakes at Saratoga. The conversion of win odds to win probabilities accounts for the public's biases (see Ziemba and Hausch [1986] and Hausch and Ziemba [forthcoming b]).

other two horses that finish in the money. If this adds up to more than the wagers we lost on the fourth and fifth place horses, then we are ahead. If, as well, the amounts we wagered on the four long-shots are such that, if Davona Dale finishes out of the money, our return covers both the bet on her and on the other out-of-the-money horse, then we have a profit regardless of the outcome of the race. A lock is clearly a very conservative betting strategy; it is consistent with a utility function that has infinite disutility for any losses. Later we will discuss the logarithmic utility function for comparison, but for now we will describe the conditions that are necessary for a lock and recommend wagers on the horses to develop a lock.

Suppose we wish to receive approximately the same profit regardless of the finish. Initially, let us assume that (1) except for the favorite, the public wagers the same amount on each horse to show, and (2) our bets do not affect the odds. With k as the fraction of the show pool on the favorite, the show bet on the favorite is kS and the equal show bets on the other horses are $(1-k)S/(n-1)$. Let x be our wager on the favorite and y be our

show bet on each of the other horses, for a total wager of $x + (n-1)y$. If the favorite is in the money then we collect five percent on x and two of the y bets, but lose $(n-3)y$ on the $n-3$ losers, for a profit of $.05(x + 2y) - (n-3)y$. If the favorite is out of the money, then, since it has been supposed that our wagers do not affect the odds, our profit is

$$3 \left(\frac{QS - \dfrac{3(1-k)S}{n-1}}{3} \right) \frac{y}{(1-k)S/(n-1)}$$
$$- x - (n-4)y.$$

The first term is the profit on the three horses that finished in the money, and it follows from expression (1). The next two terms are the losses on the favorite and on the $n-4$ other horses.

To guarantee a particular return regardless of the finish, these two profit functions must be equal. This means x and y must satisfy the following ratio:
$$x/y = -2 + Q(n-1)/[1.05(1-k)]. \qquad (3)$$
This x/y ratio yields profit of $.05yQ(n-1)/[1.05(1-k)] - (n-3)y$, and this must be positive for a lock to exist. This holds when
$$k > 1 - Q(n-1)/[21(n-3)]. \qquad (4)$$
Our example does not satisfy the first

assumption: the bets on the long shots are not equal; they range from \$3,873 to \$7,901. The second assumption will hold only if our wagers are relatively small. Despite this, $Q = .85$ and $n = 5$ for our example, so condition (4) indicates that a lock will exist if k exceeds 0.920. Davona Dale's $k = 0.955$ so we can devise a lock. Equation 3 gives $x/y = 69.96$, so with \$2,500 to bet we would wager $x = \$2,364$ on the favorite and $y = \$34$ on each of the other four horses.

If Davona Dale finished in the money, then our profit would be \$53.60. If the public's show bets on the other four horses had been the same, then the profit would also have been \$53.60 even if Davona Dale finished fourth or fifth: a guaranteed 2.1 percent return. The fact that the show bets on the four other horses are different will not affect our profit of \$53.60 if Davona Dale finished at least third. If Davona Dale finished fourth or fifth, however, our profit would depend on which of the four horses were the top three finishers. Table 2 lists the profit for the four possible permutations of these four horses being the top three finishers. If the four profits are weighted by their likelihoods, the average profit is about \$53.60.

In this analysis, we supposed that our bets do not effect the payoffs. For this example and a total wager of \$2,500, this assumption has been fine. In reality, the effect of our wages on the payoffs would reduce each of the profits in Table 2 by less than \$7.00 and have no effect on our \$53.60 profit if Davona Dale was in the money. The effect on the payoffs of a much larger total wager is more serious though. For example, betting a total of \$25,000 cannot be expected to allow a guaranteed profit of about \$536.00.

Equation 4 shows that the condition for a lock is more easily met when n is small or Q is large or both. Unlike most states which have a guaranteed minimum five percent return, Louisiana has a minimum return of 10 percent. Equation 4, revised for Louisiana, yields the less restrictive lock condition, $k > 1 - Q(n-1)/[11(n-3)]$. If $Q = .85$ and $n = 5$, as they do in our example, then $k \geq .846$ is sufficient for a lock.

Conditions (3) and (4) assumed that the public wagers the same amount on each of the nonfavorite horses. It is possible to treat the case where these amounts are different, as they are in this Alabama Stakes example. Let x continue to be our wager on the favorite, Davona Dale, but now let y_1, y_2, y_3, and y_4 be our wagers on It's in the Air, Mairzy Doates, Poppycock, and Croquis, respectively. Let R be our guaranteed return. Table 3 shows the return on each horse for each possible triplet of winners. The linear program for

Top Three Finishers (order does not matter)	Profit
It's in the Air — Mairzy Doates — Poppycock	\$ 12.60
It's in the Air — Mairzy Doates — Croquis	148.60
It's in the Air — Poppycock — Croquis	172.40
Mairzy Doates — Poppycock — Croquis	597.40

Table 2: If Davona Dale finishes out of the money, then profit depends on the identity of the top three finishers. The order in which they finish does not matter, though.

maximizing the guaranteed return to a $2,500 bankroll is

Maximize R

subject to

$$x + y_1 + y_2 + y_3 + y_4 \le 2500$$
$$.05(x + y_1 + y_2) - y_3 - y_4 \ge R$$
$$.05(x + y_1 + y_3) - y_2 - y_4 \ge R$$
$$.05(x + y_1 + y_4) - y_2 - y_3 \ge R$$
$$.05(x + y_2 + y_3) - y_1 - y_4 \ge R$$
$$.05(x + y_2 + y_4) - y_1 - y_3 \ge R$$
$$.05(x + y_3 + y_4) - y_1 - y_2 \ge R$$
$$15.60\, y_1 + 27.30\, y_2 + 28.00\, y_3 - x - y_4 \ge R$$
$$15.60\, y_1 + 27.40\, y_2 + 31.90\, y_4 - x - y_3 \ge R$$
$$15.60\, y_1 + 28.00\, y_3 + 32.00\, y_4 - x - y_2 \ge R$$
$$27.60\, y_2 + 28.30\, y_3 + 32.20 y_4 - x - y_1 \ge R$$
$$x, y_1, y_2, y_3, y_4 \ge 0.$$

The solution is $R^* = \$52.47$ with $x^* = \$2366.04$, $y_i^* = \$34.54$ for $i = 1,2,3$, and $y_4^* = \$30.34$. So, the wagers on the horses are close to those made in the equal-public-wagers case. Constraints 2, 3, and 5 have surpluses of $4.41 while constraints 10 and 11 have surpluses of $23.75 and $454.66, respectively. The remaining constraints are binding.

The lock condition is seldom met, and to get a substantial return one needs a large bankroll. However, the times when it is met should not be complete surprises; it happens when there is an extreme favorite that the crowd figures cannot be out of the money. Davona Dale allowed another lock in the 1979 Coaching Club American Oaks at Belmont. She won that race with 97.8 percent of the show pool on her! Two years after Davona Dale won that race, the entry of Heavenly Cause and De La Rose had 95.2 percent of the 1981 Coaching Club American Oaks show pool. Spectacular Bid ran 30 races and finished out of the money only once, and that was as a two-year-old. He was such a standout that he often went off at 1-20 odds, a good sign that a lock may exist. One lock on him was the 1980 Amory Haskell Handicap where he had 96.0 percent of the show pool. A more recent example is Easy Goer's impressive win in the 1989 Gotham Stakes at Belmont Park. Easy Goer had 97.1 percent of the show pool of $553,658. We thank Peter Arnold for bringing this lock to our attention.

If a horse that has virtually all the show money bet on it loses, the payoffs can be very high. In fact, show payoffs

---

The show payoff on Arbor Hoggart is thought to be the highest show payoff of any sort.

---

can exceed win payoffs. This happened when the public's favorite, Kassa Branca, finished last in the New Jersey Futurity at Freehold Racetrack on October 22, 1988. Two bettors had wagered $50,000 and $150,000 to show on him. Others brought the show bet on Kassa Branca up to $213,837. That amounted to 98.8 percent of the show pool of $216,492. The payoffs per two-dollar wager were

	WIN	PLACE	SHOW
Beta Bob	41.20	11.20	312.40
Nukes Image		7.00	113.00
Arbor Hoggart			340.80

The show payoff on Arbor Hoggart is thought to be the highest show payoff of

Top three horses	Probability of these being the top three horses	Respective show payoffs per $1 bet
Davona Dale-It's in the Air-Mairzy Doates	0.34468	1.05, 1.05, 1.05
Davona Dale-It's in the Air-Poppycock	0.25139	1.05, 1.05, 1.05
Davona Dale-It's in the Air-Croquis	0.28692	1.05, 1.05, 1.05
Davona Dale-Mairzy Doates-Poppycock	0.03589	1.05, 1.05, 1.05
Davona Dale-Mairzy Doates-Croquis	0.04109	1.05, 1.05, 1.05
Davona Dale-Poppycock-Croquis	0.02970	1.05, 1.05, 1.05
It's in the Air-Mairzy Doates-Poppycock	0.00330	16.60,28.30,29.00
It's in the Air-Mairzy Doates-Croquis	0.00376	16.60,28.40,32.90
It's in the Air-Poppycock-Croquis	0.00274	16.60,29.00,33.00
Mairzy Doates-Poppycock-Croquis	0.00053	28.60,29.30,33.20
	1.00000	

Table 3: The probability of any three horses being the in-the-money finishers is given, along with the resulting show payoffs. These probabilities were computed using the Harville [1973] formulas. The probability of an i, j, k finish, where q_i is the probability of winning, is $q_i q_j q_k/[(1 - q_i) (1 - q_i - q_j)]$. Stern [1987] discusses alternative probability models (see also the discussion in Hausch and Ziemba [forthcoming a]).

any sort. The previous harness racing record was $296.00 in 1986 at Dover Downs. This extreme example would have allowed a lock returning close to four percent. (Many thanks to Pete Asch for pointing out this race to us). Other examples are mentioned in Ziemba and Hausch [1987].

Since the lock strategy requires a guaranteed profit, it is much more conservative than the optimal capital growth strategy described in Ziemba and Hausch [1987]. The optimal capital growth strategy asymptotically maximizes the rate of growth of one's bankroll. This is achieved by maximizing, in a myopic race-by-race fashion, the expected value of log utility. A comparison of these two strategies can be made with the Alabama Stakes race. Table 3 shows possible payoffs on each horse and their likelihoods, and Table 4 gives the expected return to a show bet on each horse.

Only the show bets on Davona Dale

and It's in the Air have positive expected profits. With an initial wealth of $2,500, the optimal capital growth bets are

Davona Dale	$2,294
It's in the Air	203
Mairzy Doates	0
Poppycock	0
Croquis	3.

The entire $2,500 is wagered, and regardless of the order of finish our bettor does not go bankrupt. The certainty equivalent, that is, the certain return giving the same utility as the expected utility of the gamble using the logarithmic utility function, is $101. The wager on Croquis has an expected return of only $0.607 on the dollar. While it is unlikely that both

Horse	Expected Return on a Show Bet
Davona Dale	1.039
It's in the Air	1.090
Mairzy Doates	0.658
Poppycock	0.524
Croquis	0.607

Table 4: The expected return to show is calculated for each horse.

Top three horses	Profit
Davona Dale-It's in the Air-Mairzy Doates	$121.85
Davona Dale-It's in the Air-Poppycock	121.85
Davona Dale-It's in the Air-Croquis	125.00
Davona Dale-Mairzy Doates-Poppycock	− 91.30
Davona Dale-Mairzy Doates-Croquis	− 88.15
Davona Dale-Poppycock-Croquis	− 88.15
It's in the Air-Mairzy Doates-Poppycock	808.20
It's in the Air-Mairzy Doates-Croquis	908.20
It's in the Air-Poppycock-Croquis	908.20
Mairzy Doates-Poppycock-Croquis	−2,399.80

Table 5: Using the optimal capital growth wagers, profits are given for the possible trios of horses in the money. The profit is not affected by the order of the three, though.

Davona Dale and It's in the Air will finish out of the money, Table 3 shows that, in that case, Croquis has a better payoff than Mairzy Doates and Poppycock. Table 5 presents the possible profits from these wagers (accounting for their effect on the odds).

Tables 3 and 5 show that the chance of a loss is 10.7 percent and the expected profit can be calculated to be $106.28, or 4.25 percent on the bankroll. Clearly, this expected return is considerably higher than the lock's certain 2.1 percent return. The actual finish, Poppycock-Davona Dale-It's in the Air, yielded a profit of $121.85.

Such a lock exists maybe five to ten times a year in North America. While you are enjoying the performance of a super horse, you may make some profit as well.

Acknowledgments

Our thanks to the referees for their comments and special thanks to Bill Stein for his suggestions and for detecting an error in our numerical work.

References

Asch, Peter; Malkiel, Burton G.; and Quandt, Richard E. 1984, "Market efficiency in racetrack betting," *Journal of Business*, Vol. 57, No. 2 (April), pp. 165–175.

Asch, Peter; Malkiel, Burton G.; and Quandt, Richard E. 1986, "Market efficiency in racetrack betting: Further evidence and a correction," *Journal of Business*, Vol. 59, No. 1 (January), pp. 157–160.

Gelband, Myra, ed. 1979, "A perfect race," *Sports Illustrated*, Vol. 51, No. 11 (September 10), p. 18.

Harville, David A. 1973, "Assigning probabilities to the outcomes of multi-entry competitions," *Journal of the American Statistical Association*, Vol. 68, No. 342 (June), pp. 312–316.

Hausch, Donald B.; Ziemba, William T.; and Rubinstein, Mark 1981, "Efficiency of the market for racetrack betting," *Management Science*, Vol. 27, No. 12 (December), pp. 1435–1452.

Hausch, Donald B. and Ziemba, William T. 1985, "Transactions costs, extent of inefficiencies, entries and multiple wagers in a racetrack betting model," *Management Science*, Vol. 31, No. 4 (April), pp. 381–394.

Hausch, Donald B. and Ziemba, William T., eds. forthcoming a, *Efficiency of Racetrack Betting Markets*, Academic Press, New York.

Hausch, Donald B. and Ziemba, William T. forthcoming b, "Arbitrage strategies for cross-track betting on major horseraces," *Journal of Business*.

Leong, Siew Meng and Lim, Kian Guan 1989, "Cross-track betting: Is the grass greener on the other side?" working paper, National University of Singapore.

Seligman, Daniel 1979, "Looking for a Lock," *Fortune*, Vol. 100, No. 8 (October 22), pp. 41–42.

Stern, Hal S. 1987, "Gamma processes, paired comparisons and ranking," PhD diss., Department of Statistics, Stanford University.

Thaler, Richard and Ziemba, William T. 1988, "Parimutuel betting markets: Racetracks and lotteries," *Journal of Economic Perspectives*, Vol. 2, No. 2 (Spring), pp. 161–174.

Willis, Kenneth E. 1964, "Optimum no-risk strategy for win-place pari-mutuel betting," *Management Science*, Vol. 10, No. 3 (April), pp. 574–577.

Ziemba, William T. and Hausch, Donald B. 1986, *Betting at the Racetrack*, Norris M. Strauss, New York.

Ziemba, William T. and Hausch, Donald B. 1987, *Dr. Z's Beat the Racetrack*, William Morrow and Co., New York.

RACETRACK BETTING--AN EXAMPLE OF A MARKET WITH EFFICIENT ARBITRAGE

Jay R. Ritter*

College of Commerce and Business Administration
University of Illinois at Urbana-Champaign

Abstract. A model of racetrack betting behavior is set forward, and its implications tested, albeit with a small sample size. The model is consistent with risk-loving on the part of bettors. A simple betting rule [essentially a crude version of the Dr. Z system, described by Ziemba and Hausch (1984)] is then tested, with evidence put forward that an unexploited profit opportunity may exist. When uncertainty is introduced, however, it is found that the profits vanish, a result consistent with the joint hypothesis of market efficiency and the model of betting behavior described in the paper.

Can an intelligent bettor make money at a harness horse racetrack? Do bettors allow profitable bets to go unbet? Do all bets have the same (negative) expected return? Or, do expected returns differ between bets, although none are profitable? Using academic finance terminology, are there violations of a joint hypothesis of an efficient market and a plausible model of betting behavior? This paper will investigate the opportunities available to the casual bettor at harness horse racetracks.

One possible model of betting behavior would be that all bets have the same (negative, due to the "loading fee" that accrues to the track and state) expected return. Based upon the evidence presented by Mukhtar Ali in a 1977 paper in the Journal of Political Economy, this model can easily be rejected. Ali found that harness horse racetrack bettors consistently earn a lower average return in betting on horses with low objective probabilities of winning than they do in betting on the favorite. He concluded that bettors are risk-lovers, where risk is synonymous with the variance of return on a bet. Presumably individual bets have no covariance with anything else, so that all "systematic risk," in the traditional finance sense, can be diversified away.

A testable implication of Ali's risk-loving hypothesis is that the safest bets of all should have the highest expected payoff among all the possible bets. The safest bet is betting on the favorite to show. Only slightly more risky is betting on the favorite to place. This implication was tested using data from harness horse racetracks in Illinois.

The results, for a $1 bet, of betting on the favorite to win, place, or show are contained in Table 1. The favorite is defined as that horse in a race which has the greatest dollar amount bet on it to win, and may not correspond to the horse with the highest objective probability of winning. Using Ali's terminology, the favorite defined here is the subjective favorite.

*The author wishes to thank Mukhtar M. Ali, Melanie L. Lau, and Ronald Michener for helpful comments on an earlier draft of this paper. Thanks are due to Chuck Waller of American Totalisator Company for providing the data. This paper was originally written in 1978, while the author was a graduate student at the University of Chicago. Changes made by the author in 1993 are included in brackets [].

Table 1

Results from Betting on the Favorite to Win, Place, or Show

Bet	Average Net Gain, $	Standard Deviation of Average Gain, $
Win	(0.18)	0.08
Place	(0.06)	0.06
Show	(0.09)	0.05

N = 229 races, composed of 134 races at Sportsman's Park during 1977 and of 95 races at Hawthorne Park during 1978. It should be noted that the last race of the day is unrepresented in the sample, so that the average net gains reported here have a downward bias. This is because bettors have a strong tendency to "plunge" on long shots in the last race of the day. See McGlothlin for support for these assertions.

Betting odds are determined by the public—the bettors are betting against themselves, with the state and the track each receiving a specified percentage commission, or "takeout," of the betting dollars. This takeout, which in Illinois averages approximately 18 percent on win betting, is analogous to the "loading fee" that risk-averse individuals pay when they buy insurance.

As can be seen in Table 1, the average return is higher (less negative) for betting on the favorite to place or show than for the average bet, where the expected return is equal to the takeout, which we shall denote as α. This is consistent with the hypothesis that bettors are risk lovers. The presence of a large α, however, allows the existence of risk-loving behavior, with all expected returns still being non-positive.

Given that one would expect at least some bettors (or potential bettors) to compete away any positive expected return, the model of betting behavior set forward here is as follows: while the expected return on any bet may vary with the risk involved, all bets have non-positive expected returns. This has the following testable implication: no betting rule exists which generates a positive expected return.

Since the extremely simple strategy of betting on the favorite to place or show yields only a small negative return, it may be possible that some bettors do consistently earn a positive return. At least some professional racetrack bettors appear to earn a positive return betting on specific horses. Presumably, there are a fairly large number of amateur bettors who bet essentially at random. If this random betting is not fully arbitraged, it may be possible for even the "casual" bettor to earn a positive return. A "casual" bettor would be an individual who only uses a small subset of the publicly available information in making betting decisions, such as only the information flashed on the tote board at the racetrack before each race.[1]

For a horse with a given objective probability of winning, the less the amount bet on it as a fraction of the total amount bet on all of the horses, the higher the dollar return if it does pay off. The payoff to a $1 bet on horse h, x_h, is determined according to the formula

$$x_h = \frac{\frac{1}{n}\left[P(1-\alpha) - \sum_{i=1}^{n} X_h^i \right] + X_h}{X_h} \quad (1)$$

---

[1] The tote board contains the current win odds and the betting totals for horses in the race on which betting is currently occurring.

where P is the appropriate betting pool,[2] α is the track takeout,[3] X_h is the dollar amount bet on horse h for the specific type of bet that the pool applies to, n is the number of horses that pay off, i is the order of finish, and $\sum_{i=1}^{n} X_h^i$ is the total dollar amount bet on the horses that pay off. For the win return, n would be 1 and the numerator would reduce to $P(1-\alpha)$. For the place and show payoffs, where n would be 2 and 3, respectively, the returns are conditional upon what other horse or horses pay off. x_h is equal to 1 plus the odds when n = 1 (i.e., for the win pool).

Equation (1) describes the return, for a given type of bet, on a horse that pays off. We can now put some substance into the necessary and sufficient condition for a bet to have a positive expected net return: the return on a bet that does pay off, x_h, multiplied by the probability that it will pay off, π_h, must exceed unity. We want to see if there are any betting strategies for which $x_h \pi_h > 1$. If there are, we can reject the joint hypothesis that we have correctly specified a model of betting behavior and that the betting market is informationally efficient. Before we can test this joint hypothesis, however, we need a proxy for π_h.

Let f_{hj} be the fraction of the jth betting pool bet on horse h, with j = w,p, or s depending upon whether we are dealing with the win, place, or show pools. In general, $f_{hw} \neq f_{hp} \neq f_{hs}$. Consequently, it may be possible to use a function of f_{hw} as a proxy for π_h for place and show betting. It should be noted that while the betting is in progress, the current values of the f_{hj}'s are publicly available, since the current betting totals are available on the tote board. Support for the conjecture about the utility of using f_{hw} as a proxy for π_h comes from the fact that at some tracks as much as 70 percent of all the win, place, and show betting is in the win pool. In addition, most professional bettors concentrate on the win betting, so that f_{hw} is likely to be close to the true π_h. Hence, although hardly a sufficient statistic, one would expect that f_{hw} would be a better indicator of a horse's objective probability of paying off, π_h, than f_{hp} or f_{hs}.

In reality, only one function of f_{hp} and f_{hw}, and only one function of f_{hs} and f_{hw}, have been analyzed. The simple linear function analyzed, with the optimal parameter value included, is the basis for the following simple betting rule:
[More precise versions of this rule are contained in Ziemba and Hausch's <u>Beat the Racetrack</u>.]

<u>Rule</u>: Ignoring long shots, bet on horse h to place if $\dfrac{f_{hp}}{f_{hw}} \leq 0.7$, and bet on horse h to show if $\dfrac{f_{hs}}{f_{hw}} \leq 0.7$.

---

[2] A betting pool is the total dollar amount bet on all horses for a specific type of bet, such as a place bet. It should be noted that the relevant betting pool for the win odds is the total amount bet on horses to win only. The place pool and the show pool do not affect the win odds. A similar statement is true for the place and show odds.

[3] The takeout is actually composed of two parts, the take and the breakage. The take is a fixed percentage, such as 16 percent, while the breakage is composed of residuals due to the fact that x_h is rounded down to the nearest lower dime, subject to the constraint that x_h is always greater than one. On average, the breakage is about 2 percent for win betting, so that α is approximately 18 percent. Because the breakage is almost certainly higher for place and show betting, α is likely to be higher than 0.18 for place and show betting. The exact formulation has been used in the empirical section, although the step function described here is not as amenable to the calculus as the continuous x_h assumed in the theoretical sections.

Since there is strong evidence in Ali's paper that there are, on average, large negative returns on long shots, horses with $f_{pw} < 0.15$ were excluded from betting consideration. The 0.15 cutoff is arbitrary; no attempt has been made to find the optimal definition of "long shot."

To operationalize and test this simple betting rule, it was assumed that the f_{hj}'s--the raw data for which are reported on the tote board at the track and constantly updated until the race starts--are constant during the last few minutes of betting. This is the assumption made by Ali. In reality, they are not constant but they are highly autocorrelated. This assumption is crucial for the implementation of the simple betting rule as a profitable betting rule, and will be discussed further below. All other information available about the horses and jockeys has not been taken into account--only a small subset of the publicly available information is being used.

The 0.7 parameter in the rule was determined by calculating the net returns using different cutoff numbers. The net return was at a maximum when the cutoff was 0.7, based upon the sample of 134 races at Sportsman's Park referred to above.[4] The results of applying this simple betting rule are listed in Table 2, and the results of using other cutoff values than 0.7, not necessarily the same for both place and show betting, are described in the Appendix.

Table 2

Results of Implementing Simple Betting Rule Using 0.7 Cutoff

	Place	Show
No. of Actual Bets	9	65
Average Gain per Dollar Bet	0.47	0.17
S.D. of Average Gain, $	0.31	0.08

N = 134 races at Sportsman's Park during 1977.

Since the value of the optimal cutoff ratio (0.7) was determined from the data, "t-statistics" computed by dividing the average gain per dollar bet by the standard deviation of the average gain would be difficult to interpret. Consequently, a second sample composed of 95 harness horse races at Hawthorne Park in February, 1978, was analyzed, with the results in agreement with the profitability of the simple betting rule. The results of applying the simple betting rule to this independent sample, with the 0.7 cutoff, are contained in Table 3.

---

[4]Positive returns were also found using any parameter value less than 0.75 for both the place and show betting. Below 0.7, the lower the parameter value, the higher the average return, but the lower the total return, since with a more restrictive betting rule, fewer profitable bets would be made.

Table 3

Results of Implementing Simple Betting Rule, Independent Sample

	Place	Show
No. of Actual Bets	11	55
Average Gain per Dollar Bet	0.05	0.15
S.D. of Average Gain, $	0.37	0.14

N = 95 races at Hawthorne Park during 1978. Cutoff value used was 0.7.

Table 4 contains the combined sample of all races.

Table 4

Results of Implementing Simple Betting Rule, Combined Sample

	Place	Show
No. of Actual Bets	20	120
Average Gain per Dollar Bet	0.24	0.16
S.D. of Average Gain, $	0.24	0.08

N = 95 races at Hawthorne Park during 1978. Cutoff value used was 0.7.

In interpreting the "t-statistics" formed by taking the average net gain per dollar bet and dividing it by the sample standard deviation of the average gain, it should be kept in mind that the sample distributions are highly non-normal, discrete, and asymmetric. For example, for the show betting results in Table 4, approximately one-third of the mass of the sample probability mass function is concentrated on zero, with the non-zero mass approximating a Poisson distribution from $2.20 on up. In addition, the variance of returns on the initial sample is substantially less than that of the independent sample. This is the reason why the standard deviation of the average net gain for show betting in Table 4 (S.D. = 0.08) does not decrease from its value of 0.08 in Table 2, in spite of the number of bets increasing by more than 80 percent, from 65 to 120. Part of the reason for the larger variance in the independent sample may be that the betting pools are smaller, so that more "noise" is present.

Although the sample sizes upon which Tables 2, 3 and 4 are based are quite small, there is an indication of a profit opportunity at the track. If, for show betting, the 0.16 net return per dollar bet could occur on a $10,000 bet, this would indicate a very substantial profit opportunity. Is a profitable $10,000 bet possible?

Since the odds are affected by the size of the bet made on a horse, the optimal bet size must be computed. If the odds were not endogenous, then the optimal bet size for a risk-neutral bettor would be either zero or infinity. Given the endogeneity, however, there is a finite optimal bet size.

The optimal bet size maximizes the net dollar return, which is equal to $B(x_h\pi_h-1)$, where B is the dollar bet size, x_h is the payoff to a \$1 bet on horse h, and π_h is the (objective) probability of a horse paying off. x_h is a function of B, according to equation (1), since P is equal to the total dollar bet on all horses by the other bettors, P_o, plus B, and X_h^i for the horse to be bet on is equal to $X_{ho}^i + B$. It should be noted that $X_{ho}^i - X_{ho}$ for one i, if a successful horse is bet upon, where X_h is the horse being bet on. Consequently, equation (1) can be expressed as

$$x_h = \frac{\frac{1}{n}\left[(P_o+B)(1-\alpha) - \sum_{i=1}^{n} X_{ho}^i - B\right] + X_{ho} + B}{X_{ho} + B}. \quad (2)$$

Multiplying both sides of equation (2) by $B\pi_h$ and subtracting B from both sides, the left side becomes the objective of what should be maximized with respect to B to find the optimal bet size, given the (exogenous) parameters and variables n, P_o, α, and the X_{ho}^i. This results in

$$B(x_h\pi_h-1) = B\pi_h\left\{\frac{\frac{1}{n}\left[(P_o+B)(1-\alpha) - \sum_{i=1}^{n} X_{ho}^i - B\right] + X_{ho} + B}{X_{ho} + B}\right\} - B. \quad (3)$$

Differentiating (3) with respect to B, and setting this equal to zero, gets

$$0 = \pi_h\left\{\frac{\frac{1}{n}\left[(P_o+B)(1-\alpha) - \sum_{i=1}^{n} X_{ho}^i - B\right] + X_{ho} + B}{X_{ho} + B}\right\}$$

$$+ B\pi_h\left\{\frac{\left(1-\frac{\alpha}{n}\right)(X_{ho}+B) - \frac{1}{n}\left[(P_o+B)(1-\alpha) - \sum_{i=1}^{n} X_{ho}^i - B\right] - X_{ho} - B}{(X_{ho}+B)^2}\right\} - 1. \quad (4)$$

This is the first-order condition for the optimal bet size, B. Solving this for B results in the following quadratic equation:

$$0 = -\left[1 + \pi_h\left(\frac{\alpha}{n}-1\right)\right]B^2 + 2X_{ho}\left[1 + \pi_h\left(\frac{\alpha}{n}-1\right)\right]B$$

$$+ X_{ho}\left\{X_{ho}(1-\pi_h) - \frac{\pi_h}{n}\left[P_o(1-\alpha) - \sum_{i=1}^{n} X_{ho}^i\right]\right\}. \quad (5)$$

Table 5 contains optimal bet sizes for some arbitrary values of the exogenous parameters and variables.

Table 5

Bets and Expected Dollar Returns

		Variable and Parameter Values				Optimal	$E[B(x_h \pi_h - 1) \mid B - B_o]$
n	α	P_o	$\sum_{i=1}^{n} X_{ho}^i$	X_{ho}	π_h	Bet	
1	0.18	$25,000	$4,000	$4,000	0.40	$2,403	$970.20
2	0.18	9,500	1,500	1,500	0.50	710	549.00
3	0.18	7,600	3,600	1,600	0.77	485	40.40
3	0.20	7,600	3,600	1,600	0.77	422	31.20
3	0.22	7,600	3,600	1,600	0.77	360	23.20
3	0.22	9,500	4,000	2,000	0.77	626	56.20
3	0.22	9,500	4,500	2,000	0.77	450	29.00
3	0.22	9,500	5,000	2,000	0.77	260	9.60

While the sample values for n = 1 and 2 are totally arbitrary, those for n = 3 are representative of the sample average. Hence, one can estimate the expected returns from arbitraging show betting odds using the simple betting rule. Since there are typically 10 races in a program, and, at least for this sample, just over half of the races (120 out of 229) are bet upon, the bettor following the simple betting rule would bet several hundred dollars in each of an average of approximately five races (sitting out those races where no potential bets meet the requirements), lose approximately a quarter of the races bet upon, and walk away from the track with, on average, maybe $100 more than he or she entered with. Assuming attendance at 250 racing dates per year, this indicates a potential (presumably tax-free) income on the order of $25,000 per year.

Since each bet is an independent bet, and the returns presumably have no covariance with any other investment opportunity, it appears that a bettor with an initial capital of only several thousand dollars could earn an income of approximately $500 per week. If so, why isn't this profit opportunity being acted upon?

One possibility is that few professional bettors realize its existence, and few of them have the ability to compute several fractions using 4 and 5 digit numbers within the space of a few seconds before rushing to the betting window before it closes.

Another possibility, however, is that the final values of the f_{hj}'s are unknowable to a sufficient degree of accuracy when the bets are being made, contrary to the assumption made throughout this paper. If the f_{hj}'s did change in a random fashion within the last minute of betting, the profitable bets would usually be unknowable until after the betting window closed. To analyze the realism of the assumption that the final betting fractions are known when bets are actually made, a sample of the time sequence of these fractions has been analyzed. To the best of the author's knowledge, the time sequence of betting totals is not recorded anywhere. Consequently, the betting totals, which are posted on the tote board and updated at 45 second intervals, were recorded for 10 races at harness horse racetracks by the author.[5]

In all 10 races, the fraction of the total win bet, f_{hw}, bet on the $\underline{ex}$ $\underline{post}$ favorite increased during the last 1.5 minutes of betting, from an average of 0.318 at t_o-1.5 to an average value of 0.360 at t_o.[6] If this is a representative sample, the "best" forecast of the final fraction of the pool bet on a specific horse is the fraction at t_o-1.5 plus (0.360-0.318 = 0.042). In the sample of 10 races used, this prediction results in a forecast error standard deviation of only 0.020. Thus the coefficient of variation is only 0.05. Furthermore, at least for the initial sample, the profitability of the simple betting rule appears to be fairly robust with regard to small changes in the cutoff used (see appendix).

To explicitly analyze the effects of the stochastic nature of the ratios, further tests are required. Because of the small number of times in which the simple betting rule was implemented with place betting, analysis will be restricted to show betting. To distinguish between f_{hs}/f_{hw} at t_o, which is unknown when betting is done, and f_{hs}/f_{hw} at t_o-1.5, we will denote the ratios as $(f_{hs}/f_{hw})_{t_o}$ and $(f_{hs}/f_{hw})_{t_o-1.5}$, respectively. Ideally, to analyze the effect of differences between these two ratios on the profitability of the simple betting rule, observations of the time series of the (f_{hs}/f_{hw})'s should be used. These observations are available only by recording them at a racetrack, at a time cost of approximately 25 minutes per race--a very expensive proposition.

An alternative method would be to simulate the results by constructing a set of $(f_{hs}/f_{hw})_{t_o-1.5}$ observations by adding a "noise" term to the final ratios, betting on the basis of the $(f_{hs}/f_{hw})_{t_o-1.5}$ with noise values, and analyzing the returns based upon the $(f_{hs}/f_{hw})_{t_o}$ values. To implement this procedure, an estimate of the stochastic process determining the time series of the ratios must be used. Using the $(f_{hs}/f_{hw})_{t_o}$ observations of the 95 races in the independent sample, for all $f_{hw} \geq 0.15$, an error term drawn from a normal distribution with zero mean and variance 0.02 was added.[7] The simple betting rule, with a cutoff value of 0.63, was then applied to this simulated $(f_{hs}/f_{hw})_{t_o-1.5}$ series.[8] Whereas applying the simple betting rule with a 0.7 cutoff to the $(f_{hs}/f_{hw})_{t_o}$'s yielded 55 bets and a net return of 0.15 per dollar bet (Table 3), the simulation with

---

[5]Two races at Arlington Park on December 27, 1977 and eight races at Hawthorne Park on January 2, 1978 were analyzed. See the appendix for the actual fractions. A cassette tape recorder was used to record the approximately 25 four-digit numbers which are updated every 45 seconds.

[6]1.5 minutes has been chosen as the least amount of time during which the relevant fractions can be computed and compared and a bet actually made. The last update of the betting amounts leaves less than 45 seconds in which to accomplish these acts, which the author has found to be insufficient. The betting window closes at time t_o.

[7]The variance of 0.02 is based upon the sample variance of the differences between the ratios at t_o and t_o-1.5 for 15 horses with $f_{hw} \geq 0.15$ in the 10 races referred to in footnote 5 and in the appendix.

[8]This procedure is implicitly assuming that the relevant loss function is being minimized by using $(f_{hs}/f_{hw})_{t_o-1.5}$ as the forecast of $(f_{hs}/f_{hw})_{t_o}$.

noise yielded 39 bets and a net return of -0.01 per dollar, using the 0.63 cutoff, where the payoffs used were the actual payoffs to these 39 bets if these bets had been made.[9]

This result is quite disturbing to the earlier analysis. While the -0.01 mean return that results from the simulation with noise has a standard deviation of 0.14, it is far below the 0.15 return realized without taking account of the stochastic nature of the f_{hs}/f_{hw}'s, and casts severe doubt on the profitability of the simple betting rule in practice. In other words, the $25,000 income referred to earlier may be zero, or even negative.

In addition, I have been told by serious bettors that there are scores of bettors who try to implement the strategy of the simple betting rule, so that for low values of $(f_{hs}/f_{hw})_{t_o-1.5}$, the expected value of $[(f_{hs}/f_{hw})_{t_o} - (f_{hs}/f_{hw})_{t_o-1.5}]$ is almost certainly very positive. [i.e., the more people who use the Dr. Z system, the less attractive it is to use because the potential profits are competed away.] If this is a market in which efficient arbitrage is occurring, so that all profits are competed away, then the results of Tables 2, 3 and 4, where positive average returns appear to exist, merely represent the ex post realizations of those $(f_{hs}/f_{hw})_{t_o}$'s which ex ante did not look profitable, but which had unpredicted, and unpredictable, negative residuals from the best feasible forecasts.

Conclusions

Due to the large sampling variance, any conclusions based on a sample as small (n=229 races, n=120 show bets) as the combined sample here must be regarded as tentative. When the ratios of show betting to win betting fractions, the $(f_{hs}/f_{hw})_{t_o}$'s, are assumed known at t_o-1.5, one cannot reject the hypothesis that a profitable betting strategy may exist at some harness horse racetracks. Once uncertainty about the $(f_{hs}/f_{hw})_{t_o}$'s at t_o-1.5 is explicitly introduced, however, positive returns appear to be dissipated away. Thus, the horse race betting market would appear to be just another market which seems to exhibit what might be termed "weak form" efficiency. When the uncertainty in the $(f_{hs}/f_{hw})_{t_o}$'s is taken into account, then the joint hypothesis that risk-loving bettors have non-positive expected returns on all bets and that the betting market is informationally efficient cannot be rejected.

[Because some other bettors use the betting rule described here (the Dr. Z system), to implement it profitably one must bet only on the very most attractive bets. Otherwise, one will be consistently disappointed regarding the change in the odds between when a bet is placed and the final betting totals, upon which the payoffs are calculated.]

References

Ali, M. M. "Probability and Utility Estimates for Racetrack Bettors," J.P.E. 85, no. 4 (August 1977): 803-815.

Griffith, R. M. "Odds Adjustments by American Horse-Race Bettors," American J. Psychology 62 (April 1949): 290-294.

_____ "A Footnote of Horse Race Betting," Transactions Kentucky Academy of Science 22 (1961): 78-81.

Gruen, A. "An Inquiry into the Economics of Race-Track Gambling," J.P.E. 84, no. 1 (February 1976): 169-178.

---

[9]The 0.63 cutoff is used instead of 0.7 because of the tendency for low ratios at t_o-1.5 to increase in the last 1.5 minutes of betting. Having a lower cutoff in the stochastic world results in fewer ex post unprofitable bets being made. The 0.63 value of the cutoff is based upon the sample referred to in footnote 5.

McGlothlin, W. H. "Stability of Choices Among Uncertain Alternatives," American J. Psychology 69
     (December 1956): 604-615.
[Ziemba, W. T. and Hausch, D. B., Beat the Racetrack, New York: Harcourt, Brace Jovanovich,
     1984.]

Appendix

Description of the Data--Independent Sample

Data on the amount bet per horse for win, place, and show for horse races between
February 8, 1978 and February 20, 1978 at Hawthorne Park Racetrack were provided by Chuck Waller
of American Totalisator Co. As with the initial sample from Sportsman's Park (which, incidentally, is
located several hundred feet from Hawthorne), the last race of the day is not represented in the sample.
Otherwise, the races can be viewed as a random sample of the races occurring during February, 1978
at Hawthorne. One race was excluded from consideration because of a horse which was scratched
shortly after the betting began. Results of the races were taken from the appropriate issues of the
Chicago Sun-Times and Tribune, and verified against the calculated payoffs using equation (1).

Time Sequence of Betting Fractions

For the 10 races referred to in footnote 5, the following fractions of the win bet pool were bet
on the ex post favorite:

$$f_{bw}$$

	Final	t_o-1.5	Δ	Δ-0.042
	0.360	0.318	0.042	0.000
	0.427	0.360	0.067	0.025
	0.428	0.376	0.052	0.010
	0.415	0.392	0.023	-0.019
	0.350	0.333	0.017	-0.025
	0.290	0.253	0.037	-0.005
	0.484	0.448	0.036	-0.006
	0.339	0.322	0.017	-0.025
	0.287	0.236	0.051	0.009
	0.288	0.235	0.053	0.011
Mean	0.360	0.318	0.042	0.000

Appendix

Description of the Data--Initial Sample

Data on the amount bet per horse for win, place, and show for 136 horse races between
June 22, 1977 and August 19, 1977 at Sportsman's Park Racetrack were provided by Chuck Waller of
American Totalisator Co. Of the 136 races, which represent approximately one-third of all the races
occurring at Sportsman's Park during this time, one was excluded because of a horse numbered 1a
which paid off, which I was unable to reconcile with the betting data. Results of the races were taken

from the appropriate issues of the <u>Chicago Sun-Times</u> and <u>Tribune</u>, and verified against the calculated payoffs using equation (1).

Values of Gain Function for Sample

Place Cutoff

Value of $\dfrac{f_{hp}}{f_{hw}}$	0.85	0.80	0.75	0.70	0.65
Number of $2 Bets	97	55	23	9	4
Gross Return	185.40	107.60	53.80	26.40	11.40
Net Return	(8.60)	(2.40)	7.80	8.40	3.40

Show Cutoff

Value of $\dfrac{f_{hs}}{f_{hw}}$	0.75	0.70	0.65	0.60	0.55
Number of $2 Bets	98	65	42	26	10
Gross Return	215.60	151.60	97.20	63.60	27.00
Net Return	19.60	21.60	13.20	11.60	7.00

PART VI

EFFICIENCY OF EXOTIC WAGERING MARKETS

Introduction to the Efficiency of Exotic Wagering Markets

Donald B. Hausch, Victor S.Y. Lo and William T. Ziemba

Exotic bets are wagers involving two or more horses, and a variety of them are offered by tracks. Quinellas require one to predict the first two finishers in a race. Exactas (or perfectas) also require one to predict the first two horses in a race, but additionally one needs to predict their order of finish. The trifecta involves naming the first three finishers in the correct order. In a double (or daily double) bet, one must select the winner of two consecutive races. The double concept has been extended to predicting winners of three, four, six, and more consecutive race, providing the public extremely low probability but high payoff wagers. This section is concerned with the efficiency of these markets.

Exotic wagers tend to be very popular with the public. Part of the attraction is their risk/return tradeoff; one wins with very low probability but the payoff is high. In view of the favorite-longshot bias for win bets, it is not surprising that longer-odds events than win bets are heavily wagered. Benter (1994)[1] provides another reason for the exotics' lure. To illustrate his reason, consider double betting, which bettors can create for themselves by wagering on a horse and then, if successful, betting to win all the proceeds of the first race on a horse in the next race. This self-constructed double bet, called a parlay, does differ from a double in an important way. The double involves incurring the track take just once. The parlay, however, involves a track take on both of its win bets. Thus, the transactions costs are higher with the self-constructed exotic bets. Benter's (1994)[1] point is that in order to be a successful bettor, one needs handicapping skills exceeding those of the average bettor. For win betting, to cover the track take, one's skills must be significantly better than average. Exotic betting, however, due to its "lower" transactions costs, requires less of a skill advantage for a bettor to be profitable. Many tracks account for this difference by charging a higher track take on exotic wagers, reducing its advantage. The same phenomenon is in betting on lottos with unpopular numbers. The more numbers (races) there are the easier it is to win, see Ziemba, Brumelle, Gautier and Schwartz (1986)[2] and McLean, Ziemba and Blazenko (1992)[1].

While a parlay that is a self-constructed version of a double pays the track take twice, it does allow the bettor more information. The double itself must be made before either race; thus, a bettor sees the public's odds on the first race but has little information about the public's view of the second race (other than any information that can be gleaned from the payoffs that are offered on double combinations, but that information is usually difficult to access). The parlay allows one to wager on the second race with a better sense of the public's impression of the horses. Ali (1979)[1] found that returns of parlays and double bets were not significantly different. Thus, they are "equally priced," an implication of an efficient market. Asch and Quandt (1987)[1], however, found that doubles are statistically more profitable than parlays. When parlay payoffs are adjusted as if parlay bettors paid the track take just once, then returns on parlays and doubles are not significantly different. Lo and Busche's (1994)[1] conclusions were the same for Hong Kong data. Their results also indicate that bettors bet on double and double quinella less accurately when compared to the win bet using data from Meadowlands and Hong Kong.

Asch and Quandt (1987)[1] find some support for the notion that "smart money" is in the exotic pools. The basis for this notion is that the informational content of smart money is more difficult for the public to discern in the exotic market than it would be if it were wagered in the win market. Their analysis ignores the systematic biases of the Harville (1973)[1] model, though. Also, Dolbear (1991)[2] describes a further bias in their comparison of the theoretical and subjective probabilities of exacta outcomes. Bacon-Shone, Lo and Busche (1992)[2] address these concerns and conclude that the public's exacta betting provides more accurate estimates of ordering probabilities (the probability that i wins and by Harville (1973)[1], Henery (1981)[1] or Stern (1990)[2]). Similarly, the trifecta market provides more accurate estimates of their ordering probabilities than does the win market.

Hausch, Lo and Ziemba (1994)[1] develop a general formula for optimal betting on exotic bets. Their model allows ordering probabilities based on Harville (1973)[1], Henery (1981)[1] or Stern (1990)[1] with the help of approximations developed by Lo and Bacon-Shone (1993)[2]. The model employs the Kelly criterion and can be applied to any exotic bet. Quinella data on 369 Hong Kong races is used to illustrate the system. Benter (1994)[1] proposes another way of modifying the Harville model.

Kanto and Rosenqvist (1994)[1] develop a betting system for quinella bets (called double bets in Finland) at a Finnish racetrack. Instead of using the win odds data directly, they use maximum likelihood estimation and the Harville (1973)[1] model to estimate the win probabilities and the probabilities associated with a quinella bet by assuming that the quinella bet amounts for different combinations follow a multinomial distribution. Using the Kelly criterion for wagering and 111 races, they show some evidence of positive profits.

Post positions for horses are normally assigned in a random fashion. There are circumstances, such as for a front runner, where an inside post position can be an advantage. Canfield, Fauman and Ziemba (1987)[1] consider any post position bias in assessing the efficiency of win and exotic markets. Using a three year Canadian data set, their results indicate that inside post positions provide a winning edge and outside positions are disadvantageous. Furthermore, the bias is more pronounced the smaller the circumference of the track (because turns - where the advantage of the inside positions lies - tend to be a greater fraction of the distance covered) and the longer the race (because such races tend to involve more turns). However, after accounting for transactions costs, the public overbets the favorable bias positions to fully negate their advantage, and thus over the long run the bias provides no financial advantage. Because tracks are banked, water may accumulate near the rail when it rains. Canfield, Fauman and Ziemba found that such off track days introduce an effect that essentially neutralizes the bias in favor of post position one. The public seems not to completely appreciate this, though, as wagering on post position one on off track days generates significant losses. For off tracks, a favorable bias does persist for positions two and three, though, which seems to allow an advantage in exotic wagering.

Using more recent data from the racetrack studied by Canfield, Fauman and Ziemba (1987)[1], Betton (1994)[1] employs a probit analysis and pooled cross-sectional time series analysis to study the post position bias. She models the relationship between the odds ranking, post position, and final position of a horse. The length of the race does not significantly improve the fit of any of these models. Her results indicate that the post position significantly adds to the information available from the odds rankings in determining the probability of a horse placing in the top three positions. She also shows that as the number of horse in a race increases, the post position bias becomes a more important factor in addition to the win odds.

[1] included in this volume
[2] cited in the Annotated Bibliography

SOME EVIDENCE OF THE EFFICIENCY OF A SPECULATIVE MARKET

By Mukhtar M. Ali[1]

It is well known that the returns on various betting opportunities at a racetrack are determined by a competitive bidding of the bettors in a natural environment of their decision making. In this paper, two simple bets of unknown but identical winning probabilities are identified. An analysis of 1,089 observations shows the data are consistent with the hypothesis that both bets are identically priced, an implication of an efficient speculative market.

1. INTRODUCTION

IN MANY RESPECTS, valuation of a bet in racetrack betting is similar to the valuation of a stock in the stock market. In both cases, future earnings are unknown and investors (bettors) bid against each other to determine the prices or returns on their investments. Baumol [2, p. 46] maintains that the market of racetrack betting closely approximates the efficient market hypothesis on the ground that

> ... all opportunities for profit by systematic betting are eliminated. Bets at ten to one will in the long run come off almost as badly as bets at three to one.

However, Ali [1] has shown from an analysis of over 20,000 races that on the average, one would lose 10 cents per dollar bet by betting to win on the first favorite, the horse with the lowest odds in a race, whereas he would lose 19 cents per dollar bet by betting to win on the second favorite, the horse with the second lowest odds.

Rosett [6] claims that the relationship between the return from a bet if it succeeds, and its winning probability is consistent with the hypothesis that bettors are rational, sophisticated, and have a strong preference for low-probability-high-return bets and thus it is consistent with the efficient market hypothesis.[2] But his estimated relationship, as derived from this hypothesis, explains only part of his data and consistently over-estimates the returns of low probability bets. Although this casts doubts on the validity of his claim, it does not necessarily contradict the efficient market hypothesis.

In this paper, two distinctly different bets with identical winning probabilities[3] are identified. It is shown that the hypothesis that both bets are identically priced,

---

[1] The author wishes to acknowledge the help received from his wife Julia W. Ali in collecting the data for the present analysis. Comments from W. E. Wecker and S. I. Greenbaum are appreciated. I am highly grateful to T. Hatta for comments clarifying some fundamental issues.
[2] Bettors are rational in the sense that no one prefers a bet with a smaller winning probability and the same or lower return or with a lower return and the same or lower winning probability to what is available to him. Bettors are sophisticated in the sense that the winning probabilities of the bets are known to them.
[3] These are objective probabilities. Objective probability of an event is defined as the long run relative frequency when the experiment is repeated (the race is run) infinitely many times under the same conditions.

an implication of an efficient market, cannot be rejected. Section 2 outlines the construction of two such bets and describes the methods of analysis. Empirical findings and conclusions are in Section 3.

2. METHODOLOGY

There are various betting opportunities at a racetrack. The odds on a bet are profits per dollar bet to a successful bettor and are determined from the bets made by the public, and the track-take and breakage. A fixed proportion of the amount bet in a race is taken out by the track for maintenance costs, taxes, and profits, before it distributes the rest to the successful bettors. This proportion is known as the track-take. The breakage arises because of the following two restrictions: (a) odds cannot be below a certain prescribed minimum, and (b) odds are rounded downward except when (a) is in effect, in which case it is rounded upward. For the races analyzed, the odds are rounded to either ten or five cents and the minimum odds are also either ten or five cents depending on the particular racetrack.

Odds for different types of bets—win, place, show, daily double, quinella, etc.—are determined separately. Let us take the case of a win bet. Suppose there are H horses numbered $1, 2, \ldots, h, \ldots, H$, and let X_h be the total amount of bet for horse h to win. Then $W = \Sigma_{h=1}^{H} X_h$ is the total win bet in the race also known as the win "pool". Let the track-take and breakage be α; then the total money to be distributed to the successful win bettors is $(1 - \alpha)W$ and the odds on horse h are

$$a_h = [(1 - \alpha)W - X_h]/X_h \qquad\qquad (h = 1, 2, \ldots, H).$$

In other words, return per dollar bet to a successful bettor is $(1 + a_h)$ and the return is zero to an unsuccessful bettor. Thus, the returns per dollar bet are market determined through a competitive bidding of the bettors.

In the daily double bet one chooses a horse in the first race and another in the second race during a racing day(night) and before the first race is run. The bet is successful only when the chosen horses win their respective races. A parlay can be constructed by choosing the same two horses *before* the first race is run where a bet is made on the horse in the first race to win and if it wins, the total return from this successful bet is bet on the horse in the second race to win. It can be verified that winning probabilities of such a daily double and parlay are the same.

Let D be the return per dollar bet on a daily double and P be the return on a parlay. D can be observed directly, but P must be derived. If the odds on the win bets for the chosen horses in the first and second races are a and b, respectively, then P can be shown to be $(1 + a)(1 + b)$. Note that a, b, and D are determined from distinctly different betting pools and hence D and P are separately market determined.

The cost of a bet can be defined to be the amount that must be paid for a dollar return from a successful bet. This will be the reciprocal of the return. As a fraction of every wager made represents track-take and breakage, which can be interpreted as transaction cost, the cost of a bet has two components: transaction cost and price of the bet. Thus, the price and cost of a bet would be identical if there is no transaction cost. If the return from a bet is R when the track-take and breakage (transaction cost) is α, then the return would have been $R/(1-\alpha)$ in the absence of such a transaction cost; and as the cost of a bet is the reciprocal of its return, the cost of the bet in the absence of the transaction cost would have been $(1-\alpha)/R$ which would then also be the price of the bet. It then follows that the price of a daily double is $(1-\alpha)/D$, where D is the daily double return. If the return on a parlay is P, then $P=(1+a)(1+b)$, where $1+a$ is the return from the bet for the chosen horse in the first race to win and $(1+b)$ is the return from the bet for the chosen horse in the second race to win. Without the track-take and breakage of α, the returns on these win bets would have been $(1+a)/(1-\alpha)$ and $(1+b)/(1-\alpha)$, respectively, and therefore, the parlay return would have been $P/(1-\alpha)^2$. Hence, the price of the parlay is $(1-\alpha)^2/P$. It can be verified that the fraction of the total cost representing the transaction cost is higher for a parlay bet than for a daily double bet.

It can be seen that the daily doubles and the parlays are priced separately. Any daily double is priced in relation to all possible daily doubles involving the first two races. The parlay prices result from the prices of the win bets in the first two races. Any win bet is priced in relation to all possible win bets in that race and in that race alone.

If the market is efficient, the bets will be valued according to their intrinsic worth, i.e., their probability distributions alone, and thus, the price of a parlay bet will equal the price of the corresponding daily double bet. However, if the speculative motive of the bettors plays a significant role in pricing the bets, then this equality cannot be guaranteed; rather it is likely to be violated. If the price of a parlay differs from that of the corresponding daily double, there does not seem to exist any motivation from the mere rationality of the bettors that can equalize these prices. If the cost (price plus transaction cost) of a parlay is below the cost of the corresponding daily double, the potential bettors of the daily double will find it profitable to bet the parlay and such a betting may equalize the costs. However, as the transaction cost of the parlay bet is higher than that of the daily double bet, the price of the parlay can still be below the price of the daily double, i.e., the price inequality can exist. Utilizing the definition of odds on a bet, it can be verified that the sum of the costs of all the parlay bets involving the first two races is $1/(1-\alpha)^2$, whereas the corresponding sum for the daily double bets is $1/(1-\alpha)$. As $1/(1-\alpha)^2$ is larger than $1/(1-\alpha)$, it follows that inevitably the cost of at least some parlay will be larger than the cost of the corresponding daily double. However, mere higher cost of a parlay over the corresponding daily double cost does not guarantee the equality in their prices. The preceding discussions show that the price equality implication of the efficient market hypothesis is not an empty proposition.

The observed daily double return, D, can be different from its true value. The true daily double return, D_T, is defined to be the one which is obtained if there is no friction in the market. The true parlay return, P_T, is similarly defined. In practice, the market may not be free from frictions. For example, there is no secondary market where bettors can exchange bets already made. Further, no bet can be less than \$2 and some bets cannot be less than \$3. Moreover, the equilibrating process which adjusts D towards D_T works for only a finite length of time. There may be numerous other factors influencing the workings of the market. However, none of these seemed to have a systematic effect on the determination of D. The aggregate influence of all these factors will be called unsystematic effects on D and we will view

$$D = D_T + e_D$$

where e_D is the error in the observation due to the unsystematic effects. The error e_D can be positive or negative and can be assumed to have zero mean in repeated sampling. D_T is expected to depend upon the winning probability of the bet. As this probability differs from observation to observation, D_T will be different at different observations. However, as the unsystematic effects on D can be assumed to be independent and remain invariant in basic structure from observation to observation, e_D's can be taken as a random sample from a population with zero mean and unknown variance.

We view the observed parlay return, P, in a similar fashion so that

$$P = P_T + e_P$$

where e_P is the error in observation and the e_P's are a random sample from a population with zero mean and unknown variance.

Following our earlier discussions, the price of a daily double with a return, D_T is $(1-\alpha)/D_T$ and the price of the corresponding parlay with a return, P_T is $(1-\alpha)^2/P_T$. Thus, a test of the efficient market hypothesis can be achieved by testing the implication, $(1-\alpha)/D_T = (1-\alpha)^2/P_T$, or $(1-\alpha)D_T = P_T$.

In order to test $(1-\alpha)D_T = P_T$, we note the random variable, $(1-\alpha)D - P$ has a mean, $\mu_{D-P} = (1-\alpha)D_T - P_T$ and an unknown variance, σ^2_{D-P}. Thus, the postulated hypothesis can be tested by testing to see whether $\mu_{D-P} = 0$ or not.

From a sample of size N, an unbiased estimate of μ_{D-P} and that of σ^2_{D-P} can be obtained, respectively, as

$$\hat{\mu}_{D-P} = \frac{1}{N}\Sigma X_i = \bar{X}$$

and

$$\hat{\sigma}^2_{D-P} = \frac{1}{N-1}\Sigma (X_i - \bar{X})^2$$

where $X_i = (1-\alpha)D_i - P_i$ is the ith observation. Thus, an unbiased estimate of the variance of $\hat{\mu}_{D-P}$ is $\hat{\sigma}^2_{D-P}/N$ and the standard error of the estimate, $\hat{\mu}_{D-P}$, can be computed as

$$S.E.(\hat{\mu}_{D-P}) = \hat{\sigma}_{D-P}/\sqrt{N}.$$

For large N, it is well known that the sampling distribution of

$$Z = \frac{\hat{\mu}_{D-P} - \mu_{D-P}}{S.E.(\hat{\mu}_{D-P})}$$

is well approximated by a normal distribution with mean zero and unit variance. Thus, the hypothesis, $\mu_{D-P} = 0$, can be tested using the statistic Z, and referring it to the standardized normal distribution.

3. EMPIRICAL FINDINGS AND CONCLUSIONS

The data were collected from the race results published in *The Horseman and Fair World* during September through December, 1975. The data consist of 1,089 observations from 34 racetracks in the U.S.A. and Canada. Each observation consists of the winning daily double return, D, and the return on a \$2 bet on each of the horses finishing first in the first and second races of a racing day (night). From these returns, the respective odds on a win bet are constructed and therefrom the parlay return, P is computed. The track-take and breakage, α is taken to be 0.18 which is an average α found in Ali [1] from an analysis of over 20,000 similar races.[4,5] We find

$$\hat{\mu}_{D-P} = -0.5931$$

and

$$S.E.(\hat{\mu}_{D-P}) = 0.5673,$$

so that the computed Z when the market is efficient is -1.0455. Thus, μ_{D-P} does not differ significantly from zero and hence the efficient market hypothesis cannot be rejected.

In conclusion, two bets of equal winning probabilities have been shown to be equally priced which is an implication of an efficient market. One of the two bets is a daily double and the other is the corresponding parlay. The data are obtained from a controlled experiment conducted under a natural environment of the decision makers. This can be contrasted with various laboratory studies (Preston and Baratta [5], Mosteller and Nogee [4], Rosett [7]) to learn human decision behavior under uncertainty. The conclusion derived is independent of any behavioral assumption such as risk preference, risk aversion or risk neutrality of the individuals except that they are rational. The validity of the conclusion requires no assumption regarding the decision makers' perception of the probabilities, i.e. the bettors need not be sophisticated as assumed by Rosett [6]. Nor is it necessary to assume that decision makers are expected utility maximizers. In

---

[4] Utilizing the definition of a_h, α can be obtained from the identity,

$$\sum_{h=1}^{H} \left(\frac{1}{1+a_h} \right) = \frac{1}{1-\alpha}$$

where a_h are the odds on horse h to win a race and H is the number of horses in the race.

[5] Although the track-take and breakage vary from race to race and also from racetrack to racetrack, their variability is negligible. The coefficient of variation of α in the above study was less than one per cent.

M. M. ALI

short, the conclusion of the paper does not depend on any specific valuation process of the decision makers.[6] This is a novelty not usually found in the works dealing with market efficiency.

University of Kentucky

Manuscript received February, 1977; final revision received April, 1978.

REFERENCES

[1] ALI, M. M.: "Probability and Utility Estimates for Racetrack Bettors," *Journal of Political Economy*, 85 (1977), 803–816.
[2] BAUMOL, W. J.: *The Stock Market and Economic Efficiency*. New York: Fordham University Press, 1965.
[3] KEYNES, J. M.: *The General Theory of Employment Interest and Money*. New York: Harcourt, Brace and Co., 1935.
[4] MOSTELLER, F., AND P. NOGEE: "An Experimental Measurement of Utility," *Journal of Political Economy*, 59 (1951), 371–404.
[5] PRESTON, M. G., AND P. BARATTA: "An Experimental Study of the Auction-value of an Uncertain Outcome," *American Journal of Psychology*, 61 (1948), 183–193.
[6] ROSETT, R. N.: "Gambling and Rationality," *Journal of Political Economy*, 73 (1965), 595–607.
[7] ————: "Weak Experimental Verification of the Expected Utility Hypothesis," *Review of Economic Studies*, 37 (1971), 481–492.

[6] The discounted future earnings of a stock is often viewed as its value. There are numerous ways to make this valuation process operational. Any test for market efficiency relying on such an arbitrary valuation process is questionable because the failure of a market in the efficiency test can be ascribed to the possible invalidity of the postulated valuation model.

Efficiency and Profitability in Exotic Bets

By Peter Asch and Richard E. Quandt

Rutgers University and Princeton University

Final version received 4 September 1986. Accepted 22 September 1986.

The efficiency and profitability of exotic racetrack bets such as exactas and daily doubles are examined. Efficiency is understood to mean that above average returns cannot be made in the long run once risk is appropriately controlled for. The markets in question are found not to be efficient; the inefficiencies, however, are insufficient to permit simple strategies to show a consistent profit. Some evidence of "smart money" exists in that holders of inside information may bet on exactas rather than equivalent standard bets in order to avoid signalling their actions to the betting public.

Introduction

The efficient markets hypothesis has found considerable empirical support in studies of securities markets, where efficiency is taken to mean that consistently above-average profits cannot be gained from investing—or at least that such profits cannot be made once the differential risks of investment vehicles are taken into account.

Investigation of the efficiency of gambling markets has focused on racetrack betting. 'Efficiency' has often been interpreted as the inability to pursue a betting strategy that yields a rate of return significantly above the average loss to all bettors; this loss, which averages 17-21 per cent at US racetracks, is a function of the track take on various types of bets.[1] Some investigators find that betting markets are reasonably efficient (Dowie, 1976; Snyder, 1978; Figlewski, 1979; Ali, 1979), whereas other see significant departures from efficiency (Asch, Malkiel and Quandt, 1982, 1984; Hausch, Ziemba and Rubinstein, 1981; Ziemba and Hausch, 1984; Crafts, 1985; Zuber, Gandar and Bowers, 1985). Generally, the findings of some inefficiency seem convincing enough to suggest that complete efficiency is not the rule. Inefficiency, of course, does not necessarily mean that racetrack betting is profitable. An increase in the expected rate of return from -0.18 to -0.05, attributable to some clever betting strategy, suggests inefficiency but does not imply profitability.

In this paper we consider betting that occurs at the racetrack. Off-track betting is permitted at some locations in the United States, but is far less significant quantitatively than in Britain. Bets at US racetracks occur under a *pari mutuel* system that calculates and continually updates betting odds based on all wagers. Payoffs to all types of bets are based on the final odds that prevail at the end of each betting period. Bettors thus have no opportunity to contract for wagers at 'earlier' odds, and cannot be certain about the odds that will govern the payoff to a successful bet.

If W_i is the amount of money bet on horse i, W is the total amount bet on all horses in a race, t is the track take and p_i is the objective probability that horse i wins, the expected return to a bet is

$$R_i = p_i \frac{W(1-t)}{W_i} - 1.$$

If the bet is to have positive expected return, the proportion of all dollars bet on the horse must not exceed $p_i(1-t)$. To devise a profitable betting strategy thus requires the identification not only of winners, but of winners that the betting public does not regard as winners.

This paper investigates the efficiency of a class of wagers known as 'exotic bets'. The members of this class are compound bets, and require more complicated estimates of outcomes than do the familiar win, place and show bets. Because of the track take, all racetrack bets tend to have negative expected returns.

Even if it is difficult to explain why individuals engage repetitively in such a money-losing activity, it is interesting and important to examine whether differential avenues to losing money tend to equalize rates of return. This is the basic concern of the present paper. In Section I we characterize these bets and the data to be employed. In Section II we formulate our hypotheses concerning exactas and present our analysis of the data. Section III proceeds analogously with respect to daily doubles. Section IV presents an explanation of a discrepancy between the exacta and daily double findings. Some brief conclusions are presented in Section V.

I. Exotic Bets and Data Sources

The standard racetrack wagers are win, place and show bets. The minimum bet is usually $2, and a win bet on horse i is successful if and only if horse i wins the race. In this case, the payoff per dollar bet is $W(1-t)/W_i$, where the terms are as defined above. A place bet on horse i is successful if horse i comes in first or second; and a show bet is successful if horse i comes in first, second or third.[2]

Exotic bets involve at least two simultaneous wagers on different horses. In this paper, we investigate the exacta and the daily double. In the exacta, the bettor must pick two horses: one to win and one to come in second. The bet is successful if both picks are correct, and is unsuccessful otherwise. In the daily double, the bettor picks the winners of two consecutive races; the bet is successful if both picks win, and is unsuccessful otherwise. The racetrack designates the races in which such betting is available.

It is obvious that the exotic bets are relatively low-probability wagers. Since the track take also tends to be higher on exotic bets,[3] these bets should appeal to relative risk-lovers. Perhaps the most widely established empirical regularity in racetrack gambling concerns win bets: horses with low probabilities of winning ('long shots') are overbet; and horses with high probabilities of winning ('favourites') are underbet. That is, the proportion of all moneys bet on long shots is greater than their objective winning probability, with the reverse holding for favourites (Rosett, 1965, 1971; Snyder, 1978; Ali, 1979; Asch, Malkiel and Quandt, 1982). As a result, rates of return are higher for favourites than for long shots.

Whether a similar pattern ought to be expected for exactas or daily doubles is unclear. In straight win betting, one may easily distinguish between 'high-probability' bets (with success probabilities of perhaps $0.3-0.6$) and 'low-probability' bets (in the range of $0.02-0.05$). For exactas, however, the probabilities of success may lie largely in the range $0.05-0.001$; and we do not know whether bettors clearly discriminate among probabilities in this narrow range.

A second, and potentially more interesting, feature of exactas and daily doubles is that (unlike win, place and show betting) the amounts wagered on the various combinations are not continuously displayed at the track.[4] This may have important signalling consequences. If certain bettors are privy to inside information ('smart money' is, of course, part of the folklore of racetrack betting), it may be difficult to take advantage of this in straight win (and perhaps in place or show) betting, because one's bet is immediately revealed to the crowd, and may induce 'following' behaviour. This would drive down the odds and prospective payoffs, thereby destroying the usefulness of the inside information.

Such revelation is far less likely in exotic betting. Whereas a major discrepancy in wagering might be spotted by a few bettors, the signal conveyed to the crowd as a whole is almost always obscure. Thus, acting on inside information via exotic bets is safer. However, it also incurs the cost of having to bet on two or more horses simultaneously, while inside information may be confined to the prospects of a single animal.

The data employed in this study comprise the results of harness racing between 26 May and 11 August 1984 at the Meadowlands Racetrack. A total of 705 usable races took place (we eliminated a small number of anomalies such as the existence of two winners); exacta betting was available in 510 of these, and daily double betting in 122 pairs of races.

II. ANALYSIS OF EXACTA BETTING

In a race with n horses, there are $n(n-1)$ different exacta bets. We denote by B_{ij} the amount bet by all bettors on the exacta that horse i will win and j will run second in a given race. If B is the total amount bet on all exacta combinations in the race and t is the track take, then the payoff to the (ij)-exacta per dollar wagered, if successful, is

$$(1) \qquad A_{ij} = (1-t)B/B_{ij}.$$

If h_{ij} is the objective probability that the (ij)-exacta is successful, the expected value E_{ij} and the variance V_{ij} of return are

$$(2) \qquad E_{ij} = A_{ij}h_{ij} - 1$$

$$(3) \qquad V_{ij} = (A_{ij} - 1)^2 h_{ij} + (1 - h_{ij}) - E_{ij}.$$

The computation of these requires estimates of the objective probabilities, a problem that we consider later. We first examine average rates of return to various exactas.

Average rates of return

In analysing the profitability of straight win bets and the relationship of profitability to the 'market's' assessment of winning probabilities, investigators have often computed the actual average rates of return to horses in various odds classes.[5] In the case of exactas, there are no odds as such, but we can categorize exactas by the value of the (potential) payoff A_{ij}. The average rate of return in a class of A_{ij} values is $(A_{ij}S_{ij} - 1)/N$, where $S_{ij} = 1$ if the (ij)-exacta is successful and 0 otherwise, and N is the number of exactas over which the average is taken.

The average rates of return to exacta bets show an erratic, if not random, relationship with respect to payoff class.[6] It is not surprising that some rates of return are better than $-0{\cdot}19$ (the track take) and not even surprising that a few are positive; what is significant is that these better-than-expected rates of return do not appear to be related to payoff class. Thus, we cannot confirm the existence of a typical underbetting/overbetting bias.

Arbitrage and efficiency

The bets on various horses in the win pool define their subjective winning probabilities and thus, implicitly, define the subjective success probabilities for particular exactas. At the same time, the amounts bet on each possible exacta combination divided by the total exacta pool provide a direct estimate of the subjective success probabilities. If information is used consistently by all bettors, the two 'markets' will be fully arbitraged and the two sets of probabilities will be (statistically) the same.

To test whether this is so, we first compute the subjective winning probabilities of each horse in each race from the straight win bets (subjective probability $s_i = W_i/W$). To obtain an estimate of the objective winning probabilities p_i, we have aggregated horses into classes. We have 705 races which contained 6729 horses; these were aggregated into 20 classes, with class 1 containing the 337 horses with the lowest s_i, class 2 the 337 horses with the next lowest s_i, etc. (Some classes contain 336 horses.) For each of the 20 classes, we computed the mean of the subjective probabilities of the horses in that class $\bar{s}_1, \bar{s}_2, \ldots, \bar{s}_{20}$ and an estimate of the objective probabilities $p_1, \ldots, p_{20}$, by determining the proportion of horses that actually won in each class. These are displayed in Table 1, and confirm the existence of the familiar underbetting/overbetting bias.

The average relationship between the objective and subjective probabilities is now determined by regressing p_i on $\bar{s}_i$. The estimated equation is:

(4) $\qquad p = -0\cdot0100 + 1\cdot0959\bar{s}$

$\qquad\quad (-1\cdot9965) \ \ (31\cdot5177)$

(parenthesized values denote t-statistics throughout), and $R^2 = 0\cdot9803$.[7]

The objective probability of success in the (ij)-exacta is then obtained as follows. (1) Substitute in (4) the actual subjective probability of each horse, s_i, and compute an estimate p_i of the objective winning probability of that horse. (2) Apply the Harville (1973) approximation to these p_i to obtain an estimate that horse i wins *and* horse j is second. This approximation is:[8]

(5) $\qquad h_{ij} = \dfrac{\hat{p}_i\hat{p}_j}{1 - \hat{p}_i}.$

Finally, obtain an estimate of the implicit subjective probability of success for the (ij)-exacta by using (4): substitute $h_{ij} = -0\cdot0100 + 1\cdot0959s_{ij}^*$ and solve for s_{ij}^*.

Of course, as already noted, a direct measure of the subjective probability of success for the (ij)-exacta is given by $s_{ij} = B_{ij}/B$. The key observation is that, if the market is efficient, it must evaluate the winning and runner-up

TABLE 1

OBJECTIVE (p_i) AND THE MEAN
SUBJECTIVE ($\bar{s}_i$) WINNING
PROBABILITY ESTIMATES FOR
MEADOWLANDS DATA

p_i	$\bar{s}_i$
0·0030	0·0068
0·0030	0·0110
0·0059	0·0151
0·0119	0·0201
0·0178	0·0258
0·0445	0·0323
0·0297	0·0398
0·0386	0·0489
0·0415	0·0581
0·0804	0·0683
0·0804	0·0787
0·0923	0·0908
0·1310	0·1036
0·1161	0·1185
0·1250	0·1368
0·1637	0·1587
0·1548	0·1841
0·2024	0·2220
0·3006	0·2747
0·4554	0·4031

chances of individual horses in mathematically consistent fashion; thus the probability estimates s_{ij} and s_{ij}^* must be the same pairwise, on average. This hypothesis can be tested by regressing s_{ij}^* on s_{ij} in the equation

$$s_{ij}^* = a + bs_{ij} + u_{ij},$$

and testing H_0: $a = 0$, $b = 1$. The regression is based on 41,246 exacta pairs and is:

$$s_{ij}^* = 0 \cdot 00683 \ + \ 1 \cdot 10267 s_{ij}$$

$$(186.374) \quad (683 \cdot 350)$$

$R^2 = 0 \cdot 9190$. The null hypothesis is thus rejected at all conventional significance levels.

The 'smart money' hypothesis

Above we have compared the subjective probability estimates obtained from the exacta pool with those implicit in the win pools. We can also compare the payoffs to potential winners based on the exacta pool with those implicitly given by the s_{ij} calculated from the win pool. The actual payoffs that would accrue to (potential) winners are:

$$(6) \qquad A_{ij} = \frac{(1-t)B}{B_{ij}} = \frac{(1-t)}{s_{ij}}.$$

The implicit payoffs, denoted by A_{ij}^*, are:

$$(7) \qquad A_{ij}^* = \frac{(1-t)}{s_{ij}^*}.$$

A 'smart money' hypothesis can be tested by examining the differences $A_{ij}^* - A_{ij}$. We group the differences for all 41,246 exacta pairs as follows: group I contains the $510 A_{ij}^* - A_{ij}$ values for exacta combinations that actually won, and group II contains the values for the remaining combinations.

Suppose that there are bettors with inside information. To avoid sending clear signals, they utilize their information by betting on exactas. Whereas, in the no-inside-information case, we would expect the distribution of differences $A_{ij}^* - A_{ij}$ to be statistically the same for groups I and II, the 'smart money' hypothesis suggests that actual payoffs A_{ij} will tend to be depressed relative to A_{ij}^* for exacta winners (group I). The Kolmogorov–Smirnov test rejects the hypothesis that the two distributions of $A_{ij}^* - A_{ij}$ values were drawn from the same parent distribution.

The means and variances of the two groups are shown in Table 2. The differences in means are highly significant and are in the direction predicted by the 'smart money' hypothesis.

TABLE 2

MEANS AND VARIANCES FOR $A_{ij}^* - A_{ij}$ FOR WINNING AND
LOSING EXACTA PAIRS

	Winning	Losing
No. of pairs	510	40,736
Mean $A_{ij}^* - A_{ij}$	−33·18	−261·69
Variance $A_{ij}^* - A_{ij}$	103·84	458·85

Efficiency in a mean-variance framework?

Since racetrack betting is an unfair game, then, barring non-economic motives, racetrack bettors must be risk-lovers. Since mean return and variance are both 'goods' in this case, the indifference curves of the expected utility function must have negative slope, as in the conventional case. For equilibrium to be possible, it is necessary that the boundary of the feasible region, the 'mean-variance locus', also have negative slope. This can be examined by computing (2) and (3) for every exacta in every race and checking whether $E_{ij} > E_{k1}$ if and only if $V_{ij} \leq V_{k1}$. The facts are that in no race was the mean-variance locus uniformly of the slope compatible with efficiency.

III. ANALYSIS OF DAILY DOUBLE BETTING

The daily double provides a straightforward comparison between the compound (exotic) bet and the equivalent pair of single-horse bets. Consider betting on a horse i in the first of two races and, if it wins, betting the entire payoff on horse j in the second race. This bet, called a *parlay*, is in effect a daily double bet. Both the daily double and the parlay are successful if and only if both chosen horses win their respective races.

We now consider only winning daily double pairs. Their actual payoffs are denoted by D_k ($k = 1, \ldots, 122$). For each of these, we can also compute the equivalent parlay payoffs, denoted by L_k. If the market is efficient, it should not, on average, make any difference whether one bets a given pair of horses via the daily double or a parlay.

This is not the case, however. The average of the daily double payoffs is \$52·54 and the parlay payoffs \$41·38, suggesting that it is substantially more profitable to bet on the daily double.[9] The reason may well be that the commitment to a daily double must be made before the start of the first of two consecutive races, i.e., before the evolution of odds in race no. 2, which may play an important role in revealing relevant information.

Since winning daily double combinations appear to be underbet, one must suspect that a 'smart money' hypothesis is unlikely to be supported here. This is indeed the case, when we compute the $D_k - L_k$ differences for all daily double pairs and group them as we did for exactas. The comparison between the groups suggests the opposite of the 'smart money' hypothesis; that is, the mean $D_k - L_k$ is smaller for losing pairs than for winning pairs.[10]

V. The Discrepancy between Exactas and Daily Doubles

It is interesting to consider why the analysis of exacta and daily double betting suggests diametrically opposed conclusions with respect to the 'smart money' hypothesis. We suggest the following answer, based on considerations of signalling.

If there are n_1 horses running in the first and n_2 in the second race of a daily double pair, the total number of potential daily double bets is $n_1 n_2$. As soon as the first race is over, the number of possible winning bets is reduced to n_2, since the winner of the first race is now known. Racetracks customarily display, before the beginning of the second race, 'will pay' amounts—i.e., the sums that would be paid to ticket-holders on each of the still-possible winning combinations. These figures contain potentially important information.

Suppose that insider information has suggested to some bettors that horse i_1 in race 1 and horse j_2 in race 2 are likely to win. The combination $i_1 j_2$ will then be heavily bet by insiders in the daily double if, as we conjecture, they attempt to hide their actions from the betting crowd. After the first race is over, however, the 'will pay' figure allow the public at large to observe that horse j_2 in race 2 has been heavily bet in the daily double *relative to the odds that are evolving in the straight win betting for race 2*. The signal that insiders have hidden in the daily double betting is now revealed.

Racetrack bettors can no longer wager on the daily double; but they can follow the signal that has emerged from the 'will pay' numbers by backing horse j_2 to win in race 2. Such betting will reduce the odds and winning payoff to j_2, and also will reduce the profitability of a parlay on combination $i_1 j_2$. In effect, bettors believe the 'smart money' hypothesis that we have suggested above, and act accordingly.

There is some evidence that the following type of behaviour exists. Let f_j denote the payoff to horse j in straight win betting, m_j the payoff to j that would occur on the basis of morning line odds, and z_j some appropriate measure of the payoff to j in daily double betting *relative* to the payoff based on morning line odds such as their ratio. Since the morning line represents a professional's estimate of the intrinsic quality of a horse, it will be positively related to the final odds. If imitative (following) behaviour does occur, then a relatively low z_j will induce a relatively low value of f_j. Thus, in the regression equation

$$f_j = \alpha_0 + \alpha_1 m_j + \alpha_2 z_j + u_j$$

where u_j represents the error term, we expect, *a priori*, $\alpha_1 > 0$, $\alpha_2 > 0$. This regression yields:

$$f_j = -16 \cdot 204 + 2 \cdot 157 m_j + 10 \cdot 470 z_j$$

$$(-20 \cdot 503) \quad (35 \cdot 750) \quad (27 \cdot 000)$$

where $R^2 = 0 \cdot 706$. Qualitatively similar results emerge if z_j is defined not as a ratio of payoffs but as a difference in payoff ranks, or if the regression equation is run only over observations representing winning horses, or only over observations for losing horses.

Since the presence of imitative behaviour appears to be confirmed, we would not expect parlays to be more profitable than daily double bets. It is still not clear, however, why parlays are actually less profitable. One possibility is that the public overreacts to the signal perceived in the will-pay figures, in effect attaching too much weight to 'new' information. Such tendencies have been observed in other settings by experimental psychologists (Kahneman and Tversky, 1982); and there is plausible evidence of this sort of overreaction in stock markets (de Bondt and Thaler, 1985). The patterns of betting that we have described are clearly consistent with an overreaction phenomenon. The issue, however, deserves fuller investigation.

V. CONCLUSION

Are *pari mutuel* markets for exotic bets efficient? And is inside information ('smart money') a verifiable source of any observed inefficiency? Our analysis of two types of exotic bets, exactas and daily doubles, yields the following conclusions. (1) Both of these betting markets exhibit inefficiency in the sense that their payoffs are not statistically the same as the payoffs to the analogous bets on individual horses; in addition, the mean–variance locus of exacta combinations is inconsistent with efficiency for risk-loving bettors. (2) The differences in payoffs support the 'smart money' hypothesis for exacta betting, but not for daily double betting.[11]

ACKNOWLEDGMENTS

We are grateful to Bruce H. Garland, deputy director of the New Jersey Racing Commission, Joseph J. Malan, director of mutuels at the Meadowlands Racetrack, and Joe LaVista, New Jersey State auditor, for their help in providing the data, to the Sloan Foundation and the National Science Foundation for financial support, and to two referees for useful comments.

NOTES

1. The track take is the amount taken out of each bet by the racetrack to cover taxes, expenses and profits, plus the amount ('breakage') resulting from rounding payoffs down to the nearest 10 or 20 cents.
2. The payoffs to these bets are more difficult to calculate. The general rule is: first, the original bets of all successful place (or show) bettors are refunded; then the remaining portion of the place (or show) betting pool (net of the track take) is divided equally between the two (in the case of show bets, three) groups of individuals holding successful place (show) tickets. It follows immediately that the reward to holders of a place ticket on horse *i* depends on which other horse *j* also places. For the exact formulas, which also take breakage into account, see Ziemba and Hausch (1984).
3. The takes are approximately 0·18 for win, place and show bets, and somewhat over 0·19 for exactas at the Meadowlands Racetrack.
4. 'Probable payoff' numbers are usually displayed sequentially at certain locations. If the typical race contains ten entrants, a bettor searching for market information about a two-horse exacta must view and 'process' 90 possibilities.

5. If W_i is the amount bet on horse i, the subjective probability that i will win is W_i/W. Since the odds on horse i are defined as $D_i = W(1-t)/W - 1$, the subjective probability $= (1-t)/(1+D_i)$ and is monotone in D_i. See Asch, Malkiel and Quandt (1982).

6. We observe in the tables below that rates of return to exacta combinations in various payoff classes exhibit no systematic tendency. This is consistent with a failure of bettors to distinguish between high- and low-probability combinations (see table). The prevalence of negative returns raises the question of whether there are any 'strategies' that will yield positive profit in betting on exactas. Using the logit model of Asch, Malkiel, Quandt (1984) to predict winners and runners-up, and simulating various 'reasonable' betting strategies, indicates that on the average profits are significantly negative.

Rates of Return on Exacta Bets: Constant Class Invervals

Payoff range	No. of pairs in sample	No. of winners in category	Rate of return
0–16·00	1896	162	−0·11
16·01–32·00	3653	110	−0·30
32·01–48·00	3139	68	−0·13
48·01–64·00	2676	38	−0·26
64·01–80·00	2317	34	0·05
80·01–96·00	1866	18	−0·13
96·01–112·00	1757	17	−0·04
112·01–128·00	1469	11	−0·24
128·01–144·00	1347	5	−0·54
144·01–160·00	1188	8	0·10
160·01–∞	19938	39	−0·46

7. It is interesting to note that this regression is reasonably similar to that obtained by Fabricand (1965) on the basis of a completely different sample.

8. The simplest intuition for this is as follows: $\hat{O}_i$ is the probability that i wins; $\hat{p}_j/(1-\hat{p}_i)$ is the conditional probability that j wins in a race from which i is missing. If the two events are independent, the product is the required probability.

9. The difference is significant at the 0·05 level. It is almost precisely accounted for by the fact that daily double bettors pay a single track take of approximately 19 per cent, whereas parlay bettors pay two takes of about 18 per cent. When the parlay payoffs are adjusted as if parlay bettors paid a single 'daily double take' (multiply each L_k by 0·8091 and divide by $0·8203^2$), the mean L_k rises to $49·75, and is no longer significantly different from the mean D_k. It is as if daily double bettors did not take into account that only one (slightly higher) take is assessed against their bets.

10. We may also consider the issue of daily double profitability by predicting the winner of each component of a daily double race pair on the basis of our logit model, and simulating a betting strategy. The result of this simulation for 122 daily double pairs is an average rate of return of −0·059, substantially better than the loss implied by the track take, but not profitable. Overall, we find inefficiency and lack of profitability, as in the case of exacta betting; but we do not confirm the 'smart money' hypothesis.

11. We have reported elsewhere (Asch, Malkiel and Quandt, 1982, 1984) some evidence suggesting that 'smart money' may show up in the form of *late* betting in the 'win bet market'.

REFERENCES

ALI, MUKTAR M. (1979). Some evidence of the efficiency of a speculative market. *Econometrica*, **47**, 387-92.

ASCH, PETER, MALKIEL, BURTON G. and QUANDT, RICHARD E. (1982). Racetrack betting and informed behavior. *Journal of Financial Economics*, **10**, 187-94.

—— (1984). Market efficiency in racetrack betting. *Journal of Business*, **57**, 165-75.

CRAFTS, N. F. R. (1985). Some evidence of insider knowledge in horse race betting in Britain. *Economica*, **52**, 295-304.

DE BONDT, W. F. H. and THALER, R. (1985). Does the stock market overreact? *Journal of Finance*, **40**, 793-805.

DOWIE, J. (1976). On the efficiency and equity of betting markets. *Economica*, **43**, 139-50.

FABRICAND, BURTON F. (1965). *Horse Sense*. New York: David McKay.

FIGLEWSKI, STEPHEN (1979). Subjective information and market efficiency in a betting market. *Journal of Political Economy*, **87**, 75-88.

HARVILLE, D. A. (1973). Assigning probabilities to outcomes of multi-entry competition. *Journal of the American Statistical Association*, **68**, 312-16.

HAUSCH, DONALD B., ZIEMBA, WILLIAM T. and RUBENSTEIN, MARK (1981). Efficiency of the market of racetrack betting. *Management Science*, **17**, 1435-52.

KAHNEMAN, DANIEL and TVERSKY, AMOS (1982). Intuitive prediction: biases and corrective procedures. In Daniel Kahneman, Paul Slovic and Amos Tversky, *Judgment Under Uncertainty; Heuristics and Biases*. Cambridge: Cambridge University Press.

ROSETT, RICHARD N. (1965). Gambling and rationality. *Journal of Political Economy*, **73**, 595-607.

—— (1971). Weak experimental verification of the expected utility hypothesis. *Review of Economic Studies*, **38**, 481-92.

SNYDER, WAYNE N. (1978). Horse racing: testing the efficient markets model. *Journal of Finance*, **22**, 1109-18.

ZIEMBA, WILLIAM T. and HAUSCH, DONALD B. (1984). *Beat the Racetrack*. New York: Harcourt, Brace, Jovanovich.

ZUBER, RICHARD A., GANDAR, JOHN M. and BOWERS, BENNY D. (1985). Beating the spread: testing the efficiency of the gambling market for National Football League games. *Journal of Political Economy*, **93**, 800-6.

How Accurately Do Bettors Bet in Doubles ?

Victor S.Y. Lo and Kelly Busche[1]
The University of British Columbia and the University of Hong Kong

Previous studies in the analysis of double bets in racetrack markets compared the payoffs resulting from a double bet with a parlay contructed by the bettors. We compare the accuracy of implied winning probabilities between double bet and bets based on a single race and suggest that bettors bet on doubles less accurately. We also report the analysis based on payoff comparisons for a new data set.

1. Introduction

The racetrack betting market has long been recognised as a source for investigating market efficiency and risk behavior. The simplest bet is to win which pays off if the single horse chosen by a bettor wins the race. Many previous studies have concentrated on the favorite-longshot bias of the win bet (e.g. Ali(1977), Ziemba and Hausch (1986), Snyder(1978), Busche and Hall (1988), Bacon-Shone, Lo and Busche (1992 a)). Some researchers are interested in the efficiency of bet types which involve more than one race. A double bet pays off if the two horses chosen by a better both win in two consecutive races, and thus it is related to the win events of two races. A parlay is a way by which bettors construct low-probability-high-return bets. The simplest parlay is to bet the entire payoff from a winning horse in the first race to another horse in the next race. When both of the horses in the two races win, the parlay pays off. Therefore, the parlay is a combination of win bets in two consecutive races and it has the same probability of receiving payoff for a double bet. Ali (1979) and Asch and Quandt (1987) compare the payoffs derived from betting on double bets with parlays. Using data sets of double bets in Meadowlands and Hong Kong, we compare the accuracy of implied winning probabilities of double bets with other bet types based on single race. The Hong Kong data is utilized to support Asch and Quandt (1987)'s result that double bets have higher payoffs than parlays.

2. Accuracy of Implied Probabilities for Double Bets

The Meadowlands data has 122 pairs of races for double bets in 1984. This has been analysed by Asch and Quandt (1987) in comparing the payoffs between double bets and parlays. We now compare the accuracy of the double bet fractions with that of win bet fractions for estimating the probability of receiving profit for a double bet. Let $\pi_i^{(1)}$ and $\pi_j^{(2)}$ be the win probabilities of horses i and j in the first and second races, respectively. Betting on horses i and j to double will have the probability of receiving

profit, $\pi_{i,j} = \pi_i^{(1)}\pi_j^{(2)}$ since the win events in the two races are independent. Define $\hat{\pi}_{i,j}^{(d)}$ = the double

bet fraction (the fraction of the total double bet amounts) for i & j in the two races, and $P_i^{(1)}$ and $P_j^{(2)}$ to be the win bet fractions (fractions of money bet on the win market) of i and j in the two races. Since each of these win bet fractions is a good estimate of the associated win probability (e.g. Busche and Hall (1988), Snyder (1978)), $P_i^{(1)}P_j^{(2)}$ can be used to estimate $\pi_{i,j}$. We use the log likelihoods to compare the

---

[1] We thank Richard Quandt for his U.S. data.

accuracy of two different estimates for $\pi_{i;j}$ - Double bet fractions: $\sum_r \log \hat{\pi}^{(d)}_{[1;1],r}$; Win bet fractions

: $\sum_r \log(P^{(1)}_{[1],r} P^{(2)}_{[1],r+1})$ where the summations are over all the available pairs of races, the subscript r

denotes the race number and [1] and [1;1] indicate that the probabilities are associated with the winning horses. These values for the double bet fractions and the win bet fractions are -452.31 and -449.69, respectively. The log likelihoods for discrete events are simply logarithms of the probabilities of observing what we did observe given a particular model. Thus higher log likelihood implies higher chance of observing what we observed and hence the model is closer to reality. To compare the accuracy of the two fractions formally, we apply the Cox non-nested hypothesis test (Cox(1961,62)). To test H_f: double bet fraction reflects the true probability versus H_g: win bet fraction reflects the true probability, the test has no clear null and alternative hypotheses. We evaluate

$$T_f - \frac{(loglik_f - loglik_g) - E(loglik_f - loglik_g \,|\, H_f)}{SD(loglik_f - loglik_g \,|\, H_f)} \quad \text{and} \quad T_g - \frac{(loglik_g - loglik_f) - E(loglik_g - loglik_f \,|\, H_g)}{SD(loglik_g - loglik_f \,|\, H_g)}$$

where $loglik_i$ = the log likelihood under H_i $(i=f,g)$, and $E(.\,|\,H_i)$ and $SD(.\,|\,H_i)$ are the expectation and standard deviation respectively when H_i is true. Under H_i, $T_i \sim N(0,1)$ asymptotically. T_i will be largely negative if H_i is rejected in favor of the alternative hypothesis. In our case, $E(.\,|\,H_i)$ and $SD(.\,|\,H_i)$ can be computed straightforwardly, but their complicated forms are not shown here. Our empirical results are $T_f = -2.55$ and $T_g = -0.98$, and thus we confirm that the double bet fractions are less accurate. A reason seems to be that those bettors interested in the double bets have to place their bets before the first race and they cannot observe the evolution of odds of double bets. But for the win bet, bettors can observe the evolution of odds in both races, which may be important in revealing relevant information.

A similar analysis follows for the 1981-89 Hong Kong data. We consider the double quinella bet since the double bet in Hong Kong is complicated by consolation. Quinella is a bet that pays off if the two horses chosen by a bettor finish first and second regardless of their orders. Double quinella pays off if the four horses chosen finish first and second in the two consecutive races. We now compare the quinella bet fractions, the double quinella bet fractions and the win bet fractions for estimating the probability of making profits for double quinella by computing the relevant log likelihoods. 691 pairs of races are available, the empirical results are shown in Table 1. The log likelihoods based on win bet fractions are found by the Harville model (Harville (1973)) and the Henery model (Henery (1981)),

Table 1
Estimations of probabilities of double quinella occurring (1382 races)

Bet types	log likelihood
Double quinella bet fractions	-4844.64
Quinella bet fractions	-4777.92
Win bet fractions:	
Harville	-4805.14
Henery	-4788.26

respectively. The former assumes that the running times are exponentially distributed while the latter assumes they are normally distributed. Details are discussed in Bacon-Shone, Lo and Busche (1992 b).

Due to unavailability of data for all combinations, we can only compare the log likelihoods directly. The result that the probabilities implied by the win bet fractions are more accurate than the double quinella bet fractions is consistent with the Meadowlands result. Quinella bet fractions are most accurate among these.

3. Testing the Payoffs of Double Bets in Hong Kong

Using a simple t-test, Asch and Quandt (1987) analysed the Meadowlands data and concluded that the mean return of double bet is higher than that of parlay formed by win bet. Using the Hong Kong data, comparing the double quinella bet and the parlay derived from quinella bets in two consecutive races is trivial. To conduct a one or two-sided test H_0: Expected payoff of double quinella bet = Expected payoff of parlay, a paired comparison t-test is used and the pair of observed values are

$(1-t_{dq})/D^{(dq)}_{[12;12]}$ and $(1-t)^2/(Q^{(1)}_{[12]}Q^{(2)}_{[12]})$ where t and t_{dq} are the track take (commission) of the quinella

(0.17) and the double quinella (0.23) respectively, and $D^{(dq)}_{[12;12]}$ and $Q^{(1)}_{[12]}$, $Q^{(2)}_{[12]}$ are respectively the double quinella bet fraction and quinella bet fractions for the winning combinations in the two races. The t-statistic = 3.31 and we reject H_0 at any conventional level[2].

Another test is related to the Hong Kong double bet which involves consolation: if i wins the first race, j and k respectively finish first and second in the second race, the payoff = $0.85(1-t)/D_{i,j}$ and the consolation = $0.15(1-t)/D_{i,k}$ where $D_{i,j}$ = the double bet fraction of i & j and t = the track take = 0.17. A parlay similar to this double bet is: if i wins the first race, betting the entire payoff to place on a horse in the next race, where place bet pays off if the horse goes first or second. 108 pairs of races are available to test (one or two-sided) H_0: Expected payoff of double bet = Expected parlay payoff. The observed values are: total double bet payoff per \$1 bet = $0.85(1-t)/D_{[1;1]} + 0.15(1-t)/D_{[1;2]}$ and parlay

payoff per \$1 bet = $\dfrac{1-t}{P^{(1)}_{[1]}} \left\{ [(1-t)\sum_r Pl^{(2)}_r - Pl^{(2)}_{[1]} - Pl^{(2)}_{[2]}](\dfrac{1}{2Pl^{(2)}_{[1]}} + \dfrac{1}{2Pl^{(2)}_{[2]}}) + 2 \right\}$, where the superscript

denotes the race number (1st or 2nd), [i] denotes the horse finishing ith, [$i;j$] denote the horses finishing ith and jth in the two races, and D, P and Pl are respectively double bet fraction, win bet fraction and total amount of place bet associated with the horse(s) indicated in the subscripts. The t-statistic becomes 3.93 which also rejects H_0 at any reasonable level[3]. Thus, all the findings in this section are consistent with Asch and Quandt (1987).

---

[2]We also follow Asch and Quandt (1987) to replace $(1-t_{dq})$ by $(1-t)$ to equalise the track takes. The the t-statistic drops to 1.31. Hence, the previous signficant difference appears to be due to the effect of the difference in track takes.

[3]We have no way to equalize the track takes because of the complicated form of parlay payoff in this case.

4. Conclusions

First, the double bet fractions and the double quinella bet fractions are less accurate than the win bet fractions and the quinella bet fractions in predicting probabilities. One explanation for both findings is that bettors betting on the double or the double quinellas cannot observe the evolution of odds of any bet type or updated information for the second race which may indicate some useful information at the time they place their bets. Also, if bettors with better information bet on the double, their information would be released to the public in the second race, so it is expected that they would try to bet as late as possible in the win market. Second, we conclude that the double bet or the double quinella bet has a higher expected payoff than the self-constructed parlays. This may be because of the fact that it has less takes.

References

Ali,M.M., 1977, Probability and utility estimates for racetrack bettors, *Journal of Political Economy* 85, 803-815.

Ali,M.M., 1979, Some evidence on the efficiency of a speculative market, *Econometrica* 47, 387-392.

Asch,P. and R.E. Quandt, 1987, Efficiency and profitability in exotic bets, *Economica* 54, 289-298.

Bacon-Shone,J., V.S.Y. Lo and K. Busche, 1992a, Modelling winning probability, Research paper 10, Department of Statistics, University of Hong Kong.

Bacon-Shone,J., V.S.Y. Lo and K. Busche, 1992b, Logistic analyses of complicated bets, Research paper 11, Department of Statistics, University of Hong Kong.

Busche,K. and C.D. Hall, 1988, An exception to the risk preference anomaly, *Journal of Business* 61, 337-346.

Cox,D.R., 1961, Tests of separate families of hypotheses, *Proceedings of the Fourth Berkeley Symposium* 1, 105-123.

Cox,D.R., 1962, Further results on tests of separate families of hypotheses, *Journal of Royal Statistical Society* B 24, 406-424.

Harville,D.A., 1973, Assigning probabilities to the outcomes of multi-entry competitions, *Journal of the American Statistical Association* 68, 312-316.

Henery,R.J., 1981, Permutation probabilities as models for horse races, *Journal of Royal Statistical Society* B 43, 86-91.

Snyder,W.W., 1978, Horse racing: Testing the efficient markets model, *Journal of Finance* 33, 1109-1118.

Ziemba,W.T. and Hausch,D.B., 1986, *Betting at the racetrack.* Dr.Z Investments Inc, Los Angeles.

Pricing Exotic Racetrack Wagers

Donald B. Hausch, Victor S.Y. Lo and William T. Ziemba

School of Business, University of Wisconsin,
Madison, Wisconsin 53706

Management Science Division, Faculty of Commerce,
University of British Columbia, Vancouver BC, Canada V6T 1Z2

Management Science Division, Faculty of Commerce,
University of British Columbia, Vancouver BC, Canada V6T 1Z2

Abstract

Numerous authors have found that the win market at racetracks is essentially weak-form efficient. The relative amounts wagered at various odds levels provides a fairly accurate estimate of the true chances of winning. However, the accuracy of this estimate can be improved by adjusting for the favorite-longshot bias. This is the tendency for bettors to significantly overvalue low probability high payoff wagers on longshots and significantly undervalue high probability low payoff favorites. The resulting pricing equation coupled with a probability model for running time distributions generates accurate probabilities of all possible finishes. This allows us to price exotic wagers such as the exactor, triactor, quinella and daily double, and to identify when such bets have a positive expected return.

1. Introduction

Racetracks form interesting financial markets. They share many features of listed securities markets, such as limited liability, prices that depend upon the demand characteristics of the investors, etc. The ever-changing odds prices correspond to a constantly updated futures market with a cash settlement after the race has been run. An exotic wager involves more than one horse. Exotic wagers such as the exactor, quinella, triactor, daily double, pick three, four, six, seven and nine allow investors to construct low probability high payoff wagers much akin to highly levered options or futures positions. Important differences are higher transaction costs, the usual absence of tax effects, and the market ends with the conclusion of the race, at which time all payoff uncertainty is resolved.

Previous work on win market efficiency (see Griffith (1949), McGlothlin (1956), Ali (1977), Fabricand (1965,1979), Quirin (1979), Snyder (1978), Asch, Malkiel and Quandt (1982) and Ziemba and Hausch (1984,1986,1987)) found that bettors had a marked tendency to overvalue low probability high payoff gambles and to undervalue the high probability low payoff gambles, i.e. the favorite-longshot bias[1]. Although the bias has been found by many researchers, it is insufficient to overcome the track take

---

[1] Busche and Hall (1988) and Busche (1994) are exceptions for Hong Kong and Japan.

except for horses at odds of 3-10 or less (Ziemba and Hausch (1986)). Such extreme favorites are rare, though, so for practical purposes, the market can be considered to be weakly efficient. An apparent reason for the efficiency in the win pools is the ready availability of the prices on tote boards and TV screens and the ease with which the investors can sift through this information. Another aspect is simplicity of the win bet; it depends only on the performance of one horse.

Betting on exotic bets, however, usually requires estimates of ordering probabilities (e.g. the probability that horse i wins and j finishes second). One way is to estimate these ordering probabilities based on the estimates of the win probabilities. The win probabilities can be estimated by the win bet fractions if the win market is considered to be efficient. The stable nature of the favorite-longshot bias suggests that, by accounting for it, we can improve our win probability estimates based on the public's win odds. To estimate the ordering probabilities, Harville (1977), Henery (1981) and Stern (1990) are three alternative models. These models assume different probability distributions for the running times of horses. Bacon-Shone, Lo and Busche (1992 b) demonstrate that the Henery model appears to best fit their U.S. data and Hong Kong data. However, this does not appear to be universally true; Lo (1994) indicates that the Stern model fits the Japanese data best. In a smaller study, Stern (1990) also finds that his model fits the data better than the Harville model. Although the Harville model does not appear to have as good a fit as the others, there is a tradeoff between complexity and accuracy: the Harville model is simple to apply in practice while the other two models rely on numerical integrations. Lo and Bacon-Shone (1993) give an approximation to both models and show that it works well in practice.

With estimates of the required ordering probabilities for the exotic bet, the next question is how to bet. Maximizing expected return on a race-by-race basis is very risky, with bankruptcy likely (see Epstein (1977)). Hausch, Ziemba and Rubinstein (1981) propose using the "Kelly criterion" which was originated by Kelly (1956) - one maximizes the expected logarithm of final wealth. This method avoids bankruptcy. That study concentrated on place and show bets, that is, the bets pay off when the horses finish in the top 2 and 3 positions, respectively. Independently, Asch, Malkiel and Quandt (1984, 1986) developed logit prediction equations which indicate that place and show inefficiencies exist. This paper considers betting on exotic bets.

This paper is organized as follows. Section 2 briefly surveys the research on the win market including the favorite-longshot bias, its correction, the weak form efficiency of the win market aside from the anomaly of statistically significant profits for extreme favorites, and various probability models for predicting ordering probabilities. Then Section 3 indicates how to price a general exotic bet, with the exactor, quinella, daily double and triactor treated as particular cases. The Hong Kong quinella market is chosen to illustrate our betting system. Concluding remarks appear in Section 4.

2. Efficiency of the Win Market

Table 1 and Figure 1 summarize the data in Snyder (1978) who combined the results of Fabricand (1965), Griffith (1949), McGlothlin (1956), Seligman (1975) and Weitzman (1965) with his own data. One clearly sees the favorite-longshot bias and the weak form efficiency of the market in the sense that no trader can make profits solely on price (odds) information.

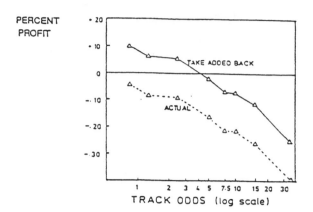

Figure 1: Expected Value Per Dollar Bet for Different Odds Levels, in 35,285 Races Run During 1947-1975; Source: Snyder (1978).

	Rates of Return on Bets to Win by Grouped Odds. Take Added Back							
	Midpoint of grouped odds							
Study	0.75	1.25	2.5	5.0	7.5	10.0	15.0	33.0
Fabricant	11.1[a]	9.0[a]	4.6[a]	− 1.4	− 3.3	− 3.7	− 8.1	− 39.5[a]
Griffith	8.0	4.9	3.1	− 3.1	− 34.6[a]	− 34.1[a]	− 10.5	− 65.5[a]
McGlothlin	8.0[b]	8.0[a]	8.0[a]	− 0.3	− 4.6	− 7.0[b]	− 9.7	− 11.0
Seligman	14.0	4.0	− 1.0	1.0	− 2.0	− 4.0	− 7.3	− 24.2
Snyder	5.5	5.5	4.0	− 1.2	3.4	2.9	2.4	− 15.8
Weitzman	9.0[a]	3.2	6.8[a]	− 1.3	− 4.2	− 5.1	− 8.2[b]	− 18.0[a]
Combined	9.1[a]	6.4[a]	6.1[a]	− 1.2	− 5.2[a]	− 5.2[a]	− 10.2[a]	− 23.7[a]

[a] Significantly different from zero at 1% level or better.
[b] Significantly different from zero at 5% level or better.

Table 1: Summary of the Six Studies Comparing the Rate of Return on Win Bets at Various Odds Levels; Source: Snyder (1978).

These data were based on pari-mutuel wagering using the totalizator method of betting in North America. Figure 2 considers wagers made head to head with bookies in England. The shape of the bias is broadly consistent with that in Figure 1. Figure 2 extends the odds range, though, and one sees positive returns for extreme favorites and extremely low expected returns for horses with the longest odds. Based on this data, plus that in Ali (1977), Asch, Malkiel and Quandt (1982), Quirin (1979) and Ziemba and Hausch (1984) one has the aggregate rate of return versus odds summary Figure 3. The graph compares California with one of the lowest track takes of 15.33% with New York which has one of the highest, 17%.

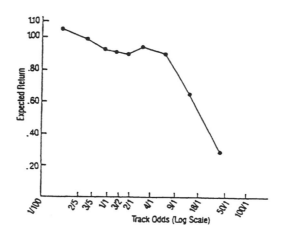

Figure 2: Expected Value Per Dollar Bet for Different Odds Levels in all British Flat Races During 1950, 1965 and 1973. Taxes of 4% on-course and 8% off-course must be subtracted to give the actual return to bettors. Source: Ziemba and Hausch (1986), using data from Figgis (1974) and Lord Rothschild (1978).

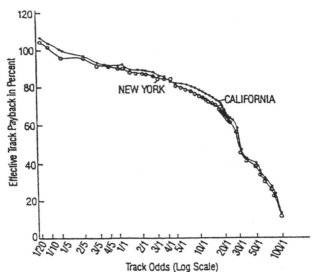

Figure 3: The Effective Track Payback Less Breakage for Various Odds Levels in California and New York. Source: Ziemba and Hausch (1986).

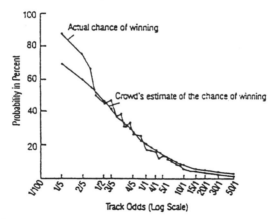

Figure 4: Actual Chance of Winning Compared with the Crowd's Estimate of this Chance at Various Odds Levels for Horses in 2,196 Races in New York in 1971. Source: Ziemba and Hausch (1986), using data from Quirin (1979).

To compute the probability that a horse at odds level O_i will win a given race we must correct for the bias. Plotting the Quirin (1979) data results in Figure 4. Similar plots for the Ali and Asch, Malkiel and Quandt data appear in Ziemba and Hausch (1984). Aggregating all the data yields the corrections in Table 2 and the formula:

Quoted Odds	Odds Range		Adjustment (Fav-Lg Shot Bias)	Effective Track Payback California	New York
1-20	0.05	0.09	20.8	104.5	104.0
1-10	0.10	0.19	20.3	104.0	102.3
1-5	0.20	0.39	18.0	101.7	100.0
2-5	0.40	0.59	14.0	97.7	96.0
3-5	0.60	0.79	10.0	93.7	92.0
4-5	0.80	0.99	9.1	92.8	91.1
1-1	1.00	1.19	8.2	91.9	90.2
6-5	1.20	1.39	7.3	91.0	89.3
7-5	1.40	1.49	6.4	90.1	88.4
8-5	1.60	1.79	6.3	90.0	88.3
9-5	1.80	1.99	6.2	89.9	88.2
2-1	2.00	2.49	6.1	89.8	88.1
5-2	2.50	2.99	6.1	89.8	88.1
3-1	3.00	3.49	4.5	83.2	86.5
7-2	3.50	3.99	3.0	86.7	85.0
4-1	4.00	4.49	1.5	85.2	83.5
9-2	4.50	4.99	0.0	83.7	82.0
5-1	5.00	5.99	-1.2	82.5	80.8
6-1	6.00	6.99	-1.9	81.8	80.1
7-1	7.00	7.99	-2.6	81.1	79.4
8-1	8.00	8.99	-3.2	80.5	78.8
9-1	9.00	9.99	-4.2	79.5	77.8
10-1	10.00	10.99	-5.2	78.5	76.8
11-1	11.00	11.99	-6.2	77.5	75.8
12-1	12.00	12.99	-7.2	76.5	74.8
13-1	13.00	13.99	-8.2	75.5	73.8
14-1	14.00	14.99	-9.2	74.5	72.8
15-1	15.00	15.99	-10.2	73.5	71.8
16-1	16.00	16.99	-11.2	72.5	70.8
17-1	17.00	17.99	-12.2	71.5	69.8
18-1	18.00	18.99	-13.2	70.5	68.8
19-1	19.00	19.99	-14.2	69.5	67.8
20-1	20.00	20.99	-15.2	68.5	66.8
21-1	21.00	21.99	-16.2	67.5	65.8
22-1	22.00	22.99	-17.2	66.5	64.8
23-1	23.00	23.99	-18.2	65.5	63.8
24-1	24.00	24.99	-19.2	64.5	62.8
25-1	25.00	29.99	-20.2	63.5	61.8
30-1	30.00	34.99	-25.2	58.5	56.8
35-1	35.00	39.99	-36.0	47.7	46.0
40-1	40.00	49.99	-39.9	43.8	42.1
50-1	50.00	59.99	-43.7	40.0	38.3
60-1	60.00	69.99	-47.5	36.2	34.5
70-1	70.00	79.99	-51.4	32.3	30.6
80-1	80.00	89.99	-55.2	28.5	26.8
90-1	90.00	99.99	-59.0	24.7	23.0
100-1	100.00	∞	-70.0	13.7	12.0

Table 2: The Effective Track Payback Less Breakage for Various Odds Levels in California and New York.* Source: Ziemba and Hausch (1986).

* The track take is 15.33% (17% New York) for an average payback of 84.67 cents per dollar wagered (83%). Breakage is to the nearest 10 cents below the true computed amount per dollar wagered. At 9-2 the breakeven point in the favorite-longshot bias, breakage amounts to about another 1% commission.

$$\pi_i = \frac{Q + D_i}{O_i + 1}, \tag{1}$$

where π_i is the probability that a horse at odds O_i will win, Q is the track payback (i.e. one minus track take) and D_i is the adjustment from Table 2 for the favorite-longshot bias of odds level O_i. Figure 5 plots equation (1) for California and New York and also gives the probabilities of placing (finishing first or second)[2] under the assumption that all horses other than the horse in question have the same odds.

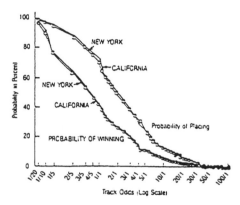

Figure 5: The Probability of Winning Odds and Placing at Various Odds Levels in California and New York. Source: Ziemba and Hausch (1986).

The data on extreme favorites consists of that of Fabricand (1965, 1979), Figgis (1974) and Lord Rothschild (1978) and Ziemba and Hausch (1984). Summarizing these data indicates, as shown in Figure 6(a) and (b), that positive profits set in at odds of about 3-10, corresponding to a win probability of about 70%.

---

[2] The probability of placing is computed by Harville (1973)'s formula. There is a systematic bias of this formula for calculating the probability that a horse finishes second. As this bias is opposite to the favorite-longshot bias, and because the probability a horse places is the sum of the probabilities that a horse finishes first and second, these two biases tend to cancel one another. However, this may not be sufficiently accurate. One way to correct for this is to use the transformation suggested in Benter (1994). This is an alternative to the models that follow.

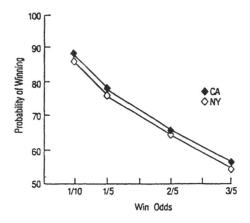

Figure 6(a): Probability of Winning on Extreme Favorites in California and New York. Source: Ziemba and Hausch (1986).

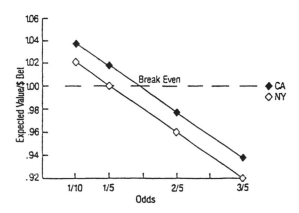

Figure 6(b): Expected Value per Dollar Bet on Extreme Favorites in California and New York. Source: Ziemba and Hausch (1986).

Equation (1) with its Table 2 correction provides an estimate of horse i's win probability, π_i. Define $\pi_{ij} \equiv P(i \text{ wins}, j \text{ finishes 2nd})$ and $\pi_{ijk} \equiv P(i \text{ wins}, j \text{ finishes 2nd}, k \text{ finishes 3rd})$. Natural estimates of these two ordering probabilities are:

and

$$\pi_{ij} = \frac{\pi_i \pi_j}{1 - \pi_i} \qquad (2)$$

$$\pi_{ijk} = \pi_i \frac{\pi_j}{1 - \pi_i} \frac{\pi_k}{1 - \pi_i - \pi_j} \qquad (3)$$

Harville (1973) first suggested the use of (2) and (3). Dansie (1983) showed that they follow from the assumption that the running times of the horses are independently exponentially distributed with different means. Henery (1981) proposed an independent normal running times model, i.e. the running time of horse i, $T_i \sim N(\theta_i, 1)$. One may argue that this model may not be realistic as the probability that $T_i < 0$ is positive. However, we are actually interested in ordering probabilities only, e.g. $P(T_i < T_j <$ others), so any monotonic increasing transform of the running times will not change the probabilities. Hence, the Henery model is equivalent to assuming that any monotonic increasing function of the running times are normally distributed. In particular, if the function is logarithmic, then the running times are lognormally distributed. With Henery's normality assumption,

$$\pi_{ij} = \int_{-\infty}^{\infty} \Phi(u + \theta_j - \theta_i) \prod_{r \neq i,j} [1 - \Phi(u + \theta_j - \theta_r)] \phi(u) \, du \quad,$$

where $\Phi(.)$ and $\phi(.)$ are cdf and pdf of the standard normal. The mean running times θ_i's are estimated by solving the following set of nonlinear equations:

$$\pi_i = \int_{-\infty}^{\infty} \prod_{s \neq i} [1 - \Phi(v + \theta_i - \theta_s)] \phi(v) \, dv \quad i = 1, 2, \ldots, n.$$

We may estimate the win probabilities π_i's by using the win bet fractions or the corrected form stated in (1). However, no closed form solutions can be found. The formula for π_{ij} is slightly more complicated.

Another probability model is due to Stern (1990). Instead of assuming independent exponential distributions for the running times, which implies the Harville model, he assumed independent Gamma distributions with predetermined shape parameter r and scale parameters that can be estimated by using the estimates of win probabilities. When $r = 1$, it is the Harville model. Using a small data set, Stern shows that $r = 2$ fits his data better than the Harville model (i.e. $r = 1$).

Bacon-Shone, Lo and Busche (1992 b) indicate that the Henery model is significantly better than the others using their U.S. and Hong Kong data. However, Lo (1994) shows that the Stern model with shape parameter, $r = 4$, fits their Japanese data best. While there does not appear to be a universal model, all the evidence points away from the Harville model. Unfortunately, the Henery and Stern models involve complicated numerical integration and solution of systems of equations, though, making their practical use questionable. However, Lo and Bacon-Shone (1993) propose the following approximation to the ordering probabilities based on the Stern model :

$$\pi_{ij} \sim \pi_i \frac{\pi_j^\lambda}{\sum\limits_{r \neq i} \pi_r^\lambda} , \tag{4}$$

and

$$\pi_{ijk} \sim \pi_i \frac{\pi_j^\lambda}{\sum\limits_{r \neq i} \pi_r^\lambda} \frac{\pi_k^\tau}{\sum\limits_{s \neq i,j} \pi_s^\tau} . \tag{5}$$

These formulas, called the discount model, include the Harville model and approximate the Henery and Stern models. Lo and Bacon-Shone suggest appropriate values for λ and τ for different r as shown in Table 3. They show that this approximation is quite accurate using their U.S., Hong Kong and Japan data. We define that when $r = \infty$, the discount model reduces to the approximate form of the Henery model.

r	λ	τ
1 (Harville)	1.00	1.00
2	0.93	0.89
3	0.90	0.84
4	0.88	0.81
5	0.87	0.80
6	0.86	0.78
7	0.86	0.77
8	0.85	0.76
10	0.84	0.75
20	0.82	0.72
30	0.81	0.71
40	0.81	0.70
∞ (Henery)	0.76	0.62

Table 3: Parameter Values for the Discount Model

3. Pricing exotic wagers

Hausch, Ziemba and Rubinstein (1981) developed a strategy for betting to place and show that utilized the "Kelly criterion" (Kelly (1956)). The properties of this criterion include the following : (i) it maximizes the asymptotic growth rate of capital; (ii) the expected time to reach a predetermined goal of capital is minimum asymptotically; and (iii) the final capital is higher than that of any other "different" strategies asymptotically. These are proved in Breiman (1960,1961). Other properties of this criterion can be found in MacLean, Ziemba and Blazenko (1992). This betting strategy in horseracing is further developed by Hausch and Ziemba (1985), and is discussed in detail in Ziemba and Hausch (1984,1987).

The system developed by Hausch, Ziemba and Rubinstein (1981) identifies the inefficiencies in the place and show markets. The simple formulas proposed by Harville (1973) in (2) and (3) are used to estimate the ordering probabilities. Lo, Bacon-Shone and Busche (1994) suggest using the discount model proposed in Lo and Bacon-Shone (1993) together with the betting strategy. Here, we concentrate on exotic markets where each bet involves more than one horse. First, we require estimates of the ordering probabilities. The formulas given in (4) and (5) with $\lambda = 0.76$ and $\tau = 0.62$ (i.e. to approximate the Henery model) appear to work quite well in the U.S. (Lo, Bacon-Shone and Busche (1994)). These formulas are more complicated than the Harville model and they require the estimates of win probabilities for all horses for each combination.

With the estimates of the ordering probabilities, the next step is to employ the "Kelly criterion" to determine the optimal bet amount on a particular combination. We now derive the general formula for an exotic bet and then discuss particular cases. To make our system more practical, we only consider the simple case of betting on one combination of horses and assume that the pool size is large enough so that the odds will not be changed by our bet. Because there will not be many good bets for longshots, we suggest screening out these horses. One way to do so is to concentrate only on those horses at odds of 8-1 or less. Suppose O is the odds (i.e. net profit per \$1 bet if we win) associated with our bet. Further, let W_o and f be the current capital and the fraction of the current capital we should bet, respectively. In addition, let the probability of winning be ρ. One may select a combination with highest positive expected return, where

$$\text{Expected Return of a \$2 wager} = 2(O+1)\rho - 2 . \qquad (6)$$

We may want to have an edge of $100E\%$, so that we bet on the combination if the expected return of a \$2 wager is at least $2E$, or

$$\text{offered price of a \$2 wager} = 2(O+1) \geq \frac{2(1+E)}{\rho} . \qquad (7)$$

Among the combinations satisfying (7), we could choose the one with highest chance of providing a return, ρ. Then we can proceed to use the "Kelly criterion" to determine the optimal bet.

The new wealth, W will be $W_0(1+fO)$ with probability ρ and $W_0(1-f)$ with probability $1-\rho$. The expected log capital is

$$E(\log W) = \rho \log[W_0(1+fO)] + (1-\rho)\log[W_0(1-f)] . \qquad (8)$$

The "Kelly criterion" maximizes this expected log capital with respect to f. The optimal f is

$$f = \frac{\rho(O+1) - 1}{O} . \qquad (9)$$

For the exactor, we must choose the horses that finish first-second to win our bet. Let O_{ij} be the exactor odds for the combination i and j. The estimate of ρ is given by (4). Using (9), the optimal wealth fraction for exacta betting on horses i and j is

$$f = \frac{\pi_{ij}(O_{ij}+1) - 1}{O_{ij}} .$$

Similar formulas hold for other exotic bets. Table 4 summarizes the odds, the relevant ordering probability and its estimate for the exactor, triactor, quinella and daily double.

With the formulas of the ordering probability for each case in Table 4, formula (9) can be easily applied[3]. In fact, formula (9) is easy to implement on a hand-held calculator or from a table plotting ρ versus A/C where A is the price being paid for the combination payoff and C is the theoretical cutoff value to achieve the edge E. See Ziemba and Hausch (1986) for details.

Exotic bet	Odds (O)	Ordering Probability (ρ)	Estimate of ρ
Exactor	O_{ij} $(i,j=1,...,n;$ $i \neq k)$	π_{ij}	$\pi_i \dfrac{\pi_j^\lambda}{\sum\limits_{r \neq i} \pi_r^\lambda}$
Triactor	O_{ijk} $(i,j,k=1,...,n;$ $i \neq j \neq k)$	π_{ijk}	$\pi_i \dfrac{\pi_j^\lambda}{\sum\limits_{r \neq i} \pi_r^\lambda} \dfrac{\pi_k^\omega}{\sum\limits_{s \neq i,j} \pi_s^\omega}$
Quinella	O_{ij} $(i,j=1,...,n;$ $i < j)$	$\pi_{ij} + \pi_{ji}$	$\pi_i \dfrac{\pi_j^\lambda}{\sum\limits_{r \neq i} \pi_r^\lambda} + \pi_j \dfrac{\pi_i^\lambda}{\sum\limits_{r \neq j} \pi_r^\lambda}$
Daily double	O_{ij} $(i=1,...,n_1;$ $j=1,...,n_2)$	$\pi_i(1) + \pi_j(2)$	sum of estimates of two win probabilities

Table 4: Formulas of the Ordering Probabilities for Various Exotic Bets.

In the case of the daily double, n_l is the number of horses in race l and $\pi_i(l)$ is the win probability of horse i in race l, $l=1$ or 2.

The system was tested using 369 Hong Kong races for which full quinella data was available. The quinella odds are shown on a big screen at the Hong Kong racetrack. The quinella pool is the largest in Hong Kong and the track take is 17%. Busche and Hall (1988) and Busche (1994) found no evidence of the favorite-longshot bias in the Hong Kong win market. This was further confirmed by Bacon-Shone,

---

[3] The formula in (9) assumes that our wagers do not influence the odds. For small wagers or large pools, this is unlikely to present a problem in practice. However, when our bets are fairly large relative to the size of the pools, then bets based on (9) are too large. Smith and Ziemba (1986) have analyzed this problem. They recommended lowering your wager by x% if you have bet x% of the total pool.

Lo and Busche (1992a). Hence, to estimate the win probabilities, the win bet fractions are used directly without any correction. For estimating the ordering probabilities, Bacon-Shone, Lo and Busche (1992b) show that the Henery model (Henery (1981)) fits the Hong Kong data better than the Harville model (Harville (1973)). The Harville model overestimates (underestimates) π_{ij} if j is a favorite horse (longshot). Nevertheless, both models are used to compute the ordering probabilities π_{ij}'s. To screen out the longshots, we concentrate only on those horses with win odds of 8-1 or less. In addition, a 15% edge is used. With initial capital of U.S.\$ 10,000, final capitals increased to \$109,532 and \$68,217 using the Harville model and approximated Henery model, respectively. The betting histories are shown in Figure 7[4].

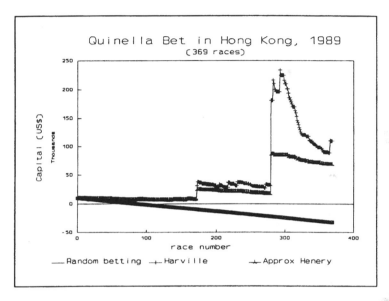

Figure 7: Application of the betting strategy in Quinella Bet in Hong Kong

The Harville model does somewhat better in this case, but this analysis is not based on a large sample. The two major combinations (the winning horses in races no. 172 and 280 in Figure 7) provided much of the profits in the betting histories. Both involved extreme favorites. With the overvaluation of ordering probabilities for favorites by the Harville model, it is clear that more will be wagered on the favorites and, thus, it is not surprising that the Harville model does better in this case. In conclusion, our system appears to be beneficial to those bettors interested in Hong Kong quinella market.

---

[4] Following HZR (1981), the straight line in Figure 7 is the approximate wealth history for random betting, where the total bet amount is the same as that under the Harville model. The total dollar bet using the Harville model is \$ 251,455 and the track take is 17%. Thus the final wealth level is 10,000 − 0.17(251,455) = -\$ 32,747.35 (which should actually be zero in practice).

4. Concluding remarks

This paper proposes a practical method for pricing exotic bet wagers. Starting with estimates of win probabilities based on the odds in the win market, the required ordering probabilities were constructed. For quinella wagering, the system is tested on a sample of Hong Kong data.

The advantages of the "Kelly criterion" are well known in the literature. However, the optimal fraction of capital bet on the horses may be very large in some cases, even when the probabilities of profits are very small, such as in triactor bets. It may take a long time to recover the loss. To reduce the risk, the fractional Kelly criterion developed by MacLean and Ziemba (1991) and McLean, Ziemba and Rubinstein (1992) is an alternative that balances the tradeoff between risk and return. This method simply invests a fixed fraction of the optimal bet amount determined by the original Kelly criterion. Risk can also be reduced by increasing the edge in (7).

References

Ali,M.M. (1977) "Probability and utility estimates for racetrack bettors." *Journal of Political Economy* 84, 803-815.

Asch,P., Malkiel,B. and Quandt,R. (1982) "Racetrack betting and informed behavior." *Journal of Financial Economics* 10, 187-194.

Asch,P., Malkiel,B. and Quandt,R. (1984) "Market efficiency in racetrack betting." *Journal of Business* 57, 165-174.

Asch,P., Malkiel,B. and Quandt,R. (1986) "Market efficiency in racetrack betting: Further evidence and a correction." *Journal of Business* 59, 157-160.

Bacon-Shone,J.H., Lo,V.S.Y. and Busche,K. (1992 a) "Modelling winning probability." *Research report* 10, Department of Statistics, University of Hong Kong.

Bacon-Shone,J.H., Lo,V.S.Y. and Busche,K. (1992 b) "Logistic analyses for complicated bets." *Research report* 11, Department of Statistics, University of Hong Kong.

Benter,W. (1994) "Computer based horse race handicapping and wagering systems: A report." In this volume, 169-184.

Busche,K. (1994) "Efficient market results in an Asian setting." In this volume, 580-581.

Busche,K. and Hall,C.D. (1988) "An exception to the risk preference anomaly." *Journal of Business* 61, 337-346.

Breiman,L. (1960) "Investment policies for expanding businesses optimal in a long-run sense." *Naval Research Logistics Quarterly* 7, 647-651.

Breiman,L. (1961) "Optimal gambling systems for favorable games." in *Proceedings of the Fourth Berkeley Symposium. Mathematical Statistics and Probability* 1, 65-78. University of California Press.

Dansie,B.R. (1983) "A note on permutation probabilities." *Journal of Royal Statistical Society B* 45, 22-24.

Epstein,R.A. (1977) *Theory of gambling and statistical logic.* 2nd edition. Academic Press, New York.

Fabricand,B.P. (1965) *Horse sense.* New York: David McKay Co.

Fabricand,B.P. (1979) *The science of winning: A random walk on the road to riches.* Van Nostrand

Reinhold.

Figgis,E.L. (1974) "Rates of return from flat race betting in England in 1973." *Sporting Life* 11 (March).

Griffith,R.M. (1949) "Odds adjustments by American horse-racing bettors." *American Journal of Psychology* 62, 290-294.

Harville,D.A. (1973) "Assigning probabilities to the outcomes of multi-entry competitions." *Journal of the American Statistical Association* 68, 312-316.

Hausch,D.B., Ziemba,W.T. and Rubinstein,M. (1981) "Efficiency of the market for racetrack betting." *Management Science* 27, 1435-1452.

Hausch,D.B. and Ziemba,W.T. (1985) "Transactions costs, extent of inefficiencies, entries and multiple wagers in a racetrack betting model." *Management Science* 31, 381-394.

Henery,R.J. (1981) "Permutation probabilities as models for horse races." *Journal of Royal Statistical Society B* 43, 86-91.

Kelly,J.L. (1956) "A new interpretation of information rate." *Bell System Technical Journal* 35, 917-926.

Lo,V.S.Y. (1994) "Application of running time distribution models in Japan." In this volume, 221-231.

Lo,V.S.Y. and Bacon-Shone,J.H. (1993) "An approximation to ordering probabilities of multi-entry competitions." Submitted to *Communications in Statistics: Theory and Methods*.

Lo,V.S.Y., Bacon-Shone,J.H. and Busche,K. (1994) "The application of ranking probability models to racetrack betting." *Management Science*, forthcoming.

MacLean,L.C. and Ziemba,W.T. (1991) "Growth-Security profiles in capital accumulation under risk." *Annals of Operations Research* 31, 501-510.

MacLean,L.C., Ziemba,W.T. and Blazenko,G. (1992) "Growth versus Security in dynamic investment analysis." *Management Science* 38, 1562-1585.

McGlothlin,W.H. (1956) "Stability of choices among uncertain alternatives." *American Journal of Psychology* 69, 604-619.

Quirin,W.L. (1979) "Winning at the races: Computer discoveries in thoroughbred handicapping" William Morrow and Co., New York.

Rothschild, Lord (1978) *Royal commission on gambling*, Vols I and II. Presented to parliament by Command of Her Majesty (July).

Seligman,D. (1975) "A thinking man's guide to losing at the track." *Fortune* 92, 91-81.

Smith,M. and Ziemba,W.T. (1985) "The effect of your bet on the odds." *Gambling Times*.

Snyder,W. (1978) "Horse racing: Testing the efficient markets model." *Journal of Finance* 33, 1109-1118.

Stern,H. (1990) "Models for distributions on permutations." *Journal of the American Statistical Association* 85, 558-564.

Weitzman,M. (1965) "Utility analysis and group behavior: An empirical study." *Journal of Political Economy* 73, 18-26.

Ziemba,W.T. and Hausch,D.B. (1984) *Beat the racetrack*. Harcourt Brace Jovanovich, San Diego.

Ziemba,W.T. and Hausch,D.B. (1986) *Betting at the racetrack*. Dr.Z Investments, Inc., Los Angeles.

Ziemba,W.T. and Hausch,D.B. (1987) *Dr.Z's beat the racetrack*. Revised edition. William Morrow, New York.

ON THE EFFICIENCY OF THE MARKET FOR DOUBLE (QUINELLA) BETS AT A FINNISH RACETRACK

Antti Kanto   and   Gunnar Rosenqvist
Helsinki School of Economics   Swedish School of Economics
Helsinki, Finland   and Business Administration,
                                            Helsinki, Finland

ABSTRACT

This paper proposes a system for double betting at a racetrack. The proposed method utilizes Harville's (1973) formulas to reduce the $\binom{k}{2}$-dimensional space of empirical odds for double bets to a $(k-1)$-dimensional space (k is the number of horses in a race). A novel feature of our betting system, compared with earlier work e.g. by Ziemba and Hausch, is that win market efficiency is not assumed since it is not supported by the Finnish data. The Kelly criterion is utilized for determining the amount of money to be wagered on each bet. The system is tested on data from 111 races at a Finnish racetrack and found to produce positive profits.

1. INTRODUCTION

Racetrack betting can be thought of as a securities market. The tote ticket constitutes a security, its price or return is determined by investors (bettors) who bid against each other. Recently the efficiency of securities markets has been the object of extensive research (surveys are provided e.g. by Fama, 1970, Schwert, 1983, Berglund, 1986, Thaler and Ziemba, 1988). The topic is important because the efficiency of the financial markets directly influences the chances for firms to obtain "rightly" priced financing, and thus accordingly has an effect on the efficiency of the whole economy and social welfare. Admittedly, the market for bets at the racetrack does not perhaps have this type of direct influence on the efficiency of the total economy. However, horse race betting markets do provide unique and interesting opportunities of research. The importance of this area of research lies in the possibilities (i) to obtain information about how securities markets function and (ii) to investigate how people make decisions under uncertainty and risk. As noted e.g. by Snyder (1978), Ali (1979), Asch, Malkiel and Quandt (1982) and Hausch, Ziemba and Rubinstein (1981), the race track shares many of the characteristics of the archetypal securities market, e.g. in listed common stocks. In both cases future earnings are not known with certainty, it is easy to enter the market, there is a large number of participants, and extensive information is available, including past data, professional advice, and information about the activities of other participants. Thus it is natural to apply the concept of an efficient market. Furthermore, as has been pointed out by the above-mentioned authors and others, the racetrack is a particularly interesting case since it is essentially a more elementary market, without several of the dynamic features which serve to make analysis of the stock market rather more complex. The tote ticket is a highly-stadardized form of security, and the bettors (investors) act in a quasi-experimental (but nevertheless natural) situation.

Recent research into the efficiency of horse racing betting markets has established the efficiency of the market for win bets, despite the so-called long shot bias (Hoerl and Fallin, 1974, Asch, Malkiel and Quandt, 1982, Snyder, 1978, and Ziemba and Hausch, 1987, chap. 3). Building on this evidence Hausch, Ziemba and Rubinstein (1981), Ziemba and Hausch (1987), and Hausch and Ziemba (1985) devised a technical system for exploiting inefficiencies in place and show betting. This system - the so-called Dr. Z system - utilizes the formulas developed by Harville (1973) for calculating from given win probabilities the probabilities of different outcomes. The principle of these formulas can be described as follows: Assume π_i is the probability that horse i wins, then the probability that i is first and j is second is

(1)
$$\frac{\pi_i \pi_j}{1 - \pi_i},$$

and the probability that i is first, j second and k third is

(2)
$$\frac{\pi_i \pi_j \pi_k}{(1 - \pi_i)(1 - \pi_i - \pi_j)}.$$

Given that the win market is efficient, good estimates of the π_i:s are obtained from the odds determined by the public in the win betting. The formulas (1) and (2) then provide "objective" odds for each possible place and show outcome. Comparing these objective odds with the empirically-observed odds provides a test of efficiency. Hausch, Ziemba and Rubinstein (1981), Ziemba and Hausch (1987) and Hausch and Ziemba (1985) also utilize a capital growth model to determine how much should be wagered when inefficiencies arise, and demonstrate significant profits.

The purpose of this paper is to explore the existence of weak inefficiencies of this type at racetracks in Finland. However, with Finnish data the market for win bets can hardly be taken as efficient. This is because (i) there is no empirical evidence of this from Finland, and (ii) the win pool is generally very small in Finland. Typically the win pool for one race is about 5-10% of that for so-called double betting, the most popular type of wager on races in Finland. Hence, we have chosen to look at double betting. To win a double bet you have to, without specifying the order, pick the first two horses. Thus, our double bet is identical to the North American quinella, or the dual forecast in Great Britain. Because the Dr. Z system is based on the win odds, which in Finland are probably unreliable, we develop an alternative, which we call the Dr. K system.

The basic idea of our approach is that, if there is a natural ranking order among the horses, the odds for the $\binom{k}{2}$ possible pairs in the double betting can not vary freely. More precisely, we assume that the probabilities of the possible outcomes in double betting should obey formula (1). In other words, if there are k horses in a race, the probabilities of the $\binom{k}{2}$ different outcomes should be determined by k parameters π_i, i =1,...,k, with the obvious restriction $\Sigma \pi_i = 1$. This gives an opportunity to fit to the empirical $\binom{k}{2}$ -1-dimensional space of odds for the double bets a (k-1)-dimensional space of theoretical probabilities, and to detect pairs of horses which are underbet by the public. In other words we reduce the $\binom{k}{2}$ -1-dimensional space of empirical odds to a (k-1)-dimensional space in order to detect "favorable" bets. Following Hausch, Ziemba and Rubinstein (1981), Ziemba and Hausch (1987), and Hausch and Ziemba (1985) we utilize the Kelly criterion (Breiman,1961) for determining the optimal wagers. The system is applied on data from Teivo Racetrack at Tampere and found to produce positive profits.

Section 2 describes the "institutional" conditions in Finland. The model is presented in Section 3 and Sections 4 and 5 contain empirical evidence. Section 6 contains a concluding discussion, and some technical details of the system are given in an Appendix.

2. HORSE RACING IN FINLAND

In Finland horse races with betting are arranged on 19 major racetracks and 25 smaller tracks. In 1985 these tracks arranged 669 races with a total of 6403 runners. All Finnish racetracks are 1000 meters long. There are no thoroughbred races in Finland, nearly all races are harness races for trotters.The only exceptions, which are rare, are "monte races", i.e. trotters with saddle. The race distances vary between 1600 and 3200 m (1 mile to 2 miles), but occasionally even longer races are arranged. About three quarters of the races include handicaps, according to the horses' personal best times or the amount of money they have earned.

In Finnish horse races there are six types of bets to be made; win, place, double, triple, V4 and V5, which are described below:

Bet	No. of horses to be chosen	You win when
Win	1	Your selection wins
Place	1	Your selection is first, second or third
Double	2	Your selections are first and second irrespective of the order
Triple	3	Your selections are first, second and third, in the correct order
V4	4x1	Your selections win four particular races
V5	5x1	Your selections win five particular races

Win, place and double betting is arranged in all races with four or more runners. Triple betting is arranged on major tracks usually once in a raceday. Triple betting began in 1987, hence experience with it is very limited. The V4 accumulator is usually arranged once a raceday, V5 betting occurs weekly at one of the major tracks and you have to make your bets no later than one or two days before the race. The most popular type of betting is the double. In 1985 the total sum of all double bets in Finland was 439 million FIM (about 100 million USD), which is nearly 10 times the sums of the corresponding win and place pools.

The smallest stake to be made is 5 FIM (≈ 1.25 USD) for win, place, and double bets, 2 FIM for triples, 1 FIM for V4 and 0.50 FIM for V5. The breakage equals 0.50 FIM. The track payback is Q=0.79 in win, place, double and triple betting, Q=0.63 in V4 and Q=0.55 in V5. Off-track betting is not allowed in Finland, except in the case of V5 which is arranged nationwide by a state-owned betting office.

3. A BETTING MODEL

3.1. Finding profitable double bets

Let x_{ij} ($i \neq j$) be a random variable corresponding to the number of bets (calculated in 5 FIM tokens) made on the pair (i,j) and $x=(k_{12},...., x_{k-1,k})$ be the vector of lenght k(k-1)/2, of all bets, where k equals the number of horses in the race in question. We assume that each pair (i,j) has a probability π_{ij} to finish first and second, in either order, with $\sum_{i<j} \sum \pi_{ij} = 1$. The players lay their bets by comparing the parimutuel odds with their subjective estimates of π_{ij}. If the market is efficient, then these subjective estimates should all be equal. However, in our view it would be unreasonable to require an efficient market to have x_{ij} exactly proportional to π_{ij} for each pair (i,j). Some reasonable stochastic variation should be allowed even within an efficient market. These considerations lead us to formulate a stochastic model for x. A simple possibility is to choose a multinomial model for x with probabilities π_{ij} and $\sum_{i<j} \sum x_{ij} = $ n, where n is the total number of bets made. The multinomial model is an approximation since the assumption of independent tokens does not hold in practice. It is, however, commonly used for discrete data and a reasonable starting point.

By Harville's formulas

(3)
$$\pi_{ij} = \frac{\pi_i \pi_j}{1-\pi_i} + \frac{\pi_i \pi_j}{1-\pi_j}$$

where π_i is the probability that horse i wins.

The likelihood function is then

$$L(\pi) = C \prod_{i<j} \prod (\pi_{ij})^{x_{ij}} = C \prod_{i<j} \prod \left[\frac{\pi_i \pi_j}{1-\pi_i} + \frac{\pi_i \pi_j}{1-\pi_j} \right]^{x_{ij}}$$

where $C = n \left[\prod_{i<j} \prod x_{ij}! \right]^{-1}$, $\sum_i \pi_i = 1$ and $0 \le \pi_i \le 1$. The model has k parameters π_i of which k-1 are independent. The π_i :s can be estimated by maximizing the likelihood function; see the Appendix for mathematical details.

Let p_i be the estimate of π_i obtained by maximizing the likelihood function above. Estimated probabilities for each pair can now be calculated as

$$p_{ij} = \frac{p_i p_j}{1-p_i} + \frac{p_i p_j}{1-p_j}$$

Let o_{ij} be the observed odds of pair (i,j), i.e. the odds determined by the public,

$$o_{ij} = \frac{Q \sum_{i<j} x_{ij}}{x_{ij}}$$

where Q is the track payback. The expected return of pair (i,j) equals $p_{ij}o_{ij}-1$ and it is profitable in the theory to wager on this pair if

$$p_{ij}o_{ij} > 1.$$

3.2. The amount to bet

Assume that (i) the bets made do not affect the odds and (ii) the sum wagered is continuous, i.e. fractions of tokens can be wagered. Let w_0 be the initial wealth, i.e. the amount of money available and w_1 be the wealth after the race. According to the Kelly criterion (Breiman, 1960 and 1961, Thorp, 1971) we assume that our utility function is logarithmic, i.e. that the utility of wealth w_1 is $\log(w_1)$. Our aim is to maximize the expected utility $E(\log(w_1))$.

Let w_{ij} be the amount of the wager made on pair (i,j). Further, o_{ij} is the odds of pair (i,j) and p_{ij} is the estimated winning probability. If $p_{ij}o_{ij} < 1$, the estimated expected return of a bet placed on the pair (i,j) is negative and thus the optimal bet equals zero. Thus we only consider pairs for which $p_{ij}o_{ij} > 1$. Then the optimal amount w_{ij} is that which maximizes

$$E(\log(w_1)) = p_{ij}\log(w_0 - w_{ij} + w_{ij}o_{ij}) + (1-p_{ij})\log(w_0 - w_{ij}),$$

which implies (Ziemba and Hausch, 1987, chap. 5)

(4)
$$w_{ij} = w_0 \left[\frac{o_{ij} p_{ij} - 1}{o_{ij} - 1} \right] \text{ if } p_{ij} o_{ij} > 1,$$

$$0 \text{ otherwise.}$$

The editors pointed out that formula (4) could be generalized by accounting for (a) one's effect on the odds, and (b) betting on several pairs in the same race. However, the equations soon become complicated. With the purpose of demonstrating an inefficiency we use equation (4) as an approximation. Also, in any case it is important that the model is simple enough to allow the calculations to be done in a short time.

3.3. Applying the procedure in practice

To sum up, practical application of our betting system at the track amounts to the following.

1. Feed the computer with the data required. This constitutes a considerable practical problem because the number of bets for all pairs are needed. Ideally the data to be utilized should be as final as possible, but in practice some time has to be reserved for data inputting, calculations and wagering. Ziemba and Hausch (1987) give evidence that profits can be made by using odds from two minutes before start. As yet our system has not been put to work in practice at the track so we lack experience of it in this respect.

2. Maximize the log likelihood function in order to get win probability estimates p_i, $i=1,...,k$

and the corresponding probabilities $p_{ij} = \frac{p_i p_j}{1 - p_i} + \frac{p_i p_j}{1 - p_j}$.

3. If $p_{ij} o_{ij} > 1$ wager the amount $w_0 \left[\frac{o_{ij} p_{ij} - 1}{o_{ij} - 1} \right]$.

Effective operation of the system requires that the data entering can be done in the last 2 or 3 minutes of wagering. It is possible that this will require either some kind of simplification of the system or some kind of hi-tech implementation of the data inputting. One possible simplification is to use data e.g. only for the two top ranked horses. A further simplification is to compare, for each of the other horses, its odds with the favorite with its odds with the second favorite. This procedure, which can even be done by eyeballing the tote board, may reveal some undervalued bets.

The perfect solution of the data inputting problem would of course be to connect ones own micro computer to the organizer's computer. As this probably is not allowed at present, other solutions must be looked for. One possibility is to devide the work between a few persons, each typing down part of the required data base, and then to merge the sub-data bases utilizing e.g. a small local net. Optical reading of the tote board or a monitor is another speculative possibility.

4. AN ILLUSTRATIVE EXAMPLE

To illustrate, consider Race 10 on the card at Killerjärvi, the racetrack of Jyväskylä, on 26th January, 1986. This race was chosen because it is one where the first author actually rode one of the horses - the number 2, Laku.

Table 4.1. The program for Race 10 at Killerjärvi, Jan. 26, 1986.

The total wagers were 1690 for win, 2085 for place, and 52880 for double. Table 4.2 gives the observed win odds and the odds obtained from win probabilities estimated from double play. (We follow the European way of announcing odds, i.e. the odds gives the amount of money payed for every mark bet. Thus, e.g. odds of 3.5 would be given as 5-2 in North America).

Musta-Hilu, horse No. 9 is the favourite both in win betting and double play. In addition both odds are similar. For other horses, the estimated and observed odds differ considerably, e.g. in the case of horse No. 3. This may be an indication of some kind of inefficiency.

Horse	observed odds	estimated odds	Horse	observed odds	estimated odds
1	26.7	34.9	8	5.9	6.8
2	53.4	58.1	9	2.7	2.5
3	53.4	12.2	10	5.1	6.3
4	29.6	19.0	11	66.7	31.5
5	17.8	22.5	12	19.0	32.2
6	9.8	13.2	13	7.6	6.8
7	20.5	18.0	14	44.5	103.6

Table 4.2. Observed win betting odds and winning odds estimated from double play betting.

The observed double odds are presented in Table 4.3. and the odds estimated from the estimated winning probabilities are in Table 4.4.

	1	2	3	4	5	6	7	8	9	10	11	12	13
2	334.2												
3	163.8	397.8											
4	225.8	225.8	88.8										
5	288.0	642.6	114.4	245.7									
6	154.7	491.4	73.9	90.8	128.5								
7	363.2	596.7	75.9	203.7	208.8	136.9							
8	177.7	363.2	54.9	108.4	134.7	73.2	70.2						
9	74.5	130.5	18.0	30.1	35.6	20.6	24.6	6.7					
10	117.6	185.6	68.4	62.3	60.9	39.5	104.4	34.0	7.4				
11	835.4	278.5	185.6	363.2	417.7	198.9	397.8	189.8	50.9	117.6			
12	759.5	1392.4	278.5	348.1	835.4	203.7	596.7	167.1	43.5	89.8	298.3		
13	174.0	464.1	51.5	141.6	112.8	73.2	65.2	24.9	7.8	22.8	126.5	149.1	
14	8355.0	—	1044.3	596.7	1193.5	464.1	1392.4	1670.9	134.7	491.4	835.4	1193.5	397.8

Table 4.3. Observed double odds. A hyphen (-) means that the pair is not played at all.

As one can see, there are considerable differences in these odds. For example, the observed double odds for the pair (1,2) is only 334.2, whereas the estimated odds is 1263.3. This shows, that the pair (1,2) is overplayed. (From experience outside the scope of this study, we believe that the pair (1,2) very often is overplayed).

	1	2	3	4	5	6	7	8	9	10	11	12	13
2	1263.3												
3	259.5	434.2											
4	407.5	681.9	140.1										
5	484.9	811.4	166.7	261.8									
6	279.7	468.1	96.2	151.0	179.7								
7	385.0	644.1	132.3	207.8	247.3	142.7							
8	140.8	235.5	48.4	76.0	90.4	52.2	71.8						
9	46.2	77.3	15.9	25.0	29.7	17.2	23.6	8.7					
10	128.6	215.1	44.2	69.4	82.6	47.7	65.6	24.0	7.9				
11	681.0	1139.6	234.1	367.6	437.4	252.3	347.3	127.0	41.7	116.0			
12	696.0	1164.7	239.2	375.7	447.0	257.9	354.9	129.8	42.6	118.6	627.9		
13	140.7	235.5	48.4	76.0	90.4	52.2	71.8	26.3	8.7	24.0	127.0	129.8	
14	2256.6	3776.0	775.5	1218.0	1449.3	836.0	1150.6	420.7	138.1	384.3	2035.5	2080.4	420.6

Table 4.4. Double odds obtained from estimated winning probabilities.

The method described earlier gives us the following betting system, where the amount is expressed as a proportion of the capital at hand.

pair		amount	odds		pair		amount	odds
1	9	0.00371	74.5		5	12	0.00057	835.4
1	14	0.00023	8354.9		6	8	0.00149	73.2
2	8	0.00060	363.2		6	13	0.00149	73.2
2	9	0.00257	130.4		7	10	0.00248	104.3
2	13	0.00120	464.1		7	12	0.00055	596.7
2	14	0.00009	8354.9		8	10	0.00357	34.0
3	10	0.00328	68.4		8	11	0.00096	189.8
3	14	0.00006	1044.3		8	12	0.00010	167.0
4	8	0.00118	108.4		8	14	0.00128	1670.9
4	13	0.00335	141.5		10	14	0.00002	491.4
5	8	0.00132	134.7					

Table 4.5. Dr. K system bets.

There are some further odds in Table 4.3 which are greater than those in 4.4, e.g. those for the pairs (1,11) and (2,12), but the difference is so small that it does not cover the track take, i.e. for these pairs the estimated expected value $p_{ij}o_{ij}$ is smaller than one. The total sum wagered equals .03010, which represents about 3% of the capital at hand. This does not seem to be a dangerous way to play.

The wagers in Table 4.5 are obtained using equation (4) on each pair one by one. Most of the bets are small and placed on pairs with high odds. One of the editors pointed out that diversification would probably lead to larger bets.

The favorite, No. 9, won with No. 1 second. With odds 74.5 the bet returned .27639 for a profit of .24630, or about one quarter of the base capital.

This example was, in fact, the first one we studied. Its results encouraged us to investigate further.

5. A LARGER STUDY

We have applied our method on ex post data from Teivo, in Tampere, the second largest racetrack in Finland. Our data consisted of 111 races over 11 days in July, August and September 1987. The development of our capital (see Figure 5.1) ends in a 46.5% increase, or about 3.5% per race day.

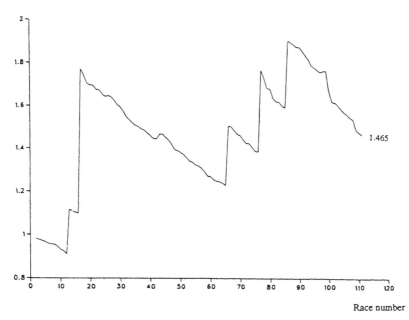

Race number

Figure 5.1. Wealth level by race.

The overall-return is built up by a few successful bets, whereas most of the bets end in a slight loss. These single losses are, however, typically so small that they are covered by wins. In an attempt to assess the reliability of the results we performed a significance test in the manner of Hausch, Ziemba and Rubinstein (1981, pp. 1450-1451). We calculate the probability of achieving a wealth level at least as high as 1.465 under a random betting model. Let π be the probability of winning a bet and X_i be the payoff,

$$X_i = \begin{cases} 1+w \text{ if the bet is won,} \\ 0 \text{ otherwise.} \end{cases}$$

It follows that $E(X_i) = \pi (1+w)$, $Var(X_i) = (1+w)^2 \pi (1-\pi)$ and the null hypothesis is that $\pi (1+w) = 0.79$, the track payback. The probability of interest ("significance level") is

$$P\left[\frac{1}{n} \sum_{i=1}^{n} X_i > 1.465\right] \approx 1 - \Phi\left[\frac{1.465 - \pi(1+w)}{(1+w)\pi \sqrt{(1-\pi)/n\pi}}\right],$$

where $n = 111$ and Φ is the cumulative distribution function of a standardized normal variable. From the data π is estimated as 0.081 and further 0.79 is, according to the null hypothesis, inserted for $\pi (1+w)$. This gives 0.0038 as the probability of obtaining by random betting a wealth level of at least 1.465 in the 111 races.

6. CONCLUSIONS

Some of the smaller Finnish racetracks have no toteboards. At these tracks, one bets with no information about the empirical odds. Even when there is a toteboard, only a few, usually 6-12, of the lowest double odds are displayed. Higher odds are available only from special monitors and, thus, it should not be surprising if some inefficiencies exist at Finnish racetracks, and especially so in double play.

This paper is the first empirical investigation into the efficiency of betting markets at Finnish racetracks. We have presented a simple model for detecting technical inefficiencies in the market for double bets. A novel feature of our approach is that no reliance is made on the assumption of an efficient win market, an assumption that would not seem to be appropriate for Finnish racing. In the data we have gathered up to now, consisting of 111 races at Teivo, our system provided significant profits. Admittedly, however, one could want further empirical evidence. Our future plans are to test this model on further data and to explore other possible inefficiencies at Finnish racetracks. We will also have to consider the practical implementability of the system. However, the present study offers a promise of possible positive expected returns. It remains to refute or sustain this hypothesis with further data.

A multinomial model was considered. Besides maximizing the multinomial likelihood, estimates of the win probabilities were also obtained by alternative methods (Poisson, normal, minimum-chi^2). However, these experiments produced win probabilities which were similar to those obtained using the multinomial approach, and hence only the latter are given here.

APPENDIX. The maximization algorithm used

The purpose is to find the maximum likelihood estimates of π_i using our model. The logarithm of the likelihood function is

$$l(\pi) = C + \sum_{i<j} \sum x_{ij} \log\left[\frac{\pi_i\pi_j}{1-\pi_i} + \frac{\pi_i\pi_j}{1-\pi_j}\right],$$

where $\pi = (\pi_1,...,\pi_k)$, $\Sigma\pi_i = 1$, and C is a constant.

A version of the steepest descent algorithm is utilized to maximize $l(\pi)$. Let $p^0=(p_1^0,...,p_k^0)$ be an initial estimate of π. A new estimate p^1 of π is obtained by correcting p_i^0, i=0,..., k-1, in the direction

$$\left(-\frac{dl}{d\pi_i} / \frac{d^2l}{d\pi_i^2}\right)\Big|_{\pi=p_0}$$

with a step size D, say, and letting $p_k^1 = 1- \sum_{i=1}^{k-1} p_i^1$. Thus

$$p_i^1 = p_i^0 - D\left(\frac{dl}{d\pi_i} / \frac{d^2l}{d\pi_i^2}\right)\Big|_{\pi=p_0}, \ i=1,...,k-1,$$

$$p_k^1 = 1- \sum_{i=1}^{k-1} p_i^1$$

The step size D has been determined in the following way (see e.g. Marquardt, 1963):

1) Set D = 1.

2) If $l(p^1) > l(p^0)$ and $0 \le p_i^1 \le 1$ the step is complete.

   Otherwise set D = D/10 and recalculate p^1.

The iteration is stopped when $D < \delta$, where δ is some predetermined quantity. We have used $\delta = 10^{-20}$.

Because one of the components of π, say π_k, is dependent of the others, we have

$$\frac{\partial\pi_k}{\partial\pi_i} = -1, \ i = 1,..., k-1$$

and thus

$$\frac{dl}{d\pi_i} = \frac{\partial l}{\partial\pi_i} + \frac{\partial l}{\partial\pi_k}\frac{\partial\pi_k}{\partial\pi_i} = \frac{\partial l}{\partial\pi_i} - \frac{\partial l}{\partial\pi_k}$$

$$= l_i - l_k, \quad i = 1,\ldots, k\text{-}1.$$

Similarly

$$\frac{d^2 l}{d\pi_i^2} = l_{ii} - 2l_{ik} + l_{kk}.$$

After some tedious but straightforward algebra the first and second derivatives of the log likelihood are found to be

$$l_i = \sum_{j \neq i} \frac{2 - 2\pi_i - \pi_j + \pi_i^2}{\pi_i(1-\pi_i)(2-\pi_i-\pi_j)} \, x_{ij},$$

$$l_{ij} = -\frac{1}{(2-\pi_i-\pi_j)^2} \, x_{ij}, \quad j \neq i,$$

$$l_{ii} = \sum_{j \neq i} \frac{4 - 12\pi_i + 10\pi_i^2 - 4\pi_i^3 + \pi_i^4 - 4\pi_j + 10\pi_i\pi_j - 4\pi_i^2\pi_j + \pi_j^2 - 2\pi_i\pi_j^2}{\pi_i^2(1-\pi_i)^2(2-\pi_i-\pi_j)^2} \, x_{ij}.$$

Acknowledgements. We are grateful to Suomen Hippos for financial support and to the editors for comments improving our presentation.

REFERENCES

Ali, M.M. (1979): Some Evidence of the Efficiency of a Speculative Market, Econometrica, 47, 387-392.

Asch, P., Malkiel, B.G. and Quandt, R.E. (1982): Racetrack Betting and Informed Behavior, Journal of Financial Economics, 10, 187-194.

Berglund, T. (1986): Anomalies in Stock Returns on a Thin Security Market, Publications of the Swedish School of Economics and Business Administration, nr 37, Helsingfors.

Breiman, L. (1960): Investment Policies for Expanding Businesses Optimal in Long-Run Sense, Naval Research Logistics Quarterly, 7, 647-651. Reprinted in Ziemba, W.T. and Vickson, R.G. (eds.): Stochastic Optimization Models in Finance, Academic Press, London, 1975.

Breiman, L. (1961): Optimal Gambling Systems for Favorable Games, Proceedings of the Fourth Berkeley Symposium on Mathematical Statistics and Probability, Neyman, J. (ed.), University of California Press, Berkeley and Los Angeles.

Harville, D.A. (1973): Assigning Probabilities to the Outcomes of Multi-Entry Competitions, Journal of the American Statistical Association, 68, 312-316.

Hausch, D.B. and Ziemba, W.T. (1985): Transaction Costs, Extent of Inefficiencies, Entries and Multiple Wagers in a Racetrack Betting Model, Management Science, 31, 381-394.

Hausch, D.B., Ziemba, W.T. and Rubinstein, M. (1981): Efficiency of the Market for Racetrack Betting, Management Science, 27, 1435-1452.

Hoerl, A.E. and Fallin, H.K. (1974): Reliability of Subjective Evaluations in a High Incentive Situation, Journal of the Royal Statistical Society, Ser. A, 137, 227-230.

Marquardt, D.W. (1963): An Algorithm for Least-Squares Estimation of Nonlinear Parameters, Journal of the Society for Industrial and Applied Mathematics, 11, 431-441.

Schwert, G.W. (1983): Size and Stock Returns, and Other Empirical Regularities, Journal of Financial Economics, 12, 3-12.

Snyder, W.W. (1978): Horse Racing: Testing the Efficient Markets Model, The Journal of Finance, 33, 1109-1118.

Thaler, R. and Ziemba, W.T. (1988): Parimutual Betting Markets: Racetracks and Lotteries, The Journal of Economic Perspectives, 2, Spring, 161-174.

Thorp, E.O. (1971): Portfolio Choice and The Kelly Criterion, Business and Economics Statistics Section, Proceedings of the American Statistical Association, 215-224. Reprinted in Ziemba, W.T. and Vickson, R.G. (eds.): Stochastic Optimization Models in Finance, Academic Press, London, 1975.

Ziemba, W.T. and Hausch, D.B. (1987): Beat the Racetrack, Harcourt Brace William Morrow, New York.

EFFICIENT MARKET ADJUSTMENT OF ODDS PRICES TO REFLECT TRACK BIASES

BRIAN R. CANFIELD, BRUCE C. FAUMAN AND WILLIAM T. ZIEMBA

Faculty of Commerce and Business Administration, University of British Columbia, Vancouver, British Columbia, Canada V6T 1Y8

Biases that reflect the economic worth of uncertain contingent claims occur in many financial markets. Parimutuel betting at racetracks is one such market with ample data to investigate such biases. The total wagering market is about $10 billion per year in North America. The configuration of racetracks leads to an advantage for horses breaking from post positions near the rail, especially for tracks with small circumferences. Can the bettor make profits with knowledge of this bias? To investigate, we utilize data from 3,345 races involving over $300 million in wagers from 1982, 1983 and 1984 on win and exotic bets at Exhibition Park in Vancouver where this bias should be strong. The results indicate that the bias exists but the prices adjust to fully negate the potential gains from the bias.
(EFFICIENT MARKETS; RACETRACK WAGERING)

Biases that reflect the economic worth of uncertain contingent claims occur in many financial markets.[1] In their surveys of this literature including the turn of the year small stock bias and the low price earnings ratio bias, Schwert (1983) and Dimson (1986) refer to these effects as "empirical regularities". This suggests that these effects are not temporary aberrations but are in fact permanent biases. The cause of such biases may have elements that are economic, psychological, institutional or physical in character. Information on these biases may be public but many if not most traders are likely to be unaware of them. This may provide a potential advantage to those who are aware of the biases. A natural question is whether or not the market prices adjust to compensate for this edge so that the potential gains are lost. In some markets the bias does indeed occur and profitable trading strategies exist to exploit these inefficiencies. MacLean, Ziemba and Blazenko (1987) have studied how one can best trade off risks and returns in such situations.

An interesting area with ample data to test such bias hypotheses is in thoroughbred racing.[2] Horseplayers are replete with such biases. Indeed the search for them is intense, see e.g., Beyer (1983), Davidowitz (1979) and Quirin (1979) for discussions of many of them.

One of the most consistent biases is that of post position. In Beyer's words:

> While most biases are due to idiosyncrasies in the racing surface, many tracks have shapes that influence the results. At tracks less than a mile in circumference, the sharp turns and short stretch almost always work to the advantage of the front runners and the horses on the inside, (Beyer 1983, p. 42).

---

[1] A typical example is the increase in volatility subsequent to the announcement of stock splits, see Bar-Yosef and Brown (1977), Ohlson and Penman (1985), and Reilly and Drzycimski (1981, 1984). The work on small firm, low PE ratio, day of the week, weekend, and turn of the year effects is also related. For a sampling of the literature, see, e.g., Brown, Kleidon and Marsh (1983), Clark and Ziemba (1988), French (1980), Gibbons and Hess (1981), Keim (1983), Lakonishok and Smidt (1986), Reinganum (1983), and Roll (1983). Surveys appear in Schwert (1983), Dimson (1986), Keim (1986), and Ziemba (1986).

[2] Racetrack betting markets have been used to study various efficiency questions by Ali (1977, 1979), Asch and Quandt (1987), Asch, Malkiel and Quandt (1982, 1984, 1986). Dowie (1976), Figgis (1974), Griffith (1949), Harville (1973), Hausch and Ziemba (1985, 1987), Hausch, Ziemba and Rubinstein (1981), McGlothlin (1956), Rosett (1965), Snyder (1978), Weitzman (1965) and Ziemba and Hausch (1984, 1986, 1987). A survey of this literature will appear in Hausch and Ziemba (1988).

TABLE 1

Winning Percentages and Rates of Return on Win Bets by Post Position in 2516 Six-Furlong Races Run over Mile Tracks. Source: Quirin (1979)

Post Position	Number of Horses	Number of Winners	Percent Winners	Rate of Return
1	2516	304	12.1	1.08
2	2516	267	10.6	0.80
3	2516	279	11.1	0.73
4	2516	276	11.0	0.71
5	2516	249	9.9	0.63
6	2481	271	10.9	0.89
7	2375	220	9.3	0.71
8	2208	204	9.2	0.79
9	2025	164	8.1	0.94
10	1886	140	7.4	0.52
11	1107	91	8.2	0.72
12	727	51	7.0	0.21
Very Outside	2516	184	7.2	0.53

The disadvantage that the outside horses have is most acute when the race starts near a turn. They are likely to be forced two or more horses wide during the turn. For each horse they race outside in a single turn they lose about one length.[3] If a horse is a front runner breaking from near the rail, then it is imperative that the horse strive to take or be near the lead before the first turn to be in a good position to win the race.[4] The inside position despite its shorter length is thought to be disadvantageous if the track surface is heavy there, especially when it is wet or drying out. Quirin (1979) has shown at mile tracks that post position 1 is the most advantageous and positions 2–6 are generally more advantageous than chance. However, these advantages do not return win bet profits because these advantages are noticed and overbet by the public; see Table 1.

1. Biases in the Win Market

The track at Exhibition Park in Vancouver, British Columbia is five furlongs and 208 feet. Since a furlong is an eighth of a mile the track is about two thirds of a mile. Horses must make two turns even in short spring races of 5 to 7 furlongs and at least three turns in longer route races of 8 to 12 furlongs. Hence there is a potential for a strong post position bias. We collected data for all the races during the 1982–1984 summer seasons. This sample is quite extensive covering 1092 races over 109 days in 1982, 1147 races over 113 days in 1983 and 1106 races over 110 days in 1984. Most of these races were either $6\frac{1}{2}$-furlong sprints around two turns or $1\frac{1}{16}$-mile routes around three turns. Table 2 indicates that there indeed is a post position bias favoring the horses on the inside, particularly positions one and two but also position three. Positions one to three have a statistically significant edge; positions seven through ten have a significant disadvantage; while positions four through six have the average number of winners. Indeed, the chance that a typical horse in post position one or two will win a given race is about double that of a horse in post position seven to ten.

---

[3] For a semicircular turn of radius r a horse on the rail travels πr feet around the turn. A horse running m horses wide must travel $\pi(r + mw)$ feet where w is the width of a horse, about 3 feet. Since a length is about 8.5 feet and πw is about 9.4 feet the horse will lose about 1.1 m length on that turn or 2.2 m lengths in a two-turn race. With two turns and $m = 3$ this is nearly seven lengths. With three turns it is over ten lengths!

[4] See Quirin (1979) for data indicating that horses in or near the lead at various points in a race have a much higher chance of winning.

TABLE 2

Winning Percentages by Post Position in 3345 Sprint and Route Races Run Over 332 Days at Exhibition Park in 1982–1984

Post Position	1982 109 days		1983 113 days		1984 110 days		Average % Winners	Avge. Diff. from Mean % Winners	t statistic***
	Number of Winners	% Winners*	Number of Winners	% Winners	Number of Winners	% Winners			
1	173	16.05	146	12.78	162	14.78	14.54	+2.56	4.09
2	192	17.88	139	12.24	149	13.72	14.61	+2.63	2.06
3	152	14.06	132	11.60	144	13.16	12.94	+0.96	2.28
4	124	11.50	141	12.29	123	11.32	11.70	-0.28	-0.46
5	123	11.45	158	13.92	132	12.10	12.49	+0.51	0.46
6	119	11.17	132	11.76	114	10.51	11.15	-0.83	-1.45
7	83	8.19	90	8.43	91	8.74	8.45	-3.53	-6.50
8	60	6.83	91	9.44	82	8.84	8.37	-3.61	-3.00
9	44	6.49	63	7.70	59	7.65	7.28	-4.70	-5.82
10	22	4.44	55	8.61	50	8.39	7.14	-4.84	-2.74
Totals and Averages	1092	12.81	1147	11.47	1106	11.67	11.98		

* These percentages do not necessarily add to one since different post positions have different numbers of starters.

** With two degrees of freedom, using the three-years data, the one-tail cut-off values are 1.89 at the 10% level and 2.92 at the 5% level. So at the 10% level, positions 1–3 have an edge, 7–10 have a disadvantage and 4–6 are average.

TABLE 3

Winning Percentages and Rates of Return on Fast Track and Off Track Days by Post Position in 1088 Sprint and Route Races Run at Exhibition Park in 1982

Post Position	Fast Track Days (72)				Off Track Days (37)				All Days (109)			
	Number of Horses	Number of Winners	Percent Winners	Rate of Return	Number of Horses	Number of Winners	Percent Winners	Rate of Return	Number of Horses	Number of Winners	Percent Winners	Rate of Return
1	712	127	17.84	1.100	366	46	12.57	0.765	1078	173	16.05	0.986
2	709	136	19.18	1.144	365	56	15.34	0.922	1074	192	17.88	1.069
3	714	99	13.87	0.943	367	53	14.44	1.036	1081	152	14.06	0.975
4	711	67	9.42	0.601	367	57	15.53	1.006	1078	124	11.50	0.741
5	709	83	11.71	0.845	365	40	10.96	0.621	1074	123	11.45	0.769
6	703	80	11.38	0.707	362	39	10.77	0.600	1065	119	11.17	0.671
7	669	54	8.07	0.500	344	29	8.43	0.650	1013	83	8.19	0.554
8	580	36	6.21	0.516	298	24	8.05	0.563	878	60	6.83	0.532
9	448	25	5.58	0.725	230	19	8.26	0.695	678	44	6.49	0.567
10	327	13	3.98	0.506	169	9	5.33	0.597	496	22	4.44	0.537
11,12	5	0	0.00	0.000	2	0	0.00	0.000	7	0	0.00	0.000
Totals and Averages	6287	720	13.35	0.829	3235	372	12.34	0.829	9522	1092*	12.81	0.829

* There were four dead heats.

TABLE 4

Profits from $2 Win Wagers on Post Positions 1–3 on Fast Track and Off Track Days at Exhibition Park, 1982–1984

Post Position	1982				1983				1984				Total Profit ($) 1982–1984	Rate of Return (%) 1982–84
	Number of Bets	Number of Wins	Total Return ($)	Profit ($)	Number of Bets	Number of Wins	Total Returns ($)	Profit ($)	Number of Bets	Number of Wins	Total Returns ($)	Profit ($)		
FAST TRACK DAYS														
1	712	127	1566.90	142.90	859	108	1364.30	−353.70	707	111	1236.40	−177.60	−388.40	−8.53
2	709	136	1622.40	204.40	855	112	1414.60	−295.40	701	109	1368.60	−33.40	−124.40	−2.75
3	714	99	1346.20	−81.80	856	95	1230.40	−481.60	706	88	1201.80	−210.20	−773.60	−16.99
OFF TRACK DAYS														
1	366	46	559.60	−172.40	283	38	485.40	−80.60	389	51	776.30	−1.70	−254.70	−12.27
2	365	56	672.90	−57.10	281	27	378.30	−183.70	385	40	538.20	−231.80	−472.60	−22.92
3	367	53	760.70	26.70	282	37	452.40	−111.60	388	56	738.00	−38.00	−122.90	−5.93
ALL DAYS														
1	1078	173	2126.50	−29.50	1142	146	1849.70	−434.30	1096	162	2012.70	−179.30	−643.10	−9.70
2	1074	192	2295.30	147.30	1136	139	1792.90	−479.10	1086	149	1906.80	−265.20	−597.00	−9.06
3	1081	152	2106.90	−55.10	1138	132	1682.80	−593.20	1094	144	1939.80	−248.20	−896.50	−13.53

TABLE 5
Profits from $2 Win Wagers on Post Positions 1–3 for Route Races on Fast Track and Off Track Days at Exhibition Park, 1982–1984

Post Position	1982				1983				1984				Total Profit ($) 1982–1984	Rate of Return (%) 1982–84
	Number of Bets	Number of Wins	Total Return ($)	Profit ($)	Number of Bets	Number of Wins	Total Returns ($)	Profit ($)	Number of Bets	Number of Wins	Total Returns ($)	Profit ($)		
FAST TRACK DAYS														
1	189	38	451.30	73.30	227	21	243.80	−210.20	195	30	454.90	64.90	−72.00	−5.89
2	189	27	381.00	3.00	227	32	394.20	−59.80	195	29	455.60	65.60	8.80	0.72 (0.57)**
3	189	21	349.00	−29.00	227	28	431.50	−22.50	195	23	347.30	−42.70	−94.20	−7.71
OFF TRACK DAYS														
1	95	14	261.80	71.80	77	13	141.20	−12.80	83	10	106.00	−60.00	−1.00	−0.20
2	95	15	179.30	−10.70	77	9	129.20	−24.80	83	5	52.10	−113.90	−149.40	−29.29
3	95	13	163.10	−26.90	77	13	160.80	6.80	83	9	134.90	−31.10	−51.20	−10.04
ALL DAYS														
1	284	52	713.10	145.10	304	34	385.00	−223.00	278	40	560.90	4.90	−73.00	−4.21
2	284	42	560.30	−7.70	304	41	523.40	−84.60	278	34	507.70	−48.30	−140.60	−8.12
3	284	34	512.10	−55.90	304	41	592.30	−15.70	278	32	482.20	−73.80	−145.40	−8.39

** With 2 degrees of freedom, this t statistic indicates that this wager does not return positive profits at any reasonable statistical level.

For a typical bet at a racetrack one is faced with two transactions costs: the track take and breakage. The track tate t is a fixed percentage commission, denoted by $t = 1 - Q$, the value of the bet, where Q is the percentage amount returned to the bettors. This track take varies from 14.8% to 22.1% on win bets at various tracks in North America. At Exhibition Park it was 17.1% in 1982, 16.4% in 1983, and 15.8% during 1984. The second transaction cost is breakage which is the rounding down of the payoffs to multiples of $0.05 or $0.10 per dollar bet. Hence a payoff that returns $7.77, on a $2 wager, is actualy paid $7.70 or $7.60 with $0.05 and $0.10 breakage, respectively. This amounts to about another 1% and 2% on win bets, respectively, see Ziemba and Hausch (1984 or 1987). Exhibition Park, like essentially all Canadian tracks, has the more favorable $0.05 breakage. Nearly all tracks in the United States use $0.10 breakage.

Hence, a typical win bet loses some 17-20% on average. This then fuels the search for winning systems using the biases. We say that the odds have fully adjusted to negate the value of knowing about the bias if bets on average return their initial stake or less so that the expected value per dollar bet does not exceed 1.00.

Quirin (1979) hypothesized that post position 1 is disadvantageous on off track days (because the banked track tends to accumulate water near the rail). Table 3 investigates this using the 1982 data. Indeed, post position 1 returns scarcely more than the average number of winners and the overbetting in light of the apparent favorable bias yields a negative 23.5% rate of return, which is lower than a random wager. Post position 2 also suffers from this difficulty on off track days though to a lesser extent. The only bets that yield potential significant positive profits are those on positions one and two on fast track days. These rates of return are 1.10 and 1.14, respectively. None of the post positions provides significant positive profits over all days. Post positions four to twelve provide progressively worse and worse returns. Horses in positions from seven or beyond return significantly less than chance on both fast and off tracks.

To look at this more carefully we investigated post positions 1-3 on fast and off track days during the three years 1982-1984. Table 4 summarizes these results. While there were small profits betting on post positions one and two on fast track days in 1982, these profits did not materialize in 1983 nor in 1984 even though the track take was lower in these years. Indeed whatever edge there may be in these positions is fully negated by the overbetting by the crowd. There are losses betting on post positions one to three on both fast track and off track days.

Longer races have more turns and tend to have a greater post position bias. This is despite the fact that strategy can be used to avoid the bias by staying off the lead in anticipation of a late charge. Therefore, even though profits are not attainable on all races for the favorable posts, they might be on the route races. Table 5 considers this. Again the post position advantage is fully negated by the crowd's overbetting on post positions one to three. In the only situation where profits are obtainable (post position 2 on fast track days), the profits are so slight that you cannot reject the hypothesis that these profits are nonpositive.

2. The Exotic Wagering Market

At Exhibition Park like many racetracks the most popular wagers are on exotic or gimmick bets. These bets are low probability high payoff gambles that combine the outcome of two, three, four or six horses and provide much entertainment for the betting public.[5] They fit in nicely with the "greed and bragging rights" of the typical bettor. The track take is usually 2 to 4 percentage points higher on exotic bets than on

---

[5] As Rosett (1965) has shown, one can construct low-probability high-payoff wagers from parlays composed sequences of win bets. Such wagers have the disadvantage of several compound commissions and the possible advantage of being underbet.

TABLE 6
Exotic Bets at Exhibition Park in 1984

Type of Bet	You Choose	You Win Only When
Quinella	Two horses in one race	These two horses finish first and second in either order.
Exacta	Two horses in one race	These two horses finish first and second in the exact order that you specified.
Triactor	Three horses in one race	These three horses finish first, second and third in the exact order that you specified.
Daily Double	One horse in one race and a second horse in another race (typically races 1 and 2)	Both horses finish first.
Win Four	One horse in each of four consecutive races	All four horses finish first.
Sweep six	One horse in each of six consecutive races	All six horses finish first.

straight win, place and show bets. However, because of the size of the payoffs, the breakage on exotic bets is much less significant. Typically it is 0.02% to 0.5% of the gross return. The total take is then 16 to 25%. At Exhibition Park, the track take on exotic wagers was 17.1% in 1982. It was raised to 18.6% on August 9, 1982 (day 70). It was raised again to 19.3% for the 1984 season.

Table 6 describes these exotic bets.

Even with our large sample, the exotic wagers cause data problems because of their low-probability high-payoff character. Hence, to look at the effect of possible post position bias in exotic wagering markets we constructed wagers that include all possible combinations of several horse numbers. These combinations are called "boxes". In the quinella a 12 box is horses 1 and 2 since the order of finish is irrelevant and a 123 box is 12, 13, and 23. In the exactor a 12 box is 12 and 21 since the order matters and a 123 box is 12, 13, 21, 23, 31 and 32.

The results from betting on two and three horse boxes in the quinella, and the exacta on fast track, off track and all days appear in Tables 7 and 8, respectively. The three-year data set is too small to adequately test such hypotheses for the daily double, triactor, Win Four, and Sweep Six. Except for two wagers, the 1-2 quinella on fast track days and the 3-1 exacta on off track days, the public is sufficiently aware of the advantage of the post position bias to overbet the favorable post positions to fully negate any advantage.

3. Conclusion

The studies summarized by Snyder (1978) and Ziemba and Hausch (1984, 1986, 1987) indicate that the win market is weakly-efficient in the sense that you cannot make profits simply by betting on particular odds horses. Bettors greatly overbet low-probability high-payoff wagers and underbet high-probability low-payoff wagers. Hence the wagers with the highest average returns are those with the lowest odds. However, even with the most extreme favorites the possible profits are slight and possibly not statistically significant anyway. Hausch, Ziemba and Rubinstein (1981), Hausch and Ziemba (1985) and Asch, Malkiel and Quandt (1984) have shown that significant mispricing inefficiencies sometimes occur in the place and show markets. These mispricings arise because of the public's greed and bragging rights that make such high probability low

TABLE 7

Profits from $2 Quinella Wagers on Post Position 1–3 Boxes on Fast Track and Off Track Days at Exhibition Park, 1982–1984

Box	1982				1983				1984				Total Profit ($) 1982–1984	Rate of Return (%) 1982–84
	Number of Bets	Number of Wins	Total Return ($)	Profit ($)	Number of Bets	Number of Wins	Total Returns ($)	Profit ($)	Number of Bets	Number of Wins	Total Returns ($)	Profit ($)		
							FAST TRACK DAYS							
12	285	24	845.40	275.40	335	15	669.80	-0.20	283	14	509.40	-56.60	218.60	12.10 (2.25)*
13	286	18	528.90	-43.10	335	15	406.60	-263.40	284	14	577.70	9.70	-296.80	-16.40
23	285	15	673.60	103.60	334	10	317.60	-350.40	283	7	318.10	-247.90	-494.70	-27.42
123	856	57	2047.90	335.90	1004	40	1394.00	-614.00	850	35	1405.20	-294.80	-572.90	-10.57
							OFF TRACK DAYS							
12	147	4	107.10	-186.90	111	3	60.70	-161.30	155	4	126.80	-183.20	-531.40	-64.33
13	147	7	203.40	-90.60	111	4	135.00	-87.00	156	7	622.00	310.00	132.40	15.99 (0.62)**
23	147	7	472.50	178.50	111	4	125.20	-96.80	155	8	334.10	24.10	105.80	12.81 (0.68)**
123	441	18	783.00	-99.00	333	11	320.90	-345.10	466	19	1082.90	150.90	-293.20	-11.82
							ALL DAYS							
12	432	28	952.50	88.50	446	18	730.50	-161.50	438	18	636.20	-239.80	-312.80	-11.88
13	433	25	732.30	-133.70	446	19	541.60	-350.40	440	21	1199.70	319.70	-164.40	-6.23
23	432	22	1146.10	282.10	445	14	442.80	-447.20	438	15	652.20	-223.80	-388.90	-14.79
123	1297	75	2830.90	236.90	1337	51	1714.90	-959.10	1316	54	2488.10	-143.90	-866.10	-10.96

* With 2 degrees of freedom, this t-statistic indicates that this wager provides positive profits at the 10% level.

** These wagers do not return positive profits at any reasonable statistical level.

TABLE 8

Profits from $2 Exacta Wagers on Post Positions 1–3 Boxes on Fast Track and Off Track Days at Exhibition Park, 1982–1984

Box	1982				1983				1984				Total Profit ($) 1982–1984	Rate of Return (%) 1982–84
	Number of Bets	Number of Wins	Total Return ($)	Profit ($)	Number of Bets	Number of Wins	Total Returns ($)	Profit ($)	Number of Bets	Number of Wins	Total Returns ($)	Profit ($)		
FAST TRACK DAYS														
12	283	8	333.40	−232.60	351	5	300.10	−401.90	285	8	343.40	−226.60	−861.10	−46.85
13	284	12	1307.20	739.20	351	9	540.60	−161.40	286	9	631.50	59.50	637.30	34.60 (1.55)**
21	283	19	897.30	331.30	351	2	226.10	−475.90	285	7	386.70	−183.30	−327.90	−17.84
23	283	9	476.40	−89.60	351	8	399.80	−302.20	285	8	501.20	−68.80	−460.60	−25.06
31	284	11	597.20	29.20	351	4	129.50	−572.50	286	11	1021.30	449.30	−94.00	−5.10
32	283	3	269.20	−296.80	351	4	252.70	−449.30	285	5	457.40	−112.60	−858.70	−46.72
123	1700	62	3880.70	480.70	2106	32	1848.80	−2363.20	1712	48	3341.50	−82.50	−1965.00	−17.81
OFF TRACK DAYS														
12	147	4	218.90	−75.10	119	1	112.90	−125.10	147	4	182.70	−111.30	−311.50	−37.71
13	147	3	89.40	−204.60	119	2	82.60	−155.40	147	4	141.30	−152.70	−512.70	−62.07
21	147	3	119.20	−174.80	119	0	0	−238.00	147	1	236.30	−57.70	−470.50	−56.96
23	147	8	417.40	123.40	118	0	0	−236.00	146	1	16.40	−275.60	−388.20	−47.23
31	147	5	562.00	268.00	119	4	324.50	86.50	147	8	457.30	163.30	517.80	62.69 (5.23)*
32	147	3	392.50	98.50	118	3	199.60	−36.40	146	3	560.30	268.30	330.40	40.19 (1.17)**
123	882	26	1799.40	35.40	712	10	719.60	−704.40	880	21	1594.30	−165.70	−834.70	−16.87
ALL DAYS														
12	430	12	552.30	−307.70	470	6	413.00	−527.00	432	12	526.10	−337.90	−1172.60	−44.02
13	431	15	1396.60	534.60	470	11	623.20	−316.80	433	13	772.80	−93.20	124.60	4.67 (0.89)**
21	430	22	1016.50	156.50	470	2	226.10	−713.90	432	8	623.00	−241.00	−798.40	−29.97
23	430	17	893.80	33.80	469	8	399.80	−538.20	431	9	517.60	−344.40	−848.80	−31.91
31	431	16	1159.20	297.20	470	8	454.00	−486.00	433	19	1478.60	612.60	423.80	15.88 (1.38)**
32	430	6	661.70	−198.30	469	7	452.30	−485.70	431	8	1017.70	155.70	−528.30	−19.86
123	2582	88	5680.10	516.10	2818	42	2568.40	−3067.60	2592	69	4935.80	−248.20	−2799.70	−17.52

* With 2 degrees of freedom, this t-statistic indicates that this wager provides positive profits at the 10% level.
** These wagers do not return positive profits at any reasonable statistical level.

payoff wagers relatively unattractive and their inability to properly evaluate the worth of such bets that require the probability and payoff information from two or more horses. Ziemba and Hausch (1986) have shown how the efficiency of the win market adjusted for the favorite-longshot bias can be used to price exotic wagers and that significant inefficiencies occur in the exacta, quinella and daily double markets all of which involve the finish of two horses. The reason for the existence of this bias seems to be the fact that the public cannot easily compute the true worth of wagers involving two or more horses hence frequently their wagers lead to mispricings. Hausch and Ziemba (1987) have shown that there are biases across pools that can be exploited for major stakes races that are simulcast to many racetracks via TV. Since most bettors are unaware of the prices at other tracks these discrepancies occur.

This paper examines the existence of a post position bias and the possible edge that might be obtained from knowledge of this information. The results indicate that the bias is present and that the low post positions have a significant edge and the high post positions have a significant disadvantage. This bias is more pronounced the smaller the circumference of the track and the longer the race. Information on the bias is regularly available to the bettors in the daily program. The public seems to react to the bias in its betting. In the win and exotic markets the results indicate that when transactions costs are considered the crowd overbets the favorable bias positions to fully negate their absolute advantage. While positive profits may be obtainable for short periods, even for a full year's duration, over the long run the bias provides no financial advantage. The data on post position bias available to the general public are aggregated over fast and off tracks. Yet the strong bias particularly for post position one does not exist on off track days. The public does not seem to fully realize this and the results from wagers on post position one on off track days lead to even more significant losses. On these off track days the favorable bias is relatively strong in positions two or three. Position two is widely known to have a significant bias and is overbet on all days. The bias in position three is less understood and on off track days it frequently leads to slight profits. However, even with the large three-year data set one cannot conclude that there are significant positive profits in the long run.[6]

[6] Without implicating him the authors would like to thank Donald B. Hausch for helpful comments on an earlier draft of this paper.

References

ALI, M. M., "Probability and Utility Estimates for Racetrack Bettors," *J. Political Economy*, 85 (1977), 803–815.

———, "Some Evidence of the Efficiency of a Speculative Market," *Econometrica*, 47 (1979), 387–392.

ASCH, P., B. G. MALKIEL AND R. E. QUANDT, "Racetrack Betting and Informed Behavior," *J. Financial Economics*, (1982), 187–194.

———, ———, AND ———, "Market Efficiency in Racetrack Betting," *J. Business*, 57 (1984), 165–175.

———, ———, AND ———, "Market Efficiency in Racetrack Betting: Further Evidence and a Correction," *J. Business*, 59 (1986), 157–160.

——— AND R. E. QUANDT, "Efficiency and Profitability in Exotic Bets," *Economica*, forthcoming, 1987.

BAR-YOSEF, S. AND L. D. BROWN, "A Re-examination of Stock Splits Using Moving Betas," *J. Finance*, 32 (September 1977), 1069–1080.

BEYER, A., *The Winning Horseplayer*, Houghton-Mifflin Co., Boston, 1983.

BROWN, P., A. W. KLEINDON, AND T. A. MARCH, "New Evidence on the Nature of Size Related Anomalies in Stock Prices," *J. Financial Economics*, 12 (1983), 33–56.

CLARK, R. AND W. T. ZIEMBA, "Playing the Turn of the Year Effect with Index Futures," *Oper. Res.*, forthcoming, (1988).

DAVIDOWITZ, S., *Betting Thoroughbreds*, E. P. Dutton, New York, 1979.

DIMSON, E. (Ed.), *Stock Market Regularities*, London Business School Conference Proceedings, 1986.

DOWIE, J., "On the Efficiency and Equity of Betting Markets," *Economica*, 43 (1976), 139–150.

FIGGIS, E. L., "Rates of Return from Flat Race Betting in England in 1973," *Sporting Life*, (1974), 11.

FIGLEWSKI, S., "Subjective Information and Market Efficiency in a Betting Model," *J. Political Economy*, 87 (1979), 75–88.

FRENCH, K. R., "Stock Returns and the Weekend Effect," *J. Financial Economics*, 8 (August 1980), 55–69.

GIBBONS, M. R. AND P. HESS, "Day of the Week Effects and Asset Returns," *J. Business*, 54 (October 1981), 579–596.

GRIFFITH, R. M., "Odds Adjustments by American Horse Race Bettors," *Amer. J. Psychology*, 62 (1949), 290–294.

HARVILLE, D. A., "Assigning Probabilities to the Outcomes of Multi-Entry Competitions," *J. Amer. Statist. Assoc.*, 68 (1973), 213–316.

HAUSCH, D. B. AND W. T. ZIEMBA, "Transactions Costs, Extent of Inefficiencies, Multiple Bets, and Entries in a Racetrack Betting Model," *Management Sci.*, 31 (April 1985), 381–394.

———, AND ———, "Cross Track Betting on Major Stakes Races," mimeo, University of British Columbia, April 1987.

——— AND ———, EDS., *Efficiency of Racetrack Betting Markets*, forthcoming, 1988.

———, M. RUBINSTEIN AND W. T. ZIEMBA, "Efficiency of the Market for Racetrack Betting," *Management Sci.*, 27 (November 1981), 1435–1452.

KEIM, D. B., "Size Related Anomalies and Stock Return Seasonality: Further Empirical Evidence," *J. Financial Economics*, 12 (1983), 13–32.

———, "The CAPM and Equity Return Regularities," *Financial Analysts J.*, (May–June 1986), 19–34.

LAKONISHOK, J. AND S. SMIDT, "Are Seasonal Anomolies Real? A Ninety Year Perspective," mimeo Johnson Graduate School of Management, Cornell University, September 1986.

MACLEAN, L., W. T. ZIEMBA AND G. BLAZENKO, "Growth Versus Security in Dynamic Investment Analysis," mimeo, University of British Columbia, April 1987.

MCCLOTHLIN, W. H., "Stability of Choices Among Uncertain Alternatives," *Amer. J. Psychology*, 63 (1956), 604–615.

OHLSON, J. A. AND S. PENMAN, "Volatility Increases Subsequent to Stock Splits: An Empirical Aberration," *Financial Ecomonics*, 14 (1985), 251–266.

QUIRIN, W. L., *Winning at the Races: Computer Discoveries in Thoroughbred Handicapping*, William Morrow, New York, 1979.

REILLY, F. K. AND E. F. DRZYCIMSKI, "Short Run Profits from Stock Splits," *Financial Management*, 10 (Summer 1981), 64–71.

——— AND ———, "Investing in Options of Stocks Announcing Splits," mimeo, Cornell University, 1984.

REINGANUM, M. R., "The Anomalous Stock Market Behavior of Small Firms in January: Empirical Tests for Tax-Loss Selling Effects," *J. Financial Economics*, (June 1983), 89–104.

ROSS, R., "Vas Ist Das? The Turn of the Year Effect and the Return Premium of Small Firms," *J. Portfolio Management*, (Winter 1983), 18–28.

ROSETT, R. H., "Gambling and Rationality," *J. Political Economy*, 73 (1965), 595–607.

SCHWERT, G. W., "Size and Stock Returns and Other Empirical Regularities," *J. Financial Economics*, 12 (1983), 3–12.

SNYDER, W. W., "Horse Racing: Testing the Efficient Markets Model," *J. Finance*, 33 (1978), 1109–1118.

WEITZMAN, M., "Utility Analysis and Group Behaviour: An Empirical Study," *J. Political Economy*, 73 (1965), 18–26.

ZIEMBA, W. T., "Security Market Inefficiencies," TIMS Workshop, Los Angeles, ORSA-TIMS, April 1986.

——— AND D. B. HAUSCH, *Beat the Racetrack*, Harcourt, Brace and Jovanovich, San Diego, 1984.

——— AND ———, *Betting at the Racetrack*, Norris Strauss, New York, 1986.

——— AND ———, *DR. Z's Beat the Racetrack*, William Morrow, New York, 1987.

POST POSITION BIAS:

An Econometric Analysis
of the
1987 Season at Exhibition Park

Sandra Betton[1]

Department of Finance, University of British Columbia

Abstract

This paper investigates the existence of a post position bias during the 1987 racing season at Exhibition Park. The econometric techniques of Probit and Pooled Cross-sectional Time Series analysis were used to analyze the relationship between the odds ranking, post position, and final position of a horse. The length of the race does not significantly improve the fit of any of the models. The results of this analysis indicate that the post position significantly adds to the information reflected in the odds rankings. The significance of the post position bias tends to increase as the number of horses taking part in the race increases.

Ever since the time of Augustus Caesar and the Roman chariot races, the betting public has been searching for a system to maximize their profits or more realistically, to minimize the losses incurred at the track and in so doing, have observed certain features inherent in horse racing. These features include, believed biases due to the length and narrowness of the track, the track conditions, race distance and the commonly held belief that the post position exerts a significant bias.

This paper examines the relationship between finishing position, odds ranking and post position. The odds rankings were found to more fully reflect the post position bias the fewer the horses involved in the race. The relationship between post position bias and abnormal profits were not examined in this paper. (See Canfield, Fauman, and Ziemba [1987] for a detailed examination of this issue).

The importance of a solid understanding of the relevant biases arising at particular race tracks is underscored by the amounts wagered and paid out at these race tracks. On average, a bettor at Exhibition Park loses between 15 and 20 percent of the amount wagered due to track commissions and taxes (Table 1). Table 2 indicates that certain uninformed betting rules may allow one to do better than average, ie, to lose less than average. Table (7) summarizes the dollars paidback for uninformed betting categorized by the number of turns in a race. Therefore, it is in the bettor's best interest to understand and, where necessary, price the results of any observed bias.

1 Causes of Post Postion Bias

The expected post position bias arises from the physical design of the modern race track. Most North American horse racing tracks are oval or elliptical and have banked turns. The oval nature of the track forces an outside horse to run further than an inside horses. Thus, if four horses started

---

[1]This paper originated as a term paper in Dr. Ziemba's course on Speculative Investments. Without implicating them, I would like to thank the editors for their help in preparing this paper for publication.

Table 1: SUMMARY STATISTICS

Exhibition Park					
Vancouver, British Columbia					
	1979	1986			
Days of racing	108	133			
Total attendance	996,573	875,336	PER PERSON BASIS		
				1979	1986
Total Wagered	$106,781,374	$128,084,885	Wagered	$107.00	$146.00
Track commission and breakage	$10,529,474	$13,611,058	Commission + Breakage	10.56	15.55
Provincial tax	$7,474,696	$8,965,941	Tax	7.50	10.24
			Payback	$88.94	$120.21

Source: Agriculture Canada: Race Track Division Annual Review, 1986

from post positions one through four and ran at the same pace, the extra 10 lengths run by the #3 horse in a three turn race would preclude him from victory.[2]

The second cause of a post position bias is the banking of the track. Banking can be advantageous as it promotes drainage, thereby, helping to prevent deep water from accumulating on the track. This provision can, however, hinder the inside horses as water would tend to accumulate next to the inside rail with the result that the outside horses are running on a relatively dry track while the inside horse's track is still wet and slow.

Banking, however, can also be a major disadvantage in dry conditions to any horse forced to run up that gradient. If only the turns are banked then the outside horse effectively gallops up and then down a hill, significantly affecting the power available from the animal. For example, a gradient of 1:100 causes a 10% reduction in a horse's pulling ability (see Marks [1941]). Unfortunately, muscular energy used running up a hill is not regained running back down the hill.

One would, therefore, expect the post position bias to be most significant on short tracks with less straights to enable one to jockey for position on the inside rail. This is particularly so on short tracks with very sharp turns.

2   The Data Set

Data were collected from the results of racing at Exhibition Park as reported in the Vancouver Sun and Province newspapers for the period from April 13, 1987 to October 18, 1987. Ninety-seven races were excluded from the sample for the following reasons:

1. two horses were reported to have the same post positions.

2. cases where normal betting would be affected due to coupled entries, late scratches and in one case, where no "show" wagering was permitted.

3. one day of racing was eliminated as only four races were run before all racing was suspended due to poor conditions.

---

[2]Canfield, Fauman, and Ziemba [1987] show that for a semicircular turn, the inside horse travels πr feet around the turn. A horse m horses outside must travel an additional $\pi m w$ feet where w is the average width of a horse.

Table 2: DOLLARS PAIDBACK PER $100 WAGERED USING UNINFORMED BETTING

Exhibition Park

April 13, 1987 - October 18, 1987

		Betting on Post Position					Betting on Odds Rank		
Post Pos'n	# of Bets	To Win	To Place	To Show	Odds Rank	# of Bets	To Win	To Place	To Show
1	1062	78.03	86.59	78.09	1	1089	85.47	93.92	97.15
2	1062	86.76	85.62	81.05	2	1068	88.70	85.68	87.44
3	1062	77.00	78.19	81.98	3	1042	79.82	83.32	87.91
4	1062	98.48	83.36	90.52	4	1062	80.66	76.69	85.93
5	1061	87.81	82.66	80.27	5	1054	78.82	79.22	78.79
6	1027	81.70	82.86	79.20	6	1037	82.83	78.23	78.30
7	935	72.70	73.50	70.73	7	935	86.27	85.79	78.22
8	790	73.66	68.77	76.68	8	787	67.05	76.59	63.42
9	641	64.80	64.21	63.57	9	641	107.25	70.28	57.18
10	484	74.68	70.18	73.78	10	483	35.53	36.46	38.39

Note: In 1986, on average Exhibition Park paid back $82 per $100 wagered.

4. cross-track betting on races which actually took place at other tracks.

One thousand and sixty-two races, representing 9239 horses, comprised the final sample. Seventeen of these races resulted in dead heats.

This paper does not examine the post position bias in conjunction with the track condition bias. The period in question was unusually dry for Vancouver and this is reflected in the distribution of track conditions. A 95% confidence interval places the average track condition between fast and good.[3]

Final win odds were ranked[4] as the actual odds quoted are functions of the relative sizes of the win pools and therefore, may not be comparable between different races and days.

Exhibition Park is an interesting track for examining post position bias as it is short, only 5/8 of a mile long. Canfield, Fauman, and Ziemba (1987) examine the issue of post position bias in relation to the length of the track and the races involved.

A second characteristic of racing at Exhibition Park that is useful in examining the post position bias is the high proportion of large numbers of horses in races. Over 70% of the races involved more than 7 horses (see Figure 1).

3 Analysis

3.1 Graphical Analysis

A graphical analysis of the results of horse races by post position was conducted to determine if a post position bias seemed likely. Figure 2 compares actual and expected results of the first 10 post positions. The expected number of finishes by each post position was determined under the assumption that, in the absence of a post position bias, each post position is equally likely to place first, second, or third. The "In Top Three" class refers to the percentage difference between expected and actual number of times the post position finished in the top three places.

[3]See Canfield, Fauman and Ziemba [1987] for a detailed analysis of the off-track effect on the post position bias.
[4]Odds ranking 1 = race favourite, $n = n^{th}$ favourite

Figure 2 indicates that the inner four to five post positions appear, on average, to do better than the outer four post positions. The significance of this apparent bias was examined using a t-test to compare the average observed post position of the top three horses to the average post position expected under the hypothesis that no post position bias exists and therefore, each post position is equally likely to finish in any position.[5]

Table 3: ODDS AND POST POSITION OF THE TOP THREE HORSES

	Post Position of Top Three Horses				Odds Ranking of Top Three Horses			
	1	2	3	All Races	1	2	3	All Races
Number of Turns								
Expected Mean	5.0424	4.74537	4.9621	4.9953	5.0310	4.9923	4.9473	4.9826
				FIRST PLACE				
Observed Mean	4.2	4.6337	4.4963	4.5829	2.5	3.0448	3.2836	3.0843
T-statistic	-2.1640	-1.1971	-3.0319	-5.2790	-8.3894	-25.1029	-12.3697	-28.8498
N	40	759	268	1067	40	759	268	1067
				SECOND PLACE				
Observed Mean	4.3095	4.6538	4.4135	4.5798	3.6667	3.8136	3.6617	3.7696
T-statistic	-1.7082	-.9487	-3.4557	-5.1279	-4.1588	-12.6667	-8.6210	-17.2009
N	42	751	266	1059	42	751	266	1059
				THIRD PLACE				
Observed Mean	5.3684	4.7871	4.8327	4.8193	4.3158	4.0815	4.1822	4.1152
T-statistic	.8541	.4458	-.8345	-2.2427	-1.6022	-10.6609	-5.5218	-12.0657
N	38	761	269	1068	38	761	269	1068

The results in Table 3 indicate that the post position bias is significant for first and second place; but, not for third place. Table 3 also summarizes similar data for odds ranking.

The post position bias is far less dramatic than the odds "bias" (see Figure 3 and Table 3). These preliminary results support the findings that at this track the win odds are useful tools in the prediction of horse race results.

If the odds were determined independently of the post position, one would expect that, on average, the odds rankings would then be equally distributed across post positions for a given size of race. The uniform distribution expectation arises due to the random manner of post position assignment.

The equal distribution expectation was examined for races having 10 horses. This size of race was selected as it comprises the largest group of races. Figure 4 indicates that the uniform distribution hypothesis is not supported by the data. The hypothesis of no post position effect, ie. that the odds rankings are uniformly distributed across post positions, was tested using the Chi-square test of goodness of fit. The results summarized in Table 4 indicate that the post position bias affects the odds ranking, and has a major impact on inside horses in these races. The hypothesis of a uniform overall distribution can be rejected at the 0.05 level for all field sizes.

---

[5]In an N horse race, each post position has a $1/N$ chance of placing in any position. The expected post position of the win, place, or show horse, in that particular race, is, therefore, $(N+1)/2$.

Table 4: TEST OF UNIFORM ODDS RANKING DISTRIBUTION

χ^2 Test of Goodness of Fit			
10 horses in the race	Observed χ^2	7 horses in the race	Observed χ^2
post position 1	17.01	post position 1	33.09
post position 5	20.18	post position 4	24.95
post position 10	53.05	post position 7	26.05
Critical values:		Critical values:	
$\alpha = .05$	$\chi^2 = 16.91$	$\alpha = .05$	$\chi^2 = 12.59$
Overall observed $\chi^2 = 187.18$		Overall observed $\chi^2 = 190.78$	
Critical $\chi^2(99, .05) = 123.22$		Critical $\chi^2(48, .05) = 65.17$	

The finding that the odds ranking distribution of the outside post position, compared to the inside position, in a ten horse race is more significantly different from uniform lends support to the notion of the odds taking into account the post position of the horse. A possible rationale for these observations is that bettors determine the probability of a horse winning before considering the post position and then adjust odds rankings for the post position drawn. Typically, the participants in a horse race are announced several days prior to the actual race. In contrast, the post position of the horses is usually announced on the day of the race. In this manner, the bettors are able to evaluate the horse's probability of winning before adjusting for the post position drawn. This hypothesis would be tested by examining the change in odds rankings due to the announcement of post positions. Unfortunately, this data is not available.

The issue of whether, or not, the odds account for the post position of the horse was further analyzed by examining the differences between the odds ranking of different post positions within a race.

The basic hypothesis is that if post positions matter, one would expect the odds rankings of horses to be more similar the closer their post positions were. The null hypothesis being tested is that the true absolute average difference in odds rankings is less than or equal to one. As we are starting from the pole position horse (post position #1) and moving outwards, we would expect the difference to be negative. Essentially, we test the joint hypothesis that odds rankings increase as we move outwards from the pole position and that the true difference between the odds rankings of two adjacent post positions is less than or equal to one.

The results of the paired t-tests are given in Table 5 The underlying assumption of normality of the differences is not significantly violated as indicated by the normal probability plots of these differences.

In Table 5, the mean difference between odds rankings in post position A and post position B (in the same races) are found at the intersection of row A and column B. The value labeled (p) refers to the probability of observing that difference under the null hypothesis that the true difference was less than or equal to minus one. For example the mean difference between the odds rankings of post position 1 and 2 is -.70 while the probability of observing that value under the null hypothesis is 1.

The null hypothesis of a difference of less than one was used in preference to the more standard null hypothesis of a difference of zero due to the odds ranking procedure. Using the "zero" null in a situation where the odds are ranked would lead to a bias toward rejection as any infinitesimal difference in the odds would result in odds rankings which differ by at least one, possibly even

Table 5: TEST OF ODDS RANKING DIFFERENCES BETWEEN POST POSITIONS IN THE SAME RACE

POST	2	3	4	5	6	7	8	9	10
1	-.7	-1.0	-1.5	-2.0	-2.2	-2.6	-3.1	-3.5	-3.7
(p)	1	.5	0	0	0	0	0	0	0
2		-.3	-.8	-1.4	-1.5	-1.9	-2.3	-2.6	-2.8
(p)		1	.95	0	0	0	0	0	0
3			-.5	-1.0	-1.2	-1.5	-2.0	-2.3	-2.5
(p)			1	.4	.04	0	0	0	0
4				-.5	-.7	-1.0	-1.4	-1.7	-1.9
(p)				1	.99	0	0	0	0
5					-.2	-.5	-.8	-1.2	-1.4
(p)					1	1	.89	.17	.04
6						-.4	-.8	-1.1	-1.3
(p)						1	.95	.22	.05
7							-.4	-.7	-.9
(p)							1	.95	.76
8								-.3	-.5
(p)								1	1
9									-0.0
(p)									1

more. Unless the odds are identical, the observed difference between odds rankings is at least one in absolute value. The minimum absolute difference in odds rankings which can be observed in practice when the underlying odds are not identical is one.

The Table 5 shows that a significant difference did not exist between the odds rankings of horses in the last five post positions. Adjacent post positions, results on the diagonal of Table 5 did not have significantly different odds rankings. As the distance between horses increases, moving toward the upper right of the table, significant differences between the odds rankings begin to appear.

One can, therefore, conclude that a post position bias appears to exist and that the odds determination process appears to consider the post position of the horse. The question of whether or not the odds rankings adequately reflect the post position bias remains unanswered.

3.2   Statistical Analysis

3.2.1   Individual Analysis

Typical econometric techniques cannot be used to estimate a dichotomous dependent variable, the horse either finishes in the top three or it does not, as one typically finds estimated probabilities greater than one. Probit and logit functions provide a technique for ensuring that the estimated probabilities lie between 0 and 1. Logit analysis is based on the logistic function while probit analysis is based on the cumulative normal probability function.

A probit analysis was undertaken to examine the impact of odds rankings, post positions and number of horses on the probability of a horse finishing in the top three positions. The number of horses was included in the analysis as the graphical evidence indicates that the strength of the post position bias is related to the number of horses in the race. As one is dealing with odds rankings the number of horses in the race would also be required to give a comparative value to the odds ranks.

Table 6: RESULTS OF PROBIT ANALYSIS

Total number of observations		9239
Number of horses in one of the top three positions		3194

Estimation One: Odds and number of horses
% correct predictions = 69.5%

	Coeff.	Asymptotic t-ratio
Odds rank	-.18123	-31.723
Numb. of horses	-.05174	-5.382
Constant	.91482	10.715

Log-likelihood function = -5320.2, LRT = 1276.21 with 2 degrees of freedom

Estimation two: Odds, post position and number of horses
% of predictions correct: 69.4%

	Coeff.	Asymptotic t-ratio
Post Position	-.01135	-2.063
Odds rank	-.18018	-31.415
Number of horses	-.04677	-4.720
Constant	.92133	10.782

Log-likelihood function = -5318.1, LRT = 1277.47 with 3 degrees of freedom

Estimation three: Odds, post position, length of race, and number of horses
% of predictions correct: 69.3%

	Coeff.	Asymptotic t-ratio
Post Position	-.01135	-2.063
Odds rank	-.18018	-31.414
Length of race	.00170	0.159
Number of horses	-.04677	-4.720
Constant	.90959	18.054

Log-likelihood function = -5318.1, LRT = 1277.49 with 4 degrees of freedom

Likelihood Ratio Test (LRT) of H0: no true relationship exists

Note: An uninformed bettor wagering that any horse would
finish in the top three would be correct on average 35% of the time.

Probit analysis was selected instead of logistic analysis as the probit analysis assumes that the probability distribution is approximately normal. This assumption was not unreasonable due to the large number of observations. The analysis estimates the cumulative probability density function:

$$F(X'\beta) = G(T \leq X'\beta)$$

where G is the standard normal distribution. Inference is carried out asymptotically in a probit model as the maximum likelihood estimator is only asymptotically normally distributed. The results of this analysis appear in Table 6 .

Estimation two, which includes the post position, is significantly better, in terms of fit, than estimation one, which does not include the post position. As Shazam (the computer code available at the University of British Columbia [White, 1978]) uses Maximum Likelihood Estimation, hypotheses can be tested using a likelihood ratio test. The test comparing estimations one and two, at the .01 significance level, had a critical value of 6.3 while the observed value was 4.2. Similar results were obtained using the odds rather than the rankings.

A bettor, randomly selecting horses, would correctly identify one of the top three horses, on average, 35% of the time. The percentage of correct horse selections rises to nearly 70% when the odds rankings and post positions are taken into account.

The length of the race, which reflects the number of turns involved in a race and therefore, the extra distance an outside horse must run, was included in the analysis in an attempt to identify the cause of the post position bias. The length of the race did not significantly add to the explanatory power of any of the models estimated. It would, therefore, appear that the information contained in the length of the race has been included in the odds rankings. The post position bias appears to be caused by something other than the extra distance run by outside horses.

It would appear that the odds do partially adjust to account for the post positions of the horses involved, as illustrated by the insignificance of the length of the race. However, there is still a significant amount of useful information available from the post position.

The results, thus far, indicate that both the strength of the post position bias and the reaction of the odds ranking, are related in some way to the number of horses in the race.

The question of whether the apparent failure of the odds rankings to fully account for the post position bias is due to an inefficiency in the market, or is due to a combination of many different probability generating processes, remains unanswered. But, it seems reasonable that races of different sizes might have different win generating processes.

3.2.2   Serial Relationship

A pooled cross section time series analysis was undertaken to examine the finish position as a function of odds rankings and post positions on the basis of race size. This analysis was also undertaken to evaluate the commonly held belief that certain post positions are "hot" during the course of a day of racing.

In this type of analysis, each race is taken to be an event, while the cross sectional units are the post positions.

The model estimated is cross-sectionally hetero- scedastic and time wise autoregressive. This analysis did not examine the probability of payoff, but, rather used the finishing positions as the dependent variables. This "pooled" time series analysis was conducted on each size of race in order to examine the relative importance of post position in different race sizes.

Pooled cross-section time-series estimates the following model:

$$Y_i = X_i \beta + \epsilon_i \tag{1}$$

With cross sectional units represented by i, with $i = 1, \ldots N$ and time series denoted by

$$Y_i = (Y_{i,1}, \ldots Y_{i,T})'$$
$$X_i = (X_{i,1}, \ldots X_{i,T})'$$

The finishing position of the horse in race t starting from post postion $X_{i,t}$ is denoted by $Y_{i,t}$. For $i = 1, \ldots N$

$$\epsilon_i = (\epsilon_{i,1}, \ldots \epsilon_{i,T})'$$
$$\epsilon_{i,t} = \rho_i \epsilon_{i,t-1} + \theta_{i,t}$$

Other conditions include:

$$E(\theta_{i,t}) \;=\; 0$$
$$E(\theta_{i,t}\theta_{j,t}) \;=\; 0$$
$$E(\theta_{i,t}\theta_{j,s}), \;=\; 0 \text{ for } t \neq s$$

Assumptions underlying the model are:

1. The coefficients are the same for each post position.

2. The errors follow a first order autoregressive process. In the estimated model, the autoregressive process is assumed to be the same for each post position.

3. The variance of ϵ can vary across different post positions.

4. The disturbances are contemporaneously correlated. (Judge et al, 1985).

The results of this estimation are summarized in Table 8 . Similar results were obtained using odds rather than the odds rankings.

Including the post position significantly improves model fit for races having 7, 9 and 10 horses. In all estimations, the autoregressive parameter, ρ, was not significantly different from zero. This finding does not support the con- tention that certain numbers become "hot" during the day; rather, that the results would lend support to the hypothesis that the results of previous races do not affect the outcome of the current race.

The general lack of serial behaviour of the post position bias is, however, consistent with the fact that horses of sufficiently variable abilities are randomly assigned to their post positions.

If all the horses are of equal ability, and there is no post position bias, we would expect, on average, that each post position would win $1/N$ times where N is the number of horses in the race. If the post position bias is significant, we would expect that certain positions would consistently perform better and that these results would be correlated over different races. If, however, the horses are not of equal ability, it is possible for a post position bias to not display a significant correlation between races while still having a significant impact on the win probability of the different horses.

4   Summary and Conclusions

1. The results of the probit analysis indicate that the post position significantly adds to the information available from the odds rankings in determining the probability of a horse placing in the top three positions.

2. We would expect, from the results of the probit and graphical analyses, that the post position would be significant for some field sizes, but, not necessarily all. The pooled cross section time series results support this contention. It is shown that if there are more than six horses in the race, the post position becomes an important factor in addition to the win odds.

The results of this analysis must be used with caution due to the very small data set and the lack of pricing analysis. See Canfield, Fauman and Ziemba [1987] for a discussion of profit making at the race track. The primary purpose of this paper was descriptive, essentially to determine if the post position bias exists and whether the odds rankings appear to have compensated for that bias. Similar results were obtained when the analysis was repeated using the odds rather than the odds ranking.

One must bear in mind that any statements generalized from this paper must take into account that the data were only from the 1987 racing season at Exhibition Park.

This paper has shown that knowledge of the post position significantly improves the information available from the odds rankings. The relatively low overall explanatory power of these models suggests that more is unknown than known in the determination of racing results.

Table 7: DOLLARS PAIDBACK PER $100 WAGERED USING UNINFORMED BETTING

Exhibition Park
April 13, 1987 - October 18, 1987

Races categorized by number of turns.

POST POSTION BETTING

	ONE TURN				TWO TURNS				THREE TURNS		
POST POS'N	# of BETS	WIN	PLC	SHW	# of BETS	WIN	PLC	SHW	# of BETS	WIN	PLC	SHW
1	40	116.5	93.00	68.38	755	79.30	84.65	79.02	267	68.58	91.12	76.93
2	40	107.88	85.88	67.13	755	86.21	83.15	81.47	267	85.15	92.55	81.95
3	40	44.25	86.25	76.13	755	74.01	80.09	81.61	267	90.36	71.61	83.89
4	40	55.88	91.38	103.88	755	102.07	84.34	88.00	267	94.72	79.36	95.64
5	40	81.13	62.88	86.50	755	85.15	83.03	79.59	266	96.35	84.57	81.26
6	39	33.97	36.92	50.77	728	79.90	81.06	73.40	260	93.90	94.81	85.73
7	37	55.00	46.49	57.57	668	78.46	72.16	70.02	230	58.80	81.76	74.89
8	31	23.71	55.16	81.13	569	65.89	68.07	76.85	190	105.08	73.08	75.42
9	26	19.81	31.92	71.73	468	73.35	70.94	65.76	147	45.51	48.50	55.17
10	21	15.95	44.05	53.57	356	70.80	72.61	77.91	107	99.11	67.20	64.02

ODDS RANKING BETTING

	ONE TURN				TWO TURNS				THREE TURNS		
ODDS RANK	# of BETS	WIN	PLC	SHW	# of BETS	WIN	PLC	SHW	# of BETS	WIN	PLC	SHW
1	40	119.13	106.50	118.50	774	87.19	93.80	98.06	275	75.45	92.44	91.49
2	41	83.54	83.05	93.78	756	91.68	85.30	88.64	271	81.18	87.14	83.14
3	39	63.46	95.64	64.23	747	76.07	85.92	91.42	256	93.24	73.89	81.29
4	41	15.73	49.51	48.54	754	81.55	75.63	80.40	267	88.13	83.86	107.30
5	39	123.72	101.41	92.26	748	74.12	79.93	77.96	267	85.45	74.01	78.73
6	41	67.44	45.61	62.32	733	87.78	81.42	79.05	263	71.43	74.41	78.69
7	35	0.00	16.43	58.71	660	79.72	88.46	79.85	229	118.32	88.69	76.81
8	31	0.00	47.90	74.68	570	67.35	79.54	65.22	186	77.31	72.34	56.02
9	26	87.31	41.54	43.65	468	112.60	61.42	54.31	147	93.74	103.57	68.74
10	21	0.00	48.10	43.33	354	32.87	26.95	36.03	108	51.16	65.37	45.14

5   References

Amemiya, Takeshi. *Advanced Econometrics*. Harvard University Press, Cambridge, Massacusetts. 1985.

Canfield, Brian R., Bruce Fauman, and William T. Ziemba. "Efficient Market Adjustments of Odds Prices to Reflect Track Biases". *Management Science*. Vol 33 #11. November 1987.

Judge, George G, W. E. Griffiths, R. Carter Hill, Helmut Lutkepohl, and Tsoung-Chao Lee. *The Theory and Practice of Econometrics*. 2nd Edition. John Wiley and Sons. New York. 1985.

Kmenta, Jan. *Elements of Econometrics*. Macmillan Company, New York. 1971.

Marks, Lionel S. editor. *Mechanical Engineer's Handbook*. McGraw-Hill Book Company, New York. p 1132. 1941.

White, Kenneth J. "A General Computer Program for Econometric Methods - SHAZAM". *Econometrica*. January 1978. PP. 239- 240.

Table 8: POOLED AND CROSS SECTIONAL TIME SERIES

Cross-sections restricted to have the same autoregressive parameter.

	Number of horses: 5 Number of races = 34		Number of horses: 6 Number of races = 92		Number of horses: 7 Number of races = 145	
	Coeffic.	T-ratio (166)	Coeffic.	T-ratio (548)	Coeffic.	T-ratio (1011)
Post position	.0889	1.21	.0307	.78	.0941	3.18
Odds	.4111	5.79	.3627	9.08	.3449	11.67
Length	.0169	0.16	-.0023	-.05	.0004	.01
Constant	1.4082	1.82	2.1493	5.27	2.2496	5.72
	Constant ρ: -.00268		Constant ρ: -.004426		Constant ρ: .011613	
	Log-Likelihood: -281.23		Log-Likelihood: -1037.80		Log-Likelihood: -2069.79	
	Coeffic.	T-ratio (167)	Coeffic.	T-ratio (549)	Coeffic.	T-ratio (1012)
Post position	.0888	1.23	.031	.79	.095	3.19
Odds	.4109	5.89	.363	9.13	.345	11.72
Constant	1.5227	4.72	2.133	10.47	2.253	13.21
	Constant ρ: -.00268		Constant ρ: -.04425		Constant ρ: .011613	
	Log-Likelihood: -281.24		Log-Likelihood: -1037.80		Log-Likelihood: -2069.79	
	Coeffic.	T-ratio (168)	Coeffic.	T-ratio (550)	Coeffic.	T-ratio (1013)
Odds	.401	5.84	.364	9.22	.351	11.91
Constant	1.799	7.77	2.237	14.61	2.595	19.67
	Constant ρ: -.00775		Constant ρ: -.00359		Constant ρ: .02085	
	Log-Likelihood: -282.100		Log-Likelihood: -1038.11		Log-Likelihood: -2074.77	

	Number of horses: 8 Number of races = 149		Number of horses: 9 Number of races = 157		Number of horses: 10 Number of races = 439	
	Coeffic.	T-ratio (1188)	Coeffic.	T-ratio (1409)	Coeffic.	T-ratio (4386)
Post Position	-.0031	-.11	.0447	1.78	.0521	3.71
Odds	.3870	14.48	.3730	14.96	.3933	28.15
Length	.0034	.06	-.0003	-.005	.0023	.07
Constant	2.7576	6.91	2.9159	7.77	3.0340	12.62
	Constant ρ: .05436		Constant ρ: .017206		Constant ρ: -.001102	
	Log-Likelihood: -2577.90		Log-Likelihood: -3236.79		Log-Likelihood: -10471.8	
	Coeffic.	T-ratio (1189)	Coeffic.	T-ratio (1410)	Coeffic.	T-ratio (4387)
Post Position	-.003	-.11	.045	1.79	.052	3.71
Odds	.387	14.53	.373	15.01	.393	28.18
Constant	2.781	15.32	2.914	15.73	3.049	28.23
	Constant ρ: .05436		Constant ρ: .01721		Constant ρ: -.001101	
	Log-Likelihood: -2577.90		Log-Likelihood: -3236.79		Log-Likelihood: -10471.80	
	Coeffic.	T-ratio (1190)	Coeffic.	T-ratio (1411)	Coeffic.	T-ratio (4388)
Odds	.387	14.59	.374	15.09	.400	28.92
Contant	2.768	20.46	3.132	22.38	3.294	38.46
	Constant ρ: .05436		Constant ρ: .0195		Constant ρ: .0002018	
	Log-Likelihood: -2577.91		Log-Likelihood: -3238.40		Log-Likelihood: -10478.70	

Figure 1.

DISTRIBUTION OF RACE SIZES
BY NUMBER OF HORSES

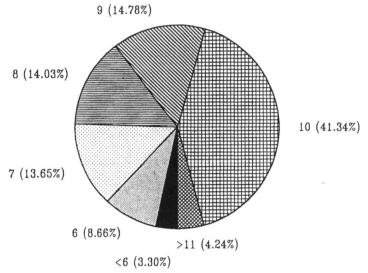

9 (14.78%)

8 (14.03%)

10 (41.34%)

7 (13.65%)

6 (8.66%)

>11 (4.24%)

<6 (3.30%)

DISTRIBUTION OF RACE LENGTHS
IN FURLONGS

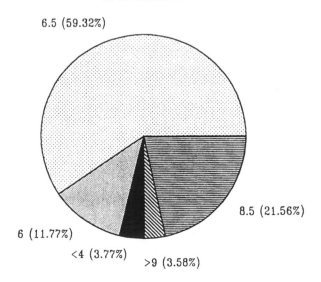

6.5 (59.32%)

8.5 (21.56%)

6 (11.77%)

<4 (3.77%)

>9 (3.58%)

Figure 2.

PERCENTAGE DIFFERENCE (POST BASIS)

ACTUAL AND EXPECTED # OF FINISHES

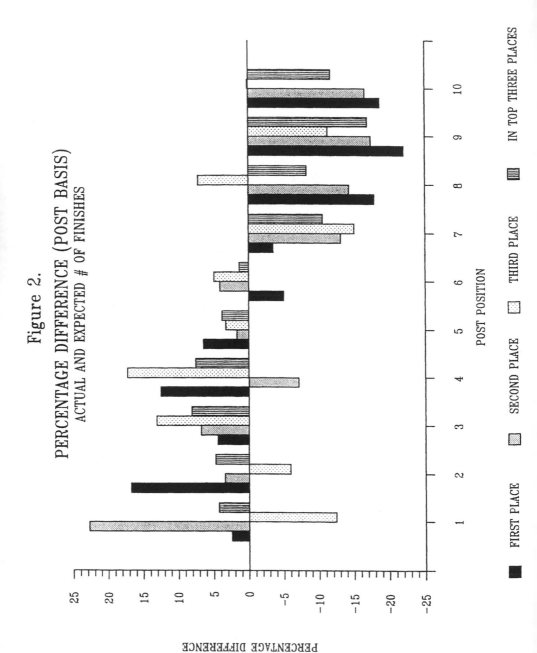

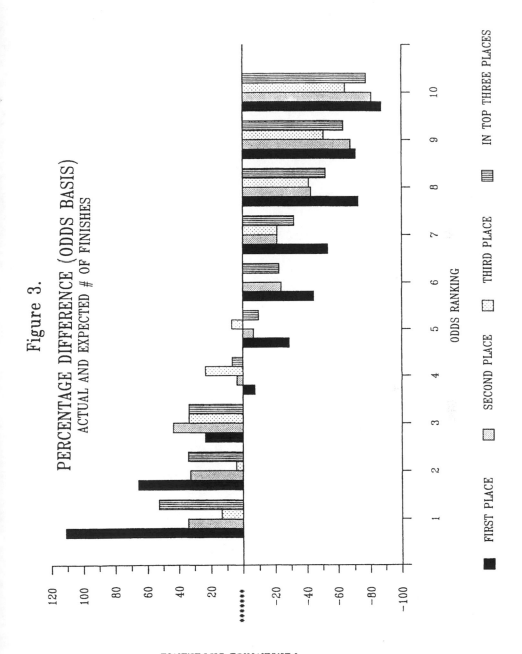

Figure 3.

PERCENTAGE DIFFERENCE (ODDS BASIS)
ACTUAL AND EXPECTED # OF FINISHES

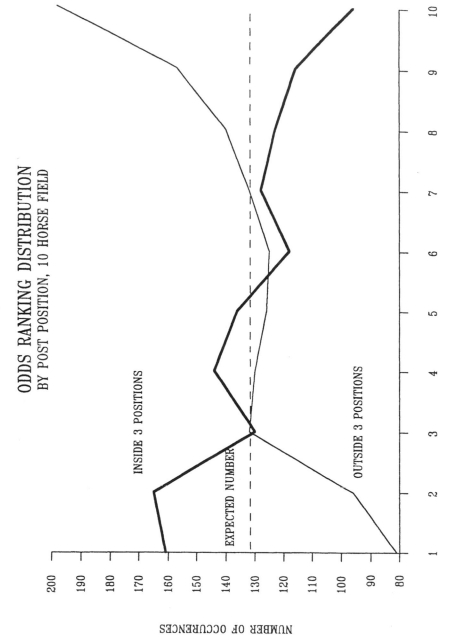

Figure 4.

ODDS RANKING DISTRIBUTION
BY POST POSITION, 10 HORSE FIELD

PART VII

RESEARCH IN THE COMMONWEALTH AND ASIA

Introduction to the Efficiency of Racetrack Betting Markets in England

Donald B. Hausch, Victor S.Y. Lo and William T. Ziemba

The parimutuel betting system utilizing electric tote boards is the dominant method of betting at North American racetracks. Under the parimutuel system, the win bet fractions may be interpreted as consensus subjective win probabilities provided by the public. In England, and in other Commonwealth countries, such as Australia and New Zealand and other European countries, such as Italy and France, the dominant system is the fixed-odds system where it is bookies who establish and offer the odds.

The fixed-odds system operates as follows. Bookies set odds that my vary throughout the betting period. Bettors lock in the odds at the level they are offered at the time of purchase, even if the offered odds changed later. Quite differently, locking in odds is not possible in the parimutuel system where odds are established only when the betting period is over. Under the fixed-odds system, changes in the odds are largely due to the bookies' efforts to balance their books to ensure a profit no matter which horse wins. Such price dynamics can allow an opportunity to hedge one's bets. Lane and Ziemba (1992)[2] analyzed how this can be accomplished for a race with two horses. Although far less popular, a parimutuel betting system using the tote board is also available in England for betting both on course and off course.

To test the market efficiency under the fixed-odds system, one can use the starting prices (SP) reported at the track. They are the bookies' final offered odds. Most off-course betting is offered at SP. Table 1 reports the rates of returns in lower odds ranges and lower returns for long odds horse. Thus, the favorite-longshot bias is present, despite the odds being the bookies' offerings.

Dowie (1976)[1] and Crafts (1985)[2] studied the efficiency of British racetrack markets. Dowie (1976)[1] considered two types of prices (odds): (i) Forecast prices (FP) - forecasts odds are estimated long before start of the race in *Sporting Life* which is widely regarded as the most authoritative racing information available in the press. Thus, it is similar to the morning line odds determined by the official handicappers in North America. (ii) Starting prices (SP) - the final odds made by the bookies as mentioned above. Dowie found that the correlation between forecast price and winning frequency is not significantly less than that between starting price and winning frequency. He considered that if insider information does exist, it will not be available to the public and only the final SP can reveal it. Therefore, he concludes that there is no insider information.

Crafts (1985)[2] disagreed with Dowie's point of view. His argument is that a gambler possessing inside information can bet earlier on more attractive odds than the final starting odds. To verify his argument empirically, he classified the win odds data using the ratio FP/SP, and then compared the expected returns in each range of FP/SP. The FP/SP ratio is an indication of the movement of the odds. For high FP/SP ratio ranges (i.e. the favorites are getting more popular), expected returns are positive and are better than the average. For low FP/SP ratios, on the other hand, the average expected returns are negative. This result is clearer for more favorite horses when the data is further classified by the SP odds levels. Hence, a gambler with insider information can bet long before the announcements of SP odds in order to gain much higher returns. According to his analyses, the market is "weakly efficient" since no strategy based on publicly available information would allow a profit at the odds available at the end of trading (SP). However, the market is also not weakly efficient because a bettor placing a bet early enough in a particular odds range can reduce expected losses (or increase expected returns) by betting

Table 1
Rates of returns for different SP odds ranges in British flat racing before taxes

SP odds Ranges	Rate of Return				
	1950	1965	1973	1975[a]	1976[a]
1-100 to 2-5	97.2	108.1	108.5	112.1	107.0
4-9 to 2-5	98.8	89.4	109.7	108.4	107.8
8-11 to 1-1	94.8	88.4	93.6		
21-20 to 3-2	96.5	87.2	88.6		
13-8 to 9-4	90.4	95.9	83.6		
95-40 to 4-1	95.5	95.0	95.5		
9-2 to 9-1	90.1	89.5	89.1		
19-2 to 18-1	64.5	64.9	66.5		
larger than 18-1	23.8	37.3	23.2		

Source: Ziemba and Hausch (1984)[2]. It is assumed that every horse in each odds range was bet to return 100 to break even.
[a] Data not available in some entries.

on those going in markedly in the betting (i.e. high FP/SP ratio). His results suggest that British racing offers considerable potential for profitable insider trading.

Crafts (1994)[1], in support of earlier work by Crafts (1985)[2], utilized the information provided by special decision rules (racing systems) which are sold to the public, with advertising reporting that they are profitable. However, his results indicate that gamblers who buy racing systems in the hope of profits do so in vain. Thus he concluded that all these advertisements of profitable systems are in fact misleading. Instead, following the results obtained in Crafts (1985)[1], he found a 55.8% return by betting on horses with a marked difference between FP and SP (FP/SP ≥ 1.5) which had been off the course for a long period. The results suggest that the British market includes insiders with potentially highly profitable information not known to the outsiders who buy racing systems which contain unprofitable methods of betting.

In Britain, the tote system is also being used as well as the fixed-odds system. Gabriel and Marsden (1990[1],1991[1]) compared the returns to starting price bets placed with bookies and pari-mutuel tote bets. They concluded that tote returns are higher than starting price returns, even though both betting forms are of similar risk and the payoffs are widely reported. This appears to reject semi-strong form efficiency. The result continued to hold as the season progressed, in spite of their argument that even if insiders exist, the final starting price and tote payments to winners should converge.

Ziemba and Hausch (1994)[1] propose a capital growth betting strategy for place wagering in Britain. This strategy is a modification of the one proposed by Hausch, Ziemba and Rubinstein (1981)[1] and further developed in Ziemba and Hausch (1984,1987)[2], because the British place bet is different from North American's.

Other articles using the U.K. racing data include Henery (1984)[2] and Henery (1985)[2]. Assuming the running times of horses are extreme-value distributed, Henery (1984)[2] tested predictive ability for the win probabilities. However, based on his U.K. fixed-odds system data, only the tail of the empirical distribution functions is consistent with the proposed extreme-value model. Henery (1985)[1], on the other hand, proposed an empirical linear regression model to explain the favorite-longshot bias in the U.K. racing market.

[1] included in this volume
[2] cited in our Annotated Bibliography

On the Efficiency and Equity of Betting Markets

By Jack Dowie

University of Kent

I

Confident in the belief that vital resource allocation decisions depend on well functioning capital markets, economists over the last two decades have spent a good deal of their time, and not a little of the computer's, studying the behaviour of stock market prices. Their main aim has been to discover the extent to which capital markets, and in particular markets in equity shares, are "efficient". Perhaps surprisingly to the layman, their answer has usually been "very efficient", and despite the amount of effort devoted to attempts at refutation, Fama was able to conclude that "the evidence in support of the efficient market's model is extensive and (somewhat uniquely in economics) contradictory evidence is sparse" (Fama, 1970, p. 416).

The definition of "efficiency" being used is not, however, the usual one and reflects the fact that stock markets are being examined as information markets, not as service industries. A stock market is "efficient" if its prices always "fully reflect" "available information". Almost all the empirical work proceeds on the assumption that the conditions of market equilibrium can be stated in terms of expected returns, and it is becoming conventional to talk of three subsets of information in relation to the determination of equilibrium expected returns: historical prices (and returns), other publicly available information (e.g. announcements of earnings, issues, etc.) and "inside" information (i.e. information to which particular groups or individuals have monopolistic access at relevant points of time). Defining "available information" as each of these in turn produces "weak", "semi-strong" and "strong" tests of efficiency.

If prices always fully reflect historical prices in the sense that price changes approximate a random walk, the market is said to be "weakly efficient". The bulk of empirical work has been done in this area. While some drift is usually found and some filters (mechanical trading rules) have been discovered that produce profits in excess of a naive buy-and-hold policy, it is generally agreed that these have little practical significance given the transaction costs (brokerage) needed to implement them.

If prices fully reflect public announcements as soon as they are made, the market is said to be "semi-strongly efficient". Only a small, though rapidly growing, volume of work has been done on this, particularly on dividend announcements and bonus issues.

The third form of test is in many ways the most interesting, because it raises the problematic link between "efficiency" and "equity" in information markets. Do corporate insiders or other in the investment community (e.g. financial journalists) have monopolistic access to information that they can exploit in such a way as to produce above-average returns and contribute to, or even produce, the semi-strong and weak efficiency ordinarily found? (Samuelson has recently shown (1973, p. 373) that "there is no incompatibility in principle between behaviour of stocks' prices that behave like random walk at the same time that there exist subsets of investors who can do systematically better than the average investors.") To the extent that such subsets exist, the phenomena of semi-strong and weak efficiency are of reduced interest, the implication being that publicly available information plays a less than crucial role in the allocation process. While Fama and others talk of "inside information"—using this as a catchall term—detracting from "strong efficiency" there seems to be a reasonable case for reserving the term "efficiency" for the publicly available information subsets (historical prices, public announcements) and using the term "equity" in place of "strong form efficiency". In other words, we will talk of a market as *efficient* to the extent that it passes the weak and semi-strong tests and *equitable* to the extent that it passes the strong test.

II

This preamble establishes the context for an excursion into another sort of market in information: betting on horse races. Even though it is quite possible that the detrimental effects on social welfare arising from inefficiency and inequity in betting markets are commensurate with those arising in capital markets, the more conventional justification for moving into this area is that

> ... institutional forms of betting are of scientific interest for two reasons. They yield behavioural implications for individual decision-making under uncertainty. Furthermore, since these betting schemes give rise to wager markets with equilibrating functions similar to ordinary security (or commodity) markets, they afford opportunities for the study of market mechanisms under a widened class of contingency and institutional conditions. (Smith, 1971, p. 242)

A brief guide to betting markets on horse-racing in Britain is required. Conceptually there are two distinct forms of betting: fixed odds betting and pool betting. In the former a bookmaker accepts bets at specific, but changing, odds throughout the betting on an event, and the return to any individual bet is unaffected by bets made subsequently. In the latter a bet is made with less-than-complete certainty as to the return if successful, the odds being determined by the weight of money on each runner at the conclusion of betting. The conceptual difference does not produce practical differences in average returns to all successful bettors to the extent that (a) bookmakers are able to adjust their odds to the weight of money on each runner; and (b) pool (totalizator, pari-mutuel) bettors are kept informed about, and act upon, the number of units invested on each runner (note that these pool bettors may include bookmakers). In any event the conceptual distinction is not preserved institutionally in the United Kingdom because most betting with bookmakers is *not* at fixed and known odds. The bulk of on-course betting certainly is, and so is quite a lot of the betting on the more important races of the year on which ante-post books are opened. But most off-course betting—95 per cent or more according to "Michael Rolfe" of the *Sporting Life* (personal communication)—is at "starting price" (SP, defined as the odds at which a "sizeable" bet could have been made *on the course* just before the " off"). This is in one sense a fixed price, but it is clearly not known with certainty at the time of making the bet. The weight of off-course SP money is reflected in on-course prices as information, and money, is relayed to the course, and the actual SP is thus determined by both off- and on-course activity. Instances of manifestly imperfect equilibration are the focus, at different times, of complaints by both smaller off-course bookmakers and off-course punters, as well as constituting the basis for many "betting coup" attempts. But it is assumed here that these instances are not significant enough to undermine the general validity of the SP as an equilibrium price.

Pool betting has one great advantage from the investigators' point of view, and the limited American analysis in this area has all been done on this basis. Since bookmaking is illegal, except in Nevada, psychologists (Griffiths, 1949; McGlothlin, 1956), economists (Weitzman, 1965; Rosett, 1965) and others (Fabricand, 1965; Harville, 1973; Hoerl and Fallin, 1974) made use of pari-mutuel data. The great advantage is that a constant percentage is taken out of the pool by the operators. As a result the prices for every horse in every race are homogeneous in the sense that each odds always represents precisely the same proportionate weight of money invested. There is, therefore, no question as to *differentially* "unfair" odds being offered against the runners in the race. In contrast, in a bookmaking setup, we *cannot* be certain that all horses starting at any particular price represent equivalent obligations on the bookmakers' part. Despite this advantage, totalizator data in the United Kingdom suffer from three major disadvantages that more than offset this particular advantage.

The first is the low relative turnover. The proportion of total turnover accounted for by totalizator betting is less than five per cent. It would be dangerous to generalize upon this limited basis. The second factor is even more crucial. The Tote is still basically a manual operation in Britain, and consequently only the dividend of the winner is published—indeed, only the dividend of the winner is calculated except when there is a photo finish. As a result we have no comprehensive record concerning the totalizator odds at which every runner started, contrary to the American situation. The third and final advantage of concerning ourselves with bookmakers' starting prices rather than totalizator data is the existence of systematic *forecasts* of the former in the daily press. It is on these forecasts that we hinge much of the argument concerning the *equity* of betting markets.

III

The basic data resource for our investigation of betting markets are sets of starting prices (SP) and forecast prices (FP). A typical example of each, along with the arithmetic of their "standardization" (to be discussed later) appears in Table 1. The examples are typical in the sense that they relate to a ten-runner event in which the SP set is about 20 per cent "over-round" and the FP set about 35 per cent "over-round", these three figures being close to the medians of their respective distributions. The column following the SP and FP in Table 1 gives the unit stakes required to yield a return (stake plus winnings) of 100 at that price. Calculated as $100/(P+1)$, these are known in racing parlance as the "percentages". The sum of these percentages indicates, under the (strong) assumption that stakes were held appropriately, the bookmakers' gross percentage margin over the bettor on the event. Naturally they do not normally add to 100 and, even more obviously, they rarely fall below 100: no instance of a "Dutch" set of SP returns was found in the 1973 flat season in Britain.

The starting prices are returned by the representatives of the two daily racing papers, the *Sporting Life* and *Sporting Chronicle*. They observe the betting

TABLE 1

EXAMPLE OF STARTING PRICES (SP) AND FORECAST PRICES (FP) AND STANDARDIZATION PROCEDURE: CHERRY HINTON STAKES, NEWMARKET, 3 JULY 1973

	SP	Percentages	Probabilities	Standardized SP	FP	Percentages	Probabilities	Standardized FP	"Life" betting report
2nd Mrs Tiggywinkle	15/8	34·8	29·0	5/2	2	33·3	24·6	3	proved a solid favourite at 7/4, 2/1 and finally 15/8
1st Celestial Dawn*	4	20·0	16·7	5	12	7·7	5·7	16	good second best, although always returning to 4 from 7/2
Caught in the Rye	6	14·3	11·9	15/2	9/2	18·2	13·4	13/2	sound business for at 7 and 6 (11/2 at times)
4th Belle Tigresse	13/2	13·3	11·1	8	7	12·5	9·2	10	sound business for, closed in at 13/2 after touching 9 from 6
Slipperty	7	12·5	10·4	17/2	3	25·0	18·5	9/2	receded from 5 to 7
Ribella	12	7·7	6·4	15	14	6·7	4·9	20	receded from 8 to 12
Wheal Jane	14	6·7	5·6	16	11/2	15·4	11·4	8	receded from 12 to 14
3rd C'est Vrai	20	4·8	4·0	25	10	9·1	6·7	14	receded from 14 to 20
Princess Donna	33	2·9	2·4	40	25	3·8	2·8	33	—
Finishing Touch	33	2·9	2·4	40	25	3·8	2·8	33	—
		119·9	99·9	(99·5)		135·5	100·0	(99·9)	

* Tote 0·60 or approximately 11/2 SP.

in the principal ring (Tattersalls), concentrating on those bookmakers who can be relied upon to lay a substantial bet:

These [bookmakers] are the men who take any money "unloaded" by bookmakers' offices through the "Blower", which usually arrives in the last few minutes and often causes rapid changes in the odds. These bookmakers co-operate with the newspaper men by allowing them to see the books if necessary, though most of the wagers can be overheard. The two journalists record the fluctuations of the odds from start to finish and immediately the "off" is signalled they meet to compare notes and fix the prices on which millions of starting-price bets will be settled. Just occasionally the market has been so stable that the two men are in perfect agreement, but this is rare and usually adjustments have to be made. One may have the favourite a sound even money chance, whereas the other can obtain 6–5. So that 11–10 is obviously a fair return. Similarly 11–4 is often a compromise between 5–2 and 3–1, while prices such as 13–2, 15–2 and 17–2 usually means that each man has "given" half a point to strike a correct balance. (O'Neill, 1971, pp. 57–58)

It is worth noting in passing that the results reported later suggest that the agreement is sometimes not reached until *after* the outcome of the race is known. Horses at compromise prices or "split odds" often tend to have a higher percentage return than those on either side—there being clearly a greater need to establish maximum agreement in the case of a winner or placed horse than in the case of a loser.

The forecast prices are produced by employees of various newspapers in the afternoon of the day preceding the race, and it must be emphasized that the forecast used in this particular analysis is simply one of several. They all differ to a greater or less extent and there is no such thing as "the" forecast. The ones we have chosen to use are those put out daily by the *Sporting Life*. These are produced by a number of individuals (about eight in all during 1973), though the forecast on any one race is produced by a single person. Two basic strategies are available and each is employed by some members of the group. The strategies differ in respect of the use made of the Press Association (PA) forecast that is sent to all subscribers soon after final declaration of runners at 11 am on the day before the race. Some papers' forecasters use this without significant change or modify it as they think fit while retaining the overall shape. Others largely

TABLE 2

	Times; Sun	Mirror	Guardian	Life	SP
2nd Prominent	9–2 Fav.		5	8	15/2
Red Power	5		4 Fav.	5 Fav.	6
1st Spring Stone	13–2		6	7	7
Ballyhot	8	7	6	6	6
3rd Malleny	8			12	14
Negus	10		8	9	13/2
Vedvyas	10				12
Offenbach	12	8	10	10	11/2 Fav.
Buss	12				14
Hardy Scot	14		10		14
Bright Fire	16			16	14
Tack On	20			14	20

ignore the PA forecast, compile a forecast of their own and, if they feel inclined, check the result against the PA. The differing practices are illustrated in Table 2 in relation to the P.T.S. Laurels Stakes run at Goodwood on 4 August (blanks indicate same odds as column (1)). *The Times* and *Sun* used the PA forecast, the *Mirror* brought a couple of middle-order horses in a bit, the *Guardian* made a lot of small changes and the *Life* (in the case of this race) clearly compiled its own forecast.

The only manipulation of the published data necessary was in the case of withdrawals. Where these occurred before the runners came under starter's orders the SP odds on all runners were adjusted in accordance with the official Tattersalls deductions, and this procedure was also followed—much more often, naturally—in the case of the forecast prices. The Tattersalls deductions are based on the odds of the withdrawn horse only, irrespective of the over-round-ness of the market, but more precise adjustment was neither practicable nor unambiguously desirable.

IV

It is of course not possible to learn much from the SP return for a single race or small number of races. A number of questions can however be posed if we aggregate the returns for a large number of races. The first question, which has direct relevance to a weak efficiency test of horse race betting markets, is "What is the expected return at each odds (and group of odds)?" Obviously the number of horses that start at each odds varies considerably—from single figures at long odds-on to thousands of runners at 20–1 and 33–1 during a season—but wherever the number is regarded as large enough the expected return calculation can be made (or rather used, given that it can always be made). Table 3 presents the results of such an analysis for the 1973 flat season, column (5) containing the percentage return at each starting price and column (6) giving the cumulative return up to that particular price (e.g. the cumulative return at 10–1 is the result

TABLE 3

1973 FLAT SEASON: RETURN AT EACH STARTING PRICE

Odds	Unit stakes required to return 100	Runners	Winning percentage	Percentage return	Cumulative return Level staking	Cumulative return Staking to return 100
(1)	(2)	(3)	(4)	(5)	(6)	(7)
1–11	91·7	1	100·0	109·0	109·0	109·0
2–13	86·7	1	0·0	0·0	54·5	56·1
1–6	85·7	2	100·0	117·0	85·8	85·8
1–5	83·3	2	100·0	120·0	97·2	96·8
2–9	81·8	4	100·0	122·0	107·1	106·7
1–4	80·0	7	85·7	107·1	107·1	106·9
2–7	77·8	2	50·0	64·5	102·6	102·6
30–100	76·9	1	100·0	130·0	104·0	103·9
1–3	75·0	8	75·0	99·8	102·8	102·9
4–11	73·3	15	66·7	90·7	98·6	98·9
2–5	71·4	18	88·9	124·4	106·2	106·0

TABLE 3—*contd.*

4–9	69·2	21	81·0	116·6	108·9	108·7
40–85	68·0	1	100·0	147·0	109·3	109·1
1–2	66·7	24	70·8	106·3	108·6	108·5
8–15	65·2	14	64·3	98·4	107·4	107·4
4–7	63·6	23	60·9	95·6	105·5	105·8
8–13	61·9	24	75·0	121·5	107·8	107·7
4–6	60·0	52	67·3	112·4	108·9	108·7
8–11	57·9	58	37·9	65·6	99·9	100·7
4–5	55·6	76	59·2	106·6	101·3	101·8
5–6	54·5	35	65·7	120·3	103·0	103·2
10–11	52·4	79	46·8	89·5	100·7	101·2
Evens	50·0	116	46·6	93·1	99·2	99·8
11–10	47·6	112	41·1	86·3	97·1	98·0
6–5	45·5	32	25·0	55·0	95·3	96·5
5–4	44·4	134	42·5	95·7	95·3	96·4
11–8	42·1	113	35·4	84·2	94·1	95·3
6–4	40·0	186	37·6	94·1	94·1	95·1
13–8	38·1	123	36·6	96·2	94·3	95·2
7–4	36·4	221	30·8	84·6	92·9	94·0
15–8	34·8	101	32·7	94·1	92·9	94·0
2	33·3	305	24·6	73·8	89·9	91·6
85–40	32·0	43	23·3	72·8	89·5	91·3
9–4	30·8	312	27·2	88·5	89·4	91·0
5–2	28·6	363	26·5	92·6	89·8	91·2
11–4	26·7	342	24·9	93·2	90·2	91·3
3	25·0	538	23·6	94·4	90·8	91·6
100–30	23·1	279	22·9	99·3	91·5	92·0
7–2	22·2	590	18·8	84·7	90·6	91·4
4	20·0	734	21·1	105·6	92·7	92·6
9–2	18·2	656	13·9	76·3	90·8	91·5
5	16·7	855	15·1	90·5	90·8	91·4
11–2	15·4	635	15·6	101·3	91·7	91·9
6	14·3	944	13·5	94·2	92·0	92·0
13–2	13·3	450	17·3	130·0	94·0	93·1
7	12·5	983	9·9	78·9	92·4	92·3
15–2	11·8	288	12·9	109·2	92·9	92·6
8	11·1	1290	8·8	78·8	91·3	91·8
17–2	10·5	36	2·8	26·4	91·1	90·5
9	10·0	621	7·6	75·7	90·3	90·1
10	9·1	1564	7·5	83·0	89·4	89·7
11	8·3	513	4·9	58·5	88·3	89·3
12	7·7	1798	4·8	62·2	85·3	88·0
13	7·1	109	5·5	77·1	85·3	85·9
14	6·7	1700	4·7	69·7	83·8	85·3
15	6·3	56	1·8	28·6	83·6	84·4
16	5·9	1501	2·5	41·9	80·3	83·2
18	5·3	124	0·8	15·3	79·9	83·1
20	4·8	3739	1·1	22·5	70·5	79·9
22	4·4	141	0·0	0·0	70·1	79·7
25	3·8	1913	1·3	34·0	67·4	78·8
28	3·4	70	1·4	41·4	67·3	78·8
30	3·2	115	0·0	0·0	67·0	78·7
33	2·9	3314	0·5	18·5	61·3	77·1
40	2·4	152	0·0	0·0	61·0	77·0
50	2·0	506	1·0	50·4	60·8	77·1
66	1·5	59	0·0	0·0	60·7	77·1
100	1·0	57	0·0	0·0	60·5	77·1
'150'	0·7	6	0·0	0·0	60·6	77·1
		29307				

of having level stake bets on all runners at 10–1 and less). From the final figure in column (6) we can see that a unit bet on every one of the 29,307 horses that ran in the 2,777 races included in the analysis would have led to a pre-tax 39·4 per cent loss. In fact, given betting turnover duty—introduced on 24 October 1966 at 2½ per cent, raised on 25 March 1968 to 5 per cent and again—as far as off-course betting was concerned—to 6 per cent on 27 April 1970—the loss would have been 43 per cent. An alternative interpretation of column (6) is that it gives the average return to "pin-pricking" selection among horses up to any particular price. This suggests a small profit even after tax for random selection up to 6–4 on, but since the 1974 budget raised the duty to 7½ per cent off-course, with bookmakers normally deducting an extra half per cent "to cover Levy Board commitments", most of this will now have been eradicated.

Given that starting prices are defined as those at which a sizeable bet could be laid, the calculations referred to above have definite policy significance, in contrast to the results of equivalent calculations made—if this were possible—on the basis of totalizator prices.

There is however another set of calculations that can be made on the basis of the SP returns. This set, which can be put to either non-controversial or controversial use, is to be found in column (7). The findings in this column are generated by assuming not that stakes wagered were identical at all odds, but that they were always such as to generate a return of 100. At any particular odds the percentage return is naturally unchanged, but in aggregating we now weight not only by number of runners but also by relative stakes. It is now assumed, for instance, that twice as great an amount of stakes is placed on each horse at even money as on each horse at 3–1. To the extent that the realized percentage returns approach 100 this different weighting system makes no difference to the cumulative return picture, but as the returns depart from 100 a discrepancy arises. In Table 3 it can be seen that the two calculations stay fairly close to-gether below 20–1, but that from this point on a substantial gap opens up. It is a gap in favour of the bettor and the non-controversial conclusion that can be drawn from this calculation is that if every horse (or random selection) had been backed so as to return 100 the pre-tax loss would have been 23 per cent rather than 40 per cent.

The controversial, but more common, use to which this calculation can be put is to estimate the bookmakers' gross margin. For this purpose it has to be assumed that the share of season total stakes placed on horses which started at x to 1 was *actually* $100/(x+1)$. This does not, of course, necessitate that in any particular race the stakes were appropriately distributed, but merely that in the long run the average amount staked at each price is in the appropriate relation, so that the cumulative return over all odds gives a reasonably accurate indication of the gross margin. The standard defence of the assumption lies in the theory of ideal bookmaking together with the empirical assertion that sufficient competition is observable to warrant our believing that *long run* practical outcomes approximate the theoretical prediction: supply and demand do determine price and hence SP returns over a reasonably long period pro-cessed in the above way will yield a fairly good indication of the gross margin. Understandably, bookmakers are always keen to deny that a reasonably accurate picture is given by this calculation—naturally they argue that it is too high rather than too low—while vigorously maintaining that bookmaking is highly competitive. We are not in a position to determine this issue, but it can be noted that the invariable reaction to an SP analysis of the present sort fails to address the central, long-run, issue.

> ... the public are not machines who bet proportionately to the estimated chance of a particular horse.... Only on a rare and lucky occasion can any S.P. layer make a perfect book.... Any data used without such information [on the volume of turnover at each SP] will produce results which at best can only express an opinion of likely profitability should the betting public follow an imaginary pattern. (letter by John Waugh, *Sporting Life*, 14 March 1974, in reaction to Figgis, 1974, pp. 128–133)

V

Our interest, however, is not in how "fair" a bet is with a bookmaker (its "expected value") but how "efficient" the betting market is: is "available information" "fully reflected" in price? The criterion used is the *pattern* of expected values throughout the odds range: weak efficiency is complete if the expected values are equal throughout the range and decreases to the extent that expected values diverge from equality.

Immediately, however, it will be obvious that the fact that the over-roundness of the SP and FP differ in any one race and that each varies over a series of races makes calculation of genuinely comparable expected values difficult. In Table 3 for instance runners at any SP are grouped irrespective of the sum of the percentages in their race: a horse at evens is a horse at evens whether there were three other horses at 3–1 (market 25 per cent over-round) or four other horses at 3–1 (market 50 per cent over-round). Our solution to this problem is to standardize the SP and FP sets for every race, turning them into hypothetical, perfectly round books, thereby producing a situation where the prices/odds are effectively probabilities. The procedure has already been illustrated in Table 1. Having found the sum of the percentages for each race, each individual percentage is divided by the sum and then allocated to the nearest available SP or FP. This procedure has the effect of producing slight deviations from perfect roundness in any set, but practical reasons favoured the preservation of this discrete basis rather than one in which the standardized SP and FP would have been allowed to take continuous values and subsequently been grouped. As a result of standardization, horses that had different SP (or FP) are grouped together and ones with the same SP (or FP) are separated, but we can now be sure of one thing—that the "probabilities" of the runners in any one race add up to 1.00.

Having generated Table 4, we are now in a position to ask the same questions of the standardized SP and FP data as we did previously about the raw data. And, more important, we are in a position to ask the crucial question in relation to the equity issue: is there any significant difference between the degree to which the SP pattern on the one hand and FP pattern on the other diverge from a strict expectation model (i.e. expected return equal at all odds and, in the case of the standardized prices, equal and zero). Why is this the crucial question?

Anyone possessing inside information will presumably exploit it continuously up to the "off", always acting so as to bring the odds on offer back into line with those suggested by the superior information they possess. Given that SP is defined as the odds at which a sizeable bet could have been placed at the "off",

TABLE 4

1973 FLAT SEASON: WINNING PERCENTAGES AT STANDARDIZED
STARTING AND FORECAST PRICES

Probability	Standardized forecast prices		Standardized starting prices	
	Runners	Winning percentage	Runners	Winning percentage
(1)	(2)	(3)	(4)	(v)
87·5	1	100·0		
85·7			1	0·0
84·6			1	100·0
81·8	1	0·0	2	100·0
80·0			2	100·0
77·8	1	100·0	2	100·0
75·0	1	100·0	6	83·3
73·3			2	100·0
71·4	1	100·0	4	75·0
69·2	5	80·0	9	66·7
68·0	3	100·0	7	71·4
66·7	4	50·0	6	66·7
65·2	5	100·0	12	83·3
63·6	5	40·0	11	81·8
61·9	9	77·8	14	71·4
60·0	8	75·0	17	70·6
57·9	11	81·8	25	68·0
55·6	13	46·2	21	76·2
54·5	17	35·3	23	47·8
52·4	23	56·5	34	70·6
50·0	26	65·4	64	45·3
47·6	38	68·4	73	61·6
45·5	38	44·7	57	57·9
44·4	44	45·5	67	47·8
42·1	49	44·9	111	45·1
40·0	58	50·0	107	43·0
38·1	84	50·0	95	43·2
36·4	87	41·4	113	36·3
34·8	102	45·1	114	48·3
33·3	91	41·8	112	34·8
32·0	114	42·1	122	32·8
30·8	215	34·9	200	30·5
28·6	237	34·6	297	29·6
26·7	307	28·0	277	24·9
25·0	344	26·2	292	25·0
23·5	204	26·5	194	27·8
23·1	162	21·0	166	32·5
22·2	414	19·6	384	20·3
20·0	723	23·5	611	21·0
18·2	757	16·5	578	23·2
16·7	766	17·2	602	20·9
15·4	719	17·1	550	16·4
14·3	675	16·2	570	15·1
13·3	635	16·4	609	17·6
12·5	562	14·8	532	10·9
11·8	611	12·1	542	16·1
11·1	595	12·4	600	11·3
10·5	506	9·7	471	13·8
10·0	695	9·8	720	11·1
9·1	985	8·5	911	12·2
8·3	866	10·9	869	8·5

TABLE 4—*contd.*

7·7	837	7·8	838	7·4
7·1	744	6·6	752	7·3
6·7	714	7·3	713	5·6
6·3	725	6·5	805	6·0
5·9	1116	4·5	1025	5·7
5·3	1266	4·7	1323	4·4
4·8	1182	4·7	1124	4·3
4·4	1360	3·5	1242	2·6
3·8	2065	2·4	1682	1·6
3·4	1858	1·4	1185	0·9
3·2	1767	1·6	1074	1·1
2·9	2465	1·5	1352	1·1
2·4	944	1·1	2403	0·8
2·0	261	1·2	1810	0·3
1·5	143	0·0	675	0·6
1·0	29	3·5	70	0·0
0·7	14	0·0	25	0·0
	29307		29307	

SP can be taken to incorporate any superior or inside information that exists in relation to the event. If inside information plays a significant role in horse race betting markets, then the correlation between the probabilities embodied in the SP returns and the realized probabilities should be significantly higher than the correlation between the latter and any other set of probabilities assigned prior to the "off" (and certainly any set assigned prior to betting on the event). If, then, the correlation between the probabilities embodied in the betting forecasts in a morning newspaper and the realized probabilities is as high as the SP correlation, we can conclude that the existence of superior "inside" information is in doubt.

The exercises reported in Table 5, and particularly the more reliable results

TABLE 5

REGRESSION OF SP AND FP PROBABILITIES ON REALIZED PROBABILITIES

	Starting prices	Forecast prices
All probabilities (no minimum number of observations at any probability)		
N	67	64
R	0·91	0·88
R^2	0·83	0·77
Regression coefficient	0·77	0·71
Std error	0·04	0·05
Constant	0·53	0·67
Probabilities 0·5 and less (gives minimum of 25 observations at any probability)		
N	48	48
R	0·98	0·98
R^2	0·96	0·97
Regression coefficient	0·84	0·78
Std error	0·03	0·02
Constant	0·16	0·22

Source: columns (1), (3) and (5) of Table 4.

in the bottom half of the table, indeed imply that the FP correlation in 1973 was as high as, if not higher than, the SP one and therefore raise serious doubts as to the significance of inside information. The main caveat that needs to be entered concerns the paucity of data in the odds-on range, where it might be argued that inside information is particularly relevant. Given the few runners in the shorter odds range, it would require many years, even decades, to get together enough observations to make the analysis feasible over this range by itself. The practical problems involved in a supplementary analysis of this sort are also quite massive but, granting the importance of the point, we are exploring ways of overcoming them.

This limitation does not however undermine the current exercise. Only 71 winners out of the total of 2,777, roughly 2½ per cent, were accorded an SP probability of more than 0·5, so that the "exploitation" involved could hardly dominate that in the more heavily populated odds ranges unless the total amount of such exploitation involved were relatively small, which is indeed the view we are canvassing. Without wishing to belittle the statistical limitations and deficiencies of the present exercise, we are fairly confident that it poses a genuine challenge to the conventional wisdom that betting markets are "strongly inefficient" while supporting, a fortiori, the more accepted/acceptable belief that they are reasonably "weakly efficient".

A final point. Even if the "outsider" has access to as good information as the "insider" it does not mean that he exploits it, or exploits it as effectively. We have shown only (with the above caveat) that outsiders (collectively) had available to them just as good information as insiders (collectively), not that they used it and so achieved equivalent returns—but then, ought not this, rather than the ex post equality of returns, be the criteria for equity?

ACKNOWLEDGMENTS

I am grateful to Ian Dallas, Ann Worthington and Elizabeth Oxborrow for programming assistance.

REFERENCES

FABRICAND, B. P. (1965). *Horse Sense*. New York: David McKay.

FAMA, E. F. (1970). Efficient capital markets: a review of theory and empirical work. *Journal of Finance*, 25, 383–417

FIGGIS, E. L. (1974). *Betting to Win*. London: Playfair.

GRIFFITHS, R. M. (1949). Odds adjustment by American horse-race bettors. *American Journal of Psychology*, 62, 290–294.

HARVILLE, D. A. (1973). Assigning probabilities to the outcome of multi-entry competitions. *Journal of the American Statistical Association*, 68, 312–316.

HOERL, A. E. and FALLIN, H. K. (1974). Reliability of subjective evaluations in a high incentive situation. *Journal of the Royal Statistical Society*, A, 137, 227–230.

McGLOTHLIN, W. H. (1956). Stability of choices among uncertain alternatives. *American Journal of Psychology*, 69, 604–615.

O'NEILL, JOHN. (1971). Why there is a double check on starting prices. In *Daily Mirror Punters' Club Guide to the Turf* 1971 (J. L. Stevenson, ed.), pp. 56–61. London: Daily Mirror Books.

ROSETT, R. N. (1965). Gambling and rationality. *Journal of Political Economy*, 73, 595–607.

SAMUELSON, P. A. (1973). Proof that properly discounted present values of assets vibrate randomly. *Bell Journal of Economics and Management Science*, 4, 369–374.

SMITH, V. L. (1971). Economic theory of wager markets. *Western Economic Journal*, 9, 242–255.

WEITZMAN, M. (1965). Utility analysis and group behaviour: an empirical study. *Journal of Political Economy*, 73, 18–26.

Winning Systems? Some Further Evidence on Insiders and Outsiders in British Horse Race Betting

N F R Crafts,
Department of Economics,
University of Warwick.

I

In an earlier paper (Crafts, 1985) I examined the relationship of the racecourse odds at the start of a horserace in Britain to the odds forecast in the morning racing press. The results showed that there is considerable potential for profitable insider trading in British horserace betting but that the adjustment of the odds to the starting prices is "weakly efficient" in the sense that bettors at starting price could not share in the exploitation of profitable opportunities. I also demonstrated the existence of "mug bets", a class of particularly poor value bets where horses listed in the morning as fancied runners lengthen markedly in the betting during pre-race trading. In other words I argued that insiders are able to profit at the expense of outsiders, a situation which appears to be acceptable to or regarded as inevitable by the authorities.

The resulted presented here give further support to this general view. It is shown that the decision rules embodied in "racing systems" on sale to the general public in Britain are distinctly unprofitable, although in many cases they have been continually advertised over a long period. I also indicate a particularly profitable form of information possessed by insiders; in this case, however, there does seem to be scope for outsiders to adopt decision rules allowing them to profit from the insiders' information, a result which involves a further departure from efficiency in the horeserace betting market. Unfortunately the investment of time involved in seeking to exploit this rule is such as to imply that this finding is unlikely to benefit many people.

II

Each week in Britain the sporting press carries advertisements for "racing systems"; the reader is encouraged to purchase a set of decision rules which it is claimed can be used to bet with profit using publicly available information. Many such systems are typically on offer and some have been advertised regularly for years. If such systems actually are profitable, this would obviously violate the conditions for an efficient market. If such systems were profitable prior to publication but not subsequently, this would suggest a less serious inconsistency with the efficient markets hypothesis but nonetheless would indicate that profitable decision rules can be discovered and that horserace betting markets are not always efficient.

Table 1 reports the results of an investigation into the profitability of three systems, researched by their proprietors in the 1970s and widely advertised in the 1980s. Each of the systems provides decision rules for betting based on readily available public information. Fineform and the Peter Smith Method are based on the recent form figures of horses in a race while Special Bets gives the formula which yields a list of horses to follow based on time and manner of a winning performance. These systems have been marketed at prices in the £10 - £15 range.

TABLE 1 : The Profitability of Three British Racing Systems

		Win Fraction	Return[a] (%)	Win Fraction [a,b]	
				To Break Even	To Make 10% Return
a)	Before Publication				
	Fineform (n = 249)	0.345	-7.1	0.372	0.409**
	Peter Smith Method (n = 349)	0.203	-18.5	0.250*	0.275*
	Special Bets (n = 254)	0.303	-12.1	0.345	0.379**
b)	After Publication				
	Fineform (n = 252)	0.333	-19.9	0.416**	0.458**
	Peter Smith Method (n = 341)	0.226	-12.9	0.259	0.285**
	Special Bets (n = 147)	0.272	-10.9	0.305	0.337*

Source: Derived using the rules proposed in Holt (1986) for maximum rated horses, in   Smith (1986) for handicap races of up to 12 runners where all horses have raced at least 5 times (his most preferred category), and in Randall (1983) for flat racing only. Results were checked using daily newspapers drawing samples from 1978-9 and 1986-7, except for Special Bets where 1978-9-80 and 1984-6-7 were used.

Notes:

[a]All calculations are made for off-course punters paying the 10% betting tax; on course punters paid 4% tax until April 1987 and now bet tax free.

[b]The required win fractions are calculated on the basis of the average odds per winning bet in the sample; * and ** indicate that this proportion is significantly different on a one-tailed test from the observed proportion at the 5 and 1% levels respectively.

The results in Table 1 are straightforward to interpret and give rise to the following points.

1.   The systems do not permit a 10% rate of return on stakes (likely to be the minimum acceptable to compensate for the time and risk involved in following such betting rules) either before or after publication.

2.   Only in the case of Fineform does there appear to be any evidence of the market responding to publication by adjusting odds to the detriment of the rule, which may have been profitable for on-course backers in the 1970s.

3.   If results before and after publication are pooled, then both Fineform and the Peter Smith Method show a win frequency below at the 1% level of statistical significance that required to break even the Special Bets is below at the 10% level.

In general, then, the results are consistent with an efficient markets hypothesis and there is no evidence to support claims that rules have been discovered which have allowed off-course bettors to make profits. Fineform, however, was a system which may have produced a small positive return for on-course punters for a time and may therefore have uncovered an inefficiency in the horserace betting market during the 1970s. The results given in Table 2 suggest that racing systems generally, as well as the "ratings services" of longstanding private handicappers such as Raceform and Timeform who sell selections for races are substantially unprofitable tools for bettors, again much as the efficient markets hypothesis would predict.

TABLE 2 : The Profitability of Other Racing Systems and Ratings Services

	Win Fraction	Return (%)	Win Fraction	
			To Break Even	To Make 10% Return
Nine Racing Systems (n = 638)	0.202	-28.5	0.283**	0.311**
Five Ratings Services	0.254	-22.6	0.306**	0.336**

Source: Derived from the results reported in Roberts and Newton (1987) using own calculations with methods as in Table 1. None of the ratings services showed a profit but one of the racing systems did - however, this method's results as given in Roberts and Newton were based on only 12 observations.

It seems reasonable to conclude that punters who buy racing systems in the hope of obtaining access to decision rules that will lead to profitable betting do so in vain - as a proponent of the efficient markets hypothesis would predict - despite the misleading claims made repeatedly in advertisements in the sporting press. The continued sale of these systems suggests that participants in British horserace betting include many gullible outsiders.

At the other extreme of the market are knowledgeable insiders, who, as my earlier paper showed

(Crafts, 1985), possess a better idea of the true odds against certain horses than do the bookmakers or the general public. These insiders have opportunities to profit in the market trading which establishes the starting prices at which offtrack bets are settled and at which the insiders' information would not be profitable. Crafts (1985, p.298) shows that where horses have a ratio of the win probability implied by the morning line odds to that of the win probability implied by the starting price (FP/SP) > 1.5 it would not be profitable to bet at SP (after tax rate of return = -12.6%) but it would have been very profitable to bet on them at FP (79.7% rate of return).

Preliminary analysis of these data suggested that there might be a category of horses where insiders were able to avail themselves of particularly valuable insider information. The category in question is that of horses racing after a long absence from the racecourse through injury, which are allowed under British rules of racing to return to racing without any previous public trial; relatively few of these horses win first time out and they are generally ignored by tipsters and the betting public. Over the five years since my initial sample I have collected data on the performance of horses who had not run since the season before last and had starting prices markedly shorter than the odds forecast in the morning press (FP/SP > 1.5). The results are shown in Table 3 for the 88 observations which have occurred in the period September 1982 to November 1987. If it had been possible to bet on these horses at FP, the rate of return would have been 261.9%!

TABLE 3 : Betting on Horses with a Marked Difference between FP and SP which had been Off the Course for a Long Period

Win Fraction	Return (%)	Win Fraction to Break Even	Win Fraction to Make 10%
0.318	+55.8	0.204**	0.225**

Source: Derived on the basis described in the text using the Sporting Life as in
                                 Crafts (1985) but including only horses with a starting price < 7/1 and
                                 making calculations on the same basis as for Tables 1 and 2.

Table 3 reveals that the information possessed by insiders in the case of horses off the course since the season before last is especially profitable as their bets would generally be struck at odds better than SP. The important difference from the results in Crafts (1985) is that backing these horses at SP is highly profitable with a 55.8% rate of return on stakes. Thus there is a possibility that outsiders, including off-track bettors, could adopt a profitable decision rule by following the insiders and backing horses at SP where they have been off the course since the season before last and market support has established an FP/SP ratio > 1.5.

The findings in Table 3 are not consistent with "weak efficiency" if that is taken to imply that SP odds are a good approximation to winning probabilities but FP odds are not. There is, however, a substantial cost to adopting the decision rule suggested by the results of Table 3, namely that on more than half of the racing days each year there is at least one runner which has not raced since the season before last and might turn out to qualify at the end of market trading as having an FP/SP ratio of > 1.5. In a period of five years only 88 bets would have resulted from a very substantial investment of time and effort in market observation in order to be able to place a bet just before the end of trading if the rule happens to be satisfied.

III

The evidence presented in this paper suggests that the British horseracing betting market is not a very equitable one and strengthens my earlier objections to Dowie's (1976) claim to the contrary. This market includes insiders with potentially highly profitable information not known to the hapless outsiders who buy racing systems which contain unprofitable methods of betting. The results do, however, offer encouragement to the view that it may be possible to generate profitable betting rules and thus that the betting market may not be even weakly efficient; if so, it may be better to look for profitable rules by selective following of insiders than by seeking to devise systems based on form figures or past times.

REFERENCES

Crafts, N.F.R. (1985), "Some Evidence of Insider Knowledge in Horse Race Betting in Britain", Economica, 52, 295-304.

Dowie, J. (1976), "On the Efficiency and Equity of Betting Markets", Economica, 43, 139-150.

Holt, C. (1986), Be a Successful Punter, Morecambe, Fineform Publications.

Randall, G. (1983), Special Bets : The Winning Formula, Cambridge.

Roberts, P.M. and Newton, B.A. (1987), The Intelligent Punter's Survey, Weymouth, T.I.P.S. Publishing.

Smith, P. (1986), The Peter Smith Method of Interpeting Racing Form, Southsea.

An Examination of Market Efficiency in British Racetrack Betting

Paul E. Gabriel

Loyola University of Chicago

James R. Marsden

University of Kentucky

The nature of the British racetrack betting market provides a distinctly different opportunity for testing market efficiency. On the basis of data from a single racing season, we compare the returns to two similar forms of betting: (1) starting price bets placed with bookmakers and (2) pari-mutuel tote bets. Our analysis indicates that tote returns are consistently higher than starting price returns, even though both betting forms are of similar risk and the payoffs are widely reported. The persistently higher tote returns suggest that the British racetrack betting market does not satisfy the conditions of semistrong efficiency. Our results also provide indirect support that the market fails to meet the conditions for strong efficiency.

I. Introduction

The notion of an efficient market was initially formalized in Fama's (1970) seminal work. A market is considered efficient if information is widely available to participants and all relevant and ascertainable information is reflected in prices. In general, studies on market effi-

We are indebted to the editor, an anonymous referee, and Arthur Walker for several helpful suggestions.

ciency have examined financial securities or commodities markets. Recently, researchers have considered whether or not the efficient markets hypothesis of Fama and others has application in different markets, including racetrack wagering (see Snyder 1978; Tuckwell 1983; Asch, Malkiel, and Quandt 1984; Crafts 1985).

Racetrack betting is a particularly interesting application for efficient market analysis because of its many similarities to financial markets. For instance, in both types of markets a large number of investors (bettors) have access to widely available information. In addition, once an investment (bet) is made, the return cannot usually be guaranteed; that is, the return on an investment is uncertain. Further, the possibility of "insider information" closely parallels that in the stock market. Such similarities make racetrack wagering a logical choice for further considering the efficient market analysis begun utilizing stock market data.

In a paper on betting market efficiency and equity, Dowie (1976) suggested three types of market efficiency tests differentiated by the definition of "available information" utilized. In his approach, tests of weak efficiency equate available information with historical prices and returns. Tests of semistrong efficiency add public announcements to the set of available information and focus on whether prices fully reflect them as soon as they are made. The third type of test, that of strong efficiency, considers the existence of specific subsets of market participants possessing monopolistic access to or control over specific information. Dowie chose to refer to this last set of tests as dealing with *equity*, apparently because of the differentiation in information accessibility: "In other words, we will talk of a market as *efficient* to the extent that it passes the weak and semi-strong tests and *equitable* to the extent that it passes the strong test" (p. 140). In the present paper we utilize a key aspect of the racetrack betting market in England, the ability to place either a bookie starting price or pari-mutuel tote bet, to analyze what Dowie refers to as semistrong efficiency. In doing this, we develop information relevant to, but not directly testing, the equity of this market. An additional test of market efficiency based on the work of de Leeuw and McKelvey (1984) and Zuber, Gandar, and Bowers (1985) is also provided.

Section II briefly describes the English thoroughbred betting market. Section III analyzes data on winning payoffs made during the 1978 racing season and discusses possible implications for market efficiency. As explained in more detail below, the 1978 season was chosen since it represented a period when electronic boards providing rapid odds updates for tote wagering were still rare in England. Section IV summarizes our analysis.

II. The Racetrack Betting Market in England

A typical English bettor ("punter") can utilize two general betting mediums. First, a bet can be placed on a horse through the totalizator—the "tote." The tote is similar to American racetracks' parimutuel betting systems. If a horse wins, the payoff made on a tote bet is based on the amount of money bet on the winning horse relative to the total bet in the winning pool. That is, once the takeout (taxes, track cut, owners' and trainers' cut) is removed, the winning pool is proportioned to those wagering on the winning horse. Each bet on a horse thus affects the odds paid if that horse should win. Though for small bets the impact may be almost negligible, large bets can have great marginal impacts on winning payouts. Because of the general absence of tote boards (electronic boards that update wagering and provide current odds) in England during the late 1970s (see the next section), tote bettors were betting with only minimal information (morning line odds or hand-calculated and "chalked" updates) about likely final odds for their betting choice.

A bettor also has the option of placing bets with a bookmaker. Bets with the book can be placed either at fixed odds, offered at the time of the bet, or at what is termed "starting price" odds. In the starting price bet, the odds are determined as the average of a set of the largest (or "ring") bookmakers at the racecourse just before the race starts. Starting price odds represent the "odds at which a 'sizeable' bet could have been made *on the course* just before the 'off' " (Dowie [1976, p. 140]; see also O'Neill [1971] for a more complete explanation). Starting price bets can be placed anytime prior to the race and remain available up until the start of the race. Though there is an almost infinite array of bets that can be made with a bookmaker (ante post, bar the fave, etc.), we needed to choose one that was sufficiently recorded to provide the data necessary for our analysis. Further, we wished to choose a bet that was not an insignificant part of the market.[1] At the time a bet is placed at the starting price, the bettor is uncertain about the amount to be received if a bet wins. Thus this type of betting shares a similarity with tote betting. A critical difference, however, is that the starting price odds reflect betting trends and bets made with bookmakers, while the tote odds reflect an aggregate measure of bettors' subjective winning probabilities based on a separate pool of funds. The final tote odds are determined by the total amounts in the pool bet on each horse, while the starting price

---

[1] Starting price betting constitutes a large amount of off-course betting. Dowie notes that "most off-course betting—95 per cent or more according to 'Michael Rolfe' of the *Sporting Life* (personal communication)—is at 'starting price'. This occurs despite the readily available option of off track TOTE wagers" (p. 140).

odds represent the last odds offered and reflect trends and the attempts of bookmakers to "round their books." Under either betting form, however, each bettor is uncertain of the exact odds he or she will receive until after the race starts. In terms of risk, starting price and tote bets are essentially equivalent. That is, both are defaultless, and the true likelihood that a particular horse will win and, thus, the odds of receiving a return for betting on that horse are identical. The difference in return from each is the focus of our analysis. Since the differing bets are two options for purchasing exactly the same item (a bet to win on a specific horse), we would expect the odds to converge. If, over time, tote odds (and thus payouts for a given bet on a winning horse) repeatedly exceed those of the corresponding starting price bet, we would expect bettor shifts to occur, leading to the tote and starting price odds being driven closer. If the betting market is semistrongly efficient, then historical returns (payouts) from each betting type and differences in those payouts (published announcements following the outcome of the race) should be included in the market prices. Since the objects compared are identical (as explained below, a 10-pence bet on a winning race horse), the odds or returns from that object should be identical. If the market is equitable (or strongly efficient), then monopolistic or "insider" information should be absent. In the next section we investigate whether the returns on the two bet types show evidence of systematic differences and what, if any, implications exist concerning market equity.

III. Empirical Analysis

The data analyzed in this study are drawn from the 1978 thoroughbred racing season in England. We chose the year on the basis of a critical factor: the general absence of mechanical or electronic tote boards (as suggested by a letter from the Horserace Totalisator Board, London).[2] This placed the tote bettors in the position of having limited, if any, information on betting patterns or likely final odds for their betting choice. The sample consists of the first 1,427 flat races run in the 1978 season.[3] Information on the following variables was compiled for each race from the weekly race results published in the *Sporting Chronicle Handicap Book:* (1) the number of horses in each race, (2) the total (winning) purse in each race, (3) the type of race (e.g., maiden, claiming, stakes, etc.), (4) the track conditions, (5) the

[2] The letter, dated September 26, 1978, was received in response to an inquiry concerning the existence of electronic tote boards and the pool deductions (percentage) that affected tote wagering.
[3] The 1978 season includes a few scattered races occurring in late 1977. Our data run through June 1978.

TABLE 1

AVERAGE WINNING PAYOUTS PER 10-PENCE BET

	Number of Observations	Tote	Starting Price	Difference
All races	1,427	108.21 (270.27)	63.50 (52.33)	44.28* (251.27)

NOTE.—Standard deviations are in parentheses.

* The difference in average tote and starting price payouts is significant at the 1 percent level using a Wilcoxson matched-pairs signed-ranks test.

length of the race (in furlongs), (6) the totalizator payoff on the winning horse per 10-pence bet, (7) the date of the race, and (8) the starting price payoff per 10-pence bet.

As noted earlier, if the betting market is semistrongly efficient, then the expected returns to two bets of similar risk should be equal. However, if market inefficiencies exist, they may lead to a divergence between overall tote and starting price payouts. Thus a rough test of such market efficiency is simply to compare the average tote and starting price payoffs across races.

However, if the market is not equitable (not strongly efficient), some bettors would have inside information. We would expect them to place early fixed-odds bets with bookmakers. Since tote is parimutuel wagering in which odds are determined by the amount bet on individual horses, bettors with inside information would tend to avoid tote bets in order to place fixed-odds bets with bookmakers. If substantial accurate inside betting occurred as tote betting, tote payouts would be lowered. At the same time, if significant insider betting (based on good information) occurs with bookmakers, winning starting price payouts are driven downward. For example, if insider information is favorable on a horse that opens betting at 10 to 1 odds, bookmakers would receive substantial wagering. "Tick-tack" signaling (interbookmaker signaling) would tend to push the subsequent starting price odds below the early odds locked in by inside bettors. Hence, it is conceivable that insider betting may actually lower observed starting price payouts relative to tote payouts because the former are driven down as bookmakers react to the flow of insider funds.

Table 1 shows the mean payoffs and accompanying standard deviations from each of the two bet types and the average difference in payout for all winning horses in our sample period.[4] The table also includes the results of a Wilcoxson matched-pairs signed-ranks test

---

[4] The tote winnings are reported after the customary track takeout has been deducted. No similar deduction is made for starting price winnings.

for the difference in mean tote and starting price payouts. As indicated in the table, the average difference for a 10-pence bet (the common payout reporting base used in England) is 44.28 pence. That is, on average, over all races in the entire sample, the tote return exceeded the starting price return by 44.28 pence on a 10-pence bet. The average tote payoff was 108.21 (approximately 10 to 1) and the average starting price payoff was 63.50. Further, the Wilcoxson test indicated that the differences were significant at an extremely high level. These results indicate that the tote yielded a higher average payout, suggesting that the market fails to satisfy semistrong efficiency conditions. Further, the results are consistent with a lack of market equity and the presence of insider information, though, as noted earlier, the inferences here are only indirect.

It is still possible that the market really is efficient and that the results we find are due to factors that represent either aberrations or initial market conditions that get worked out over time. These possibilities lead us to consider the following two questions: (1) Since bookmakers generally limit offered odds (to avoid bankruptcy perhaps) and the tote odds are virtually unlimited (e.g., if only 10 pence in a £1,000,000 pool is bet on a horse that wins, the payoff rate would be rather large), could the findings of differences in average return simply reflect a few very large tote payouts? (2) Since bettors gain information over time (e.g., past race performances or past betting payouts), could it be that the differences are due to early race insider information? Once the season is well under way and more information is available on horses, jockeys, trainers, and betting returns, does the observed difference begin to disappear?

To address the first question, we ran our comparisons for a variety of subsets of our original data. Three new data sets were constructed by deleting observations that included tote payouts above specified odds level as follows:

Data Set	Observations Deleted if Tote Payout Odds Greater than
1	20 to 1
2	15 to 1
3	10 to 1

Even with observations deleted on the basis of the tote payout level, the tote payout average continued to exceed the starting price payout average. As detailed in table 2, the differences between the tote payouts and starting price payouts averaged 9.81 for data set 1, 6.86 for data set 2, and 3.13 for data set 3. Though, as we might expect, the

TABLE 2

AVERAGE WINNING PAYOUTS PER 10-PENCE BET

Races	Number of Observations	Tote	Starting Price	Difference
Tote ≤ 20 to 1	1,283	62.73	52.93	9.81*
		(43.5)	(35.22)	(32.33)
Tote ≤ 15 to 1	1,225	56.96	50.09	6.86*
		(35.13)	(31.18)	(27.43)
Tote ≤ 10 to 1	1,106	48.80	45.68	3.13*
		(25.64)	(25.89)	(21.89)

NOTE.—Standard deviations are in parentheses.
* Significant at the 1 percent level.

average difference declined as the level of payout declined, all three of the Wilcoxson matched-pairs signed-ranks tests were significant at the 1 percent level.

The issue raised by the second question suggested above concerns whether the observed differences tend to disappear over time. To address this issue, we performed our calculations and the Wilcoxson test for three different racing periods: (1) races up to and including April (533 races), (2) races run during May (509 races), and (3) races run during June (374 races). Dividing the periods up in this chronological order enabled us to examine whether the differences noted above diminished over time and whether the returns on the two betting types converged as bettors observed horse performance and betting returns. Tables 3, 4, and 5 present the tote and starting price return comparisons for each of the three periods. Each table includes mean payout comparisons for all races during the period and for

TABLE 3

AVERAGE WINNING PAYOUTS PER 10-PENCE BET FOR RACES PRIOR TO MAY 1

Races	Number of Observations	Tote	Starting Price	Difference
All	533	147.10	68.00	79.10*
		(415.43)	(55.86)	(393.80)
Tote ≤ 20 to 1	460	65.19	53.21	12.04*
		(45.37)	(34.47)	(33.12)
Tote ≤ 15 to 1	439	59.42	50.58	8.88*
		(35.59)	(31.91)	(29.55)
Tote ≤ 10 to 1	385	48.92	45.06	3.86**
		(26.05)	(25.28)	(22.66)

NOTE.—Standard deviations are in parentheses.
* Significant at the 1 percent level.
** Significant at the 5 percent level.

TABLE 4

AVERAGE WINNING PAYOUTS PER 10-PENCE BET FOR RACES RUN DURING MAY

Races	Number of Observations	Tote	Starting Price	Difference
All	509	91.79	63.03	28.76*
		(124.15)	(48.39)	(92.67)
Tote ≤ 20 to 1	468	62.09	53.68	8.41*
		(41.33)	(33.07)	(31.08)
Tote ≤ 15 to 1	448	56.65	51.03	5.62*
		(32.89)	(29.13)	(26.26)
Tote ≤ 10 to 1	409	49.71	47.44	2.27**
		(24.73)	(25.25)	(21.91)

NOTE.—Standard deviations are in parentheses.
* Significant at the 1 percent level.
** Not significant.

races during the period in which the tote odds were no greater than 20 to 1, no greater than 15 to 1, and no greater than 10 to 1.

The results presented in tables 3–5 suggest some mild degree of learning over time by bettors; that is, there is some apparent reaction to information announcements. These mild adjustments, however, fall far short of immediate (and complete) reactions required for semistrong efficiency to hold. For example, there were generally lower average tote and starting price payoffs on winning horses during May and during June than in the earlier period running through the end of April. Further, there were generally lower differences in tote versus starting price payoffs in each of the later periods (May and June) compared with the earlier periods. As we might expect, additional information (e.g., announcements of results of previous races

TABLE 5

AVERAGE WINNING PAYOUTS PER 10-PENCE BET FOR RACES RUN DURING JUNE

Races	Number of Observations	Tote	Starting Price	Difference
All	374	75.51	58.26	17.26*
		(81.37)	(52.00)	(53.74)
Tote ≤ 20 to 1	353	60.52	51.77	8.75*
		(43.85)	(39.17)	(32.96)
Tote ≤ 15 to 1	336	54.28	48.39	5.90*
		(34.62)	(32.83)	(26.07)
Tote ≤ 10 to 1	310	47.57	44.30	3.27**
		(26.39)	(27.39)	(20.98)

NOTE.—Standard deviations are in parentheses.
* Significant at the 1 percent level.
** Significant at the 5 percent level.

in a given season) did tend to reduce the average payouts. What is surprising, however, is that, except for the "all race" category, this was not true when we compare May racing with June racing. In the all race category the difference went from 28.76 to 17.26. But for categories with high-tote-payoff races removed, this narrowing of difference failed to occur. In fact, the differences in June remained almost identical to those in May (see tables 4 and 5) with even actual slight increases (8.75 vs. 8.41, 5.90 vs. 5.62, and 3.27 vs. 2.27) in June. The one instance of a nonsignificant Wilcoxson matched-pairs signed-ranks test occurred in the May category for races with tote odds of 10 to 1 or less for the winning horse. All other such tests were significant at the 5 percent level or higher.

These results are consistent with a continued lessening of the occurrence of extremely high tote payoffs and with an ongoing presence of insider information leading to continuing differences between the returns on the two betting forms. Though there was a lessening of the difference from the early period of racing to the later two periods considered, the differences for races with winning tote odds of 20 to 1 or less stayed roughly the same from May to June. Further, even deep into the flat racing season in June, the differences remained statistically significant with no evidence of convergence.[5]

The statistically significant differences in average tote and starting price payouts presented in tables 1–5 cast doubt on the semistrong efficiency of British racetrack betting. We can extend our analysis further to incorporate an additional test based on recent literature in market efficiency. The formulation of this additional test is based on the following notion: If British gamblers use available information efficiently, then the starting price payout should be the best unbiased forecast of the tote payout (and vice versa). Following de Leeuw and McKelvey (1984) and Zuber et al. (1985), we can test this notion of efficiency by estimating the parameters of the following linear expression:

$$\text{TOTE}_i = \alpha_0 + \alpha_1 \text{SP}_i + \mu_i, \tag{1}$$

where TOTE_i is the totalizator payout in the ith race, SP_i is the starting price payout in the ith race, and μ is the error term.[6] The test for market efficiency based on equation (1) is the null hypothesis that

---

[5] Conventional t-tests were also conducted for the differences in average tote and starting price payouts. In all comparisons, tote payouts were significantly different from starting price payouts at conventional levels. The results of the Wilcoxson test for differences in means are reported since this test makes no underlying assumption about the distributions of tote and starting price payoffs.

[6] We are grateful to an anonymous referee for suggesting this alternative test for market efficiency.

$\alpha_0 = 0$ and $\alpha_1 = 1$ jointly (Zuber et al. 1985, pp. 800–801). As Zuber et al. note, ordinary least squares estimation of the parameters in (1) allows the null hypothesis to be tested using a standard F-test.

Table 6 presents the regression estimates for equation (1) for all the subsets of races presented in tables 1–5. The F-values indicate that the null hypothesis is rejected at conventional levels for every subset. Hence, this second statistical test provides additional evidence against the efficiency of the segment of the British racetrack betting market being studied here.

The presence of insider knowledge is consistent with our observation of lower starting price payouts: bookmakers attempt to balance their books on horses whose odds are firming at race time because of insider betting. However, if insider information exists, the market should assimilate this fact and agents in the market should integrate this information into their decision-making process. Rational starting price bettors should realize that insiders are benefiting at their expense. Hence, individuals who normally bet at starting price odds should tend to switch to the tote. This adjustment in betting patterns should lower tote payments because increased betting on a given horse will lower the odds. The reverse should occur for starting price odds as funds move out of this form of betting. Therefore, even with insider information, the starting price and tote payments to winners should converge. Yet, our empirical observations indicate otherwise.

One additional point adds further emphasis to our analysis. Tote payoffs represent "after-tax" payoffs. On the other hand, the starting price bettors must choose to pay a tax on the bet or pay a tax on the winnings, thereby escaping a tax on losing bets. In either case, the actual starting price payoff is less than the figures we utilized. Since we had no way to calculate how winners might have chosen to pay the tax, we ignored this in our calculations. Including this tax would further strengthen our results since this operates to reduce the true starting price return and increases the differences in payoffs we observed.[7]

IV. Summary

Under both types of tests utilized, our analysis of British racetrack betting provides evidence that conditions for semistrong efficiency are not satisfied. Tote payoffs were consistently greater than identical bets made at starting price odds. This result held true even when large tote payoffs (and even moderately high tote payoffs) are elimi-

---

[7] Including the 5 percent tax on either starting price bets or winnings did not alter our conclusions about higher average tote payouts.

TABLE 6

Regression Estimates for Equation (1)

	Intercept	Slope	R^2	F[a]
	Whole Season			
All races	−36.30	2.27	.193	77.02*
	(10.15)	(.12)		
Tote ≤ 20 to 1	18.29	.84	.465	81.11*
	(1.60)	(.03)		
Tote ≤ 15 to 1	14.50	.75	.441	96.54*
	(1.42)	(.02)		
Tote ≤ 10 to 1	19.88	.63	.408	141.77*
	(1.20)	(.02)		
	Races through April			
All races	−63.34	3.09	.170	37.31*
	(25.72)	(.29)		
Tote ≤ 20 to 1	17.04	.90	.474	33.15*
	(2.81)	(.04)		
Tote ≤ 15 to 1	20.68	.76	.424	36.42*
	(2.55)	(.04)		
Tote ≤ 10 to 1	20.54	.63	.375	46.74*
	(2.14)	(.03)		
	May Races			
All races	−30.70	1.94	.574	111.52*
	(5.88)	(.07)		
Tote ≤ 20 to 1	17.04	.84	.451	24.64*
	(2.70)	(.04)		
Tote ≤ 15 to 1	19.33	.73	.419	33.00*
	(2.39)	(.04)		
Tote ≤ 10 to 1	21.09	.60	.379	56.56*
	(2.05)	(.04)		
	June Races			
All races	6.18	1.19	.579	26.44*
	(4.11)	(.05)		
Tote ≤ 20 to 1	20.51	.77	.477	27.18*
	(2.81)	(.04)		
Tote ≤ 15 to 1	18.44	.74	.494	29.50*
	(2.40)	(.04)		
Tote ≤ 10 to 1	17.85	.67	.485	39.48*
	(2.05)	(.04)		

Note.—Standard errors are in parentheses.
[a] F-values for null hypothesis $\alpha_0 = 0$, $\alpha_1 = 1$, jointly.
* Significant at the 1 percent level.

nated from the data set. Further, the result continued to hold as the season progressed, despite increased information availability on horse performance and on differing tote and starting price payoffs early in the season. Although we did find a mild narrowing of the differences in tote and starting price payoffs when comparing early-season races with May or June races, the differences remained at absolute levels approximately equal to those earlier in the season.

Both types of bets are accessible to bettors, and the outcomes are published daily with weekly summaries of all race and payoff results. The fact that these differences persist leads us to a quandary. Are we observing an inefficient market or simply one in which the tastes and preferences of the market participants lead to the observed results? Is all relevant and ascertainable information reflected in the "prices," the market odds? In fact, is it ever possible to differentiate between the two, to separate an inefficient market from one in which the participants are pursuing the satisfaction of nonmonetary preferences?

References

Asch, Peter; Malkiel, Burton G.; and Quandt, Richard E. "Market Efficiency in Racetrack Betting." *J. Bus.* 57 (April 1984): 165–75.

Crafts, Nicholas F. R. "Some Evidence of Insider Knowledge in Horse Race Betting in Britain." *Economica* 52 (August 1985): 295–304.

de Leeuw, Frank, and McKelvey, Michael J. "Price Expectations of Business Firms: Bias in the Short and Long Run." *A.E.R.* 74 (March 1984): 99–110.

Dowie, Jack A. "On the Efficiency and Equity of Betting Markets." *Economica* 43 (May 1976): 139–50.

Fama, Eugene F. "Efficient Capital Markets: A Review of Theory and Empirical Work." *J. Finance* 25 (May 1970): 383–417.

O'Neill, J. "Why There Is a Double Check on Starting Prices." In *Daily Mirror Punters' Club Guide to the Turf,* edited by J. L. Stevenson. London: Daily Mirror Books, 1971.

Snyder, Wayne W. "Horse Racing: Testing the Efficient Markets Model." *J. Finance* 33 (September 1978): 1109–18.

Tuckwell, R. H. "The Thoroughbred Gambling Market: Efficiency, Equity and Related Issues." *Australian Econ. Papers* 22 (June 1983): 106–18.

Zuber, Richard A.; Gandar, John M.; and Bowers, Benny D. "Beating the Spread: Testing the Efficiency of the Gambling Market for National Football League Games." *J.P.E.* 93 (August 1985): 800–806.

An Examination of Efficiency in British Racetrack Betting: Errata and Corrections

Paul Gabriel

Loyola University of Chicago

James R. Marsden

University of Kentucky

Shortly after the publication of our paper in this *Journal* (Gabriel and Marsden 1990), we received a kind letter with several informational questions from Paddy Waldron, a graduate student at Wharton and long-time "punter" (bettor) in his native Ireland. One of Waldron's questions caused us to recheck the original source (*The Sporting Chronicle Handicap Book*, 1978) from which our data were gathered by research assistants. As his questions suggested to us, we discovered that a footnote had been overlooked and that the tote returns on Irish races included in the flat season (about 27 percent of all such races) are quoted as returns on a 20-pence bet (their minimum bet) rather than as returns on the 10-pence minimum used for reporting the other flat races. Thus we found that we had overstated the tote returns. We completely recollected all our data, making the necessary adjustments for Irish races and adding in eight races that had been reported out of sequence in the *Handicap Book*.

Using the corrected data, we still find that tote returns exceed starting price returns for all races in our period and for each subperiod (races through April, races in May, and races in June), though, of course, the differences are somewhat smaller than those reported earlier (see table 1). Further, all joint-hypothesis regression equation tests suggest the same conclusions as earlier, though three of the significance levels do differ, two going from .01 to .05 and one from .01 to .1.

We are indebted to Paddy Waldron for his interest and the questions that he raised. We are also indebted to the editor for this opportunity to remedy the error in our earlier computations. For brevity, only part of the revised figures are provided here. A complete set of revised tables is available from either author on request.

TABLE 1

Average Winning Payouts per 10-Pence Bet

	Number of Observations	Tote	Starting Price	Difference
All races	1,435	81.55	63.37	18.18*
		(150.47)	(52.15)	(127.41)
Pre-May	535	102.51	68.17	34.3*
		(219.54)	(55.92)	(194.60)
May	514	74.78	62.76	12.0*
		(96.55)	(48.27)	(63.81)
June	386	61.52	57.53	4.0**
		(63.15)	(51.20)	(43.54)

Note.—Standard deviations are in parentheses.
* Significant at the 1 percent level.
** Significant at the 10 percent level.

One part of our earlier results, that dealing with restricting long-shot payoffs, needs revision. To this end, our original table 2 has been expanded to include restrictions using tote odds and restrictions using starting price odds. Originally we reported only the former since these were the stronger results. With the corrected data set, however, comparisons of restricted samples (tote < 20:1, < 15:1, and < 10:1) yield mixed results, though the "after-tax" nature of tote returns suggests that if starting price returns were adjusted for taxes, the earlier reported results might still hold. *Simply put, the corrected results are not as strong as those reported earlier.* Restricting observa-

TABLE 2

Average Winning Payouts per 10-Pence Bet

Races	Number of Observations	Tote	Starting Price	Difference
Tote ≤ 20:1	1,339	55.34	54.24	1.1**
		(40.21)	(36.09)	(23.74)
Tote ≤ 15:1	1,296	51.02	51.64	−.6
		(32.91)	(32.05)	(19.62)
Tote ≤ 10:1	1,202	44.70	46.92	−2.2*
		(24.45)	(26.02)	(16.18)
Starting price ≤ 20:1	1,408	74.85	59.10	15.7*
		(140.17)	(41.78)	(122.93)
Starting price ≤ 15:1	1,353	63.87	53.71	10.2*
		(117.35)	(32.49)	(106.55)
Starting price ≤ 10:1	1,271	52.69	48.40	4.3*
		(42.86)	(25.48)	(29.85)

Note.—Standard deviations are in parentheses.
* Significant at the 1 percent level.
** Significant at the 10 percent level.

tions based on tote values eliminates those with high tote values and low starting price values (e.g., tote odds of 40:1 and starting price odds of 5:1 would be eliminated). Further, sorting by tote odds is based on unobservable values. Bettors can observe information on starting price odds through trends and firming in bookie odds. We resorted our restricted samples based on starting price odds (< 20:1, < 15:1, and < 10:1) and include the results for all races in table 2. *In all cases (table 2's entire period and the pre-May, May, and June groupings not shown here), the results of the Wilcoxon matched-pairs signed-ranks tests indicate that a bettor about to make a winning bet is, on average, better off making a tote wager (11 at a .01 or higher and one at a .05 level of significance).* Complete recalculations are available on request.

We apologize for any inconvenience that our oversight has caused. We hope that these corrections prove helpful and offer our thanks to Paddy Waldron for his inquiry, which led to the discovery of the data collection error on a segment of our observations.

Reference

Gabriel, Paul E., and Marsden, James R. "An Examination of Market Efficiency in British Racetrack Betting." *J.P.E* 98 (August 1990): 874–85.

The Dr.Z Betting System in England[1]

William T. Ziemba and Donald B. Hausch

Management Science Division, Faculty of Commerce,
University of British Columbia, Vancouver BC, Canada V6T 1Z2

School of Business, University of Wisconsin,
Madison, Wisconsin 53706

Abstract

The betting strategy proposed in Hausch, Ziemba and Rubinstein (1981) and Ziemba and Hausch (1984,1987) has had considerable some success in North American place and show pools. The place pool is England is very different. This paper applies a similar strategy with appropriate modifications for places bet at British racetracks. The system or minor modifications also applies in a number of other countries such as Singapore with similar betting rules. The system appears to provide positive expectation wagers. However, with the higher track take it is not known how often profitable wagers will exist or what the long run performance might be.

I. Introduction

At North American racetracks, the parimutuel system of betting utilizing electric totalizator boards in the dominant method of betting. Las Vegas and the other legal sports books may set odds on particular betting situations, but these fixed odds are not available at racetracks. In England and in other Commonwealth countries, such as Italy and France, odds betting against bookies is the dominant betting scheme. This fixed-odds system is introduced in Hausch, Lo and Ziemba (1994). They also indicate a tendency of higher (lower) returns for lower (higher) odds ranges. Thus, the favorite-longshot bias appears to exist in England (see e.g. Ali (1977), Busche and Hall (1988)).

This paper applies the betting system proposed by Hausch, Ziemba and Rubinstein (1981) and Hausch and Ziemba (1985) in England. This strategy is also called the Dr.Z system in the trade books Ziemba and Hausch (1984,1987) who discuss it more fully. The strategy utilizes the Kelly criterion (Kelly (1956)) which maximizes the expected logarithm of wealth. The Kelly criterion has several advantages. First, it maximizes the capital growth asymptotically. Second, it prevents bankruptcy. Third, the expected time to reach a specified goal is minimum when the goal increases. Fourth, it is superior to any different strategy in the long run. These properties were proved in Breiman (1961). See McLean, Ziemba and Blazenko (1992) for discussion of these properties.

II. The System

Instead of the North American parimutuel system of win, place, and show, the bets in England are to win and place. By "place" the British mean "finish in the money." This is what North Americans call show except for one important difference. The number of horses that can place in a particular race is dependent on the number of starters.

---

[1] Modified from an Appendix in Ziemba and Hausch (1987).

Table 1. Relationship between number of horses that place and number of starters

Number of Horses that Place	Number of Starters
one: the winner	four or less
two: winner and second	five, six, or seven
three: winner, second, and third	eight to fifteen
four: winner, second, third, and fourth	sixteen or more

The place pools are not shown on the tote board, but the current payoffs for place bets for each horse are flashed on the screen. Bookies, on the other hand, simply pay a percentage of the win odds, as shown in Table 2.

There are many types of exotic bets as well. The tote jackpot corresponds to what North Americans call the pick six or sweep six. The tote placepot bet has no analogue in North America. The average rates of return on various bets on and off course against a bookmaker or the tote are listed in Table 3. The track take is 5% larger in the place pool than in the win pool. The tote take is larger than what the bookies make on average, and on-course betting takes are much less than off-course takes.

The races in England are on the turf for distances of generally at least a mile, except for some shorter races for two-year-olds. The season in southern England is unique in that races are run for about three days at each race course. The jockeys, trainers, and so forth then move on to a new course. After a month or so they return to the same course. Handicapping is very sophisticated in England. It has to be, with little information easily accessible (they have no analogue of the *Daily Racing Form*, although some past performances are available in newspapers) and all that moving from course to course.

The method of computing the place payoffs in England differs from that used in North America. In both locales, the net pool is the total amount wagered minus the track take. In North America, the cost of the winning in-the-money tickets is first subtracted to form the profit. This profit is then shared equally among the in-the-money horses. Holders of winning tickets receive a payoff consisting of the original stake plus their proportionate share of the horse's profits. This means that the amount of money wagered on the other horses in the money greatly affects the payoff. In England, the total net pool is divided equally among the horses that finish in the money. This means that the payoff on a particular horse depends upon how much is bet on this horse to place but not on how much is bet on the other horses. Since the minimum payoff is £1 per £1 wager, management is able to keep a control on betting for particular favorites. Once this minimum level is reached, it does not pay to wager on a given horse. This occurs whenever the percentage of the place pool that is bet on a given horse becomes as large as Q_p, which is the track take for place, divided by m, which is the number of in-the-money horses. In a race with 8-15 starters, if Q_p is about 0.735, and $m=3$, the just-get-your-money-back point is reached when the bet on a particular horse to place becomes 24.5% of the total place pool: $0.735/3 = 24.5\%$. Hence in England you will often see horses whose place payoffs are £1 or just slightly higher. This method of sharing the place pool tends to favor longer-priced horses at the expense of the favorites.

Table 2. Bookmakers payoff for place bets

Number of Runners	Type of Race	Fraction of Win Odds Paid on Place Element	Horses Regarded
Two to five		No place betting	
Six or seven	Any	$\frac{1}{4}$	First and second
Eight or more	Any except handicaps involving twelve or more runners	$\frac{1}{5}$	First, second, and third
Twelve to fifteen	Handicaps	$\frac{1}{4}$	First, second, and third
Sixteen to twenty-one	Handicaps	$\frac{1}{5}$	First, second, third, and fourth
Twenty-two or more	Handicaps	$\frac{1}{4}$	First, second, third, and fourth

Source: Rothschild (1978).

Table 3. Rates of return on different types of bets in England on thoroughbred and greyhound racing

Type of Bet	Rate of Return (%)
On-course bookmaker	90
Off-course bookmaker	81
Single bet to win with off-course bookmaker	85
Double bet to win with off-course bookmaker	78
Treble bet to win with off-course bookmaker	72
ITV Seven bet to win with off-course bookmaker	70–75
Computer straight forecast with off-course bookmaker	65
Greyhound forecast with off-course bookmaker	76
Greyhound forecast double with off-course bookmaker	58
Place element of each-way with off-course bookmaker	80
Ante-post betting with off-course bookmaker	96
Horse race tote win pool (on course)	80
Horse race tote win pool (off course)	77
Horse race tote place pool (on course)	75
Horse race tote place pool (off course)	72
Horse race tote daily double pool (on course)	74
Horse race tote daily double pool (off course)	71
Horse race tote daily treble pool (on course)	70
Horse race tote daily treble pool (off course)	67
Horse race tote daily forecast pool (on course)	70
Horse race tote daily forecast pool (off course)	67
Horse race tote jackpot pool (on course)	70
Horse race tote jackpot pool (off course)	67
Horse race tote placepot pool (on course)	70
Horse race tote placepot pool (off course)	67
Greyhound tote pool betting, average	83.5

Source: Rothschild (1978).

The current track take to win is about 20.6% and to place is 26.5%, and the breakage is of 10¢ variety, or more properly 10p, for pence[2]. These track takes are much higher than those in North America. Since the track paybacks to win and place are different, we call the former, $Q_w=0.794$, and the latter, $Q_p=0.735$.

It is easy to apply the Dr.Z system in Great Britain, although with its much higher track takes, there may not be many Dr.Z system bets, see Mordin (1992) for a discussion of this. We utilize the substitution that $q_i=Q_w/O_i$, where O_i are the odds to win on the horse under consideration.

The expected value per pound bet to place on horse i is

$$EXPlace - \text{(probability of placing) (place odds)} - (Prob) (PO_i). \qquad (1)$$

In (1), PO_i refers to the odds to place on horse i. Prob, the probability of placing is determined as follows.[3][4]

---

[2]We can calculate these track takes as follows: The payoff on horse i if it wins is $Q_w W_i/W$, where Q_w is the track payback to win, and W_i and W are the bet amount of horse i and the total bet amount, respectively. So let $q_i = W_i/W$, the efficient-market assumption. Let B be the average breakage, namely, 4.5p. Since breakage can be 0,1,2,...,9 pence, its average is 4.5p. Then the payoff on i is Q_w/q_i-B, which equals the odds O_i, since the odds are based on total return (not return plus original stake as in North America). So $q_i = Q_w/(B+O_i)$. Summing over all n horses gives

$$\sum_{i-1}^{n} q_i - 1 - Q_w \sum_{i-1}^{n} (\frac{1}{B+O_i}),$$

since some horse must win. Hence

$$Q_w - \frac{1}{\sum_{i-1}^{n}(B+O_i)}.$$

For place, there are one, two, three, or four horses that are in the money, depending upon the number of starters. So

$$Q_p - \frac{m}{\sum_{i-1}^{n}(B+O_i)},$$

where $m=1,2,3$ or 4.

With the above formulas, $Q_w \approx 0.794$ and $Q_p \approx 0.745$.

[3]These equations were developed using the 1981-1982 Aqueduct data to relate probability of in-the-money finishes to q, the probability of winning and n, the number of horses. Equations (2), (3) and (4) had R^2 of 0.991, 0.993 and 0.998, respectively. These equations are valid when q ranges from 0 to 0.6

With $n=5$ to 7 horses, the first 2 horses place and

$$\text{Prob} \; - \; 0.0667 + 2.37q - 1.61q^2 - 0.0097n. \tag{2}$$

With $n=8$ to 15 horses, the first 3 horses place and

$$\text{Prob} \; - \; 0.0665 + 3.44q - 3.47q^2 - 0.0049n. \tag{3}$$

With $n=16$ or more horses, the first 4 horses place and

$$\text{Prob} \; - \; 0.0371 + 4.47q - 6.29q^2 - 0.00164n. \tag{4}$$

Figures 1,2, and 3 determine Prob directly using only O_i, the win odds on the horse in question. Figure 1 applies when there are five, six or seven horses. Figure 2 corresponds to equation (3) and applies when there are eight to fifteen horses. Finally, Figure 3 corresponds to equation (4) and applies when there are sixteen or more horses.

The optimal Kelly criterion bet is to wager (Prob PO_i-1)/(PO_i-1) percent of your betting wealth[5]. We can determine the optimal fraction of your wealth to bet indicated by equation (5) using Figure 4.

---

for (2), from 0 to 0.45 for (3), and 0 to 0.3 for (4), which should be the case in most instances. However, Figures 1,2 and 3 are valid for any q.

[4]In a race with $n=2,3$, or 4 horses, only one horse places, the winner. Such races are rare. Also, it is unlikely that the win and place pools would then become so unbalanced as to yield a Dr.Z system bet. However, one would occur when PO_i/O_i was at least 1.44, for a track payback of 0.794 and an expected-value cutoff of 1.14, since 1.14/0.794 is 1.44. In such a case, one would have a good bet.

[5]We have assumed that your bets will be small and hence will not affect the odds very much. Thus to determine the optimal bet b for betting wealth w_0, you maximize Prob $\log[w_0 + (PO_i\text{-}1)b] + (1 - \text{Prob})\log(w_{0\text{-}b})$, whose solution is equation (5).

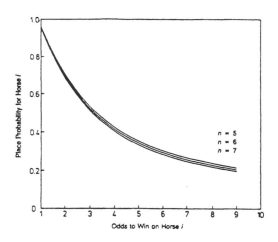

Figure 1. Probabilities of placing for different odds horses when the race has five to seven starters

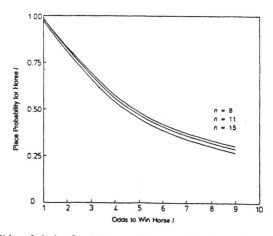

Figure 3. Probabilities of placing for different odds horses when the race has eight to fifteen starters

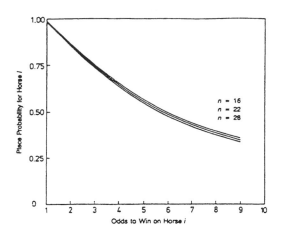

Figure 3. Probabilities of placing for different odds horses when the race has sixteen or more starters

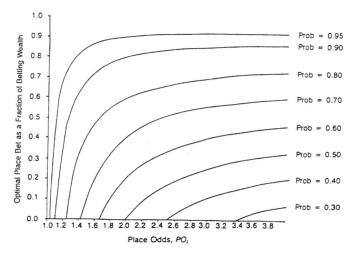

Figure 4. Optimal bets when the probability of placing is Prob and the place odds of the horse in question is PO_i

References

Breiman,L. (1961) "Optimal gambling systems for favorable games." in Proceedings of the Fourth Berkeley Symposium. Mathematical Statistics and Probability 1, 65-78. University of California Press.

Figgis,E.L. (1974) "Rates of return from flat race betting in England in 1973." *Sporting Life* 11 (March).

Hausch,D.B., Ziemba,W.T. and Rubinstein,M. (1981) "Efficiency of the market for racetrack betting." *Management Science* 27, 1435-1452.

Hausch,D.B. and Ziemba,W.T. (1985) "Transactions costs, extent of inefficiencies, entries and multiple wagers in a racetrack betting model." *Management Science* 31, 381-394.

Kelly,J.L. (1956) "A new interpretation of information rate." *Bell System Technical Journal* 35, 917-926.

MacLean,L.C., Ziemba,W.T. and Blazenko,G. (1992) "Growth versus security in dynamic investment analysis." *Management Science* 38, 1562-1585.

Mordin,N. (1992) "Grab your place in the pool!" *Sporting Life*.

Rothschild, Lord (1978) Royal commission on gambling, Vols I and II. Presented to parliament by Command of Her Majesty (July).

Ziemba,W.T. and Hausch,D.B. (1984) Beat the racetrack. Harcourt Brace Jovanovich, San Diego.

Ziemba,W.T. and Hausch,D.B. (1987) Dr.Z's beat the racetrack. Revised edition. William Morrow and Co. Inc. New York.

THE EFFICIENCY OF RACETRACK BETTING MARKETS: AUSTRALIAN EVIDENCE

R. Bird* and M. McCrae**

Introduction

Several Australian empirical studies have addressed issues that are the focus of some of the North American studies in this volume. In Australia, as in England and other Commonwealth countries, both bookmaker and Totalisator (parimutuel) betting is used. The availability of bookmaker odds enables researchers to conduct tests that are unavailable to those whose data sets are restricted to totalisator odds. The relative importance of racetrack gambling in Australia is discussed in Section One and to summarise the major findings of the empirical tests relating to this market are presented in Section Two. Section Three contains a brief discussion of New Zealand evidence. Outstanding issues are discussed in Section Four.

1 : Gambling in Australia

The major gambling activities in Australia are restricted to betting on races (thoroughbreds, harness racing and dogs) which operate on a regular basis in all States, acquiring tickets in lotteries which again operate nationwide[1], and gambling on poker machines which is only legal in New South Wales (N.S.W.), Victoria and the Australian Capital Territory (A.C.T.). Gambling at casinos has recently been legalised in a number of states. The latest census data on turnover for each of these forms of gambling is presented in Table 1 on a state-by-state and Australia-wide basis. Although the greatest turnover is on racing, by far the most popular form of gambling on the basis of per capita turnover is gambling on poker machines - the per capita turnover being in excess of $1,000 in N.S.W. and S800 in the A.C.T. As a consequence, N.S.W, is by far the greatest gambling state with per capita turnover being nearly three times that of the other states.

State gambling laws have recently been liberalised in Australia to allow the operation of casinos. Tasmania, South Australia, Western Australia, the Northern Territory have well established casinos. The Australian Capital Territory recently introduced a casino. While New South Wales and Victoria are about to follow suit.

Opinion is divided about on the horse racing industry. Some claim that casinos may cause a short-term decline in the proportion gambling expenditure spent on horse racing, but that the long term impact will be minimal. Others predict a more permanent decline in racing's share of total gambling expenditure (Australian Financial Review, [AFR],1993). In America, Thalheimer and Ali (1993) found that the demand relationships for horse race attendance and wagering were price and income elastic to competitive products. Competition from state lotteries and other professional sports (baseball, basketball, football) are significant determinants of demand relationships. The existence of state lotteries can result in a substantial loss in both attendance and wagering revenue. If these elasticities are also present in Australia then the continued growth of competing products may present a substantial threat to racetrack attendance and wagering volumes.

* TPF&C Ltd. Consulting Actuaries, Melbourne, Australia

** Associate Professor, Dept. of Accountancy, University of Wollongong, NSW, Australia

The latter argument is supported by recent data which show that total gambling expenditure in Australia grew at an average of 15.5 percent per year between 1973 and 1992, compared to the Consumer Price Index increase of 9.1 per cent. However, racing increased at only 10.5 per cent, well below the industry growth rate (AFR. 1993). The implication is that while gambling expenditure has increased faster than inflation, the introduction of casinos into the States may cause a long-term decline in the racing industry share of the gambling dollar, unless the industry re-examines the appeal of its product, especially to the younger generation.

The available means for betting on horse races in Australia are the totalisator, which operates both on-course and off-course, the on-course bookmakers and the illegal off-course (SP) bookmakers. The relative importance of the three forms of legal betting is indicated by the per capita expenditure on each which is presented [2] in Table 2. The expenditure of off-course gamblers is about double that of on-course gamblers.[3] Further, there is relatively little disparity between the per capita expenditure on racing across the more populated states.

Tuckwell (1984) isolated those factors that most strongly influenced the level of betting with the totalisator and bookmakers. The main influences on totalisator betting were the level of real wages, unemployment and lottery turnover. This suggests an association between the level of totalisator betting and the level of disposable income. Further, there is some substitution between the level of expenditure on lotteries and totalisator gambling. For bookmakers, the findings were not as conclusive. Tuckwell found only a weak association between turnover and real wages. Bookmaker betting may be somewhat insulated from changes in the level of per capita disposable income by the higher preponderance of wealthy gamblers who use this betting medium. He also found a persistent decline in real per capita turnover over time. This may be a result of the general decline in the attendance at race meetings, which decrease bookmakers' turnover.

2: Australian Evidence

The major difference between placing a bet with a bookmaker or on the totalisator is that with the former the odds are determined at the time that the bet is placed while with the latter the odds are not determined until the end of betting. The implications of this difference include (i) the realised take is fixed for totalisator but not for the bookmakers, (ii) optimal strategies for betting with bookmakers and on the totalisator may differ, and (iii) the characteristics of gamblers that will be attracted to each form of betting may differ. The papers that follow provide a detailed discussion of various peculiarities of the two forms of betting. We now summarise the Australian empirical evidence which are keyed to the relevant section headings for the other papers in this volume.

Utility Preferences of Racetrack Bettors

Bird, McCrae and Beggs[1987] use information derived from the racetrack to evaluate the risk-preferences of gamblers. They find the usual favourite-longshot bias which suggests that gamblers are willing to accept a lower return in order to participate in a high risk investment - ie. they are risk-takers in the usual two parameter sense, see also Kanto et al (1992), Busche and Hall (1988) for Hong Kong evidence. However, with the addition of the third moment (skewness) as an explanatory variable of gambler behaviour, the authors find that the gamblers are *averse to risk* as measured by the variance of the return distribution, but they have a strong preference for positive skewness. The authors then address the question of why a risk-averse investor would choose to participate in a negative expected return activity and conclude that they probably gain psychic income from the activity.

Efficiency of Win Markets and the Favourite-Longshot Bias

Several Australian studies, Bird and McCrae (1987b) and Tuckwell (1983), have addressed the efficiency of the market for gambling on racehorses to win. The availability of a series of bookmaker odds throughout the course of betting facilitates tests that are particularly appropriate for testing market efficiency. The tests have largely concentrated on testing market efficiency with respect to three types of information - the movement in bookmaker odds, the information supplied by tipsters and insider information. The co-existence of a market framed by bookmakers and by the totalisator provides the opportunity for punters to arbitrage between the two markets (Gabriel and Marsden [1990]). In the sequel we consider some evidence on these arbitrage opportunities and the implication of the existence of these two markets for the betting strategies of gamblers.

The Movements in Bookmaker Odds

The movements in bookmaker odds are readily available to on-course punters during the course of betting on each race. Bird and McCrae (1987b) applied filter tests to determine whether this information could be used as the basis for a profitable betting strategy. They concluded that such profits could not be realised. In contrast, Tuckwell (1983) concluded the odds movements could generate profits from betting, although he used a much smaller sample. He supported this finding by demonstrating that odds movements were not random. A result which may suggest a possible inefficiency in the case of capital markets with financial securities but not in the case of racetrack markets.

Tipsters Information

Another source of information that has received attention in Australian studies of racetrack efficiency is that supplied by newspaper tipsters. Anderson, Clarke and Ziegler (1985) and Bird and McCrae (1987) evaluated the extent to which this information was incorporated in starting prices and/or the ability of a gambler to use this information to generate positive returns. The former study used a two stage regression technique and concluded that the tipsters information is impounded into the bookmaker odds. The latter confirmed this finding and also concluded that the information cannot be used to earn positive returns. In contrast, Tuckwell (1983) evaluated a strategy which involved backing horses whose actual odds departed from published odds prepared by an "expert". He found that this strategy could be used to generate positive returns which suggests a possible inefficiency in the market.

Insider Information

Information held by a very limited number of market participants is often described as inside information. An example is where the trainer and owner are aware that a horse has been suffering from a cold and is likely to perform below expectations in a particular race. As a result, the probability of the horse winning as reflected by the odds, is likely to exceed its true winning chance and those with such inside information are in a position to use it to their advantage. Both Tuckwell (1983) and Bird and McCrae (1987b) used odds movements as evidence of the existence of inside information and evaluated whether prior knowledge of this information could be used to generate positive returns. Both studies find that an inefficiency exists with respect to inside information. Tuckwell used this finding to support a case for attempting to minimise inside information as he argues that its demise would result in bookmakers being able to offer better odds to all gamblers.

Two Markets, Bookmakers and the Totalisator

Bird and McCrae (1987a) calculated the bookmaker 'take' implied by the odds at various times during the course of betting. They found that this take was around 26 per cent at the commencement of betting but reduced to, and settled, around 17 per cent about half-way through betting. Further, the totalisator take is approximately 17 per cent, which is consistent with competition between the two modes of betting. The higher take required by bookmakers in the early part of betting is probably due to a desire by bookmakers to afford themselves some protection during a period when they are at an information disadvantage.

Bird and McCrae also consider the optimum strategy for an informed gambler betting with bookmakers. Asch, Malkiel and Quandt (1982) point out that an informed gambler betting on the totalisator should delay placing a bet until the very last moment and so minimise any information passed on to other gamblers. The same incentive does not exist when betting with the bookmakers as the odds for a particular gamble are determined at the time the bet is placed. Further, we have already seen that the implied take by bookmakers decreases during betting which would suggest that a gambler could expect the odds on a favourite to lengthen during betting. Bird and McCrae found that the greater volume of movements in odds occurred in the first half of betting. This suggests that most informed gamblers choose to utilise any information advantage early in betting, perhaps because they feel they do not have a monopoly on the insider information.

Bird and McCrae also examined the opportunity to arbitrage between the two markets. They found that any difference between bookmaker and totalisator odds tends to disappear towards the end of betting. However, the starting prices offered by bookmakers tend to be higher (lower) than those offered by the totalisator for those horses whose odds shorten (lengthen). Bird and McCrae found that this knowledge could not be used to generate a profitable gambling strategy. In contrast, Gabriel and Marsden's (1990) British study found that totalisator returns were consistently higher than starting price returns. From this result, they concluded that the racetrack betting market does not satisfy semi strong efficiency conditions.

Prices vs. Handicapping: Place and Show Anomalies

A number of North American studies reported in this volume evaluate the performance of strategies for backing horses for a place and show. Tuckwell (1981) used Australian data to conduct such a test. He evaluated whether the calculation of the show odds implied by the win odds could be used as the basis for a profitable gambling strategy. The basis for this strategy was to back a horse for a show where the actual show odds exceeded the implied show odds. A situation often known as a 'lock'. Tuckwell found evidence of an inefficiency in that the gambling strategy generated a positive return of 20 per cent. This result is subsequently supported by Hausch and Ziemba's (1990) work on locks in North America.

3: New Zealand Evidence

The racetrack betting market in New Zealand is similar to that in the United States in that the only medium for betting is on the on- and off-course totalisator. The scant research based upon New Zealand data are interesting replications of previous American studies. For example, van Zijl (1984) and Snyder (1981) conclude that the expected return from backing horses for either a win or place where the horses are grouped by their odds is negative and becomes more negative for horses at higher odds. These findings have been taken as an indication of a market inefficiency, but as van Zijl points out, this conclusion is only a valid when risk-neutrality is assumed. McCulloch and van Zijl (1986) used the same data set to evaluate the Harville model previously used by both Hausch, Ziemba and Rubinstein (1981) and Tuckwell (1981). They obtained estimates of a horse's show probability given its win odds, and found that this model gave a small and variable under-estimation of the show probabilities. They concluded that the extent to which this finding could be generalised to studies in other countries was open to question.

4: Other Issues

There is little discussion in the Australian literature on other important issues in the market efficiency of racetrack betting. These include the following: the need to allow for post position and other biases at individual tracks (Canfield, Fauman and Ziemba [1987]); the possibility of arbitrage opportunities for cross-track betting on major races (Hausch and Ziemba [1990]); tests of pricing effects predicted by the divergence of opinion' hypothesis given heterogeneous expectations (Lusht and Saunders [1989]). Furthermore, there is no Australian evidence which tests market efficiencies for more exotic bets such as daily doubles and trifectas as treated by Asch and Quandt (1988). Finally, there is still scope for further work on insider trading, such as that by Crafts (1985) and Shin (1992) in Britain.

Endnotes

1   Includes the turnover on lotto, pools and bingo.

2.   Estimates of the extent of gambling with SP bookmakers indicates a level of turnover at least equal to that of on-course bookmakers.

3.   In contrast, the turnover for on- and off-course gambling is almost equal.

References

Anderson, D., R. Clarke and P. Ziegler (1985). " Information, Equilibrium, Efficiency in Betting Markets", *University of Queensland Working Paper*

*Asch, P., B. Malkiel and R. Quandt (1982), "Racetrack Betting and Informed Behavior". J. *Financial Economics*, July, 187-194.

Asch, P., B. Malkiel and R. Quandt, (1988), "Betting Bias in 'Exotic Bets' ", *Economic Letters*, 28(3), 215-19, Australian Financial Review (1993), Wednesday, September 8, J. Fairfax Ltd, Melbourne.

Bird, R. and M. McCrae, (1987a), "Battling the Books: The Australian Experience", Published Proceedings, 7th International Conference on Gambling and Risk Taking, Reno, Nevada, August.

*Bird, R. and M. McCrae (1987b), " Tests of the Efficiency of Racetrack Berting Using Bookmaker Odds", *Management Science*, December.

Bird R., M. McCrae and J. Beggs, (1987), " Are Gamblers Really Risk Takers?", *Australian Economic Papers*, December, 237-253.

*Busche, K. and C. Hall, (1988), "An Exception to the Risk Preference Anomaly (Horse Race Betting)", *J. of Business*, July, 337-46.

Canfield, B.., Fauman, B. and Ziemba, W., (1987), "Efficient Market Adjustment of Odds Prices to Reflect Track Biases", *Management Science*, November, 1428-39.

*Crafts, N., (1985), "Some Evidence of Insider Knowledge in Horse Race Betting in Britain", *Economica*, August, 295-304.

*Gabriel, P. and Marsden, J., (1990), "An Examination of Market Efficiency in British Racetrack Betting", *J. of Political Economy*, August, 874-85.

*Hausch D., W. Ziemba and M. Rubinstein (1981), "Efficiency of the Market for Racetrack Betting", *Management Science*, December, 1435-1452.

*Hausch, D. and W. Ziemba, (1990), "Locks at the Racetrack (hedging opportunities)", Interfaces May-June, 41-8.

*Hausch, D. and W. Ziemba, (1990), "Arbitrage Strategies for Cross-Track Betting on Major Horse Races, *J. of Business*, January, 61-78.

*Kanto, A. and G. Rosenqvist, (1994), "On the Efficiency of the Market for Double (Quinella) Bets at a Finnish Racetrack, in Hausch, D., V. Lo and W. Ziemba (eds), "Efficiency of Racetrack Betting Markets", Academic Press.

Lusht, K. and Saunders, E., (1989), "Direct Tests of the Divergence of Opinion Hypothesis in the Market for Racetrack Betting", *J. of Financial Research*, 12(4), Winter, 285-91.

McCulloch B. and A. van Zijl (1986), " Direct Test of Harville's Multi-Entry Competitions Model on Race-track Betting Data", J. *of Applied Statistics*, May, 213-219.

*Shin, H., (1992), "Prices of State Contingent Claims with Insider Traders, and the Favourite-Longshot Bias", *Economic Journal*, March, 426-35.

*Snyder, W., "Horse Racing: The Efficient Markets Model", *J.* Finance, September 1109-1118.

Thalheimer, R. and Ali, M. (1993) "The Demand for Parimutuel Horse Race Wagering and Attendance", Research Project, Dept. of Equine Administration, University of Louisville, Kentucky.

*Tuckwell R. (1981), "Anomalies in the Gambling Market", *Australian J. Statistics*, December, 287-295.

Tuckwell R., (1983) The Thoroughbred Gambling market: Efficiency, Equity and Related Issues", *Australian Economic Papers*, June, 106-118.

Tuckwell R, (1984), "Determinants of Betting Turnover", *Australian J.Management*, December.

van Zijl A. (1984), Returns and Weak Form Efficiency: Betting Markets", Victoria University of Wellington Working Paper.

* Article included in this volume

TABLE 1: Turnover by Different Forms of Gambling, 1985/86

State	Total racing $m	Lotteries $m	Poker Machines $m	Casinos $m	Total Turnover $m	Turnover per capita $
Aust. Capital Terr.	94.40	20.55	213.05	-	328.49	1,242.56
Victoria	2,200.49	618.10	-	-	2,818.59	676.78
Queerisland	1,200.08	445.42	-	-	1,645.51	634.69
Tasmania	141-58	66.19	-	120.71	349.36	781.76
Northern Terr,	50.19	13.42	-	106.85	170.46	1,151.14
Western Aust.	530.88	154.35	-	530.88	789.42	547.98
South Aust.	520.76	178.93	-	178.93	832.18	606.04
New South Wales	3,298.87	481.89	5,890.19	-	9,670.65	1,744.52
Australia	89036.95	1,978.85	6,103.24	937.37	16,604.66	1,039.49

TABLE 2: Expenditure on Racing. 1985/86

State	Off-course Parimutuel $m	On-course Parimutuel $m	On-course Bookmakers $m	Total Expenditure $m	Expenditure per capita $
Aust. Capital Terr.	7.67	0.64	2.55	10.85	41.04
Victoria	194.76	40.82	39.54	275.11	66.05
Queensland	100.86	19-19	29-32	149.37	57.61
Tasmania	13.26	1.22	2.77	17.25	38.60
Northern Terr.	2.37	0.13	1.23	3.73	25.18
Western Aust.	59.84	8.63	8.49	76.60	53.17
South Aust.	37.03	8.65	11.18	55.85	40.67
New South Wales	254.86	54.61	80.55	390.02	70.35
Australia	670.65	133.89	175.63	978.78	61.27

ANOMALIES IN THE GAMBLING MARKET[1]

R. H. TUCKWELL

Macquarie University

Summary

Apparent irregularities between the win and the place betting markets in Australian horseracing are examined. Win odds are used to predict win probabilities from which place probabilities are estimated and compared with the place odds on offer. It is concluded that anomalies do in fact exist and are capable, in theory at least, of profitable exploitation.

1. Introduction

This paper is concerned with what appear to be anomalies between the win and the place betting markets in Sydney and Melbourne horseracing. The relationship between the odds quoted for a win and those quoted for a place frequently appear to be inconsistent. For example, two horses in the same race may both be quoted at seven to one for a win but at markedly different odds for a place. One horse may be even money for a place while the other is two to one. Alternatively, two horses may have the same place odds but quite different odds for a win. It is posible that these apparent inconsistencies are justified by the fact that some horses are more reliable, or consistent, than others and that this superior reliability raises their chances of running a place relative to their winning chances. The paper examines whether the "anomalies" are justified or not, concludes that they are not and then proceeds to determine whether the inconsistencies are sufficiently large to be capable of profitable exploitation. It was found that a profitable betting strategy is possible but that its effective implementation is inhibited by one or two, not insuperable, practical problems.

In more detail, the approach employed took the following lines. A large sample of bookmakers' starting price odds was collected and the proportion of winners at each of a number of odds ranges noted. Estimates of the probability of a horse winning at each odds range were then derived. On the assumption that the apparent anomalies are not justified and that the win probabilities of all horses in a race are all

---

[1] Manuscript received August 4, 1980; revised July 20, 1981.

that is needed to determine the probabilities of running a place, the place probabilities were computed for 3,849 horses in 286 races in Sydney and Melbourne over the period August–November 1975. These estimated place probabilities were then tested for accuracy by comparing the proportion of horses running a place with the predicted probability over a number of categories of predicted probabilities. The estimates were found to be reasonably accurate, suggesting that the apparent anomalies are in fact anomalies. The scope for profitable exploitation was examined by comparing the odds-equivalent of the estimated place probabilities with the odds for a place on the totalisator at starting time. Of the sample of 3,849 horses, 874 (almost 23 per cent) started at place odds yielding a positive expected return. The expected rate of return on outlay was approximately 17 per cent and the actual rate of return, if the strategy had been effectively implemented over the sample period, would have been somewhat higher at approximately 20 per cent.

2. The Gambling Market

There are two betting mediums on Australian courses, bookmakers and the totalisator (the "tote"). Bookmakers usually only bet for a win and/or each-way, while the tote runs separate pools for win and place only betting. Describing the tote operation first, the sums wagered on each horse in a race are pooled, a certain percentage deducted (approximately fourteen per cent, fifteen per cent in some States) and the remainder is distributed to those who bet on the winner. In the case of place betting, the pool, net of tax, is distributed equally between the first three place-getters. The tote odds cannot be determined with any certainty until all betting ceases just prior to race-start. The tote board, however, does show how the odds are progressing.

In addition to the tote, there are a large number of bookmakers who operate on course. Each bookmaker has his own board which lists his odds on offer for each horse. As in the case of the tote, these odds continually change up until starting time depending on the relative weight of money and also, in this case, on bookmakers' opinions. Unlike the tote operation, punters betting with a bookmaker receive the odds currently quoted on his board—not the starting price odds as in the case of the tote. Most serious punters consequently bet with bookmakers as they know precisely what odds they are getting and can take advantage of favourable fluctuations in the odds during the course of betting. If, however, a punter finds himself in the position of still considering a bet very close to starting time, the odds on the tote will be known reasonably accurately, and if the bet is not of sufficient size to unduly depress the odds, he would of course bet on the tote if the odds are better than with bookmakers. This type of activity ensures that the tote odds are brought into reasonably close line with the starting price odds offered by bookmakers. In this respect, observation suggests that operators exist whose specific objective it is to take advantage of discrepancies between the two sets of odds. Similar operators also take advantage of place odds on the tote which are regarded as high in relation to the corresponding win odds. But the criterion of what is high is more difficult to determine in this case and the conclusions reached here suggest that they are by no means exhausting all the opportunities for profit.

3. Starting Price Odds and Winning Probability

In estimating win probabilities, bookmakers' starting price odds were used, since an examination of several hundred odds movements suggested that the final odds, as opposed to any earlier quotes, provide the more accurate indication of a horse's chances. The published starting price odds (SPO) for all horses running in Sydney and Melbourne metropolitan meetings during the calendar year of 1974 were collected. In all, there were just under sixteen thousand observations. The proportion of winners at each level of SPO was noted. When there were very few runners at a particular level of SPO (usually the very short odds) and where the proportion of winners was zero, or close to it (some of the very long odds), neighbouring odds categories were aggregated. Win probabilities were estimated by first estimating the percentage loss associated with each level of SPO, or its probability equivalent (p^*), and substituting these estimates into (1), which relates the probability of winning to the percentage loss.

$$\hat{p} = p^*(1 - \hat{L}/100), \tag{1}$$

where

$\hat{p}$ = estimated win probability,

p^* = probability equivalent of SPO = $1/(1 + \text{SPO})$,

$\hat{L}$ = estimated percentage loss.

The estimated percentage loss was derived by regressing the sample percentage loss (L) on p^*. The sample percentage loss is the percentage of outlay that would have been lost by betting one unit on each runner at the particular level of SPO in the sample. L was regressed on p^* using alternative degrees of polynomial and weighting each observation by the standard error of L. A degree three polynomial was found to be the highest warranted. The estimated relation appears below with t-values in parentheses. Sydney and Melbourne races were aggregated as no significant difference was found when the relationship was estimated separately for each city.

$$\hat{L} = 50 \cdot 3 - 362 \cdot 7 p^* + 978 \cdot 8 p^{*2} - 777 \cdot 3 p^{*3}. \qquad \bar{R}^2 = 0 \cdot 44 \tag{2}$$
$$(7 \cdot 1) \quad (3 \cdot 1) \qquad (3 \cdot 1) \qquad (2 \cdot 6)$$

Substituting (2) into (1) gives

$$\hat{p} = 0\cdot497p^* + 3\cdot627p^{*2} - 9\cdot788p^{*3} + 7\cdot773p^{*4}. \tag{3}$$

Relation (3) was used to predict the win probabilities for the sample of 3,849 over the period August–November 1975. The estimates were adjusted, where necessary, to ensure that the win probabilities in each race summed to unity.

The predicted percentage losses at different levels of SPO (from relation (3)) are of interest for their own sake and are reproduced in Table 1.

TABLE 1

Predicted percentage loss and starting price odds

SPO	$\hat{L}$	SPO	$\hat{L}$
0·41	8·24	5·00	13·44
0·67	17·15	5·50	14·84
0·73	17·55	6·00	16·20
0·80	17·61	6·50	17·50
0·90	17·21	7·00	18·74
1·00	16·49	7·50	19·91
1·11	15·50	8·00	21·02
1·25	14·20	9·00	23·04
1·38	13·09	10·00	24·84
1·50	12·09	11·00	26·43
1·63	11·21	12·00	27·84
1·75	10·46	14·00	30·24
1·88	9·85	15·00	31·27
2·00	9·37	16·00	32·20
2·25	8·73	20·00	35·17
2·50	8·45	25·00	37·76
2·75	8·45	32·91	40·44
3·00	8·66	40·00	42·03
3·25	9·03	50·00	43·56
3·50	9·51	66·00	45·10
3·75	10·07	98·73	46·76
4·00	10·70	187·30	48·40
4·50	12·04		

From a level of approximately 16 to 17 per cent for SPO in the odds-on categories, the percentage loss gradually falls as SPO rises to be a minimum of approximately 8.5 per cent when SPO is in the vicinity of three to one. Thereafter the percentage loss rises with SPO, until the loss is almost 50 per cent with SPO of the order of one hundred to one and above. Scott (1978), page 142, obtained a similar type of relationship using Sydney data for the 1950's.

4. Predicting Place Probabilities from Win Probabilities

If there are n horses in a race, let k_1, k_2 and k_3 represent the probabilities of the kth horse coming first, second and third, respectively, where k takes the values $1, 2, 3, \ldots, n$.

The probability of the kth horse running a place is $(k_1 + k_2 + k_3)$. k_1 has been estimated. Estimates of k_2 and k_3 are required. Dealing with k_2 first, if the jth horse wins the race, the probability that the kth horse runs second is the probability of it beating the remaining horses. This is assumed to be $k_1/(1 - j_1)$. The weighted sum of these probabilities over all j $(j \neq k)$, where the weights are j_1, gives the required k_2.

$$k_2 = \sum_{j=1}^{n} j_1(k_1/(1 - j_1)) - k_1(k_1/(1 - k_1)).$$

Turning to k_3, if the ith and jth horses occupy the first two places, the probability of the kth horse coming third is the probability of it beating the remaining horses, which is assumed to be $k_1/(1 - i_1 - j_1)$. The probability of the ith and jth horses occupying the first two places is assumed to be $i_1(j_1/(1 - i_1)) + j_1(i_1/(1 - j_1))$, which simplifies to $i_1 j_1/(1 - i_1) + i_1 j_1(1 - j_1)$. The probability of the kth horse coming third is the weighted sum of $k_1/(1 - i_1 - j_1)$ over all possible pairs of i and j $(i \neq k$ and $j \neq k)$ where the weights are $i_1 j_1/(1 - i_1) + i_1 j_1/(1 - j_1)$.

$$k_3 = \sum_{i=1}^{n-1} \sum_{j=i+1}^{n} \left(\frac{i_1 j_1}{1 - i_1} + \frac{i_1 j_1}{1 - j_1}\right)\left(\frac{k_1}{1 - i_1 - j_1}\right)$$
$$- \left[\sum_{j=1}^{n} \left(\frac{k_1 j_1}{1 - k_1} + \frac{k_1 j_1}{1 - j_1}\right)\left(\frac{k_1}{1 - k_1 - j_1}\right) - \left(\frac{2k_1^2}{1 - k_1}\right)\left(\frac{k_1}{1 - 2k_1}\right)\right]$$

The term subtracted in square brackets eliminates those cases where $i = k$ and $j = k$.

The derivation of the above formula assumes that the win probabilities provide sufficient information for the assignment of probabilities to other ranks (second and third in this case). This may not be so. For example, reliable horses may have a higher chance of running a place relative to their winning changes than unreliable horses. For the formula to be correct, it has to be assumed that the probability of a horse beating any subset of horses in the race is the same irrespective of where the other horses (those not in the subset) finish. Some idea of the extent to which this assumption is transgressed may be gained by applying the formula to the sample 3,849 horses over the August–November period 1975. The estimated place probabilities were subdivided into ten different ranges and the average probability of placing within each range was compared with the proportion of horses running a place. This was done for all horses within each probability range and also for two classes, (1) where the odds-equivalent of the predicted place probability was exceeded by the place odds on the tote (which were noted just prior to race start at Sydney meetings) and the expected return was positive and (2) where the odds-equivalent of the estimated place probability exceeded the tote place odds and the expected return was negative. The results appear in Table 2. At the

TABLE 2

Comparison of estimated place probabilities with proportion placing

Range of estimated place probability		Positive expected return	Negative expected return	All
0·0 and under 0·03	x	0·019	0·017	0·017
	y	0·0	0·025(0·006)	0·025(0·005)
	n	30	829	859
0·03 and under 0·05	x	0·042	0·039	0·040
	y	0·087(0·059)	0·054(0·015)	0·057(0·014)
	n	23	239	262
0·05 and under 0·08	x	0·066	0·064*	0·064*
	y	0·136(0·073)	0·108(0·017)	0·110(0·017)
	n	22	324	346
0·08 and under 0·13	x	0·100	0·102	0·102
	y	0·114(0·048)	0·127(0·018)	0·125(0·017)
	n	44	355	399
0·13 and under 0·20	x	0·165	0·163	0·163
	y	0·181(0·045)	0·191(0·020)	0·190(0·018)
	n	72	397	469
0·20 and under 0·30	x	0·240	0·248	0·248
	y	0·231(0·044)	0·291(0·025)	0·278(0·022)
	n	91	323	414
0·30 and under 0·40	x	0·352	0·343	0·346
	y	0·337(0·046)	0·360(0·034)	0·352(0·027)
	n	104	200	304
0·40 and under 0·60	x	0·497	0·484	0·492
	y	0·438(0·032)	0·456(0·033)	0·447(0·024)
	n	235	226	461
0·60 and under 0·80	x	0·698*	0·667*	0·690*
	y	0·604(0·037)	0·452(0·063)	0·564(0·032)
	n	174	62	236
0·80 and under 1·00	x	0·888	0·867	0·884
	y	0·873(0·037)	0·750(0·097)	0·849(0·036)
	n	79	20	99

x = average estimated place probability in range
y = proportion of horses running a place (standard errors in parentheses)
n = number of horses
* denotes a significant difference at the 5 per cent level

five per cent significance level the estimated place probability differs significantly from the proportion placing in only five cases out of a possible thirty. There appears, nevertheless, to be a consistent tendency for the formula to under-estimate the probability of placing for the low probabilities and to overestimate it for the high probabilities. This tendency, however, appears to be shared by both of the classes and the apparent lack of difference between the results for the two classes suggests that the assumptions made in estimating the place probabilities may well be justified; that is, that differences between horses in reliability, or consistency, are relatively unimportant for present purposes. If reliability is an important factor in determining place probabilities, the formula will over-estimate the place probabilities of the unreliable horses and under-estimate the place probabilities of the more reliable ones. This will tend to push the unreliable horses into class (1), where the expected return is positive, and the reliable horses into class (2), where the expected return is negative. The fact that there is no obvious and consistent over-estimation of the place probabilities in class (1) and under-estimation in class (2) lends support to the view that reliability is a relatively unimportant factor.

Regressing the proportion placing on the predicted place probability, and weighting by the standard error of the proportion placing, for each of the two classes yielded the following results.

Expected Return Positive

$$\hat{y} = 0 \cdot 0052 + 0 \cdot 955\, x \qquad\qquad \bar{R}^2 = 0 \cdot 984$$
$$(0 \cdot 006)(0 \cdot 029)$$
(standard errors in parentheses)

Expected Return Negative

$$\hat{y} = 0 \cdot 0151 + 0 \cdot 928\, x \qquad\qquad \bar{R}^2 = 0 \cdot 752$$
$$(0 \cdot 008)(0 \cdot 067)$$

The constant terms are both positive and the coefficients of x less than unity which reflects the tendency for the low probabilities to be under-estimated and the high probabilities to be over-estimated. However, at the five per cent level, neither of the constant terms differs significantly from zero and neither of the coefficients of x differs significantly from unity. Perhaps more importantly, the regressions do not differ significantly from each other. Neither the intercepts nor the slopes differ significantly.

5. Feasibility of Place-betting Strategy

One can only conclude from the above that the apparent inconsistencies between the win and place betting markets are not justified and that they are in effect genuine anomalies, capable, in theory at least, of profitable exploitation as evidenced by the fact that 874 (almost 23 per cent) of the sample of 3,849 horses fall in the category where the expected return is positive. The proportion of horses showing a positive expected return is considerably lower than the average of 23 per cent for the low probabilities and considerably higher for the high probabilities. In the highest place probability range (0·8 to 1·0) almost 80 per cent of horses show a positive expected return which compares with only 12 per cent for horses with estimated place probabilities of less than 0·2. This may be partly due to the tendency to underestimate the place chances for the low probabilities and to overestimate them for the high probabilities. However, it seems more likely that it is a consequence of place bettors, not renowned for their expertise, putting relatively too much money on the long-shots in the hope of a high reward. Operating the place-betting strategy in retrospect over the sample period, a bet of one unit on each of the 874 selections would have yielded a profit of 177 units, equivalent to a rate of return on outlay of just over 20 per cent. The expected rate of return was somewhat lower at approximately 17 per cent.

In view of the fact that the majority of bets came from the high probability categories and there was a tendency to overestimate these probabilities, it might be expected that the actual rate of return would be below the expected rate of return, not above it. The high actual rate of return appears to have been due to some unusually favourable results in the low probability categories. Table 3 gives the details for each of the ten probability ranges.

In general, the expected rate of return exceeds the actual rate of return for the high probabilities and is below the actual rate of return for the low probabilities. In the overall figures, however, the numerical dominance of the high probabilities is outweighted by the extremely high actual returns, in relation to expected returns, for the low probability categories 2, 3 and 4. The number of observations in these categories is not large and the very high returns appear to have been the result of a combination of chance occurrences, (1) an unusually high proportion of place-getters in categories 2 and 3, even allowing for some under-estimation (see Table 2) and (2) an unusually high return on the place-getters in categories 3 and 4, which is evident from a comparison of the place odds, or their equivalent probabilities, with the estimated place probabilities in Table 3.

In the longer run the actual rate of return can be expected to fall below the expected rate of return. However, the effective operation of such a place betting strategy is inhibited by two factors. First, bookmakers' starting price odds are required to estimate place probabilities. By definition, starting price odds are not known until starting time. What is required is a programmable hand-operated calculator capable of performing the calculations in a matter of seconds. At the present

TABLE 3
Details of return for each probability range

Category	x	N	N_P	ARR	ERR	PO_P	$P - PO_P$
1	0·019	30	0	−100·0	17·1	—	—
2	0·042	23	2	135·7	19·5	26·1	0·037
3	0·066	22	3	297·3	19·3	28·1	0·034
4	0·100	44	5	108·6	23·3	17·4	0·054
5	0·165	72	13	25·6	23·6	6·0	0·143
6	0·249	91	21	6·8	21·1	3·6	0·217
7	0·352	104	35	22·9	14·7	2·7	0·270
8	0·497	235	103	0·4	15·2	1·3	0·435
9	0·698	174	105	0·7	15·8	0·7	0·588
10	0·888	79	69	16·0	6·9	0·3	0·769
		874	356	20·3	16·8		

x = average estimated place probability in range
N = number of horses showing a positive expected return
N_P = number of place-getters
ARR = actual percentage rate of return
ERR = expected percentage rate of return
PO_P = average place odds for place-getters
$P - PO_P$ = probability equivalent of average place odds for place-getters

time the best of them takes at least several minutes, but no doubt the continuing rapid technological advance in this area will soon ensure the solution to this problem.

The second factor which may restrict the effective operation of the strategy is that a bet of any size on the tote will depress the odds and reduce the rate of return. However, with present-day place pools in the vicinity of $100,000 and upwards, this should not present much of a problem if bets are restricted to less than four figure amounts.

6. Conclusion

Genuine inconsistencies do appear to exist between the win and place gambling markets in metropolitan Australian horseracing. In approximately twenty to twenty-five percent of cases these inconsistencies are of sufficient magnitude to be capable, in theory at least, of profitable exploitations on the place tote. The continued existence of these anomalies can only be explained by a combination of ignorance on the part of everyday place punters, a lack of expertise on the part of operators attempting to exploit irregularities in the market and certain practical difficulties in the 'fine-tuning' of such an exploitative strategy.

References

SCOTT. D. (1978). *Winning*. Sydney: Wentworth Press.

TESTS OF THE EFFICIENCY OF RACETRACK BETTING USING BOOKMAKER ODDS

RON BIRD AND MICHAEL McCRAE

Commerce Department, Australian National University, Canberra, ACT 2601, Australia

The objective of this study is to evaluate the informational efficiency of the market for betting on horse races. Whereas the price data used in previous studies have been drawn largely from totalizator (parimutuel) odds, the data used in this study are derived from bookmaker odds. The availability of a series of prices throughout betting facilitates the use of filter tests to evaluate market efficiency. We conclude that this gambling market is efficient in the use of information supplied via both the movements in odds during the course of betting and the selections of newspaper tipsters. However, there is evidence to suggest that those with access to private information can earn positive returns from gambling.
(EFFICIENCY; GAMBLING MARKETS; BOOKMAKER ODDS)

Fama (1970) was the first to formalize the concept of informational efficiency. Since that time many authors have evaluated the efficiency of markets with respect to numerous types of information. The majority of these studies have concentrated on the informational efficiency of markets for financial assets, and particularly the stock market. However, there has been a small number of studies that have reported on the informational efficiency of gambling markets—specifically the market for gambling on horse races. The objective of this paper is to extend the analysis of these studies, particularly by taking advantage of data that enable the application of more appropriate statistical techniques. The source of the price information used in most of these studies has been totalizator (parimutuel) odds which only become available at the completion of betting. The data set used in this paper consists of bookmaker odds with a series of odds being available for each horse in each race. Implications from using bookmaker, as opposed to totalizator, odds are discussed in §1. §2 reviews the literature on the efficiency of this gambling market. The tests of market efficiency undertaken in this study are reported in §3. Finally, §4 summarizes our findings and suggests possible extensions.

1. The Data

There are basically two forms of betting on horse races—one where bets are placed with a bookmaker and the other where bets are placed on a totalizator. Further, the opportunity is often available to place bets both at the racetrack (on-course betting) and in person and/or by phone at venues away from the racetrack (off-course betting). In Australia, the legal forms of betting are provided by on-course bookmakers and by a totalizator which operates both on-course and off-course.[1] Although totalizator betting exceeds one-half of all betting by volume, approximately 80 percent of all on-course betting is with bookmakers. The data used in this study are derived from the odds on offer from on-course bookmakers. This is the only data which provide a meaningful series for prices throughout the course of betting. In addition, bookmaker odds provide

---

* Accepted by William T. Ziemba; received May 1986.

[1] In addition to the legal forms of betting, illegal betting takes place with off-course bookmakers who are referred to as SP bookmakers. Estimates of the volume of bets placed with SP bookmakers relative to the total volume of all betting vary between 25 percent and 40 percent.

the best means to gauge the advent of information onto the market since the largest and most knowledgeable gamblers concentrate their betting activities via bookmakers.[2]

The actual source of the data used is the bookmaker odds prepared for transmission from three Melbourne racetracks to other racetracks where bookmakers are operating on the Melbourne meeting (referred to as the interstate betting service). These odds are transmitted throughout the course of betting and typically are revised eight to ten times at intervals of three to five minutes. Each bookmaker establishes his own odds throughout the course of betting on each race. The odds used in this study represent a consensus of the odds on offer by the five largest bookmakers at particular points in time during betting. To reduce the recording, storing and analyzing of data, we chose to use the odds from the four points of time during betting at which there is a complete call of the odds. The betting period for most races is approximately 30 minutes and the times chosen were at the commencement of betting (t_1), five minutes later (t_2), approximately 13 minutes before the start of the race (t_3) and just prior to the start of the race (t_4). Information was collected over the 1983 and 1984 calendar years and resulted in usable data on 1026 races.[3]

The odds obtained on a totalizator bet can only be determined at the completion of betting. At that time the statutory take is deducted from the total pool and the number of units bet on the winner is divided into the remaining pool to determine the odds (dividend) of the winner.[4] In contrast, a gambler placing a bet with a bookmaker obtains the odds showing on the bookmaker's board at that time.[5] Therefore, a series of odds for each horse in a race can be obtained which provide the necessary data to conduct filter tests to evaluate the efficiency of this gambling market. The results of these tests are reported in §3. Note that, unlike the totalizator, bookmakers cannot guarantee a fixed return on turnover although they do establish their odds at each point of time during betting to imply a particular take where horses perform to expectations.[6]

2. Previous Efficient Market Studies

The majority of studies of the efficiency of the market for gambling on horse races have been conducted in America and so are based on totalizator odds.[7] Snyder (1978) evaluated a betting strategy based solely on the odds for each horse and concluded that such a strategy could not earn positive returns. Further, he found that the odds set by "experts" (i.e. handicappers and newspaper tipsters) provided no better prediction of a horse's winning chance than those provided by the totalizator odds. On the basis of

---

[2] In most of North America the only legal forms of betting are with the on-course and off-course totalizator. In England, there is on-course betting with both bookmakers and the totalizator and off-course betting with "legal" SP bookmakers.

[3] A small number of races were deleted from the sample when there was the withdrawal of a horse after the commencement of betting on the race.

[4] A win pool on a race of $1 million and a statutory take of 15 percent would result in $850,000 being available for distribution to those who back the winner of the race. If there are 140,000 $1 units bet on the winner, then the maximum dividend would be $6.07 (i.e. $850,000/140,000). However, dividends are usually paid in increments of 10 cents and so the dividend paid in this case would be $6.00 which represents odds equivalent to 5/1 (i.e. five to one).

[5] Each bookmaker has a betting board on which he displays the odds on offer for each horse in a race. For example, odds for a particular horse of 4/1 indicates that someone betting $100 on that horse will receive $500 if it wins (i.e. winnings of $400 and a return of the $100 outlay). The bookmaker changes the odds on offer for each horse during the course of betting, largely on the basis of the volume of betting on each horse.

[6] For a discussion of the calculation of the bookmaker's implied take, see Dowie (1976) and Bird and McCrae (1985).

[7] For a detailed discussion of the studies of the efficiency of this gambling market, see Ziemba and Hausch (1984).

these findings, Snyder concluded that this gambling market was efficient. The Snyder study was extended by Losey and Talbot (1980) who evaluated a gambling strategy based upon the difference between the totalizator odds and those set by the "experts". They found that the return on a strategy of backing horses whose actual odds exceeded their "experts" odds was very poor and became worse the greater the disparity between the actual and "experts" odds.

Asch, Malkiel and Quandt (1982) found that the bets placed in the second half of betting provided a much better prediction of outcomes than those placed in the first half of betting. They took this as an indication of the existence of an "informed" class of gambler but they could not determine whether this class could earn positive returns.[8] In a subsequent study, Asch, Malkiel and Quandt (1984, 1986) used a multinomial logit probability model to develop various gambling strategies based upon the information contained in a combination of the odds set by "experts" (i.e. handicappers) with those established by gamblers either over the total period of betting or over the second half of betting. They found that the gambling strategies generated outperformed a random betting strategy but failed to achieve positive returns. Figlewski (1979) used a similar logit model to determine the information content of selections by "experts" (in this case, newspaper tipsters) and the extent to which this information is incorporated into the market-determined odds. He found that the "experts" selections prove superior to a random selection strategy and so inferred that the "experts" do possess some relevant information. Further, he found that this information is incorporated into the odds, especially in relation to on-course betting.

In Britain, Dowie (1976) investigated the relationship between race results and two sets of odds—those set by "experts" and presented in the morning paper and the starting price (SP) odds of bookmakers. He found no evidence to suggest that the correlation between SP odds and actual results was greater than that between "experts" odds and actual results. He took this as evidence of market efficiency as it suggests an absence of gamblers with private information. Craft (1985) extended the Dowie study in a similar way to the Losey and Talbot extension of the Snyder study. Craft used the relationship between Dowie's "experts" odds and SP odds as the basis for generating a gambling strategy. In contrast to the Losey and Talbot results, Craft found that a strategy of backing each horse whose odds decreased (i.e. "experts" odds exceed SP odds) at its initial "experts" odds gave rise to positive returns while the reverse strategy of backing horses whose odds drifted gave rise to negative returns. When he applied the same strategies but with the bets now being placed at SP odds, he found that the returns were typically close to zero. On the basis of this evidence Craft concluded that those with private access to information can use it to generate positive returns but those whose sole information source is the movement in odds cannot earn positive returns. In an Australian study, similar results were obtained by Tuckwell (1983) who investigated the returns from a gambling strategy based upon the difference between the "experts" and SP odds.

The above papers on market efficiency have concentrated on three types of information—information contained in the horses' odds; information supplied by "experts"; information that is privately held. We evaluate these three types of information in §3. However, in our brief summary we have failed to mention a number of studies that have evaluated backing horses for a place and a show. Hausch, Ziemba and Rubinstein (1981) calculated the place and show odds implied by the win odds. They tested a strategy of backing a horse whose actual place and/or show odds exceeded those implied by the win odds on offer for that horse. The strategy proved highly profitable,

---

[8] Using the data on which this paper is based, Bird and McCrae (1985) found that the odds do not become a significantly better predictor of outcomes during the course of betting.

suggesting an inefficiency since it is based solely on price information which is available at little or no cost.[9] Similar inefficiencies have been found in studies conducted in Australia (Tuckwell 1981) and New Zealand (McCulloch and van Zijl 1984).

3. Empirical Tests of Market Efficiency

In the typical test of capital market efficiency, an inability to utilize a piece of information to earn risk-adjusted abnormal returns is taken as evidence of market efficiency with respect to that piece of information. As there is no reason to undertake a risk-adjustment when evaluating gambling markets, this suggests that efficiency should be judged against a benchmark of a zero rate of return (i.e. where the total returns equal the total amount outlaid). However, a number of studies have proposed that a gambler's willingness to bet when the expected returns are negative may be due to their deriving consumption benefits from gambling.[10] This suggests that consumption benefits should be incorporated into the benchmark against which efficiency is judged.

The problem remains as to how consumption benefits should be incorporated into the analysis. In a number of capital market studies using filter tests similar to those used in this study, the rate of return on a randomly selected portfolio is used as a benchmark. We found that a strategy of placing a $1 bet on a number of randomly selected horses realized a rate of return of −35 percent (i.e. on average the return from every $1 bet was 65 cents).[11] An alternative staking strategy where each randomly selected horse was backed to win $1 realized a rate of return of −19 percent (i.e. on average the return on every $1 outlayed was 81 cents).[12] The problem with using such a benchmark in this study is that it attributes "average" consumption benefits to all betting strategies when it is unlikely that consumption benefits are independent of the strategy being pursued. In particular, the consumption benefits are related to a preference of punters for high return-low probability outcomes and such outcomes are associated with backing horses at long (high) odds. Therefore, the consumption benefits associated with a particular bet are directly related to the odds at which the bet is placed. One test used in this study attempts to control for consumption benefits by comparing the rate of return on two strategies which are matched on the basis of the odds of the horses included in each strategy group.

As mentioned previously, the availability of a series of odds for each horse in a race enables us to test for efficiency using filter tests similar to those used in a number of studies which evaluate aspects of the efficiency of capital markets (e.g. Fama and Blume 1966). The filter tests used in this study are expressed in terms of a specified movement in the "sum to one" probabilities over the course of betting. The movement was measured in terms of probabilities rather than odds, because the use of the latter would have resulted in picking up a preponderance of relatively small price movements in

[9] For a confirmatory extension of this paper, see Hausch and Ziemba (1985).

[10] For a discussion of these studies, see Bird and McCrae (1986).

[11] All the rates of return reported in this paper are calculated by dividing the net return from bets placed under a particular strategy by the total amount outlaid on all bets placed under the strategy. For example, a gross return of $420 from a betting strategy that results in a $1 bet on 350 horses produces a net return of $70 and a rate of return of +20 percent. It should be noted that the minimum rate of return on any strategy is −100 percent but there is no theoretical upper bound.

[12] Under the staking system of backing each horse to win $1, the size of the bet on each horse will vary according to the odds on offer for that horse. For example, it will involve a 2 cent bet on a horse whose odds is 50/1 and a 40 cent bet on a horse whose odds is 5/2. Therefore, this staking system results in relatively larger bets on short-priced horses viz-a-viz long-priced horses. A result which when combined with the well-documented phenomenon that long-priced horses are poorer bets explains why the returns from this staking system are less negative than those from a staking system of backing each horse to win $1 (i.e. −19 percent versus −35 percent).

long-priced horses. The first step in calculating the "sum to one" probabilities is to convert the odds on offer to the implied probabilities. For example, odds of 3/1 imply that a horse has a chance of one in four of winning and so the implied probability is 0.25. The odds of each horse in a race at a point of time during betting are converted to implied probabilities which are then aggregated. This sum of the implied probabilities is always greater than one.[13] The "sum to one" probability for each horse is then obtained by dividing the implied probability for each horse by the sum of the implied probabilities for all horses in the race. For example, if the implied probabilities sum to 1.25, then the "sum to one" probabilities of a horse whose odds are 3/1 would be 0.20 (i.e. 0.25/1.25).

The "sum to one" probabilities were used in order to abstract from the general drift in odds that occurs during the course of betting.[14] The positive and negative filters tested were 0.025, 0.05, 0.075, 0.10, 0.125 and 0.15. These filters were applied to movements in "sum to one" probabilities over the four time periods utilized in our study and a \$1 bet was placed on a horse at the time it first satisfied a particular filter. A horse could satisfy a number of filters over the course of betting. For example, a horse might satisfy both a +0.025 and +0.05 filter between t_1 and t_2 and a 0.075 filter between t_1 and t_3. In this case a \$1 bet would be placed on the horse at its t_2 odds as part of the +0.025 filter, at its t_2 odds as part of the +0.05 filter and at its t_3 odds as part of its +0.075 filter. The returns under this strategy are reported in Table 1 where it is described as a "bet at end" strategy.

As we have only used data on odds on offer at four points of time during betting, it is likely that a horse might have more than satisfied a particular filter before it is identified as having done so. For example, a horse's odds might move from 10/1 to 6/1 between t_1 and t_2 resulting in its "sum to one" probability increasing by 0.04. Assume the +0.025 filter strategy would have been satisfied when the horses odds had fallen to 7/1. In this case, it is likely that a gambler pursuing a +0.025 filter strategy would back this horse at 7/1 rather than the 6/1 which would be assumed in our analysis given our data restrictions. In order to overcome this deficiency we simulated a continuous series of prices by calculating the rate of return for each filter strategy where a \$1 bet was placed at "as soon as possible" odds. The "as soon as possible" odds being the first odds at which the horse would have satisfied the particular filter. The returns from backing at the "as soon as possible" odds under each filter are also reported in Table 1.

A particular problem associated with the "as soon as possible" odds is that they would often imply that a bet is placed at odds that are not offered by bookmakers. For example, a horse whose odds are shortening may satisfy a particular filter at odds of 3.1/1 but no such odds are offered by bookmakers and the first odds at which the bet could be placed would be 3/1. For the positive (negative) filters the rate of return at the "as soon as possible" odds is likely to be better (worse) than what could be achieved by a gambler using a particular filter; while the rate of return at "bet at end" odds is likely to be less (greater) than that obtainable by the same gambler. Therefore, the actual rate of return achievable under each filter will lie between that obtained using the "bet at end" odds and that obtained using the "bet as soon as possible" odds. The rates of returns calculated using both "bet at end" odds and "bet as soon as possible" odds prove to be almost always negative. Further, the application of t-tests to these returns failed to

---

[13] If the sum of the implied probabilities was less than one it would be possible for a gambler to ensure a win by backing each horse in the race to win a fixed amount. In over 4,000 observations we failed to find one instance where the sum of the implied probabilities was less than one.

[14] The average sum of the implied probabilities across all races in the sample was 1.351 at t_1, 1.282 at t_2, 1.207 at t_3 and 1.210 at t_4. This is consistent with a general upward drift in odds between t_1 and t_3 and odds remaining fairly stable between t_3 and t_4.

TABLE 1

Rates of Return from Placing a $1 Bet on Horses Identified as having Satisfied Various Filter Strategies

Filter (no. of bets)	Bet at beginning %	Bet at end %	Bet "as soon as possible" %	Filter (no. of bets)	Bet at beginning %	Bet at end %	Bet "as soon as possible" %
+0.025 (1923)	+8.16	−11.85†	−8.30	−0.025 (1532)	−46.03†	−23.21†	−38.04†
+0.05 (696)	+18.41†	−8.28	−5.52	−0.05 (277)	−43.74†	−10.11	−30.48†
+0.075 (243)	+28.01†	−4.33	−2.56	−0.075 (54)	−47.50†	−9.96	−32.09
+0.1 (77)	+42.66‡	0.52	2.73	−0.1 (9)	−80.00†	−66.67	−75.56
+0.125 (26)	+52.31	−15.50	−6.46	−0.125 (3)	−100.00	−100.00	−100.00
+0.15 (10)	+196.10	47.40	62.30	−0.15 (0)	—	—	—

† Significant at 0.05 level.
‡ Significant at 0.10 level.
N.B. Significance tests only applied where the sample size exceeds 30.

identify a single instance of a significant positive return. These findings strongly suggest that a strategy based upon backing horses whose odds move by a specified amount is unlikely to prove profitable.

We constructed a control group to incorporate consumption benefits into the analysis. This matched each bet placed under the various filter strategies with a $1 bet on the horse in the same race whose "bet at end" odds, at the time the filter strategy bet is placed, is closest to that of the horse backed under the filter strategy (e.g. if the horse backed under the filter strategy is 4/1, then a bet is placed as part of the control group on a different horse in the same race whose odds at that time is closest to 4/1).[15] Since the horses in both groups have approximately the same odds, it is likely that any consumption benefits associated with the bets placed on the horses in each group will be similar. The distinguishing feature between the two groups is that the odds of the horses included in the filter strategy have moved in a particular direction whereas this is not true for the horses included in the control group. The returns for the filter strategy and the control group are reported in Table 2. A *t*-test failed to reveal a single instance of a significant difference between the rate of return for each filter strategy group and that for the corresponding control group. Further, there is no pattern in the relationship between the rates of return for the two groups with five instances of the filter strategy group outperforming the control group and six instances where the opposite occurs. These findings are consistent with the proposition that past movements in odds cannot be used as the basis for a profitable gambling strategy.

We saw previously that a staking system where each randomly selected horse is backed to win $1 outperforms a staking system where a $1 bet is placed on each horse. In order to evaluate whether our findings are sensitive to the staking system utilized, we repeated the previous filter tests using the alternative staking system where each horse is backed to win $1. The returns as reported in Table 3 prove to be very similar to those obtained under the previous staking system (see Table 1) with no filter strategy earning a positive return.[16] This evidence confirms that the previous findings on the efficiency of the market with respect to past movements in odds and indicates that this result is not sensitive to the staking system being utilized.[17]

[15] Ball (1977) criticized the methodology employed in many of the filter tests applied to capital markets (e.g. Fama and Blume 1966) and suggested an approach analogous to the control group utilized in this study.

[16] One factor that would mitigate against finding any difference between the returns under the two staking systems is where the horses included in each filter group are at similar odds.

[17] The rates of return were also calculated for a control group under a staking system of backing each horse to win $1. The results were consistent with those reported in Table 2 and so are not presented.

TABLE 2

Rate of Return from Placing a $1 Bet at "End" Odds on Horses that Satisfy Various Filter Strategies and Horses Designated to Control Groups

Bet at End

Filter (no. of bets)	Filter strategy	Control group	Filter (no. of bets)	Filter strategy	Control group
+0.025 (1923)	−11.85	−16.52	−0.025 (1532)	−23.21	−11.06
+0.05 (696)	−8.28	−6.94	−0.05 (277)	−10.11	−28.07
+0.075 (243)	−4.33	−5.10	−0.075 (54)	−9.96	−21.30
+0.1 (77)	0.52	18.83	−0.1 (9)	−66.67	83.33
+0.125 (26)	−15.50	14.42	−0.125 (3)	−100.00	208.33
+0.15 (10)	47.40	−12.50	−0.15 (0)	—	—

N.B. The differences between the rates of return for the filter strategy and the control group for all positive and negative filters are not significant at either the 0.05 and 0.10 level.

Little or no market exists to establish the odds for each horse in a race until the on-course betting commences approximately 30 minutes before the starting time of each race. The initial odds are set by the bookmakers who have access to certain information including that provided in the daily newspapers and specialist racing publications. However, it is likely that some relevant information relating to the winning chances of horses in each race is not available to bookmakers at the time of the commencement of betting. Therefore, one would expect a movement in odds as a more informed market is established. Our previous analysis establishes that these price movements cannot be used as the basis for a profitable strategy. However, the question remains as to whether a prior knowledge of these price movements would give rise to positive returns? This question is of particular interest as the price movements reflect the private information in existence at the commencement of betting.

The approach used in this study to address the above question involves applying the same filter tests as described above. The only difference being that now it is assumed that bets on horses identified as satisfying a particular filter strategy are placed prior to the price movement (i.e. "bet at beginning" odds). The rate of return for the various filters where a $1 bet is placed at "bet at beginning" odds is reported in Table 1. The positive filters evaluated all results in positive rates of return with three of them being

TABLE 3

Rates of Return from Placing a Bet to Win $1 on Horses that have Been Identified as Having Satisfied Various Filter Strategies

Filter (no. of bets)	Bet at beginning %	Bet at end %	Bet "as soon as possible" %	Filter (no. of bets)	Bet at beginning %	Bet at end %	Bet "as soon as possible" %
+0.025 (1923)	+3.07	−8.11‡	−6.93‡	−0.025 (1532)	−38.22†	−16.21†	−30.66†
+0.05 (696)	+10.31	−7.24	−6.03	−0.05 (277)	−42.24†	−12.94	−30.59†
+0.075 (243)	+18.81‡	−3.32	−2.82	−0.075 (54)	−46.70†	−10.48	−32.16‡
+0.1 (77)	+20.64	−5.49	−4.77	−0.1 (9)	−68.97	−47.55	−60.60
+0.125 (26)	−2.65	−29.54	−27.95	−0.125 (3)	−100.00	−100.00	−100.00
+0.15 (10)	+42.52	+42.71	+53.72	−0.15 (0)	—	—	—

† Significant at 0.05 level.
‡ Significant at 0.10 level.
N.B. Significance tests only applied where the sample size exceeds 30.

TABLE 4

Rate of Return from Placing a $1 Bet at "Beginning" Odds on Horses that Satisfy Various
Filter Strategies and Horses Designated to Control Groups

Bet at Beginning

Filter (no. of bets)	Filter strategy %	Control group %	Filter (no. of bets)	Filter strategy %	Control group %
+0.025 (1923)	+8.16†	−25.62	−0.025 (1532)	−46.03†	−21.47
+0.05 (696)	+18.41‡	−15.19	−0.05 (277)	−43.74†	−16.79
+0.075 (243)	+28.01†	−52.16	−0.075 (54)	−47.50	−20.37
+0.1 (77)	+42.66†	−37.34	−0.1 (9)	−80.00	−38.89
+0.125 (26)	+52.31	−27.88	−0.125 (3)	−100.00	−100.00
+0.15 (10)	+196.10	−100.00	−0.15 (0)	—	—

Difference between rate of return on filter strategy and control group:
† Significant at 0.05 level.
‡ Significant at 0.10 level.
N.B. Significance tests only applied where the sample size exceeds 30.

statistically different from zero. In addition, the returns increase with the size of the filter, which is consistent with the size of the price movement reflecting the magnitude and quality of the private information relating to particular horses. The negative filters all generate negative rates of return which are statistically different from zero.

In order to control for consumption benefits we again formed control groups in the same way as previously described. The rates of return for the control groups and filter strategies are reported in Table 4. For the positive filters, the rates of return under the filter strategy outperforms those for the control group in all instances and these differences are significant in all cases where it is appropriate to apply a *t*-test. For the negative filters, the filter strategy performs worse than the control group in all instances. The conclusion that can be drawn from the above evidence is that a prior knowledge of price movements can be used as the basis for earning positive returns. This finding suggests the existence of private information which is not impounded in the odds on offer at the commencement of betting.

We repeated our test of the informational efficiency of the market with respect to private information applying the alternative staking system of backing each horse that satisfies a particular filter to win $1. Our findings are reported in Table 3 where the relevant column is headed "bet at beginning" odds. Overall, the findings for the positive filters as reported for the two staking systems in Tables 1 and 3 are consistent for the "bet at beginning" odds with all filters considered generating positive returns. However, these positive returns are higher in every case and more significant in a statistical sense where the staking system of placing a $1 bet on each horse is used. This suggests that the informed gambler would be advised to use this staking system whereas previous evidence suggests that the uninformed gambler will minimize losses by using the alternative staking system of backing horses to win $1. The results for the negative filters under both staking systems are also consistent for the "bet at beginning" odds with all filters generating negative returns.[18] However, in this case there is no clear evidence to suggest an informed gambler should favor either staking system.

[18] The rates of return were also calculated for a control group under a staking system of backing each horse to win $1. The results were consistent with those reported in Table 4 and so are not presented.

As mentioned earlier a number of writers have evaluated the market reaction to information supplied by "experts" in gambling markets (i.e. handicappers and newspaper tipsters). In this study we also evaluate information supplied by "experts" with the objective of determining whether this information could be used to make positive returns and whether the information is incorporated into the bookmaker odds. The "experts" used in this study are ten tipsters whose selections are published in a Melbourne newspaper on the morning of the race meeting. Each "expert" makes a first, second and third selection in each race. Each first selection is awarded three points, each second selection is awarded two points and each third selection is awarded one point and the points awarded each horse are aggregated in order to calculate an "experts" poll for each race. The analysis was based on the consensus of the selections of the ten "experts" as reflected in the "experts" poll, rather than the selections of each individual "expert". The rationale for using this "experts" poll is that it provides an ungarbled signal based on the aggregate of the information available to all "experts" (Grossman 1976; Verrecchia 1979).

The horses in each race were ranked according to the "experts" poll and the rate of return calculated for a betting strategy of placing a $1 bet on all horses with the same tipsters ranking (i.e. all horses ranked first, all horses ranked second, etc). Our results are reported in Table 5. The first point worth noting is that none of the strategies evaluated yield a positive rate of return. On the basis of this evidence it would appear that the gambling market is efficient in its utilization of information supplied by the "experts".

In order to evaluate whether the information from the "experts" is incorporated into the horses' odds, we also ranked horses on the basis of their odds (i.e. the horse with the shortest odds being the first favorite, the horse with the next shortest odds being the second favorite, etc). We then calculated the rate of return that would be realized from placing a $1 bet on all horses ranked at a particular level of favoritism. These returns are also reported in Table 5. Using a t-test, we found no evidence of any significant differences between the rates of returns where the horses were ranked on the basis of the tipsters' poll and on the basis of the level of favoritism. This result was confirmed when a sign test was applied to the number of times the rate of return based upon the "experts" selections outperformed the rate of return based upon the odds. These find-

TABLE 5

Rates of Return from Placing a $1 Bet on Horses Ranked Both on the Basis of the "Experts" Poll and Their Level of Favoritism.

	Period							
	t_1		t_2		t_3		t_4	
Ranking	Tipsters %	Favoritism %	Tipsters %	Favoritism %	Tipsters %	Favoritism %	Tipsters %	Favoritism %
1	−12.29	−15.28	−7.18	−9.00	−4.50	−5.40	−5.50	−7.38
2	−26.45	−26.57	−22.29	−20.61	−18.06	−13.33	−18.78	−10.81
3	−16.62	−13.38	−12.67	−17.80	−7.79	−11.54	−9.66	−17.99
4	−27.66	−24.40	−25.54	−14.93	−19.18	−22.61	−19.61	−24.45
5	−20.71	−35.72	−15.91	−36.20	−9.23	−33.64	−9.62	−29.16
6	−53.27	−39.95	−52.98	−33.55	−50.05	−20.08	−40.46	−27.12
7	−46.09	−42.49	−44.65	−38.56	−36.92	−41.63	−38.40	−38.96

N.B. None of the differences between the returns where the horses are ranked on the basis of the tipsters' polls and on basis of the level of favoritism prove to be significant at the 0.10 level.

ings suggest that the information supplied by the "experts" is incorporated into the bookmaker odds.

4. Conclusions

The objective of this study has been to evaluate the informational efficiency of the market for betting on horse races. Two types of public information were evaluated—one being the movement in bookmaker odds during the course of betting on each race and the other being the selections of newspaper tipsters. We were unable to identify a betting strategy based on either of these types of information that could generate a positive rate of return. Therefore, we concluded that the gambling market is efficient in its utilization of both these sources of information. We did find that those with prior knowledge of movements in odds during the course of betting could use this knowledge to earn significant returns. This finding suggests that all information is not impounded into the odds on offer at the commencement of betting on each race. Further, it suggests that those with access to private information can earn positive returns although one would have to identify the costs of this information before suggesting this indicates a definite market inefficiency.

One possible extension of this paper is to control for a number of parameters which might be relevant to the betting behavior of punters. Examples of these include the type of race, the quality of the participating horses and the state of the track. A second extension would be to utilize a more formal approach when choosing the filters to apply in our tests. It is usual practice to arbitrarily choose the filters to be evaluated. However, a knowledge of the distribution of price movements would enable a choice of filters that have a strict statistical interpretation. Finally, we have stressed in this paper the need to formally model the benefits to punters from gambling, in order to make a judgement on the efficiency of the gambling market. In capital market studies, an asset pricing model is used as a basis for estimating expected returns. No such model yet exists for estimating the expected returns from gambling and the derivation of such a model would be an important advance towards gaining a better understanding of the efficiency of this gambling market.[19]

[19] The authors would like to acknowledge the helpful comments of Mark Tippett, Graeme Rankine and the anonymous referees.

References

ASCH, P., B. MALKIEL AND R. QUANDT, "Racetrack Betting and Informed Behavior," *J. Financial Economics*, (July 1982), 187–194.

———, ——— AND ———, "Market Efficiency in Racetrack Betting," *J. Business*, (April 1984), 165–175.

———, ——— AND ———, "Market Efficiency in Racetrack Betting: Further Evidence and Correction," *J. Business*, (January 1986), 157–160.

BALL, R., "Filter Rules: Interpretation of Market Efficiency, Experimental Problems and Australian Evidence," *Accounting Education*, (May 1977), 1–17.

BIRD, R. AND M. MCCRAE, "Battling the Books: The Australian Experience," Australian National University Working Paper, Canberra, Australia, (November 1985).

——— AND ———, "Attitudes Towards Risk in Gambling Markets," Australian National University Working Paper, Canberra, Australia, (May 1986).

CRAFT, N., "Some Evidence of Insider Knowledge in Horse Race Betting in Britain," *Economica*, (August 1985), 295–305.

DOWIE, J., "On the Efficiency and Equity of Betting Markets," *Economica*, (May 1976), 139–151.

FAMA, E. AND M. BLUME, "Filter Rules and Stock Market Trading," *J. Business*, (January 1966), 226–241.

FIGLEWSKI, S., "Subjective Information and Market Efficiency in a Betting Market," *J. Political Economy*, (February 1979), 75–88.

GROSSMAN, S., "On the Efficiency of Competitive Stock Markets Where Traders Have Diverse Information," *J. Finance*, (May 1976), 573–585.

HAUSCH, D. AND W. ZIEMBA, "Transaction Costs, Extent of Inefficiencies, Entries and Multiple Wagers in a Racetrack Betting Model," *Management Sci.*, 31 (April 1985), 381–417.

———, ——— AND M. RUBINSTEIN, "Efficiency of the Market for Racetrack Betting," *Management Sci.*, (December 1981), 1435–1452.

LOSEY, R. AND J. TALBOT, "Back on the Track with the Efficient Markets Hypothesis," *J. Finance*, (September 1980), 1039–1043.

McCULLOCH, B. AND A. VAN ZIJL, "A Direct Test of a Multi Entry Competition Probability Model," Victoria University Working Paper, Wellington, New Zealand, January 1984.

SNYDER, W., "Horse Racing: The Efficient Markets Model," *J. Finance*, (September 1978), 1109–1118.

TUCKWELL, R., "Anomalies in the Gambling Market," *Australian J. Statistics*, (December 1981), 287–295.

———, "The Thoroughbred Gambling Market: Efficiency, Equity and Related Issues," *Australian Economic Papers*, (June 1983), 106–118.

VERRECCHIA, R., "On the Theory of Market Information Efficiency," *J. Accounting and Economics*, (March 1979), 77–90.

ZIEMBA, W. AND D. HAUSCH, *Beat the Racetrack*, Harcourt, Brace and Jovanovich, San Diego, 1984.

Kelly Busche
Christopher D. Hall
University of Hong Kong

An Exception to the Risk Preference Anomaly

I. Introduction

We report here an anomaly of an anomaly. The general definition of risk, involving variability in wealth or consumption, implies risk aversion.[1] Yet, the generality of risk aversion seems to stop at horse racing's betting window and perhaps at less easily studied portals, too.[2] In this and other journals, investigators marshal massive data sets that reject the proposition of generalized risk aversion.[3] Study after study confirms this anomaly of risk-preferring bettors. In light of this aberrant tradition, bettors in Hong Kong are perverse: they provide no evidence for the assertion that bettors prefer risk.

The data are examined using two techniques borrowed from previous studies. We have also borrowed the general model, hypothesis, or view of these studies. With these new data viewed in the old manner, however, we cannot replicate previous results.

The theory of risk bearing implies risk aversion. In every published study of horse race betting known to us, however, investigators reject this implication in favor of "risk-loving" behavior. Using the techniques of these studies, we examine a new data set from Hong Kong and find a rather different result: Hong Kong bettors seem to be either risk neutral or risk averse. A striking difference between the Hong Kong data and the previously studied North American data is the much larger betting volume per race.

1. If risk is a general and costlessly transacted economic good, each of us would consume so much of it as to leave us risk averse at the margin. (See, e.g., Bailey, Olson, and Wonnacott 1980; Friedman 1981.)

2. Do criminals prefer risk? (See Becker 1968, p. 178.)

3. The most familiar papers reporting risk preference at race tracks include Rosett (1965), Weitzman (1965), Harville (1973), Ali (1977), Snyder (1978), Losey and Talbot (1980), Hausch, Ziemba, and Rubinstein (1981), and Asch, Malkiel, and Quandt (1982, 1984, 1985). References to similar studies are available in the psychology literature and may be found in the works cited above.

II.   Risk Preference?

We will discuss the methods of the previous studies more fully in the next sections. Nevertheless, it may be helpful to note two troublesome issues first. Given that betting is a negative sum game, are risk preferrers the only people willing to bet? And, are bettors homogeneous with respect to risk attitudes?

Race tracks "take" a commission and collect betting taxes from the pari-mutuel pools such that the average bettor loses about 15%–25% of each dollar wagered, depending on the type of bet, the track, and the year considered.[4]

One rationalization of horse betting, in light of this fact, is that for a representative bettor the expected utility of the gamble exceeds the utility of the certain and greater wealth of not gambling. A utility function consistent with these assumptions and with the North American betting data has been estimated by Ali (1977): gambling is explained as a means to the end of reallocating consumption possibilities in the face of rising marginal utility of wealth.

An alternative assumption is that horse betting is like going to the opera or owning a boat: some days are better than others, but it is a hobby for a representative consumer. Horse betting then enters the utility function directly. Although the function's convexity in wealth remains a possibility, this is not a requirement for such rational gamblers: gambling is fun.

A third possibility brings us to the second question. How useful is it to envision a representative bettor? Perhaps differences of opinion between people with identical utility functions are an important cause of horse betting: gambling aggregates diverse predictions. Or, diverse opinions may be held together with rising marginal utilities or gambling as a good itself.

The literature assumes gambling is a means, not an end in itself. A significant advantage of this approach is its simple modeling. It also places horse betting in a traditional context of risky portfolio decisions. Although our "prior" assigns a low probability to this hypothesis, it is our intention to see how far it takes us in explaining these new data. We do not, therefore, offer a theory of why horse betting exists. Our task here is to report on the consistency, persistence, or replicability of findings that heretofore seemed destined for consecration as the "Law of Underwinning Long Shots."

---

4. See National Association of State Racing Commissioners (1983). Track take for Hong Kong win betting is as high or higher than takes at tracks studied in the United States. In New York (Ali's 1977 data, covering 1970–74), the take was 15%; take in Illinois (Asch, Malkiel, and Quandt 1982) is 17%; in California (Hausch, Ziemba, and Rubinstein 1981), take is about 16.8%. In Hong Kong, the take is 17%. Brokerage fees in commodity markets are lower.

To conserve space and assist in comparability of the present and past results, we adopt the terminology and techniques used by others. Ali (1977), for example, speaks of a "representative" bettor and analyzes his data accordingly. Other studies do the same: expected utility is assumed to be equalized on the margin, and observed betting patterns are examined for consistency. This study may be viewed as an attempt to replicate this representative-bettor approach.

III. The Data

This study is based on data from racing at Happy Valley and Shatin racecourses in Hong Kong for the racing seasons of 1981–82 through the middle of the 1986–87 season. It is useful to describe two main features of our data.[5] First, the data are from 2,653 races providing more than 26,000 pari-mutuel dollar totals for the horses competing. The largest data sets (Weitzman 1965; Ali 1977) include about seven-and-a-half times as many races, but studies using smaller numbers of races are common. For example, Snyder (1978) examined 849 races; Ali (1979), 1,089 races; Figlewski (1979), 189 races; Hausch, Ziemba, and Rubinstein (1981), 627 and 1,065 races; and Asch, Malkiel, and Quandt (1982), 729 races.

The second feature of the data relates to the (legal) betting volume. In Hong Kong, the "handle," or total value of all wagers handled by the track, is enormous by North American standards. The average handle for 1985–86 (thoroughbred racing) in Hong Kong was $5.6 million (U.S.) per race. The U.S. per-race average (1983) for all racing was $87,000 (thoroughbred alone was $152,000). The average win pool per race in Hong Kong during the most recent season was about $1.2 million (U.S.). Win pools at U.S. tracks that have been studied are much smaller. Ali (1977), for example, reports win pools ranging from $24,000 to $235,000 (in 1970–74 dollars), and win pools at Longacres, a thoroughbred racetrack in Washington State, averaged around $100,000 in 1982.

IV. Revealed "Risk Attitudes" Assuming True Win Probabilities

The pari-mutuel odds for a given horse is the total money bet on the other horses racing divided by the total bet on the given horse. Legal bookmakers in England and Australia, in contrast to the United States and Hong Kong, offer fixed betting odds. Fixed odds reveal marginal valuations of bets directly, for the standard reason that quantities ad-

5. The data and a good deal of thoughtful consulting on matters of racetrack betting were generously provided by Alan Woods, Mark Lemieux, Walter Simmons, and William Benter.

just until marginal gains per dollar are equated. The interpretation of pari-mutuel odds is more complicated because the odds respond to the amount of money bet by each person: the odds on a horse fall mechanically as the bet increases, and the actual odds are only determined once the race starts. The existing literature, nevertheless, assumes that pari-mutuel odds reveal expected marginal utilities directly.

Each person is, by definition, indifferent toward alternative bets at the margin in any betting market, but how and when will these marginal valuations be revealed under the constraint of pari-mutuel betting? The literature provides one answer. Assuming the final odds are anticipated correctly, and that each one of the many bettors places a small, fixed amount on each race, the final or equilibrium pari-mutuel odds reveal these marginal utilities. These assumptions characterize a price- or odds-taking market, just as in the case of a competitive fixed-odds betting market.[6]

Risk attitudes are assessed by posing an alternative hypothesis based on risk neutrality in the context of an odds-taking market with a representative individual. Risk-neutral bettors simply minimize losses for a set of given true winning probabilities. Bettors may be considered as having utility functions where gambling itself is consumption. This implies the equalization of marginal returns across competing horses. A sufficient condition for this is that the proportion of the win pool bet on a given horse is equal to the proportion of expected wins in a repeated sample context.

The hypothesis actually tested assumes further that bettors form unbiased predictions of horses' true winning probabilities.[7] The (joint) alternative hypothesis, then, is that the expected fractions of money bet equals the winning probabilities *given* a set of expectations or true probabilities. Since neither of the classification variables, expectations, or true probabilities are observable, a proxy variable must be created to implement the test. The proxy researchers use for the expected win probability of a given horse is the frequency of wins for "similar" horses. "Similar" is clarified in the next section.

If bettors prefer risk, this will be revealed by "overbetting" ("underbetting") long-shot (favorite) horses. That is, long-shot horses will attract a percentage of the win pool that is greater than the percentage of races they will win. The opposite will be true for favorites.

6. Formal derivations of the maximization conditions are available in many sources, including Rosett (1965), Ali (1977), Hausch, Ziemba, and Rubinstein (1981), and Asch, Malkiel, and Quandt (1982). An added but implicit assumption is that the change in wealth and change in utility of wealth is sufficiently large to be measured. Bettors cannot, e.g., bet on enough races to render the random variation trivially small.

7. An anonymous referee of this journal suggests that our study measures risk preference at the margin, whereas previous studies measure an average risk preference. The larger betting volume in Hong Kong, if largely exogenous, makes this likely.

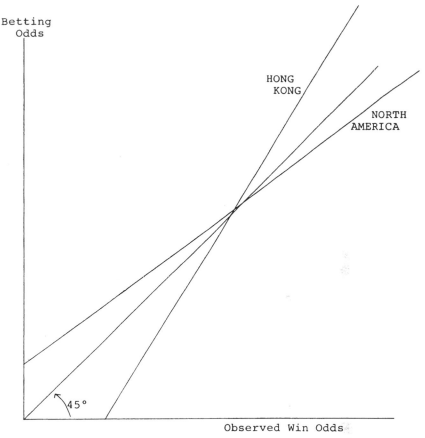

FIG. 1.—Betting and win odds: Hong Kong and North America

V. The Empirical Case For and Against Risk Preference:
Inferred Win Probabilities

Are horses bettors generally risk lovers? The case for risk preference has been presented with data as in figure 1. Hong Kong bettors, however, defy the North American convention.

Suppose the horizontal axis represents the spectrum of true win probabilities while the vertical axis records the fractions of money bet. The slope of a line fitted to the North American data would have a slope less than unity: horses with below average win fractions receive more than a proportionate share of the betting volume. In Hong Kong, horses with below average win fractions receive less than a proportionate share of the betting volume. Bettors in Hong Kong seem to be risk averse.

Interpreting such a figure, and a similar representation of the betting data that will be discussed shortly, requires understanding the corre-spondence between the true win probabilities, or odds, and the proxy variable used in the regression analysis.

The present study, together with all of those cited, defines this proxy as the observed fraction of wins *given* the betting odds. Notice that this conditionality is opposite to that sought for a test of the risk neutrality hypothesis, which asks, instead, "What are the betting odds (propor-tions bet) given the expected win frequencies?" Employing this proxy amounts to asking "What are the win frequencies (returns) given the betting odds?" To appreciate this reversal of positions it may be useful to examine two methods of estimating win probabilities.

A. Grouping by Pari-mutuel Odds

Figure 1 employs the most common method. First, the betting data are classified by the fractions of money bet on each horse in a race. For example, this classification rule puts all of those horses with one-ninth of the money in the betting pool (eight-to-one odds horses) in the same group. The fraction of races won by the group is then interpreted as the group's representative win fraction, and both the bet and win fractions are expressed as odds. The win odds are defined *given* the betting odds.

Following the studies cited, we have also grouped the horses accord-ing to betting odds. Both the numbers of groups used and the range of fractions included within a group vary from study to study. Subject to the constraint that no group may have zero winners, the groupings are chosen to maximize between-group (minimize within-group) variation. Our grouping is similar to the typical study and includes equal numbers of horses in each group. Along the horizontal axis in figure 1, then, are the observed-win odds. Along the vertical axis are the betting odds, which are first used to find the winning proportions.[8]

It is then argued that, if bettors are risk neutral and make unbiased and accurate predictions, returns must be equated across horses of differing win odds: the regression line through the scatter will have a slope of one.[9] Risk preference, however, is described by a regression with slope less than unity: the return from a random bet on a favorite is higher than one on a long shot. In Hausch, Ziemba, and Rubinstein (1981), and in six other studies cited there, where data from more than 30,000 races are classified by betting odds, returns on favorites were

---

8. These odds are adjusted by adding the track's commission and taxes to them so that the fractions of money bet and the suggested probabilities sum to unity.

9. A further condition is that a bettor's prediction error is "small." The larger is the prediction error, the greater the upward bias in classifying data by betting odds.

higher than on long shots. Using these combined results, the implied regression (standard errors[10] in parentheses) is

$$\text{betting odds} = 1.144 + 0.747 \text{ win odds}, \quad R^2 = 0.993. \quad (1)$$
$$(0.403) \quad (0.023)$$

Hong Kong betting data from 2,653 races (from 1981 to the end of 1986), when interpreted in this manner, fail to repeat the risk-preference results. That is, if one accepts the general method employed in previous studies of asking whether returns on random bets across odds categories are equated, Hong Kong is different. These returns seem to be higher for long-shot horses: risk *aversion* is suggested by *this* technique with *these* data.

These Hong Kong data are different from that reported typically for North America. The slope of the regression line here is 1.251, as reported below (with standard errors in parentheses):

$$\text{betting odds} = -2.908 + 1.251 \text{ win odds}, \quad R^2 = 0.99. \quad (2)$$
$$(1.40) \quad (0.036)$$

This slope estimate is seven standard errors larger than unity, which is significant beyond the 0.001 level. Using the technique of pooling by pari-mutuel odds magnitudes, bettors "reveal" risk aversion.

B. Grouping by Favorite Position

A problem with the classification method used for figure 1 leads to an alternative method of looking at these data. With the above classification it is possible that two or more horses from the same race might end up in the same odds group, while only one horse per race actually wins. The win total, then, for such a category may be unrepresentative of a typical horse with these betting odds. If, instead, all of the most-favored horses are grouped to compute an average betting odd and win odd for the group, and this is repeated for each rank of relative betting volume, there will never be two horses from the same race in the same odds category.[11]

Table 1 presents the data classified by rank order of relative betting volume in column 1. Column 5 is the difference between bet and win fractions per standard error where the latter is derived assuming that the fraction of wins is a binomial statistic, that is, the difference per standard error $= (\psi - \zeta)[N/\zeta(1 - \zeta)]^{.5}$. When the sign of the value in

10. Since the sampling distribution cannot be bivariate normal, standard errors are reported in deference to tradition only.

11. This method is used by Ali (1977). Asch, Malkiel, and Quandt (1982) also use rank but put ties together in the same odds group. To prevent horses from the same race being grouped together in figure 1, we could delete all but one horse from each race, but at the cost of statistical efficiency.

TABLE 1   Hong Kong Betting Data Classified by Favorite Position

Rank (1)	No. of Races (N) (2)	Win Fraction (ζ) (3)	Fraction of Money Bet (ψ) (4)	Standard Errors (ψ − ζ) (5)	Ali 1977 (6)	Asch, Malkiel, and Quant 1982 (7)
1	2,653	.276	.284	.92	− 10.29	− 2.119
2	2,653	.190	.187	− .39	.99	− .903
3	2,651	.151	.142	− 1.29	− .52	− 1.972
4	2,647	.0985	.1104	2.05	3.45	− .961
5	2,638	.0835	.0862	.46	3.49	.074
6	2,598	.0627	.0662	.74	3.01	− .279
7	2,509	.0483	.0498	.35	5.80	.480
8	2,346	.0473	.0370	− 2.35	6.20	1.096
9	1,992	.0336	.0277	− 1.46	. . .	2.095
10	1,620	.0209	.0209	.0	. . .	. . .
11	969	.0227	.0175	− 1.09	. . .	. . .
12	726	.0138	.0133	− .12	. . .	. . .
13	411	.0098	.0103	.10	. . .	. . .
14	233	.0129	.0078	− .69	. . .	. . .

column 5 is positive, there has been "overbetting." In contrast, columns 6 and 7 show the measures of betting "bias" found by Ali (1977) and Asch, Malkiel, and Quandt (1982). Whereas their data show a tendency to overbet long shots, there is no such tendency in the Hong Kong data.[12]

VI. The Classification Problem

Since either method of classifying data, by fractions bet or by favorite position, defines the win odds *given* the betting odds, we may not ask directly our initial question, Do bettors prefer risky horses? Instead, we ask, Are returns equated across betting odds? Or, is the average return from a random bet on a long shot lower than one on a favorite odds category (rather than a favorite horse)? In a classic regression model this reversal of questions corresponds to running the regression backward and biases the test in favor of accepting risk aversion versus risk preference. Such a bias makes the North American data all the more striking, and, in this model, the coefficient of betting odds given win fractions may be deduced readily in any event.

In practice, the appeal of this question is reinforced by the impossibility of asking directly the alternative question, Are returns equated across actual win probabilities of horses? Tackling this head-on re-

12. The Asch, Malkiel, and Quandt (1982) results are similar to those in Ali (1977), but for a much smaller sample, i.e., specifically 729 vs. 20,247 races, respectively. There is more variety in the number of horses running in Hong Kong than at the tracks studied by Ali (1977) and Asch, Malkiel, and Quandt (1982). Standard errors for races with nine or more horses are not provided by Ali on the grounds that there are too few observations.

quires horses to be built like roulette tables or in some manner where their objective probabilities are independently verifiable.[13] Until the genetics of this problem are solved, we can only classify data by betting odds and remembering which question we have addressed in examining the results.

Unfortunately, bias in these data is unlikely to be as simple as noted above. There is certainly an error in measuring true win frequencies of representative horses. When added to the classic regression model, such as error, if independent of the error in determining the fraction bet, biases the data toward accepting a risk-preference hypothesis: such a bias opposes the bias of classifying by betting odds. The central problem, however, is that the underlying error process surely includes correlations between errors in measuring both variables. Rain, for example, may affect both the betting and the win probability measurements. The sudden scratching of a horse, imperfectly known jockey skills, or handicap consequences, and so forth, may well affect both measurement errors systematically. If such correlations are significant, it is not possible to measure risk attitudes without first estimating the structure of measurement errors.

The larger betting volume per race in Hong Kong may be relevant to the question of classification errors. Suppose, for example, that each dollar bet is placed according to the true but unobservable win probabilities of the competing horses. The binomial distribution then might characterize the correspondence between observed or sample bet fractions and win probabilities. In such a world, this aspect of the error in observed bet fractions would decline as betting volume increases. In turn, the error in classifying wins by betting fractions would also decline.

Frankly, we cannot provide an appropriate sampling distribution for the situation described above, nor do we yet have any data to assess the empirical significance of sampling errors. Our intention, instead, is to show that the resolve of long shots to win less frequently as dictated by the fractions of money bet on them may yet yield to larger and more diverse data collection. Further, inasmuch as the sums wagered are typically trivial in comparison to the wealth of the bettors, and variations in consumption alternatives are cheaper to obtain at casinos, commodity markets, or by spending less on automobile brake repairs, our "prior" is that risk attitudes should not be important in racetrack betting.

13. Kevin MacKeown has pointed out to us that physicists working in an area known as "confinement" face similar problems. He drew our attention to an interesting footnote from Lyons (1985, p. 226): "Two physicists working on confinement were asked to predict the result of a horse race involving N horses. The first did extensive analytic and Monte Carlo calculations, and came back months later with the solution to the problem, provided the horses were spherical. The second had found a unique solution, but only in the limit that $1/(1 - N)$ tended to infinity."

VII.   A Final Note

Every study of horse-race betting known to us is presented as rejecting the proposition of risk aversion. Data from Hong Kong racetracks viewed in the same way tell a different story. Although the literature in economics and psychology abounds with speculations regarding the special nature of the gambler's psyche, the inscrutable risk preference among North American gamblers, may, perhaps, be tempered by further study.

References

Ali, M. M. 1977. Probability and utility estimates for racetrack bettors. *Journal of Political Economy* 85 (August): 803–15.

Ali, M. M. 1979. Some evidence on the efficiency of a speculative market. *Econometrica* 47 (March): 387–92.

Asch, Peter; Malkiel, B.; and Quandt, R. 1982. Racetrack betting and informed behavior. *Journal of Financial Economics* 10 (June): 187–94.

Asch, Peter; Malkiel, B.; and Quandt, R. 1984. Market efficiency in racetrack betting. *Journal of Business* 57 (April): 165–74.

Asch, Peter; Malkiel, B.; and Quandt, R. 1985. Market efficiency and racetrack betting: Further evidence and a correction. *Journal of Business* 59 (January): 157–60.

Bailey, Martin J.; Olson, M.; and Wonnacott, P. 1980. The marginal utility of income does not increase: Borrowing, lending, and Friedman-Savage gambles. *American Economic Review* 70 (June): 372–79.

Becker, Gary. 1968. Crime and punishment: An economic approach. *Journal of Political Economy* 76 (March): 169–219.

Figlewski, S. 1979. Subjective information and market efficiency in a betting model. *Journal of Political Economy* 87 (February): 75–88.

Friedman, David. 1981. Why there are no risk preferrers. *Journal of Political Economy* 89 (June): 600.

Harville, D. A. 1973. Assigning probabilities to the outcomes of multi-entry competitions. *Journal of the American Statistical Association* 68 (June): 312–16.

Hausch, D. B.; Ziemba, W. T.; and Rubinstein, M. 1981. Efficiency of the market for racetrack betting. *Management Science* 27 (December): 1435–52.

Losey, Robert, and Talbot, John C., Jr. 1980. Back at the track with the efficient markets hypothesis. *Journal of Finance* 35 (September): 1039–43.

Lyons, L. 1985. Quark search experiments at accelerators and in cosmic rays. *Physical Reports* 129 (December): 225–89.

National Association of State Racing Commissioners (NASRC). 1983. *Parimutuel Racing, 1983*. Lexington, Ky.: NASRC.

Rosett, R. N. 1965. Gambling and rationality. *Journal of Political Economy* 73 (December): 595–607.

Snyder, Wayne W. 1978. Horse racing: Testing the efficient markets model. *Journal of Finance* 33 (September): 1109–18.

Weitzman, Martin. 1965. Utility analysis and group behavior. *Journal of Political Economy* 73 (February): 18–26.

Efficient Market Results in an Asian Setting*

Kelly Busche
School of Economics
University of Hong Kong

Researchers concerned with efficiency of racetrack betting seem agreed on the existence of market inefficiency or risk preference among gamblers. However, Busche and Hall (1988) found an anomaly: they analyzed 2,653 races run in Hong Kong from 1981 to 1986, and could not reject a hypothesis of equal average returns across groups of horses. I present evidence that the above result is not unique. I find virtually identical results when I examine a sample of 2,690 new Hong Kong races, pool all 5,343 Hong Kong races (Table 1), or examine 1,738 races run in Japan during 1990 (Table 2).

Previous results suggesting a systematic inefficiency at race tracks find no support in these new data.

Comments

Clearly, more research is needed on the question of how gambling markets work. The large betting volumes at Hong Kong and at most Japanese tracks perhaps provide a clue. It is possible that gambling markets are composed of some gamblers who act as money maximizers and some who treat their days at the races as consumption. Expected returns from privately produced information are limited by the size of the betting pool, so potential money maximizers must limit their attendance to tracks with large pools. Excitement at the races will be the return at small volume tracks and application of investment theory may be misplaced there.

References

[1] Busche, Kelly, and Hall, Christopher D., "An exception to the risk preference anomaly". *Journal of Business*, 61 (July 1988):337-46.

*I thank Junji Shiba for translation of Japanese data sources

Table 1: Hong Kong- 1987-92 and 1981-92

fav. pos'n	races	betf	winf	std dev winf	1987-92 Zvalue	1981-92 Zvalue 5343races
1	2690	.2854	.2766	.0086	1.0238	1.3565
2	2690	.1821	.1885	.0075	-0.8443	-0.8339
3	2690	.1374	.1368	.0066	0.0872	-0.8433
4	2688	.1074	.1004	.0058	1.2068	2.2830
5	2673	.0845	.0860	.0054	-0.2804	0.1602
6	2644	.0659	.0692	.0049	-0.6709	0.0141
7	2571	.0511	.0521	.0044	-0.2376	0.0661
8	2413	.0391	.0390	.0039	0.0282	-1.6467
9	2014	.0306	.0357	.0041	-1.2519	-1.9319
10	1730	.0234	.0277	.0039	-1.0978	-0.8643
11	1067	.0197	.0150	.0037	1.2573	-0.0019
12	859	.0155	.0198	.0048	-0.9039	-0.7936
13	622	.0123	.0080	.0036	1.1796	0.9613
14	410	.0090	.0049	.0034	1.2090	0.2314

Table 2: Japan- 1990

fav. pos'n	races	betf	winf	std dev winf	Z value
1	1738	.3447	.3493	.0114	-0.3994
2	1738	.1846	.1974	.0095	-1.3354
3	1738	.1282	.1410	.0083	-1.5326
4	1738	.0938	.0834	.0066	1.5618
5	1738	.0698	.0673	.0060	0.4190
6	1721	.0529	.0494	.0052	0.6705
7	1663	.0409	.0421	.0049	-0.2473
8	1580	.0316	.0285	.0042	0.7360
9	1420	.0249	.0254	.0042	-0.1104
10	1216	.0200	.0164	.0036	0.9779
11	986	.0167	.0071	.0027	3.6037
12	792	.0137	.0114	.0038	0.6154
13	542	.0120	.0074	.0037	1.2525
14	427	.0098	.0070	.0040	0.6859

Japanese volumes are highly variant but sometimes approach $500 million US. for a single race. The track take for win bets averages 26%, but the breakage is very low: winning bets are paid to the lower 10 Yen on a 100 Yen ticket, or .05% of winnings on a 10:1 ticket.

The Z-value measures, in standard deviation (of win fraction) units, the difference between returns on horses in this group and average returns across all horses. Detailed values are shown only for the new 1987-92 sample. See Busche and Hall for methods and exact values from 1981-87. betf is the average fraction of the win pool on horses in the listed favorite position and winf is the corresponding win frequency;

Hong Kong betting volumes are enormous compared to U.S. volumes which are seldom over a few million per day. Current daily handles in Hong Kong approach $100 million US. The track take on win bets is 17.5%, which was recently raised from 17%. There is zero average breakage in Hong Kong. Winning tickets of $10 pay to the nearest (higher or lower) ten cents.

Cross-Track Betting: Is the Grass Greener on the Other Side ?

Siew Meng Leong and Kian Guan Lim

Abstract

This paper reports the results of an empirical study on cross-track betting based on a sample of about 10,000 horses from 867 races in Singapore and Malaysia. Statistical test shows that the usual favorite-longshot bias exists at both the home and cross tracks. Arbitrage profit was possible in 13 races. The paper also shows that by employing a stylized version of the Kelly criterion as suggested by Hausch and Ziemba (1990a), using simultaneous home and cross track information on dividends, positive returns can be made. Implications of these results are discussed and directions for future research suggested.

1. Introduction

The use of racetrack betting data to test market efficiency and rationality has become an accepted and prevalent practice. Racetrack efficiency has been analyzed from such theoretical perspectives as economics (e.g., Ali 1977, Rosett 1965), psychology (Griffith 1949), mathematics (Isaacs 1953), and statistics (Harville 1973, Henery 1981). Various types of betting including win (Quandt 1986, Snyder 1978), place and show (Asch and Quandt 1986; Hausch and Ziemba 1985; Hausch, Ziemba, and Rubinstein 1981), and exotic wagering (Ali 1979, Asch and Quandt 1987, Hausch and Ziemba 1986) have also been subject to scrutiny.

Research has spanned the globe from the dirt tracks of North America to the grass tracks of Australia, Britain, Hong Kong, and New Zealand (Busche and Hall 1988; Dowie 1976; Tuckwell 1981, 1983). Bookmakers' odds (Bird and McCrae 1987) and data from harness races (Ali 1977) have been employed apart from the more commonly used information from parimutuel or totalisator betting and flat races.

In synthesizing this literature, Thaler and Ziemba (1988) concluded that while the racetrack may be "surprisingly efficient," there is "substantial evidence" that various forms of market efficiency have been violated (p.163).[1] The most robust empirical anomaly has been the favorite-longshot bias in which the expected returns per dollar bet increase monotonically with the probability of the horse winning. In short, favorites win more often than their subjective probabilities[2] imply whereas longshots win less frequently (see e.g., Ziemba and Hausch 1986). A more recent line of inquiry has been the uncovering of other empirical regularities such as post position bias (Canfield, Fauman, and Ziemba 1987) and the development of profitable betting strategies to exploit such inefficiencies or biases particularly in the place, show, and exotic wagering markets (Asch, Malkiel, and Quandt 1984, 1986; Hausch and Ziemba 1985, 1989, 1990b; Hausch, Ziemba, and Rubinstein 1981; Ziemba and Hausch 1987).

Continuing this stream of research is the focus on formulating an optimal betting model for exploiting the apparent inefficiencies in cross-track betting (Hausch and Ziemba 1990a). This recent development allows bettors to wager at their track (a cross track) on races run at another track (the home track). Thaler and Ziemba (1988) have noted that this form of betting raises new and interesting questions about market efficiency. Rational expectations would predict that the odds at every track would be approximately the same despite the absence of pure arbitraging opportunities and lack of rapid communication access across tracks. However, anecdotal evidence suggests that they "frequently vary dramatically" (Thaler and Ziemba 1988, p.167). One notable case involved Ferdinand in the 1986 Kentucky Derby, which offered win payoffs ranging from $13.20 to $90.00 for a $2 bet. The empirical evidence in Hausch and Ziemba (1990a) indicated that the weak efficiency of the win market at a single track did not seem to be maintained across several win markets with cross-track betting. Their analysis was based on 10 Triple Crown races involving 106 horses. This small

---

[1] See Snyder (1978) for definitions of market efficiency in the racetrack betting context.

[2] The proportion of the money in the win pool bet on any given horse can be interpreted as the subjective probability of that horse winning the race.

sample was used because cross-track betting is mainly confined to major stakes races in North America, although it is growing rapidly. From this analysis, Hausch and Ziemba (1990a) developed an optimal betting model for cross-track wagering founded on the assumption that the home track odds are accurate (after correcting for the favorite-longshot bias).

Our research augments the thus far scant literature on cross-track betting by providing the first large-scale empirical study of this new form of wagering. Unlike the races in North America, cross-track betting is the norm rather than the exception in Singapore and Malaysia where all races are so wagered. Thus, the data is voluminous and available for analyses. We assess the efficiencies of the home- versus cross-track of the win markets in Singaporean and Malaysian races. We also attempt to ascertain if the risk preference anomaly exists among Singapore and Malaysian bettors. In another Asian setting, Busche and Hall (1988) and Busche (1994) have found that Hong Kong bettors do not have a favorite-longshot bias and seem to be either risk neutral or risk averse. Section 2 of this paper briefly describes the betting markets in Singapore and Malaysia and the basic data employed in our study. Section 3 contains the empirical findings. Section 4 concludes with a discussion of the implications of our results and furnishes some directions for future research in cross-track betting.

2. Research Context and Data

Horse racing in Singapore and Malaysia is conducted under the auspices of the Malayan Racing Association. There are four grass tracks which form the racing circuit - one in Singapore and three in Malaysia at Kuala Lumpur, Ipoh, and Penang. The "market" at these centers convenes for about 30 minutes, during which participants place bets on any number of generally 10 to 12 horses in the ensuing race. In a typical race, participants can bet on each horse, either to win or place. All participants who have bet a horse to win realize a positive return on that bet only if the horse comes in first, while a place bet realizes a positive return if the horse is placed first, second, or third.[3] Regardless of the outcome, all bets carry limited liability. The payoffs are jointly determined by all the participants and a rule governing transactions costs.[4] The totalisator system in Singapore and Malaysia operates in a manner identical to the North America parimutuel system for the win pool (for a description, see Hausch, Ziemba, and Rubinstein 1981).

While there are four racing locations, only two totalisators (totes) are operated - one in Singapore and the other in Malaysia. The one in Malaysia aggregates bets from the three Malaysian racetracks. The reasons for this separation effected in 1981 are partly geographical, partly political, and partly financial given the exchange rate differences in the two currencies (the Singapore dollar and Malaysian dollar). For all races, cross betting occurs simultaneously and there is "live" simulcasting of races from the home to the cross tracks. However, betting in Singapore and Malaysia are independent in that the payoffs reflect only the betting in one country as no payoff information is transmitted between the two countries at the time the study was conducted.

Bets are placed in $5 multiples in both countries in each country's currency and payoffs range from a statutory minimum of $6 to several hundreds in $1 increments.[5] The exchange rate between the Singapore dollar and the Malaysian dollar has been fairly constant and highly predictable. In 1986, the year under study, some S$268 million was wagered in Singapore alone with a similar amount in each of the three Malaysian centers.[6] There were 867 races (269 and 598 races in Singapore and Malaysia respectively) involving 9,839 horses over

---

[3] Unlike the races in North America, there is no show betting in Singapore and Malaysia. Only the win market is of concern in this study. The place totalizator operates along the lines of the English system (see Ziemba and Hausch 1987, Chapter 17 for details).

[4] At the time this study was conducted, this was 30% of the value of the win and place tickets in Singapore (20% for government tax and 10% for the track "take" or commission), and 29.25% in Malaysia (19.25% tax and 10% take). These are much higher takes than those in North America, and similar to the 30-35% in South America.

[5] This represents a relatively high "20c breakage" as payoffs declared are rounded down to the nearest dollar for each $5 bet. Estimates of the amount of breakage are not available. In November 1988, the minimum payoff was reduced to $5. In 1991, both the Singapore and the Malaysian tracks readjusted this to $6.

[6] S$1.70 and S$1.50 are approximately equal to US$1 and C$1. Interestingly, ignoring exchange rate differences, this may be compared with the reported C$300 million wagered in 3,345 races over three years at Exhibition Park in Vancouver (Canfield, Fauman, and Ziemba 1987).

52 weekends of racing.[7] While the number of races and horses are significantly lower than the classic studies by Fabricant (1965), McGlothlin (1956), and Weitzman (1965), they are nevertheless comparable to more contemporary research efforts (e.g., each of the two data sets used by Hausch, Ziemba, and Rubinstein (1981) and that employed by Asch, Malkiel, and Quandt (1984)). Thus, the number of observations appears to form a sufficiently large sample so that more detailed analyses are possible and useful particularly in respect of cross-track betting. Further, the ``handle," or total value of all wagers on a particular track on an average race is large by North American standards (US$500,000 against US$87,000 in 1983) although much less than the US$5.6 million bet in Hong Kong in 1985-86.

3. Empirical Results

In this section, we first assess the risk preferences of Singapore and Malaysian bettors. The methodology follows that of Busche and Hall (1988). No distributional assumption is made in this case and the regression is based on odds. We then ascertain if arbitrage possibilities existed in cross-track betting in Singapore and Malaysia. Following that we implemented the optimal capital growth model in wagering and verified if positive returns could have been made. In all cases, home-track data involve horses running in Singapore and Malaysia and bet by Singapore and Malaysian bettors respectively collapsed into one group; while cross-track refers to horses running in Singapore and bet in Malaysia and vice versa aggregated in another group.

3.1 Risk Preference Assessment

The ex-ante win probability for a given horse i,

$$ P_i = \left(x_i \sum_j^n \frac{1}{x_j} \right)^{-1} $$

where x_i is the payoff on horse i when it wins and n is the number of horses in that race. The betting odds is then $\frac{1}{P_i} - 1$ to 1.

Following Busche and Hall (1988), we regress the ex-ante (betting) odds on the win odds. The win odds are computed by dividing the total number of horses into 30 groups classified by ex-ante odds.[8] The win ratio in each group j is W_j = number of horses winning / total number of horses. The win odds is then $\frac{1}{W_j} - 1$ to 1. The ex-ante odds used in the regression are the average odds of the horses in each of the 30 odds groups.

The cross-sectional betting odds regressed on the win odds using OLS produced the following results (t-statistics in parentheses):

Ex-ante odds (home track) = 3.54 + 0.45 win odds
(3.39) (-9.86)
R^2 = 0.696

---

[7] The data for this study were obtained from Punters' Way, a popular racing guide for bettors in Singapore and Malaysia.

[8] Sample sizes were roughly equal in each group. A further constraint was that no group may have zero winners as per Busche and Hall's (1988) criterion.

Ex-ante odds (cross track) =   4.91 + 0.40 win odds
                                (4.26) (-11.42)
                                R^2 = 0.668

The t-statistics for the slopes are based on the null hypotheses of unit slopes. The slope coefficients are significantly less than one in both the home track and the cross track cases. The intercepts are also significantly positive in both cases. The results indicate risk preference, and are consistent with prior research with the exception of the Hong Kong data.

We also plot the payoffs against each odds group. The payoff is the effective track payback less breakage. This is computed as the total dividends paid on all winners in each odds group divided by the total wager in each group, using one-ticket wager per horse. This is shown in Figure 1. From Figure 1, the favorite-longshot bias is observed. There was a tendency to overbet longshots and underbet favorites. This confirms our earlier findings.

FIGURE 1

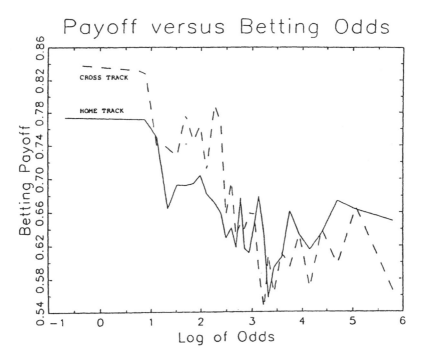

3.2 Arbitrage Analysis

We now replicate the arbitrage strategies for cross-track betting suggested by Hausch and Ziemba (1990a). The idea is that where betting on both tracks is possible simultaneously, bettors can achieve the least cost of ensuring \$1 certain payoff by betting $\dfrac{5}{x_i}$ on each horse in a race where x_i^* denotes max $[x_i^{II}, x_i^C]$ and the betting is done on the track with the higher payoff. The least cost is

$$\theta = \sum_i^k \frac{5}{x_i^*}$$

where k is the number of horses in a race.[9] The results are tabulated below, showing the number of races with varying levels of θ.

TABLE 1

Least Cost of Wagering in 1\$ certainty payoff
in Cross-Track Betting

$\theta <$	0.8	0.9	1.0	1.1	1.2	1.3	1.4	1.5	≥ 1.5
No. of Races	7	1	5	53	192	425	176	6	2

There were 13 out of 867 races (1.5%) in 1986 where arbitrage profit could have been made. In these cases, wagers of less than a dollar brought a certain \$1 payoff. However, in Singapore and Malaysia, the broadcasting of odds to other tracks was not permitted. Hence, it may not be too surprising that such arbitraging does not occur to eliminate the arbitrage potential in equilibrium. In other words, bettors can access only publicly available payoffs at their track and cannot act on those prevailing at the other track. Another possibility is that the other track's information is available privately, but that market imperfections such as communications delay and transactions costs do not allow arbitrage profit to take place.

3.3 Kelly Wagering

To determine whether this cross-track inefficiency arises out of lack of the other track's information and not market imperfection, we employ the Kelly wagering system suggested in Hausch and Ziemba (1990a).[10] In Hausch and Ziemba's approach, only information on the home track is required for wagering at the cross track. There is no need for physical transaction at the home track, thus avoiding the issue of market imperfection raised above. Kelly betting in the stylized fashion that Hausch and Ziemba demonstrated is practicable if home-track dividend information is available at the cross track. An advantage of using this wagering model is that the effect one's bets have on the aggregate payoffs is explicitly taken into account. The wagering attempts to follow the Kelly criterion, and one chooses the amounts of bets in any race by solving the following optimization problem. The program is

---

[9] It does not matter if the payoffs in different tracks are in different currencies as long as the amounts wagered are geared up to ensure a common unit payoff. The problem of currency risk is minimal as the S\$/M\$ exchange rate is very stable.

[10] See MacLean, Ziemba, and Blazenko (1992) for a discussion of the advantages of the Kelly criterion.

$$\max_{\theta_i} \sum_i^n q_i \ln\left(W_0 + Q\left(W + \sum_i^n \theta_i \right)(\theta_i / (W_i + \theta_i)) - \sum_i^n \theta_i \right)$$

subject to $\sum_i^n \theta_i \le W_0$, and $\theta_i \ge 0$, $l = 1, \ldots, n$

where W_0 is the initial wealth for betting, q_i is the ex-ante probability of a win by the i [th] horse suitably adjusted for any favorite/longshot bias using the regression equations in section 3.1, and n is the number of horses in the race.

Following Hausch and Ziemba (1990a), we assume an initial wealth of S$2,500 for each race wagered. This stylized version of Kelly wagering without updating the wealth level after each subsequent race is due to Hausch and Ziemba and we think is a more natural way in betting than to pool the entire wealth of a bettor each time.[11] W_i is the total bet on horse i, W is the total bet in a race, Q is the ratio one minus the track-take, in this case, 0.7, and θ_i is the optimal amount that should be wagered on horse l. As in Hausch and Ziemba (1990a), restrictions are placed on the optimal bets: only horses with expected returns above a pre-determined level are considered for wagering. In our study, we show results with levels of 1.1, 1.5, 2.0, and 2.5. The wagering is done on the cross track where q_i is determined using dividend payoffs from the home track.[12] Of the 867 races in 1986, 776 races are Kelly-wagered when the expected return is fixed at a minimum of 1.1. This number decreases to 223 when the minimum expected return is 2.5. The various statistics associated with applying Kelly-wagering are shown in Table 2. In Tables 3 and 4, we show the frequency breakdown of total money wagered in the races, and the frequency breakdown of total actual payoff to the wagers in the races.

From Table 3 it is seen that Kelly wagering total bets were mostly below $1000. Total high-end bets were less frequent. From Table 4, it is clearly evident that bettors lose "small" but win "big", i.e. they lose more often, but make very large amounts when they win. For example, there were 6 races with actual payoff of over S$5000 in the case with minimum expected return of 2.0. In the case with minimum expected return of 2.5, "big" winners can be seen from the results in Table 5. In all cases the wagers placed did not materially influence the payoffs given the large win pools in question, although such adjustments have been taken into account.

TABLE 2

Kelly-Wagering Characteristics

Expected Return	> 1.1	> 1.5	> 2.0	> 2.5
No. of races wagered	776	558	357	223
No. of races with winner	69	21	12	6
No. of winning races with 1 bet	14	7	5	4
No. of races with no bet	91	309	510	644
No. of races with one bet	212	175	198	204
No. of races with two bets	530	376	159	19
No. of races with three bets	34	7	0	0

[11] After all, the optimal growth model as propounded by financial economists was meant to be an investment model for lifetime consumption - unless we live in a world in which securities markets do not exist, but only the race track exists for risk-taking and "investment", then perhaps it makes sense to wager based on total wealth updated as if no other investments existed. Moreover, even if updating was done, it would never be an exactly optimal growth model - as Hausch and Ziemba (1990a) put it, (p.72), "...however, that due to the approximations in the model it was more profitable to wager only if the expected return was well in excess of 1.00. For large tracks and races of the quality of these Triple Crown races, a minimum expected return of 1.10 is reasonable." The minimum expected return cutoff would result in suboptimal wagering, not exactly the optimal growth model intended.

[12] See Hausch and Ziemba (1990a) for the reasons why the home track information is used.

TABLE 3

Frequencies of Total Wagering Amounts in S$'s

Expected Return	> 1.1	> 1.5	> 2.0	> 2.5
Total Wager < 250	376	438	313	209
Total Wager < 500	171	58	21	7
Total Wager < 750	98	42	14	3
Total Wager < 1000	78	10	2	0
Total Wager < 1250	28	5	3	0
Total Wager < 1500	13	1	0	0
Total Wager < 1750	7	1	0	0
Total Wager < 2000	1	0	1	1
Total Wager < 2250	2	1	1	1
Total Wager < 2500	2	2	2	2

TABLE 4
Frequencies of Actual Payoff Amounts in S$'s

Expected Return	> 1.1	> 1.5	> 2.0	> 2.5
Payoff ≤ −2300	1	1	1	1
Payoff ≤ −2000	2	1	1	1
Payoff ≤ −1500	5	1	0	0
Payoff ≤ −1000	31	6	3	0
Payoff ≤ −500	142	44	14	3
Payoff ≤ −400	61	16	3	1
Payoff ≤ −300	64	26	14	4
Payoff ≤ −200	78	35	10	9
Payoff ≤ −100	114	93	47	24
Payoff ≤ 0	211	315	252	174
Payoff ≤ 100	0	0	0	0
Payoff ≤ 200	0	0	0	0
Payoff ≤ 300	0	0	0	0
Payoff ≤ 400	1	0	1	1
Payoff ≤ 500	1	1	1	1
Payoff ≤ 1000	2	0	0	0
Payoff ≤ 1500	1	2	2	0
Payoff ≤ 2000	3	0	0	0
Payoff ≤ 2500	14	0	0	0
Payoff ≤ 5000	40	11	2	1
Payoff ≤ 10000	4	5	5	2
Payoff ≤ 20000	1	1	1	1

TABLE 5
Winners in the Case of Expected Return > 2.5

Winning Odds	$ Dividends	$ Bet Amount	$ Payoff	Rate of Return
0.4 : 1	22	110	374	2.4
19 : 1	267	80	4192	51.4
26 : 1	414	85	6953	80.8
5 : 1	77	35	504	13.4
24 : 1	551	100	10920	108.2
32 : 1	596	75	8865	117.2

From the actual payoffs in the races wagered, a wealth series is constructed for each case of the minimum expected return. The series is an accumulation of the payoffs due to the season's betting. For convenience of presentation, the starting point of the series is chosen as S$50,000. The 4 series corresponding to the minimum expected returns of 1.1, 1.5, 2.0, and 2.5, are shown in Figures 2a, 2b, 2c, and 2d respectively.

FIGURE 2a

Kelly Wagering Wealth Series 1

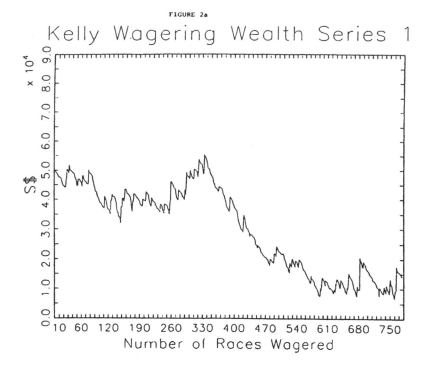

Number of Races Wagered

FIGURE 2b

Kelly Wagering Wealth Series 2

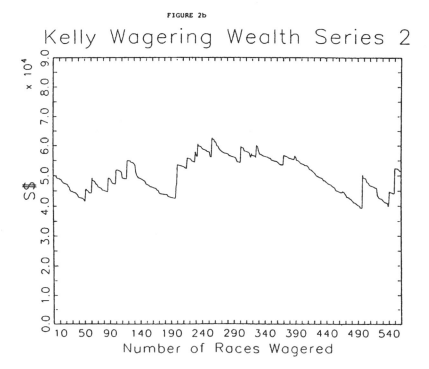

Number of Races Wagered

FIGURE 2c

Kelly Wagering Wealth Series 3

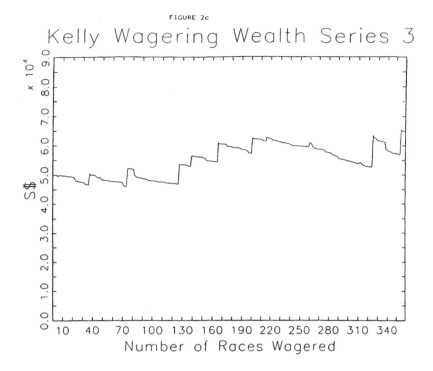

FIGURE 2d

Kelly Wagering Wealth Series 4

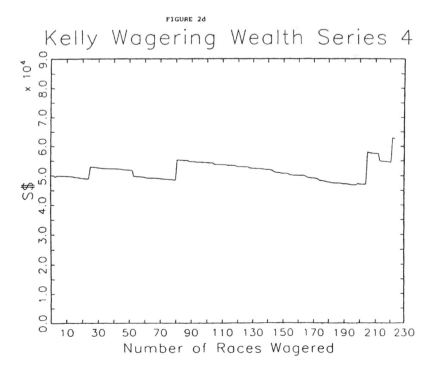

From the Figures, it is clear that when Kelly wagering in the stylized version used by Hausch and Ziemba (1990a) is used subject to bets only on horses with a minimum expected return of 1.5 or more, some profits can be made. When the expected return restriction is 1.1, there was a loss of -S$33,370 (-11%) based on the total bet across all the wagered races. For a minimum expected return of 1.5, there was a small profit of S$3880 for a 3.7% return. However, for minimum expected returns of 2.0 and 2.5, the gains were S$17,400 and S$15,140 respectively. These posted returns of 37% and 65% respectively, which are impressive. This is particularly so considering the high track take of 30%. Hausch and Ziemba (1990a) report returns ranging from 9.2% to 15% with lower North American track takes of 15-18%. For the case of minimum expected return of 2.5, the average payoff per race is $68. Using the 223 races wagered, the t-statistic of this average payoff based on a null hypothesis of zero expected payoff per race is 0.77, or a p-value of 0.22. However, since the payoff distribution is non-stationary across races, the t-statistic is not applicable. We use Monte Carlo technique to generate the probability distribution of the average payoff and, based on the same null, found the p-value to be smaller at 0.14. Thus, the profits from the Kelly wagering scheme during this period are seen to be statistically greater than zero at the 15 percent significance level. Considering the track take, a naive betting scheme would yield an expected return of -30 percent; the Kelly wagering scheme is statistically significantly superior to the naive betting scheme at 10 percent significance level (a p-value of 0.07).

The Kelly criterion is seen to produce high returns, as suggested by Hausch and Ziemba (1990a). The wagering scheme essentially exploits the difference in the dividends of a horse on home versus cross track. This difference is of course due to a lack of information transfer across both tracks, which was also enforced by the authorities at the time of this study. If we assume correctly that the home track dividends, after any necessary favorite-longshot bias adjustment, reflect an accurate assessment of the win probabilities, then the availability of this home track information for betting on the cross track is a highly profitable proposition. The Kelly wagering scheme in essence suggests betting heavily on those horses with high win probabilities based on home track information, and as well having high dividends or low win probabilities on the cross track. The profits that we show imply that the betting market in Singapore and Malaysian race tracks is inefficient because of lack of the other track's information, not so much because of market imperfection. This argument could be elaborated as follows. Earlier in section 3.2 we showed the existence of arbitrage profit which could be due to informational inefficiency or due to high transactions costs in simultaneous betting across markets, which we termed market imperfection. In the stylized Kelly wagering here, only betting in the cross track is necessary, so there is no cost of market imperfection in the way we define. However, there is possibility of informational inefficiency. Since some large profits appeared to be possible, we suggest that this was likely due to informational inefficiency, not market imperfection. However, if no significant profits were observable, then we would suggest that the arbitrage profits would then be due to market imperfection, since there appeared to be informational efficiency at least in the Kelly wagering scheme.

4. Discussion

4.1 Summary of Findings

There are three major findings from our research on cross-track betting in Singapore and Malaysia. First, bettors in Singapore and Malaysia were found to also underbet favorites and overbet longshots. Second, significant arbitrage opportunities were available in 1.5% of the races using a simple strategy of betting the higher payoff for each horse offered by the home and cross track, assuming simultaneous betting is permissible and possible at the home and cross tracks. Third, it was also shown that where home track information is available to cross-track bettors, the astute bettors may make abnormal profits by utilizing the information at the home track while wagering at the cross track employing Kelly betting. However, economically significant profits were possible only when restrictions to bets on horses with high expected returns were imposed. Our study suggests that information alone without access to betting on the home and cross tracks simultaneously would seem adequate for profit-making. It is interesting to note that the betting market inefficiency observed was due largely to information rather than transaction costs.

4.2 Implications

Several implications of these findings are noteworthy of discussion. First, the robust empirical anomaly of the favorite-longshot bias was found in this study. This indicates that the Busche and Hall (1988) ``anomaly of an anomaly'' may be isolated in Hong Kong and not generalizable to other Asian racing centers.

Our results provide an initial estimate of the existence of "locks" (or races in which it is possible to construct a risk-free hedge) in cross-track betting. In our study, 13 races (or 1.5% of the total number of races) could be profitably arbitraged using the simple betting rule in Hausch and Ziemba (1990a). We also show that if home track information is accessible to cross-track bettors, using Kelly wagering on the cross-track would yield some profits. However, astute bettors need be highly selective in implementing Kelly wagering. Specifically, although a lower expected return cut-off may be set for Triple Crown events, our findings suggest that a much higher threshold may be required for races of more typical quality and at tracks with higher takes.

Nonetheless, that these potential profits were available indicated that there must be some strong barriers to information transfer between the two tracks. Moreover, there must have been equally strong incentives to overcome the blockage. Indeed, racetracks in both countries have strictly denied any communication devices on their premises, even though cross-track betting is not strictly illegal. Police enforcement is active in raiding and shutting down illegal communications centers operated by unauthorised parties or individuals. In addition, as the telecommunication networks of both countries are state-owned, close monitoring is feasible with regard to their use for illegal gambling (bookmaking is outlawed in both countries). Payoff information from the other track was also not available to bettors in Singapore and Malaysia by mutual agreement of the race tracks. This made it very difficult for the small-time bettor to employ information from the other track to make abnormal profits.

It is difficult to ascertain the reasons for the blackout policy. Clearly, the lifting of the blackout of payoff information would enhance market efficiency in the race tracks of both countries. The welfare of small-time bettors would thus be raised, although this would also eradicate any possibility of successfully employing arbitraging strategies of the type developed here.

4.3 Future Research

Several suggestions for future research in cross-track betting can be furnished given the encouraging findings reported here. First, more data can be collected for added longitudinal insights both in Singapore and Malaysia as well as for other locations where this form of betting is available. In Singapore and Malaysia, several recent developments (e.g. reduction of government betting taxes to differing degrees in both countries) may produce a different pattern of findings than those reported here.[13] In addition, the place totalisator may be analyzed for cross-track inefficiency. Research may also be extended to exotic wagers including forecast and tierce betting[14] which may yield additional insights into cross-track betting efficiency.

---

[13] The current total take is 23% (15% tax and 8% commission) for all types of bets in Singapore.

[14] Forecast and tierce betting are equivalent to quinella and trifecta betting in North America. The former involves selecting two horses per race and payoffs are made should both finish first and second in any order in a race. The latter involves selecting three horses per race and payoffs are made should all three finish in first, second, and third positions in the order specified by the bettor. In 1986, the takes for these exotic wagers were 29.25% in Malaysia and 40% in Singapore. They are currently 23% and 29.25% in Singapore and Malaysia respectively.

REFERENCES

ALI, M. M., Probability and Utility Estimates for Racetrack Bettors, J. Political Economy, 85 (August 1977), 803-815.

_____ , Some Evidence of the Efficiency of a Speculative Market, Econometrica, 47 (March 1979), 387-392.

ASCH, P., B. G. MALKIEL, AND R. E. QUANDT, Market Efficiency in Racetrack Betting, J. Business, 57 (April 1984), 165-175.

_____ , Market Efficiency in Racetrack Betting: Further Further Evidence and A Correction, J. Business, 59 (January 1986), 157-160.

_____ AND R. E. QUANDT, Racetrack Betting: The Professors' Guide to Strategies, Dover, MA: Auburn House, 1986.

_____ , Efficiency and Profitability in Exotic Bets, Economica, 59 (1987), 278-298.

BIRD, R. AND M. McCRAE, Tests of the Efficiency of Racetrack Betting Using Bookmaker Odds, Management Science, 33 (December 1987), 1552-1562.

BUSCHE, K., Efficient Market Results in an Asian Setting, (1994), in this Volume.

_____ AND C. D. HALL, An Exception to the Risk Preference Anomaly, J. Business, 61 (March 1988), 337-346.

CANFIELD, B. R., B. C. FAUMAN, AND W. T. ZIEMBA, Efficient Market Adjustment of Odds Prices to Reflect Track Biases, Management Science, 33 (November 1987), 1428-1439.

DOWIE, J., On the Efficiency and Equity of Betting Markets, Economica, 43 (1976), 139-150.

FABRICANT, B. F., Horse Sense, New York: David McKay Co., 1965.

GRIFFITH, R. M., Odds Adjustments by American Horse Race Bettors, Amer. J. Psychology, 62 (April 1949), 290-294.

HARVILLE, D. A., Assigning Probabilities to the Outcomes of Multi-Entry Competitions, J. Amer. Statist. Assoc., 68 (June 1973), 213-216.

HAUSCH, D. B. AND W. T. ZIEMBA, Transactions Costs, Extent of Inefficiencies, Multiple Bets and Entries in a Racetrack Betting Model, Management Science, 31 (April 1985), 381-394.

_____ Arbitrage Strategies for Cross-Track Betting on Major Horseraces, Journal of Business, 63 (No. 1, Pt. 1, 1990a) 61-78.

_____ Locks at the Racetrack, Interfaces, 20 (May-June 1990b), 41-48.

_____ AND M. RUBINSTEIN, Efficiency of the Market for Racetrack Betting, Management Science, 27 (December 1981), 1435-1452.

HENERY, R. J., Permutation Probabilities as Models for Horse Races, J. Royal Statistical Society B , 43 (No 1, 1981), 86-91.

ISAACS, R., Optimal Horse Race Bets, Amer. Mathematical Monthly, (1953), 310-315.

MacLEAN, L.C., W.T. ZIEMBA, AND G. BLAZENKO, Growth versus Security in Dynamic Investment Analysis, Management Science, 38 (November 1992), 1562-1585.

McGLOTHLIN, W. H., Stability of Choices Among Uncertain Alternatives, Amer. J. Psychology, 69 (December 1956), 604-615.

QUANDT, R. E., Betting and Equilibrium, Quarterly J. of Economics, 101 (February 1986), 201-207.

ROSETT, R. N., Gambling and Rationality, J. Political Economy, 73 (1965), 595-607.

SNYDER, W. W., Horse Racing: Testing the Efficient Markets Model, J. Finance, 23 (September 1978), 1109-1118.

THALER, R. H. AND W. T. ZIEMBA, Parimutuel Betting Markets: Racetracks and Lotteries, J. Economic Perspectives, 2 (Spring 1988), 161-174.

TUCKWELL, R. H., Anomalies in the Gambling Market, Australian J. Statist., 23 (December 1981), 287-295.

_____ The Thoroughbred Gambling Market: Efficiency, Equity and Related Issues, Australian Economic Papers, 22 (June 1983), 106-118.

WEITZMAN, M., Utility Analysis and Group Behavior: An Empirical Study, J. Political Economy, 73 (February 1965), 18-26.

ZIEMBA, W. T. AND D. B. HAUSCH, Betting at the Racetrack, New York: Norris Strauss, 1986.

_____ , Dr. Z's Beat the Racetrack, New York: William Morrow, 1987.

Annotated Additional Bibliography

In the following descriptions, those references marked with superscript [1] are included in this volume while those marked with superscript [2] are on this additional bibliography.

1. Algoet, P.H. and Cover, T.M. (1988) "Asymptotic Optimality and Asymptotic Equipartition Properties of Log-Optimum Investment." *The Annals of Probability* 16, 875-898.

   Breiman's (1961)[2] theoretical results for the "Kelly criterion" are generalized in this paper.

2. Asch, P. and Quandt, R. (1988) Betting Bias in 'Exotic' Bets." *Economics Letters* 28, 215-219.

   This paper tests the unbiasedness of the probabilities implied by exacta and trifecta bet fractions. Using simple linear regressions on the relationship between the objective probabilities (estimated by the corresponding win frequencies) and the average bet fractions (or the subjective probabilities), they draw different conclusions for the two pools. For the exacta pool, the implied probabilities appear to be unbiased. However, the trifecta bet fractions appear to be a weaker approximation of the objective probabilities and some clear over/underbetting pattern appears to exist.

3. Arvesen, J.N. and Rosner, B. (1971) "Optimal Pari-Mutuel Wagering." in Gupta, S.S. and Yackel, J., *Statistical Decision Theory and Related Topics*. Academic Press, 239-254.

   This paper analyzes the betting problem in a Bayesian framework where information and prior win probabilities are combined to form posterior win probabilities. The objective is then to minimize the Bayes risk with different sets of prior probabilities. The paper is an early example in the decision analysis literature of combining the opinions of experts.

4. Arvesen, J.N. (1973) "Multivariate Analysis of Horse Race Betting." in Proceedings - American Statistical Association : Business and Economic Statistics.

   Using a Bayesian decision rule derived in Arvesen & Rosner (1971)[2] and a small data set, the author develops a betting rule that is shown to be profitable.

5. Aucamp, D.C. (1993) "Number of Plays to Achieve Superior Performance with Geometric Mean Strategy." *Management Science* 39, 1163-1172.

   In theory, the Kelly criterion will "eventually" beat any different strategy almost surely in the long run. This paper determines the number of plays required to ensure that the Kelly criterion will beat any different strategy with a high probability. It demonstrates that "eventually" can be quite long in risky situations.

6. Bacon-Shone, J.H., Lo, V.S.Y. and Busche, K. (1992a) "Modelling Winning Probability." Research Report 10, Department of Statistics, University of Hong Kong.

   A simple logit model is proposed to relate win probabilities and win bet fractions. This model is shown to outperform some previous models. Empirical results are obtained for racetracks in the U.S., Japan, Hong Kong and China. No strong conclusions for bettor risk preference can be supported.

7. Bacon-Shone, J.H., Lo, V.S.Y. and Busche, K. (1992b) "Logistic Analyses for Complicated Bets." Research Report 11, Department of Statistics, University of Hong Kong.

   With data sets from the U.S. and Hong Kong, logit models are applied to compare the accuracies of different models for predicting ordering probabilities. The empirical results suggest that the model based on independent normal running times is superior.

8. Bird, R., McCrae, M. and Beggs, J. (1987) "Are Gamblers Really Risk Takers?" *Australian Economic*

Papers 26, 237-253.
     Using fixed-odds data from Melbourne racetracks, the authors analyze the subjective probabilities derived from different fixed odds at four different times during the betting period. In a two moment model, they find bettors to be risk seeking. In a three moment model, however, they find that bettors are averse to variance but seek skewness.

9. Bochonko, R. and Rosenbloom, E.S. (1992) "An Arbitrage Model for the Racetrack." Working paper. University of Manitoba.
     A linear programming model is proposed to create a risk-free hedge by betting on different pools: win, exacta and double.

10. Breiman,L. (1960) "Investment Policies for Expanding Businesses Optimal in a Long-Run Sense." *Naval Research Logistics Quarterly* 7, 647-651.
     Brieman considers the asymptotic properties of the "Kelly criterion." The Kelly criterion is shown to minimize the expected time to achieve a fixed wealth level asymptotically, and the asymptotic wealth exceeds that of any other different strategy.

11. Breiman,L. (1961) "Optimal Gambling Systems for Favorable Games." in *Proceedings of the Fourth Berkeley Symposium, Mathematical Statistics and Probability* 1, 65-78. University of California Press.
     This paper contains more theoretical results on the properties of the "Kelly Criterion." In addition to asymptotic properties, the author also considers the finite case for his coin-tossing example. The Kelly criterion's wide acceptance owes much to this influential paper.

12. Cover, T.M. (1984) "An Algorithm for Maximizing Expected Log Investment Return." *IEEE Transactions on Information Theory* IT-30, 369-373.
     An algorithm for maximizing the expected log return of any portfolio is developed. The algorithm monotonically improves the expected log return.

13. Crafts, N.F.R. (1985) "Some Evidence of Insider Knowledge in Horse Race Betting in Britain." *Economica* 52, 295-304.
     Crafts argues that the test used by Dowie (1976)[1] is inappropriate. He develops a new test and concludes that the British fixed-odds system enables profitable arbitrage at prices different from the starting price (SP), that is, profitable insider betting exists that is not available to SP bettors.

14. Dansie, B.R. (1983) "A Note on Permutation Probabilities." *Journal of Royal Statistical Society B* 45, 22-24.
     This paper shows that the Harville model (Harville (1973)[1]) is implied by assuming that the underlying running time distributions are independent exponential.

15. Dolbear, F.T. (1991) "Is Racetrack Betting on Exactas Efficient?" Working paper, Department of Economics, Brandeis University.
     It has long been thought that "smart money" is attracted to the exotic market because the informational content of a bet made in that market is not easily discerned by the public. Asch and Quandt (1987)[1] find evidence of this phenomenon but Dolbear shows a bias in their comparison of theoretical and subjective probabilities for exacta bets, a bias that may vitiate their result.

16. Eisenberg, E. and Gale, D. (1959) "Consensus of Subjective Probabilities : The Pari-Mutuel Method." *The Annals of Mathematical Statistics* 39, 165-168.
     It is shown that bet fractions can be considered a consensus of subjective probabilities of the

public in equilibrium.

17. Epstein, R.A. (1977) *Theory of Gambling and Statistical Logic*. 2nd ed. Academic Press
   This classic book is a definitive source of information on the mathematics of gambling.

18. Fama, E. (1970) "Efficient Capital Markets: A Review of Theory and Empirical Work." *Journal of Finance* 25, 383-417.
   A survey of the research on stock market efficiency.

19. Fama, E. (1991) "Efficient Capital Markets II." *Journal of Finance* 46, 1575-1617.
   An update of Fama's 1970 paper. Surveyed is recent work on return predictability for time-varying processes, cross-sectional return predictability, event studies, and tests for private information.

20. Friedman, M. and Savage, L.J. (1948) "The Utility Analysis of Choices Involving Risk." *Journal of Political Economy* 56, 279-304.
   A particular form of utility function is proposed which can accommodate both gambling and insurance behavior of the public. The utility function is concave below our current wealth. Beyond our current wealth it is initially convex but eventually concave.

21. Hakansson, N.H. (1971) "Capital Growth and the Mean-Variance Approach to Portfolio Selection." *Journal of Financial and Quantitative Analysis* 6, 517-557.
   The mean-variance approach is compared with the capital growth criterion (i.e. the "Kelly criterion"). For an example, it is found that the two criteria are not close and the mean-variance criterion may lead to ruin in the long run.

22. Hausch, D.B. and Ziemba, W.T. (1994) "Efficiency of Sports and Lottery Betting Markets." In *Finance Handbook*, eds., Jarrow, R.A., Maksimovic, V. and Ziemba, W.T. North Holland Series, forthcoming.
   A survey of the efficiency of various gambling markets. Included are horseracing, baseball, football, basketball and lotteries.

23. Henery, R.J. (1984) "An Extreme-Value Model for Predicting the Results of Horse Races." *Applied Statistics* 33, 125-133.
   Henery examines the potential of an extreme-value running time model to predict the win probabilities. Using U.K. fixed-odds system win bet data, only the tail of the empirical distribution function is consistent with the model.

24. Henery, R.J. (1985) "On the Average Probability of Losing Bets on Horses with Given Starting Price Odds." *Journal of Royal Statistical Society A* 148, 342-349.
   An attempt at explaining the favourite-longshot bias in the U.K. market.

25. Hoerl, A.E. and Fallin, H.K. (1974) "Reliability of Subjective Evaluations in a High Incentive Situation." *Journal of Royal Statistical Society A* 137, 227-230.
   Compares the subjective win probabilities (average win bet fractions) and the relative frequencies of winnings conditional on the race size. Subjective win probabilities are shown to undervalue the larger win probabilities and overvalue the smaller win probabilities, consistent with the favorite-longshot bias.

26. Kahneman, D. and Tversky, A. (1984) "Choices, Values and Frames." *American Psychologist* 39, 341-350.

This paper discusses how framing decision problems in different but equivalent ways can lead to different choices. "Mental accounting" explains some anomalous behavior.

27. Kallberg, J.G. and Ziemba, W.T. (1983) "Comparison of Alternative Utility Functions in Portfolio Selection Problems." *Management Science* 29, 1259-1276.
This paper examines the effect of alternative utility functions and parameter values on the composition of the optimal investment portfolio.

28. Kelly,J.L. (1956) "A New Interpretation of Information Rate." *Bell System Technical Journal* 35, 917-926.
The paper that proposed the "Kelly criterion."

29. Kendall, M.G. (1953) "The Analysis of Economic Time-Series, Part I: Prices." *Journal of the Royal Statistical Society* 96 (Part I), 11-25.
Examines the behavior of industrial share prices and spot prices for cotton and wheat. After extensive analysis of serial correlations, prices appear to follow a random walk.

30. Lane, D. and Ziemba, W.T. (1992) "Jai Alai Hedging Strategies." Working paper, University of British Columbia.
A hedging strategy is developed for betting in the game of team Jai Alai. Like the fixed-odds system for horseracing in Britain, fixed-odds bets may be placed before every point is played.

31. Lo, V.S.Y. and Bacon-Shone, J.H. (1993) "Approximating the Ordering Probabilities of Multi-Entry Competitions by a Simple Method." Working Paper, Department of Statistics, University of Hong Kong.
Proposed is a simple approximation to the Henery and Stern models for computing ordering probabilities. The accuracy of the approximation is empirically tested on different data sets and the approximation itself is applied in Lo, Bacon-Shone and Busche (1994)[2].

32. Lo, V.S.Y. and Bacon-Shone, J.H. (1994) "A Comparison Between Two Models for Predicting Ordering Probabilities in Multi-Entry Competitions." *The Statistician*, forthcoming.
Using Hong Kong data, conditional logistic analysis is applied to show that the Henery model is superior to the Harville model for predicting ordering probabilities. It is also proved that if running times of horses are normally distributed, the probabilities produced by the Harville model systematically overestimate (underestimate) probabilities associated with favorites.

33. Lo, V.S.Y., Bacon-Shone, J.H. and Busche, K. (1994) "An Application of Ranking Probability Models to Racetrack Betting." *Management Science*, forthcoming.
In the betting system of Hausch, Ziemba and Rubinstein (1981)[1], the Harville model for ordering probabilities is replaced by the approximation method developed by Lo and Bacon-Shone (1993)[2]. Using three data sets from Meadowlands, Hong Kong and Japan, the empirical results generally show improved profit. To use this modified system at racetrack, however, we need all the win odds. Hence, a scanner or other computer facility is needed in order to make the system operational.

34. MacLean, L.C. and Ziemba, W.T. (1991) "Growth-Security Profiles in Capital Accumulation Under Risk." *Annals of Operations Research* 31, 501-510.
This paper focuses on the "fractional Kelly criterion." It shows how growth of capital can be traded for security with simple strategies generated from the optimal growth and optimal security problems.

35. Maier, S.F., Peterson, D.W. and Weide, J.H.V. (1977) "A Monte Carlo Investigation of Characteristics of Optimal Geometric Mean Portfolios." *Journal of Financial and Quantitative Analysis* 12, 215-233.
Some characteristics of the optimal geometric mean portfolios are studied using simulations. The optimal portfolio depends on assumptions of the market and opportunities of borrowing and lending. The optimal portfolio can be approximated by heuristic portfolio building rules.

36. Markowitz, H. (1952) "The Utility of Wealth." *Journal of Political Economy* 60, 151-158.
Amends the utility function proposed by Friedman and Savage (1948)[2] to explain more general risk behavior.

37. McCardell, M. (1992) "Easy Money." *Equity*, Vancouver.
Describes the Dr. Z system (Ziemba and Hausch (1984,1987)[2]).

38. McCulloch, B. and Van Zijl, T. (1986) "Direct Test of Harville's Multi-Entry Competitions Model on Race-Track Betting Data." *Journal of Applied Statistics* 13, 213-220.
The authors utilize the fact that the payoff to a show bet in New Zealand is independent of which other horses finish in-the-money and thus the show bet fraction is used as a direct estimate of the probability of showing. This estimate is compared to that given by the Harville (1973)[1] model, assuming that the show bet fraction is a good proxy for the probability of showing. The results indicate that the Harville model overestimates (underestimates) the probabilities of showing for high (low) values.

39. Mordin, N. (1992) "Grab Your Place in the Pool!" *The Sporting Life*.
A short article reviewing the Dr. Z system in England.

40. Quandt, R.E. (1992) "On the Distribution of Horse Qualities at Racetracks: An Analysis of Cournot-Nash Equilibria." Working paper, Princeton University.
A game theoretic approach is used to explain the observed pattern of win bet amounts. A simulation study with particular parameters generates data similar to two data sets from Atlantic City and Meadowlands.

41. Roberts, H.V. (1959) "Stock Market Patterns and Financial Analysis: Methodological Suggestions." *Journal of Finance* 14, 1-10.
Following Kendall (1953)[2], Roberts underlined the implications of the random walk model for stock market research and financial analysis.

42. Rosenbloom, E.S. (1992) "Picking the Lock: A Note on 'Locks at the Racetrack'." *Interfaces* 22, 15-17.
Provides counter-examples, albeit highly unlikely ones (e.g., a race where most of the horses fail to finish the race), to show that the risk-free hedging possibilities described in Hausch and Ziemba (1990)[1] are not precisely risk-free.

43. Rosner, B. (1975) "Optimal Allocation of Resources in a Pari-Mutuel Setting." *Management Science* 21, 997-1006.
Proposes an algorithm for betting in a pari-mutuel situation when true win probabilities are known. The objective is to maximize the expected log return.

44. Shin, H.S. (1991) "Optimal Betting Odds Against Insider Traders." *The Economic Journal* 101, 1119-1185.

A game is constructed with a bookmaker setting odds, an informed punter (the Insider), and a continuum of uninformed punters. The race is assumed to be between two horses. The optimal odds and the value of the insider's information are shown to depend on the insider's wealth and the true winning probabilities of the horses. The general case of n horses is considered in Shin (1992)[1].

45. Shin, H.S. (1993) "Measuring the Incidence of Insider Trading in a Market for State-Contingent Claims." *The Economic Journal.*
    Assuming the existence of insiders who know with certainty which horse will win, a relationship is derived between the objective and subjective probabilities under the U.K. fixed-odds system. The relationship exhibits the favorite-longshot bias. An iterative regression estimation procedure for the fraction of insiders is also proposed. Empirically, the author finds that this fraction is significantly different from zero.

46. Skinner, M. (1989) "Easy Money." *OMNI*, May, 42-49.
    Discusses the Dr. Z system. Also summarizes the history of the development of this betting system and describes the dosage theory of breeding.

47. Snyder, W. (1978) "Decision-Making with Risk and Uncertainty: The Case of Horse Racing." *American Journal of Psychology* 91, 201-209.
    Confirms the favorite-longshot bias by plotting returns versus win odds. Moreover, the author finds that expert handicappers tend to increase the favorite-longshot bias. Neither the public's nor the experts' predictions can be used to make a profit. This paper is similar to the one by the same author in this book.

48. Stein, W.E. and Mizzi, P.J. (1991) "Estimation of the Degree of Market Inefficiency in Place and Show Betting at the Racetrack." *Mathematical and Computer Modelling* 15, 55-63.
    Bounds are derived on the percentage bet to win, place and show for each of the first three finishers where the bet amount is based on the system of Hausch, Ziemba and Rubinstein (1981)[1]. Bounds on the estimated expected profit are also obtained. These bounds are useful in retrospective studies of wagering.

49. Stern, H. (1988) "Gamma Processes, Paired Comparisons and Ranking." in *Computer Science and Statistics: Proceedings of the 20th Symposium on the Interface*, 635-639.
    This paper makes use of the gamma distribution for two situations: paired comparisons and ranking. The latter leads to prediction of ordering probabilities proposed by Stern (1990)[2].

50. Stern, H. (1990) "Models for Distributions on Permutations." *Journal of the American Statistical Association* 85, 558-564.
    The Gamma distribution with a predetermined shape parameter r is proposed to model the running times of horses. Harville (1973)[1], with r = 1, is a special case. While the resulting ordering probabilities for general r are complicated, a small empirical analysis shows that r = 2 fits the data better than r = 1. A modified version of this paper appears in this volume.

51. Thalheimer, R. and Ali, M.M. (1994) "The Demand for Parimutuel Horse Race Wagering and Attendance." *Management Science*, forthcoming.
    The demand for parimutuel horse race wagering and attendance is statistically modelled for both thoroughbred and standardbred tracks, using Ohio and Kentucky data from 1960-87. Several covariates are considered: track take, racing quality, personal income, number of racing days, and competition from a state lottery and from professional sports. Most of these factors are significant in determining both

wagering and attendance at the track. For example, a state lottery has a substantial negative impact on the demand for parimutuel betting.

52. Tuckwell, R.H. (1983) "The Thoroughbred Gambling Market: Efficiency, Equity and Related Issues." *Australian Economic Papers* 22, 106-118.

Using bookmakers' odd data from Sydney and Melbourne, Tuckwell showed that if bettors wager on horses that are not initially offered at short-odds but that eventually over the betting period become favorites, then they can achieve profits.

53. Tversky, A. and Kahneman, D. (1986) "Rational Choice and the Framing of Decisions." *Journal of Business* 59, S251-S278.

Prospect theory is proposed as an explanation of the betting behavior of the public. For example, to compute an expectation, they use a set of decision weight instead of probabilities to reflect the framing and evaluation of the bettor.

54. Vannebo, O. (1980) "Horse Racing : Testing the Efficient Markets Model: Comment." *Journal of Finance* 35, 201-202.

Regarding Snyder's (1978)[1] analysis, this paper comments that the skewness of rate of return is also important in addition to the mean and standard deviation of the return. The author suggests that the biased expected returns is a result inherent in rational behavior towards risk and will be exhibited even if the betting market is efficient.

55. Vergin, R.C. (1977) "An Investigation of Decision Rules for Thoroughbred Race Horse Wagering." *Interfaces* 8, 34-45.

Tests several filter rules for betting.

56. von Neumann, J. and Morgenstern, O. (1944) *Theory of Games and Economic Behavior*. Princeton, N.J.: Princeton University Press.

A classic book on decision theory under risk and game theory.

57. White, E.M., Dattero, R. and Flores, B. (1992) "Combining Vector Forecasts to Predict Thoroughbred Horse Race Outcomes." *International Journal of Forecasting* 8, 595-611.

An approach is offered for combining information from different sources to forecast the outcome of a horse race. Sources considered are the final win odds, handicappers' predictions, and the past racing histories of the horses.

58. Wills, K.E. (1964) "Optimum No-Risk Strategy for Win-Place Pari-Mutuel Betting." *Management Science* 10, 574-577.

A risk-free betting system is proposed for win and place betting. Linear programming provides the maximum profits regardless of which horses win and place. For a basic feasible solution to exist, though, it is necessary that the fractions bet on a horse to win and place be quite different.

59. Ziemba, W.T. (1994) "World Wide Security Market Regularities." *European Journal of Operations Research*, April.

Survey of systematic violations of stock market efficiencies in the U.S., Japan and other world wide equity markets. Included are the January small firm effect, and the turn-of-the-year and turn-of-the-month effects.

Annotated Bibliography of Handicapping Books

Ainslie,T. (1978) *Ainslie's Encyclopedia of thoroughbred handicapping*. New York: Morrow.
This book contains brief abstracts about many horseracing topics.

Ainslie,T. (1986) *Ainslie's complete guide to thoroughbred racing*. 3rd edition. New York: Simon & Schuster.
A revised edition of an early guide to handicapping fundamentals.

Asch,P. and Quandt,R. (1986) *Racetrack betting: The professor's guide to strategies*. Dover, MA: Auburn House.
This book proposes some betting strategies. It covers statistical models for place and show betting and how to apply them.

Beyer,A. (1975) *Picking Winners*. New York: Houghton Mifflin Co.
A good source of information on speed handicapping.

Beyer,A. (1978) *My $50,000 year at the races*. New York: Harcourt, Brace, Jovanovich, Inc.
This book contains a story of this famous handicapper's attempts to beat the races over a full season.

Beyer,A. (1983) *The winning horseplayer*. New York: Houghton Mifflin Co.
An introduction to trip handicapping written in an entertaining style.

Beyer,A. (1993) *Beyer on speed*. Houghton Mifflin.
This book shows explains how to use speed figures in wagering.

Fabricand,B.P. (1965) *Horse sense*. New York : David McKay Co.
This trade book presents an interesting theory of horse-race betting based on the Rule of Similarity.

Fabricand,B.P. (1979) *The science of winning : A random walk on the road to riches*. Van Nostrand Reinhold.
This book discusses the possibility of making profits in different gambling and risky situations such as horse-racing and stock market. By exploiting the favorite-longshot bias in horse-racing, the author suggests some rules (the Rule of Similarity) that may make profits by betting on favorites.

Herbert,I. (1980) *Horse racing: The complete guide to the world of the Turf*. New York: St. Martin's Press.
Lavish pictorial book with useful information concerning the sport of kings.

Mitchell,D. (1985a) *A winning thoroughbred strategy*. Los Angeles: Cynthia Publishing Co.
This book discusses strategies for successful play at the trade and contains a number of useful handicapping computer programs.

Mitchell,D. (1985b) *Myths that destroy a horseplayer's bankroll*. Los Angeles: Cynthia Publishing Co.
A discussion of many racing fallacies and why you should avoid them.

Mitchell,D. (1993) *Commonsense handicapping*. William Morrow.

This book introduces the simple idea that bettors should choose bets that carry a positive expectation.

Quinn,J. (1986a) *The handicapper's condition book*. 2nd ed. New York: Morrow.
A useful book on thoroughbred class with special information on international racing, dosage indices, and two-year-olds and claiming races at minor tracks.

Quinn,J. (1986b) *High tech handicapping in the information age*. New York: Morrow.
A look into the world of computers in racing by one of the game's most talented handicappers.

Quinn,J. (1987a) *The best of thoroughbred handicapping: 1965-1986*. New York: Morrow.
This contains useful essays on major handicapping ideas.

Quinn,J. (1987b) *Class of the field*. New York: Morrow.
A good discussion of the subjective and quantitative approaches to ascertain class.

Quinn,J. (1992) *Figure handicapping*. New York: Morrow.
This is a guide to the interpretation and use of speed and pace figures in spotting the winning horses.

Quirin,W.L. (1979) *Winning at the races: Computer discoveries in thoroughbred handicapping*. New York: Morrow.
A seminal work with much data analysis of various betting strategies.

Quirin,W.L. (1984) *Thoroughbred handicapping: State of the Art*. New York: Morrow.
A followup to the 1979 book emphasizing recent handicapping developments.

Quirin,W.L. (1986) *Handicapping by example*. New York: Morrow.
Forty-one races are analyzed, each to bring out a handicapping point or angle.

Scott,W.L. (1984) *How will your horse run today?* Baltimore: Amicus.
A handicapping theory developed by careful study of 433 races at four Eastern racetracks in 1981.

Ziemba,W.T., Brumelle,S.L., Gautier,A. and Schwartz,S.L. (1986) *Dr.Z's 6/49 lotto guidebook*. Dr Z Investments, Inc., Los Angeles.
This book proposes betting on lotto scientifically with unpopular numbers and discusses many aspects of various types of lottery games.

Ziemba,W.T. and Hausch,D.B. (1984) *Beat the racetrack*. Harcourt Brace Jovanovich, San Diego.
This trade book is a detailed treatment of theory and applications of the Dr.Z system to place and show bets. The system was originally proposed by Hausch, Ziemba and Rubinstein (1981)[1]. There are many simulations showing that the system would have yielded profits at many racetracks over entire seasons and some real betting results at Exhibition Park and the Kentucky Derby over a number of years.

Ziemba,W.T. and Hausch,D.B. (1986) *Betting at the racetrack*. Dr Z Investments, Inc., Los Angeles.
This is a trade book for betting on exotic pools such as quinella, exacta, daily double and trifecta as well as show and place pools. Included are strategies for pricing wagers on different pools.

Ziemba,W.T. and Hausch,D.B. (1987) *Dr.Z's beat the racetrack*. Revised edition. William Morrow and Co. Inc., New York.

    Revised edition of Ziemba and Hausch (1984)[2], this book includes more empirical evidence, some new results, betting in England, dosage and the Kentucky Derby analysis, and further ideas using the Dr.Z place and show betting system.

Index

Printed in the United States
By Bookmasters